London
and the
Reformation

London
and the
Reformation

Susan Brigden

CLARENDON PRESS · OXFORD

Oxford University Press, Walton Street, Oxford OX2 6DP
Oxford New York Toronto
Delhi Bombay Calcutta Madras Karachi
Petaling Jaya Singapore Hong Kong Tokyo
Nairobi Dar es Salaam Cape Town
Melbourne Auckland
and associated companies in
Berlin Ibadan

Oxford is a trade mark of Oxford University Press

Published in the United States
by Oxford University Press, New York

First published 1989
First issued in paperback (with corrections) 1991
Hardback reprinted 1992

British Library Cataloguing in Publication Data
Brigden, Susan
London and the Reformation.
1. London. Christian church. Reformation.
I. Title   274.21
ISBN 0–19–822774–4
ISBN 0–19–820256–3 (Pbk)

Library of Congress Cataloging in Publication Data
Brigden, Susan.
London and the Reformation / Susan Brigden.
p.   cm.
Includes bibliographical references.
1. London (England)—History—16th century.   2. Reformation—
England—London.   I. Title.
DA680.B86   1989
942.1'05—dc19   89-16260   CIP
ISBN 0–19–822774–4
ISBN 0–19–820256–3 (Pbk)

Printed in Great Britain by
The Ipswich Book Company Ltd., Suffolk

for
My Parents

Acknowledgements

THE DEBTS I have incurred in writing this book are, as the sands by the sea shore, innumerable. My most evident scholarly debts are to the legion of historians of the English Reformation, from John Foxe onwards, and are recorded in the notes. Yet many works upon other times and other places are, lest I make a long book longer, too little acknowledged.

It is my pleasure to record my gratitude to my teachers: to Miss Margaret Powell, to Dr Christopher Haigh and the late Professor J. K. Hyde, and to Professor Sir Geoffrey Elton. Perhaps none but Geoffrey Elton's other research students can imagine quite how constant and how great his support and inspiration have been.

Blair Worden has seen this book in its many incarnations. Without his wise counsel and his unfailing encouragement it would have been quite different, and might not have been at all.

I have been more than fortunate in the institutions which have harboured me, and the colleagues I have found. For the beneficence of the University of Newcastle upon Tyne, which awarded me a Sir James Knott Fellowship, I am very grateful, as I am for the chance of working in the Departments of History and Classics there. The friendship and support of my fellow historians at Lincoln College—Vivian Green, Paul Langford, Gervase Rosser, Jane Garnett, Christopher Cunliffe, and several generations of undergraduates—have meant a great deal.

The research for this book has been made much easier by the helpfulness and patience of the librarians and archivists at the libraries and record offices where I have worked. I remember in particular the kindness of Mr Christopher Cooper, Miss Janet Foster, Miss Susan Hare, Miss Jean Imray, Dr Christopher Kitching, Miss Betty Masters and Miss Anne Sutton.

I gratefully acknowledge the permission granted to me by the Worshipful Companies of the Mercers, Goldsmiths, Drapers and Cloth-workers to consult and cite their records. The

Master and Scholars of Balliol College, Oxford allowed me to read and quote from the commonplace book of Richard Hill.

I am greatly indebted to all those whose dissertations I have consulted. Some of these dissertations have since been published: would that they all had been. Dr Richard Wunderli generously allowed me to publish the results of his research in Table 2.

The errors and infelicities in this book are regrettable, and all my own. There would have been more had it not been for the generosity of Ian Archer, Peter Marshall, and Gervase Rosser who read the manuscript. I am most grateful to them. The reader at the Oxford University Press—anonymous, but I have my suspicions—made sagacious criticisms, for which I am thankful.

Ivon Asquith has been the most patient of publishers, and has taken the long view.

To my husband, Jeremy Wormell, who most wanted me to finish this book, and has helped me to finish it, an infinity of thanks.

S.B.

Lincoln College, Oxford
April 1989

Contents

List of Illustrations

List of Tables

Abbreviations

AHR	*American Historical Review.*
Anglica Historia	*The Anglica Historia of Polydore Vergil, 1485–1537*, ed. D. Hay (Camden Society, 3rd series, lxxiv, 1950).
APC	*Acts of the Privy Council of England*, ed. J. R. Dasent, vols. i–ix (1890–1907).
ARG	*Archiv für Reformationsgeschichte.*
BIHR	*Bulletin of the Institute of Historical Research.*
BJRL	*Bulletin of the John Rylands Library.*
BL	British Library.
Brinklow, 'Complaynt'	Henry Brinklow, 'Complaynt of Roderyck Mors', ed. J. M. Cowper (EETS, extra series, xxii, 1874).
Brinklow, 'Lamentacyon'	Henry Brinklow, 'The lamentacyon of a Christen agaynst the cytye of London', ed. J. M. Cowper (EETS, extra series, xxii, 1874).
BRUO	A. B. Emden, *A Biographical Register of the University of Oxford, AD 1501 to 1540* (Oxford, 1974).
Cavendish, *Life*	George Cavendish, *The Life and Death of Cardinal Wolsey*, ed. R. S. Sylvester (EETS, original series, 243, 1959).
Chronicle of Calais	*Chronicle of Calais*, ed. J. G. Nichols (Camden Society, xxxv, 1846).
Chronicle of Queen Jane	*The Chronicle of Queen Jane, and of two years of Queen Mary*, ed. J. G. Nichols (Camden Society, xlviii, 1850).

CLRO	Corporation of London Record Office.
Constantine, 'Memorial'	'A memorial from George Constantine', ed. T. Amyot, *Archaeologia*, xxiii (1831).
Cooper, *Chronicle*	*Coopers chronicle . . . unto the late death of Queene Marie* (1560; *RSTC* 15218).
Correspondance de Marillac	*Correspondance politique de Castillon et de Marillac*, ed. J. Kaulek (Paris, 1885).
Coverdale, *Remains*	*Remains of Myles Coverdale*, ed. G. Pearson (Parker Society, Cambridge, 1846).
CPR	*Calendar of Patent Rolls, Edward VI, Philip and Mary* (1924–).
CSP Mil.	*Calendar of State Papers, Milan, 1385–1618*, ed. A. B. Hinds (1912).
CSP Sp.	*Calendar of State Papers, Spanish*, ed. G. A. Bergenroth *et al.*, 13 vols. (1862–1964).
CSP Ven.	*Calendar of State Papers, Venetian*, ed. Rawdon Brown *et al.*, 9 vols. (1864–98).
CW	*The Yale Edition of the Complete Works of St Thomas More* (New Haven and London, 1963–).
Diary of Machyn	*The Diary of Henry Machyn, Citizen and Merchant Taylor of London, 1550–1563*, ed. J. G. Nichols (Camden Society, xlii, 1848).
DKPR	*Reports of the Deputy Keeper of the Public Records*, III & IV (1840, 1847).
DNB	*Dictionary of National Biography*, 63 vols. (1885–1900).
Documentary Annals	*Documentary Annals of the Reformed Church of England: being a collection of injunctions, declarations, orders & c . . 1546 . . 1716,*

	ed. E. Cardwell, 2 vols. (Oxford, 1839, 1844).
Edward VI Chronicle	*The Chronicle and Political Papers of King Edward VI*, ed. W. K. Jordan (1966).
EETS	Early English Text Society
EHD	*English Historical Documents*, V, 1485–1558, ed. C. H. Williams (1967).
EHR	*English Historical Review.*
Ellis, *Original Letters*	*Original Letters Illustrative of English History*, 3rd series, 11 vols. (1824–46).
Elton, *Policy and Police*	G. R. Elton, *Policy and Police: The Enforcement of the Reformation in the Age of Thomas Cromwell* (Cambridge, 1972).
Elton, *Reform and Renewal*	G. R. Elton, *Reform and Renewal: Thomas Cromwell and the Common Weal* (Cambridge, 1973).
Elton, *Studies*	G. R. Elton, *Studies in Tudor and Stuart Politics and Government: Papers and Reviews* (Cambridge, 1974, 1983).
Erasmus, *Correspondence*	*The Correspondence of Erasmus*, in *The Collected Works of Erasmus* (Toronto, 1974–).
Fish, 'Supplicacyon'	Simon Fish, 'A supplicacyon for the beggers', ed. J. M. Cowper (EETS, extra series, xiii, 1871).
Formularies of Faith	*Formularies of Faith put forth by authority during the reign of Henry VIII*, ed. C. Lloyd (Oxford, 1825).
Foxe, *Acts and Monuments*	John Foxe, *Acts and Monuments*, ed. S. R. Cattley and G. Townsend, 8 vols. (1837–41).
Gee and Hardy, *Documents Illustrative*	*Documents Illustrative of English Church History*, ed. H. Gee and W. J. Hardy (1896).
GLRO	Greater London Record Office.

Gorham, *Gleanings*	G. C. Gorham, *Gleanings of a few scattered ears during the period of the Reformation in England* (1857).
Grafton, *Chronicle*	*Grafton's Chronicle or History of England*, ed. H. Ellis, 2 vols. (1809).
Great Chronicle	*The Great Chronicle of London*, ed. A. H. Thomas and I. D. Thornley (1938).
Grey Friars Chronicle	*Grey Friars Chronicle of London*, ed. J. G. Nichols (Camden Society, liii, 1852).
Guildhall	Guildhall Library, London.
Hale, *Precedents*	*A Series of Precedents and Proceedings in Criminal Causes, 1475–1640, extracted from act books of ecclesiastical courts in the diocese of London*, ed. W. H. Hale (1847).
Hall, *Chronicle*	Edward Hall, *The Union of the two Noble and Illustre Famelies of Lancastre and Yorke*, ed. H. Ellis (1809).
Harpsfield, *Life*	Nicholas Harpsfield, *The life and death of Sir Thomas Moore*, ed. E. V. Hitchcock and R. W. Chambers (EETS, original series, clxxxvi, 1932).
Hennessy, *Repertorium*	G. Hennessy, *Novum Repertorium ecclesiasticum parochiale Londinense* (1898).
HJ	*Historical Journal.*
HL	Huntington Library, San Marino, California.
HLQ	*Huntington Library Quarterly.*
HMC	*Historical Manuscripts Commission.*
Holinshed, *Chronicles*	Raphael Holinshed, *Chronicles*, ed. H. Ellis, 6 vols. (1807–8).
House of Commons, 1509–1558	*The House of Commons, 1509–1558*, ed. S. T. Bindoff, 3 vols. (1982).
HSL	*Proceedings of the Huguenot Society.*
HSR	*Harleian Society Registers.*

Inner Temple	Inner Temple Library, London.
JBS	*Journal of British Studies.*
JEH	*Journal of Ecclesiastical History.*
Journal	Journal of the Court of Common Council.
Knowles, *Religious Orders*	M. D. Knowles, *The Religious Orders in England*, 3 vols. (Cambridge, 1959).
Lambeth	Lambeth Palace Library.
L&P	*Letters and Papers, Foreign and Domestic, of the Reign of Henry VIII*, ed. J. S. Brewer, J. Gairdner, and R. S. Brodie, 21 vols. (1862–1932).
Latimer, *Remains*	*Sermons and Remains of Hugh Latimer*, ed. G. E. Corrie (Parker Society, Cambridge, 1845).
Latimer, *Sermons*	*Sermons by Hugh Latimer*, ed. G. E. Corrie (Parker Society, Cambridge, 1844).
LCCW	*London Consistory Court Wills, 1492–1547*, ed. I. Darlington (London Record Society, 3, 1967).
Letters of Gardiner	*Letters of Stephen Gardiner*, ed. J. A. Muller (Cambridge, 1933).
LFMB	*The Lay Folks' Mass Book*, ed. T. F. Simmons (EETS, original series, 71, 1879).
Lisle Letters	*The Lisle Letters*, ed. M. St. C. Byrne, 6 vols. (Chicago and London, 1981).
LJ	*Journals of the House of Lords*, vol. I.
LMCC	*London and Middlesex Chantry Certificates, 1548*, ed. C. J. Kitching (London Record Society, 16, 1980).
Lyndwood	*Lyndwood's Provinciale*, ed. J. V. Bullard and H. Chalmer Bell (1929).

More, *Answer to a Poisoned Book*	*The Answer to a Poisoned Book*, *CW* 11, ed. S. M. Foley and C. H. Miller (1985).
More, *Apology*	*The Apology*, ed. J. B. Trapp, *CW* 9 (1979).
More, *Confutation*	*The Confutation of Tyndale's Answer*, ed. L. A. Schuster, R. C. Marius, J. P. Lusardi, and R. J. Schoeck, *CW* 8, 3 vols. (1973).
More, *Correspondence*	*The Correspondence of Sir Thomas More*, ed. E. F. Rogers (Princeton, 1947).
More, *Debellation*	*The Debellation of Salem and Bizance*, ed. J. A. Guy, R. Keen, and C. H. Miller, *CW* 10 (1988).
More, *De Tristitia Christi*	*De Tristitia Christi*, *CW* 14 ed. C. H. Miller (1976).
More, *Dialogue concerning Heresies*	*A Dialogue concerning Heresies*, ed. T. M. C. Lawler, G. Marc'hadour, and R. C. Marius, *CW* 6, 2 vols. (1981).
More, *Dialogue of Comfort*	*A Dialogue of Comfort, against Tribulation*, ed. L. L. Martz and F. Manley, *CW* 12 (1976).
More, *Richard III*	*The History of King Richard III*, *CW* 2, ed. R. S. Sylvester (1963).
More, *Treatise on the Passion*	*Treatise on the Passion, Treatise on the Blessed Body, Instructions and Prayers*, *CW* 13, ed. G. E. Haupt (1976).
More, *Utopia*	*Utopia*, ed. E. Surtz and J. H. Hexter, *CW* 4 (1965).
Myrc, *Instructions*	John Myrc, *Instructions for Parish Priests*, ed. E. Peacock (EETS, original series, 31, 1868).
Narratives of the Reformation	*Narratives of the Days of the Reformation*, ed. J. G. Nichols (Camden Society, lxxvii, 1859).
NLS	National Library of Scotland.

Original Letters	*Original Letters relative to the English Reformation*, ed. H. Robinson, 2 vols. (Parker Society, Cambridge, 1846–7).
P&P	*Past and Present.*
PCC	Prerogative Court of Canterbury Wills.
PPC	*Proceedings and Ordinances of the Privy Council of England, 1386–1542*, ed. N. H. Nicolas, 7 vols. (1834–7).
PRO	Public Record Office.
P. V.	*Historical Narrations . . . written by P. V.* (1865).
'Recollections'	'Religion and Politics in mid Tudor England through the eyes of an English Protestant Woman: the Recollections of Rose Hickman', ed. M. Dowling and J. Shakespeare, *BIHR* lv (1982), pp. 94–102.
Repertory	Repertory of the Court of Aldermen.
Roper, *Lyfe*	*The Lyfe of Sir Thomas Moore, Knighte* (EETS, original series, cxcvii, 1935).
RSTC	*A Short-Title Catalogue of Books Printed in England, Scotland and Ireland, and of English Books Printed Abroad, 1475–1640*, 2nd edn., ed. W. A. Jackson, F. J. Ferguson, and K. F. Pantzer, 2 vols. (1976 and 1986).
St Mary at Hill	*The Medieval Records of a London City Church: St Mary at Hill, 1420–1559*, ed. H. Littlehales, 2 vols. (EETS, original series, 125, 128, 1904–5).
State Papers	*State Papers . . . King Henry VIII*, 2 vols. (1830–52).

State Trials   A Complete Collection of State Trials,
                                          ed. W. Cobbett, T. B. Howell,
                                          et al., 42 vols. (1816–98).

Statutes of the Realm   The Statutes of the Realm,
                                          ed. A. Luders et al., 11 vols.
                                          (1810–28).

Stow, Annales   John Stow, The Annales, or generall
                                          chronicle of England . . . ,
                                          ed. E. Howes (1615).

Stow, Survey   John Stow, A Survey of London,
                                          ed. C. L. Kingsford, 2 vols.
                                          (Oxford, 1908).

Strype, Ecclesiastical Memorials   John Strype, Ecclesiastical
                                          Memorials, relating chiefly to religion,
                                          and the reformation of it . . . under
                                          King Henry VIII, King Edward VI,
                                          and Queen Mary I, 3 vols. (Oxford,
                                          1822).

Strype, Memorials of Cranmer   John Strype, Memorials of the most
                                          reverend father in God, Thomas
                                          Cranmer, 2 vols. (Oxford, 1840).

TCD   Trinity College, Dublin.

TED   Tudor Economic Documents,
                                          ed. R. H. Tawney and E. Power,
                                          3 vols. (1924).

TLMAS   Transactions of the London and
                                          Middlesex Archaeological Society.

TRHS   Transactions of the Royal Historical
                                          Society.

TRP   Tudor Royal Proclamations,
                                          ed. P. L. Hughes and J. F.
                                          Larkin (1964–9).

Tudor Tracts   Tudor Tracts, 1532–1588,
                                          ed. A. F. Pollard (1903).

Two London Chronicles   Two London Chronicles from the
                                          collections of John Stow,
                                          ed. C. L. Kingsford (Camden
                                          Society, miscellany, xii, 1910).

VCH London   The Victoria History of the Counties
                                          of England. A History of London,
                                          i (1909).

Venn, Alumni	*Alumni Cantabrigienses,* J. and J. A. Venn, 4 vols. (Cambridge, 1922–7).
Visitation Articles	*Visitation Articles and Injunctions of the period of the Reformation,* ed. W. H. Frere and W. M. Kennedy, 3 vols. (Alcuin Club Collections, xiv, xv, xvi, 1910).
'Vita Mariae'	'The *Vita Mariae Angliae Reginae* of Robert Wingfield of Brantham', ed. D. MacCulloch (Camden Miscellany, xxviii, 1984).
Walters, *London Churches*	H. B. Walters, *London Churches at the Reformation, with an account of their contents* (1939).
Wilkins, *Concilia*	*Concilia Magna Britanniae et Hiberniae . . . 446–1718,* ed. D. Wilkins, 4 vols. (1737).
Wriothesley, *Chronicle*	Charles Wriothesley, *A Chronicle of England during the Reigns of the Tudors, 1485–1559,* ed. W. D. Hamilton, 2 vols. (Camden Society, new series, xi and xx, 1875–7).

Reference, unless otherwise stated, is to document numbers throughout. Place of publication, unless otherwise stated, is London throughout.

Prologue

IN THE Tower of London in 1534 Sir Thomas More thought upon the passion of Christ and his own passage out of the world. He had faced a test not far different from that put to every other citizen of London: whether to swear an oath, 'without scrupulosity of conscience', as the King commanded. All the citizens swore, but this political test had spiritual consequences. More alone refused it. In the following year England was sundered from Catholic Christendom, and More was executed for refusing to follow the laws of one realm, against his conscience. Yet almost every Londoner continued to hold with More a belief in the universality of the Catholic Church, was obedient to its authority, and was convinced of the power and ineluctability of its seven sacraments. A shared faith bound the community, and common religious observance marked the rites of passage of the citizens and of their City. Few saw then how desperately the Church was in danger, or from which quarters. But there were signs.

Some sought reform in the Church, before it was too late. John Colet, Dean of St Paul's, had warned the Catholic clergy in 1511 that their failings left them defenceless against attack. Thomas More feared that heresy would undermine and pervert the foundations of the faith, and imperil the society which harboured it. Again and again, he urged 'good Catholic men' never to imagine their Church, seemingly adamantine, to be unassailable. For More, moving in many London worlds, saw heresy pervading his own parish, his own company, even his own family. The revival of an old heresy, Lollardy, and the arrival of a new one, more dangerous yet, threatened terrene disorder. Once, Catholics had cast out apostates in their midst, refusing to talk or trade with them: would they still if these heretics grew in strength and numbers, and what would then be the consequences for this community, once bound by a common faith? No one could have foretold that the threat, when it came, would not be from the heretics alone, but from that King whom More served, and who had vaunted himself *defensor fidei*: Henry VIII. Knowingly, or unknowingly, he had gathered evangelicals around him, in the highest places of all, and for a time it served his purpose to listen to them.

The people of England found themselves caught up in a Reformation, not at first of their making, but in time made by them. Religious choices were demanded most immediately from the Londoners, because in London the English Reformation began, and the capital was 'the common country of all England.' Maybe as many as one in twenty of the population of England found their home in London in the mid-sixteenth century, at least for a time. The power of the City's religious example was immense. There the new faith was first and most powerfully evangelized; there, under the eyes of government, conformity to the royal will was most imperative. The cause and experience of conversion, or of resistance to it, were different for every individual and, because they were private, remain largely hidden. Yet in London it was not always easy to hide convictions and allegiances.

Though the citizens of London boasted of their world-wide trading connections, they lived within a City of one square mile, bound still by its ancient walls. Within the City of London there were worlds within worlds. In their parishes, wards, precincts, companies, and fraternities the citizens met, worshipped, and feasted together. Great distinctions of wealth and status divided them, but the rich and poor were neighbours, worshipping in the same parish churches, knowing each other as fellow communicants, as givers and receivers of charity. As Londoners met daily in the markets or company halls, passed each other in the streets, attended the same Church services, kept watch together, much of City life was public. Deponents before the courts of the Church and City vividly attested how closely the citizens knew and observed each other. The wills of Londoners reveal how widely the networks of social relationship extended: from kin, to friends, to neighbours, to fellow parishioners, to godchildren, apprentices, servants, members of the same company, to debtors, to the poor they aided, and—until the Reformation—to all those dead souls of friends and family who awaited release from Purgatory, 'all Christian souls'. True, a great metropolis affords the best chance for fugitives to disappear, and in

London there was a growing population of vagrants and mi-
grants, yet there were probably few in the capital with whom
no one claimed relationship. In such a society private affairs
were hard to hide, and memories were long. So, when Andrew
Boord fled the rigours of the London Charterhouse his Catholic
neighbours did not easily forgive him. They called him 'apostate', and remembered his ancient failings; that 'twenty years
agone' he had been, though in religious orders, 'conversant
with women'.[1] The religious convictions of individuals might
be public knowledge. For to conceal faith, to worship one way
outwardly, while believing another way privately, was condemned. Hypocrisy and Nicodemism, though prudently practised, were widely held to damn the soul, even though they
might seem to heal divisions within the community. It was
because beliefs were known that Londoners notoriously came
to taunt each other: 'thou papist', 'thou heretic'; 'he is a
papist, he is a gospeller, he is of the new sort, he is of the old
faith'. And contention was to be expected, and even necessary,
in a greater cause. Bishop Latimer counselled his Protestant
brethren: 'where as is quietness . . . there is not the truth'.[2]
When one faith was evangelical, determined that the Word
should go forth, whatever the risk, and the other rested upon
authority, giving all power in the determination of doctrine to
the Church, there could hardly be peace.

Did this religious crisis destroy the sense and fact of community in London, or was the crisis in religion itself in part a
symptom of fracture having other causes? The new divisions in
faith might reflect and exacerbate divisions within society older
and deeper still, and one path to salvation be chosen rather
than another for reasons not of faith alone. Maybe the poor
and dispossessed could find a special appeal in the liberating
doctrines of Protestantism. Had not Christ and His disciples
been poor too, and had not Christ pointed out the special
difficulties for the rich to enter the Kingdom of Heaven? So
reforming preachers reminded their audiences. Certainly the
rich, the governors, feared the consequences of giving the
socially irresponsible the freedom to read the Gospel in English,
and the oppressors of the poor and the persecutors of the

---

[1] *L&P* xi. 297.   [2] Foxe, *Acts and Monuments*, vii, p. 508.

gospellers appeared to be one and the same. The richest and the poorest in England lived in London, side by side; the rich in great houses on the street fronts, the poor in alleys behind. As the Reformation progressed the distinctions in wealth became greater; not because of the Reformation, but because of the social and economic transformation created by the rise in population coeval with it. These demographic changes would be of the greatest consequence for the religion and politics of London.

But the Reformation was made by individuals, not by social forces. In this first generation of Reformation everyone was faced, for the first time, with a choice in religion. Though the Reformation was first imposed upon the English people, who were unknowing, unwilling, it became their particular creation. For many—perhaps most—acquiescence to royal command was the course of least resistance, but the experience of conforming, against conscience, might be bitter. In every parish and street in London families, friends, and neighbours who had once shared faith and ideals, as well as much else, could find their ways dividing. When some people converted, and others did not, there would be personal consequences. When in 1531 Thomas Crispe made a will, providing for his children if they chose to enter the religious life, he naturally made his cousin, a fellow mercer, his executor.[3] But his cousin was Robert Packington, who was already a committed evangelical, and for him the monastic vocation was an absurdity. What would his duty be: to follow his friend's wishes or his own religious inclination? The pious Catholic benefactors of the monastic houses might become in time ardent gospellers, denouncing Purgatory and the prayers for souls to which monastic life was devoted. Conversely, those who had once called for the translation of Scripture could become its fiercest opponents when they realized that defence of the Church was more desperately needed than reform. The Reformation, like other revolutions, created enemies. Though Protestants and Catholics might in the end find in their faiths more to unite than to divide them, they both seemed then to see only the perversion and traduction of the truth by the other.

[3] PRO, PCC, Prob. 11/24, fo. 90^r.

At the Reformation the Christian found himself forced to choose between true and false images, between free will and predestination, between private faith and public conformity. All that follows is an enquiry into how the citizens of London made those choices and what were the consequences.

I   The Catholic Community

THE LAITY

The City Churches

WITHIN THE ancient walls of the City of London on the eve
of the Reformation were just over a hundred parish churches, a
cathedral, and thirty-nine religious houses. The panoramas of
sixteenth-century London show their spires and towers and
steeples dominating the skyline.[1] Among secular buildings in
late medieval London only the Guildhall and the Tower could
compare in beauty or grandeur with the churches. No layman
aspired to build to rival the Church: those who did might be
punished for their vainglory. When Sir John Champneys, Lord
Mayor in 1534, added a 'high tower of brick' to his house in
Tower Street ('the first . . . in any private man's house to
overlook his neighbours in this City'), 'this delight of his eye
was punished with blindness'.[2] Generation after generation of
Londoners gave of their wealth and labour towards edifying the
churches where they celebrated. If concern to build and decorate
churches, and determination to adorn their City with religious
artefacts, were marks of faith, then the citizens of London were
of a remarkable piety. The very street names were religious:
Ave Maria Lane, Rood Lane, Creed Lane, Pater Noster Row.

In London 'there was in every corner a cross set'.[3] Most
famous of these was the Eleanor Cross in Cheapside which was
gilded and adorned with images—of the Virgin and Child, of
Christ's Resurrection, with the arms of its donor 'embraced by

---

[1] See Antony von den Wyngaerde's panorama: Ashmolean Museum, Oxford,
Sutherland Collection: *St Mary at Hill* (frontispiece). For the Agas and Braun and
Hogenburg maps, see *A to Z of Elizabethan London*, compiled by A. Prockter and R. Taylor
(Lympne Castle, Kent, 1979). *VCH London*, i, pp. 407–585. M. Honeybourne, 'The
extent and value of property in London and Southwark occupied by the Religious
Houses, Inns of Bishops and Abbots and churches and churchyards before the
Dissolution' (University of London, MA thesis, 1948).

[2] Stow, *Survey*, i, p. 133; cf. also p. 151. For the Earl of Surrey's castigation of
London's 'proud towers', see below, pp. 340–2.

[3] Guildhall, MS 9531/10, fo. 135ᵛ.

angels'. For each great occasion of public ceremonial—the entry of Emperor Charles V or Philip of Spain, the coronations of Anne Boleyn and Edward VI—this cross was specially regilded.[4] Every parish church had a cross on its spire or tower—save St Michael Cornhill where an image of the patron stood—and a preaching cross in its churchyard. St Paul's Cross was the first pulpit of the realm, and it was from this cross that the changes of the Reformation were expounded. These parish crosses were venerated. At St Mary Magdalen Milk Street in 1449 there stood a cross in the churchyard which was 'worshipped by the parishioners there as crosses be commonly worshipped in other churchyards'.[5] The 'Rood of Northern' at the north door of St Paul's, allegedly carved by St Joseph of Arimathea and discovered during a flood of the Thames in the days of the fabled King Lucius, became a favourite object of devotion because miracles were performed there.[6] John Paston wrote to his mother in 1465: 'I pray you visit the Rood of North door and Saint Saviour at Bermondsey, while ye abide in London, and let my sister Margery go with you to pray to them that she may have a good husband ere she come home again.'[7] Crucifixes were everywhere as a remembrance of Christ's passion, and the strongest defence against the temptations of man's 'ghostly enemy', the Devil. They were not to be worshipped for themselves, but venerated just as the King's seal was venerated: not for love of the seal itself but for love of the man who owned it. The rood, so the fourteenth-century homilist John Mirk insisted, was 'the King's seal of Heaven'.[8]

Images and paintings of saints were 'lewd men's books', for there were 'many thousands of people' who could not imagine in their hearts how Christ died on the cross to redeem mankind until they learnt this truth from pictures of His passion.[9] At the opening of his *Dialogue concerning Heresies* Sir Thomas More appealed to the legend that Christ inspired St Luke to paint the

---

[4] Stow, *Survey*, i, pp. 26, 266–7.

[5] C. Pendrill, *Old Parish Life in London* (Oxford, 1937), pp. 4–6.

[6] A. Vallance, *Greater English Church Screens* (1947), pp. 75–6.

[7] Cited in Pendrill, *Old Parish Life*, p. 9.

[8] *Mirk's Festial: A Collection of Homilies*, ed. T. Erbe (EETS, extra series, 96, 1905), p. 171 (de solemnitate Corporis Christi).

[9] Ibid. M. Aston, 'Devotional Literacy', in *Lollards and Reformers* (1984), pp. 115–16; G. R. Owst, *Literature and the Pulpit in Medieval England* (Oxford, 1961).

'lovely visage of Our Blessed Lady' as proof that images are pleasing to God.[10] More knew by his own experience that miracles were performed at the shrines of saints to reveal divine power to the faithful. He had seen how Sir Roger Wentworth's twelve-year-old daughter, 'tormented by our ghostly enemy the Devil, raving with despising and blasphemy of God and hatred of all hallowed things', was saved by visiting the shrine of the Virgin at Ipswich.[11] In his youth More had seen at Barking Abbey kerchiefs reputedly sewn by Our Lady; 'as clean seams to my seeming as ever I saw in my life'. These relics, hidden in the back of a golden tabernacle which had been shut away for four or five hundred years since 'the abbey was burned by infidels', had remained unknown, 'till now that God gave that chance that opened it'.[12] If relics and shrines were venerated by the most learned of Londoners, how much more emotive they might be to the ignorant and credulous. More (Moria) saw the folly of superstitious excess. Writing to Erasmus of the life of the courtier, he compared the King's suitors to those London wives who, praying to an image of the Virgin by the Tower, 'gaze upon it so fixedly that they imagine it smiles upon them'.[13]

The Virgin Mary, 'mother and maiden', 'Queen of Heaven, lady of the world and Empress of Hell', was ceaselessly invoked as mediator to God for men. The saints, her courtiers in Paradise, might act for their perpetual suitors as 'holy patrons' in the 'blessed Court of Heaven'.[14] The saints were believed to have power to ward off the disasters which might befall their supplicants, and their favours were daily called upon.[15] The saints could also be angered, and must be placated. In 1533 'a letter written by Mary Magdalen's hand' was delivered to a London widow, warning that 'if she did diminish any part of the gold hidden by her husband . . . and bestowed it not entirely in the ornaments of the church', it would be 'to her husband's

---

[10] More, *Dialogue concerning Heresies*, i, p. 39.

[11] Ibid., p. 93. She may indeed have been deranged; see *L&P Addenda* i/1. 522.

[12] More, *Dialogue concerning Heresies*, i, p. 222.

[13] Cited in R. C. Marius, *Thomas More* (1984), p. 201.

[14] See, for example, PRO, PCC, Prob. 11/20, fos. 39ᵛ, 177ᵛ; 11/24, fos. 104ᵛ, 121ʳ.

[15] Keith Thomas, *Religion and the Decline of Magic: Studies in Popular Beliefs in Sixteenth- and Seventeenth-century England* (1971), ch. 2; *The Oxford Dictionary of Saints*, ed. D. H. Farmer (Oxford, 1978).

utter damnation and hers both'.[16] Saints were glorified
through their images and at their shrines. In every City church
candles burned before the images of favourite saints set up by
their faithful votaries. The narrator in *The ymage of loue*, published in 1525, told of the 'many good men that be nowadays
which honour the temples of God with many goodly images of
great cost of silver and of gold set with pearl and stone'.[17]

Men and women called especially upon their own patrons,
remembering their images particularly in their wills; 'St John
Evangelist', for example, 'whom I have always worshipped
and loved'.[18] To St Margaret Pattens Margaret Sale had given
an image of Our Lady, and when she made her will in 1527 she
bequeathed her funeral tapers to burn thereafter before the
images she had honoured during her life: before Our Lady, St
Katherine, St Anne, St Sythe, and the new rood at her parish
church, before Our Blessed Lady at Barking, and St Gabriel at
St Gabriel Fenchurch.[19] The citizens of London often wished
to be buried beside their own patron saints in their churches:
before the image of St James the Apostle at St Michael Bassishaw; before the image of St John the Evangelist; at St Christopher le Stocks by the image of Jesus; before 'Our Lady of pity'
at St Antholin; in the Grey Friars before St Francis; in Savoy
Chapel before the image of St George.[20] Londoners provided
candles to burn forever to light the images. At St Mary at Hill
John Causton bequeathed money for two tapers to burn

afore the image of Our Lady at high altar on Sundays and holy days,
and two tapers burning before the Angel's Salutation of the image of
Our Lady every evening at the time of singing of Salve Regina, and
one taper should burn at the south altar between the figures of St
Thomas and St Nicholas.[21]

The saints looked down from the windows of the churches and
from the walls and altars inside. In the windows of St Mary at

[16] PRO, SP 1/82, fo. 93ʳ (*L&P* vii. 72).
[17] John Ryckes, *The ymage of loue* (7 Oct. 1525; *RSTC* 21471.5), sig. B iiiʳ. For
The ymage of loue, see below, pp. 80-2.
[18] PRO, PCC, Prob. 11/24, fo. 110ᵛ.
[19] Guildhall, MS 9171/10, fo. 96ʳ.
[20] See *inter alia* PRO, PCC, Prob. 11/20, fos. 5ᵛ, 109ᵛ, 199ʳ, 217ʳ, 219ᵛ; 11/24,
fos. 25ᵛ, 33ᵛ, 121ʳ; Guildhall, MS 9171/10; fos. 53ʳ, 55ᵛ, 74ʳ, 137ʳ, 160ᵛ, 231ʳ.
[21] *St Mary at Hill*, i, p. 6.

Hill were images of the Trinity, of the seven works of mercy, and of St John.[22] St Dunstan in the West and St Thomas Acon had windows enshrining St Thomas à Becket.[23] The figures were often clad in special coats, and wearing silver shoes, which their faithful votaries kissed.[24] The image of St George at St Peter Cornhill and St Mary Woolnoth was a martial figure, on horseback, with 'coat armour of sarcenet' and 'a headstall for a horse'.[25] Still in 1538 George Robinson found a woman kneeling before the image of St Uncumber (whose aid unhappy wives invoked when they wished to be rid of their husbands), who was 'in her old place and seat with her grey gown and silver shoes'.[26]

The manner in which the citizens chose to build and adorn the churches where they worshipped is revealing of the nature of popular devotion. The late medieval parish churches of London were ornate and adorned, gilded and painted. Every space was covered with 'gay outward things', sacred paintings, tabernacles, banners, and veils. The churchwardens' accounts of every City church reveal the parishioners' continuing devotion to their special saints.[27] In All Hallows Staining in 1528, for example, the accounts record the repainting of the image of St Luke:

bringing of St Luke's tabernacle home from painting	2d.
for plaster, mortar and brick, for setting up St Luke	16d.
pulley and cord and a strap to pull up and down the shrine	6d.[28]

At St Peter Cornhill were cross banners 'with the image of St Peter with keys and swords', 'of the baptizing of Jesu Christ

[22] *St Mary at Hill*, i, pp. 19, 252, 313.

[23] PRO, SP 1/92, fo. 128ʳ (*L&P* viii. 626); Guildhall, MS 2968/1, fo. 94ʳ.

[24] Walters, *London Churches*, pp. 51, 91, 351, 432, 447, 457, 554, 566, 602, 646.

[25] Ibid., pp. 51, 467. Cf. Erasmus, *Praise of Folly* (Harmondsworth, 1971), p. 126.

[26] PRO, SP 1/134, fo. 183ʳ (*L&P* xiii/1. 1393).

[27] The contents of the City churches were listed for the Edwardian commissioners of 1552: Walters, *London Churches*. For printed churchwardens' accounts, see *Churchwardens' accounts of the parish of Allhallows, London Wall, 1455-1536*, ed. C. Welch (1912); *St Mary at Hill*; *Annals of St Helen's Bishopsgate*, ed. J. E. Cox (1876); C. W. F. Goss, 'The parish and church of St Martin Outwich', *LMAS*, new series, vi (1933), pp. 65-9, 81-90; *The accounts of the churchwardens of the parish of St Michael Cornhill*, ed. W. H. Overall (1871); T. Milbourn, 'The church of St Stephen Walbrook', *TLMAS* v (1881), pp. 327-402. Extracts from the accounts of many of the churches appear in J. P. Malcolm, *London Redivivum*, 4 vols. (1802-7).

[28] Guildhall, MS 4956/1, fo. 126ʳ.

and St John Baptist', 'of the coronation of Our Lady and assumption', 'of the transfiguration of Our Lord', 'with the image of St Peter written with Sancte Petre orate pro nobis'. There were Lenten veils there to hang before the rood until Easter, 'stained with divers stories', with the Scriptural text 'surrexit dominus vere', and with the Passion.[29] In the lady chapel at St Stephen Walbrook there were seven wooden images; in the choir eight; in the chapel of St Nicholas and St Katherine were more, of St Nicholas, St Katherine, St Michael, St Margaret, St James, and St Mary Magdalen; in the cloister were two images of 'Our Lady holding God Almighty in her lap, called the Pity', and images of St Anne, St Laurence, St Vincent, St Peter, and St Paul. There were pictures too in St Stephen Walbrook: of Christ crucified, of the twelve apostles and four doctors of Holy Church; a painted hand—Manus Meditationis—with thumb and fingers inscribed with verses; tables inscribed with the ten commandments, the seven works of mercy, and the seven deadly sins.[30]

Hidden behind screens of wood and stone, shut off from the nave and chancel, was a series of chapels; chapels for gilds, chapels for chantries, each with its own altar, guarded by its own saint, with its own priest singing perpetual masses. The late medieval churches were mysterious; one secret sanctuary followed after another.[31] The incessant invocation of the saints suggests the belief that the soul could never reach God without ceaseless intercession and meditation. At the altars enclosed around the sides of the church the people prayed privately to the Virgin and the holy company of Heaven, but during the public worship of the parochial Mass their attention should be drawn always upwards towards the east end.

The great rood—the figure of Christ crucified, with images of the Blessed Virgin Mary and St John Evangelist at either side—dominated the interior of every parish church. The rood, set above the congregation, separated the parishioners in the nave from the priest celebrating the sacred mysteries in the

---

[29] Walters, *London Churches*, pp. 572–86.

[30] Pendrill, *Old Parish Life*, pp. 10–11, 16.

[31] For what follows, see G. W. O. Addleshaw and F. Etchells, *The Architectural Setting of Anglican Worship* (1948); A. Heales, *Archaeology of the Christian Altar* (1881); *English Church Furniture, Ornaments and Decorations, at the Period of the Reformation*, ed. E. Peacock (1866).

chancel. Before the rood, candles burnt constantly to illumine
the figures, painted and with golden diadems. Painted upon
the tympanum which filled the space between the rood loft and
the chancel arch was the doom, the portrayal of the General
Resurrection and the Last Judgement. Beyond the rood, and
hidden from the laity by the chancel screen, were the reredos
and high altar, where the priest celebrated. Elevated above the
high altar during the Mass, and left hanging there perpetually
reserved in a pyx, was the consecrated Host, which the faithful
came to gaze upon and adore.

The Mass and Salvation

For every Catholic the path to salvation was found only by
following the teachings of the Church, and by receiving divine
grace through the seven sacraments. None of the faithful could
challenge the Church nor abandon its practices without con-
signing his immortal soul to eternal perdition. Of all the seven
sacraments—baptism, confirmation, matrimony, extreme
unction (anealing), ordination, penance, and the Mass—it was
the Mass, the sacrament of the altar, which played the most
sacred part in the religious life of the faithful.[32] The lives of the
people were marked by the rites of passage of the Church:
birth, marriage, and death by the sacraments of baptism,
matrimony, and extreme unction; but these occurred but once
a lifetime. The mystery of the Mass, performed daily in each of
the City churches, was the perpetual affirmation of Christ's
incarnation, passion, and ascension. It was also the unifying
bond of the Christian community. No Catholic should doubt
the power and sanctity of this sacrament. At every celebration
of the Mass a miracle occurs: the consecrated elements of the
bread and the wine are transformed, by the working of God's
grace, into the very body and blood of Christ. So powerful was
the sacrament of the altar that merely to be in the church and
watch it performed might bring the faithful closer to Heaven.

---

[32] This account of the Mass on the eve of the Reformation is drawn from J. A.
Jungman, *The Mass of the Roman Rite: its Origin and Development*, 2 vols. (New York,
1951–5); F. E. Brightman, *The English Rite*, 2 vols. (1915); J. Bossy, 'The Mass as a
Social Institution, 1200–1700', *P&P* 100 (1983); B. L. Manning, *The People's Faith in
the Time of Wyclif* (Cambridge, 1919).

Special benedictions were granted 'to all such as be in clean life with reverence and good devotion seeth the sacrament':

Venial sin and idle oaths are forgiven, and sight shall not fail; they shall die no sudden death; they shall not wear old in sight; and all the steps that they make coming there to be numbered of an angel for their merit.[33]

To know the 'Manner and Mede of the Mass', to see the parts played and the words spoken by the priest and people, is to begin to understand the nature of popular devotion on the eve of the Reformation. Though public worship might not reflect private faith; though everyone in the congregation might conceive the soul's journey to God in a different way; nevertheless the Mass was at the centre of their religious world. The words of the rite, and the directions for the ritual, which we know, still cannot express the essence of the Mass: that it is a mystery, not given to men to understand. The full, public, sung rite of the Latin Mass took the following form: a confession of sin by the priest and people, and absolution for the penitent; the promise of earthly peace and heavenly bliss in the hymn, *Gloria*; a declaration of the faith in readings from the Epistles, Gospels, and Nicene Creed; the offertory, whereby the priest prepared the bread and wine for sacrifice; the canon, the consecration by the priest of the bread and wine, by which they were transformed into the very body and blood of Christ; the communion, the reception of the eucharist by the priest and, sometimes, by the people; the postcommunion, the blessing of the people by the priest. All this while the priest was at the altar, usually speaking low, praying secretly, and always in Latin.

The Mass had once been more evidently the communion of the faithful, with priest and congregation celebrating in union, but by the eve of the Reformation the Mass had become a clerical service, performed by the priest alone, with the people looking on and worshipping apart.[34] Even Princess Mary, most

---

[33] W. Harrington, *In thys boke are conteyned the comendacions of matrimony* (c.1517; *RSTC* 12798.7), sig. Dv^v. William Harrington was a canon of St Paul's and vicar of St Giles Cripplegate.

[34] This is one aspect of the reformers' attack upon the Mass. See Thomas Becon, 'The Displaying of the Popish Mass' and 'A Comparison between the Lord's Supper

devout of Catholics, had wondered why the laity's part in the Mass was so circumscribed; why they must not pray themselves, only listen. As a girl, she had asked her almoner:

In my God, I cannot see what we shall do at the Mass, if we pray not.

Ye shall think to the mystery of the Mass and shall harken to the words that the priest say.

Yea, and what shall they do which understand it not?

They shall behold, and shall hear, and think, and by that they shall understand.[35]

Yet if it seemed as though the Catholic laity were excluded from the rite, as if they were but 'vain gazers', they thought quite otherwise. When they attended Mass the people took their own part in the service: the Mass did involve participation and propitiation for them too.[36] While the priest went to the altar, the people knelt to pray (on one knee).[37] The laity could not hear the words of the Mass, whispered by the priest; nor could most of them have understood them anyway, for they were in Latin, a secret language; but they could follow the rite.

For devout and literate lay men and women there were books to read to guide their worship.[38] One fifteenth-century guide for the spiritual direction of the laity advised, 'while the clerks are singing, look at the books of the church; and on every feast day look at the Gospel, and the exposition of it, and at the Epistle'.[39] *The Lay Folks Mass Book* of the late fourteenth century explained something of the form of the rite, while not revealing matters too sacred for the laity to know. Only the priest knew the exact words of the office, which he celebrated

and the Pope's Mass', in *Prayers and Other Pieces of Thomas Becon*, ed. J. Ayre (Parker Society, Cambridge, 1844), pp. 278–81, 374.

[35] *LFMB*, p. 158.

[36] Bossy, 'The Mass as a Social Institution', p. 36; *Christianity in the West, 1400–1700* (Oxford, 1985), pp. 66–70; Manning, *The People's Faith in the Time of Wyclif*, pp. 14–16; Becon, 'Displaying of the Popish Mass', pp. 257–8.

[37] *LFMB*, pp. 162–3.

[38] H. C. White, *The Tudor Books of Private Devotion* (Westport, Conn., 1951), pp. 53–86; J. Hartham, *Books of Hours and their Owners* (1977); M. Aston, 'Devotional Literacy', in *Lollards and Reformers*, ch. 4.

[39] W. A. Pantin, 'Instructions for a Devout and Literate Layman', in *Medieval Learning and Literature: Essays presented to R. W. Hunt*, ed. J. J. G. Alexander and M. T. Gibson (Oxford, 1976), p. 399.

on behalf of the congregation, apart from them; 'so that his communication is to God, and not to the people'.[40] But there were ways for the laity to follow the rite. Langforde's *Meditations for spiritual exercise in the time of the Mass*, intended to 'move souls to the devotion of the Mass', gave instructions for the people to accompany 'all the secret prayers and gestures of the priest' which they witnessed.[41] *The Lay Folks Mass Book* told the people to:

> Behold the elevation reverently,
> Such prayer then thou make,
> As liketh thee best for to take.[42]

That prayer was most likely to be chosen from the Primer— *Horae Beatae Mariae Virginis*. This lay folks' prayer book contained usually the calendar, the hours or office of the Virgin, the penitential psalms, the litany, the office of the dead or suffrages of saints, and numerous prayers. On the eve of the Reformation dozens of editions were printed, mainly still in Latin, but increasingly with English explications. This book of devotion was used by devout laymen to guide their daily prayer and meditation and to help them bear their part in the services of the Church.[43] In 1529 Cavendish discovered Thomas Cromwell, weeping at his master's fall, in the window of the great hall at Esher, with his primer in his hand, reading Our Lady's matins. But 'this had been since a very strange sight'.[44] In the capital the people were more literate than elsewhere, and there in 1500 an Italian visitor saw at Mass daily 'the women carrying long rosaries in their hands, and those who can read taking the office of Our Lady with them, and with some companion

---

[40] John Standish, *A discourse wherin is debated whether the scripture should be in English* (1554; *RSTC* 23207), sig. Kviii^v; cited Aston, *Lollards and Reformers*, p. 123.

[41] *Tracts on the Mass*, ed. J. Wickham Legg (Henry Bradshaw Society, xxvii, 1904), pp. 17–29; Aston, *Lollards and Reformers*, p. 123.

[42] *LFMB*, p. 39.

[43] See *RSTC* 15867–15985a (editions of the primer printed before Marshall's 1534 English edition). E. Hoskins, *Horae Beatae Virginis or Sarum and York Primers . . . of the Reformed and Roman Use* (1901); C. C. Butterworth, *The English Primers, 1529–1545* (Philadelphia, 1953), ch. 1. A general indulgence was granted in 1499 to 'who so ever being in a state of grace that devoutly will say the psalter of Our Lady': *RSTC* 14077c.148. Londoners bequeathed primers in their wills: *LCCW* 12, 40, 108, 177.

[44] Cavendish, *Life*, p. 104. Foxe condemned 'Our Lady's Matins' as 'full of popish blasphemy': *Acts and Monuments*, vii, pp. 129–38.

reciting it in the church verse by verse, in a low voice, after the manner of churchmen'.[45]

Even those Londoners who could not read—still most Londoners—could nevertheless participate. As the Mass began, the priest and people confessed together, the people kneeling and saying their paternosters.[46] At the reading of the Gospel 'the people stand up and make courtesy when they hear the name of Jesus'.[47] 'Thousands', said Tyndale, crossed themselves all the while when the priest read from St John's Gospel, 'a legion of crosses, behind and before; and . . . pluck up their legs, and cross as much as their heels and the very soles of their feet', 'that there shall no mischance happen them that day'.[48] During the *secreta*, while the priest made private prayers for the sacrifice to be offered, the people too knelt and prayed.[49] At the *Sanctus* a bell rang, warning of the sacrifice to come: the people stood. Before the canon they knelt, and during the bidding prayers the congregation commemorated the living and offered prayers for their salvation; they prayed too for the intercession of the saints, for souls in Purgatory.[50]

The elevation of the Host was the most sacred moment of the sacrament: as the sacring bell rang the priest displayed to the wondering laity the body of Christ in form of bread.

> A little bell he will to us ring,
> Then is reason that we do reverence
> To Jesus Christ's presence,
> That may loose of all baleful bonds;
> Therefore kneeling, hold up thine hands.[51]

Devout Londoners left money to light candles in dark City churches at that very moment, 'at the time of the elevation of the most glorious sacrament of the altar at high Mass', so that

[45] *A Relation, or rather a True Account, of the Island of England . . . about the year 1500*, ed. C. A. Sneyd (Camden Society, old series, xxxvii, 1847), p. 23. Cf. More, *Debellation*, p. 7.

[46] *LFMB*, pp. 6, 181–4.

[47] Becon, 'Displaying of the Popish Mass', p. 264.

[48] *An Answer to Sir Thomas More's Dialogue . . . by William Tyndale*, ed. H. Walter (Parker Society, Cambridge, 1850), p. 61; cited Thomas, *Religion and the Decline of Magic*, p. 36.

[49] *LFMB*, pp. lxii–lxiii.

[50] Ibid., pp. 40–4.

[51] Ibid., p. 39.

everyone could witness and be blessed by the miracle.[52] The people must mark their worship of the Host: they ought to 'reverently and meekly kneel on their knees & hold up their hands at the lifting up of the most blessed sacrament in the Mass, saying devoutly some prayer in the honour and reverence of that blessed sacrament'.[53] Even to see the sacrament was to have grace conferred thereby: but the parishioners rarely received it themselves. The priest must counsel them to receive the eucharist three times in the year: at Easter, Whitsun, and Christmas. In fact, the Church required only that the laity confess and receive once a year: at the 'blessed season' of Easter.[54] Anyone who failed to take the Easter houseling 'ought to be excluded out of the church and company of Christian people. And after that they be dead for to want Christian men's burial.'[55] Not to attend was to risk the worst of stigmas, suspicion of heresy. 'Thou art an heretic for thou tookest not thy rights at Easter', so one woman defamed another in 1521.[56]

The Host, created by a miracle, might have miraculous powers itself. Into his commonplace book in the 1530s a Londoner, John Colyns, copied an account of a 'great miracle'. A Cornish knight had travelled on pilgrimage to the Holy Land in quest of a splinter from the Holy Cross. This he found, and the splinter cured and remained in a wound on his thigh. Returning home after a perilous voyage, the knight fell into a deep sleep. In a church nearby, just as the priest was elevating the blessed sacrament, a white dove flew down and carried away the Host to where the sleeping knight lay. The dove laid the Host upon the knight's thigh, whereupon the splinter of the Holy Cross was revealed, to the wonderment of the parishioners.[57] Tudor Londoners could look, too, to miracles which

---

[52] For example, PRO, PCC, Prob. 11/20, fos. 45ʳ, 69ᵛ, 107ᵛ, 171ʳ.

[53] Harrington, *Comendacions of matrimony*, sig. Dvᵛ; Lyndwood, III, tit. 23 c. iv; Becon, 'Displaying of the Popish Mass', p. 267; Myrc, *Instructions*, p. 9.

[54] *LFMB*, p. 121; T. N. Tentler, *Sin and Confession on the Eve of the Reformation* (Princeton, 1977), pp. 20–2; H. Maynard Smith, *Pre-Reformation England* (1938), pp. 93–6.

[55] Harrington, *Comendacions of matrimony*, sig. Dv.

[56] GLRO, DL/C/207, fos. 7ʳ, 10ʳ, 24ᵛ.

[57] BL, Harleian MS 2252, fos. 50ᵛ–51ʳ. At least two Londoners kept relics of the Holy Cross in their houses: Guildhall, MS 9171/10, fo. 75ᵛ; PRO, PCC, Prob. 11/24, fo. 86ʳ.

happened nearer home and during their own lifetimes: when the rood and altar at Rickmansworth church were set on fire by 'wretched heretics', though the pyx was melted the 'blessed body of Our Lord Jesus Christ in form of bread was . . . nothing perished', 'through the might . . . of Our Saviour'.[58] Only the power which the blessed sacrament held in the popular imagination could explain the consternation in the City when in 1532, even on Good Friday, the reserved host was stolen away from the church of St Botolph Aldersgate. When it was recovered the whole parish restored it to the church in solemn procession.[59]

The Mass was a work of great power, and it was a symbol of peace. This sacrament might reconcile and bring the faithful 'into charity', and unite the Christian community as nothing else could.[60] Many might have been slain in a 'great skirmish in Fleet Street' in April 1459 had not, 'as God would', the Bishops brought 'crosses and our Lord's body' to pacify the rioters.[61] At the centre of the service of the Mass was a ceremony of reconciliation: the kiss of peace or pax. No one was allowed to receive the sacrament unless he was 'in charity', reconciled with God and man. The priest was admonished to warn all partakers in the Mass 'in God's name', 'that none of you come thus to God's board but if ye be in perfect love and charity, and be clean shriven and in full purpose to leave your sin'.[62] Anyone who remained unrepentant and unreconciled was sundered from the Church and the community of the faithful, dead as well as living. The priest himself must say as he took the cup: 'O Lord Jesu Christ, let not the sacrament of thy body and blood which I receive (though unworthy) be to my judgement and damnation.'[63] Many may have received the eucharist casuistically—even in Bonner's *Homilies* it was suggested that the threat of damnation was not to be taken abso-

[58] BL, C 18 e 2. 96.

[59] *Grey Friars Chronicle*, pp. 35–6.

[60] Bossy, 'The Mass as a Social Institution'; M. James, 'Ritual, Drama and Social Body in the late medieval English Town', *P&P* 98 (1983), pp. 9–10. The reformers thought, on the contrary, that the Mass was divisive: Becon, 'Displaying of the Popish Mass', p. 279.

[61] R. Flenley, *Six Town Chronicles of England* (Oxford, 1911), p. 146.

[62] *LFMB*, p. 121.

[63] Foxe, *Acts and Monuments*, vi, p. 366. More, *Answer to a Poisoned Book*, pp. 73–7.

lutely literally[64]—but there were some who did have scruples about asking for divine forgiveness without giving or deserving human forgiveness; about receiving the host whilst 'out of charity'. Anne Williamson alleged as her reason for not attending her parish church throughout 1554 her belief that the good might receive the sacrament of the altar to their salvation, but the unworthy to their damnation.[65]

Catholics must pacify their social enemies as well as the wrath of the Almighty before they be pure enough to receive the sacraments. The sacrament of penance could restore concord within the Church: contrition by itself would restore the sinner to God, but only contrite repentance expressed before the priest in confession could restore him to the body of the faithful.[66] In the beginning of Lent the priest 'should exhort the people to come shortly to confession, and also at all other times when they fall to any deadly sin lest that one sin be occasion of another greater'.[67] Richard Hill described in his commonplace book the qualities of the confessor and the form of confession.[68] As the sinner knelt at the 'shriving pew', the priest gave a blessing, the sign of the cross. The confessor should comfort the penitent, showing him first that Christ died for our sins; reminding him that he was not the first in the world to sin; that Peter himself had denied Christ; recounting the sins of Mary Magdalen, and of the woman of Canaan taken in adultery. All these sinners were 'sanctificati', because they had confessed their faults. Much better to confess sins in this world than to come to universal judgement in the next and be damned eternally. After general confession the sinner was questioned in great detail concerning his failings: of the five senses, of the seven deadly sins, against the twelve articles of the faith, the seven sacraments of the Church, the seven works of corporal

---

[64] *Homelies sette fourth by the righte reuerende father in God, Edmunde, Byshop of London* (1555; *RSTC* 3285.2), fo. 70ᵛ.

[65] GLRO, DL/C/614, fo. 22ʳ. Cf. A. G. Dickens, *The Marian Reaction in the Diocese of York*, 2 vols. (York, 1957), ii, pp. 139–40.

[66] B. Poschmann, *Penance and the Anointing of the Sick* (1964); Tentler, *Sin and Confession on the Eve of the Reformation*; J. Bossy, 'The Social History of Confession in the Age of the Reformation', *TRHS* 5th series, xxv (1975), pp. 22–3; H. C. Lea, *A History of Auricular Confession and Indulgences in the Latin Church*, 3 vols. (1896).

[67] Harrington, *Comendacions of matrimony*, sig. Dv.

[68] Balliol College, MS 354, fos. 362 ff. I am grateful to the Master and Scholars of Balliol College for allowing me to consult and quote from this manuscript.

and spiritual mercy. No one who had not confessed, was not penitent, had not made satisfaction, who remained knowingly in sin, who harboured rancour against another, who was doubtful in faith, should presume to receive the eucharist, or be allowed to receive it. Owning up to sins to another, in confession, was embarrassing; it was, therefore, more penitential in character than private contrition; even so, the confessional was still secret; supposedly only the priest knew the sin and the shame. The London community understood penance in a wider sense still, believing that the forgiveness of the aggrieved party and public expiation were needed as well as the forgiveness of God. This was why Joanna Carpenter challenged Margaret Chambers at Mass in St Michael Queenhithe in 1529. As Margaret knelt at the altar, preparing to receive the host, Joanna took her by the arm, insisting: 'I pray you, let me speak a word with you, for you have need to ask my forgiveness before you receive your rights.'[69] Margaret Chambers was unworthy of divine forgiveness if she had not sought her neighbour's; she must show her penitence and be 'in charity' before she be pure enough to receive the Mass.

Never was the need to be 'in charity', with God and man, so urgent as when death approached. All Catholics prayed that 'death with his unavoidable dart' would not take them suddenly—they prayed especially to St Barbara, patron saint of gunpowder makers, for this grace—so that they might have time to repent and to confess, when the Kingdom of Heaven was most imminently at hand.[70] People were warned to reflect incessantly upon the eternal verity that 'every man in his transitory life is mortal and that the end of this universal flesh is death'.[71] Londoners were reminded of death, immanent and ineluctable, in sermons and in pictures. On the walls of the north cloister of St Paul's was 'richly painted the dance of Machabray, or dance of death', with death leading every estate, and with verses by Lydgate of death's speeches and his victims' replies.[72] The 'loathly figure of our dead bony bodies'

[69] Hale, *Precedents*, cccxl.

[70] Erasmus, *Praise of Folly*, p. 126; *Oxford Dictionary of Saints*; PRO, PCC, Prob. 11/36, fo. 33^v.

[71] Ibid., Prob. 11/27, fo. 132^r; cf. also fos. 5^r, 16^r, 28^r, 223^r, 224^r, 249^v, 250^v.

[72] Stow, *Survey*, i, pp. 109, 327-9; ii, p. 346.

should stir passers-by to remembrance of the Last Judgement, when their souls would be consigned to Hell after death, where they became subjects of the Devil, condemned perpetually to suffer the torments which he and his evil spirits had prepared for them. These torments were terrifyingly portrayed in the dooms in the London churches.[73]

The Devil was believed to be immanent, powerful, and persuasive. Since he had once been one of God's angels he knew all the mysteries of the natural world and the weaknesses of men. Christians who fell from grace could be recruited easily to his diabolical band.[74] The temptations of the Devil were vividly described: 'the image of carnal love' had 'harlot's lips as sweet as a honey comb', and 'a little from her was there death and Hell's mouth gaping'.[75] In the air all around the Devil and his demons waited to tempt the weak: 'I am not able to tell how many thousand be here amongst us', said Latimer.[76] Men expressed in their wills their fears of their 'ghostly enemy', the 'horrible fiend'.[77] John Stow had heard 'oft' as a boy the story of bell-ringers in the steeple of St Michael Cornhill on St James's eve pealing the bells, when suddenly 'a tempest of lightning and thunder did arise, an ugly shapen sight appeared'. It was the Devil. The stones around the north window were 'as if they had been so much butter, printed with a lion's claw': Stow had seen the Devil's mark himself, and put a feather into the holes in the wall.[78] St Augustine had advised that 'the mind of Christ's passion is the best defence against temptations of the fiend'.[79]

'Ever the image of the crucifix' must be had in the sight of the dying. Those who were about to die should have with them 'a special friend', who would read 'some story of saints or the seven psalms with the litany or Our Lady psalter', and cast holy water 'for avoiding of evil spirits'. The sign of the cross

---

73 Doom stones stood in the churchyards of St Leonard Foster Lane and St Christopher le Stocks, and in the window at the west end of St Mary Woolnoth was the figure of St Michael weighing souls: Walters, *London Churches*, pp. 49, 77, 341, 426; Pendrill, *Old Parish Life*, p. 280.

74 Thomas, *Religion and the Decline of Magic*, pp. 469–77.

75 *The ymage of loue*, sig. Bii.

76 Latimer, *Sermons*, p. 493; cited Thomas, *Religion and the Decline of Magic*, p. 471.

77 For example, PRO, PCC, Prob. 11/20, fo. 89ᵛ.

78 Stow, *Survey*, i, p. 196, also p. 97.

79 *Mirk's Festial*, p. 171.

armed the Christian against his 'ghostly enemies', and marked him as God's 'child of salvation'. The friend should ask the dying these questions:

Be ye glad that ye shall die in Christian belief? Let him answer: Yea. Know ye that ye have not so well lived as ye should? Yea. Have ye will to amend if that ye should live? Yea. Believe ye that Jesus Christ God, son of Heaven, was born of Blessed Mary? Yea. Believe ye also that Jesus Christ died upon the cross to buy man's soul on Good Friday? Yea. Do ye thank God therefore? Yea. Believe ye that ye may not be saved but by His passion and death? Yea. As long as the soul is in your body thank God for His death, and have a sure trust by it and His passion to be saved.[80]

At the Mass a part of the Host was reserved as a *viaticum* for the sick. Londoners who saw the sacrament 'borne by the streets to any sick person' were to kneel and do reverence to it.[81] The desperation of good Catholics to die reconciled with the Almighty and with their fellows is revealed in death-bed confessions to priests hurriedly summoned. Londoners confessed at last sins upon their consciences, debts spiritual and worldly that they wished to repay.[82] Christopher Payne's dying wish in 1546 was to make restitution to his poor neighbours; 'and divers parishioners being poor folk resorted to Payne and he asked them of charity to forgive him'.[83] As Alice Grisby lay upon her deathbed in Aldermanbury in 1538 her curate and women friends sat anxiously about her, imploring her to 'look upon the sacrament', 'to remember the passion of Christ'. They 'knocked her upon the breast', and pleaded, 'What, will ye die like a hell hound and a beast, not remembering your maker?' At the last Alice did 'knock herself upon the breast', and looked up at the sacrament, and 'so continued until the extreme pains of

---

[80] *Here begynneth a lytyll treatyse . . . called ars moriendi . . . the craft for to deye for the helthe of mannes sowle* (1491; *RSTC* 786), sigs. Ai[r]–iii[r]. In primers there was often a prayer asking Christ to put His passion and death between judgement and the sinner in the hour of death: White, *Tudor Books of Private Devotion*, pp. 60, 63; M. C. O'Connor, *The Art of Dying Well: the development of Ars Moriendi* (New York, 1942).

[81] Harrington, *Comendacions of matrimony*, sig. Dv[v]; Lyndwood, III, tit. xxvi.c.2. In St Andrew Hubbard they paid a man 'to bear the torch with the housel': Guildhall, MS 1279/1, fo. 40[v]. C. Harris, 'The Communion of the Sick', in W. K. Lowther Clarke (ed.), *Liturgy and Worship* (1932), pp. 541–615.

[82] See, for example, CLRO, Journal 15, fos. 177[v], 232[r], 287[r]; Hale, *Precedents*, cccli, ccclx, ccclxxxii; PRO, PCC, Prob. 11/20, fo. 209[v].

[83] CLRO, Journal 15, fos. 177[v]–8[r].

death'.[84] So she died a good Catholic. The relief of her friends and neighbours about her way of dying (*ars moriendi*) says much about the collective anxiety of the community for the Christian life of its members.

Private Faith and Public Worship

Communal religious observance marked the autonomy of the City of London, as of every city in the early sixteenth century, and faith might bind the citizens as nothing else could. Londoners worshipped together on Sundays and festivals, and even daily. They walked in parish processions on holy days, and processed communally through the City whenever there was a special reason for thanksgiving to the Almighty or need for propitiation.[85] Most of the hundred parish churches of London had been founded in the twelfth century when the laity wished to worship in small congregations, close to their priest.[86] Even on the eve of the Reformation, as the City expanded, the wish remained. Parishioners not only worshipped together, but fêted and celebrated together, and there were parish entertainments. Edifying their parish churches was a communal enterprise. In the 1520s St Andrew Undershaft was 'new builded' by the parishioners, 'every man putting to his helping hand, some with their purses, others with their bodies'. At All Hallows London Wall and St Katherine Creechurch stage plays were performed in 1528 and 1530 to raise money for rebuilding, and every year at St Andrew Holborn the parishioners held shooting matches and church ales to collect money for their church.[87] At St Margaret Pattens there was a bowl, inscribed on the outside, 'Of God's hand blessed be he that taketh this cup and drinketh to me', and on the inside, 'God that sitteth in Trinity,

---

[84] Guildhall, MS 9065/A/1, fos. 19ᵛ–21ᵛ; see also Wriothesley, *Chronicle*, i, p. 76; Cavendish, *Life*, pp. 175, 181.

[85] For what follows see Brigden, 'Religion and Social Obligation in early sixteenth-century London', *P&P* 103 (1984), pp. 67–112.

[86] Balliol College, MS 354, fos. 202ʳ–203ʳ (lists 101 parishes); BL, Harleian MS 2252, fos. 10ʳ⁻ᵛ (97 within the City walls); Egerton MS 1995, fos. 82ʳ–84ʳ (98 within the walls): C. N. L. Brooke and G. Keir, *London 800–1200: The Shaping of a City* (1974), pp. 122–47.

[87] Stow, *Survey*, i, p. 145; *Churchwardens' accounts of Allhallows, London Wall*, pp. 56, 57, 59–60; CLRO, Letter Book O, fo. 164ʳ; *St Mary at Hill*, i, pp. 157, 291; Edward Griffith, *Cases of supposed exemption from Poor Rates . . . with a preliminary sketch of the ancient history of the parish of St Andrew Holborn* (1831), pp. ii, v, vi, viii–ix, xi.

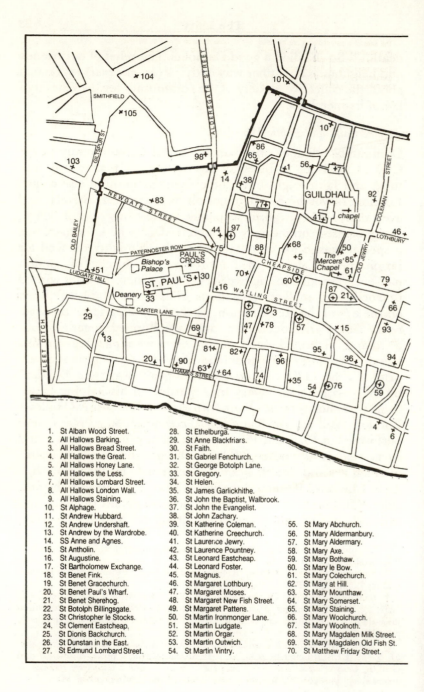

The sites of parish churches are indicated by a square cross +, the numbers appended referring to the accompanying list.
Those belonging to the Deanery of Bow (peculiars of the Archbishop of Canterbury) are marked by a circle round the cross ⊕

Outside the area shown in the map, but within the bars, were the parish churches of St Andrew Holborn, St Dunstan in the West, and St Bride.

Moor Fields

LONDON WALL

The Dutch Church

The French Church

CORNHILL

LOMBARD ST

GRACECHURCH STREET

BISHOPSGATE STREET

LEADEN HALL STREET

ST MARY AXE

HOUNDSDITCH

FENCHURCH STREET

MARK LANE

MINORIES

TOWER STREET

THAMES STREET

LONDON BRIDGE

TOWER HILL

THE TOWER

71. St Michael Bassishaw.	83. Christchurch Newgate.	95. St Thomas Apostle.
72. St Michael Cornhill.	84. St Olave Hart Street.	96. Holy Trinity the Less.
73. St Michael Crooked Lane.	85. St Olave Jewry.	97. St Vedast Foster Lane.
74. St Michael Queenhithe.	86. St Olave Silver Street.	98. St Botolph Aldersgate.
75. St Michael le Querne.	87. St Pancras Soper Lane.	99. St Botolph Aldgate.
76. St Michael Paternoster Royal.	88. St Peter Westcheap.	100. St Botolph Bishopsgate.
77. St Michael Wood Street.	89. St Peter Cornhill.	101. St Giles Cripplegate.
78. St Mildred Bread Street.	90. St Peter Paul's Wharf.	102. St Peter in the Tower.
79. St Mildred Poultry.	91. St Peter le Poor.	103. St Sepulchre.
80. St Nicholas Acon.	92. St Stephen Coleman Street.	104. St Bartholomew the Great.
81. St Nicholas Cole Abbey.	93. St Stephen Walbrook.	105. St Bartholomew the Less.
82. St Nicholas Olave.	94. St Swithin.	106. Holy Trinity Minories.

'The City Churches at the Time of the Reformation'

send us peace and unity'. This bowl was surely for festive, not sacramental, purposes.[88] The neighbourhood loyalties growing up around the parish churches may have done much to sustain friendships between families which otherwise moved in different social spheres.[89]

At special masses and at annual festivals the citizens celebrated their communal good fortune. Even the great Midsummer pageants of London had a spiritual as well as a secular function, as the Grocers' Company recognized by wearing coats embossed with crosses.[90] These festivals were celebrations also of a communal spirit, of reconciliation, a spirit which was to be appreciated the more at the losing of it. John Stow recalled the great summer festivals of his youth in the 1530s. There were, he remembered, bonfires in the streets; the 'wealthier sort' set out tables before their houses, replenished with 'sweet bread and good drink', and invited their neighbours and passers-by to sit, and 'be merry with them in great familiarity, praising God for his benefits bestowed upon them'. These were called '*bon fires*' because of the 'good amity amongst neighbours' they engendered, and because of the way they 'made of bitter enemies, loving friends'.[91] But in the 1590s Stow was remembering this past social unity with the nostalgia of one who thought it forever lost.

Londoners promised constantly that they would in all their dealings with their fellow citizens behave in accordance with Christian virtues. 'Forasmuch as amongst all things most pleasant to Our Lord God in this transitory world, after due love had unto him, is the love, amity, and good accord to be had amongst Christian people', the Leathersellers, incorporating with the Glovers Pursers in 1500, promised to be 'knit together in very true amity'.[92] A clear association was made between breaching the laws of the Church and offending against the moral code of the City. Those who broke the City's

[88] Pendrill, *Old Parish Life*, p. 161.

[89] S. L. Thrupp, *The Merchant Class of Medieval London, 1300–1500* (Ann Arbor, 1948), pp. 38 ff.

[90] Guildhall, MS 11571/5, fo. 21[r]. For the midsummer pageant, see *CSP Ven.* iii. 244.

[91] Stow, *Survey*, i, p. 101.

[92] CLRO, Journal 10, fo. 258[v].

rules were accused of 'not dreading God nor shame of the world', of acting to the 'displeasure of God', and public penance was ordered by the City, as by the Church.[93] All the rites of passage in City life—apprenticeship, freedom, holding office—were sanctioned by the taking of oaths or by religious services. Swearing oaths signified both assent to the duty, and the belief that divine favour had been bestowed upon the office holder.[94] Excerpts from the holy evangelists were transcribed in the oath books of the City and its companies, and upon these books solemn vows were made.[95] Of assessors to the subsidy in 1497 it was required, for example, that they should impartially assess every citizen, 'sparing no man for favour, nor grieving no man for hate, and this ye shall do as God you help, and all saints, and by this book'.[96] That in 1516 the Mayor and Aldermen chose to make up the full value of the subsidy themselves rather than, as Wolsey commanded them, swear an oath that every citizen had truly assessed himself, reveals both the force of oaths and the sense of religious and social obligation implicit in civic service.[97] For breach of faith was a mortal sin, as well as subversion of legal and secular arrangements. This was why Church Courts as well as secular courts could insist upon the performance of promises, and why it was that the Commissary Court punished John Pinchbeke in 1476 for failing to keep his oath to the Bakers' Company.[98] Violation of oaths had terrifying consequences, for 'Dame Perjury' led her followers to Hell. So when Robert Church, a known perjurer, died before he could perform penance and be absolved, this note was

[93] Ibid., Letter Book N, fo. 233^r; Journal 15, fo. 30^r; Repertory 10, fo. 285^r; Repertory 12, fo. 454^r; Stow, *Survey*, i, pp. 188–9; Thrupp, *Merchant Class*, p. 18.

[94] See, for example, the City oaths recorded in the sixteenth-century oath book in CLRO, and Journal 10, fos. 83^r, 132^r, 155^r, 200^v, 269^v, 270^v; BL, Lansdowne MS 762, fos. 37^v, 40^r, 41^r, 42^r. C. Phythian-Adams has shown the importance of oaths for Coventry civic life: 'Ceremony and the citizen: the communal year at Coventry', in P. Clark and P. Slack (eds.), *Crisis and Order in English Towns, 1500–1700: Essays in Urban History* (1972), pp. 59–62; *Desolation of a City: Coventry and the Urban Crisis of the Late Middle Ages* (Cambridge, 1979), p. 138.

[95] Bodleian, MS Gough 10; Guildhall, MSS 5197, 2890, 6350, 7114, 5070, 5385A, 4645.

[96] CLRO, Journal 10, fo. 83^r.

[97] Ibid., Repertory 3, fos. 110^r, 116^{r-v}.

[98] Lyndwood, v, tit. 15 c.1; B. L. Woodcock, *Medieval Ecclesiastical Courts in the Diocese of Canterbury* (Oxford, 1952), pp. 82–92; Hale, *Precedents*, lxviii.

added in the Commissary Court book: 'Deus Rex celestis, miserere anime sue'.[99]

To appear religious was in early Tudor London conventional, at the very least, and piety was to be vaunted rather than hidden. Letters were very often headed 'Jesu' or 'IHS' or with the sign of the cross, and correspondents usually ended by commending their friends or patrons to Jesus, 'who long continue your ladyship in honour', or 'to have you in His blessed keeping'.[100] When George Monoux, draper and Alderman, began a new ledger in 1507 he dedicated it to Father, Son, and Holy Ghost, beseeching 'mercy and grace and good conveyance in all my works', and he signed it with a cross. Such a dedication was conventional: divine favour was required for business ventures.[101] People were exhorted, 'see that you do nothing only of custom', and urged to follow God rather than man, conscious of the Day of Judgement when 'no man of law may speak for us, nor any excuse may serve us'.[102] But such injunctions are hard to follow. On Passion Eve 1528 Stephen Vaughan wrote to Cromwell of his search through London to recover a debt from Alderman Munday. At last Vaughan found him, as befitted the day, at evensong at St Faith's, but Munday was 'otherwise disposed to serve God', and declined to discuss money. So Vaughan was 'bold to answer him' that there was no better way to serve God than by 'restoring the right to his brother whom he had wrongfully defrauded'.[103] True piety lay in righteous conduct rather than ritual observance. Hypocrisy was a vice which was particularly condemned as the Reformation proceeded. So deeply was faith a part of the mental world of Tudor Londoners that their very doodles might be religious.[104]

Real religious beliefs, personal convictions about the soul's journey to God, are likely to remain secret and inward. Yet

[99] Thomas Becon, 'An Invective against Swearing', in *The Early Works of Thomas Becon*, ed. J. Ayre (Parker Society, Cambridge, 1843), p. 369; Hale, *Precedents*, cclxxii.

[100] See the variety of commendations expressed throughout the *Lisle Letters*.

[101] BL, Additional MS 18783, fo. 2ʳ. (The dedication to the Virgin was later excised.) Thrupp, *Merchant Class*, pp. 174–80.

[102] *A dyurnall: for deuoute soules: to ordre themselfe therafter* (1532?; *RSTC* 6928), sig. Bi; Richard Whitford, *A Werke for Housholders, or for them that haue the gydynge or gouernaunce of ony company* (1530; *RSTC* 25422), sig. Aii.

[103] PRO, SP 1/47, fo. 149ʳ (*L&P* iv/2. 4107).

[104] See, *inter alia*, CLRO, Journal 10 (endpapers); Clothworkers' Company, Court Orders, 1 (endpapers).

Tudor Londoners did express at the last their private hopes for salvation and their fears of damnation in a public document: the will. Wills were testaments of individual faith as well as mundane instructions for the disposal of goods after life. There are doubts about the writing of wills; suspicions that *in extremis* people vaunted a piety they had never manifested during life, and that they did not even write these pious affirmations themselves, but had thoughts expressed for them by professionals, priests and scribes.[105] People may have wished to be guided; however, as death approached there were compelling reasons also to tell the truth. Londoners usually did not make their wills until there was little time remaining for repentance and atonement. Many spoke of mortal mutability and frailty, and wrote of man's mortal plight; 'wandering and labouring in this wretched world whose end is death temporal the which is very certain and nothing more uncertain than is the hour of death'.[106] Tudor Londoners were not complacent in their final settling of accounts, acknowledging that fortune in this life was not of their own deserving; that 'all men be lent and not given this world, be stewards here and not owners'; that wealth upon earth might not promise treasure in Heaven.[107] For those 'goods that my Lord God of his bountiful goodness hath lent me in my life time', a Londoner should make restitution, for the wealth of his soul. According to the ancient custom of the City, after debts were duly repaid and funeral expenses discharged, the testator's estate was divided into three equal parts: the first and second to be assigned to the widow and children, and the third to be spent in works of 'pity and charity', for the beneficiary to pray for the testator's soul and all Christian souls.[108]

The first bequest was always spiritual, that of the soul to God, from Whom it came and to Whom, it was hoped, it would return. Those who were about to die considered their belief in

[105] Professor Dickens pioneered the study of wills to examine faith: A. G. Dickens, *Lollards and Protestants in the Diocese of York, 1509–1558* (Oxford, 1959), pp. 171–2, 220–1. M. L. Zell, 'The use of religious preambles as a measure of religious belief in the sixteenth century', *BIHR* l (1977), pp. 246–9.

[106] PRO, PCC, Prob. 11/20, fo. 39ᵛ.

[107] For example, ibid., and Guildhall, MS 9171/10, fo. 130ʳ.

[108] The custom is described by J. A. F. Thomson in 'Piety and Charity in late medieval London', *JEH* xvi (1965), pp. 181–2.

the way of salvation they hoped presently to find. Almost all bequeathed their souls to Almighty God, to His son, Jesus Christ, Saviour and Redeemer, to the Blessed Virgin Mary and the saints in Heaven.[109] Some expressed that dedication in more complex ways than others. In 1522 Ralph Shelton, an apprentice, commended his soul

into the hands of the merciful and indivisible Trinity, the Father, the Son and the Holy Ghost, three persons and one very God, and to the protection of the immaculate and most beautiful Virgin Mary, heartily beseeching her to be my protectrix from the horrible fiend at the separating of my soul from my body, and desiring heartily that it may please her to be as a mediatrix for my soul to the high majesty of God for mercy and grace, and also I commend it to the custody and petitions of all the holy company of Heaven that they will vouchsafe to pray for remission of mine offences that I may by the mercy and will of God and by the mediation of Our Lady and by their inter-cessions be partaker of everlasting bliss.[110]

Three-quarters of those Londoners whose wills were proved in the Commissary and Consistory Courts of the City on the eve of the Reformation expressed their belief in the Catholic doctrine of mediation when they bequeathed their souls to the glorious company of Heaven. The rest committed their souls to God or Christ alone, as their 'maker, creator and redeemer', with no dedication to the Virgin or the saints, but since they often added to that formula an '&c' the intercession of saints may have been taken for granted.[111] Those images of saints which they had worshipped during life the citizens often re-membered at their deaths, leaving money to adorn them. Their own houses, too, were often hung with tapestries of religious scenes, and images of favourite saints looked down from the walls protectively. Joan Rice of St Sepulchre left to a friend her hanging of the passion, and Margaret Finch of St Dunstan in the West left a picture of St James, and in John Porth's house

---

[109] The following discussion is based upon wills proved in the early sixteenth century in the Prerogative Court of Canterbury (PRO, PCC, Prob. 11/20, 24, 27, 28), the Commissary Court (Guildhall, MSS 9171/10 & 11), and the Consistory Court (*LCCW*). See also *The Fifty Earliest English Wills*, ed. F. J. Furnivall (EETS, original series, 78, 1882).

[110] PRO, PCC, Prob. 11/20, fo. 89ᵛ.

[111] Guildhall, MS 9171/10; *LCCW*.

at St Mary at Hill was an 'old pageant of Jesus' and images of Our Lady.[112] Thomas Cromwell (or perhaps his mother-in-law) had in his house at Austin Friars images of the Virgin and saints in almost every room, and took with him to his new house at St Peter le Poor in 1534 an image of the Child Jesus, and pictures of Our Lady of Pity and the Passion of Our Lord.[113] Thomas Wells of St Martin Orgar left a silver crucifix, a little badge of Our Lady, and a brooch of the Salutation of Our Lady.[114] Much of Tudor jewellery was decorated with images of the saints: Dame Maud Parr left her daughter a chain with an image of St Gregory, and others left rings decorated with representations of the five wounds of Christ.[115] The saints in Heaven were trusted to protect their votaries in this vale of tears, and their images were carried as talismans; still more urgently needed was their intercession for the souls of their supplicants in the hereafter.

Countless souls awaited final redemption in Purgatory; languishing, 'sleepless, restless, burning and broiling in the dark fire one long night of many days, of many weeks . . . of many years . . .'.[116] Their release might be speeded by the prayers of those in Heaven and on earth. Only those without sin could avoid penance in Purgatory, and all good Catholics tried to reduce the days and years of durance there. In their lives they purchased pardons and indulgences to procure release.[117] Printed indulgences were sold, assuring, for example, that 'whosoever devoutly beholdeth these arms of Christ's passion hath 6,755 years of pardon'.[118] Cardinal Wolsey promised 'one hundred days of pardon releasing of penance in Purgatory' to anyone who gave towards the re-building

---

[112] Guildhall, MS 9171/11, fo. 99ᵛ; MS 9171/10, fo. 34ᵛ; *St Mary at Hill*, i, pp. 37, 38, 39, 45.

[113] *L&P* iv/2. 3197; xv. 1029(6).

[114] Guildhall, MS 9171/10, fo. 82ʳ.

[115] PRO, PCC, Prob. 11/24, fo. 86ʳ; 11/20, fo. 69ᵛ.

[116] Thomas More, *The Supplicacyon of soulys* (1529; *RSTC* 18092); cited in A. G. Dickens, *The English Reformation* (1964), pp. 5–6.

[117] The printed indulgences are catalogued in *RSTC* 14077c.5–14077c.148. For the doctrine of Purgatory, see H. Rondet, *Le Purgatoire* (1948); E. H. Plumtre, *The Spirits in Prison and other studies on the Life after Death* (1884); J. le Goff, *The Birth of Purgatory* (1984).

[118] Bodleian, MS Rawl. D 403 (c.1500; *RSTC* 14077c.22).

of Rickmansworth church.[119] In 1532 William Kirkby willed his letters of pardon and indulgence to be redelivered to the religious houses which had granted them, where he was lay brother: to the Charterhouse of Sheen, and the nunneries of St Helen's and of the Minories.[120] Expiation for sins could be sought from beyond the grave. John Banester willed a priest to sing for three years 'for all the trespasses and offences that I have done to any man'.[121] The Catholic community was not only among the living. Those on earth still owed a duty to their fellow Catholics departed: to remember them in their prayers. This is why the dying conventionally ordered plaques on their graves for themselves and their families, asking to 'be had in memory in the prayers of good people passing by', or that 'of your charity you shall pray for the souls of . . .'.[122] To their friends Londoners would leave rings, inscribed with messages urging remembrance: 'miseremini mei saltem vos amici mei' ('Have mercy upon me, you, at least, my friends').[123] In September 1521 John Garrard added a postscript to his will, addressed to his wife: 'Margaret, I pray you if I die go to Our Lady of Walsingham . . . and to Our Lady of Willesdon to pray.'[124] Those who were left behind must work for the salvation of dead friends and family, as they would that others should do for them, when they were 'in like case'.

In every City church every day there were masses sung for the souls of the dead. The first time that masses were sung was, of course, at the funeral. Arrangements for funerals on the eve of the Reformation were usually as elaborate as the testator or his friends could afford, for the repose of the departed soul might be hastened by this last rite of passage. As the body went to its final resting place dirige, placebo, and requiem were sung by attendant priests, monks, and friars, knells were rung, and

---

[119] BL, C 18 e 2. 96.

[120] PRO, PCC, Prob. 11/24, fo. 104ᵛ. Perhaps one of these indulgences, issued by Sheen, survives: Bodleian, Arch. G. fo. 14ʳ (*RSTC* 14077c.15).

[121] PRO, PCC, Prob. 11/20, fo. 48ʳ.

[122] Bossy, 'The Mass as a Social Institution', pp. 37, 42–3; N. Z. Davis, 'Ghosts, Kin and Progeny: Some features of Family Life in Early Modern France', *Daedalus*, cvi (1977), pp. 87–114. PRO, PCC, Prob. 11/20, fos. 89ᵛ, 107ᵛ, 110ʳ, 145ᵛ, 227ᵛ; Prob. 11/28, fos. 9ʳ, 271ᵛ–2ʳ; Guildhall, MS 9171/10, fos. 158ʳ, 168ʳ, 193ᵛ, 269ʳ, 305ʳ; 9171/11, fo. 93ʳ.

[123] PRO, PCC, Prob. 11/28, fo. 90ʳ; 11/20, fo. 182ᵛ.

[124] PRO, PCC, Prob. 10/2.

the churches were lit by candles and tapers. Special rituals might have special efficacy: Maud Gowsell willed in 1524 that at her funeral there should be 'fifteen tapers of wax borne by fifteen children', and that 'five priests say five masses of the five wounds of Our Lord in the honour of His glorious passion that He suffered for the redemption of me and all mankind'; John Jones left 15*d.* for 'fifteen poor men and women in the honour of the fifteen pains Our Lady suffered for all mankind and to the intent that they shall pray for my soul and all Christian souls', and seven 'King Henry pence' for 'seven children in honour of the seven joys of Our Lady' to pray; thirteen men were to hold torches at Gerard Danet's funeral in 1520 'in honour of Our Lord Jesus Christ and twelve apostles'.[125] Some citizens already renounced such elaborate ceremonial—William Porter desired in 1522 that no more than £10 be spent on his burial, which should be accomplished 'without any pompous manner or much ringing of bells, not having above the number of eight torches and without calling of any common beggars to my burial and without any great tapers'.[126] Porter's puritanism was doubtless prompted by the ostentation of some of his fellow citizens' funerals, which were used to vaunt their wealth and standing as much as to celebrate their piety. Detailed accounts remain of City funerals which evinced 'pomp and pride of the world'.[127] John Hosier, citizen and mercer, and Merchant of the Staple at Calais, left £100 for his funeral expenses: for sixteen torches to burn at his hearse, and thereafter to be lit at 'times of the levation of the blessed sacrament' in seven City churches, for 'placebo and de profundis and requiem' to be sung 'by note' by the five orders of friars in London; and for £20 to be distributed among the poor who came to his burial, 'every such poor body to have thereof a penny a piece'. Hosier may well have been boasting his status in City society, but the prayers of 4,800 poor were believed to confer a special blessing, and to win the soul remission in Purgatory.[128]

---

[125] Guildhall, MS 9171/10, fos. 47ʳ, 75ʳ, 97ʳ, 116ʳ, 224ʳ, 321ʳ; PRO, PCC, Prob. 11/20, fos. 7ᵛ, 48ʳ.

[126] Ibid. 11/20, fo. 177ᵛ.

[127] Thrupp, *Merchant Class*, pp. 153–4, 250.

[128] PRO, PCC, Prob. 11/20, fo. 45ᵛ; cf. 11/40, fo. 83ʳ; £10 in 'two penny dole'. B. Tierney, *Medieval Poor Law: a sketch of Canonical Theory and its Application in England* (Berkeley, 1959).

Late medieval society expended a considerable part of its resources for the spiritual welfare of its dead members. The masses continued long after death; for the wealthy, sometimes perpetually.[129] In March 1521 Dame Elizabeth Thurston willed money 'for keeping daily forevermore Mass of the name of Jesu on the Fridays and Mass of Our Lady on the Saturday, and every day of the week two Salves . . . one before the Rood and another before Our Lady' at the churches of St Vedast and St Magnus.[130] Almost every other testator asked for prayers, and many would endow as many masses as they could afford, for their own soul, for their father's and mother's souls, for 'all those souls that I have fared better for',[131] for all Christian souls. In half the wills proved in the Prerogative Court of Canterbury between 1520 and 1522 there were bequests for chantry priests to sing for souls, for one year or more; in wills proved between 1529 and 1531 a fifth endowed chantries for an average period of three years. Of usually poorer Londoners, whose wills were proved in the Commissary Court, between 1522 and 1530 9 per cent bequeathed money for a chantry priest to sing for them.[132] A month after the funeral thirty memorial masses were celebrated—a trental—and on the anniversary of the testator's death masses were sung in remembrance—obits. In September 1527 Thomas Gibson, a mercer of St Olave Hart Street, willed that a trental of masses be sung for his soul: 'of the five wounds of Our Lord . . . other five of the joys of Our Lady and other five to the Holy Ghost, and other five of the resurrection of Our Lord, other five of the eleven virgins and other five of requiem'.[133]

Catholics were reminded constantly to remember the spiritual plight of the departed. Every church kept a bead-roll, a list of benefactors for whose souls the people ought to pray, with the dates of their anniversaries. At the 'bidding of beads' at Mass every Sunday the priest would read out the bead-roll from the pulpit.[134] In 1521 John Lapham of St Stephen Walbrook left

---

[129] K. L. Wood-Legh, *Perpetual Chantries in Britain* (Cambridge, 1965); A. Kreider, *English Chantries: The Road to Dissolution* (Cambridge, Mass., 1979).

[130] PRO, PCC, Prob. 11/20, fos. 183^v–185^r.

[131] Ibid., fo. 127^r.

[132] Guildhall, MS 9171/10.

[133] Ibid., fo. 97^v.

[134] Pendrill, *Old Parish Life*, pp. 150–1.

his parish priest 3*s*. 4*d*. and his spectacles to be remembered in the bead-roll of his church, and in 1517 Hugh Fenne left 20*s*. to be remembered every week for sixty years at All Hallows the Less.[135] The chantry priest who sang for John Wilkinson at St Mary Abchurch was enjoined to turn 'from time to time . . . towards the people saying in this wise; of your charity you shall pray for the soul of John Wilkinson, late Alderman of this City'.[136] William Roye left a bequest for a priest to come to his tomb daily after Mass, and sing the psalm *De profundis*, and 'cast holy water upon my grave and upon the people there standing about'.[137]

To some of the departed the living owed special devotions: to their families first of all. But other organizations, other brotherhoods in the City had created artificial kinships, and owed obligations to their dead brethren: the confraternities and the seventy-five trade gilds of London. Spiritual brotherhood had been the first reason for the existence of the trade gilds, and in the sixteenth century the first reason still mattered, or at least it was piously invoked. The City companies kept their titles as religious fraternities: 'the gild . . . of the Skinners of London, to the honour of God, and the precious body of Our Lord Jesus Christ', 'the Merchant-Taylors of the fraternity of St John the Baptist', the 'fraternity of the Blessed Mary the Virgin, of the mystery of the Drapers'. Gild members worshipped together on the day of their patronal feast, and maintained lights in the City churches.[138] Citizens made special bequests for worship within their companies to continue.[139] Company members attended the funerals and obits of their fellows. On the eve of the Reformation the Goldsmiths still attended twenty-five obits a year, the Merchant Taylors twenty-seven, the Mercers at least sixteen, the Grocers twelve. Feasts, 'drinkings', and 'recreations' were conventionally held after the funerals. To attend funerals and anniversary masses

---

[135] PRO, PCC, Prob. 11/20, fos. 145[v], 227[v].

[136] Ibid., fo. 109[v].

[137] Ibid., fo. 30[r].

[138] W. Herbert, *The History of the Twelve Great Livery Companies of London*, 2 vols. (1836). Expenditure upon obits, funerals, lights, priests' wages, church ornaments is evident in all the accounts of the City companies.

[139] See, for example, amongst many other such bequests, some made between 1517 and 1521: PRO, PCC, Prob. 11/20, fos. 107[v], 141[r], 141[v], 221[r], 227[v], 231[r].

may have been the action of a friend, or a mark of respect from a colleague; it was also one of the first duties of the gildsmen of Tudor London.[140] Yet although the religious life of the trade gilds was still important, it was not now the first reason for their organization as it had been at their foundation. Indeed, their economic function may often have been incompatible with their spiritual purpose.

Londoners who sought to bind particular members of their community to care for their spiritual welfare joined religious gilds. The purpose of the lay confraternities was to support living brethren through friendship and charity, and their dead brethren through their prayers. The religious gilds, established within the City parishes but often transcending parish loyalties, were of central importance in the devotional lives of the citizenry.[141] In fifteenth-century London there had been perhaps as many as 176 such gilds; and on the eve of the Reformation Londoners still remembered over eighty gilds in half of the parishes of the City in their wills.[142] Pardons and indulgences granted to London confraternities in the early sixteenth century still survive.[143] New gilds were being founded on the eve of the

---

[140] Herbert, *History of the Twelve Great Livery Companies*, ii, p. 33; Guildhall, MSS 11571/5, fos. 129^v, 152^v; 298/2.4; 12071/1. fos. 107^v, 132^r. G. Unwin, *The Gilds and Companies of London* (4th edn., 1966), ch. xiii.

[141] Brigden, 'Religion and Social Obligation', pp. 94–102; H. F. Westlake, *The Parish Gilds of Medieval England* (1919); *English Gilds*, ed. L. Toulmin Smith (EETS, original series, xl, 1870); J. A. F. Thomson, 'Clergy and Laity in London, 1379–1529' (University of Oxford, D.Phil. thesis, 1960), pp. 82–125; Thrupp, *Merchant Class*, pp. 34–8; *Parish Fraternity Register: Fraternity of the Holy Trinity and SS Fabian and Sebastian in the Parish of St Botolph without Aldersgate*, ed. P. Basing (London Record Society, 1982); A. G. Rosser, 'The Essence of Medieval Urban Communities: the Vill of Westminster, 1200–1540', *TRHS* 5th series, 34 (1984); C. M. Barron, 'The Parish Fraternities of Medieval London', in C. M. Barron and C. Harper-Bill (eds.), *The Church in Pre-Reformation Society* (Woodbridge, 1985), pp. 13–37; J. J. Scarisbrick, *The Reformation and the English People* (Oxford, 1984), ch. 2. Cf. J. Bossy, 'The Counter Reformation and the People of Catholic Europe', *P&P* 47 (1970), pp. 51–70; J. Toussaert, *Le Sentiment religieux en Flandre à la fin du Moyen Age* (Paris, 1963), pp. 478–93; A. N. Galpern, *The Religions of the People in Sixteenth-Century Champagne* (Cambridge, Mass., 1976).

[142] Thomson, 'Clergy and Laity', p. 86; C. L. Kingsford, *Prejudice and Promise in Fifteenth-Century England* (Oxford, 1925), p. 141; Barron, 'The Parish Fraternities of Medieval London', p. 13. The sixteenth-century gilds of the City are discovered in churchwardens' accounts and in the bequests which citizens made to them in their wills.

[143] BL, C 18 e 2, nos. 33, 91, 122; C 37 h 8; C 54 d 4, no. 2; 1A 55480; Bodleian, Arch. A b 8, nos. 3, 16, 35 a & b, Vet. A 1 b 12; Trinity College, Oxford, I.2.13.

Reformation: two were established in honour of the Visitation of the Blessed Virgin Mary at the Crossed Friars and the Austin Friars, another for her conception at the Black Friars, one for St Barbara at the Friars Preachers, and a gild of St Christopher of the Waterbearers at the Austin Friars and a new fraternity of St Katherine.[144] Brothers and sisters in the confraternities maintained lights at the gild altars in the churches and friaries of London, celebrated Mass on special festivals and saints' days, and held feasts. Dying brothers had the solace of knowing that their funerals would be attended by their friends and brethren in the gild. William Roye left money to the brothers and sisters in his gild of St James in the Friars Preachers 'for a recreation to pray devoutly for his soul'.[145] At the gilds of the Visitation of the Blessed Virgin Mary at the Austin Friars and the Crossed Friars 'when a brother or sister . . . is departed out of this world then every brother is bound to come at the Mass of requiem and to offer a penny, and shall tarry and abide there until the body be put into the grave'.[146] If brothers and sisters happened to die outside London they still had a requiem sung for them.[147] At the gild of St Barbara at the Friars Preachers four poor men would be found to attend the departed.[148] Very nearly a quarter of the wills proved in the London Commissary Court between 1522 and 1539 contained bequests for confraternities, and a quarter of Londoners whose wills were proved in the Prerogative Court of Canterbury between 1520 and 1522 remembered their gilds.[149] As long as the prayers of the living could avail the souls of dead brothers awaiting release from Purgatory, then fraternity extended to the hereafter. This was why William More left money to 'be remembered as a dead brother in their devout prayers' at Our Lady and St Thomas confraternity in St Magnus, and why Christiane Smith willed to be made 'a dead sister' of Our Lady and St Stephen at St Sepulchre.[150]

[144] Guildhall, MS 9531/9, fos. 9ᵛ–10ʳ, 27ʳ–29ᵛ, 142ʳ–43ᵛ; MS 9168/8, fos. 98ʳ–99ʳ, 138ʳ–139ʳ; 'The Ordinances of Some Secular Gilds of London', ed. H. C. Coote, *TLMAS* iv (1871), pp. 1–59; *L&P* v. 766(7); CLRO, Repertory 5, fo. 66ᵛ.

[145] PRO, PCC, Prob. 11/20, fo. 30ʳ.

[146] Guildhall, MS 9531/9, fos. 10ʳ, 28ʳ, 143ʳ.

[147] Ibid., fo. 29ʳ.   [148] Ibid., fo. 28ᵛ.

[149] Guildhall, MS 9171/10; PRO, PCC, Prob. 11/20.

[150] Guildhall, MS 9171/10, fo. 311ᵛ; PRO, PCC, Prob. 11/20, fos. 171ʳ, 211ʳ.

Membership of most City fraternities was open to anyone 'known for a good true body or person . . . received by the more party of the brotherhood', and who could pay the entrance fee and quarterages. If a dispute arose about the admission of a new brother the master of the gild must make 'good peace, for . . . the old shall not be forsaken for the new without great lawful cause'.[151] Sisters in the gilds, unlike women in any other association within the City, had more or less the same status as brothers.[152] This openness of entry itself points to the charitable and religious aims of the gilds, for everywhere else in London strict hierarchies were observed. In the parish churches they set 'rich and poor . . . in the pews it belongeth every man to have': there were pews for men, pews for married women and for maidens, pews for the poor, and pews for individual families, and still people squabbled for place.[153] In the City processions there was often contention 'by reason of fond courtesy and challenging of places'.[154] The fraternities may have transcended the social boundaries which usually circumscribed the lives of Londoners: when entrance fines of only 20*d*. were demanded, and quarterages were only 4*d*. or 8*d*. the fraternities might have been open to almost all comers, for such charges were beyond the pockets only of the labouring poor of the City.[155] Grander citizens might choose grander gilds. John Thurston, Sheriff in 1516, willed that his executors 'cause knowledge to be given to all such gilds and places that I am brother of, and especially the gilds of St Audrey of Lynn, Our Lady of Boston, Houndslow, Burton Lazar, Our Lady in the Sea and Jesus in Paul's', that he might have their 'prayers and suffrages . . . as a brother ought to have'.[156] Some London gilds counted Henry VIII himself as a member.[157] Grandest of all the City gilds was that of the Name of Jesus in the Shrouds of St Paul's founded in the

[151] Guildhall, MS 9531/9, fos. 10ʳ, 143ʳ.

[152] For sisters who remembered their brotherhoods in their wills, see Guildhall, MS 9171/10, fos. 3ʳ, 24ʳ, 32ʳ, 35ʳ, 48ᵛ, 53ᵛ, 58ᵛ, 62ʳ, 74ʳ, 75ᵛ, *passim*.

[153] BL, Harleian MS 2252, fo. 22ʳ; *St Mary at Hill*, i, pp. 215, 225, 130, 198, 251, 252, 255, 323, 328; Guildhall, MS 1279/1, fos. 22ᵛ, 101ᵛ, 19ᵛ, 80ᵛ; Walters, *London Churches*, p. 581; GLRO, DL/C/208, fos. 6ᵛ–7ʳ.

[154] *Inivnccions geven by the Moste Excellent Prince Edward the Sixte* (1547; *RSTC* 10088), sig. Ciʳ.

[155] Guildhall, MS 9531/9, fos. 10ʳ, 143ʳ.

[156] PRO, PCC, Prob. 11/20, fo. 181ᵛ.

[157] *L&P* v, pp. 308, 318, 324, 325.

reign of Henry VI, which had among its members some of
London's richest citizens.[158] To be a member there signified
success in City life. While other fraternities in the City sheltered
their brothers fallen into poverty, or stricken by life's vicissi-
tudes, this was an association of another sort: membership here
was part of London's *cursus honorum*, but the brothers also came
to be marked by their attachment to the old forms of religion as
the Reformation proceeded.[159] This was a gild which collected
oblations from all over England; it had an income of £385. 3s. 4d.
in 1533, and was wealthy enough to have daily masses sung in
its chapel, to have 'masses of Jesu' sung every Friday by the
subdean, peticanons, six vicars choral, ten choristers, and a
chantry priest, to pay every Sunday the preachers at Paul's
Cross and St Mary Spital, 'that in their beads shall remember'
the brothers and sisters of the fraternity, to pay for musical
waits with banners to process on the vigil of the feast of the
Name of Jesus, and to hire cloth of Arras on 'Jesus day'.[160]
The gild maintained four almsmen: not their own members
fallen on hard times, but poor men drawn from the community
to attend at the celebration of the Mass.[161] In the minds of
Catholic Londoners there was an ancient conviction that the
poor were somehow blessed.

God's answer to a 'certain creature that desired to wit what
thing was most pleasure to him in this world' was thus—so an
anonymous Londoner recorded in his commonplace book—
'give thy alms unto poor folk whilst thou livest for that pleasureth me more than thou cravest a great hill of gold after thy
death'.[162] Almsgiving might aid the soul of the donor in after-
life, if he were contrite, as much as it relieved the plight of the
poor on earth, for it was the doctrine of the Church that 'good
works', including acts of charity to the poor, promised to those
who performed them a means of salvation. Doles to the poor,
'Christ's poor', were called 'devotions', and the crumbs
gathered from the rich man's table 'Our Lady's bread', because there was for the Catholic donor no essential distinction

---

158 Its account book, with its ordinances, survives: Bodleian, Tanner MS 221.
159 See Brigden, 'Religion and Social Obligation', pp. 100–1.
160 Bodleian, Tanner MS 221, fos. 18^r, 6^v–8^v, 18^v, 19^r.
161 Ibid., fos. 19^r, 19^v, 96^v; PRO, PCC, Prob. 11/27, fo. 13^v.
162 BL, Lansdowne MS 762, fo. 9^r.

between pious or spiritual benefactions and charitable gifts:
for both were intended to prosper the soul in the world to
come.[163] The Church Courts ordered donations to the poor
and to the debtors in London's prisons as a kind of spiritual
penance.[164] Even in London's market-places, with every
bargain made, the merchants pledged 'God's penny', a token
not included in the price of the goods, nor returnable, but given
in alms to the poor. 'God's penny' was an earnest, regarded as
giving divine sanction to the transaction: though, as Italian
visitors and Protestant moralists alike pointed out, this did not
prevent merchants from breaking their promises.[165] No one will
ever know how many Londoners were moved to give alms to
the wretched poor who stood in 'alleys and lanes innumerable'
beseeching the devotions of passers by, but when sixteenth-
century London children learned to count, like twentieth-
century children, they answered problems drawn from everyday
life. This was one such:

As I went by the way I met with a poor man, and he prayed me to
give him a penny, and I bade him to pray to God that that was in my
purse might be doubled and I would give him a penny. So he did, and
I gave him a penny. And so I met with another, and I served him the
same. And also with the third man. And when I had given to each of
these three men I had nothing in my purse. What had I at the first?[166]

The City poor knew well the best time and place to stir the
consciences and evoke the good will of the devout: at church
doors on Sundays and holy days.[167] Good Catholic men and
women gave alms to beggars and the distressed in the streets,
and daily remembered their poor neighbours, like 'John with
the sore arm' and others.[168] From Calais Lady Lisle in London

---

[163] This point is made by Thomson in 'Piety and Charity', pp. 180–1. Pendrill, *Old
Parish Life*, p. 163.

[164] See, for example, Guildhall, MS 9168/9, fo. 109ᵛ; GLRO, DL/C/330, fo. 59ᵛ;
Hale, *Precedents*, cccxviii, cccxxx, cccxxxiii.

[165] C. Barron, C. Coleman, and C. Gobbi (eds.), 'The London Journal of
Alessandro Magno, 1562', *London Journal*, ix (1983), p. 146; *Calendar of Plea and
Memoranda Rolls of the City of London, 1364–81*, ed. A. H. Thomas (Cambridge, 1929),
p. 2; Becon, 'Invective against Swearing', in *Early Works*.

[166] Balliol College, MS 354, fo. 212ʳ.

[167] CLRO, Repertory 8, fo. 274ᵛ; Letter Book P, fo. 9ᵛ; Letter Book Q, fo. 9ʳ;
Journal 15, fos. 48ʳ, 213ᵛ; Bodleian, Tanner MS 221, fo. 113ʳ.

[168] See, for example, *LCCW* 38, 171; PRO, PCC, Prob. 11/20, fo. 185ᵛ.

in November 1538 received a plea: 'all your ladyship's poor neighbours in your street desireth God to send you shortly home'.[169] In a window at St Mary at Hill were depicted the seven works of mercy to remind the parishioners of their Christian duty. The records of almsgiving there suggest that they followed the exhortation.[170]

At death devout Londoners always remembered the poor, for the prayers of the poor might bring a special benediction. Doles were customarily given at funerals to poor men and women who would pray for the wealth of the departed soul. Testators left gifts to paupers in a kind of litany: five pence for five poor men for the five wounds of Christ; to 'nine poor men . . . a groat a piece in the honour of God and the nine orders of angels'.[171] Gifts of money, ale, bread, and coals were distributed among the poor on the day of burial, usually to households the testator knew to be down on their luck in his own parish. These gifts often continued after the benefactor's death: Randall Egerton left 2d. a day for a year to two poor men or women of St Peter Cornhill to hear the Mass said there daily for his soul; a poor man was bequeathed a penny every Sunday for a year if he would say 'five aves five paters and a creed' for Thomas Stowe's soul; Jasper Shuckburgh gave 52s. 'to some poor honest man that can perfectly read and say David's psalter in the honour of God and all saints . . . every day' for a year.[172] Gifts to the poor were very common—a third of testators whose wills were proved in the Prerogative Court of Canterbury on the eve of the Reformation remembered the poor: of those whose wills were recorded in the Consistory Court register, 1522–39, 13.4 per cent left money to relieve poverty.[173]

The gifts took many forms: some forgave 'all such poor persons which upon their conscience' owed them money; others insisted that executors 'in no wise do sue, vex, or trouble any person being indebted . . . which be fallen in poverty and

[169] *Lisle Letters*, 5. 1277.

[170] *St Mary at Hill*, i, pp. 19, 1, 12, 14, 17, 20, 89, 97, 141, 154, 155, 167, 168, 179, 180, 190, 191, 202–4, 211, 284, 289, 290, 299, 295, 386.

[171] Guildhall, MS 9171/10, fos. 75^v, 97^r, 116^r, 224^r, 321^r; PRO, PCC, Prob. 11/37, fo. 222^r.

[172] Guildhall, MS 9171/10, fos. 19^v, 89^v, 116^r.

[173] PRO, PCC, Prob. 11/20, 27 & 28 (Will registers Maynwaryng, Dingeley, and Alenger); Guildhall, MS 9171/10.

decay and of none power to pay'.[174] This was because the
penalty for debt was prison, whence few returned. Between a
quarter and a third of testators whose wills were proved in the
Prerogative Court of Canterbury between 1520 and 1525 left
money to redeem or relieve prisoners.[175] Nicholas Worley left
a bequest to the 'prisoners of God' at Bedlam, the insane.[176]
Some left money to excuse their poor fellow parishioners from
the charges of the parish: John Wilkinson left an income of £20
to St Mary Abchurch in 1521 that the poor should be forever
discharged from paying towards the sepulchre light.[177] One
woman left a torch to burn at the burials of those too poor to
afford one. It was quite usual for the rich to pay for the burials
of the very poor.[178] Most onerous of all the parish charges were
the offerings and tithes paid to the clergy, and nothing was
more likely to cause trouble in the parish. So in 1521 Robert
Whited, mercer and Merchant of the Staple, left £12 a year for
three years to the curate of All Hallows Barking to pay the
Easter offerings for poor parishioners, on condition that the
poor 'upon Easter day, of their charity . . . do say for my soul
and all Christian souls afore the receiving of their communion
five pater nosters and five aves and one creed'.[179]

At their death-beds, as they made their wills and thought on
last things, the citizens had their families, neighbours, and
friends about them. They also summoned their priests. Priests
were on hand to write the wills (sometimes), to offer spiritual
consolation, and minister the last rites. They might also be sent
as emissaries to seek forgiveness for the dying and to make
restitution. As Ralph Hyde lay dying of the plague in 1540, a
divine visitation, he implored his priest:

Sir William, I desire you heartily to go into Lombard Street and to
enquire for Thomas Curtis the Irishman, a hosier there, and I desire
you to ask of him that he will forgive me that I have born false witness
of late and am perjured in so doing . . . Jesus have mercy upon me.[180]

---

[174] See PRO, PCC, Prob. 11/20, fos. 28ᵛ, 55ʳ, 65ʳ.
[175] Ibid., Prob. 11/20 & 21; Thomson, 'Piety and Charity', p. 185.
[176] PRO, PCC, Prob. 11/20, fo. 38ᵛ.
[177] Ibid., fo. 109ᵛ.
[178] *St Mary at Hill*, i, pp. 1, 129, 171, 284. Pendrill, *Old Parish Life*, pp. 153, 172.
[179] PRO, PCC, Prob. 11/20, fo. 69ᵛ.
[180] CLRO, Journal 15, fo. 111ᵛ.

Half of the Londoners making their wills on the eve of the Reformation summoned their priests, 'ghostly fathers', to act as witnesses.[181] A generation later this was no longer so. Family, friends, and neighbours were called upon, but the clergy more rarely. For at the Reformation the bond between the clergy and the laity was profoundly challenged, and transformed.

The relationship between the people and their priests had always been, by its nature, ambivalent. The secular clergy lived among their 'ghostly children', in the world and of it. Yet they were set apart from the lay society into which they had been born by the vocation they had chosen; by their ordination. Through that sacrament they were empowered to celebrate all the others: by the sacrament of penance they could bind and loose; in the sacrament of the altar they could perform a miracle. For the laity to challenge the authority of the clergy might have the direst spiritual consequences, but that challenge would soon come.

THE CLERGY

Priests were 'mediators and means unto God for men', ordained to a sacred vocation, and endowed with secret knowledge which separated them from the laity. 'The dignity of the priesthood is equal with the dignity of angels . . . for unto them said Our Saviour: You are the light of the world', so Dean Colet of St Paul's preached in 1511. The cure of souls, the guidance of spiritual and moral life, demanded that the priest should eschew the world, which was to be found in 'devilish pride, in carnal concupiscence, in worldly covetousness, in secular business'.[182] In his commonplace book Richard Hill, a London grocer, noted the qualities of the good priest:

Sacerdos debet esse almus/ a peccatis segregatus/ Rector et non Raptor/ Speculator & non spiculator/ dispensator & non dissipator/ pius in ludicro/ iustus in ecclesia/ sobrius in cena/ tacens in choro/ prudens in leticia/ purus in conscientia/ assiduus in oracione/ humilis

---

[181] Guildhall, MS 9171/10, fos. 1ʳ–200ᵛ (46 per cent of 350 wills); PRO, PCC, Prob. 11/20 (43 per cent).

[182] Dean Colet's sermon before Convocation, 1511; printed in *EHD* v, pp. 652–60.

in adversitate/ levis in prosperitate/ Dives in virtutibus/ misericors in actibus/ sapiens sermone/ verax in predicacione.[183]

If priests should break their ordination vows and fail to live according to their sacred calling they would endanger those whose spiritual lives they were supposed to guide, as well as falling into sin themselves. Thomas More warned that 'undoubtedly if the clergy be nought, we must needs be worse'. He had once heard Colet preach of the clergy as 'the light of the world, and if the light, saith he, be darked, how dark will then the darkness be, that is to wit, all the world beside?'[184]

By the sacrament of ordination the priest held the power of absolution and could reconcile the contrite sinner to God. Every Catholic must confess his sins to his priest penitently: the priest could impose penance and demand amendment of life, and he could absolve from sin.[185] To the Crossed Friars in Lent 1536 came John Stanton to confess to his 'ghostly father', George Rowland. The sins and agonies of conscience divulged in confession were almost always hidden, for the priest was bound never to break the seal of the confession, but John Stanton had special reasons for revealing the secrets of this confession:

John Stanton said 'Benediciti', and the priest said 'Domini'. John said 'confiteor', and afterward rehearsed the seven deadly sins particularly, and then the misspending of his five wits. And the priest said, 'Have you not sinned in not doing the seven works of mercy?'. Said John . . . 'Yea, forsooth; for the which and all other I cry God mercy, and beseech my ghostly father of forgiveness, and give me penance of my sins'. Then the priest asked whether John were married or not, and John said 'Nay'. And then the priest gave him for his penance to eat neither fish nor flesh two Wednesdays between Easter and Whitsuntide . . . and further, ere he went out of the church to say five pater nosters, and five aves with a credo.

Just as the priest was about to lay his hand upon John's head and give him absolution John asked a question which troubled his conscience; a question which gave the priest pause, as we

---

[183] Balliol College, MS 354, fo. 6ʳ.

[184] More, *Dialogue concerning Heresies*, i, p. 298.

[185] Certificates of confession were printed: see one given at St Thomas Acon in 1517: *RSTC* 14077c.5.

shall see later.[186] Only the priest could give absolution, but he might withhold it. Rowland did, and so, too, did the vicar of Twickenham at Easter 1524. He could not offer communion to manifestly unrepentant parishioners; like Gerard Stockard who had 'openly said . . . that he would not be in charity for him nor for no man living'. Roger Hampton could not have his 'rights' because he had not confessed that Lent, and when ordered to search his conscience, Hampton retorted, 'priest, it is better than thine'.[187]

The priest could ask soul-searching questions because he was ordained to be the parish peace-maker, the arbitrator of quarrels. Yet such delicate intervention would be better received if the priest was thought to be above reproach himself. This was not always so. The wife of Peter Fernandez, a physician, came often to her confessor, Sir Thomas Clerk, seeking his comfort and counsel, when her husband 'put her in great fear'. Clerk tried to 'treat a peace and concord between them', and admonished the errant husband:

Master Peter, ye remember that, by your own confession, ye have lived in adultery, wherefore it shall be well done that ye leave that use of your life in such adultery, and to entreat your wife like an honest Christian woman. If Peter would not reform himself . . . he would not meddle between them.

Peter retorted that he would 'never be bound nor subject to his wife'. The wife left him, taking £5 of her husband's property with her, and sued for divorce. Peter promptly retaliated by bringing a bill of complaint in Star Chamber against Clerk for counselling his wife to leave him, alleging in turn that Clerk's own 'detestable living and demeanour the more part of the City of London doth abhor, as the common fame of the City doth sound'. Clerk's neighbours, Fernandez alleged, were particularly appalled, and Clerk threatened that if they dared to testify against him he would 'undo them by expenses of law'.[188] Instead of bringing reconciliation the priest could be at the centre of parish hostilities.

---

[186] PRO, SP 1/102, fos. 73^r–4^r (*L&P* x. 346). See pp. 206–7, 243, 266–7 below.
[187] GLRO, DL/C/330, fo. 75^v.
[188] PRO, Sta Cha 2/15/45–48. The case came before Wolsey in Star Chamber in April 1528. As usual, no conclusion is recorded. Clerk was steward of St Antony's school in 1522; by 1528 curate of St Mary Abchurch.

The London clergy were not always peaceable. Some left weapons in their wills, and sometimes used them.[189] During the Pilgrimage of Grace in November 1536, when the allegiance of the clergy was particularly suspect, every London friar and secular priest between the ages of sixteen and sixty had all his weapons confiscated, save his meat knife.[190] Some may have needed to take to arms, for despite their supposed sacrosanctity priests were sometimes attacked by the citizenry for sport or as scapegoats. In 1543 as he was walking through the City Sir William Gravesend was set upon and 'evil entreated' by a group of young apprentices and servants who were playing football. They had halberds, but he had 'only a little dagger to defend himself'.[191] Some priests were guilty of attacking their fellow clergy, while in clerical dress, and even in church.[192] But to assault a priest meant instant excommunication from the Church and its sacraments. 'Such as maliciously kill or maim . . . any parson or vicar or other priest . . . never after shall be able, nor his successors within the third degree, to receive any profit of the Church.'[193] When two priests at All Hallows Bread Street attacked each other and drew blood in their church they did public penance before the general procession and the church was suspended.[194] But the clerical caste could hardly be perfect when it was so large: 'it is not well possible to be without many very nought of that company whereof there is such a many multitude'.[195]

In More's Utopia the priests were of an extraordinary sanctity: it followed that they were very few. In Utopia there were only thirteen priests in each city.[196] In London, where More lived (the Amaurotum of Utopia), they were countless. His moral was clear. In the capital there was, or there seemed to

---

[189] See, for example, PRO, PCC, Prob. 11/28, fo. 243ʳ; E 159/315 m. 21. C. Harper-Bill, 'Bishop Richard Hill and the Court of Canterbury, 1494–6', *Guildhall Studies in London History*, iii (1977).

[190] CLRO, Repertory 9, fo. 200ʳ.

[191] CLRO, Journal 15, fo. 8ʳ; see also GLRO, DL/C/207, fos. 193ᵛ, 194ᵛ, 197ᵛ, 211ʳ–212ᵛ. E. G. Ashby, 'Some Aspects of Parish Life in the City of London, 1429–1529' (University of London, MA thesis, 1951), ch. 7.

[192] GLRO, DL/C/207, fos. 211ʳ–212ᵛ, 219ʳ, 229ʳ.

[193] Harrington, *Comendacions of matrimony*, sig. Divᵛ.

[194] *Two London Chronicles*, pp. 6–7.

[195] More, *Dialogue concerning Heresies*, i, p. 301.

[196] More, *Utopia*, pp. 226–7.

be, 'an infinite number of priests'.[197] At the general procession through the City in 1535 a chronicler counted 718 regular and secular clergy.[198] The hundred parishes of London were served not only by the beneficed clergy with cure of souls, but also by a burgeoning ecclesiastical proletariat of curates, chantry priests, morrow mass priests, fraternity and stipendiary priests. In 1546 a survey found 317 chantry priests in London.[199] There were, too, the hundreds of monks, friars, and nuns of the capital's thirty-nine religious foundations. Like Chaucer's chantry priest of St Paul's, the clergy came to London seeking employment, and would advertise their services. In 1479 William Paston asked his servant if he knew 'any young priest in London that setteth bills upon Paul's door'.[200] Many served in the City as personal chaplains: More complained that 'every mean man must have a priest in his house to wait upon his wife, which no man almost lacketh now'.[201] The clerical aristocracy of the realm congregated in London: the Bishops with their households, chaplains at the royal Court and the Cardinal's court ('gnatonical elbowhangers'); canon lawyers and Church diplomats, country incumbents, bored with the shires and fled to London to seek the company of the like-minded at Doctors' Commons.[202] Over them all for a decade was Thomas Wolsey, Archbishop of York, Lord Chancellor, papal legate; at once the Church's greatest patron and the personification of its worst abuses. Well might the Church's critics think that London was priest-ridden.

During the episcopate of Bishop Fitzjames (1506–22) 840 men presented themselves for ordination in the diocese of London.[203] Why so many? Spiritual vocation should have been the first reason for seeking ordination, and so it was, for some. Yet the priesthood had long been a worldly as well as a spiritual

---

[197] So the Venetian ambassador wrote in 1531: *CSP Ven.* iv. 694.

[198] *Two London Chronicles*, pp. 11–12; Foxe, *Acts and Monuments*, v, p. 102. In 1381 the secular clerical population of the City had numbered nearly 700: 102 beneficed clergy, 513 chaplains, and 75 clerks: Thrupp, *Merchant Class*, p. 186.

[199] PRO, E 301/88.

[200] Cited in W. Jenkinson, *London Churches before the Great Fire* (1917), p. 20.

[201] More, *Dialogue concerning Heresies*, i, p. 301.

[202] 'A Supplication of the Poore Commons', ed. F. J. Furnivall (EETS, xiii, 1871), p. 77. G. D. Squibb, *Doctors' Commons: A History of the College of Advocates and Doctors of Law* (Oxford, 1977).

[203] Guildhall, MS 9531/9, fos. 156 ff.

calling, and 'they that lewd be desireth it for worldly winning'.[204] The clerical caste enjoyed privileges and exemptions
denied to the laity. The priesthood provided a way to wealth, to
constant employment, to social advancement which no lay
office could bring, especially for the low-born. This Thomas
Cromwell had learnt in Wolsey's service, and that lesson would
have consequences for the future. At his fall in 1529 the
Cardinal regretted that he had nothing left to give to his loyal
lay retainers; thereby provoking Cromwell into a furious tirade
against his master's pampered chaplains:

> Why, Sir . . . ye have no one chaplain within all your house . . . but
> he may dispend at the least well by your procurement or preferment
> 300 marks yearly, who had all the profits and advantages at your
> hands. And these [lay retainers] take much more pain for you in one
> day than all your idle chaplains hath done in a year.[205]

The Church's orthodox critics lamented the pride and the
covetousness which perverted the spiritual vocation. John
Colet observed, 'how much greediness and appetite of honour
and dignity is nowadays in men of the Church? How run they,
yea, almost out of breath, from one benefice to another?'[206]
'Dionysius Carthusianus' attacked those who 'run headlong
into holy orders without any reverence or consideration'.[207]
Certainly there were magical methods of divination whether a
man should take a benefice or enter the religious life.[208]
Richard Whitford, monk of Syon Abbey, and Henry Gold,
rector of St Mary Aldermary, blamed parents who, urging
children to take the vows, warned that otherwise 'thou shalt go
to the plough; thou shalt fare hardly'.[209]

The City clergy did not seem to 'fare hardly'. Complaints
were made in the 1520s that 'priests and all spiritual persons
waxed so proud, that they wore velvet and silk'.[210] So defensive
had many of the London clergy become about their income

---

[204] More, *Dialogue concerning Heresies*, i, p. 301.

[205] Cavendish, *Life*, pp. 106, 109.

[206] *EHD* v, p. 653.

[207] *The Lyfe of Prestes* (Dionysius Carthusianus? 1533; *RSTC* 6894), fo. 13ʳ.

[208] *Narratives of the Reformation*, p. 327.

[209] *L&P* vii. 523 (8iii). For Whitford's criticisms of contemporary religious life, see
J. Rhodes, 'Private Devotion on the Eve of the Reformation' (University of Durham,
Ph.D. thesis, 1974), pp. 27 ff.

[210] Hall, *Chronicle*, p. 593.

that in 1525 they pleaded that laymen should not assess them for the subsidy.[211] The City livings were unusually wealthy. It has been estimated that three-quarters of English livings were worth less than £15 per annum and half less than £10 per annum, but in London less than one-third of benefices had an income of less than £15, almost half were worth more than £20 annually, and some considerably more.[212] True, the unbeneficed clergy, who did most of the pastoral work in the London parishes, were poorer, and grew poorer still as prices rose sharply from the second decade of the sixteenth century.[213] Bishop Stokesley faced a violent throng of angry 'curates and stipendiary chaplains' in September 1531, who complained that 'twenty nobles [£6. 13s. 4d.] a year is but a bare living for a priest, for now victual and everything in manner is so dear'.[214] The faithful were bound to make an offering of a tenth of the 'lively gift of God's grace' that was their earthly income to their priest, God's minister. In the country the priest's prosperity would rise or fall with that of his parishioners, for his income was a tenth—a tithe—of their year's produce, but this was not so in London.

In the capital, since the income could no longer be of the land, tithe had long been commuted for a cash payment.[215] The tithe settlement which stood at the time of the Reformation was one ratified in 1453 by a bull of Pope Nicholas V, which had fixed tithe at the rate of 3s. 5d. in every £ paid in house rent. One farthing would be paid to the Church for every offering day in the year for every 10s. rent paid by the tenant. Those who owned rather than rented houses owed tithe at the same rate, based on the putative rental value of the house. All

---

[211] Ibid., p. 646.

[212] These figures for the London clergy are taken from *Valor Ecclesiasticus*, iii, pp. 370–8. In the diocese of Canterbury only 14.5 per cent of benefices were worth more than £20 p.a.; M. L. Zell, 'Church and Gentry in Reformation Kent, 1533–1553' (University of California, Ph.D. thesis, 1974), pp. 161–2.

[213] The survey of 1522 shows City curates to have had an average annual salary of £8. 2s., and stipendiary and chantry priests to have had an average of £7. 3s. 4d.: BL, Harleian MS 133.

[214] Hall, *Chronicle*, pp. 783–4; PRO, SP 1/67, fos. 8ʳ–12ʳ (*L&P* v. 387); Sta Cha 2/ii/172.

[215] For what follows, see J. A. F. Thomson, 'Tithe Disputes in Later Medieval London', *EHR* lxxviii (1963), pp. 1–17; *VCH London*, i, pp. 247–52; S. E. Brigden, 'Tithe Controversy in Reformation London', *JEH* xxxii (1981), pp. 285–301.

communicants who were not householders—the poor, servants, apprentices, and women—were to make token payments of 2*d*. for their four offering days at Easter. 'Personal' or 'privy' tithes were also due from the income of wage and salary earners, craftsmen and merchants. The bull of 1453 had left payment of 'privy' tithe solely to the 'good devotion of the parishioner', with the consequence that this income was virtually lost to the clergy—until the parishioner salved his conscience at death. On the eve of the Reformation seven out of ten testators still left money 'for my tithes and oblations by me forgotten or negligently withholden, if any such be, in discharging of my soul'.[216] But some London clergy were not prepared to wait for this unreliable bonus. In 1519 a series of sermons were preached at Paul's Cross by leading City clerics, warning Londoners that they stood in 'great danger to God' for withholding personal tithe.[217]

Tithe was only the first among many other dues which the citizens owed their clergy. In 1513 or 1514 the Londoners were provoked to send a bill before Star Chamber to protest against exorbitant exactions:[218]

At weddings: First at every wedding the . . . curates will have of every man laid upon the Book after their custom 8*d*., and in the two tapers at Mass 2*d*. . . . Also the parson will have the whole offering at the Mass . . . sometimes 2*s*. or 3*s*. or 6*s*. 8*d*. or more, as the curate perceiveth they be friended . . . Moreover, if any person will be married before the high Mass the curate will have 20*d*. or 40*d*. or 5*s*. . . . Also if that the man do dwell out of the parish where his wife is he must agree or give to his curate for a certificate . . . 12*d*. or 20*d*. or 40*d*. . . .

for Burials: At burials where any man woman or child of any reputation is departed the parson . . . will covenant beforehand with their friends for his fee . . . 12*d*. or more, and every priest in the church 8*d*., or else they will not sing him to his burial, & that at every month's mind the curate will have for his part 8*d*. or 12*d*., and every priest in his church 4*d*. or 6*d*. Moreover, they will have all tapers and wax brought into the church with the corpse that be under the 1lb. . . . and at the death of every man they demand for privy tithes 20*d*., 40*d*.,

[216] Guildhall, MSS 9171/10 & 11; *LCCW*; PRO, PCC, Prob. 11/27 & 28.

[217] CLRO, Repertory 5, fo. 154[r].

[218] Lambeth, Carte Miscellane, viii/2d. (This bill may be the one calendared in *L&P* i. 5725, now missing from PRO.)

5s. or 20s. or 40s. or more . . . If any will be buried in the chancel
or high choir the parson or curate will have for place at least 10s.,
13s. 4d., 20s., 40s. or more.

Churchings: When women do lie in childbed the curate hath every
Sunday of some women for the saying of the Gospel one penny or 2d.,
and at her purification they demand of custom 1d., with the chrism
over, and besides the offerings at the Mass.

Bead roll: If any man will have his friends prayed for in the bead roll
the curate will have of every man by the year 4d. or 8d. or more.

Men's devotions on divers days: Item, at all principal feasts in the year, as
Candlemas day, All Souls' day and Creeping to the Cross on Good
Friday & Easter day and in confessions in Lent . . . men and women
did give and offer in the church to the parson or curates . . . And if
there be any saints in the church standing without the high choir, &
have a brotherhood as in some churches the wardens of the brotherhood must . . . compound with his parson . . . 3s. 4d., 5s., 6s. 8d. or
more every year, or else the parson will not let no such brotherhood to
be kept within their church.

The churchwardens' accounts detailing the rectors' incomes
show just how often the citizens did make offerings to their
priests.

Nothing put the clergy and laity at odds so much as money.
Quarrels over the tithe provide the background against which
any hostility between Londoners and their parish priests must
be seen. Since the thirteenth century the citizenry had engaged
in periodic disputes with the City clergy over the assessment
of tithe, but at the Reformation the quarrels became intractable. Between 1520 and 1546 (when a final settlement for
London tithe was reached) over a third of the City parishes are
known to have been involved in disputes prolonged and acrimonious enough to take them to court to be settled. There were
many reasons why a parishioner might decide to withhold tithe
from his priest: doubt about the incumbent's right to tithe or,
more likely, whether he deserved it, or resentment of new and
excessive demands by the cleric. The disputes invariably
involved personal grievances as much as arguments about
money. Certainly tithe legislation was the quickest way to
poison relations within a parish.

Very few laymen denied the principle of paying tithe to their
priest. Scripture, sermons, and tracts proclaimed men's duty

to render to God's ministers a tenth of the wealth that God lent them. Refusal to pay was an act of ingratitude to God and would endanger the eternal salvation of the culprit. For those who found it difficult to envisage that handing money to the priest was tantamount to making an offering to God the Church emphasized the duty by making non-payment a sin punishable by greater excommunication. But opposition to paying tithe might become part of a wider criticism of the Church. If the quarrel was not with the principle of paying tithe, it was with making any payment to undeserving clergy: those who were venal, immoral, or negligent. Between 1529 and 1533, as the English clergy sustained unprecedented assaults upon their privilege and independence, four City rectors made exorbitant demands upon their parishioners and provoked tithe suits which stirred contention through the City. These rectors of All Hallows Lombard Street, St Leonard East-cheap, St Benet Gracechurch, and St Magnus were notoriously non-resident, 'all things to themselves uncharitably taking and from nothing departing'.[219] In 1543 bills were set up against Dr Weston and Dr Wilson, men whose conservatism had enraged London radicals, and whose negligence in serving their cures despite their presence in London was blatant.[220] The most intractable tithe disputes in the whole of the period were in St Dunstan in the East and St Magnus: not only were these the richest benefices in London, but they were held in plurality by John Palsgrave, who was always in the country, and Maurice Griffith, who held another City benefice but was resident in neither.[221] Into his commonplace book an anonymous Londoner copied the twenty-four articles of reform demanded by the German peasants in 1525. The second article had a special resonance for London citizens: 'that no parson shall have the profits of his parsonage but if he do serve it himself, and he so doing to have a competent and reasonable living'.[222]

[219] PRO, SP 2/M, fos. 200ʳ–201ʳ (L&P v. 1788); Lambeth, Carte Miscellane, viii/2 a–e. John Colyns recorded in his commonplace book that the rector of St Benet Gracechurch had incurred praemunire by appealing to Pope Nicholas's bull; BL, Harleian MS 2252, fos. 32ᵛ–33ʳ.

[220] CLRO, Repertory 10, fo. 322ᵛ.

[221] Brigden, 'Tithe Controversy', pp. 299–300; LMCC 21, 24, 50.

[222] BL, Lansdowne MS 762, fo. 76ʳ.

There was a story told of one of the King's chaplains who came upon a 'terrestrial paradise' on his journeys. To his servant he said, 'Robin, yonder benefice standeth very pleasantly, I would it were mine.' 'Why, sir . . . it *is* your own benefice.'[223] Such tales were commonplace in the early sixteenth century, for so was the practice of pluralism, and its consequence, absenteeism. Colet lamented, 'we care not how many, how chargeful, how great benefices we take, so that they be of great value. O covetousness!'[224] It was the opinion of many that as the wealth of the clergy increased, so their responsibilities to their parishioners diminished. Clement Armstrong, a parishioner of St Martin Orgar and an aspirant member of Cromwell's reforming circle in the 1530s, wrote a tract on 'The Reformation of Causes', complaining of the City clergy who 'hath all the profits [of the parishes] and never dwelling upon them nor never helpeth no parishioner therein but ready rather to hinder and hurt them, as knoweth God'.[225] Many London parishioners were aggrieved that pastoral interests seemed, at best, to come second to self-seeking clergy.

> They gasp and they gape
> All to have promotion
> There is all their devotion.[226]

The clergy's failings would, ominously, come to be used against them by those who challenged their authority outright, as Colet had warned. John Parkyn of St Andrew Hubbard reasoned ironically that if a priest could have two benefices, surely a layman could have two wives.[227]

Pluralism was among the most scandalous abuses in the early sixteenth-century Church, with potentially serious consequences for the cure of souls, but it was a perennial problem

---

[223] 'A Supplication of the Poore Commons', p. 78.

[224] *EHD* v, p. 654.

[225] PRO, SP 6/2, fo. 44ᵛ (*L&P* viii. 453); see also SP 2/M, fos. 200ʳ–201ʳ (*L&P* v. 1788). Elton, *Reform and Renewal*, pp. 62–5, 69–70.

[226] BL, Harleian MS 2252, fo. 147ʳ; cf. 'Of Double Beneficed Men', in *Select Works of Robert Crowley*, ed. J. M. Cowper (EETS, extra series, xv, 1872), pp. 27–8.

[227] GLRO, DL/C/330, fo. 138ʳ; cf. John Hig's and Christopher St Germain's statements to the same effect: Foxe, *Acts and Monuments*, iv, p. 178; J. A. Guy, *The Public Career of Sir Thomas More* (Brighton, 1980), p. 111.

which the medieval Church had never solved.[228] Any system of
dispensation was open to abuse, and abused it was. There were
many dispensations, for pluralism was sometimes a necessary
evil.[229] Some clerics were obliged to combine livings simply to
provide a sufficient income, but this was hardly the case in
London, where in 1535 only 16 per cent of benefices were
worth less than £10 annually.[230] The clergy had to be maintained while they were being educated. Yet while William
Longford, curate of Whitechapel, was away at Oxford improving his mind, an illiterate parishioner was preaching heresy to
his flock.[231] Since no adequate stipend was provided for
diocesan officials nor for servants of the Crown and nobility,
these clerics were rewarded by their masters with livings on
which they might never set foot.[232] Because the masters usually
lived in London, so did their chaplains. Of twenty-six City
clergy given dispensations for plurality between 1534 and 1546,
eighteen were chaplains to the King, or to leading nobles and
ecclesiastics; twenty-two men holding London livings between
1521 and 1546 were royal chaplains, and six at least were the
luxurious chaplains of the Cardinal.[233] Thirty-seven London
incumbents between 1521 and 1546 also held diocesan or episcopal appointments, and were paid for service to the Church by
sinecures which were meant to provide service for souls. An
extreme example of a pluralist was Richard Gwent: at the time
of his death in 1544 he was chaplain to Henry VIII, and held
the Archdeaconries of London, Huntingdon, and Brecon, as

---

[228] M. Bowker, *The Secular Clergy in the Diocese of Lincoln, 1495–1520* (Cambridge,
1968), pp. 85 ff.; P. Heath, *The English Parish Clergy on the Eve of the Reformation* (1969),
pp. 49 ff.

[229] See GLRO, DL/C/330, fos. 33ᵛ, 51ᵛ, 66ᵛ, 174ᵛ, 187ᵛ, 189ᵛ, 199ʳ, 208ʳ. The
power of dispensation in the pre-Reformation Church lay with the Pope: in 1534 the
dispensing power passed to the King. It seems that before 1534 papal collectors in
England had power to issue some dispensations locally, and Wolsey as legate also
provided in England some dispensations previously available only in Rome. No
catalogue of dispensations to be found in the Vatican Archives for this period is
available. After 1534 the Faculty Office could dispense in individual cases: D. S.
Chambers, *Faculty Office Registers, 1534–49* (1966).

[230] *Valor Ecclesiasticus*, iii, pp. 370–8.

[231] PRO, E 36/120, fos. 133ʳ–5ʳ (*L&P* xiv/2. 42); see below, pp. 273–4.

[232] See the complaints of the reformers: Fish, 'Supplicacyon' and 'Supplication of
the Poore Commons', pp. 30–1, 76–7.

[233] *Faculty Office Registers.*

well as three prebends, six rectories, and the Deanery of Arches.[234]

The extent of pluralism in London on the eve of the Reformation was, and is, extremely difficult to discover. Certainly at least 112 of the 326 men who held London benefices between 1521 and 1546 were pluralists; more may have been. Two-thirds of City parishes had pluralist incumbents for some part of this period. At least 16 per cent of incumbents of City parishes during the same period are known to have held an additional benefice; a further 11 per cent held an additional two simultaneously; and 3 per cent held more.[235] Some of these men simply collected livings, serving neither Church nor state, but only themselves. The Church was not unconcerned by the prevalence of this abuse, and sometimes, but rarely, dispensations were refused. Thus Christopher Worsley vacated his Lincoln cure for the wealthy and prestigious City parish, St Laurence Jewry, because he was not allowed to hold both.[236] Dispensations were often granted because the supplicant was of high birth, or had powerful friends. Alan Percy, rector of St Mary at Hill between 1521 and 1560, and of at least three other benefices at the same time, was the son of the fourth Earl of Northumberland. In 1535 he was granted the right to benefices to the value of £100 annually without residing personally. He did at least live in London.[237] George Neville, non-resident and pluralist rector of St Mary Woolnoth, was the seventh son of Richard Neville, Lord Latimer.[238] Roger Townsend held at least four benefices, despite his reforming and humanist ideas which should have made him abhor such a practice. In 1537 he was granted a dispensation to hold any benefice, as long as his total annual income did not exceed £300.[239] To those that had it was given: rich benefices went to men of influence, men

234 G. Williams, 'Two neglected London-Welsh Clerics, Richard Whitford and Richard Gwent', *Transactions of the Honourable Society of Cymmrodorion* (1961), p. 41.

235 S. E. Brigden, 'The Early Reformation in London, 1522–1547: the Conflict in the Parishes' (University of Cambridge, Ph.D. thesis, 1979).

236 *BRUO*.

237 Venn, *Alumni*; *L&P* viii. 802(31); PRO, C 1/1145/32; CLRO, Letter Book N, fo. 288[r]; Letter Book Q, fo. 123[r].

238 Venn, *Alumni*.

239 Ibid.; *Faculty Office Registers*, p. 105. He was resident at St Mary Woolchurch. He left Bibles to four of his parishes: PRO, PCC, Prob. 11/27, fos. 169[v]–170[r].

powerful enough to win other prizes; thus wealthy pluralists gathered significant appointments to themselves. It was in London and at the Court that such prizes were to be sought and found. Such was the competition for City benefices that vacancies might be filled overnight. So it was in 1518 when four leading clerics were candidates for St Michael Cornhill.[240]

In the provinces pluralism very often had the consequence of non-residence. In Lancashire, for example, there were rich benefices to attract the ambitious, but social conditions to dissuade them from residence. Many of these errant provincial clerics removed to London. Twenty-two of the absentee clerics of the diocese of Lincoln between 1495 and 1520 were to be found in the capital.[241] To the Bishop of London and his officials such truants were an embarrassment: between 1523 and 1526 ten men were prosecuted before the Vicar-General for leaving their cures in the country to serve as assistant clergy in London.[242] John Oterburn was celebrating at St Clement Danes in 1529, though a vicar at Farleigh in the diocese of Winchester;[243] Thomas Allen was a chantry priest at St Dunstan as well as parson of Gunthorpe in Norfolk; in 1543 William Dingley chose a chantry in London rather than a benefice in the shires.[244] The advantages of living in the capital were evident: here were the main sources of clerical preferment, the wealthiest patrons, and the best chances to meet with the like-minded. At the Doctors' Commons, an association of civilians and canonists linked with the Court of Arches, those high in legal and ecclesiastical circles congregated. The same attractions which drew clerics from the provinces to London might induce London priests to stay in their cures.

Were the City clergy resident? Although so many London clerics were blatant pluralists, very few men were called to answer for absence. William Vicars, chantry priest at St

---

[240] Drapers' Company, Repertory 7/i. p. 65.

[241] C. A. Haigh, *Reformation and Resistance in Tudor Lancashire* (Cambridge, 1974), p. 29; Bowker, *Secular Clergy*, appendix iv (I have used additional biographical information for City clergy to make this deduction).

[242] GLRO, DL/C/330, fos. 43ᵛ, 44ʳ, 51ᵛ, 52ᵛ, 53ʳ, 84ᵛ, 113ʳ. In January 1543, for instance, Leonard Barneston, curate of St Andrew Undershaft, was ordered to return to his cure in Essex: ibid. DL/C/3, fos. 206ᵛ, 208ᵛ.

[243] Ibid. DL/C/330, fo. 33ᵛ; Heath, *English Parish Clergy*, p. 55.

[244] *LCCW* 95; Mercers' Company, Acts of Court, ii, fos. 163ᵛ, 208ᵛ; Venn, *Alumni*.

Michael Queenhithe, was alone deprived for non-residence.[245]
In July 1529 John Grigell, vicar of All Hallows Barking, was
asked to show cause why he should not reside personally,
according to the oath he had taken at his institution to reside or
to suffer deprivation.[246] Some men were certainly away for
part of the time; like the diplomats in royal service—William
Knight, William Gwynn, Thomas Lupset, Richard Layton,
Thomas Paynell, Laurence Taylor, and Bishops Tunstall and
Bonner—but then the reason for absence was clear, and the
absence temporary. Two surveys made in the City reveal something about the limits to the assiduity of the beneficed clergy.
When the City clergy were assessed in the general survey of
1522 beneficed clergy were not assessed in 21 of 88 parishes:
presumably because they were absent. Three were specifically
said to be non-resident. In ten of the twenty-one parishes the
cure of souls was left to a curate and in the other eleven to
assistant clergy.[247] In the 1548 chantry survey reports were
made of the pastoral care in the parishes: it appears that in a
quarter of City parishes the incumbent was habitually absent,
but that half the London benefices were usually served by their
rector or vicar.[248] London clergy were no more negligent than
their Canterbury or Lincoln colleagues,[249] but the difference
was that many clergy who were known to be in London in
other capacities were seen to be neglecting the cure of souls.

The clergy were set above the laity by their superior wisdom
and learning; at least they should have been: the very word
'clergy' signified an élite. Only a learned clergy, *litteratura
sufficiens*, well versed in the sacred writings of the Church,
could instruct the people. Being able to read had originally set
them apart from the laity and even gave them benefit of clergy
against legal prosecution.[250] The education of the clergy was of

[245] GLRO, DL/C/330, fos. 174ᵛ, 175ʳ.
[246] Ibid., fo. 179ʳ. Grigell was later licensed to hold three benefices and to be non-
resident. He was in London to sign the Oath of Supremacy in 1534: *L&P* vii.
1498(25); PRO, E 36/64, fo. 4ʳ (*L&P* vii. 1025).
[247] BL, Harleian MS 133.
[248] *LMCC.*
[249] Non-residence in Lincoln diocese 1514–21 ran to 25 per cent; in Kent the
incidence of pluralism and non-residence between 1500 and 1550 was between 15 and
25 per cent: Bowker, *Secular Clergy*, pp. 73, 90; Zell, 'Church and Gentry', pp. 139–41.
[250] L. C. Gabel, *Benefit of Clergy in England in the Middle Ages* (Northampton, Mass.,
1928–9); J. G. Bellamy, 'Benefit of Clergy in the Fifteenth and Sixteenth Centuries',

perennial concern to the Church, but became a more urgent desideratum on the eve of the Reformation as old complaints of clerical ignorance multiplied while lay standards and expectations rose,[251] and because the faith had never needed defending more urgently. Colet warned in 1511 that the priests had become the warriors 'rather of this world than of Christ', while their 'warring is to pray, to read, and study scriptures, to preach the Word of God, to minister the sacraments of health, to do sacrifice for the people, and offer hosts for their sins'.[252]

Did Londoners have cause to complain that their clergy lacked learning? The beneficed clergy of the capital were the best educated of the realm, and their standard of education was rising. A survey of London beneficed clergy between 1429 and 1479 found a quarter to be graduates; by 1479–1529 60 per cent were graduates.[253] Of 326 incumbents in London cures between 1521 and 1546 more than 60 per cent were certainly graduates and even more may have been.[254] By the time that Mary came to the throne, 55 of 79 City incumbents were graduates.[255] In 1522 seven of 66 London clergy assessed in the general survey were doctors, 44 were masters of arts, and only fifteen styled 'dominus', which probably signified that they held no degree.[256] Some London clerics were not only well educated, but contributed to learning by their own writings: works of spiritual edification, guides to conduct, treatises on grammar, mathematics, or surveying, and, increasingly, tracts of theological controversy.[257]

in his *Criminal Law and Society in Late Medieval and Tudor England* (Gloucester, 1984), pp. 115 ff.; J. D. M. Derrett, 'The Affairs of Richard Hunne and Friar Standish', *CW* 9, appendix B.

[251] D. Cressy, *Literacy and the Social Order* (Cambridge, 1980); J. Simon, *Education and Society in Tudor England* (Cambridge, 1966); H. S. Bennett, *English Books and their Readers, 1475–1557* (Cambridge, 1952).

[252] *EHD* v, p. 654.

[253] Ashby, 'Some Aspects of Parish Life in the City of London', p. 88.

[254] This information derives primarily from an analysis of the alumni of the Universities of Oxford and Cambridge: Venn, *Alumni* and *BRUO*.

[255] E. L. C. Mullins, 'The Effects of the Marian and Elizabethan Religious Settlements on the Clergy of London' (University of London, MA thesis, 1948), p. 74.

[256] BL, Harleian MS 133.

[257] See *inter alia* Thomas Abell (*RSTC* 61); John Bradford (*RSTC* 3480.5–3504); William Harrington (*RSTC* 12798.7); Edward Laborne (BL, Royal MS 7F xiv); Thomas Lancaster (*RSTC* 15188); Thomas Lupset (*RSTC* 16932–16941); John Palsgrave (*RSTC* 19166); Thomas Paynell (Hatfield House, Salisbury MS 474; *RSTC*

Few could doubt the learning of the first estate of London clergy: it was the way they chose to use it—rather, not to use it—which angered their parishioners. The charge of vainglory had long been levelled at the learned; that is, at the clergy;[258] and many London priests particularly deserved it. It was

> A great abomination
> To see prelates & doctors of divinity
> Thus to be blinded with pride and iniquity.[259]

Having risen by an education which was more or less reserved for those who would be ordained, many of the clergy refused to share their knowledge with their flock. Spiritual knowledge was too precious to be given to the laity, who were theologically ignorant, jealous priests supposed, and should remain so. The prayers which the priest said at the altar during the Mass were called the 'secrets'; in 1514 the chantry priest of St Mary Woolnoth bequeathed his 'secrets of the Mass written in text hand . . . and limned with gold'.[260] Henry Gold, rector of St Mary Aldermary, humanist, and chaplain to Archbishop Warham, prepared a sermon which declared the priest's trust to protect the Word. In this sermon a parishioner challenged his vicar:

Vicar, I say why speak thou not? . . . Why goest thou into the pulpit if thee cannot say nothing, but goest and look upon us . . . and again say thy cross row?

For I am so afraid, as it were a child that could not tell what it should say or speak.

Vicar, why art thou thus afraid?

I am in a great fear and doubt whether I may . . . preach unto the people that be nowadays His holy doctrine or gospels or not, for He

19492-6); William Peryn (*RSTC* 19784–19787); Edward Powell (*RSTC* 20140); Lancelot Ridley (*RSTC* 21038–21043); Nicholas Ridley (*RSTC* 21046–21053); Robert Ridley (CUL, MS Dd. v 27; *RSTC* 11892); John Rogers (*RSTC* 17799); Richard Smith (*RSTC* 22815–22824); John Standish (*RSTC* 23207–23211); Thomas Starkey (*RSTC* 23236); Cuthbert Tunstall (*RSTC* 24318–24323.5); John Veron (*RSTC* 24676–24687); Hugh Weston (*RSTC* 25291.5); John Wymmesley (*RSTC* 12794).

258 A. Murray, *Reason and Society in the Middle Ages* (Oxford, 1978), pp. 244–51.
259 BL, Harleian MS 2252, fo. 26ʳ.
260 Guildhall, MS 9531/9, fo. 13ᵛ.

had given strait commandment . . . See that you do not give that
thing which is holy unto dogs, nor cast [pearls] to hogs.[261]

Gold's admonition to protect Scripture from the people came
just at a time when they were most eager to know it. Time was
when prelates had preached. Into his commonplace book
Richard Hill copied this poem:

> Peter at Rome sometime Pope was;
> Our Lord's law he kept truly;
> He preached the Gospel, through God's grace,
> That many a soul was saved thereby.[262]

Concern that the sacred Word should not be sullied was not
the only reason for keeping knowledge secret: the laziness and
ignorance of some who were charged with the cure of souls in
the parishes meant that safe recourse to easy fables and prepared sermons too often served as instruction. In the savagely
anticlerical tract, *Rede me and be nott wrothe* ('The beryeng of the
Masse'), 'Watkin' and 'Jeffrey' discuss the intellectual torpor
and credulity of 'Sir John' and his fellows:

> WAT. Thou never sawest miracle wrought?
> JEF. I: no, by Him that me bought,
>   But as the priests make rehearsal.
> WAT. Canst thou rehearse me now one?
> JEF. No I cannot, but our sir John
>   Can in his English Festivall.
> WAT. Give they to such fables credence?
> JEF. They have them in more reverence
>   Than to the Gospel a thousand fold.[263]

John Bale, too, mocked 'Sir John', whose idea of preaching
was to read out his prepared 'sermones dormi secure' (so called
because the priest might sleep untroubled on Saturday
night).[264] Perhaps he was unfair. Seventeen of fifty City priests

---

[261] PRO, SP 1/83, fo. 147^r (*L&P* vii. 523(4)). Matt. 7:6. For Gold's troubles later,
see below, pp. 215, 222, 224.

[262] Balliol College, MS 354, fo. 210^r.

[263] 'Rede me and be nott wrothe' (1528), in *English Reprints*, ed. E. Arber (1871),
p. 63; A. Hume, 'English Protestant Books Printed Abroad, 1525–1535', *CW* 8/ii,
appendix B, p. 1070.

[264] John Bale, *Yet a course at the romysh foxe* (1543; *RSTC* 1309), sig. Hi^r; G. R. Owst,
Preaching in Medieval England (Cambridge, 1926), pp. 237–8; More, *Correspondence*,
p. 54.

whose wills were proved in the London Consistory Court between 1514 and 1520 left books; as did eleven of twenty London priests whose wills are recorded in Bishop Fitzjames's register. True, most of these books were the tools of the priestly trade—breviaries, mass books, antiphoners, hymnals, missals —but there were also the Dialogues and Homilies of St Gregory the Great, the sermons of Philip Repyngdon, *Bonaventura de Vita Christi*, Peter Lombard's *Sententiarum*, the Epistles and Gospels, the *Expositions* of Nicholas de Lyra, the confessor's manual *Summa Baptistiana*.[265] William Lambert, rector of All Hallows Honey Lane, left a collection of theological works: 'Hugonem upon the whole bible; sermones opuscula; Epistolas beati Augustini; Valentinum super psalterum; sermones Vincentii; sermones Magdalene; Iustinus historicus; sermones Augustini . . . with all my quires of preaching'. He bequeathed his books to Pembroke Hall, Cambridge, which soon became the academy of the Reformation's leading evangelists.[266] William Lichfield, rector of All Hallows the Great, had left 3,083 manuscript sermons at his death in 1448.[267]

Though the beneficed clergy of the capital were well educated, the assistant clergy they found to serve their parishes hardly represented the flower of learning. More than a thousand (1,035) beneficed priests serving in the parish churches of London between 1520 and 1546 are known by name; only 74 are known to be graduates, though in the 1522 survey seventeen of a hundred assistant clergy were titled 'Master'.[268] Perhaps many had attended university at some time; like John Milner, celebrant at St Laurence Jewry, who had no degree, but had spent seven years at Cambridge. Maybe he ran out of money, or maybe, like William Stoddard, the reprobate chantry priest of St Michael Queenhithe, who went up to New College from Winchester in 1522, he left 'quia non habuit animum amplius studendi'.[269] Of 124 lay ordinands admitted by Bishop Tunstall

---

265 *LCCW*; Guildhall, MS 9531/9, fos. 1 ff.

266 PRO, PCC, Prob. 11/20, fo. 187ʳ. Lambert was appointed as arbiter of orthodoxy, see below, p. 160. For other substantial libraries belonging to clergy living or beneficed in London, see *BRUO*, appendix B (Henry Cole, Gabriel Dunne, Thomas Martyn, Thomas Paynell, David Pole, Robert Talbot). See also PRO, PCC, Prob. 11/20, fos. 8ᵛ, 185ᵛ, 228ᵛ.

267 Stow, *Survey*, i, p. 235.

268 BL, Harleian MS 133.

269 GLRO, DL/C/207, fos. 205ᵛ–206ʳ; *BRUO*.

only twenty-four were graduates.[270] When John Stokesley, a renowned scholar himself, became Bishop of London in 1530 he determined to have clergy worthy of their office, who could teach the truth and defend the faith. Observing that the cure of souls lay effectively with the curates, he conducted personally an examination into their 'letters and in their capacity and suitability for those things which pertain to the cure of souls'. His findings must have made him fearful: of the 58 curates examined only fourteen could be freely admitted. Sixteen curates were suspended, forbidden from celebrating because of their (unspecified) unsuitability; one resigned; and twenty-one were banned because of their ignorance and lack of letters (including two masters and two bachelors of arts). Others were told to go away, to study, and return again.[271] Yet the Bishop could afford to dismiss wretched curates only if he could be sure of finding educated and worthy men to replace them. Half of the lay ordinands admitted by Stokesley were indeed graduates, but he admitted only twenty-eight altogether.[272] Twelve of those curates suspended in 1530 were still serving, and in the City, in the 1530s, despite the Bishop's ban.[273] For the parishes must be served, and unsatisfactory priests were better perhaps than no priests at all. To such intellectually feeble priests fell the task of expounding the royal supremacy from the pulpits in the 1530s. In 1535 Thomas Bedyll, Archdeacon of London and a pluralist himself, despaired: he had perforce altered certain passages in the form setting out the King's new title of Supreme Head, because the City curates were so 'brute' that they would read the order verbatim, and 'say of themselves in the pulpit "they shall preach and declare"'.[274] Not, of course, that education has ever precluded stupidity.

Colet had warned of the dangers to the people if the clergy were ignorant and blinded by love of the world. Reformers uttered the same warning, and in the same biblical imagery:

A little learning in the curates doth soon corrupt and make sour all the parish. And if their light be darkness how great will the darkness

---

[270] Guildhall, MS 9531/10, fos. 152^r–163^r.

[271] GLRO, DL/C/330, fos. 265^v–266^r; Heath, *English Parish Clergy*, pp. 73–4 (my figures differ slightly from Mr Heath's).

[272] Guildhall, MS 9531/11, fos. 128^r–136^v.

[273] Some signed the Oath of Supremacy; others were witnessing wills.

[274] BL, Cotton MS Cleo. E vi. fo. 255^r (*L&P* ix. 29).

be of all other? Thus the blind guides shall soon bring all the parish to follow them into the ditch of destruction.[275]

Yet ignorance was not the worst of clerical vices: not at all. To behave badly, while knowing better, was particularly reprehensible: 'for without virtue the better they be learned the worse they be'.[276] Colet counselled that 'it is not enough for a priest . . . to construe a collect, to put forth a question, or to answer to a sophism', better by far to live 'a good, a pure, and a holy life'.[277]

Each priest was vowed to chastity. Celibacy had long been demanded by the Church as essential to the religious life.[278] Whenever the vow was broken every good Catholic should be shocked, not only because of the perjury, but also because the priesthood was set apart from and above the laity, eschewing the world, the flesh, and the Devil, to be a purer channel of divine grace. Among a series of horrors foretold for 1525, when Nature would turn in her course, was the spectre that 'religious men shall go out of their cloister. . . . Chastity shall be broken.'[279] That the priestly fall from grace usually involved the deflowering of a daughter, the cuckolding of a husband, or procuring a prostitute made the offence the greater. Complaints of clerical concupiscence were doubtless as old as the vows themselves. In 1297 Edward I had forbidden the practice of citizens who 'upon mere spite do enter in their watches into clerks' chambers, and like felons carry them to the Tun', the prison for night walkers.[280] Such was Londoners' 'filthy delight of evil communication', Thomas More insisted, that one defaulting priest would be talked about far more than many worthy ones.[281] Yet beyond the natural love of scandal lay the real hope of Londoners that their priests should provide them with an example of lives more blameless than their own.

---

[275] PRO, E 36/120, fo. 108ʳ (*L&P* xiii/1. 1111(2)). For the learning and leanings of Master Laborne, against whom this rhyme was directed, see below, pp. 281–2.

[276] More, *Dialogue concerning Heresies*, i, p. 301.

[277] *EHD* v, p. 657.

[278] Dionysius Carthusianus adduced nine reasons why chastity was a precondition of the religious life: *The Lyfe of Prestes*. Whitford wrote *The Pype, or tonne, of the lyfe of perfection* (1532; *RSTC* 25421) to encourage the religious to keep their vows.

[279] BL, Lansdowne MS 762, fo. 64ᵛ.

[280] Stow, *Survey*, i, p. 189.

[281] More, *Dialogue concerning Heresies*, i, p. 296.

To call someone 'priest's whore' or 'priest's bawd' was deeply insulting but the insult was extremely common,[282] reflecting the contemporary obsession with this sin and the contempt in which it was held.

The Church was principally responsible for clerical discipline, and incontinent priests were brought before the Church Courts. In theory, no priest could be retained once found guilty of fornication;[283] in practice no London priests seem to have been deprived for this offence until Mary's reign, when priests legally married in Edward's reign were ejected. The City authorities had long taken it upon themselves to guard the morality of the citizens: before the wardmote enquests came the City's bawds on St Thomas Eve, and among their clients were the City's priests.[284] The wardmote enquest was bound to ask if any priest 'hath been within the Tun of Cornhill'?[285] When two priests spent the night with prostitutes in St Paul's churchyard in 1529, they were rudely awakened by the Alderman's deputy, a constable, and a large crowd of morally outraged or prurient citizens.[286]

Many charges of incontinence made against the City clergy arose through malice, but the choice of this particular charge is significant: it was so damaging because the laity hoped for a virtuous priesthood. When John Roo, curate and farmer of the wealthy parish of St Christopher le Stocks, faced a paternity suit in 1529 many conflicts within that parish were revealed.[287] Roo maintained that the 'honest, good and substantial wives' of St Christopher's gave no credence to the scandalous stories which were circulating, but it was the jealous chantry priest and various 'poor folk' who sought to expel him. Before the

---

[282] See, for example, Guildhall, 9064/11, fos. 273^r, 300^r, 308^v; GLRO, DL/C/208, fos. 16^v, 83^v, 160^v; CLRO, Repertory 5, fo. 101^r. William Stoddard, chantry priest at St Michael Queenhithe, called his own mother 'priest's whore': DL/C/330, fos. 172^r, 173^r. For sexual offences and defamation, see R. M. Wunderli, *London Church Courts and Society on the Eve of the Reformation* (Cambridge, Mass., 1981), chs. 3 and 4.

[283] Lyndwood, III, tit. 3 c.1; III, tit. 12; III, tit. 13 c.1; Heath, *English Parish Clergy*, pp. 104 ff.

[284] See, for example, CLRO, Repertory 5, fo. 52^r; Repertory 9, fo. 229^r; Letter Book O, fo. 91^v. The cases were very many.

[285] BL, Lansdowne MS 762, fo. 38^r.

[286] GLRO, DL/C/330, fo. 179^v, also fo. 188^v.

[287] This account of the feud is drawn from a miscellany of records, too confused to cite separately: GLRO, DL/C/208, fos. 1^r, 3^r, 6^r, 10^r; PRO, C 1/600/46; SP 1/40, fos. 176^r–88^r (*L&P* iv/2. 2754, 2854).

Bishop of London's Commissary, Roo was accused of fathering the child of Emma Singleton. The case was dismissed. Roo's accusers, ascribing the failure to the Commissary's prejudice, brought the case before the wardmote enquest. They failed to gain a conviction there, or from the Court of Aldermen, or from the Bishop himself. But Singleton, incited by the chantry priest, continued to plague Roo, by leaving her child in his house, or even crying before the altar during divine service. Finally, Singleton sued a writ of trespass against Roo in the Sheriff's court, claiming that Roo had promised to maintain her and her child. She now demanded £800 (*sic*) in maintenance. In desperation, Roo appealed to Wolsey in Chancery, but whether he was protected there or not we cannot tell. In another case John Man, a chantry priest, had bribed a woman to say that she had slept with John Turney of St John Zachary, whose chantry he wanted.[288] Allegations of sexual impropriety have always been one of the best ways to bring down those in authority, who are expected to show higher standards of rectitude than ordinary mortals. Even while 'the preacher is preaching of the holy Gospels, Word of God in the temple of God', so Henry Gold alleged, 'such mad dogs, such common obstinate . . . condemners of other men's lives do slander you, tear and rend your good name and fame with their venom teeth and poison tongues'.[289] Some accusations were undeserved, but too many were just for the Church to be complacent.

While the laity provided priests for themselves, to sing for their souls and all Christian souls, the same condition was very often expressed: that the priest be 'of good name and fame', of 'virtuous life'.[290] In 1529 Thomas Cromwell willed that a priest should pray for his soul; 'being an honest person of continent and good living'.[291] If priests were mediators for men to God they should be beyond the world's temptations. In St Andrew Hubbard in 1532 Rowland Kendall was taunting William Lawles with the cuckold gesture of horns when Sir Thomas Kirkham approached, a priest deprived by Stokesley

---

[288] GLRO, DL/C/330, fo. 188ᵛ.

[289] PRO, SP 1/83, fos. 147ʳ–148ʳ (*L&P* vii. 523(4)).

[290] See, for example, PRO, PCC, Prob. 11/20, fos. 23ᵛ, 30ʳ, 45ʳ, 46ᵛ, 64ʳ, 79ᵛ, 80ʳ, 89ᵛ, 93ʳ, 99ʳ, 107ᵛ, 109ᵛ, 142ʳ, 150ʳ, 171ʳ, 180ᵛ, 209ᵛ, 227ᵛ.

[291] R. B. Merriman, *The Life and Letters of Thomas Cromwell*, 2 vols. (Oxford, 1902), i, p. 61.

as 'indignus and indoctus'. Kendall promptly challenged Kirkham: 'knave priest, it were more fit for thee to sing Mass underneath a hedge than in a church'; for all the parish knew that 'since Christmas thou didst not lie five nights in thine own chamber', because he was staying with Lawles's wife. [292]

Much of the anger against immoral priests derived from the doubt as to whether an unworthy priest could be the channel of divine grace in the seven sacraments. In 1510 Elizabeth Sampson reviled the holy bread administered by priests 'when they be not in clean life',[293] and, later, an unnamed London priest himself maintained that 'malus sacerdos non consecrat nec conferat baptismum nec ceteras sacramentas'[294]: but such opinions were heretical and condemned. The Church maintained that the unworthiness of the minister did not affect the validity of the sacraments, since, as Augustine had insisted, their true minister was Christ. However unworthy the priest, his ordination gave him the power to perform the sacrament of the altar. Even the best and wisest layman must always yield place to the most ignorant and venal priest, for only the priest could celebrate the sacred mystery. Thomas More wrote of the 'special prerogative that we have by a priest, be he never so bad, in that his naughtiness cannot take from us the profit of his Mass'.[295] The Church might be criticized, its clergy found wanting, but for Catholics there was no salvation outside the Church, and without the priesthood admitting the laity to the sacraments their immortal souls were lost.

While the dependence of the people upon their priests was so fundamental there could be no possibility, no question, of concerted lay opposition to clerical. privilege nor of united attacks upon a worldly clergy unworthy of its other worldly duties. Some resented their parish priests (and often with reason), scorned the clerical potentates of the City, despised the unlearned or unchaste, but this was not to denigrate the essential authority of the clergy, to doubt sacerdotal power to bind and loose or to perform the miracle of the Mass. Though there

---

[292] GLRO, DL/C/208, fos. 263ʳ–64ʳ; see also DL/C/330, fo. 265ʳ.
[293] Guildhall, MS 9531/9, fos. 4ʳ⁻ᵛ.
[294] PRO, SP 1/49, fo. 46ʳ (*L&P* iv/2. 4444).
[295] More, *Dialogue concerning Heresies*, i, ch. xii, p. 299. J. N. D. Kelly, *Early Christian Doctrines* (1958), pp. 409–11; W. II. C. Frend, *The Donatist Church* (Oxford, 1952).

might be more talk of the vices of priests than ever there was of their virtues, still—despite assertions to the contrary, then and since—the evidence of Londoners universally hating their priests is nugatory.[296] The attitude of the citizens to their clergy was complex and ambivalent, for the relationship was personal, and, like other personal relationships, subject to the vagaries of personality, the strains of proximity, and complications of financial obligation. Resentment of clerical wealth and privilege there certainly was, but it was at some times more profound than at others, as we shall see, and prevailingly the relationship between clergy and citizenry was rather harmonious than hostile. While the Catholic Church seemed so adamantine and unassailable, criticism of the many failings inevitable in an institution which was, after all, human were many and usually beneficial. The intent was to reform not to undermine. Still in 1533 Thomas More did not doubt that '(God be thanked) the faith is itself as fast rooted in this realm as ever it was before'.[297] If ever this changed, as soon it would, the Church's critics might rally to its defence.

Dean Colet had been prescient when he had warned in 1511 that the resurgent Lollard heretics were 'not so pestilent and pernicious unto us and the people as the evil and wicked life of priests'.[298] For the failings of the clergy gave the Lollards and, imminently and more dangerously, the reformers the opportunity to claim that the sacraments were vitiated by the corruption of the clergy. Even the Mass could be portrayed as an invention of priests to beguile the faithful into giving tithe to support their indolent and worldly lives. In the virulent satire *Rede me and be nott wrothe* (1528) the Mass, slain in Strasburg by Scriptural truth, was brought home for burial at St Thomas's shrine, bewailed by Catholic priests, who lost with the Mass their livelihood.

> What availeth now to have a shaven head
> Or to be apparelled with a long gown.
> Our anointed hands do us little stead.

---

[296] For the controversy regarding this subject, see C. A. Haigh, 'Anticlericalism and the English Reformation', *History*, lxviii (1983), pp. 391–407; A. G. Dickens, 'The Shape of Anti-clericalism and the English Reformation', in *Politics and Society in Western Europe*, ed. E. I. Kouri and T. Scott (1987).

[297] More, *Answer to a Poisoned Book*, preface, p. 4. [298] *EHD* v. p. 656.

.   .   .   .

Seeing that gone is the Mass,
Now deceased. Alas, Alas.[299]

This was the true anticlericalism, the antisacerdotalism of
heresy, denying the essential place and function of the clergy.

HUMANISM AND REFORM

Criticism of the Church was not simply a negative spirit. There
was on the eve of the Reformation a pious and fervently orthodox desire among influential laity and spiritualty in London
for reform. The New Learning—a concern to study the sources
of the faith in a humanist manner and to purify spiritual life in
imitation of Christ—had touched an élite in the City. At
Court, in literate lay households, in the austere and influential
religious houses in and about the capital, at the Inns of Court,
in Doctors' Commons, a group of devout laymen and clerics
were spreading the spiritual message of Erasmus and his
friends. Christian humanism, in which the ideal Christian life
was contrasted with the flawed reality, was a moving force in
the reception of the Reformation: while its spirit was essentially
and unimpeachably orthodox, it prepared the way for a more
radical vision.[300]

In London at the turn of the sixteenth century were gathered
some of Europe's leading humanists; there Erasmus found five
or six men so learned that he doubted whether even Italy had
such scholars. In December 1499 he wrote 'tumultuarie' from
London:

When I listen to Colet it seems to me that I am listening to Plato
himself. Who could fail to be astonished by the universal scope of
Grocyn's accomplishments? Could anything be more clever or profound or sophisticated than Linacre's mind? Did Nature ever create

---

[299] 'Rede me and be nott wrothe', p. 36.

[300] For a bibliography of English humanism and a useful consideration of the
problems of definition, see A. Fox, 'Facts and Fallacies: Interpreting English
Humanism', in *Reassessing the Henrician Age*, ed. A. Fox and J. Guy (Oxford, 1986),
ch. 1. See also M. Dowling, *Humanism in the Age of Henry VIII* (1986); J. K. McConica,
English Humanism and Reformation Politics under Henry VIII and Edward VI (Oxford, 1968).

anything kinder, sweeter, or more harmonious than the character of Thomas More?[301]

John Colet returned to his native London from Oxford in 1505, as Dean of St Paul's. Grocyn and Linacre were there already.[302] At St Laurence Jewry, Grocyn's parish, Thomas More lectured upon Augustine's *De Civitate Dei* 'to the great admiration of all his audience', including 'the chief and best learned men' of the City.[303] More wrote to Colet, his mentor, in November 1504; 'I shall spend my time with Grocyn, Linacre and Lilly. The first . . . is the director of my life in your absence; the second, the master of my studies; the third, my most dear companion.'[304] William Lilly, the grammarian, became the first High Master of St Paul's school, which Colet founded on humanist principles in 1509, and thereby the tutor of a generation of new humanists and reformers.[305] There were suspicions among the more conservative that the school was 'a temple of idolatry . . . because the Poets are to be taught there', and worse.[306] More told Colet:

I am not surprised that the school of Jesus excites the envy and anger of dissolutes and obdurates. These perverse people can only contemplate with fear this crowd of Christians who, like the Greeks from the Trojan horse, spring from that academy to destroy their ignorance and disorder.[307]

The New Learning did indeed come to be seen as a threat by the old order; even as akin to heresy. Certainly the Catholic reformers and the heretical opponents they would condemn had common spiritual ends: both desired above all purification and reform within the Christian community, and religious life spent in emulation of the sanctity and simplicity of Christ and His apostles. Colet's sermons at St Paul's in 1505 followed the

---

[301] Erasmus, *Correspondence*, 1, pp. 235–6.

[302] J. H. Lupton, *A Life of Dean Colet* (1909); F. Seebohm, *The Oxford Reformers: John Colet, Erasmus and Thomas More* (1867), pp. 137–40.

[303] Roper, *Lyfe*, p. 6.

[304] More, *Correspondence*, pp. 8–9; cited in Seebohm, *Oxford Reformers*, p. 149.

[305] For the foundation of the school, see *EHD* v, pp. 1039–45; J. H. Rieger, 'Erasmus, Colet and the Schoolboy Jesus', *Studies in the Renaissance*, 9 (1962), pp. 187–94; Dowling, *Humanism*, pp. 113–17.

[306] Seebohm, *Oxford Reformers*, p. 252.

[307] Cited in Dowling, *Humanism*, pp. 114–15.

spirit of humanist reform: he did not take a discrete and isolated text and preach a detailed discourse to prove a particular point of faith, in the way of the schoolmen; rather he followed one argument, drawn not from the Fathers but from the Bible. Colet's study moved increasingly away from the apostolic Epistles which had preoccupied him at Oxford, and his thoughts were always and ever upon the 'wonderful majesty of the Gospels', upon Christ himself. At St Paul's he preached of 'Gospel history', of the Apostle's Creed and the Lord's Prayer.[308] Erasmus praised his dedication: as 'father . . . to all your fellow citizens' children and indeed all your fellow-citizens . . . you devote your entire energies to winning them for Christ'.[309] Yet Colet's sermons came to touch upon matters which were politically controversial or doctrinally unsafe: he preached against war, just as Henry VIII launched grandiose expeditions to France; he preached against image-worship and against the temporal possessions of the Church; and he preached against those who could give only 'bosom sermons', reading from prepared scripts. Bishop Fitzjames took the last barb personally—he was meant to—and 'would have made . . . Colet . . . a heretic'. In 1513 Colet was banned from the pulpit.[310]

Colet 'read carefully heretical works', wrote Erasmus in his friend's obituary in 1519, 'and said he often got more profit from them than from those which are employed in endless definitions and servile adulation of certain doctors'.[311] In turn, heretics came to learn from him; like Thomas Geffrey, who said that 'true pilgrimage was, barefoot to go and visit the poor, weak and sick; for they are the true images of God'. He may have learnt that message from the humanist Dean as much as from his Lollard brethren.[312] In one of his *Colloquies*

[308] *L&P* iii/1. 303; Erasmus, *Correspondence*, 7, p. 163; Seebohm, *Oxford Reformers*, pp. 141–2.

[309] Erasmus, *Correspondence*, 2, p. 227.

[310] Foxe, *Acts and Monuments*, iv, p. 247; P. S. Allen, 'Dean Colet and Archbishop Warham', *EHR* xvii (1902), pp. 305–6; J. W. Blench, *Preaching in England in the late fifteenth and sixteenth centuries* (Oxford, 1964), pp. 79–80, 118, 216; J. A. F. Thomson, *The Later Lollards, 1414–1520* (Oxford, 1965), p. 252; H. C. Porter, 'The Gloomy Dean and the Law: John Colet, 1466–1519', in G. V. Bennett and J. D. Walsh (eds.), *Essays in Modern English Church History* (1966), pp. 18–43.

[311] *L&P* iii/1. 303.   [312] Foxe, *Acts and Monuments*, iv, p. 230.

Erasmus wrote of Colet's contempt for idle pilgrimages to saints.[313] Stephen Vaughan and Thomas Cromwell remembered Colet later, and may even have learnt reforming lessons for the future from his sermons.[314] But William Tyndale, who would have heard Colet's sermons in Oxford, said that it was because of the Dean's translation of the Pater Noster into English that he was suspected of heresy.[315] For the people to have the Scripture in English was thought to be dangerous, and had long been forbidden.

Erasmus published in 1516 his new Greek text of the New Testament, with a Latin translation. His hope was that laymen would be inspired to know the Bible; that one day every ploughboy at his plough would read the Gospel. Yet the learned language left this translation only for the educated. It was this first edition which Cromwell took with him to Italy and learnt by heart on his journey; and by this edition that Thomas Bilney was inspired in Cambridge.[316] To meddle with the Vulgate at all was thought, by some, doubtful and dangerous. In 1520 a Scotist Spanish friar challenged John Stokesley, a royal chaplain, to defend Erasmus, and attacked Erasmus for translating 'Word of God' in St John's Gospel as *sermo* rather than the traditional *verbum*. Friar Standish had already attacked Erasmus on the same point in a sermon at St Paul's, and was now effectively routed in argument by Thomas More.[317] Stokesley and More were both still champions of humanist reform, but later, as Bishop of London and as Lord Chancellor, they saw the urgent necessity to defend the Church against any novelty, and by the most savage means.

The brightest and best of the English humanist community were gathered in London on the eve of the Reformation. Were they infusing the ideal of orthodox renewal among the London citizenry? Colet was, and his influence was felt long after his

313 Seebohm, *Oxford Reformers*, pp. 288–9.

314 PRO, SP 1/47, fo. 149ʳ (*L&P* iv/2. 4107).

315 Foxe, *Acts and Monuments*, iv, p. 247; Tyndale, *Answer to Sir Thomas More's Dialogue*, ed. H. Walter (Parker Society, Cambridge, 1850), p. 168. John Gough would print Colet's English Pater Noster in his *Myrrour or lokynge glasse of lyfe* (1532?; *RSTC* 11499).

316 Foxe, *Acts and Monuments*, v, pp. 363–5; iv, p. 635.

317 Dowling, *Humanism*, p. 22.

death.[318] More's model for the missionary Bishop in Utopia was Rowland Phillips, one of the most eloquent of English preachers.[319] Phillips denounced the scandalous behaviour of pilgrims to shrines to the Virgin Mary, and so too did Friar Donald of the reformed order of Franciscan Observants, for they knew that such abuses undermined the sanctity of the practice. Friar Donald was probably the 'scotus quidam doctor theologus' who in January 1508, as Colet before him, expounded the Pauline Epistles at Paul's Cross. More had heard the friar's sermons in his youth and revered him as saintly.[320] Richard Ridley and Henry Gold, redoubtably orthodox critics of William Tyndale, left notes for sermons which urged reform in the life of the Church.[321] Bishop Tunstall, who was in a position to effect reform, ordered all dedication days of City churches to be held upon a single day from 1523, to prevent the abuses which had marred the festivals.[322] Such men may have been creating among Londoners an expectation of reform. Before the Reformation, and its emphasis upon a learned preaching ministry, there were already citizens who looked for edifying sermons. James Wilford, Sheriff in 1499, left money to provide for a 'doctor of divinity, every Good Friday forever, to preach a sermon of Christ's passion' at St Bartholomew's.[323] In 1520 Sir John Thurston, Sheriff, bequeathed £40 for exhibitions for two 'scholars priests and students . . . studying holy divinity . . . whereby the faith of Christ may be increased', and his widow left exhibitions to 'scholars priests students

---

[318] In 1533 Martin Tyndale offered Cromwell his translation of Erasmus's life of Colet: Elton, *Reform and Renewal*, p. 18.

[319] More, *Correspondence*, p. 80. The gild of St Mary Rounceval paid large sums in 1521 to attract Phillips to preach the Annunciation day sermon there: Westminster Abbey muniment room, account of the Guild of St Mary Rounceval (unnumbered). I owe this reference to the kindness of Gervase Rosser. For Phillips's preaching tour in 1511–12, see *Kentish Visitation of Archbishop Warham and his Deputies, 1511–12*, ed. K. L. Wood-Legh (Kent Records, xxiv, 1984), pp. xii, 38, 134, 143, 191.

[320] Bernard André, 'Annales Henrici VII', ed. J. Gairdner in *Memorials of King Henry VII* (Rolls series, x, 1858), pp. 105–6; More, *Dialogue concerning Heresies*, i, p. 100.

[321] BL, Cotton MS Cleo. E v, fo. 362ʳ (*L&P* iv/2. 3960). Gold's sermon notes were seized upon his attainder: *L&P* vii. 523. Cambridge University Library, MS Dd. v. 27.

[322] CLRO, Repertory 4, fo. 159ʳ; Repertory 6, fo. 50ʳ; Journal 12, fo. 243ʳ; Letter Book N, fos. 246ʳ–47ʳ, 263ʳ; C. Sturge, *Cuthbert Tunstal* (1938).

[323] Stow, *Survey*, i, p. 185; see also p. 198.

being pulpit men'.[324] Their desire was for sermons of orthodox tenor.

The religious houses in and around the capital were held in special reverence by the citizens. When Londoners founded new gilds at the turn of the sixteenth century they often chose to establish them in the City friaries, marking thereby their admiration for the spirituality of the friars.[325] In the 1520s the house of the Crossed Friars was being rebuilt with the aid of the citizens, and Sir John Milborne built fourteen alms houses there.[326] The nobility and the wealthy often chose to be buried in the friaries rather than in their parish churches.[327] The dowager Duchess of Buckingham asked that her heart be buried before the image of St Francis at the Grey Friars.[328] On the eve of the Reformation bequests to the monks and friars and nuns of the City were very common: Londoners sought their prayers and asked that the friars attend their funerals and sing masses. Nearly half the Londoners whose wills were proved in the Prerogative Court of Canterbury between 1520 and 1521 and between 1523 and 1525 remembered London's religious houses, and a third of those whose wills were proved between 1529 and 1530 did so. But very significantly, a large proportion of their bequests went to the mendicant rather than the possessioner orders, and the comparative popularity of the friars increased as the Reformation drew nearer. The Londoners, impressed by the austerity and spirituality of the Carthusians and Observant Franciscans, remembered them especially in their wills.[329] Leading Londoners had special links with the Carthusians of London and Sheen, the Bridgettines of Syon, and the Franciscan Observants of Greenwich and Richmond. Some became lay brothers of the houses.[330] Thomas More had withdrawn to the London

---

[324] PRO, PCC, Prob. 11/20, fos. 181ʳ, 183ᵛ–85ʳ. The will of John Thurston was not performed, it seems: *L&P* Addenda, i/1, 447.

[325] See above, p. 37.

[326] Stow, *Survey*, i, pp. 112, 147–9; Guildhall, MS 9171/10, fos. 36ʳ, 107ᵛ, 134ʳ; PRO, PCC, Prob. 11/20, fos. 69ᵛ, 211ᵛ. Letters of confraternity were issued to benefactors: BL, I A 55480.

[327] Stow in his *Survey* lists the memorials.

[328] PRO, PCC, Prob. 11/24, fo. 25ᵛ.

[329] PRO, PCC, Prob. 11/20, 21 & 24 (Registers Maynwaryng, Bodfelde, and Thrower). Thomson, 'Piety and Charity', pp. 189–90.

[330] See, for example, PRO, PCC, Prob. 11/20, fo. 215ʳ; 11/24, fo. 104ᵛ.

Charterhouse: for four years he lived there 'religiously' but 'without vow', contemplating entry to the religious life.[331] Sir Thomas Exmewe, Mayor in 1517, had a stepson, John West, who became a Franciscan Observant, and a nephew(?) William Exmewe, who entered the London Charterhouse and died a martyr.[332] Sir John Aleyn, member of the King's Council and Mayor in 1535, had contacts with Sheen, and Humphrey Monmouth, Sheriff in the same year, was a benefactor of Denny Abbey.[333] The devout laity admired the orthodoxy of these houses, but some monks provided more than a model of the religious life: some, notably Richard Whitford, John Fewterer, and William Bonde of Syon, were writing devotional tracts in English for the edification of 'simple souls'. Until the 1530s the Catholic conduct books had been directed primarily to the nobility and religious, but in the early 1530s works dwelling upon the medieval themes of confession and prayer, the life and passion of Christ, and tribulation, were becoming best sellers.[334] The capital, the centre of the book trade, with a populace more literate and prosperous than elsewhere, probably provided the most eager market for these works.

The great popularity of tracts such as Whitford's *Werke for housholders*, reprinted ten times between 1530 and 1537, and a *Pomander of Prayer*, written by a religious of Sheen Charterhouse, and in four editions between 1528 and 1532,[335] and the appearance of other treatises addressed to householders suggests the emergence of a new type of literate lay reader. Men were urged in the *Werke for housholders* that 'it should . . . be a good pastime and much meritorious for you that can read to gather your neighbours about you on the holy day, specially the young sort, and read to them this poor lesson'.[336] The Christian faith could be taught within the family; children and servants instructed, in the medieval catechical tradition, how to live well. The devout laity were seeking instruction in private devotion, beyond their public worship. Simon, the anchorite dwelling in

---

[331] Roper, *Lyfe*, p. 6.

[332] *L&P* iv/3. 5275. The relationship with William Exmewe is surmise.

[333] PRO, SP 1/85, fo. 145ʳ (*L&P* vii. 1091); Strype, *Ecclesiastical Memorials*, i/2, p. 365.

[334] J. Rhodes, 'Private Devotion in England on the Eve of the Reformation'.

[335] *RSTC* 25421.2–6; 25421.8–25426.

[336] Whitford, *A werke for housholders*, sig. Biiʳ.

London Wall, compiled *The fruyte of redempcyon* in 1514, a treatise which recounted, chapter by chapter, the life and passion of Christ. It was written 'in English for your ghostly comfort that understand no Latin', and went forth with the blessing of Bishop Fitzjames, who approved it to be 'read of the true servants of sweet Jesus, to their great consolation and ghostly comfort'.[337]

Christian conduct books adumbrated the ways of living well and avoiding sin. Whitford's *Werke for housholders* suggested that the day should begin with spiritual exercises: on waking 'make one whole cross from your head unto your feet, and from the left shoulder unto the right, saying all together In nomine patris et filii et spiritus sancti'. Every day must be spent as if it were the last: 'The wise man sayeth . . . in all thy works remember thine ending day.'[338] The anonymous author of *A dyurnall: for deuoute souls: to ordre themselfes therafter* advised 'perseverant exercise' to follow good and avoid evil. Upon waking the Christian must ask himself if anything other 'hath occupied your heart than the rule of perfection requireth'. If so, he must admit his fault and desire to amend: 'begin anon with humble prostration both of spirit and also of body (if ye be alone) to praise the glorious Trinity'. Then 'when ye do on your clothes, see that your mind be occupied in the praise of God, thanking Him that He hath so plenteously provided for you . . . and pray Him to move your heart to relieve his poor people'. Before leaving his chamber the Christian should say a prayer 'before some image of our Lord Jesus Christ, first thanking Him for your creation and redemption'. On the way to church 'have mind of your good angel and other saints', 'preparing your heart to prayer by the remembrance of some part of their life'. In church give thanks for your 'leisure to ensue the spiritual life', and remember those who are not free to come to church even though 'peradventure . . . much more fervent in the love of God'. 'Departing from the church, beware that ye fall not anon to idle speech.' Returning home, 'on your knees salute Our Saviour and his mother'. 'Apply yourself to some profitable occupation . . . and say to yourself, and wouldest thou be thus occupied if thou shouldst die this

337 *The fruyte of redempcyon* (1514; *RSTC* 22557), sig. Diiii^r.
338 Whitford, *A werke for housholders*, sig. Aii^v.

day?', and consequently 'spend your time at every hour to the most profit of your soul'. At the end of the day the Christian should confess his faults to Jesus, and consider what 'saints in special ye have served that day', counting the day lost in which 'ye do not obtain some friendship of the citizens of Heaven'.[339] Citizens of London bought such books, certainly, aspiring to follow these counsels of perfection.

The personal and devotional lives of the mass of Tudor Londoners remain inscrutable, but there were some who left writings which provide glimpses of their spiritual aspirations, and show them touched by the spirit of reform. John Colyns and Richard Hill, men of some standing in the City, and an anonymous Londoner kept commonplace books in the 1520s and 1530s—collections of poems, songs, fables, prophecies, miracles, *exempla*, jokes, aphorisms, recipes, remedies, City chronicles, devotional instructions, descriptions of London and its customs, accounts of contemporary and past political events —which reveal more about the religious and political attitudes of individual citizens than almost any other source.[340] John Colyns was churchwarden at St Mary Woolchurch, a mercer and like other mercers, involved in the new printing trade.[341] Richard Hill became a Merchant Adventurer and was made free of the Grocers' Company in November 1511. He married Margaret, daughter of Henry Wyngar, haberdasher, and they had seven children ('God make them all his servants').[342] Colyns died a good Catholic. At his death Colyns willed that 'there shall be said for my soul seven masses at scala celi whilst I do lie in extremis, and other four as soon after my decease as may be done'. He was a brother of the fraternity of Our Lady and St Anne in his parish.[343] Yet there were also signs that

---

[339] *A dyurnall: for deuoute soules: to ordre themselfe therafter*, sigs. Aii^r–Ci^v.

[340] John Colyns's commonplace book: BL, Harleian MS 2252. The verse is catalogued in B. L. Besserman, G. Gilman and V. Weisblatt, 'Three Unpublished Middle English Poems from the Commonplace book of John Colyns', *Neuphilogische Mitteilungen*, lxxi (1970), pp. 212–38. Richard Hill's commonplace book: Balliol College, MS 354, part of which is printed in *Songs, Carols and other Miscellaneous Poems*, ed. R. Dyboski; EETS, extra series, 101, 1908). BL, Lansdowne MS 762 (anonymous).

[341] *Acts of Court of the Mercers' Company*, p. 509. He was made free of the company in 1498. GLRO, DL/C/330, fo. 123^r; BL, Harleian MS 2252, fo. 22^r.

[342] Balliol College, MS 354, fos. 17^r, 107^r.

[343] Guildhall, MSS 9171/11, fo. 56^v; 9168/9, fo. 180^v.

Colyns, and Hill too, had come into contact with reforming circles in the City.

What religious aspirations and expectations led these men to transcribe poems of exemplary Catholic piety and anticlerical tirades side by side in the pages of their commonplace books? Colyns collected anticlerical poems: not only John Skelton's famous jibes, 'Colin Clout' and 'Speke Parrot', but more virulent condemnations of the Cardinal.[344] The attacks upon the priesthood were meant to be more than simply destructive. In certain of the poems the moral intent of criticizing the clergy was clear:

> He that sweat both water and blood
> Amend our priests and make them good.[345]

Of all the poems the most impassioned was 'The Ruin of a Realm', in which the author nostalgically lamented the moral, political, and above all the religious decadence of the country. The cause of the decay was manifest:

> It is apparent to every man's eye
> That spiritual men undoubtedly
> Doth rule the realm brought to misery.

The remedy was harder to find than the reason. 'The spiritualty is disguised like men in a play', the prelates so 'blinded with pride and iniquity', that no one dared to reproach or reform them, save one:

> A famous divine,
> Which to these vices will not incline,
> As by his preaching perceive we may,
> Saying this realm beginneth to decay.

This paragon was surely Colet, who stood apart from the self-seeking of the rest, 'alleging Scripture for every sentence, a profound man of learning and sapience'.

> Christ would we had many of this sort
> In living and preaching from vice us to guide.[346]

---

344 BL, Harleian MS 2252, fos. 133ᵛ–40ʳ, 147ᵛ–154ᵛ.
345 Ibid., fo. 156ʳ.
346 Ibid., fos. 25ʳ–28ʳ.

While deploring the current state of the Church, Colyns headed every page 'Jesus', or 'IHS', and side by side with the anticlerical poems were stories of miracles, and *exempla* urging pious charity. 'O Mortal Man call to remembrance' was a prayer for souls in Purgatory, urging personal mediation rather than the proliferation of masses conventionally ordered.[347] In the anonymous commonplace book the most pious invocations of divine grace were included among the prophecy of Skelton to 'let Colin Cloute alone' with its implied criticism of the clergy. And that unknown author transcribed, too, the articles of reform demanded by the German Peasants in 1525:

first, that no manner priest shall be suffered to take cure of souls but if he be of honest living and have good and sufficient learning, and forty years of age or more . . .
   That all manner of sacraments of the Church to be ministered unto every person at times and as often as shall be desired without taking any duty therefore.[348]

These demands had a special resonance for Londoners in the 1520s. Richard Hill's book contained poems of undoubted orthodoxy, instructions for the priest in confessional, descriptions of the virtues of the good priest, but also criticisms of the worldliness of the clergy:

> I was with Pope and Cardinal,
> And with Bishops and priests great and small,
> Yet was never none of them all
> That had enough and could say whoa.[349]

Criticism of the clergy and conventional Catholic piety could, indeed should, coexist in pre-Reformation London. A good Catholic should hope for renovation within the Church, deplore its current state, and yearn for a purity and sanctity which had once existed in an apostolic golden age. Reform was needed, and urgently, but whence should it come?
   'The way whereby the Church may be reformed into better fashion is not to make new laws. For there be laws many enough and out of number', so Colet and other Catholic

[347] BL, Harleian MS 2252, fos. 23ʳ–24ᵛ.
[348] BL, Lansdowne MS 762, fos. 7ᵛ, 9ʳ, 9ᵛ, 16ᵛ, 21ᵛ–22ʳ, 51ᵛ, 76ʳ–77ʳ.
[349] Balliol College, MS 354, fo. 210ᵛ.

reformers admitted.[350] The people were so bound by a multiplicity of laws that the simplicity of religion might be lost, and while the faithful were so little reminded of love and compassion the mercy of Catholicism might be forgotten.[351] So Erasmus lamented, and so Skelton mocked the typical cleric:

> And yet he will mell
> To amend the Gospel,
> And will preach and tell
> What they do in Hell.
> And he dare not well neven
> What they do in Heaven
> Nor how far Temple bar is
> From the seven stars.[352]

The castigation of the Catholic Church for the 'blockish burden of their constitutions, laws and statutes', and the insistence instead upon the compassion of Christ's teachings and promises would be at the centre of reformist teaching soon.[353] Yet Catholic reformers insisted too that Christ was to be worshipped not by custom and ceremonies, but in charity, and by meditation upon His passion. God's answers to 'a certain creature that desired to wit what thing was most pleasure to him in this world' were thus, recorded in the anonymous Londoner's commonplace book:

Give out tears for thy sins and for my passion, for that pleasureth me more than thou wept for worldly things as much water as is in the sea.

Me only love, and all other for me, for that pleaseth me more than if thou every day go upon a wheel sticking full of nails that should prick thy body through.[354]

The late medieval Catholic Church was not monolithic; it was capable of self-criticism and adaptation. But the critics within might go too far, demanding change too radical, and challenging practices too sensitive to be questioned. When this

---

[350] *EHD* v, p. 656.

[351] This point is made in M. Bowker, 'Some Archdeacons' Court Books and the Commons' Supplication against the Ordinaries of 1532', in *The Study of Medieval Records: Essays in Honour of Kathleen Major*, ed. D. A. Bullough (Oxford, 1971), pp. 314–16.

[352] John Skelton, *The Complete English Poems*, ed. J. Scattergood (Harmondsworth, 1983), p. 267.

[353] See below, pp. 110, 119–20.   [354] BL, Lansdowne MS 762, fos. 9ʳ⁻ᵛ.

happened those critics, though thinking themselves orthodox, might be charged with heresy.

In 1525 an Observant friar wrote a devotional treatise which he offered in humility as a New Year's gift to the 'good ladies' of Syon Abbey.[355] He wanted to present them with an 'image of love', but this gift was hard to find. Dame Nature, the World, and artificers all showed him tempting images of love, but these were of the world's love, and like the world, transitory, 'and if ye set your love upon the world . . . the charity of God Our Saviour Christ that is the very image of love cannot be in you'. The only true image of love is discovered in Holy Scripture; seen in a glass darkly, but tantalizing in its beauty. The more the beholder aspires to a life of purity, of charity, of love towards God, neighbours, and even enemies, the more clearly he will perceive the image, for God will be in him, and the image of love will be reflected. The treatise ends with an exposition of Hugh of St Victor's celebration of charity. The spirit of *The ymage of loue* may seem admirably orthodox, yet two months after its printing (7 October) the book's printer and translator were before the Vicar-General of London and charged to reclaim the New Year's gift from sixty nuns of Syon, and to retrieve all copies sold.[356]

For in his ardent insistence that religious life should return to a simpler, other-worldly state the author had attacked some of the most cherished forms of Catholic devotion. In the shop of the artificers the author had almost abandoned his search for the true image of love when he was shown 'many goodly images which I thought should stir a man to devotion and to the love of God'.[357] But a 'holy devout doctor' appeared to rebuke him: 'why dost thou cast away thy money upon these corruptible and vain things?' 'Seest not thou the goodly living image of God [the poor] most pitifully fade and decay every day in great multitude?'[358] When the author protested that the

---

[355] *The ymage of loue*. The tract, its authorship and significance are discussed in E. Ruth Harvey, 'The Image of Love', *CW* 6/ii, appendix A, pp. 729–59. For Syon and its patronage of spiritual writings, see P. Hodgson, 'The *Orchard of Syon* and the English Mystical Tradition', *Proceedings of the British Academy*, l (1964), pp. 229–49.

[356] GLRO, DL/C/330, fo. 103ᵛ; A. W. Reed, *Early Tudor Drama* (1926), pp. 166–8.

[357] *The ymage of loue*, sig. Biiᵛ.

[358] Ibid., sig. Biiiʳ.

doctor thereby condemned the 'many good men that be now-adays which honour the temples of God with many goodly images', the doctor made a rueful comparison between the simple worship of the early Church and the corrupt present: 'then were treen chalices and golden priests now be golden chalices and treen priests or rather earthen priests' (a criticism so like Colet's that the doctor might almost be the late Dean). 'We find not', said the doctor, 'that Christ commandeth us to have so costly ornaments in His Church, but He commandeth us many times to nourish His poor people.'[359] The author then began to question some of the traditional forms of Catholic piety. 'Thus I perceived that charity might lack, for all these gay outward things'; that these might hinder 'charity and contemplation, stirring the mind to elation and vainglory'. The doctor knew that such questioning was dangerous: 'why do I labour in vain? What need me to stir all the world against me?' Fear of the world would not stop the reformers, who would adopt the doctor's message and dare to evangelize it.

[359] Ibid., sig. Biiii^{r-v}.

II The Heretical Community

THE YMAGE of loue could not be suppressed. In 1532 it was reprinted by its first translator, John Gough, one of London's most fervent evangelicals.[1] Two years earlier the Pewterers' Company had hired John Gough to bind their account book. Seizing any and every chance to spread the essential reforming message, he inscribed on the endpaper of the book: 'Heaven and earth shall pass away, but my Word will remain forever; the Word of God shall last for eternity, saith Christ.'[2] The 'brethren and sisters of the false fraternity of heresy', like Gough, had a mission which was urgent and dangerous: the Word, hidden from the faithful for a thousand years, must go forth by whatever means, and whatever the risk. Thomas More warned Catholics, complacent in their ancient faith, of the threat from this 'new broached brotherhood',[3] spawned by the Devil, and with all his wiles.

Like as a few birds always chirking and flying from bush to bush, many times seem a great many; so these heretics be so busily walking, that in every ale house, in every tavern, in every barge, . . . as few as they may be a man shall find some, and there be they . . . so fervent and importune . . . that between their importune praising, and the diligence or rather the negligence of good Catholic men, appeareth often times as great a difference, as between frost and fire.[4]

More was right that the new heretics were few, but formidable. They had everything to gain for their faith, and in persecution, for Christ's sake, there was a special glory. Thomas Arthur implored his audience at St Mary Woolchurch on Trinity Sunday 1527 to pray for those 'that now be in prison for

---

[1] More found it necessary to refute in his *Dialogue concerning Heresies* the teachings of *The ymage of loue* which had been adopted by the Messenger. This suggests that copies were still available between 1525 and 1531. *RSTC* 21472.

[2] Guildhall, MS 7086/1, fo. 220ʳ. Gough printed the biblical assurance: 'Heaven and earth shall fail, but my word and promise shall never fail'; John G, *A lytell treatyse called, the myrrour or lokynge glasse of lyfe* (1532? *RSTC* 11499), sig. A viiiᵛ.

[3] For More's scornful references to the brethren, see *Apology*, pp. 7, 9, 14, 15, 29; and for their naming of themselves, More, *Debellation*, pp. 24 ff.

[4] More, *Apology*, pp. 159–60; *Answer to a Poisoned Book*, p. 3.

preaching the true Gospel of God', and made this plea for himself in tears:

if I should suffer persecution for the preaching of the Gospel of God, yet there is seven thousand more that shall preach . . . therefore good people, good people . . . think not you that if these tyrants and persecutors put a man to death . . . that he is an heretic therefore, but rather a martyr.[5]

To be accused of heresy outraged Arthur and his fellows: it was not they who had betrayed the teachings of Christ, but the priesthood, who had so far traduced their sacred charge that now 'Of Christ's word they make heresy'.[6]

Wilfully to doubt or to deny cardinal doctrines of the Catholic faith was heresy. Only the Church, which had been disobeyed, could define what constituted this darkest of sins.[7] But before the Reformation, and after, heresy might take many forms. No single and adamantine code of heretical belief existed in England at the turn of the sixteenth century; indeed, for Sir Thomas More, 'among heretics there be as many divers minds almost as there be men'.[8] In November 1511 Andrea Ammonio wrote to Erasmus from London that his servant's dullard brother 'has instituted a sect on his own, if you please, and has followers too'.[9] Before the new creed arrived from Germany, the heretics who were discovered were those who scandalized good Christians, and who offended against the mores of the community in which they lived. They were reported lest their crimes bring down the vengeance of God, whom they had mocked, upon the society which sheltered them. Yet people might say all sorts of unorthodox things, which might not be called heresy, and the laity might find heresy in unlikely actions. In March 1518, just before the 'blessed season' of Easter, four men and two women 'well deserved to be punished as *heretics*'. They had 'dissembled with God Almighty and with

---

[5] Guildhall, MS 9531/10, fo. 135ᵛ.

[6] 'Rede me and be nott wrothe', ed. E. Arber, *English Reprints* (1871), p. 69.

[7] Lyndwood, v, tit. 5. c.1–4; G. Leff, *Heresy in the Later Middle Ages*, 2 vols. (Manchester, 1967); *The Concept of Heresy in the Middle Ages*, ed. W. Lourdaux and D. Verhelst (Louvain, 1976); J. A. Guy, 'The Legal Context of the Controversy: The Law of Heresy', in More, *CW* 10, pp. xlvii ff. I am most grateful to Dr Guy for allowing me to read his introduction to *The Debellation* in advance of its publication.

[8] More, *Dialogue concerning Heresies*, i, p. 191.

[9] Erasmus, *Correspondence*, 2, p. 189.

the world', seeking to extort money from the credulous faithful by pretending the 'fearful visitation of God called the falling sickness'.[10] When in 1514 a man left his City company for another he was accused of being 'worse than an heretic' for breaking his oath.[11]

With challenges to the priesthood many of the London laity might often have sympathized, but to the clergy any denial of their God-given power of the keys smacked of heresy. So when John Steward of St Mary Magdalen Old Fish Street scorned the sentence of excommunication against him in 1493: 'I set nothing by curse if I be once on horseback, and my feet within the stirrups', he was accused 'de crimine heresim sonante et tangente', for he had spoken in gravest contempt of the Church.[12] Elena Dalok's utterances in the same year might have seemed more heretical by far to her neighbours, certainly more insulting to God: she had boasted that none who received her malediction ever lived to tell the tale; said that to have Heaven in this life would be to be denied it in the world to come; claimed that she had the power to sunder the hooks of Hell which bound the dead John Gibbs. Elena Dalok never lived in a godly manner, but always 'diabolice'; she would not confess to her priest at St Mary Abchurch, for he knew nothing of the spiritual life, she said; she could make it rain at her command, and had a book which foretold all future events. Yet for all this, she was accused not of heresy, but of being an enchantress, an anathematizer, a slanderer of her neighbours; all lesser sins than heresy.[13]

The history of heresy is often only the history of persecution; heretical enclaves are discovered only when the authorities seek them, and find them; the nature of their dissent revealed only in the light of the questions which the persecutors ask. There were individual dissidents in London, with beliefs so bizarre that no one else shared them, but there was also a distinct heretical community. Beleaguered and persecuted, it was perforce secret, composed of 'privy men'. The City's heretical underground remains indistinct, because those who sought it out

---

[10] CLRO, Journal 11, fo. 333ʳ; cf. Repertory 3, fo. 197ʳ.
[11] Ibid., Repertory 2, fo. 187ʳ.
[12] Hale, *Precedents*, cli.
[13] Ibid. cxxxvii.

could never find all its members, nor discover its movements, but the names of the 'known men' of Lollardy, and of the 'brethren' of early Protestantism (as they called themselves), with their families, fellows, and friends, will appear and reappear throughout the history of London's Reformation, and their activities may be glimpsed as a picture seen in a fire. This chapter will attempt to tell something of the history of heresy in London, rather than the history of persecution.[14] Yet the evidence will often come from the persecutors: from Bishop Fitzjames's discovery of thirty-nine and maybe more Lollard conventiclers in the capital between 1510 and 1518, from the *magna abjuratio* after Bishop Longland's quest through his great diocese of Lincoln in 1521 when at least twenty-five City Lollards were found out, and from the searches made by Bishop Tunstall in his own diocese of London.[15] The Bishops' initiatives against pestilential heresy, their plan of campaign, their success in finding offenders will be the story of the next chapter.

[14] Every historian of the Reformation depends upon the great work of the first historian of the Lollards and the Protestants: John Foxe, *Acts and Monuments*. For later histories of Lollardy and the Reformation, see James Gairdner, *Lollardy and the Reformation in England*, 4 vols. (1908–13); A. G. Dickens, *Lollards and Protestants in the Diocese of York, 1509–1558* (Oxford, 1959); *The English Reformation* (1964); 'Heresy and the Origins of the English Reformation', in *Reformation Studies* (1982), pp. 363–82; J. F. Davis, 'Heresy and Reformation in the South-East of England, 1520–1559' (University of Oxford, D.Phil. thesis, 1968) and *Heresy and Reformation in the South-East of England, 1520–1559* (1983); 'The Trials of Thomas Bylney and the English Reformation', *HJ* xxiv (1981), pp. 775–90; 'Joan of Kent, Lollardy and the English Reformation', *JEH* xxxiii (1982), pp. 225–33; Margaret Aston, *Lollards and Reformers: Images and Literacy in Late Medieval Religion* (1984), chs. 5, 6, 7; J. A. F. Thomson, *The Later Lollards, 1414–1520* (Oxford, 1965); J. Fines, 'Heresy Trials in the Diocese of Coventry and Lichfield, 1511–1512', *JEH* xiv (1963), pp. 160–74; A. Hope, 'Lollardy: the Stone the Builders Rejected?', in P. Lake and M. Dowling (eds.), *Protestantism and the National Church in sixteenth-century England* (1987), pp. 1–35. As this book goes to press, a classic study of Lollardy and the Reformation appears: Anne Hudson, *The Premature Reformation: Wycliffite Texts and Lollard History* (Oxford, 1988). This work is of fundamental importance to the subject, but is published too late for it to be sufficiently cited here. I was fortunate to read part of the book in typescript, for which I am most grateful to Dr Hudson.

[15] Bishop Fitzjames's Court book containing abjurations for heresy made between 1509 and 1518 (numbering at least 285 folios) is now lost, but before it disappeared Archbishop Ussher transcribed a number of cases, and Foxe used it too: TCD, MS 775, fos. 122ᵛ–125ʳ; Foxe, *Acts and Monuments*, iv, pp. 173–214. Ussher's transcription of the articles against Richard Hunne is printed in J. Fines, 'The Post-Mortem Condemnation for Heresy of Richard Hunne', *EHR* lxxviii (1963), pp. 528–31. For Longland's persecution of 1521, see Foxe, *Acts and Monuments*, iv, pp. 221–43. For Tunstall's drive against heretics in his diocese in 1527–8, see BL, Harleian MS 421, fos. 7ʳ–35ʳ; Strype, *Ecclesiastical Memorials*, i/1, pp. 113–34; i/2, pp. 50–65.

THE KNOWN MEN

London's first heretical community was that of the 'Lollards', the name their enemies gave to the followers of the teachings of John Wycliffe. Joan Boughton, an 'old cankered heretic' of eighty, born in the year of Oldcastle's ill-fated revolt, went to the stake on 28 April 1494. So convinced a 'disciple of Wycliffe' was she that all the divines in London could not turn her from even one of her heretical beliefs. Warned that she would be burned for her obstinacy and false belief, she defied them: 'she said she was so beloved with God and his holy angels, that all the fire in London should not hurt her'. Soon it appeared that she left some of her disciples behind her, for the night following her burning her ashes were stolen away in an earthen pot, as a 'precious relic'.[16] She had died a martyr for a cause, and for a sect which recognized itself as marked by special providences. Maybe Joan Boughton had been a heretic all her life, since Lollardy's very beginnings, but maybe she was a recent convert, for the hidden heresy was beginning to revive. In 1499 the Milanese ambassador reported home the appearance of 'a new sect of heretics', holding bizarre and scandalous opinions: 'that baptism is not necessary to those born Christians, that marriage is superfluous and copulation suffices, and that the sacrament of the altar is not true.' The ambassador was mistaken to think this a new sect, but right that the Lollards were on the increase and that the worried 'prelates have begun to persecute them'.[17]

Lollards had long congregated in the capital. It was from London that the Lollard armies, under the captaincy of Sir John Oldcastle, had in 1414 attempted to capture the King, to seize political power by force, to dispossess the Church, and to divide the kingdom among themselves. The rebellion was a spectacular failure. The Lollards gained martyrs for their cause, but lost forever any chance of support from the political orders, or of conciliation with Church or state.[18] Throughout

---

[16] *The Great Chronicle*, p. 252; *The Chronicles of London*, ed. C. L. Kingsford (Oxford, 1905), p. 200. Boughton's case is discussed in Thomson, *Later Lollards*, pp. 156–7.

[17] *CSP Milan*, i. 627 (13 July 1499).

[18] K. B. McFarlane, *John Wycliffe and the Beginnings of English Nonconformity* (1952), ch. 6, pp. 160 ff.; Aston, *Lollards and Reformers*, ch. 1.

the fifteenth century the 'known' men and women of the sect
kept the faith in secret, intermittently persecuted. Certain in
their adherence to fundamental Lollard tenets, guarding the
cherished tracts which enshrined their lore, and reading from
their vernacular Bibles, they were nevertheless completely
without political leadership, and without theological guidance
to ensure theological orthodoxy and spiritual regeneration.[19]
The Lollard community of fifteenth-century London was
shadowy. The sect sustained itself in ways which were hard to
discover, but there were suspicions: was it true, asked Church
officials in 1521, 'that such as were of that sort did contract
matrimony only with themselves, and not with other Christians?'[20]

Lollardy was a faith practised in homes, not in churches. To
understand the nature of the Lollard conventicles is to understand the way in which the faith evolved, for Lollard doctrine
was not an immutable canon, unchanged since Wycliffe wrote
and his first followers proselytized. Lollard belief developed as
diverse dissident opinions emerged from private speculation
and were exchanged in discussion with the like-minded. At one
'night school' in London 'heretics were wont to resort to their
readings in a chamber at midnight'.[21] Every Lollard meeting
like this took on the character of a conspiracy, a confederacy,
the very secrecy strengthening the comradeship of the 'privy'
men and women.[22] From the testimony which Bishop Fitzjames
and his officials prised from a group of some forty City Lollards,
who were detected and who abjured between 1510 and 1518,
something of the nature of their fellowship appears.[23] Richard
Wolman, a draper from the parish of St Ethelburga, would
often go to John Calverton, a cooper in St Bartholomew the
Little, to borrow his copies of the *Passion of Nicodemus* and of
Antichrist, and would sit reading them in the window of John's
house. William Mason and John Stilman, survivors of a daring
escape from the Bishop's prison, would read the visionary
Apocalypse together companionably in William's house for three

[19] Thomson, *Later Lollards*, ch. 6; Aston, *Lollards and Reformers*, ch. 6.
[20] Foxe, *Acts and Monuments*, iv, pp. 222–3.
[21] More, *Dialogue concerning Heresies*, i, p. 328.
[22] *De Haeretico Comburendo*; *Statutes of the Realm*, II, p. 126; Aston, *Lollards and Reformers*, pp. 198–9.
[23] TCD, MS 775, fos. 122ᵛ–125ʳ.

hours at a time.[24] Thomas Austy, Thomas Vincent, Wolman, Mason and his wife, and Thomas Goodred used to assemble to recite the Lollard laws, tell their stories, and expound Scripture. Roger Heliar and William Sweeting 'magnum inter se habebant familiaritatem et societatem'.[25] It was natural that Lollards should often be 'in company' with each other, because they would avoid, and should be banished from friendship with 'other Christians', for all the Church's attempts to reconcile them.[26]

These Lollards met together not to pray: vocal prayer was but 'lip labour'. Reading the Bible was the beginning of wisdom. The Lollards were story tellers, and the stories they told were from Scripture. The adolescent John Forge remembered hearing James Brewster, a Colchester carpenter, telling his father and Thomas Goodred the story of Moses and the making of the brazen calf, of the Salutation of the Blessed Virgin Mary by Archangel Gabriel, and of Tobias and the angel. William Sweeting, 'le Cowherd' of Chelsea, told Thomas Forge the story of the conversion of St Paul; of how Paul, while still a persecutor of the Christians, set forth with his troops against the city of Damascus, of how three claps of thunder threw Paul to the ground, whence he was raised up by the Angel of the Lord who told him that he must go to the house of Simon to hear what he must do.[27] Elizabeth Blake, the thirteen-year-old daughter of a Lollard living by St Anthony's school, already knew by heart and could recite the Epistles and Gospels of the Evangelists.[28] Knowledge of the Scriptures, and in English, was the touchstone of the Lollard faith: 'it seems to them that they need nothing into the School of God's law and service save only the Holy Scripture alone'. It was the Lollard obsession with vernacular Scripture which had outlawed the English Bible, not only to them but also to all others, since Archbishop Arundel's constitutions of 1409.[29]

[24] TCD, MS 775, fo. 124^r.

[25] Ibid., fo. 123^r.

[26] Foxe, *Acts and Monuments*, iv, pp. 206, 209, 219, 223, 226, 227.

[27] TCD, MS 775, fo. 123^r.

[28] Ibid., fo. 124^r.

[29] Reginald Pecock, *The Repressor of Over much Blaming of the Clergy*, ed. C. Babington (1860), i, p. 129; M. Deanesley, *The Lollard Bible* (Cambridge, 1920); Lyndwood, v, tit. iv. c.3.

Images were the books of poor and illiterate Catholics, but poor and illiterate Lollards had real books. Possession of the Lollard Bible and of the other treasured texts was always the mark of the 'known man', and the first interest in the heresy trials, for the Lollards' books sustained the movement.[30] In 1495 penitent at Paul's Cross 'stood four Lollers with the books of their lore hanging about them, which books were, at the time of the sermon, there burnt with the faggots that the said Lollers bore'. In 1521 John Phips of Hichendon would perforce burn his books, lest the Bishops find them and the books burn him, but he was severely chastised by his fellows because those books were worth a hundred marks.[31] Poor Lollards would make huge financial sacrifices to buy their texts, which were the more expensive because they were in manuscript.[32] Robert Benet, a wool racker and humble water carrier from the parish of St Mary Somerset, had already been detected for heresy in 1496 but still in 1504 or 1505 he had sold his looms and shears in order to pay 3s. 4d. for a little manuscript copy of the Four Evangelists. Even so, Benet could not read or understand his treasure, but kept it secret and safe in his belt. Thomas Capon, the stationer of Pater Noster Row who had sold Benet the book, came to stay in Benet's house and there taught him its truths. When Capon died Benet sold the book to Thomas Austy of St Mary Matfelon for a 'horse load of hay'. Austy owned a notable Lollard library already. A volume containing 'the Epistles of James, the Apocalypse and a book called Mathew' was 'left with him at Bristol when he was keeper of the prisoners there'. His wife Joan had brought a copy of Wycliffe's 'Wicket' with her when she married him, as a Lollard dowry. Her late husband, John Redman, had entrusted the treasure to her upon his death-bed.[33] Henry Hert's copy of the Ten Commandments had passed from John Woodruff, a net knitter of All Hallows the Great, to Elizabeth Bate, the wife of a poor bladesmith in St Andrew Hubbard, and back to Hert

---

[30] Hudson, *The Premature Reformation*; *Heresy Trials in the Diocese of Norwich, 1428–31*, ed. N. P. Tanner (Camden Society, xx, 4th series, 1977).

[31] *Chronicles of London*, ed. Kingsford, p. 211; Foxe, *Acts and Monuments*, iv, pp. 237, 218.

[32] Hudson, *The Premature Reformation*, pp. 200–8; Aston, *Lollards and Reformers*, p. 200.

[33] TCD, MS 775, fos. 123^{r-v}, 124^v.

again. Small wonder that the book was 'senex et vetusta et aliquantulum obfuscata'.[34] Robert Raskell confessed that he had copied out 'Wycliffe's Wicket and heard it read and taken heed thereof'. This book was the *vade mecum* of the conventicle he attended with his master Robert Cook, with Austy, John Wyly, Thomas Blake, and Robert Quick. For John Stilman, who confessed in September 1518 and was martyred, the Wicket was a 'good and holy book'.[35] Almost all of these detected during Fitzjames's quest were found to be reading, owning, or listening to the same forbidden works: the book of the Four Evangelists, the Ten Commandments, the Apocalypse, the Epistles of St James and St Paul, and Wycliffe's Wicket.[36]

The Gospels and the Epistles were the inspiration of the movement, for it was the life of Christ and his apostles, before the Church became corrupted, that the Lollards aspired to emulate. St James's Epistle especially provided the model of simple piety whereby 'known men' should live. John Barret, a City goldsmith detected in 1521, 'was heard in his own house to recite the Epistle of St James'. He had his entire household for an audience: his wife Joan, his maid Jude, John Newman, and his apprentices. Their copy of the Gospels of St Matthew and St Mark his wife lent to John Scrivener.[37] Reading the Bible aloud, and expounding its sacred truths, was the basis for any Lollard assembly. It mattered little that most of the 'known men' were unlearned, even illiterate, for those who could not read would listen, and have knowledge of the Word 'by the spirit'.[38] John Harrydance, a Whitechapel bricklayer detected for his evangelical zeal in 1537, admitted that he 'hath the New Testament ever about him', and that 'he all these thirty years hath endeavoured himself to learn the Scripture, but he cannot

---

[34] TCD, MS 775, fos. 122ᵛ, 123ᵛ. Foxe describes this book as 'of an old writing almost worn for age'; *Acts and Monuments*, iv, pp. 215–16. William Bate, bladesmith of St Andrew Hubbard, died a pauper in 1511: Guildhall, MS 9168/3, fo. 57ᵛ.

[35] TCD, MS 775, fo. 125ʳ.

[36] Ibid., fos. 122ʳ–125ʳ. Cf. Foxe's summary of the books most commonly found during Fitzjames's persecution: 'the book of the four evangelists, a book of the epistles of Paul and Peter, the epistle of St James, a book of the Apocalypse and of Antichrist, of the Ten Commandments, and Wickliff's Wicket': *Acts and Monuments*, iv, p. 207.

[37] E. G. Rupp, *Studies in the Making of the English Protestant Tradition* (1947), p. 8; Foxe, *Acts and Monuments*, iv, pp. 228, 224.

[38] Aston, *Lollards and Reformers*, ch. 6; A. Duke, 'The Face of Popular Religious Dissent in the Low Countries, 1520–1530', *JEH* xxvi (1975), pp. 41–67.

write or read as he saith'.[39] Thomas Man had 'heard say, the Word of God and God to be all one, and that he that worthily receiveth the Word of God, receiveth God'.[40] For the Lollards, as for their spiritual heirs the Puritans, to hear the Word was a kind of sacrament.

Catholic devotion was, for Lollards, superstition; Catholic veneration, idolatry. In four cardinal points the Lollards stood against the Catholic Church: 'in pilgrimages, in adoration of saints, in reading of Scripture-books in English, and in the carnal presence of Christ's body in the sacrament'.[41] Those beliefs which the orthodox held most sacred the Lollards refuted, even vilified. For one heresy above all Lollardy was feared and persecuted, because it was among the gravest heresies of all: the Lollards impugned the sacred mysteries of the Mass. The Lollards dared to doubt the divine miracle of transubstantiation saying that Christ, one and indivisible, could not be both on earth with men and in Heaven with the Father, that Our Lord's body could not be made by the workings of corrupt priests.[42] Thomas Austy and his Lollard friends spoke together of the sacrament of the altar, and insisted that the body of Christ was not there, only material bread.[43] At Aldermanbury in 1510 Elizabeth Sampson defiantly avowed: 'I will not give my dogs that bread that some priests doth minister at the altar when they be not in clean life, and also said that thy self could make as good bread as that was and that it was not the body of Our Lord, for it is but bread, for God cannot be both in Heaven and earth.'[44] Experiencing such doubts themselves, the Lollards presumed to enlighten the faithful. As Rivelay came from Mass at the Grey Friars in 1520, asserting that he had just seen 'his Lord God in form of bread and wine over the priest's head' he met John Southwick who contradicted him: 'Nay, William! thou sawest not thy Lord God, thou sawest but bread, wine and the chalice . . . but only a figure or

---

[39] PRO, SP 1/124, fo. 155ʳ (*L&P* xii/2. 624).

[40] Foxe, *Acts and Monuments*, iv, p. 209.

[41] Ibid. iv, p. 218.

[42] Fines, 'Heresy Trials in the Diocese of Coventry and Lichfield', pp. 166–7; Thomson, *Later Lollards*, ch. 12; Hudson, *The Premature Reformation*, pp. 281–90.

[43] TCD, MS 775, fo. 123ʳ; Foxe, *Acts and Monuments*, iv, p. 175.

[44] Guildhall, MS 9531/9, fo. 4ʳ.

sacrament' of Christ.[45] Lollard rejection of the Mass was not just materialist and rational but spiritual. Elizabeth Stamford of London had been taught in 1506 that 'the holy Sacrament, Christ's own body: this is not received by chewing of teeth but by hearing with ears, and understanding with your soul, and wisely working thereafter'.[46]

The Lollards did not share with their Catholic neighbours the certainty of the Mass to bring them closer to Heaven. Neither can they have shared the same vision of Heaven, nor a conviction that they would enjoy eternal bliss together. The Lollards speculated about the soul's path to God, and came to diverse conclusions. For some it was simple: holding a candle in his hand, John Bowkyn asserted in 1493 that 'as this candle doth fade and goeth out, likewise my soul shall go and ascend to Heaven'.[47] This was to deny the existence of Purgatory, as many Lollards did. Elizabeth Sampson of Aldermanbury doubted the general resurrection, sure that 'more souls than is in Heaven already shall never come to Heaven'.[48] Free to find the truth as they saw it, the Lollards might develop heresies so esoteric that even most of the 'known' men and women might have disowned them. In 1535 a man declared that 'Christ never died nor shed His blood for us, but only for them that be in limbo patrum'.[49] Foxe, who claimed the Lollards as spiritual ancestors of the persecuted Protestants, refrained from printing all the details of the heresies imputed to William Pottier, claiming that he had been slandered by his enemies. Well might the Protestant historian have been embarrassed by Pottier's belief that 'thou cannot tell what we be relieved by the passion of Christ for before he suffered we were but damned, and now that notwithstanding if we commit deadly sin we likewise be damned'. Pottier held that there were six gods: the Father, the Son, and the Holy Ghost ('iii gods for the which we say three times Sanctus Sanctus Sanctus'), the Devil, 'that thing that a man setteth his mind most upon', and 'a priest's concubine being in his chamber twelve months and more'.[50]

---

[45] Foxe, *Acts and Monuments*, iv, pp. 206–7.
[46] Ibid. iv, p. 205.
[47] Hale, *Precedents*, cxxxv.
[48] Guildhall, MS 9531/9, fo. 4ʳ.
[49] PRO, SP 1/94, fo. 227ʳ (*L&P* viii. 1129).
[50] Guildhall, MS 9531/9, fo. 26ᵛ; Foxe, *Acts and Monuments*, iv, p. 175. I find J. F. Mozley's explanation for Foxe's omission of the articles against Pottier uncon-

'Known men' were marked by their hatred of priests. Lollards were convinced that the Catholic Church and its priesthood had traduced and perverted the teachings of Christ. 'All priests since our Saviour were heretics', said Thomas Underwood in 1476, especially his own at All Hallows the Great.[51] 'In Christ's time there were no priests', and many Lollards wished that there were still none.[52] Lollardy was the only truly anticlerical movement on the eve of the English Reformation. John Stilman declared that the Pope was Antichrist, and all other priests and prelates 'the synagogue of Satan'. The 'doctors of the Church', for all their learning, had 'subverted the truth of Holy Scripture . . . and therefore their works be nought, and they in Hell'.[53] The reasons for the corruption of the Church were not hard to find: 'the Church was too rich'. In April 1518 George Browne abjured his assertion 'that when land was first given to the Church a holy man determined that the Church was then . . . corrupted by the same saying these words: Jam effusum est venemum in Ecclesia'.[54] It was in London that the wealth of the clergy was most vaunted, and in London the Lollard complaints were most forcibly expressed. 'Pilgrimages were nothing worth, saving to make the priests rich', thought William and Alice Cowper; John Household called the Pope 'a strong strumpet', who with his pardons had 'drowned in blindness all Christian realms; and that for money'. Richard Wolman termed the 'church of Paul's a house of thieves'. Images, Joan Baker said, were yet another clerical device to extort money from the deluded, 'set up but of covetousness of priests and to make them rich'. Tithes were another clerical confidence trick, 'never ordained to be due, saving only by the covetousness of priests'.[55]

Lollards held aloof from popular superstition and detested the external works of the Church. Images to saints and pilgrimages to their shrines, to which the orthodox accorded almost

vincing: J. F. Mozley, *John Foxe and His Book* (1940), pp. 213–16. P. Collinson, 'Truth and Legend: the Veracity of John Foxe's Book of Martyrs' in *Clio's Mirror: Historiography in Britain and the Netherlands*, ed. A. Duke and C. A. Tamse (Zutphen, 1985), pp. 31–54.

51 Hale, *Precedents*, lx.
52 Foxe, *Acts and Monuments*, iv, p. 230.   53 Ibid. iv, p. 208.
54 TCD, MS 775, fo. 124ᵛ; Foxe, *Acts and Monuments*, iv, p. 177.
55 Ibid. iv, pp. 177, 175, 183; Guildhall, MS 9531/9, fos. 25ʳ⁻ᵛ.

magical powers, the Lollards reviled as idolatry. Yet some
Lollards had their own saints: 'John Wycliffe was a saint in
Heaven', thought John Stilman; Thomas Man said, perhaps
mockingly, that 'if he were taken again of the pilled knave
priests . . . he wist well that he should go to the Holy Angel;
and then be an angel in Heaven'.[56] At the stake Joan Boughton
cried out to God and Our Lady;[57] she believed that the saints
might intercede for her, but not through totems set up for them.
But most Lollards denied utterly the power of those in Heaven to
intercede for those in this vale of tears. Thomas Wasshingburn
said in 1482 that the blessed Virgin was a 'false queen', and St
Paul and St Bartholomew were false murderers of men.[58]
While good Catholics filed to the shrines of the Blessed Virgin
Mary at Willesdon, Crome, and Walsingham, Lollards outraged by the idolatry and the hypocrisy would insult the
Mother of Heaven in the worst ways possible, impugning more
than her intercessory powers. Elizabeth Sampson admitted
uttering this repeated blasphemy:

Our Lady of Willesdon was a burnt arse elf and a burnt arse stock, and
if she might have helpen men and women which go to her of pilgrimage she would not have suffered her tail to have been burnt, and what
should folk worship Our Lady of Willesdon or Our Lady of Crome
for the one is but a burnt arse stock and the other is but a puppet.

Our Lady at St Saviour's Bermondsey she dubbed 'Sym Sawyer
wth kyt lyppe'. Joan Baker of St Mary Magdalen Milk Street
wished that she had never gone on pilgrimage, so she said in
1510, for the images at the shrines were 'but mawments and
false gods', 'idols and not to be worshipped or honoured'.[59]

Lollard criticism of images was essentially moral and sceptical. Sampson thought it folly to go on pilgrimage to stocks and
stones; rather 'better it were for people to give their alms at
home to poor people'.[60] 'True pilgrimage', avowed Thomas
Geffrey in 1521, was to go barefoot to visit the poor, weak and

[56] TCD, MS 775, fo. 125[r]; Foxe, *Acts and Monuments*, iv, p. 213.
[57] *Great Chronicle*, p. 252.
[58] Hale, *Precedents*, xxxv.
[59] Guildhall, MS 9531/9, fos. 4[r-v], 25[r-v]; Foxe, *Acts and Monuments*, iv, p. 175. To be
'burnt' meant to be afflicted with venereal disease.
[60] Guildhall, MS 9531/9, fo. 4[r].

sick, 'the true images of God'.[61] Catholic preachers implied
that women on pilgrimages did not always emulate the Virgin
they worshipped: the Scottish friar Donald warned 'ye men of
London, gang on your self with your wives to Willesdon in the
Devil's name, or else, keep them at heme [home] with you with
sorrow', and John Hewes, a draper, 'had heard the vicar of
Croydon thus preach openly, that there is much immorality
kept up by going on pilgrimage to Willesdon or Mouswell,
&c'.[62] The Lollards were not slow to mock this weakness.

Crosses were everywhere in Tudor London as a remembrance
of Christ's sacrifice, but Lollards despised such reminders.
Why should the cross be worshipped, asked George Browne,
when it was but 'a hurt and pain unto our Saviour Christ in the
time of His passion?'[63] As the crucifix was carried to the Lollard
Thomas Blake as he lay on his death-bed, Joan Baker protested
to their parish priest in St Mary Magdalen Milk Street that
'the crucifix was not to give confidence nor trust in but as a
false god'.[64] Opposition to images sometimes went beyond
words. Iconoclasm was the most shocking of Lollard outrages
against the Church: it was meant to be. At Lincoln's Inn in
1512 the image of Pety John was destroyed during Mass.[65]
Even one of Sir Thomas More's servants 'blasphemed' an
image of Our Lady at London Bridge 'with diabolical words',
and broke the neck of the Christ child in her arms.[66] Lollards
taunted the images, challenging them to use their power to
defend themselves and sometimes, by a miracle, they did.[67]

The Lollards broke the unity of the Catholic community: their
heresies outraged their neighbours, and the flagrantly anti-social
behaviour of some of their number put them outside society.
When the host was carried through the City streets in 1482 as
a spiritual solace for the sick and dying, Thomas Wasshingburn taunted the priest: 'where goeth the costermonger?'[68]

---

[61] Foxe, *Acts and Monuments*, iv, pp. 229–30.

[62] More, *Dialogue concerning Heresies*, i, p. 100; Foxe, *Acts and Monuments*, v, p. 34.

[63] Foxe, *Acts and Monuments*, iv, p. 177.

[64] Guildhall, MS 9531/9, fo. 25[v]. Probate was granted for his will in 1511; ibid.,
MS 9168/3, fos. 43[r], 77[r].

[65] R. M. Fisher, 'The Inns of Court and the Reformation, 1530–1580' (University
of Cambridge, Ph.D. thesis, 1974), p. 123.

[66] More, *Debellation*, p. 16.

[67] BL, C 18 e 2. 96. See above, p. 18.   [68] Hale, *Precedents*, xxxiv–xxxv.

Catholic women called upon the Virgin to aid and comfort them in childbirth, but Lollard women scorned such imprecations. The woman who implored Our Lady to help Joan Sampson in her labour Joan spat upon and set away, and at another woman's childbed she 'contumeliously spoke against the invocators'.[69] Sixteenth-century society was bound by oaths, and civic office sanctioned by religious vows, but many Lollards, as fundamentalists, refused ever to swear. Thomas Walker, a fuller of St Clement Danes, rebuked his wife for swearing an oath. In April 1516 William Ramsay refused to serve as a constable in his ward, 'affirming that he made a vow to make no manner oath during his life which he would keep for any man'.[70] John Sampye, merchant taylor of St Mary at Hill, announced 'precisely and presumptuously that he would not be sworn' to the wardmote enquest in 1520, and when asked 'how he keepeth his obedience and oath which he made to this City, he answered . . . that his coming hither at the Mayor's commandment was enough'.[71] But it was not, for refusal to swear that oath threatened instability. Three years later an unrepentant Sampye was found inveighing publicly at the Crossed Friars against the Mayor who had incarcerated him for his disobedience.[72] In his parish of St Mary at Hill, Sampye had also rebelled against prevailing customs, this time of devotion. 'For the recovery of the duties of Saint Anne' (perhaps payments for the lights at her altar) his fellow parishioners brought him before the spiritual court in 1512, and for his failure to conform would do so again in 1529.[73] Of Sampye's religious nonconformity there is no doubt, but whether he was a 'known man' of the Lollard sect is less than sure. What is certain is that Sampye was one of those men in the leading London companies who had been touched by heresy.

Radical sectaries of the later middle ages were usually of

---

[69] Foxe, *Acts and Monuments*, iv, p. 206.

[70] S. E. Brigden, 'Religion and Social Obligation in early sixteenth-century London', *P&P* 103 (1984), pp. 86–92; Thomson, *Later Lollards*, p. 157; CLRO, Repertory 3, fo. 80ᵛ.

[71] Ibid., Repertory 5, fo. 43ʳ.

[72] *L&P* iv/1. 245.

[73] *St Mary at Hill*, i, pp. 278–9, 347; GLRO, DL/C/330, fos. 178ʳ⁻ᵛ.

artisan status, and so most City Lollards were. In 1523 a disgruntled London curate could allege that 'these weavers and millers be naughty fellows and heretics many of you'.[74] But not all London Lollards were poor. Lollardy was entering the upper echelons of London society by the turn of the sixteenth century, and this made the spread of the heresy the more disturbing. 'I could bring my Lord of London to the doors of heretics in London, both of men and women, that be worth a thousand pound',[75] alleged the Bishop's summoner in 1514; it seems with some truth. Joan Boughton had been mother-in-law to no less than the Mayor of London, Sir John Yonge, and her 'daughter, as some reported, had a great smell of an heretic after the mother'. It appears that Lady Yonge suffered too, for Joan Baker avowed in 1510 that Lady Yonge 'died well', 'a martyr before God'.[76] If ever he went to London, confided a 'known man' to John Whithorn, the rector of Letcombe Basset, in 1508, he would find there rich heretics ('locupletes hereticos'), who had books which he might wish to read.[77] Alice Ray's apprentice son had told her (sometime before August 1511) how his master Richard Wright, 'vir honestus dives ac bonus', owned heretical books which he read often, and in secret. Wright was a 'privy man', but for him the spirit of heretical brotherhood did not transcend the customary social barriers between the City rich and the City poor: 'he will deal with no poor folks': rather he was 'great with my master Barret and his wife'. John Barret was a goldsmith of Cheapside and a Merchant of the Staple at Calais, but his faith made him more egalitarian. It was to Barret's house that Robert Benet, the illiterate Lollard water carrier, had gone for shelter during the battle called 'le Blakehethe' (the Cornish rising of 1497).[78] Lollard masters may well have chosen Lollard apprentices. Two at least of Barret's six apprentices in the plutocratic Goldsmiths' Company were also Lollards—Robert Wigge and William

---

[74] GLRO, DL/C/330, fo. 46[r].

[75] Foxe, *Acts and Monuments*, iv, p. 193.

[76] *Great Chronicle*, p. 252; Guildhall, MS 9531/9, fo. 25[v]; TCD, MS 775, fo. 122[v]. There is no corroborative evidence, however, that Lady Yonge was burned. Her husband seemed unimpeachably orthodox: Thomson, *Later Lollards*, pp. 156–7.

[77] Ibid., pp. 85–6. Thomson prefers to translate 'locupletes' as 'trusty', while allowing that 'rich' might have been intended.

[78] TCD, MS 775, fos. 123[v], 124[r].

Tilseworth.[79] Heretics of wealth and standing took far greater risks than their humbler brethren, for they had more to lose. Thomas Grove, a London butcher, thought it worth offering a £20 bribe in 1521 to be excused the shame and the risk to trade of public penance for heresy.[80] But some dared all. When Joan Baker abjured her shocking heresies in January 1511, a fellow member of her husband's company, the Merchant Taylors, and a fellow parishioner at St Margaret New Fish Street, boasted that 'he would defend her and her opinions, if it cost him five hundred marks'.[81] Her defender was Richard Hunne, and his stand against the clergy would be remembered and spoken of in London 'long after his days'.[82]

Hunne's challenge to the clergy, as reported by Foxe, was unequivocal: 'her [Joan Baker's] sayings be according to the laws of God: wherefore the Bishop and his officers are more worthy to be punished for heresy than she is'.[83] Within the year Hunne became involved in two further disputes with the London clergy. Such a misfortune came not through carelessness. In St Mary Matfelon, Whitechapel, in March 1511 Hunne's baby son Stephen died. As was the immemorial custom, the rector demanded the child's only property, his winding sheet, as a mortuary fee for burying him. It was a small price, and a usual exaction, but Hunne refused to pay it: the

---

[79] Goldsmiths' Company records. Robert Wigge abjured during Henry VIII's reign: Foxe, *Acts and Monuments*, iv, p. 585. For Tilseworth, see below, pp. 104–5.

[80] Foxe, *Acts and Monuments*, iv, p. 227.

[81] Ibid. iv, p. 184.

[82] More was right that Hunne's challenge to the clergy would be called 'Hunne's case', and that he would be long remembered. The evidence upon which discussion of Hunne's case rests is considered by E. J. Davis, 'The Authorities for the Case of Richard Hunne', *EHR* xxx (1915), pp. 477–88. The case has been written about often: see Thomson, *Later Lollards*, ch. 7; A. F. Pollard, *Wolsey* (paperback edn., 1965), pp. 31–42; J. D. M. Derrett, 'The Affairs of Richard Hunne and Friar Standish', in More, *Apology*, appendix B, pp. 213–46; P. Gwynn, 'Cardinal Wolsey and the "affairs" of Richard Hunne and Henry Standish' (I am extremely grateful to Mr Gwynn for allowing me to read this chapter in advance of its publication); A. Ogle, *The Tragedy of the Lollards' Tower: The Case of Richard Hunne with its Aftermath in the Reformation Parliament, 1529–1532* (Oxford, 1949); R. M. Wunderli, 'Pre-Reformation London Summoners and the Case of Richard Hunne', *JEH* xxxiii (1982), pp. 209–24; S. J. Smart, 'John Foxe and "the Story of Richard Hun, Martyr"', *JEH* xxxvii (1986), pp. 1–14.

[83] Foxe, *Acts and Monuments*, iv, p. 184. Foxe dated Baker's trial to 1510, but Ussher's transcript of Fitzjames's lost register dates it to January 1511: ibid. iv, p. 174; TCD, MS 775, fo. 122[v].

cloth was his, not his dead son's.[84] That summer a fire in the house of William Lamberd, a salter, in Wood Street had threatened the lives and property of Lamberd's neighbours. In August, while the circumstances of the fire and the damage caused were being investigated, it was Hunne who, with John Gerrard, a scrivener, stood surety for Lamberd. By November 1511 the dramatis personae of the dispute had altered: now Hunne and Lamberd were bound in £100 to abide by the arbitration made in the suit between them and the parson and churchwardens of St Michael Cornhill over the burnt tenement in Wood Street. That one of the arbitrators was John Munday, the goldsmith, who was himself one of Lamberd's outraged neighbours, does not seem to accord with the strict impartiality of the law. By February 1512 the recognizance was void:[85] why, we do not know, but Hunne's tangles with the law, and with the clergy, were far from over.

Thomas Dryffeld, the rector of St Mary Matfelon, at last had recourse to the spiritual courts to extract the mortuary payment from Hunne in April 1512: not to the Bishop of London's courts, but to the Archbishop's Court of Audience at Lambeth. Tried before Cuthbert Tunstall, the Archbishop's Chancellor, on 13 May 1512, the mortuary was found—rightly, given the evidence—for the rector, against Hunne.[86] But Hunne had no intention of paying, and planned further confrontation. On 27 December 1512, a time of special peace and goodwill, Hunne crossed the City, leaving his own parish to attend evensong at St Mary Matfelon. There, as he must have foreseen, Hunne met with no charitable welcome. Marshall, the rector's chaplain, ringingly denounced him: 'Hunne, thou art accursed and thou standest accursed, and therefore go thou out of the church,

---

[84] This we know from the pleadings in King's Bench, discovered by S. F. C. Milsom: PRO, KB 27/1006 m. 37; Milsom, 'Richard Hunne's "Praemunire"', *EHR* lxxvi (1961), p. 80. Cf. in 1523 the rector of St Margaret Pattens received 1*d.* as half the value of a dead infant's shirt: Guildhall, MS 4750/1, fo. 91[v].

[85] CLRO, Journal 11, fos. 138[r], 144[r]. The dispute also came before the Court of Aldermen; ibid., Repertory 2, fo. 122[r]: see Wunderli, 'Pre-Reformation Summoners', p. 218.

[86] Milsom, 'Richard Hunne's "Praemunire"', p. 80; Derrett, 'The Affairs of Richard Hunne and Friar Standish', p. 222. On the eve of the Reformation the Court of Arches became the arena for the City's most intractable tithe disputes: S. E. Brigden, 'Tithe Controversy in Reformation London', *JEH* xxxii (1982), pp. 289, 294.

for as long as thou art in this church I will say no evensong nor service.'[87] Any excommunicate was banned from worship, from association with the faithful, and from the services and sacraments of the Church. The stigma was an awesome one. But Hunne was not excommunicate—yet.[88] In order to restore his good name and fame, and his credit within the City merchant community, Hunne instituted a suit for slander against Marshall on 25 January 1513. The defendant backed down, and Hunne soon sought more expedient ways to attack Dryffeld and Dryffeld's abettors. In Hilary term 1513 he dared to bring a praemunire action in the King's Bench.[89]

What brought Hunne to dare to impugn the authority of the clergy and the spiritual law? 'The spirit of pride' thought Sir Thomas More, then Undersheriff of London, who witnessed the proceedings against Hunne and knew the case from 'top to toe'. More described Hunne later as

a man high minded and set on the glory of a victory, which he hoped to have in the praemunire, whereof he much boasted, as they said, among his familiar friends that he trusted to be spoken of long after his days, and have his matter in the years and terms called Hunne's case.[90]

But for all his boasting that he would humble the clergy, for all the support he may have received from his friends, Hunne's action may have been defensive as well as offensive. More thought so: he insisted that Hunne 'was detected of heresy before the praemunire sued or thought upon', suggesting that Hunne's action brought in the King's Bench was a desperate ploy to allay proceedings against him for heresy.[91] So terrible were the penalties against even a suspect heretic—prison, trial, penance—that any means to prevent investigation might well

---

[87] Milsom, 'Richard Hunne's "Praemunire" ', pp. 81–2.

[88] For excommunication, see F. D. Logan, *Excommunication and the Secular Arm in Medieval England* (Toronto, 1968), and Brigden, 'Religion and Social Obligation'. Dr Wunderli pointed out that Hunne, as an excommunicate, could not have brought a suit before a royal court: 'Pre-Reformation Summoners', p. 219, n. 39.

[89] Milsom, 'Richard Hunne's "Praemunire" ', pp. 80–1; Derrett, 'The Affairs of Richard Hunne and Friar Standish', pp. 223–4; Dickens, *The English Reformation*, pp. 127–8.

[90] More, *Apology*, p. 126; *Dialogue concerning Heresies*, i, pp. 318, 326.

[91] More, *The supplycacyon of soulys*, in *Workes* (1557), p. 297; cited in Milsom, 'Richard Hunne's "Praemunire" ', p. 80.

be sought. London opinion disagreed with More—Wriothesley
wrote later that 'Hunne . . . was made an heretic for suing a
Praemunire'—but London opinion may have been wrong.[92]
While the two cases were pending in King's Bench the clergy
brought their own action against Hunne—for heresy. On
14 October 1514 Hunne was brought to the Lollards' Tower.
On 2 December he was examined by Fitzjames.[93] Two days
later he was found hanging in his cell, dead. The clergy maintained that he had committed suicide, but the citizens of
London thought otherwise, especially when the Bishop's summoner fled on 10 December.[94] Before the coroner's jury
brought its verdict the clergy brought theirs: Hunne was condemned posthumously as a heretic, and on 20 December his
body was exhumed and burnt.[95] The citizens of the coroner's
jury found in February that Hunne had not taken his own
life.[96] If Hunne was innocent of *felonia de se*, then someone had
killed him, and everyone suspected the clergy.

To be a Lollard, murdered by the clergy, was to become a
martyr. It was for the Church to judge what was apostasy; and
the Church had condemned Hunne as a heretic, and handed
his body over to the secular arm for burning. Yet there are different degrees of error: some of the opinions imputed to Hunne
were also held by Dean Colet, and even while Hunne was
under suspicion, so too was Colet.[97] Why was Hunne condemned? Articles against him survive, and these provide the
evidence against him.[98] In defending Joan Baker, Hunne was

[92] Wriothesley, *Chronicle*, i, p. 9; see Simon Fish, 'Supplicacyon', pp. 9, 12.

[93] Richard Arnold, *The Customs of London*, ed. F. Douce (1811), p. xlix. Dr Fines
discovered an account of Hunne's examination for heresy—questions put to Hunne
himself during his life, and to witnesses after his death—in Ussher's transcript of
Fitzjames's lost Court book: J. Fines, 'The Post-Mortem Condemnation for Heresy of
Richard Hunne', *EHR* lxxviii (1963), pp. 523–31.

[94] Foxe, *Acts and Monuments*, iv, p. 184; Hall, *Chronicle*, pp. 573–80; Wunderli, 'Pre-
Reformation Summoners', p. 221.

[95] Pollard, *Wolsey*, p. 34; PRO, C/85/126/26 (Hunne's signification for heresy,
16 Dec. 1514).

[96] Foxe, *Acts and Monuments*, iv, pp. 190–7.

[97] The doubts cast upon Colet's orthodoxy are discussed in ch. 1 above.

[98] Investigation of Hunne's putative heresy rests upon two accounts: Ussher's
transcript, and *The enquirie and verdite of the quest panneld of the death of Richard Hune wich
was founde hanged in Lolars tower* (c.1536): the first printed by Fines, 'Post-Mortem
Condemnation', the second used by Foxe: *Acts and Monuments*, iv, pp. 183 ff. That
some of the articles transcribed by Ussher are the same verbatim as those transcribed
by Foxe gives more credence to the tract.

seen to condone the most extreme beliefs, yet not even she was accused of sacramentarianism, as Hunne was.[99] Most of the charges against Hunne record his invective against priestly power: that priests and Bishops were 'the scribes and pharisees that did crucify Christ'; that they were 'teachers and preachers, but no doers . . . all things taking, and nothing ministering'.[100] But most convincing and credible of all the charges against him was that he 'hath in his keeping diverse English books prohibit and damned by the law: as the Apocalypse in English, Epistles and Gospels in English, Wycliffe's damnable works, other books containing infinite errors in the which he hath been long time accustomed to read, teach, and study daily'.[101] Against the Church's proscriptions, Hunne defended vigorously the translation of the Bible into English, and the layman's right to read it. As it said in the prologue of his Bible, 'poor men and idiots have the truth of the holy Scriptures, more than a thousand prelates, and religious, and clerks of the schools'.[102]

Witnesses testified to Hunne's devotion to the principle of vernacular Scripture, to his assiduous reading of his own 'pulcherrimam Bibliam in Anglicis', openly in the doorway of his house. With him when he died were four books—'the Bible in English, a book of the four Evangelists, a book of the prick of conscience, and a book of the ten commandments': these he had asked his servant to bring to him in prison for his comfort. Hunne's own copy of the English Bible was 'wont to lie in St Margaret church in Bridge Street sometimes a month together' for the edification of all who would read it.[103] The spirit which lay behind the desire to have the Scripture in English need not have been heretical: Hunne may have wished for the Bible in English while disavowing the wilder Lollard heresies. There were others high in City circles who began to 'smell the Gospel'

---

[99] Guildhall, MS 9531/9, fos. 25ʳ⁻ᵛ; Foxe, *Acts and Monuments*, iv, p. 186; More, *Dialogue concerning Heresies*, i, pp. 327, 330.

[100] Foxe, *Acts and Monuments*, iv, p. 184.

[101] Fines, 'Post-Mortem Condemnation', p. 530 (verbatim in Foxe, *Acts and Monuments*, iv, p. 184).

[102] Fines, 'Post-Mortem Condemnation', p. 530; Foxe, *Acts and Monuments*, iv, p. 186.

[103] Fines, 'Post-Mortem Condemnation', pp. 530–1.

at this time, though yet untouched by heresy.[104] 'Familiar friends' had encouraged Hunne to challenge the clergy. Their identity can only be suspected, but it seems that Hunne had close connections with the 'privy' men of the Lollard sect through marriage. He had married a woman named Anne Vincent.[105] Thomas Vincent was a Lollard apostle: it was he who taught 'Father Hacker' 'all and single his errors and heresies', and gave him St Matthew's Gospel in English. Vincent had two Lollard sons-in-law—'him that was burnt for heresy' in about 1513, and Thomas Austy—and maybe, he had a third, Richard Hunne.[106]

Nothing daunted by the martyrdoms of Hunne, of Thomas Man, James Brewster, John Stilman, and William Sweeting, a conventicle of 'secret favourers' dared to meet in the heart of the City throughout the 1520s to rehearse and evangelize their Lollard convictions.[107] They usually gathered in Bird's Alley beside St Stephen Coleman Street in a house belonging to William Russell, a tailor. Of 'the learning and sect' were three founders from the neighbouring parish of St Margaret Lothbury —John Gator, Stere, and Knight—and another disciple was John Tewkesbury, a haberdasher from St Martin Ludgate. Tewkesbury had been studying Scripture since 1512: 'as he may see the spots of his face through the glass, so in reading the New Testament he knoweth the faults of his soul'.[108] At the new school of St Anthony's by the Austin Friars, Cony, the clerk, and his wife owned a copy of the Lollard classic, *The Bayly*. 'A chief reader and teacher' was Thomas Phillips, a pointmaker,

---

104 See below, pp. 107–8.

105 Drapers' Company, Deeds, A.vi, 266, 7, 4. Anne Vincent's property passed to her daughter, Margaret, who was married to the draper, Roger Whaplode.

106 BL, Harleian MS 421, fo. 12ʳ (*L&P* iv/2. 4029). Vincent was a member of that conventicle which abjured before Fitzjames in 1511: Foxe, *Acts and Monuments*, iv, p. 176; TCD, MS 775, fo. 123ʳ.

107 PRO, C 85/126/19 (significavit for William Sweeting of Chelsea and James Brewster of Colchester, 14 Sept. 1511); Foxe, *Acts and Monuments*, iv, pp. 180–1, 207–16. Foxe's own transcription of part of a register (now lost) of Bishop Tunstall, or his Vicar-General, made during the visitation of 1527–8 provides the evidence for this conventicle: BL, Harleian MS 421, fos. 11ʳ–14ʳ (*L&P* iv/2. 4029). See Dickens, 'Heresy and the Origins of English Protestantism', *Reformation Studies*, pp. 370–2; *The English Reformation*, pp. 28–9; Davis, *Heresy and Reformation*, pp. 57–8.

108 Foxe, *Acts and Monuments*, iv, p. 690.

who had had 'the New Testament of . . . Saint Jerome's translation' since about 1514, and who would 'sometime read in a
book of St Paul and sometime in a book of the Epistles'.[109] His
own house at the Little Conduit in Cheap was an occasional
meeting house. Leader and chief teacher of the community was
John 'Father' Hacker. It was he who revealed the activities of
the Coleman Street conventicle in 1527.

Many of these 'known men' and women had long been of
the sect, and some had already recanted their heresy, on pain
of death if they relapsed. John Household of All Hallows the
Less had abjured in 1517 his denunciations of the Mass, the
Pope, and the saints: a decade later he denied that he had
strayed from the path of orthodoxy since.[110] But others had
defied the persecutors. Laurence Swarffer, in 1527 a tailor in
Shoreditch, had been detected before, in Berkshire in 1514 for
holding Lollard opinions and for possessing the Gospels of St
Luke and St John in English; Thomas Geffrey, now a tailor in
Coleman Street, had been detected and had abjured in the great
heresy quest in 1521.[111] For all her past promises to reconcile,
Goodwife Bristow of Wood Street still found 'delectation' in
Lollard teachings, and owned St Luke's Gospel.[112] Most steadfast of all was Thomas Austy: having abjured in July 1511, he
was found in 1527 to be a follower of Hacker's 'learning'. For
his obdurate refusal to wear the mark of the faggot he was condemned to perpetual custody in St Bartholomew's priory (in
1530?), but he managed to escape.[113] The Tilseworths were
already, and would remain, a heretical dynasty. In 1511
'Dr' William Tilseworth, one of the 'four principal readers'
of the Amersham Lollards, went to the stake. Others of the
family continued the tradition: in 1521 Emma Tilseworth
was condemned to wear the sign of the faggot forever, and
William Tilseworth, John Barret's apprentice, had been discovered at Thomas Man's conventicle at Amersham. In 1527
Robert and Thomas Tilseworth, both tailors, of Abchurch

---

[109] PRO, SP 2/P, fo. 11^r (*L&P* vii. 155).

[110] Foxe, *Acts and Monuments*, iv, pp. 176–7: BL, Harleian MS 421, fo. 14^r.

[111] Ibid., fo. 13^r; Foxe, *Acts and Monuments*, iv, pp. 229, 230, 232, 585; Thomson, *Later Lollards*, p. 89.

[112] BL, Harleian MS 421, fo. 14^r; Foxe, *Acts and Monuments*, iv, p. 239.

[113] TCD, MS 775, fos. 123^r, 124^v; Foxe, *Acts and Monuments*, iv, p. 175; v, p. 29.

Lane and of Budge Rowe, were detected as members 'of the sect'.[114]

In the Coleman Street conventicle they read together from *The Bayly*, the devotional *Prick of Conscience*, the Ten Commandments, and St Matthew's Gospel, and there Hacker taught the cardinal Lollard doctrines. He spoke against images and pilgrimages, and taught that 'the sacrament of the altar was not the very body of God but a remembrance of God that was in Heaven'.[115] After the martyrdom of Thomas Man in 1518 it was to Hacker that the Lollards looked as the sect's foremost apostle. Later, Mistress Dolly of Stanton Harcourt in Oxfordshire testified to Hacker's evangelical charisma: he was 'very expert in the Gospels . . . and could declare it and the Paternoster in English as well as any priest, and it would do one good to hear him'.[116] From Essex to Gloucestershire, and always back through the capital, Hacker travelled to read the Scriptures, to prophesy, to instruct in the Lollard lore, strengthening 'known' men in their convictions, and winning converts to the cause.[117] Thomas Man had on his missions, so he was accused in 1518, 'turned 5, 6, or 7 hundred people to your law and opinions', and Hacker continued his task. In 1514 the Berkshire Lollards claimed Hacker as their teacher.[118] Thomas More wrote of one such 'ancient heretic' (maybe Man, maybe Hacker), who had 'in diverse countries spread about almost all the heresies that any lewd heretic holdeth'. Asked this heretic's name, More could not tell; for he had 'in every diocese a diverse name'.[119] Hacker had changed his name to Richardson, and this at the behest of Elizabeth Newman, a girl with special reason for caution. A Londoner named Newman had been a Lollard teacher too, even after his abjuration in 1496.[120] John Stacy and Laurence Maxwell,

[114] Foxe, *Acts and Monuments*, iv, pp. 123–4, 214, 222, 226, 228, 229, 232, 581; BL, Harleian MS 421, fo. 13ʳ; Thomson, *Later Lollards*, p. 87.

[115] BL, Harleian MS 421, fos. 11ʳ–12ʳ.

[116] Foxe, *Acts and Monuments*, iv, p. 582. Foxe dated this testimony to 1532.

[117] Foxe, *Acts and Monuments*, iv, pp. 226, 234, 236, 239, 582, 585; Strype, *Ecclesiastical Memorials*, i/2, pp. 52, 65.

[118] TCD, MS 775, fo. 124ᵛ; Foxe, *Acts and Monuments*, iv, pp. 208–14; Thomson, *Later Lollards*, p. 89.

[119] More, *Dialogue concerning Heresies*, i, pp. 268–9.

[120] BL, Harleian MS 421, fo. 12ᵛ; TCD, MS 775, fo. 123ʳ; Thomson, *Later Lollards*, pp. 89, 158.

wardens of the Bricklayers' Company, were Hacker's disciples and followed his evangelical path. They 'converted many men and women, both in London and the country; and once a year, of their own cost, went about to visit the brethren and sisters scattered abroad'.[121] Stacy kept a Lollard scriptorium in his house in Coleman Street. It was to Stacy that the Collins of Ginge had come for their manuscript copy of the Bible, for which they paid 20s., and there in 1527 a man called John was copying the Lollard Apocalypse.[122] Lollards had long come to London to find their forbidden texts, because hidden in the capital the copyists could work undetected.[123] But by 1527 country Lollards were looking for different works: Thomas Hilles, servant to a Lollard master in Essex, 'hath a book of the New Testament in English printed which he bought at London'.[124] This was Tyndale's translation, and it contained a message yet unknown to the Lollards. Hacker, like other radical sectaries, lived in hopes of an imminent millennium in which the oppressed would triumph and the forces of Antichrist be vanquished: he foresaw a 'battle of priests, and all the priests should be slain . . . because they hold against the law of holy Church, and for making a false gods'. And he foretold a 'merry world'.[125]

THE EVANGELICAL BRETHREN

In July 1523 a young priest, William Tyndale, came to London, seeking the patronage of Bishop Tunstall. Believing with Martin Luther that the truth of Christ's saving passion was to be discovered in Scripture alone, Tyndale was determined to translate the New Testament into English, according to the humanist principles of Erasmus.[126] The Wycliffite English Gospels could provide no model: he could not be 'helped with

[121] BL, Harleian MS 421, fo. 12ᵛ; Foxe, *Acts and Monuments*, iv, p. 681.
[122] BL, Harleian MS 421, fo. 12ᵛ; Foxe, *Acts and Monuments*, iv, pp. 236, 237. The copyist's expenses were paid by a neighbouring grocer, John Sercot.
[123] Aston, *Lollards and Reformers*, pp. 198, 200, 205.
[124] BL, Harleian MS 421, fo. 11ʳ.
[125] See, for example, R. Lerner, 'Medieval Prophecy and Religious Dissent', *P&P* 72 (1976), pp. 3–24; N. Cohn, *The Pursuit of the Millennium* (paperback edn., 1970). Foxe, *Acts and Monuments*, iv, p. 234.
[126] For Tyndale's life and work, see J. F. Mozley, *William Tyndale* (1937); R. Demaus, *William Tindale* (1871).

English of any that had interpreted the same or such like thing in the Scripture beforetime'.[127] While awaiting the Bishop's answer to his suit, Tyndale preached a series of sermons at St Dunstan in the West, and these sermons were doubtless animated by the New Learning he had acquired at Cambridge. In his congregation was Humphrey Monmouth, a wealthy draper of London. When Tunstall answered Tyndale that 'he had chaplains enough', Monmouth took Tyndale into his own house at All Hallows Barking, and there Tyndale remained, immersed in his studies, until his departure for Hamburg and exile early in 1524.[128]

Monmouth had long been 'a Scripture-man'. 'Even at that time when Dr Colet was in trouble', suspected of heresy, and while Hunne mounted his challenge against the clergy, Monmouth, like them, 'had begun to smell the Gospel'.[129] The desire to have the Scripture in English gave Monmouth an affinity with the 'Scripture-men' in London Lollard circles. 'Known men—like Wolman and Mason—were to be found within his own company, the Drapers. But not all 'Scripture-men' were Lollards. Thomas Downes, Hunne's fellow parishioner at St Margaret New Fish Street, had owned the Bible which condemned Hunne before Hunne owned it. Yet when Downes died in 1508 his bequests for his soul evinced an orthodox piety; he asked to be buried before the image of the Virgin in his church; bequeathed torches to burn 'at sacring time . . . in honour of the blessed sacrament'; and left 100 lb. of wax to burn before the crucifix.[130] If Tyndale's sermons and treatises had at first seemed of dubious orthodoxy to his patron, the devout and pietistic impulse which animated them may have excused all, and Monmouth may have thought to have the Scripture in English while remaining resolutely orthodox, if disobedient. Monmouth kept the texts of Tyndale's sermons at St Dunstan's, his letters from abroad, and had his servant copy Tyndale's treatises. 'And all those books . . . lay openly in my house for the space of two years or more, that every man might

---

[127] W. A. Clebsch, *England's Earliest Protestants, 1520–1535* (New Haven and London, 1964), p. 142.

[128] Strype, *Ecclesiastical Memorials*, i/2, p. 364; Mozley, *William Tyndale*, pp. 44–50.

[129] Foxe, *Acts and Monuments*, iv, p. 618.

[130] Fines, 'Post-Mortem Condemnation', p. 531; PRO, PCC, Prob. 11/16, fo. 36^r.

read on them that would'—just as Hunne's Bible had lain open in his church. But Monmouth kept secret his books of the English New Testament, and when Tyndale was outlawed, with his translation, in 1526, Monmouth burnt his hidden copies.[131]

When in May 1528 Monmouth wrote to Wolsey, seeking to excuse and explain his support for the heretic Tyndale, he staunchly protested his orthodoxy. Had he not 'given more exhibitions to scholars in my days, than to that priest'; among them Tunstall's own chaplain, and the King's; 'and if any of those other chance to turn, as that priest hath done, as God forbid, were I to blame . . .?' Was not Erasmus's *Enchiridion*, which Tyndale had translated, orthodox enough? After all, Bishop Fisher possessed the work; the Father Confessor of Syon Abbey and other learned priests around London had seen it, and 'never found fault in him to my knowledge'. Indeed, their only reservation about Luther's *Liberty of a Christian Man* was that 'there was in him things somewhat hard, except the reader were wise'. Devout religious, like the Abbess of Denny and an Observant friar of Greenwich, sought Tyndale's translations from Monmouth. Yet while they were attracted by Erasmus's programme of spiritual reform, they remained unimpeachably orthodox. Monmouth, too, could vaunt irreproachable proof of his orthodoxy:

If I had broken most part of the Ten Commandments of God, being penitent and confessed, (I should be forgiven) by reason of certain pardons that I have, the which my company and I had granted when we were at Rome, going to Jerusalem, of the holy Father the Pope, *a poena* and *a culpa*.

Everyone, or almost everyone, converted to a more radical vision of the path to salvation had once been of the old faith. However he might protest before his accusers in 1528, Monmouth knew well enough that to shield Tyndale was to admit complicity with his spiritual ideals. Monmouth's confession shows how, for men of his generation, orthodoxy and heterodoxy might well exist side by side. Yet events would move

---

[131] Strype, *Ecclesiastical Memorials*, i/2, pp. 364–7. What follows rests on Monmouth's confession.

Monmouth, and others with him, away from the old faith to a new.

The message of the new Christianity—that through faith alone, in Christ alone, discovered through Scripture alone, men might find their salvation—reached England fast. The incendiary tracts of Martin Luther were being imported by 1518, and without restriction. A correspondent writing from London noted in 1520, 'as for news there is none but of late there was heretics here which did take Luther's opinions'.[132] Luther's ideas probably spread first among his own compatriots, living in the expatriate communities in London like the German Steelyard. Here German merchants read and passed around the forbidden tracts brought from home, and may have sent them too to their contacts in the English merchant community.[133] It was a merchant of the Steelyard, Hans Collenbeke, who took money from Monmouth to Tyndale in Hamburg.[134] Some of Luther's ideas may have touched those who did not yet adopt all their consequences. In 1520 Gerard Danet, a gentleman and foreigner of St Faith's parish, made his will: it was a will which was in many ways orthodox and traditional, but the bequest of his 'sinful body' looked back to the Lollard knights, and the bequest of his 'sinful soul' evinced a new emphasis. Danet trusted 'faithfully and humbly', 'without any presumption by the merits of Christ's passion to come to the endless bliss which He bought me unto and all mankind upon the mount of Calvary with His most precious blood.'[135] Every Catholic would hold this belief central to his faith, but the reformers insisted, as Danet did not quite, that Christ's passion *alone* could justify them.

Some in the Catholic Church suffered spiritual agony, doubting that man's own striving, his own works, could ever merit redemption or bring his sinful soul to God. Catholic writers on the eve of the Reformation were much concerned with this

---

[132] Preserved Smith, *Luther's Correspondence*, I, ep. 235, p. 295; cited in C. S. Meyer, 'Henry VIII burns Luther's Books, 12 May 1521', *JEH* ix (1958), p. 178.

[133] PRO, SP 1/37, fos. 147^r-57^r (*L&P* iv/1. 1962). The discovery of a cell of German Lutherans at the Steelyard in 1526 is discussed by A. G. Chester, 'Robert Barnes and the Burning of the Books', *HLQ* xiv (1951), pp. 216–19.

[134] Strype, *Ecclesiastical Memorials*, i/2, p. 364.

[135] PRO, PCC, Prob. 11/20, fos. 7^{r-v}. K. B. McFarlane, *Lancastrian Kings and Lollard Knights* (Oxford, 1972), pp. 207–20.

problem of doubt, of scrupulosity.[136] With reason, for it was through just such a crisis in faith that Luther discovered his theology of grace. In the Inns of Court, William Roper, a young lawyer, became tormented too by a 'scruple of his own conscience'. With 'immoderate fasting and many prayers', Roper 'did weary himself even usque ad taedium' (until sick at heart). In his spiritual despair he found writings to comfort him. Down the river from the Inns of Court was the German Steelyard, and there Lutheran merchants, whom Roper knew, had copies of Lutheran tracts, *The Liberty of a Christian Man* and *The Babylonian Captivity*, which 'bewitched' him. Roper found a message of profound hope and mercy for man's salvation, and was soon

fully persuaded that faith only did justify, that the works of man did nothing profit, and that, if man could once believe that our Saviour Christ shed His precious blood and died on the cross for our sins, the same only belief should be sufficient for our salvation. Then thought he that all the ceremonies and sacraments in Christ's Church were very vain.[137]

The experience of grace must be brought to others by the proclaiming of it. William Roper, 'so well and properly liked of himself and his divine learning', became the model for the evangelical Messenger in the *Dialogue concerning Heresies* of Sir Thomas More, his own father-in-law. Not content to whisper 'Luther's new broached religion' 'in hugger mugger', Roper and his fellows 'thirsted very sore to publish' it.[138] Those still in spiritual bondage must be brought the liberating message. Yet in England from 1521 Luther and his works were outlawed: his followers must keep the faith in secret.

An underworld of 'evangelical brethren' soon emerged in London; loyal to each other and united in their purpose. Reformed clerics, won to the new doctrines at the universities, especially at Cambridge, and City merchants sheltered and sustained each other under persecution, converts bound together lastingly in common cause. The networks are shadowy,

---

[136] J. Rhodes describes the writings of Erasmus, Bonde, and Whitford upon scrupulosity in her 'Private Devotion in England on the Eve of the Reformation' (University of Durham, Ph.D. thesis, 1974), pp. 134 ff.

[137] Harpsfield, *Life*, pp. 85–6.

[138] More, *Dialogue concerning Heresies*, ii, pp. 444, 491–2; Harpsfield, *Life*, p. 84.

sometimes only revealed much later. In 1554 Rowland Taylor admitted how, thirty years before, he had married 'non in facie ecclesie'. This marriage took place in the house of John Tyndale and his wife at the Well with Two Buckets by the Austin Friars with the Tyndales and Benet, a priest, as witnesses.[139] John Tyndale was perhaps a real as well as an evangelical brother of William Tyndale, and certainly imported his Testaments; Benet, Thomas Bilney's friend from Cambridge, was martyred at Exeter for his beliefs.[140] Twenty years on, when John Gough made his will it was John Tyndale whom he chose as his executor.[141] Sir Thomas More advanced a conspiracy theory of the Reformation: he was right that a few prime movers led a revolutionary movement.

The passage of the truth was not left to chance. The faithful were fired and organized to proselytize, and they made of London the storm-centre of their mission. For in the City was the largest audience to convert, and there the hunted brethren might be 'privily cloaked'.[142] To the reformers, preaching was the way whereby the Word might be illumined for the people, and though risk was acute, preach they did. 'In preaching to the people they make a visage as though they came straight from Heaven to teach them a better way and more true than the Church teacheth, or hath taught this many hundred years', More complained.[143] The most daring evangelists were those divines who left academe for a wider arena. Thomas Arthur preached on Whit Sunday 1527 that all who knew the 'Gospel of God should go forth and preach in every place', and that summer he, with Thomas Bilney, embarked upon a preaching campaign in London and East Anglia.[144]

---

[139] W. H. Frere, *The Marian Reaction in its Relation to the English Clergy* (1896), p. 67.

[140] Foxe, *Acts and Monuments*, v, pp. 18 ff., 29, 803–4. For the relationship between John and William Tyndale, see G. R. Elton. *Reform and Renewal: Thomas Cromwell and the Common Weal* (Cambridge, 1973), p. 18 n. 30; H. C. Porter, *Reformation and Reaction in Tudor Cambridge* (Cambridge, 1958), p. 41.

[141] Guildhall, MS 9171/11, fo. 133^r (1543).

[142] PRO, SP 1/47, fo. 10^r (*L&P* iv/2. 3962).

[143] More, *Dialogue concerning Heresies*, i, p. 399.

[144] Guildhall, MS 9531/10, fo. 136^r. At the churches of St Magnus, St Mary Woolchurch, Willesdon, Kensington, and Newington the incumbents were either sympathetic enough to the preachers' cause to lend them their pulpits, or, more likely, they were unaware until too late of the radicalism of what they would preach. The campaign is discussed by J. F. Davis: 'Trials of Thomas Bylney'.

Bilney and Arthur were convinced of the sole power of Christ in mediation and salvation; both held a *theologia crucis* in which 'man in no wise can merit by his own deeds'.[145] Yet this belief they expressed in vehement attacks upon the idolatry and superstition of popular worship, sounding for all the world like Lollard invective, rather than by simple avowal of the doctrine of faith alone. The veneration of images, they said, detracted from the worship of God; people were 'fools' to set up candles before images; 'ye make them your gods and they can do nothing for you'. Even before the famous rood at St Magnus Bilney urged that 'likewise as Ezechias destroyed the brazen serpent . . . even so should Kings and princes nowadays destroy and burn images of the saints'.[146] Invocation of saints blasphemed and compromised Christ's power as sole mediator, and the Church's use of pardons was even to the 'injury of Christ's passion'.[147] This was no mere attack on outward ceremony: More recognized it as an attempt to stop the devotions of the faithful, and arrant heresy.[148] For Bilney and Arthur, nurtured in the new Erasmian tradition at Cambridge, the emphasis was upon personal faith, faith found in Scripture. 'Lo, here is the New Testament, and here is the Old. These be the two swords of Our Saviour Christ which I will preach and show to you and nothing else.'[149] Bilney was in some ways still the Church's loyal son: he allowed that Luther's condemnation was just, never denied confession, nor transubstantiation, nor papal primacy, held that the Councils of the Church should be universally observed, for the Church could not err. Yet *his* Church was 'the congregation of the elect, and so known only unto God'.[150] Bilney and Arthur would perforce abjure, in December 1527, but they did so only 'upon hope of preaching again'.[151]

---

[145] There are two sets of interrogatories put to Bilney in Tunstall's register, and one put to Arthur: the responses to the questions were copied into a book, which is now lost, but Foxe saw it: Guildhall, MS 9531/10, fos. 131^r–136^r; Foxe, *Acts and Monuments*, iv, pp. 623–6; Davis, 'Trials of Thomas Bylney', pp. 776–8.

[146] Guildhall, MS 9531/10, fos. 133^v, 134^v.

[147] Foxe, *Acts and Monuments*, iv, p. 626.

[148] More, *Dialogue concerning Heresies*, i, pp. 28, 36–7, 255–8.

[149] Guildhall, MS 9531/10, fo. 134^v.

[150] Davis, *Heresy and Reformation*, pp. 30–3, 46–53; Foxe, *Acts and Monuments*, iv, p. 626.

[151] Harpsfield, *Life*, p. 85. For the veracity of Bilney's recantations, see E. G. Rupp, 'The recantations of Thomas Bilney', *The London Quarterly and Holborn Review*,

John Skelton and, so he said, many of the Londoners were shocked by this casuistry:

> One of you there was
> That laughed when he did pass
> With his faggot in procession.[152]

Bilney went to the stake in 1531, but he left converts behind him.

At All Hallows Honey Lane the rector created a stronghold for 'the secret sowing and setting forth of Luther's heresies'. Dr Robert Forman was already of known reforming sympathies when he arrived in London early in 1525, down from Cambridge, for there he had 'concealed and kept Luther's books when sought for to be burnt'.[153] Perhaps this was why he was chosen for the City cure by its patrons, the Grocers, because there were in that company already some who favoured the new doctrines, including John Petyt, 'one of the first that . . . caught a sweetness in God's Word', and perhaps Geoffrey Lome.[154] To Forman's sermons in Honey Lane the brethren came. Tunstall, too, sent 'some to hear', but he 'could never know that he hath preached otherwise than well *openly*'. Secretly, Forman may have taught differently, 'being learned and using to hear confessions'.[155] The secret disputation held on 19 March 1528 between Forman and Tunstall, with London's most learned clerics in attendance, was recorded by More (as truly 'as his story of Utopia', wrote Tyndale sourly).[156] For More, Forman was one of those 'that bear two

clxvii (1942); Davis, 'The Trials of Thomas Bylney'; J. Guy, *The Public Career of Sir Thomas More* (Brighton, 1980), pp. 167–71.

[152] Skelton, 'A Replycacion Agaynst Certayne Yong Scolers Abjured of Late, &c', in *John Skelton: The Complete English Poems*, ed. J. Scattergood (Harmondsworth, 1983), p. 379, ll. 186–8. Skelton himself was witness at the abjuration of Thomas Bowgas of Colchester in May 1528; ibid., p. 19.

[153] More, *Dialogue concerning Heresies*, i, p. 379; Porter, *Reformation and Reaction in Tudor Cambridge*, p. 46.

[154] Guildhall, MS 9531/10, fo. 10^v; *Narratives of the Reformation*, p. 25; PRO, PCC, Prob 11/27, fo. 110^r.

[155] Strype, *Ecclesiastical Memorials*, i/2, p. 64; PRO, SP 1/47, fo. 109^r (*L&P* iv/2. 4073) (my italics); More, *Dialogue concerning Heresies*, i, p. 379.

[156] BL, Harleian MS 421, fo. 19^r (*L&P* iv/2. 4175). In the eleventh chapter of *The Dialogue concerning Heresies* More described the examination of a reformer, whom he does not name. That this reformer, whose disputation the clergy 'ruffleth up in darkness', was Forman is revealed by Tyndale in his *Answer to Sir Thomas More's Dialogue*, ed. H. Walter (Parker Society, Cambridge, 1850), p. 193.

faces in one hood'; a heretic while among heretics; but dissembling before churchmen.[157] Yet Forman's heresy, as More related it, was fundamental. Adducing Scripture, Forman avowed:

that all our salvation came of faith, as Abraham was justified by faith and not by his works. And that if our good works should be the cause of our salvation then, as Saint Paul saith, Christ died for nought.

All the works of man . . . be stark nought, as things all spotted with sin.

Moreover (but here More admitted 'that my remembrance may partly miss the order'), Forman and his fellows

believed that only God worketh all in every man, good works and bad. Howbeit such as He foreknoweth to be damned, no manner works be profitable to them. For God taketh them for nought, be they never so good. But on the other side, in those He hath chosen from the beginning and predestinate to glory, all works be good enough.[158]

Such predestinarian heresies Forman planned to broadcast, and he had a sermon prepared 'if the world change so that the time would serve it'. But preaching became ever more dangerous: in June 1528 Robert West, priest of St Andrew Undershaft, claimed that only 'flatterers and dissemblers' were allowed to preach at Paul's Cross, 'for they that say truth are punished as Bilney and Arthur was'.[159] So, another way was found.

Forman was the master mind of a contraband book trade. He had bought up quantities of books by Luther, Wycliffe, Hus, and Zwingli, and these he planned to deliver among 'young scholars . . . such as he thought of youth and lightness most likely to be soon corrupted'.[160] On Christmas Eve 1527 Thomas Garrett, Forman's chosen curate at All Hallows Honey Lane, returned to his old university, Oxford. By Lent 1528 Garrett 'hath do much hurt' among susceptible undergraduates, for with him he had brought two fardels of books,

---

157 More, *Dialogue concerning Heresies*, i, p. 399.
158 Ibid., pp. 391, 394, 398.
159 GLRO, DL/C/330, fo. 175ᵛ.
160 More, *Dialogue concerning Heresies*, i, p. 379.

containing 'many privy heresies'.[161] John Goodale, Garrett's old pupil from Oxford, now in Forman's service too, carried the 'very heavy' parcels of books back and forth; claiming later that he had no idea what was in them. But then, he also denied knowing that Garrett held any 'sinister opinion'.[162] Garrett abjured late in 1528, and with him abjured another in Forman's service and confederacy, Geoffrey Lome, usher of St Anthony's school. Both denounced the cult of images, and the doctrine of Purgatory, and both were convinced of the cardinal Lutheran doctrine that 'faith only, without good works, will bring a man to Heaven'.[163] This truth Lome had discovered perhaps from Forman, but also by reading Luther's works for himself. He would 'advance' it by translating Lutheran tracts from Latin, and circulating them through the universities, and through the dioceses of London, Ely, Norwich, and Lincoln.[164] It was from Lome that Robert Necton acquired forbidden works when he came down from Norwich to hear his countryman, Dr Forman, preach.[165] The books which Necton bought —Tyndale's English New Testament, *Unio Dissidentium*, and *Oeconomica Christiana*—were among those that Garrett had taken to 'infect' Oxford.[166] Gough was suspected of supplying the books, but when he protested his innocence, Tunstall was innocent enough to believe him. Tunstall had another suspect: Theodorycke, a 'Dutch man' from Antwerp, living now in London, who had already imported and sold *Oeconomica Christiana*, *Precationes Piae* (a work which Lome had translated), and

---

[161] PRO, SP 1/47, fos. 10^r, 54^r, 109^{r-v} (*L&P* iv/2. 3962, 4004, 4073). Thomas Cromwell happened to be in Oxford on Wolsey's service at this time: did he have contacts with this group? R. B. Merriman, *The Life and Letters of Thomas Cromwell*, 2 vols. (Oxford, 1902), i, pp. 318–19.

[162] PRO, SP 1/47, fos. 54^r, 109^{r-v} (*L&P* iv/2. 4004, 4073).

[163] Lome's abjuration and the articles against him are undated, but he abjured at the same time as Thomas Garrett, in 1528: Guildhall, MS 9531/10, fos. 136^v–137^r; Foxe, *Acts and Monuments*, v, p. 26, appendix vi.

[164] Guildhall, MS 9531/10, fo. 136^v.

[165] Strype, *Ecclesiastical Memorials*, i/2, p. 64. Strype calls this man Geffrey Usher, but surely he meant Geoffrey Lome, the usher. Dr Davis assumes the existence of two different men; *Heresy and Reformation*, pp. 56, 59.

[166] Strype, *Ecclesiastical Memorials*, i/2, p. 64; PRO, SP 1/47, fos. 10^r–11^r; Foxe, *Acts and Monuments*, v, appendix vi. The circle read not only Luther, but Melanchthon, Hus, Brentius, Bucer, Bugenhagen, François Lambert, Wycliffe, and Zwingli: Clebsch, *England's Earliest Protestants*, p. 80.

many of Tyndale's Testaments 'of the little volume'.[167] All these works were circulated 'so secretly that all the town should have sought them long ere they should have found them out'.[168]

London had no university, but it had the Inns of Court. Entering the Inns in the early 1520s were some of the prime movers of the English Reformation—Simon Fish, James Bainham, William Collins, Robert Packington, Christopher Brittayn, Nicholas Arnold, Francis Denham, and, most influential of them all, Thomas Cromwell.[169] Fraternal bonds were forged between young men, away from home and thrown together in the metropolis to study, and never more so than if they had a common experience of conversion. In 1524, two young men, 'marvellously well learned' '*Latine* and *Graece*', were admitted to the Inns: Francis Denham to the Middle Temple, and John Corbett to Lincoln's Inn. (Humphrey Monmouth was married to Margery Denham: was he a relative, and a recruiter?)[170] These two young scholars were soon recruited to use their talents to further the Gospel. By 1525 or 1526 'Dynamus' was at the Staple of the English merchants of Calais, which was already the headquarters of the brethren overseas. There Denham conversed with the chaplain of poetry and of Luther, and persuaded him to abandon the old interpreters of Scripture for the new: 'Dimitte delirum illum Lyram, and take this new preacher of the Gospel', Luther.[171] From Calais and from Antwerp Denham acquired new reforming texts. At the requests of his friends and mentors in London—Bilney, Fish, and George Constantine—Denham, still only nineteen, translated for the brethren François Lambert's *De Causis Excaecationis*, and a letter from Pomeranus to the faithful in England. By 1527 Denham was living in Paris in Constantine's house, collecting there a library of Lutheran tracts and works of the French

[167] PRO, SP 1/47, fos. 54ʳ, 109ʳ⁻ᵛ (*L&P* iv/2. 4004, 4073); Guildhall, MS 9531/10, fo. 136ᵛ.

[168] More, *Dialogue concerning Heresies*, i, p. 269.

[169] *A Calendar of the Inner Temple Records*, ed. F. A. Inderwick, 5 vols. (1896–1937); *Register of Admissions to the Honourable Society of the Middle Temple*, ed. H. A. C. Sturgess, 3 vols. (1949); *The Register of Admissions to Gray's Inn, 1521–1889*, ed. J. Foster, 2 vols. (1889); *Records of the Honourable Society of Lincoln's Inn*, 2 vols. (1896).

[170] *L&P* iv/2. 4326, 4493. Corbett is, however, once described as a monk (*L&P* iv/2. 4328). Strype, *Ecclesiastical Memorials*, i/2, p. 371.

[171] *L&P* iv/2. 4407, 5094.

Reformation.[172] These were surely not for him alone. His capture, and death from the plague in 1528, was a grave loss to the cause.[173] His fellow member of the Middle Temple, William Collins, was in prison by 1526 for speaking against the Church of Rome, and in favour of clerical marriage (it seems), and would remain there 'frantic', but still an evangelist: he 'lasheth out Scripture in Bedlam' as fast as Luther and Tyndale did in Germany.[174] Daring to take the part of Wolsey, and mocking the Cardinal in the Gray's Inn Christmas play of 1526, made it politic for Simon Fish to lie low, but his book running made it sure, sooner or later, that he must leave England. Fish joined Tyndale in exile, there to translate the *Sum of Scripture* from the Dutch, and to write his vitriolic satire against a venal priest-hood which invented 'Purgatory pin-fold' only to claim power to bind and loose the deluded faithful from its terrors, *The Supplicacyon for the Beggers*.[175]

One by one the brethren sought refuge 'beyond the sea' from the persecution at home, and freedom to further God's Word. In the Low Countries, France, and Germany, Tyndale, Frith, Barnes, Roye, Joye, Barlow, Denham, Constantine, Fish, and others provided inspiration for their brethren in England and writings to sustain the cause.[176] Not that exile was safe. Tyndale would be betrayed there. In 1528 Richard Akyrston, morrow mass priest of St Botolph's, who had already born a faggot in London, was imprisoned in Antwerp at the Emperor's pleasure, for his heresy.[177] Nevertheless the reformers held themselves ready for flight. Latimer trusted God to help him, so he said in 1531, 'which if I had not, the ocean sea, I think, should have

---

[172] Ibid. iv/2. 4396. Denham's books are listed: Mozley, *William Tyndale*, pp. 54–5, 349.

[173] *L&P* iv/2. 4326–8, 4338, 4359, 4374, 4390, 4394–6, 4407, 4493, 5094. Was it from Denham that Anne Boleyn received the *Epistles and Gospels for 52 weeks*? Denham certainly owned it (ibid. iv/2. 4396).

[174] More, *Dialogue concerning Heresies*, i, pp. 433–4; ii, pp. 724–5; PRO, SP 1/242, fo. 229^r (*L&P Addenda*, i/2. 1407).

[175] Foxe, *Acts and Monuments*, iv, pp. 656–8; Clebsch, *England's Earliest Protestants*, pp. 88–9, 240–51; W. W. Haas, 'Simon Fish, William Tyndale, and Sir Thomas More's "Lutheran Conspiracy"', *JEH* xxii (1972), pp. 125–36.

[176] A. Hume, 'English Protestant Books Printed Abroad, 1525–1535'; More, *Confutation*, *CW* 8/ii, pp. 1063–92; Mozley, *William Tyndale*; C. Butterworth and A. Chester, *George Joye* (Philadelphia, 1962).

[177] *The Letters of Sir John Hackett, 1526–1534*, ed. E. F. Rogers, 2 vols. (Morgantown, 1971), pp. 155, 156, 157, 169, 170, 171, 178, 201.

divided my Lord of London and me by this day'.[178] They had friends prepared to secure their flight to the 'sea side', and to hide them if and when they returned. While all Wolsey's agents on the Continent were hunting Friar Roye in 1528, 'Roye hath been in England with his mother in Westminster'.[179] In exile the brethren looked always for the day when they could return; when 'the King's pleasure is that the New Testament in English should go forth'.[180] That that day would come they were certain, for 'Heaven and earth shall fail', but God's Word never. In 1530 Stephen Vaughan signed a letter to Thomas Cromwell with the biblical message, 'Behold the signs of the world which be wondrous and again fare ye well'.[181]

In London the brethren anxiously awaited consignments of smuggled books, and sent news from home. 'This is the news in England now, Richard Harman,' wrote John Sadler, a draper, in September 1526, 'none other but that the New Testament should be put down and burnt, which God forfend.' A City ironmonger, Richard Hall, wrote to Harman in Antwerp, asking for 'two new books of the New Testament in English'.[182] 'Mr Fish, dwelling by the White Friars', commissioned Testaments from Harman, so Constantine told Robert Necton early in 1527, and from Fish Necton bought 'now five and now ten and sometime more, and sometime less, to the number of twenty or thirty in the great volume'. These Necton sold to eager London merchants; to 'young Elderton' of St Mary at Hill, William Gibson of St Margaret Pattens, and to five or six others, whose names he conveniently forgot. At Christmas 1527 Necton was offered a shipment of two or three hundred Testaments for himself by a 'Dutch man', and agreed, 'Look what Mr Fish doth, I will do the same'.[183] But by then Fish was fled to exile.

---

[178] *L&P* vi. 607.

[179] PRO, SP 1/47, fo. 10ʳ; SP 1/53, fo. 201ʳ (*L&P* iv/2. 3962; iv/3. 5462).

[180] *L&P* iv/3. 6385.

[181] PRO, SP 1/68, fo. 57ʳ (*L&P* v. 533).

[182] PRO, SP 1/50, fos. 75ʳ–76ʳ (*L&P* iv/2. 4693). A cache of letters to Harman from England was discovered in the autumn of 1528: *Letters of Sir John Hackett*, pp. 155–6, 161–2, 164, 167–70, 173–5, 178–81, 183–4, 188, 199–201, 207.

[183] Strype, *Ecclesiastical Memorials*, i/2, pp. 63–5; Mozley, *William Tyndale*, pp. 72–3, 205, 349–50. Anthony Elderton was churchwarden at St Mary at Hill between 1527 and 1530, but perhaps this was the father, rather than the son? *St Mary at Hill*, i, pp. 342–9.

CONVERSION

Who were the evangelists winning to the new doctrines by their preaching? Who was reading the books which the brethren ran such risks to distribute? Why should men abandon an old faith and an old obedience for a new and persecuted faith and discipline? The most essential questions are hardest to answer. Religious conversion occurs for reasons which are private, and usually secret; but never more secret than when the new faith is forbidden. Yet Catholic writers found these questions easy to answer: the desire for 'parasite liberty' lay at the heart of all religious disobedience. Richard Whitford was persuaded in 1532 to publish an answer to 'these new fangled persons, which in deed been heretics although they will not be so called', who questioned religious vows. [184] The heretic argument, the more 'jeopardous' became it 'done seem to be surely grounded upon Scripture', was simple:

Our sweet Lord and saviour Jesu by . . . marvellous pain and passion . . . hath bought us out of the course of the law of thraldom and bondage, and a law of fear and dread. And hath put us into the liberty and freedom of the Gospel; which is a law . . . of liberty, and a law of love. . . . But alas, alas, for most deep sorrow, that we now have slipped . . . from that sweet and most pleasant Christian liberty: unto more thraldom and bondage than ever were the children of Israel in Egypt.

'What enchantments', the heretics asked, 'have thus bewitched us', that the people should have turned 'from faith unto mistrust, from hope unto doubt, from love unto dread'? [185] And they, too, found a simple answer: the Church had deluded Christ's people, and the clergy hidden His Word. The Church bound the people to it by laws, not love, so the reformers averred. John Lambert remembered a divine preaching this warning, which was too little heeded: 'Adam, being in paradise, had but one law to observe, and yet he brake it: What

---

[184] Richard Whitford, . . . *the Pype/ or Tonne/ of the lyfe of perfection* (1532; *RSTC* 25421), sig. Ai^v.

[185] Ibid., sigs. Aiiii–Bi. Cf. the Messenger's argument in More, *Dialogue concerning Heresies*, i, p. 105.

other thing then shall this multitude do . . . but multiply transgression? For when a faggot is bound over-strait, the bond must break.'[186] Men like Roper and Bilney, 'before wounded with the guilt' of their sins and 'almost in despair', found in the Gospel Christ's promise of salvation to those who believed, and then were free to ignore the 'multitude of laws', 'so many snares of constitutions' of the Church, too many for any man to keep.[187] Headstrong evangelicals sanctioned the law-breaking. Thomas Arthur told the Londoners in 1527 that the laws of the Church were like the crosses set in their City walls: when there was only one cross, or very few, 'men would not piss there' but when there were many . . .[188] Evangelism brought many freedoms.

Early Protestantism was a 'religion of novelty and protest'.[189] Dangerous enough to be adopted only by the daring and determined, the new faith soon took on the character of a rebellion, of a protest movement. Destroying images, posting bills, singing seditious ballads, spreading forbidden books, hiding those on the run from the authorities, taunting priests, meeting in secret conventicles, planning daring escapes, preaching in spite of persecution, the first followers of the new doctrines marked themselves as rebels as well as heretics. All these activities well suited the fraternal and subversive mentality of many young people. Since more than half the population of early modern England was under twenty years old, and since young men had always flocked to the capital seeking their fortunes, it may prove little to show the youth of many of the new faith's early converts in the City. Yet Catholic opponents of reform saw it as the movement of radical youth and blamed 'lewd lads', 'these younglings' for the spread of 'ungracious heresy'. Soon Catholic mothers wrote despairingly to their cuckoo Protestant children who shamed their families by turning apostate and abandoning the old ways, and threatened to cut them off without a penny. Certainly young people were known to be especially ready to

---

[186] Foxe, *Acts and Monuments*, v, p. 202.

[187] Ibid. iv, pp. 635, 625.

[188] Guildhall, MS 9531/10 fo. 135[v]. Foxe's Victorian editors bowdlerized the text: *Acts and Monuments*, iv, p. 623.

[189] P. Collinson, 'The English Reformation and the English People: the Rise and Fall of Protestantism as a Popular Cause' (3rd Birkbeck lecture, 1982). I am most grateful to Professor Collinson for showing me this essay before its publication.

mock priests and scorn their authority.[190] The young Duke of Richmond, the King's bastard son, taunted his tutor, Richard Croke, laughing in his face: 'Praeceptor, si tu me verberes, ego te verberabo.' Ever since the boy had listened to the 'indecent songs' of 'buffoons', abusing the clergy, his tutor could 'do nothing with him'.[191] Despising priests was not heresy, but many feared that it was one step on the way.

London's apprentices were thought to be perennially unstable and easy to rally to a cause, especially if that cause were forbidden. The turbulence of 'prentices' and 'younglings' during the riots of Evil May Day 1517 gave reason to fear the consequences if they should adopt reform.[192] But apprentices became masters themselves in time, and once they joined the City establishment the infiltration of their heresy would become of even greater concern. Even within the City's first companies—the Mercers, the Grocers, the Drapers—secret cells of brethren were formed. In the Drapers' Company the older generation of 'Scripture men', like Humphrey Monmouth, was joined in the early 1520s by a younger generation— John Sadler, Thomas Mekins, John Hewes, Roger Whaplode, Thomas Patmore, John Parnell.[193] Why they all became favourers of the Gospel will remain a matter for speculation, but their fellowship within the company may well have made them susceptible to the evangelism of one of their number. At the 'Bachelors' supper' in the company in 1519 these young men were entertained by a conventional satire upon the clergy, it seems: 'to him that played the friar' they paid 8d.[194] Anticlericalism need not be a heretical spirit, but sometimes it came to be one. Few in London had better reason to resent the clergy than Roger Whaplode, for he was Richard Hunne's son-in-law, and few would express that resentment more consistently than he.[195] In St Mary Woolchurch Thomas Mekins was involved in parish affairs with his anticlerical neighbour, the

[190] *L&P* vii. 667; S. Brigden, 'Youth and the English Reformation', *P&P* 95 (1982), pp. 37–67.   [191] *L&P* iv/2. 3135.

[192] More, *Apology*, p. 156. For Evil May Day, see below, ch. 3(i).

[193] Monmouth was made free of the company in 1503, Mekins in 1516, Whaplode in 1519, Patmore in 1530, Parnell in 1509: Drapers' Company records.

[194] Drapers' Company, Repertory 7, p. 101.

[195] See below, pp. 172–3, 558, 568. He joined the riot of City priests against Stokesley in 1531.

mercer and bookseller John Colyns; but while Colyns remained orthodox, Mekins did not. In September 1528 he stood surety in £100 for Necton, the colporteur of Lutheran books.[196] John Sadler, like Humphrey Monmouth, commissioned New Testaments from the exiles.[197] Hewes was troubled for speaking against pilgrimage and the veneration of images in 1531.[198] John Parnell's apprentice abjured his heresy at Paul's Cross in November 1532, and Parnell himself married the widow of the Lutheran John Petyt, keeping the new faith within the family. We shall hear later of Thomas Patmore's heretical career.[199] Maybe the brethren kept their views hidden from the Catholic majority in their companies; certainly the Drapers seem to have thought that heresy touched others, not them. When, early in 1528, the Drapers purchased a proclamation against forbidden books, mostly works of the Reformation, their accounts recorded: 'Item, for copies of the proclamations for *lollers'* books'.[200]

Among the first enthusiasts for the new heresy were indeed the adherents of an older one, the Lollards. Lollards and reformers had much in common.[201] At the heart of both faiths lay the conviction that Scripture alone enshrined all religious truth, and that to every layman belonged the right to find that truth for himself. From the freedom to read God's Word would follow another: the liberation from priestly authority. When Thomas Man asserted that 'all holy men of his sect were only priests' he was anticipating the Lutheran doctrine of the priesthood of all believers, a personal faith in which 'every layman is priest'.[202] Pardons, confession, penance, 'Purgatory pinfold'— the whole penitential system whereby the clergy held the laity

---

[196] BL, Harleian MS 2252, fo. 165ʳ; PRO, Sta Cha 2/31, fragments. The bond was discovered by Dr Guy: J. Guy, *The Public Career of Sir Thomas More*, p. 108. The other sureties were Thomas Power, grocer of St Mary at Hill, and Thomas Necton of Norwich, the Sheriff who 'was Bilney's special good friend': Foxe, *Acts and Monuments*, iv, p. 652.

[197] See above, p. 118.

[198] Foxe, *Acts and Monuments*, v, p. 34.

[199] *Two London Chronicles*, p. 5; *House of Commons, 1509–1558* (Petyt); see below, pp. 205–7.

[200] Drapers' Company, Repertory 7, p. 317.

[201] See Dickens, 'Heresy and the Origins of the English Reformation'; Davis, *Heresy and Reformation in the South-East of England*.

[202] Foxe, *Acts and Monuments*, iv, p. 209; Guildhall, MS 9531/10, fo. 135ᵛ.

in thrall—could be discarded. Denial of the doctrines and authority of the Catholic Church gave Lollards and reformers a shared destiny, and a shared expectation of suffering under persecution.

To hear the reformers denounce the excesses of Catholic worship and inveigh publicly against the idolatry and superstition which the Lollards themselves had long decried in their conventicles, strengthened and dignified the Lollard cause. In his sermons Bilney urged his listeners, 'pray you only to God and to no saints'; insisted that 'Christian people should set up no light, and the images have no eyes to see'; that if priests were ignorant and venal 'rather let them starve then give them any penny'; alleged that priests took the people's offerings from the necks of images to bedeck their own whores; said that it was better to 'tarry at home and give somewhat in alms' than to go on pilgrimage; called Our Lady of Willesdon a 'common bawd', and St Mary Magdalen 'a stewed whore'. All this was almost indistinguishable from the Lollards' own invective.[203] Yet, although Bilney owed much to the Wycliffite tradition, it was not Lollard belief that Bilney longed to evangelize. Through his humanist study of Erasmus's Latin New Testament Bilney had 'chanced upon this sentence of St Paul . . . "It is a true saying, and worthy of all men to be embraced, that Christ Jesus came into the world to save sinners; of whom I am the chief and principal"'.[204] Did the Lollards hear behind Bilney's attacks upon images the new doctrine of faith? How would Lollards become Protestants?

At Michaelmas 1526 two Lollard companions, John Tyball and Thomas Hilles, came down to London from their home in Steeple Bumpstead in Essex in search of the Lutheran Friar Barnes, 'because they would have his counsel in the New Testament, which they desired to have of him'.[205] They found Dr Barnes in his chamber at the Austin Friars, reading aloud from St Paul, the *locus classicus* of justificatory doctrine, to City merchants who sat at his feet, one grand 'young gentleman . . . having a chain about his neck'. Here the old heresy and

---

[203] Davis, 'Trials of Thomas Bylney'.

[204] Foxe, *Acts and Monuments*, iv, p. 635.

[205] BL, Harleian MS 421, fo. 35ʳ; Strype, *Ecclesiastical Memorials*, i/2, pp. 54–5; Aston, *Lollards and Reformers*, pp. 231, 240.

the new—'known men' from the shires and wealthy, sophisticated City brethren—confronted each other. Tyball and Hilles
proffered their treasured 'four Evangelists, and certain epistles
of Peter and Paul in English', bought eight years back, and
studied 'in continuance of time'. Barnes 'made a twit of it, and
said, "A point for them, for they be not to be regarded towards
the new printed Testament . . . for it is of more cleaner
English" '. Back to Essex the Lollards took their Tyndale
Testament, warned to 'keep it close'. But though the learned
Barnes might scorn the barbarous texts of the Lollards, it was
not his intellectual Lutheranism which prevailed in the creation
of the theology of the English Reformation. In 1532 Tyndale
warned John Frith: 'Of the presence of Christ's body in the
sacrament meddle as little as you can, that there appear no
division among us. Barnes will be hot against you.'[206] But
already there was division. Soon those who had adopted a
purer form of Lutheranism, like Barnes, would be yesterday's
men. More radical teachings on the Mass, akin to Lollardy,
stemming from Zwingli and the Swiss reformers, with Frith as
their most eloquent exponent, were spreading.[207]

In the 1520s an advance guard of the brethren formed a
secret society, the Christian Brethren.[208] The organization and
membership of this clandestine fraternity were shadowy, to all
but the members, but its purposes were clear. The Christian
Brethren held a belief which had set the Lollards apart, and
they determined to proselytize it: 'the sacrament of the altar
after the consecration is neither body nor blood but remaineth
bread and wine as it did before'. They sponsored the publication
of sacramentarian books; perhaps even of the first Zwinglian
work in English, Roye's *Brefe Dialogue bitwene a Christen Father
and his stobborne Sonne* of August 1527. Their conspiracy to
convert was financed by a network of men (and women?) who
were prepared to subscribe to support their heretical convic-

[206] *L&P* vi. 403.

[207] See More, *Apology*, p. 5; Clebsch, *England's Earliest Protestants*, p. 80; A. Hume,
'William Roye's "Brefe Dialoge" (1527): An English Version of a Strassburg
Catechism', *Harvard Theological Review*, lx (1967), pp. 307–21; More, *CW* 11,
pp. xvii–xxxvii.

[208] The activities of the Christian Brethren are discussed by Professor Rupp and
Dr Davis: *Studies in the Making of the English Protestant Tradition*, pp. 6–14; Davis,
'Heresy and Reformation in the South-East of England, 1520–1559', pp. 250–67.

tions, just as members of Catholic confraternities paid quarterages. A London mercer, Thomas Keyle, revealed that

There was made for the augmentation of Christian Brethren of his sort auditors and clerks within this City. And that every Christian brother of their sort should pay a certain sum of money to the aforesaid clerks which should go into all quarters of the realm, and at certain time the auditor to take account of them.

The subscription lists of members never fell into the hands of the authorities. While we may suspect more, we know the names of only five of the Christian Brethren—Thomas Keyle, Shreve, a barber surgeon of Coleman Street, and three priests, Parker, Patmore, and Marshall.[209] The three priests were of that advanced coterie who proclaimed God's Word, whatever the danger. George Marshall had told his Essex congregation in 1527, 'I will show you the Gospel'.[210]

Thomas Keyle was one of London's most prominent citizens. In 1523 he was warden of the City's first company, the Mercers.[211] The will made by his apprentice, Ralph Shelton, in 1522 was original enough in form to suggest that Keyle's was a household in which theology was earnestly debated, if not yet in an openly unorthodox way.[212] Keyle, like other mercers, had links with the Continent, and in 1524 and 1527 he was granted protections to go in the retinues of Berners and Wingfield to Calais; like other mercers, too, Keyle was involved in the printing trade (if he was that Thomas Kellys summoned before Tunstall in October 1524 and warned against selling heretical books). Connections with Calais, with Antwerp, and with the printers, made Keyle specially susceptible to reforming ideas, and in a unique position to further them.[213] Risking all, Keyle dared to propagate contraband literature, and, like the Lollard goldsmith John Barret, seems to have suffered for the cause. Fallen into poverty, Keyle would receive charity from the Mercers, as Barret had done from the Goldsmiths.[214]

---

209 PRO, SP 1/237, fo. 78ʳ (*L&P Addenda*, i/1. 752).
210 GLRO, DL/C/330, fos. 139ʳ⁻ᵛ; Davis, *Heresy and Reformation*, pp. 47–8.
211 *Acts of Court of the Mercers' Company*, pp. 385, 490, 507, 511, 528, 661, 664–80, 683, 727, 753.
212 PRO, PCC, Prob. 11/20, fo. 89ᵛ; see above, p. 30.
213 *L&P* iv. 787(18), 3008(9), 4594(20); GLRO, DL/C/330, fo. 83ᵛ.
214 John Barret was a company almsman in 1516: Keyle was dismissed an almsman in 1543, because 'he is of better ability to live than he hath been heretofore', but was

So urgent was the mission that the brethren did not count the cost. Asked 'how he and his other fellows would do, seeing the King's grace and those great lords of the realm were against them?', George Parker answered defiantly that 'they had already two thousand books out against the blessed sacrament in the commons' hands, with books concerning diverse other matters, affirming that if it were once in the commons' heads they should have no further care'.[215]

Yet, for all their confidence and determination, the brethren were still so few and so beleaguered that their chances of revealing the light of God's grace to a whole people blinded by superstition might have seemed hopeless—if not to them. They were winning converts to their cause, and steadily; they were now a fifth column within the English Church; but, persecuted and on the run, they seemed likely to remain a Church under the cross. It was true that heretics, of whatever kind, were now growing in number, and to be found in such positions of influence that the whole of London's Catholic community could no longer ostracize them, if business were to continue as usual. Sir Thomas More warned good Catholics against fraternizing with the enemy, who were encouraged because they were no longer shunned; 'the greater hope have they . . . for they see that it beginneth almost to grow in custom that among good Catholic folk, yet be they suffered boldly to talk unchecked'.[216] But though the brethren might secretly infiltrate London society, the new faith could only become established if it were protected. Only the King, persuaded by those who had his counsel, could call a halt to the persecution. Without a voice at Court the brethren would remain outlawed, and the Word silenced.

But heresy had already entered Henry VIII's Court. On 8 November 'one Foster', 'a gentleman of the Court bare faggots at Paul's for heresy'.[217] This was Richard Foster,

---

again receiving £6 p.a. by 1553: Goldsmiths' Company records; Mercers' Company, Acts of Court, ii, fos. 174ʳ, 264ʳ.

[215] PRO, SP 1/237, fo. 78ʳ (*L&P Addenda*, i/1. 752).
[216] More, *Apology*, p. 158; *Answer to a Poisoned Book*, pp. 4–5.
[217] *Two London Chronicles*, p. 3.

yeoman usher of the Crown, who that summer had been granted a fee of 6*d*. a day.[218] To be a daily waiter at Court was to have propinquity to those around the King, which gave some influence. Foster had been before the Bishops already, it seems. In December 1527, accused with Bilney and Arthur, he had abjured. He admitted that he had 'accompanied with certain of the manner of living of Martin Luther and his sect' in eating flesh on forbidden days, but his heresies were more extreme than the Lutherans'. Accused of denying that Christ's very body was in the sacrament of the altar, he confessed to believing that 'a priest could not consecrate the body of Christ'.[219] There was, however, one at Court who had been touched by reform, who was more influential by far than any yeoman usher; one whose power over the King by 1527 was unrivalled; one who had bewitched him, so many said: Anne Boleyn.

Anne Boleyn was deeply committed to the promotion of the Gospel, and early reputed to have evangelical sympathies.[220] An encomium of Queen Anne, written for her royal daughter by William Latimer, a reforming London rector, recalled Anne's devotion to vernacular Scripture. She was 'continually . . . reading the French Bible, and other French books of like effect, and conceived great pleasure in the same'. Once Queen, she would order the chaplains in her household 'to be furnished of all kinds of French books that reverently treated of the Holy Scripture'.[221] William Locke, a wealthy London mercer, and patriarch of a reforming dynasty, 'when he was a young merchant and used to go beyond sea the Queen Anne Boleyn . . . caused him to get the Gospels and Epistles written in parchment', so his daughter remembered.[222] This book was not proscribed—yet—but to own it was to admit an evangelical impulse. In Francis Denham's heretical library in Paris the Epistles and Gospels in French had been found side by side

[218] *L&P* iv/2. 4445(4).

[219] BL, Harleian MS 421, fo. 7ʳ; Guildhall, MS 9531/10, fo. 130ᵛ; Davis, *Heresy and Reformation*, pp. 25, 46.

[220] M. Dowling, 'Anne Boleyn and Reform', *JEH* xxxv (1984), pp. 30–46; E. W. Ives, *Anne Boleyn* (1986), chs. 13, 14.

[221] William Latimer, 'Treatyse' on Anne Boleyn; Bodleian, MS Don. C. 42.

[222] 'Recollections', p. 97.

with radical reforming works in the summer of 1528.[223] Somehow, Simon Fish in exile knew that if he sent Anne Boleyn a copy of his *Supplicacyon for the Beggers* his offering might be welcomed. So it was, and so impressed was Anne that—so it was said—she showed the work to the King. In 1529 Anne knew how to find Fish in his hiding place among the City brethren, not a mile from the Court.[224] Clearly, she had links with reforming circles in London.

Before ever she was Queen, Anne dared to use her influence to protect those who were persecuted for the Gospel's sake. When Dr Forman's book-running operation was discovered at All Hallows Honey Lane in 1528 she interceded with the Cardinal for him: 'I beseech your grace with all mine heart to remember the parson of Honey Lane for my sake shortly.'[225] Later she pleaded for Richard Harman, who had been expelled from the house of the English merchants at Antwerp in 1528, 'but only for that, that he did both with his goods and policy, to his great hurt and hindrance in this world, help to the setting forth of the New Testament in English'.[226] Her plea was made to Thomas Cromwell, whose commitment to the Gospel was even deeper than hers, as he would prove. His protection of the brethren was more daring, for among them were his old friends and acquaintances in London.

The brethren were sure that the time would come, and soon, when their cause would prevail. 'And out of question that day they not only long for, but also daily look for.' Into Tunstall's palace in London they cast a letter, which promised:

There will once come a day.[227]

---

[223] Anne's own copy of 'The Pistellis and Gospelles for the LII Sondayes in the Yere, with an Exhortation to Each in English' survives in the British Library: Harleian MS 6561: M. J. C. Dowling, 'Scholarship, Politics and the Court of Henry VIII', (University of London, Ph.D. thesis, 1981), p. 133.

[224] Foxe, *Acts and Monuments*, iv, pp. 657–8. Professor Ives casts doubt upon the story that Anne showed Henry *The Supplicacyon*: *Anne Boleyn*, p. 163 n.

[225] BL, Cotton MS Vespasian F iii, fo. 15ᵛ (*L&P* iv/3, appendix 197).

[226] Dowling, 'Anne Boleyn and Reform', p. 43.

[227] More, *Apology*, p. 158.

III City and Church

THE CITY of London was small enough for news to travel fast, for individual disputes soon to be noticed and causes to be swiftly followed. It was also large enough for a formidable volume of mass resentment to grow and fearsome numbers to mobilize. As the seat of government, the centre of national trade and commerce, London's security was vital. Thomas Bedyll warned Thomas Cromwell: 'London is the common country of all England from which is derived to all parts of the realm all good and ill occurrant there.'[1] But if London's compliance was essential, it was never easily achieved. Edmund Grindal, Elizabethan Bishop of London, advised Secretary Cecil of London's dissidence: 'It was a port and overmuch populous, which Aristotle doth disallow', and for his own part he lamented that 'the Bishop of London is always to be pitied'.[2] Within the capital there were many unruly elements, and the causes which the Londoners followed might sometimes be quite different from those which the security of the realm required.

EVIL MAY DAY

In the spring of 1517 there were signs in the City of disturbance to come. London's governors feared disorder, and with reason.[3] There had been foreign merchants in the City almost as long as there had been trade there, but the citizens resented their presence; at some times more than others, and especially in times of dearth.[4] In 1517 the resentment of 'strangers' was almost universal. Londoners were convinced that strangers stole their work, their markets, and sometimes their wives. In April 1516 bills were posted on the doors of St Paul's and All

[1] BL, Cotton MS Cleo. E iv, fo. 217ʳ (*L&P* xii/2. 91).
[2] PRO, SP 12/24, fo. 42ʳ; P. Collinson, *Archbishop Grindal, 1519–1583* (1979), p. 108.
[3] For the problems of disorder at this time, see J. Kennedy, 'The City of London and the Crown, *c*.1509–*c*.1529' (University of Manchester, MA thesis, 1978), ch. 5.
[4] *A Relation, or Rather a True Account of the Island of England*, ed. C. A. Sneyd (Camden Society, xxxvii, 1857), pp. 23–4; Hall, *Chronicle*, p. 586.

Hallows Barking, attacking the King and Council for lending money to foreigners, with which they 'bought much wool which was to the undoing of Englishmen'.[5] A promise was made in March 1517 to 'subdue all strangers that be breakers of the privileges' of the City, but so widespread was the anger against aliens that it might easily turn to violence.[6] John Lincoln, broker and surveyor of 'goods foreign bought and sold', began to incite the citizens: 'Englishmen want and starve', while the aliens lived 'abundantly in great pleasure'. He prevailed upon Dr Bele, canon of St Mary Spital ('you were born in London'), to read a 'pitiful' bill before a huge audience at the Spital sermon that Easter Tuesday (14 April). Bele preached that 'as birds would defend their nest so ought Englishmen to defend themselves and to hurt . . . aliens for the common weal'. Urging 'God's law' that men should fight for their country, 'ever he subtly moved the people to rebel against the strangers'.[7] At Court a mercer threatened the 'whoreson Lombards . . . by the Mass, we will one day have a day at you'.[8]

The young men of London were notoriously easy to rally to a cause. Soon, by the persuasion of Lincoln and two apprentices, malcontents were incited to band together and a conspiracy prepared against the strangers for May Day coming, the traditional day for apprentice misrule. On 28 April 'divers young men' set upon aliens in the City streets, and the culprits' master, William Daniell, was bound upon pain of death that neither he nor his servants should hurt any 'Frenchman', nor venture abroad on May Day 'in maying or in may running'.[9] Rumours reached Cardinal Wolsey of a confederacy among 'young and riotous people', and on May eve he commanded the City rulers to keep the peace, 'to keep the young men asunder', and to order a curfew—rather than an armed watch—for May morning. Too late. Alderman Munday, finding two young men in Cheapside playing at 'bucklers',

---

[5] *L&P* ii/1. 1832. Aldermen were ordered to check the handwriting of all those in their wards who could write.

[6] *Acts of Court of the Mercers' Company, 1453–1527*, ed. L. Lyell and F. D. Watney (Cambridge, 1936), pp. 443–4; CLRO, Repertory 3, fo. 133[r].

[7] Ibid., fo. 55[v]; Hall, *Chronicle*, pp. 586–7.

[8] Ibid., p. 588.

[9] Ibid.; CLRO, Repertory 3, fos. 141[v], 142[r].

with a crowd looking on, attempted to arrest them. They resisted, crying 'prentices and clubs'. By 11 o'clock six or seven hundred rioters were thronging through Cheapside, and three hundred more through Paul's churchyard.[10] In St Martin's sanctuary they broke up the shops of foreign workmen. Orders to desist from Thomas More, Undersheriff, and the Sergeant at Arms went unheard. A vigilante mob surged to the house of John Meautas, the French secretary and harbourer of 'pick-purses', and 'would have stricken off his head', had he not escaped up the belfry of a neighbouring church. Thwarted, the young men, and 'certain young priests' with them, turned to looting. No lives were lost; only property. But by 3 o'clock that afternoon the riot was over; three hundred of the two thousand rioters were arrested, and they were reportedly mostly 'poor prentices'.[11]

The King was 'highly displeased and offended'. The 'sub-stantial persons' of his City 'did wink at the matter'; the royal credit abroad was damaged for he had failed to keep order among his own subjects or to protect foreign residents. A com-mission of Oyer and Terminer sat in London on 4 May. The Duke of Norfolk rode into the City with an armed retinue of 1,300 men, 'whose opprobrious words to the citizens . . . grieved them sore'. So volatile was the mood in the capital that men were ordered to keep their wives at home, for their 'babble and talk' was disturbing. Nearly three hundred prisoners were brought through the City, roped together, some only children. Thirteen were found guilty of treason, and the utmost penalty enjoined.[12] Fearing the loss of the City's liberties, the Mayor and Aldermen sent delegations seeking the royal pardon, but the King was in no forgiving mood. Relenting finally, maybe at the Queen's behest, he came to Westminster on 22 May to preside over a theatrical scene of forgiveness. Four hundred men and eleven women were pardoned.[13] Yet the old griev-ances against the aliens remained. Leading companies prepared 'a book . . . humbly beseeching a redress of certain things . . .

[10] Hall, *Chronicle*, pp. 588–9; More, *Apology*, p. 156.

[11] Stow, *Survey*, i, p. 152; *L&P* ii/2. 3204; Hall, *Chronicle*, p. 589.

[12] Ibid., pp. 589–90.

[13] CLRO, Repertory 3, fos. 142ᵛ, 143ʳ, 143ᵛ, 144ᵛ; Hall, *Chronicle*, p. 591; *Acts of Court of the Mercers' Company*, p. 445.

greatly against the common weal'.[14] Even while the dismembered corpses of the rioters hung in the City, new conspiracies were rumoured. In St Martin Ludgate there were complaints that the Chamberlain, the Mayor, and Aldermen 'marred all and lost the liberties', and threats that it 'shall not be well till we have a new Aldermanbury'.[15] As the sweating sickness raged in September another confederacy was formed to make 'as good a skirmish with strangers as was on May Day last', and, in the King and Cardinal's absence, to murder the aliens and sack their houses. Three thousand householders promptly armed themselves for a week and set a curfew for their servants.[16] In that month too, 'simple persons' were imprisoned for seditious words about an 'insurrection to be made upon the strangers'.[17]

Evil May Day proved to be the most serious riot of the sixteenth century in London, but the King and the City's governors could not be comforted by future knowledge. They lived in dread of worse to come, and the riots of 1517 were long remembered. In 1527 there was trouble over the Mayoral election: 'for the commons would not have had Seymour for because of Ill May Day'.[18] On May Day 1527 a foreign visitor observed how apprentices were kept at home 'for fear of the London artisans, who go in arms "guerir le May", and sometimes attack foreigners'. When in 1529 some apprentices, 'under pretence of playing a cudgel game', took up arms against foreign merchants, the Mayor made sixty arrests.[19] The riots of 1517 left Henry convinced that all London was 'traitor to his grace'. More's investigations of the confederacy of two apprentices, who 'compassed between them twain' the May Day rising, persuaded him how easily the citizens—especially turbulent youth—might be won to seditious causes.[20] Few could raise many; rumour could turn to action; threats to

[14] *L&P* ii/2. 3259; CLRO, Repertory 3, fos. 145^v, 165^r; Mercers' Company, Register of Writings, ii, fos. 153^r, 162^r. Thomas Howell was bound in £400 not to talk of the 'grudge borne by the citizens against the merchant strangers'.

[15] CLRO, Repertory 3, fos. 151^r–153^r, 154^v, 155^r, 158^v, 159^r.

[16] *L&P* ii/2. 3571–2, 3641, 3697, 3747; CLRO, Repertory 3, fo. 164^v.

[17] Ibid., fos. 165^r–166^r.

[18] *Grey Friars Chronicle*, p. 33.

[19] *L&P* iv/2. 3105, p. 1413; *CSP Ven.* iv. 569.

[20] More, *Apology*, p. 156.

violence. To survey the size and diversity of the population of Tudor London is to discover a potential for disorder which the authorities constantly feared and lamented.

THE POPULATION OF LONDON

'The population of London is immense and comprises many artificers', observed a Venetian visitor in 1531; another estimated that 70,000 people lived there.[21] The capital's population was growing so rapidly that in 1557 a visitor could imagine that 185,000 people lived in London, its liberties and Westminster.[22] He exaggerated. Historians estimate that the population of the City in 1500 was about 50,000, and that by the death of Elizabeth in 1603 it had trebled, so that 141,000 lived in the City, now overspilling its ancient walls. In the mid-sixteenth century there were perhaps 70,000 inhabitants in London, and by 1565 85,000.[23] The metropolis may have been home for one in twenty of the English people at the accession of Elizabeth. Contemporaries saw London's great growth as almost pathological: by the early seventeenth century it would be called the 'great wen': 'soon London will be all England'. The capital drew upon the rest of England and upon Europe for its inhabitants, because in the City the death rate continually exceeded the birth rate. Without the 'great resort and confluence' there London's population would have declined.[24] Londoners 'congregated there from all parts of the island, from Flanders and from every other place'.[25] Most of the City's inhabitants had not been born there, as the depositions of witnesses before the

---

[21] *CSP Ven.* iv. 682, 694.

[22] Ibid. vi/2. 884, p. 1045.

[23] R. Finlay, *Population and Metropolis: The Demography of London, 1580–1650* (Cambridge, 1981); C. Creighton, 'The Population of Old London', *Blackwood's Magazine*, cxlix (1891), pp. 477–96; G. D. Ramsay, *The City of London in International Politics at the Accession of Elizabeth* (Manchester, 1975), p. 33. Sylvia Thrupp estimated that the minimum figure for 1500 should be 33,000 'exclusive of the ecclesiastical, official and floating population of the City'—a considerable exception: *The Merchant Class of Medieval London, 1300–1500* (Ann Arbor, 1948), pp. 51–2.

[24] E. A. Wrigley, 'A Simple Model of London's Importance in Changing English Society and Economy', in P. Abram and E. A. Wrigley (eds.), *Towns in Societies: Essays in Economic History and Historical Sociology* (1978); R. Finlay, *Population and Metropolis*, ch. 3; More, *Apology*, p. 116.

[25] *A Relation, or Rather a True Account*, p. 43.

courts invariably show: for example, of 189 witnesses before the London Consistory Court between 1520 and 1524 only thirteen were certainly London-born.[26] People had long migrated from the country to the City, as Dick Whittington had done, but that movement increased in the sixteenth century.

London's population was as diverse as it was large, and this diversity had social and political consequences. Perhaps half the people in early modern England were under the age of twenty, but in the capital the youth of the inhabitants was even more obvious than elsewhere.[27] Adolescents were the most restless, the most untrammelled of the population, and most evidently on the move. Many a youth thought that the streets of the metropolis were paved with gold, and came up from the shires to learn a trade and to seek work. G. D. Ramsay analysed the origins of 1,088 young men admitted to the freedom of the City in 1551–3, and found that less than a quarter were from London, only eighty from the Home Counties, and the rest from much further afield, including 168 from Yorkshire.[28] Young men and women also came to serve, for almost every household probably supported a servant.[29] Most difficult to chart were the vagrant poor, overwhelmingly children, adolescents, and young adults.

London was the hope of the indigent and vagrant poor who came in search of work or alms.[30] Vagrants presented a spectre of instability. The complaint of the parishioners of St Ewen in 1540 that 'idle young men, vagabonds and masterless men have their common haunt unto the said parish to the great fear

[26] GLRO, DL/C/207. Forty-two per cent of deponents before the Commissary Court in the 1490s had lived in the capital for less than five years; half a century later only one-eighth of the deponents before the same court were London-born: Guildhall, MSS 9065; 9065/1A.

[27] D. V. Glass and D. E. C. Eckersley (eds.), *Population in History: Essays in Historical Demography* (1965), pp. 207, 212; Ramsay, *The City of London in International Politics*, p. 32; Finlay, *Population and Metropolis,* pp. 16, 64–5, 129–30.

[28] G. D. Ramsay, 'The Recruitment and Fortunes of Some London Freemen in the mid-sixteenth century', *Economic History Review*, 2nd series, xxxi (1978), pp. 536–40.

[29] D. M. Palliser, *The Age of Elizabeth: England under the Later Tudors, 1547–1603* (Harlow, 1983), pp. 39–40, 61–2; C. Phythian-Adams, *The Desolation of a City: Coventry and the Urban Crisis of the Late Middle Ages* (Cambridge, 1979), pp. 204–17.

[30] The best account of the problems of poverty and vagrancy in London remains that of the late Dame Kitty Anderson: 'The Treatment of Vagrancy and the Relief of the Poor and Destitute in the Tudor period, based upon the local records of London to 1552 and Hull to 1576' (University of London, Ph.D. thesis, 1933).

and jeopardy of all' was neither untypical nor exaggerated.[31] It may have been that the unfree population of London was growing faster than the citizenry: certainly the vagrant population was then, and is now, most difficult to estimate, for these men and women were masterless, without place or status, against all good order.[32] It was beyond the powers of the City governors to count London's poor exactly, and so it is for the historian. In February 1518 821 tokens were given to the Aldermen to distribute among the deserving poor in their wards—those who were 'so impotent, aged, feeble or blind that they be not able to get their livings by labour and work', who 'may not live but only by alms and charity of the people'— so that they might freely beg.[33] But there were many more who, though not deemed deserving, were nevertheless destitute because there was no work for them. There was little sympathy for their plight. In September 1517, when disorder was especially feared, those who were taken as 'vagabonds being mighty in body and able to get their living . . . but live by begging and other men's labour, contrary to the pleasure of God and to diverse acts of Parliament' were ordered to wear the sign of V in yellow cloth, and banished from the City 'with a basin ringing afore them' to their 'own countries [counties] where they were born', there to work in the harvest.[34] In London the penalties against vagrancy were more severe and more rigorously enforced than was required by government proclamation.[35] Between November 1514 and 15 March 1515 there were three searches through the City wards in 'all hostelries and all other suspicious places' for vagabonds and 'mighty beggars'.[36] In the summer of 1516 restrictions were placed on the owners

[31] CLRO, Repertory 10, fo. 174ᵛ. See also the vagabonds prosecuted in the King's Bench: for example, PRO, KB 9/548/185.

[32] P. Corfield, 'Urban Development in England and Wales in the sixteenth and seventeenth centuries', in D. C. Coleman and A. John (eds.), *Trade, Government and Economy in pre-industrial England* (1976), p. 246 n. 53.

[33] CLRO, Repertory 3, fo. 194ʳ; Journal 11, fos. 337ʳ–338ʳ; Letter Book N, fo. 76ᵛ.

[34] Ibid., Repertory 3, fo. 164ʳ; Journal 11, fo. 304ʳ.

[35] R. W. Heinze, *The Proclamations of the Tudor Kings* (Cambridge, 1976), pp. 118, 255.

[36] CLRO, Repertory 2, fos. 200ᵛ–202ʳ, 205ʳ; Repertory 3, fo. 11ᵛ. The Cardinal ordered further searches in 1519 and 1525: PRO, SP 1/18, fo. 225ʳ; SP 1/19, fo. 76ʳ; SP 1/33, fos. 156ʳ–157ʳ (*L&P* iii/1. 365, 484; iv/1. 1082).

of bowling alleys and tennis courts, and watches and searches were ordered for the apprehension of vagabonds.[37] Constant watch was kept in the City for the vagrant poor, especially in times of political crisis.

In the capital, as nowhere else in England, aliens formed a significant element of the community. Most of the strangers were 'Dutch' coming from Germany and the Low Countries, but there were also immigrants from France, Italy, and Spain, who were settled in their own communities.[38] Aliens probably constituted about 4 per cent of the City's population in the fifteenth century; by the mid-sixteenth century they numbered perhaps five or six thousand. There were more than two thousand French denizens in 1544, and probably at least as many Netherlanders.[39] Aliens appeared more numerous than they really were because of their natural tendency in a foreign city to congregate in ghettoes; in poorer areas less favoured by Londoners themselves. Most of the aliens were huddled around the periphery of the City in the East, by the waterfront and in the City's liberties.[40] London aliens were subject to many restrictions upon their activity: they might not buy or sell directly among themselves.[41] More important, though, than the real size and presence of the alien communities was the feeling prevalent among traditionally xenophobic Londoners that foreigners were everywhere, that they had too many privileges, and funny ways. Certainly some were not Christians, and rewards were offered to those who would convert. In 1532 a Portuguese woman renounced her Judaism.[42] Before the King's Bench in 1540 a family of Spanish silk weavers were indicted for holding firm to their 'Jewish and heretical faith',

---

[37] CLRO, Repertory 3, fos. 86ʳ, 88ʳ, 88ᵛ, 90ʳ, 95ᵛ, 96ʳ, 97ʳ, 97ᵛ.

[38] W. Cunningham, *Alien Immigrants to England* (1897); G. A. Williams, *Medieval London: From Commune to Capital* (1963), ch. 1; S. L. Thrupp, 'Aliens in and around London in the fifteenth century', in A. E. J. Hollaender and W. Kellaway (eds.), *Studies in London History* (1969), pp. 251–72; M. E. Bratchell, 'Alien Merchant Communities in London, 1500–1550' (University of Cambridge, Ph.D. thesis, 1975); A. Pettegree, 'The Foreign Population of London in 1549', *HSL* xxiv (1984), pp. 141–6.

[39] A. Pettegree, *Foreign Protestant Communities in Sixteenth-Century London* (Oxford, 1986), pp. 15–18.

[40] Pettegree, 'The Foreign Population of London in 1549', p. 142.

[41] Pettegree, *Foreign Protestant Communities*, pp. 14–15.

[42] *L&P* v. 1649.

for observing the Jewish sabbath and breaking the Christian one.[43] The Hanse merchants of the Steelyard proved all the worst fears about aliens' dangerous views to be true when they abandoned the Mass at All Hallows the Great in the 1520s, and evangelized the teachings of Martin Luther.[44] 'Scottish', 'Welsh', 'Irish', 'Lombard' were all terms of abuse in Tudor London. Thomas Melmerby brought a defamation case against his rector at St Antholin who had publicly and maliciously called him a Scot.[45] This was in 1513 when England and Scotland were at war. The greatest hostility was always reserved for the enemy of the moment, so Londoners would attack Netherlanders, Frenchmen, or Scots in turn.

Migration to the cosmopolitan capital could never remove the distinctions of origin and status which governed social attitudes. A defamation case of 1521 explains something of the social casting in Tudor London. Royden called Edith Stokker 'bawdy queen [prostitute] and foreign drab', to which she retorted disdainfully, 'though that I be a foreign my master is a freeman'.[46] There were legal distinctions between Londoners: the first distinction was between the citizens sworn to the freedom, and the country- or City-born 'foreigns' and overseas aliens. Only citizenship—possession of the prized 'freedom'—allowed full participation in the economic, political, even social life of the City.[47] The citizens alone owed the duties of paying taxes, serving in the City courts, and as its officers, but to them alone belonged the privileges granted to the Communalty. The trade gilds of London possessed the power to determine who obtained citizenship. They had won that right when Edward II granted a charter to the City in 1319 which ordered that 'every freeman shall be of some mystery or trade', thereby forging a lasting link between gild membership and citizenship.[48] Since the gilds determined admission to the

---

43 PRO, KB 9/547/45–6.

44 See above, pp. 109–110, and below, pp. 158–9.

45 R. M. Wunderli, *London Church Courts and Society on the eve of the Reformation* (Cambridge, Mass., 1981), p. 79. See, for example, Guildhall, MS 9064/11, fo. 274ʳ ('Lombard's whore'), 275ʳ ('whoreson Scot').

46 GLRO, DL/C/207, fo. 18ʳ.

47 P. E. Jones, *The Corporation of London: its Origin, Constitution, Powers and Duties* (1950); G. Norton, *Commentaries upon the History, Constitution and Chartered Franchises of the City of London* (1869).

48 Williams, *Medieval London*, pp. 264–84.

freedom, they controlled participation in London's economy and political life.

Freemen accounted for only half of the adult male population of Tudor York, Bristol, and Norwich, and in medieval London only one in four was of the freedom. But in the early sixteenth century a remarkable change occurred in London. Because there were growing signs of strain and discord between the free and unfree, the City had made a prudential decision to widen the freedom: to offer the privileges of economic independence in exchange for the right of the City and its companies to regulate their economic activities. By the mid-sixteenth century, Dr Rappaport estimates, two-thirds to three-quarters of the adult male population of London was sworn to the freedom.[49] Yet although a majority of Londoners possessed the rights and privileges of the City in which they lived, the social and economic conditions of the 1520s made it difficult precisely to enjoy them.

This decade was a time of unusual economic deprivation. Sharp price increases following bad harvests in 1520-1 and 1527-8 had the effect, so Dr Rappaport has shown, of pushing prices in London during the entire decade up to an average level which was 11 per cent higher than the average price index for 1510-20. The price level in 1520-1 was higher than at any point since 1490, and 24 per cent higher than the average level of the previous decade.[50] By the summer of 1527 people in London 'did starve daily for bread', and there were people crushed to death at the press to the bread carts.[51] At the end of 1526 people had been 'sore troubled with poverty for the great payments of money that were past'. More seriously, great rains through September, November, and December 1526 were followed by floods in January, and it rained still, continuously, from 12 April until 3 June. A disastrous harvest followed that summer, and corn 'sore failed'.[52] Special measures were taken

[49] S. Rappaport, 'Social Structure and Mobility in Sixteenth-Century London', Part I, *London Journal*, ix (1983), pp. 109–14.

[50] I have benefited very greatly from reading Dr Rappaport's thesis, 'Social Structure and Mobility in Sixteenth-Century London' (University of Columbia, Ph.D. thesis, 1983), which will be published in amended form as *Worlds within Worlds: the Structures of Life in Sixteenth-Century London* (Cambridge, 1989).

[51] Hall, *Chronicle*, p. 736; *Two London Chronicles*, pp. 2–3.

[52] Hall, *Chronicle*, p. 721.

to provision the capital, not least because bread riots were feared. 'Substantial men' dared not lay up stores of wheat in their houses lest they be ransacked. The war with the Emperor in 1528 seriously disrupted trade with Spain and Flanders, for the markets were closed to English merchants.[53] Throughout the 1520s, the people suffered failed harvests and recurrent plague and sweating sickness; more harrowing than all the religious and political disturbances.

Plague, even more than other misfortunes, was seen as a divine visitation. The descriptions of the symptoms revealed the analysis of their cause: the widow of Sir William Coffin wrote to Cromwell that her husband had died of the 'great sickness, and full of God's marks all over his body'.[54] In these years Londoners were more than usually afflicted. The plague of the winter of 1517–18 was followed by another as serious in the autumn of 1521, and by another still in the winter of 1525–6.[55] Brian Tuke thought that the 'sweat' proceeded from fear, and came 'by report'; 'it came in this way from Sussex to London, and a thousand fell in a night after the news was spread'.[56] So devastating was the sweating sickness in the summer of 1528 that the Midsummer watch was abandoned, and that July Bishop Tunstall ordered a general procession to invoke divine mercy.[57] In August 152 Londoners died within a week.[58] It may have been some consolation to the authorities that the sweat was decimating the most talented of the 'brethren' too—Francis Denham, Henry Summers, and others died of it.[59] Afflicted persistently by epidemic, and undermined by economic deprivation, the citizens of London might have been less likely to riot or revolt. Somehow the forces for disruption within London society were contained, for throughout the

[53] *TRP* i. 116; Hall, *Chronicle*, p. 736; *Acts of Court of the Mercers' Company*, pp. 759–62; CLRO, Journal 13, fo. 34ᵛ; Repertory 7, fos. 228ᵛ, 265ᵛ; *L&P* iv/2. 3572, 4398.

[54] *Lisle Letters*, 5, p. 409. P. Slack, *The Impact of the Plague in Tudor and Stuart England* (1985).

[55] CLRO, Repertory 3, fos. 110ᵛ, 184ᵛ–185ʳ, 191ʳ, 192ʳ; Repertory 4, fo. 89ʳ; Repertory 5, fo. 217ʳ; Hall, *Chronicle*, pp. 628–9, 707.

[56] *L&P* iv/2. 4510.

[57] *CSP Ven.* iv. 320; Hall, *Chronicle*, p. 750; *L&P* iv/2. 4409, 4417, 4418, 4440, 4453, 4489.

[58] *L&P* iv/2. 4633.

[59] Ibid. iv/2. 4493, 4690.

sixteenth century the capital remained remarkably calm, stable, and well ordered, compared with the great cities of Europe. How so? London's stability had much to do with the nature of its government.

GOVERNMENT

London's population was large and growing larger; it was diverse and contained unstable elements. Then as now, the metropolis afforded a certain anonymity, and outlaws sought shelter and sanctuary there.[60] Yet the City was tiny, bounded still by its ancient walls and by the river. The walk between the Temple in the West and the Tower in the East took less than an hour. This tiny area was divided and subdivided into hundreds of separate territories. The twenty-six wards of the City contained still smaller units called precincts.[61] Within the walls were over a hundred parishes, most of them only a few hundred yards square and with only a few hundred parishioners.[62] In St Anne and St Agnes there were only 60 households in 1546.[63] At the turn of the sixteenth century there were 78 gilds in London,[64] and within each gild was a hierarchy of five status groups, each with its own rights and responsibilities: apprentices, journeymen, householders, liverymen, and assistants. The gilds were then of the first importance in the City, but they only had few members. In 1501, before the great expansion of the freedom, even the twelve great companies had perhaps only about two hundred members each.[65] The workshops, too, were small, with each craft master allowed no more than two apprentices from the mid-sixteenth century.[66] Each little

---

[60] F. D. Logan, *Excommunication and the Secular Arm in Medieval England* (Toronto, 1968), pp. 93–4; I. Thornley, 'The Destruction of Sanctuary', in *Tudor Studies*, ed. R. W. Seton-Watson (1924), pp. 182–207.

[61] Valerie Pearl, 'Change and Stability in Seventeenth-Century London', *London Journal*, v (1979), p. 15.

[62] The Chantry Certificate of 1549 lists the number of houseling people in the parishes of the City: *LMCC*.

[63] PRO, C 1/1330/39; C 1/1330/65.

[64] Richard Hill listed them in his commonplace book: Balliol College, MS 354, fo. 107^v.

[65] CLRO, Journal 10, fo. 373^r; Thrupp, *Merchant Class*, p. 43.

[66] CLRO, Repertory 12, fos. 57^v, 99^r, 369^r, 528^r; Repertory 13, fos. 20^r, 24^r, 36^v, 105^v; Guildhall, MSS 7090/1, fo. 50^v; 11588/1, fos. 7^r, 9^r, 12^v; 5177/1, fo. 34^r;

community in London had its own institutions and officers. Every Londoner was daily overseen in his working, religious, and moral life. Such a proliferation of civic officers and the diffusion of collective responsibility for order might earth conflict and bring stability.[67]

London was controlled by a political and mercantile cartel. All power was in the hands of those who governed the City companies.[68] By Henry VIII's reign high civic office was restricted to those who belonged to one of the twelve great livery companies—Mercers, Grocers, Drapers, Fishmongers, Goldsmiths, Skinners, Merchant Taylors, Haberdashers, Salters, Ironmongers, Vintners, and Clothworkers. Their precedence, long disputed, was at last settled by Billesdon's order in 1512. There was a recognized path to success in the City; from yeomanry to livery within the gild, from membership of a smaller to a greater company, to office within that company, and thence to election as Alderman, Sheriff, or even Lord Mayor. The Mayor was elected annually from among the Aldermen, and for the year of his office his company had precedence. The Mayor and Aldermen enjoyed a status commensurate with their power, dressed in scarlet splendour, and were attended with pomp. Religious services and religious oaths marked the rites of passage in civic life, giving thereby supernatural sanction to civic office. Any derogation of the authority of the most 'worshipful' in London was regarded seriously as a sin as well as a misdemeanour, and was severely punished.

Yet contempt for Aldermen was often enough manifested. In

Mercers' Company, Acts of Court, ii, fo. 246ᵛ; Goldsmiths' Company, Court Book I, fo. 34ᵛ. Aliens were forbidden to indenture apprentices on pain of disenfranchisement; *TED* i, pp. 293–5.

[67] G. Unwin, *The Gilds and Companies of London* (4th edn., 1963); S. Rappaport, 'Social Structure and Mobility in Sixteenth-Century London', Part II, *London Journal*, x (1984), pp. 107–34; Pearl, 'Change and Stability', and 'Social Policy in Early Modern London', in H. Lloyd-Jones, V. Pearl, and B. Worden (eds.), *History and Imagination* (1981), pp. 115–31.

[68] F. F. Foster, *The Politics of Stability: A Portrait of the Rulers of Elizabethan London* (1977); V. Pearl, *London and the Outbreak of the Puritan Revolution: City Government and National Politics, 1625–43* (1964), ch. 1; Thrupp, *Merchant Class*, pp. 18–19. For the history of the Aldermen of London and a chronological list of those elected, see A. B. Beaven, *The Aldermen of London*, 2 vols. (1908–13). For the order of precedence, see CLRO, Journal 10, fo. 371ʳ; Journal 11, fo. 68ᵛ; Repertory 3, fo. 66ᵛ.

1510 Christopher Markham attacked Alderman Copynger with a knife, and told another, 'men have lost their heads that have not done so much harm to this City as ye have done'.[69] Another man threatened that the Mayor should be 'pulled out of his house and his head shall be stricken off at his own threshold'.[70] Some said that those who served as Sheriffs lived by 'polling' (plundering).[71] So many were the threats against the Mayor and his brethren in the early years of Henry VIII's reign that the Court of Aldermen sought precedents for how to punish such sedition.[72] Time was when those who spoke disparagingly of the City's sworn officers were made to walk barefooted and bareheaded through the streets, bearing lighted torches, like religious penitents.[73]

The sovereign council of the City was the Court of Aldermen, presided over by the Lord Mayor.[74] The Court of Common Council, of 212 chosen citizens, 'the wisest and most wealthy' of each ward, was the supreme legislative body, but too unwieldy to take executive action. Maintenance of public order, regulation of trade and industry, and all civic administration were controlled by the Court of Aldermen. The Aldermen, elected for life, exercised both legislative and executive functions, and constituted as the Mayor's Court they also had judicial authority. The constant and growing activity of the Court is clear from the Repertories which record its deliberations and decisions (Repertories which grow thicker as the decades pass). The Aldermen met often: four times a week in the 1520s, sometimes sabbath-breaking. Small wonder that election to aldermanic office was not invariably regarded as an honour to be accepted eagerly.[75] To be an Alderman, and particularly Mayor, involved considerable time and expense, the expense compounded by the time spent away from business at the Court of Aldermen and at endless ceremonial functions. Great wealth was a prerequisite of office: the Aldermen were

---

[69] CLRO, Repertory 2, fo. 99^v.   [70] Ibid., Repertory 2, fo. 155^v.

[71] Ibid., Repertory 2, fos. 168^v–169^r, 183^v.

[72] Ibid., Repertory 2, fo. 195^r; see also fos. 72^r, 73^v, 96^r, 169^r, 187^r–189^r, 190^r, 192^v; Repertory 3, fos. 110^v, 121^v, 151^r–154^r, 173^r, 184^r, 203^v, 213^r, 246^v, 248^r, 258^v–259^v.

[73] Thrupp, *Merchant Class*, p. 18.

[74] For the government of the City, see Foster, *Politics of Stability*, pp. 15 ff.; P. E. Jones, *Corporation of London*.   [75] *L&P* viii. 208.

London's most 'substantial' citizens. Men often did refuse office; not least at this time because the City's rulers were bound to enforce the King's will, and Henry's wrath, should they fail, was a terrifying prospect.[76]

The original tribunal of London was the wardmote enquest, presided over by the Alderman of the ward.[77] As the whole ward assembled, the wardmote appointed the ward officials—the constable, beadle, scavenger—and through the enquest acted as a body of accusation. It met annually on St Thomas the Apostle's day (21 December). Twelve men from every ward were appointed as an indicting jury to hear accusations, to examine and present offenders. First they were to inquire whether the King's peace was kept, whether there were traitors or felons in the ward; whether any man gathered 'evil company' or disturbed his neighbours.[78] Many of the offenders they sought sinned against the morals of the City: 'common strumpets', concupiscent priests, keepers of brothels where 'young men' might resort, usurers. The secondary faults sought by the wardmote were of public nuisance: casting rubbish, keeping unsavoury animals, blocked drains, fire hazards. Offenders against the City's trading regulations were punished; those who sold for short measure, who took the privileges of London without paying its charges. The deliberations and accusations before the quest were supposed to be confidential—'whatsoever is uttered unto them ought not to be opened and declared again'—but in so intimate a society this was a counsel of perfection, and there were tale bearers.[79]

The wardmote presentments for Reformation London are lost—save for those of Portsoken ward, for Bread Street ward for a single year, and a fragment for Farringdon Without for 1522.[80] In the 'book of the wardmote' of Bread Street—drafted by no less a secretary than Thomas Cromwell—the verdicts were

---

[76] See below, pp. 164, 239–41.

[77] *Calendar of Plea and Memoranda Rolls, 1413–1437*, ed. A. H. Thomas (Cambridge, 1943), pp. xxiv–xxx; Foster, *Politics of Stability*, p. 37.

[78] Rules governing the procedure of the wardmote enquests were transcribed in a contemporary commonplace book: BL, Lansdowne MS 762, fos. 37^v–39^r, 43^r–45^r.

[79] See, for example, GLRO, DL/C/209, fos. 61^{r-v}; DL/C/628 (without folio numbers).

[80] For Portsoken ward, see Wunderli, *London Church Courts*, pp. 33–5. PRO, SP 1/26, fo. 224^r; SP 1/29, fos. 115^r–120^r (*L&P* iii/2. 2736; 3657).

presented against offenders parish by parish.[81] In St Mildred
and St Mary Woolchurch the goosehouse of George Betyson
was found to be noisome, pavements were dangerous, and
Richard Bradshaw was presented for 'scolding, fighting and
brawling with his neighbours'; in St Christopher le Stocks John
Peyrson 'will not come to the enquest of wardmote as a good
citizen ought to do', and, contrary to his oath as freeman,
would not 'bear such office as he is elect unto', and was found
to be 'perjured' thereby, the worst of civic crimes. In St Bartholomew, St Benet Fink, St Margaret's, and All Hallows London
Wall the offences were either against public hygiene or safety,
or against the moral order. Wardmote indictments were turned
over to the Mayoral court for prosecution. Many of these
offences could also have been treated by other tribunals, over
which no layman had any authority, which owed the final
loyalty not to the King but to the Pope: the Courts spiritual.

The Church governed the spiritual lives of the people, and
by declaring the rewards of Heaven and the pains of Hell
guided moral life. What it could not prevent it punished.
Through the Courts Christian the Church exercised enormous
powers of detection and judgement; not least in cases which
could only by special pleading be held to pertain to the cure of
souls.[82] Where the boundaries of their jurisdiction were unclear—regarding moral offences such as bawdery, prostitution,
and defamation—the royal courts, the civic courts, and the
London Church Courts contended to charge citizens. Breaking
the peace was a civil offence, punished by the secular courts in
serious cases, but the Church Courts too judged marital quarrellers, scolds, and barrators, and those who were 'out of
charity', for they disturbed the Christian community.

The system of ecclesiastical justice in Reformation London
was complex in the extreme. Dr Wunderli has explained its
structures and its procedures.[83] The Court of Audience, where
the Bishop of London sat in judgement, was the highest court
of the diocese. In practice the Bishop delegated his judicial

---

[81] PRO, SP 1/29, fos. 115ʳ–120ʳ (*L&P* iii/2. 3657).

[82] For the operation of the Church Courts, see *inter alia*, B. L. Woodcock,
Ecclesiastical Courts in the Diocese of Canterbury (Oxford, 1952); R. Houlbrooke, *Church
Courts and the People during the English Reformation, 1520–1570* (Oxford, 1979); Wunderli,
London Church Courts.

[83] Wunderli, *London Church Courts*, ch. 1, especially fig. 1.

authority to his Official Principal or Chancellor, who judged
suits from the whole diocese in the Consistory Court.[84] The
Consistory convened weekly in the Long Chapel of St Paul's
cathedral. The Commissary Court was the Bishop's lower
court, a City court restricted mostly to instance cases and
probate, and differing from the Consistory Court in procedure.
This court, presided over by the Commissary General, heard an
enormous volume of business and sat daily, save for Sundays
and festivals.[85] Each of the four archdeaconries of London
diocese—London, Essex, Middlesex, and Colchester—had a
court, but their records have not survived. The jurisdiction of
the London archdeaconry overlapped with the Commissary
Court, but this does not seem to have caused conflict. Within
the City at least three other Church courts had limited jurisdiction under the Bishop of London: the Dean and Chapter of St
Paul's, and until the Dissolution, the priory hospitals of St
John of Jerusalem and St Katherine by the Tower.[86] The
Bishop of London did not have exclusive ecclesiastical control
in the City: thirteen City parishes were subject to the Archbishop of Canterbury, and their judge was the Dean of the
Peculiars. The Archbishop's Court of the Peculiars had early
become identified with the appellate court for Canterbury
province, the Court of Arches. Important cases in London
were taken to the Court of Arches in this period. It became the
main arena for the City tithe disputes in the 1520s. Yet the
extent of its significance can never be known because its
records were consumed in the Great Fire.[87] Bishop Stokesley
would complain bitterly of the Archbishop of Canterbury's
rival and superior jurisdiction in the capital,[88] but this would
be the least of conservative Stokesley's contests with the reformist Cranmer in the 1530s.

[84] Ibid., pp. 7–10. For the records of the court, see below, p. 642.
[85] Ibid., pp. 10–15. Dr Wunderli has analysed the great volume of cases recorded
in the Acta quoad correctionem delinquentium: Guildhall, MSS 9064–11. Commissions to the Commissaries General survive: Thomas Benet (Guildhall, MS 9168/7,
fo. 44r); William Clyff (ibid., MS 9171/10, fo. 6r); Geoffrey Wharton (ibid., 9531/10,
fo. 146r); John Longe (ibid., MS 9168/10, fo. 146r); Nicholas Wotton (ibid., MS
9168/10, fo. 193v); John Story (ibid., MS 9171/11, fo. 1); John Croke (ibid., MS
9531/12, fo. 122r).
[86] Wunderli, *London Church Courts*, pp. 15–18.
[87] D. Slatter, 'The Records of the Court of Arches', *JEH* iv (1953), pp. 139–53.
[88] See below, p. 225.

The Church Courts were consistently and immensely active. The Commissary Court sat several days a week, and heard hundreds of cases a year (980 in 1471; 1,305 in 1472; 764 in 1484; 351 in 1485; 1,061 in 1493). Dr Wunderli has counted and analysed the cases. Finding a marked decline in the number of cases before the Commissary Court in the early years of the sixteenth century (547 in 1502; 294 in 1512; 303 in 1513; 328 in 1514), he postulated that litigants in London had chosen to move away from the haphazard and dilatory processes in the Church Courts to the more efficient secular courts of the City.[89] If this were so, London would have been quite unlike other places where the Church Courts have been shown to be rigorously and ruthlessly efficient. Maybe London was different, but the loss of Commissary Court *acta* after 1516, the almost complete absence of Shrieval and Mayoral court records, the loss of the proceedings of the Court of Arches, the lack of wardmote presentments, the fragmentary survival of other Church Court records, makes such a conclusion speculative. Moreover, Dr Wunderli shows the Commissary Court to have been relentless and thoroughgoing in its correction of moral laxity in the last years of the fifteenth century. Was there so great a change, in so short a time? The absence of so many records, both of the spiritual and secular courts of the City, makes it impossible to know for certain whether the amount of litigation was rising or falling, or which courts plaintiffs chose to use. London's courts were unlikely to be less efficient than those elsewhere, and Londoners surely shared the fears which were to be expressed in the Commons' Supplication against the Ordinaries in 1532.[90]

Litigation in Church Courts arose either when plaintiffs sought justice from judges and initiated proceedings by asking that the defendants be cited to appear (instance cases), or when the judges themselves proceeded *ex officio*. There was a distinction in office cases between those heard on their own merit— *ex officio mero*—and those heard through the promotion of another—*ex officio promoto*. In either case suspects were reported

---

[89] Wunderli, *London Church Courts*, pp. 4, 18–23, 62, 137–9.
[90] See below, pp. 198–208.

to the court.[91] The distinction between litigation brought by the laity against the laity in the courts and that brought by clerical initiative against lay offenders is one of prime importance on the eve of the Reformation.[92] The Church Courts were not empowered to seek out suspects, rather their purpose was to judge delicts presented before them. By whom?

There was a small army of lowlier court officials, men not in Holy Orders—the summoners, apparitors, and mandatories. They were sent to cite wrongdoers, and as bearers of unwelcome summonses they were often vilified, sometimes worse.[93] Drawing his sword, Thomas Banester threatened the apparitor who came to find him in St Mary Woolchurch in 1522: 'Thou whoreson knave, without thou tell me who set thee a work to summon me to the court, by God's wounds, and by this gold I shall break thy head.'[94] But, as Dr Wunderli discovered, hardly any criminal charges originated with summoners' inquisitions. The Church did have a procedure for searching out offenders against its laws: the episcopal visitation. Parish by parish, the churchwardens came before the Bishop or his official to report failings in priests or parishioners, and defects in the church fabric. Though the visitations were regular enough in Reformation London, no visitation book is extant before 1561: only a fragment from St Magnus in 1498 survives.[95] From the churchwardens' presentments stemmed the *ex officio* cases in the courts. Those who were summoned before the Church Courts had been reported by their own neighbours and workmates.

'Publica fama', public fame, revealed the offence, and it was

[91] The procedure is explained by H. Conset, *The Practice of the Spiritual or Ecclesiastical Courts* (1700); Woodcock, *Medieval Ecclesiastical Courts*, chs. 4–6; Houlbrooke, *Church Courts and the People*, pp. 38–54.

[92] M. Bowker, 'Some Archdeacons' Court Books and the Commons' Supplication against the Ordinaries of 1532', in *The Study of Medieval Records: Essays in Honour of Kathleen Major*, ed. D. L. Bullough and R. L. Storey (Oxford, 1971), pp. 289–93.

[93] R. M. Wunderli, 'The Pre-Reformation London Summoners and the Murder of Richard Hunne', *JEH* xxx (1982), pp. 209–24.

[94] Hale, *Precedents*, cccxv. For other attacks on summoners, see ibid. cccxi; GLRO, DL/C/330, fos. 73ᵛ, 184ʳ; DL/C/207, fos. 107ʳ–110ᵛ; DL/C/208, fo. 181ʳ; PRO, C 1/401/34. Yet Dr Wunderli found only forty-three attacks upon summoners prosecuted before the Commissary Court, 1471–1514; 'Pre-Reformation London Summoners', p. 215.

[95] Richard Arnold, *The Customs of London*, ed. F. Douce (1811), pp. 273–6; Guildhall, MS 9537/2 (visitation book, 1561).

upon parish gossip that conviction and judgement ultimately rested. The system of trial involved more still of the parish, for the guilt or innocence of the victim was often proved by compurgation. A defendant was charged to swear to his innocence, and then his friends or neighbours were called upon, as compurgators to swear in their turn to the truth of the defendant's oath.[96] In some cases many oath-helpers were summoned: John Roo, curate of St Christopher le Stocks, was ordered to find twelve lay and twelve clerical compurgators.[97] That the overwhelming number of defendants before the London Commissary Court on the eve of the Reformation successfully purged themselves suggests either that they were truly innocent, or that they did not stand in fear of perjuring themselves.[98] Their universal innocence would seem unlikely, but violation of a solemn oath was a mortal sin, one so serious that it could only be confessed to the Bishop.[99]

The Church had sanctions and penances to impose upon the guilty. More shameful than being the subject of open or covert gossip was 'the peril of open penance'. City and Church Courts alike would punish by humiliating the wrongdoer. In the City, marital quarrellers and fornicators were shamed by the ringing of pans and basins. In 1537 Agnes Hopton, who had slept with a married minstrel, and had, against nature, dressed in 'man's rayment', was ordered to do public penance. The minstrel was to ride, facing the horse's tail, playing a musical instrument.[100] Of course, not everyone who did penance was penitent. Thomas Wiggins shocked the City in 1519 by taking his penance 'rather in derision and scorn than for any lamenting or sorrow'.[101] The usual penance ordered by the Church Courts was for the offender to proceed before the Sunday procession, clad in a white sheet, barefoot and bare-

---

[96] For compurgation, see Lyndwood, v. 14; Wunderli, *London Church Courts*, pp. 41–4, 47–8.   [97] See above, pp. 64–5.

[98] Wunderli, *London Church Courts*, pp. 48–9. Dr Houlbrooke found that between two-thirds and three-quarters of those known to have attempted compurgation succeeded in clearing their names: *Church Courts and the People*, pp. 45–6. Yet in Lincoln canonical purgation was ordered more rarely, and was even more rarely successful: *An Episcopal Court Book for the Diocese of Lincoln, 1514–1520*, ed. M. Bowker (Lincoln Record Society, lxi, 1963), pp. 6, 112, 122, 130, 136, 139, 140.

[99] Myrc, *Instructions*, p. 54.

[100] CLRO, Repertory 9, fo. 241[r].

[101] Ibid., Repertory 3, fo. 247[r]; Journal 11, fos. 387[v]–388[r].

headed, carrying a lighted candle; this candle he would present
kneeling at the high altar, to the celebrant at Mass.[102] Restitution was a condition of penance, so the penitent was required to
seek the forgiveness of those he had wronged. In 1544 Dionise
Constable was ordered to confess her fault publicly in these
words: 'Neighbours, whereas I have offended God and commonwealth, in that you have had me in suspicion of ill living
with George Wakefield, I cry God mercy, and am sorry for it;
and I pray you be in love and charity with me.'[103] Such was the
shame of public penance that some sought to have their penance
commuted to a gift of charity. In 1555, Gregory Newman was
so 'contrite of heart . . . vehemently knocking upon his breast',
that his penance was commuted instead to a payment to poor
prisoners, to poor scholars, and to edify his parish church.[104]
The judges could and did cancel penance if the offender
seemed contrite. Church Courts would often order gifts to
'Christ's poor' and to prisoners as a form of penance, particularly for the more 'honest', 'substantial', and 'worshipful' of
London's citizens whose humiliation as penitents would be the
greater perhaps, and who could offer larger gifts of charity.[105]
Those who were not penitent, who remained obdurately disobedient—'contumacious'—and would not obey the Courts'
mandates were excommunicated, cast out from the Church
and from the community of the faithful.

'First we accurse all them that break the peace of holy
Church or disturb it'; so went the curse of major excommunication pronounced by the parish priest four times in the
year.[106] Those who were so cursed were sundered from the
Church, from its sacraments, and from the Christian community, and the solemnity of the anathema was marked by the
ringing of bells and extinction of candles.[107] How many were
excommunicated in pre-Reformation London is hard to tell,

[102] See, for example, Guildhall MS 9064/11, fos. 1ᵛ, 2ʳ, 2ᵛ, 4ʳ, 277ᵛ.

[103] Hale, *Precedents*, cccxcii.

[104] GLRO, DL/C/331, fos. 188ᵛ–189ʳ; see also fos. 255ᵛ, 297ʳ.

[105] See, for example, Hale, *Precedents*, cccxviii, cccxxx, cccxxxiii; GLRO,
DL/C/330, fo. 59ᵛ; Guildhall, MS 9168/9, fo. 109ᵛ; Wunderli, *London Church Courts*,
pp. 51–3; Woodcock, *Medieval Ecclesiastical Courts*, pp. 98–9, 102.

[106] Myrc, *Instructions*, pp. 21–4; Strype, *Ecclesiastical Memorials*, i/2, pp. 188–93.

[107] T. N. Tentler, *Sin and Confession on the Eve of the Reformation* (Princeton, 1977),
pp. 302–4; R. Hill, 'The Theory and Practice of Excommunication in Medieval
England', *History*, xlii (1957).

for the sentence was sometimes given all too lightly. At St Botolph Aldersgate at Easter 1523 the curate, Robert Shoter, guilty of the gamut of clerical abuses, refused the sacrament to some of his parishioners. They could not, said Shoter, be 'in charity' and so 'take their rights' until they had asked the forgiveness of his concubine for reporting her and Shoter to the wardmote enquest. Against his enemies in the parish Shoter imposed an awesome sanction: 'Ye be accursed and I pronounce you accursed.'[108] When the clergy could utter so grave a sentence so lightly, and for crimes which were not dishonourable, then excommunication might lose its horror. The 'terrible censures of excommunication' could be reinforced by writs of *significavit* in Chancery to enforce obedience upon the contumacious, but such writs were much rarer than the sentence: only 33 survive from the whole diocese between 1470 and 1516, and only fourteen in the City between 1520 and 1532.[109] But there were other ways of making the ban real. The Church punished any who associated with excommunicates, and forbade services to be celebrated until those accursed were banned from the church. So the vicar of Wansted, Essex, who in 1523 had pronounced (selfishly), 'All ye be accursed that stop up the ways by which the tithes should be carried', could instruct a parishioner to throw out the tithe-dodgers: 'I pray you go to John Philip and desire him to go out of the church that ye and the parish may have evensong, for he is accursed.'[110] More compelling were the sanctions against heretics. The parish priest at Whitechapel had outlawed Richard Hunne.[111] In the same church, twenty years on, another priest cried out 'Blood and fire' upon John Harrydance, a hot gospeller.[112] The Church had special punishments for heretics in this world too, as well as damning them in the world to come.

'The detestable crime and sin of heresy' was the worst offence against God and Christian society. It was the only offence for which the Church could order a layman's detention. The statute *De Haeretico Comburendo* (2 Hen. IV c. 15) empowered

---

[108] GLRO, DL/C/330, fos. 45ᵛ–47ʳ.

[109] PRO, C 85/125–6; Logan, *Excommunication and the Secular Arm*, pp. 67–8.

[110] GLRO, DL/C/207, fos. 137ʳ–139ʳ; see also Hale, *Precedents*, cccxli, Dcxix, Dccxi.

[111] S. F. C. Milsom, 'Richard Hunne's "Praemunire"', *EHR* lxxvi (1962), p. 81.

[112] PRO, SP 1/124, fo. 155ʳ (*L&P* xii/2. 624).

Bishops to arrest suspected heretics on the basis of common fame alone, and to detain them until they had purged themselves or abjured heretical doctrines. Obdurate heretics, who refused to abjure, or who, having abjured, relapsed, were handed over to the secular arm, to the Sheriffs for burning.[113] Heresy could be judged by the Church alone. Against the freedom of the clergy to accuse and judge any layman for heresy, simply upon suspicion, the laity had no defence.[114] We shall see that prevalent and genuine dread of this ultimate judicial power of the Church lay behind all the recrimination and suspicion between clergy and citizenry as the Reformation Parliament opened in 1529.

HERESY

The citizens' resentment of the power and privileges of the clerical estate grew in the 1520s, for the clergy became so overweeningly arrogant that they would brook no criticism, so it was said.[115] For this breakdown in relations one man bore particular responsibility, for throughout the 1520s one man held sway in England in both secular and spiritual spheres—Thomas Wolsey, Lord Chancellor, Archbishop of York, Cardinal and *legatus a latere* to Rome. Two episodes in London on consecutive days in May 1521 illustrate something of the deleterious consequences for politics and religion of his unbounded authority.

On Sunday, 12 May 1521 Wolsey, accompanied by a retinue of Bishops, ambassadors, and English nobility, rode in solemn procession, under a golden canopy, through the City to St Paul's 'as if the Pope in person had arrived'. In Paul's churchyard the Cardinal mounted a high platform; at his right foot sat the Archbishop of Canterbury and the Papal nuncio, at his left, the Bishop of Durham, and ranged below the Bishops and nobility. A great crowd (of 30,000 the Venetian ambassador thought) looked on. The occasion for this spectacle was to deliver England's anathema against Martin Luther, the heretic excommunicated by the Pope and outlawed by the Emperor.

---

113 J. A. Guy, 'The Legal Context of the Controversy: The Law of Heresy', *CW* 10, pp. xlvii–lxvii.

114 See pp. 204–7, 218–19 below.

115 Hall, *Chronicle*, p. 593.

Archbishop Warham made a 'laudatory oration, praising the Cardinal vastly', and there followed a two-hour sermon by John Fisher. 'The spirit of Christ is not in Martin Luther'; he 'hath stirred a mighty storm and tempest in the Church'. Luther's doctrine of justification by faith alone Fisher denounced. The King himself, Fisher said, condemned Luther, and in the Cardinal's hand was a manuscript copy of Henry VIII's attack on Luther: *Assertio Septem Sacramentorum*. Wolsey pronounced the sentence of excommunication against Luther and his followers, and as he did so Lutheran works were ritually burnt.[116] Thus England declared its orthodoxy, its obedience to the Pope, and unity with Catholic Christendom.

The morning following the book burning England's premier noble, Edward Stafford, Duke of Buckingham, was brought by river from the Tower to trial by his peers at Westminster.[117] The crime alleged against him was treason; that he had compassed the deposition and death of the King and claimed the throne for himself. The Duke had, it was alleged, listened to and believed the treasonable prophecies of a monk of the Charterhouse of Henton, and was prepared to act upon them. In April 1512 the monk had first shown the Duke that he should 'have all': this the monk knew by revelation and by the grace of God. In July 1517 he foretold the time for fulfilment of the prophecy: 'before Christmas following there shall be a change and the Duke should have rule of all England'. In his City manor of the Red Rose the reckless Buckingham was declaring by May 1520 that 'neither the King nor his heirs should prosper'. This treachery was revealed by the Duke's chancellor Gilbert, and by his confessor Delacourt (who was thereby breaking the seal of the confessional).[118] Whether the Duke was guilty or not, to an insecure monarch, with the succession unsettled, he was too dangerous to live.[119] The Duke of

[116] BL, Cotton MS Vitellius B iv. fo. 111^r (*L&P* iii/1. 1274); *L&P* iii/1. 1273; *CSP Ven.* iii. 208, 210, 213; *The sermon of John the bysshop of Rochester made agayn the pernicious doctryne of Martin Luther* (*RSTC* 10894). C. S. Meyer, 'Henry VIII burns Luther's books, 12 May 1521', *JEH* ix (1958), pp. 173–87.

[117] Hall, *Chronicle*, pp. 622–4.

[118] PRO, KB 8/5; *DKPR 3rd report*, pp. 230–4; *L&P* iii/1. 1284(1); BL, Harleian MS 283, fos. 70^r, 72^r (*L&P* iii/1. 1284 (2 & 3)).

[119] J. J. Scarisbrick, *Henry VIII* (1968), pp. 120–3; G. R. Elton, *Reform and Reformation* (1977), pp. 81–2; C. Rawcliffe, *The Staffords, Earls of Stafford and Dukes of Buckingham, 1394–1521* (Cambridge, 1978), pp. 43–4.

Norfolk delivered the inevitable guilty verdict, weeping.[120] Buckingham was led back to the Tower on foot, the axe turned towards him. That same night an outrage occurred: on the bull posted on the door of St Paul's, declaring the condemnation of Luther, an unknown hand had scrawled these words:

> Bulla bullae ambae amicullae.
> Araine ante tubam.[121]

Mocking the Pope and his legate on the day of Buckingham's trial may have had telling and unforeseen consequences for the citizens' perception of Luther and his error. For the Duke was a popular hero, 'guiltless', and Wolsey was blamed for his fall. Perhaps this other enemy of the Cardinal, Luther, might arouse the people's sympathy, or at least their curiosity.

In 1519 there had already been reports that Buckingham was 'extremely popular'; 'it is thought that were the King to die without heirs male he might easily obtain the Crown'.[122] The Charterhouse monk had urged the Duke 'that he should get the favour of the commons and he should have rule of all'.[123] The events in London at Buckingham's fall revealed his popular following. Upon his arrest the people 'much mused what the cause might be, and among them was much speaking'.[124] The Duke was brought for trial 'under a strong escort of armed men lest he should be rescued; by reason of his numerous followers in London'. At his execution a guard of five hundred was in attendance, and as the Duke recited the penitential psalms men wept. His death 'grieved the City universally'.[125] Still on 20 June Buckingham's mourners went daily to the scaffold at Tower hill, and to his grave at the Austin Friars, where they would 'lament and wail' his death, 'reputing him as a saint and holy man', 'saying that he died guiltless'. On the eve of St Peter and St Paul very strict watch was kept throughout the City.[126]

---

[120] Hall, *Chronicle*, p. 624.
[121] *CSP Ven.* iii. 213.
[122] *L&P* iii/1. 402 (p. 143).
[123] BL, Harleian MS 283, fo. 72ʳ (*L&P* iii/1. 1284(2)). By this deposition was marked 'notandum'.
[124] Hall, *Chronicle*, p. 622.
[125] *CSP Ven.* iii. 213. The memory of this scene long remained: see William Shakespeare, *Henry VIII*, Act II, scene 1.
[126] CLRO, Repertory 5, fos. 199ᵛ, 204ʳ.

Five men were discovered to have spoken and written sedition 'concerning the King's most noble person, his blood royal, and certain of the noble men of the King's most honourable Council'.[127]

Why and when the Duke had gained such popularity in the City is less than clear. Many City merchants had reason to regret his fall, for he was one of their best customers, but they hardly constituted a popular following.[128] John Colet had been among the inner circle of friends and counsellors of the Duke, and John Lincoln, the stirrer of the Evil May Day riots, was of his household.[129] They had the influence to raise popular support. But if Buckingham did follow the monk's advice to win over the common people, there is no sign of when or how he did so. The sorrow at his fall revealed the hatred of the Cardinal and of his pride in bringing down a Duke. After Buckingham's death 'there were libels scattered and set up in London against the Cardinal as author thereof, calling him butcher and filius carnificis, etc.'.[130] Wolsey's critics came to judge the fall of Buckingham as among the worst of his crimes. In his commonplace book John Colyns recorded a poem for Buckingham which ended:

> Jesu reward them both bodily and ghostly
> From all adversity and great distress
> That will pray for the soul of the Duke of Buckingham,
> That late was exiled remedyless.[131]

Another poem in his book called upon the City stones to speak—'lapides loquentur'—against Wolsey for his treatment of the 'Buck':

> For as men say by thee it was done
> That since had the land no good luck.[132]

[127] CLRO, Journal 12, fo. 113ᵛ.

[128] *L&P* iii/1. 1285 (5, 9, 13, 14, 21, 25, 27).

[129] Rawcliffe, *The Staffords*, p. 97 n. 39. Colet arbitrated in a dispute between the Duke and his secretary: PRO, Sta Cha 2/23/iii; 26/386. Kennedy, 'City and the Crown', pp. 121–2.

[130] Bodleian, Jesus College MS 74, fo. 126ᵛ.

[131] BL, Harleian MS 2252, fos. 2ᵛ–3ʳ.

[132] Ibid., fo. 158ᵛ.

John Skelton often lamented the death of the 'Swan', Buck-ingham:

> So many swans dead, and so small revel.[133]

And *Rede me and be nott wrothe*, which gave the Cardinal no quarter, insistently alleged his implacability against England's greatest noble.[134]

Londoners had often supported those whose interests were quite contrary to those of the Crown.[135] Though the citizens were conscious of the loyalty demanded of them they did not always give it. Henry's awareness of this political truth, raised at the time of Evil May Day, became certainty in 1521. None of the 'lewd persons' who so publicly lamented a traitor's death were 'taken nor arrested nor punished'. The 'King's grace is sore and highly displeased . . . and will not forget'. The City scribe nervously headed this page of the Repertory 'lex domini immaculata vivat Rex currat lex', to proclaim the City's loyalty, after the event. The Mayor and Aldermen went, penitent, to Wolsey to beg forgiveness.[136] Watches were held throughout the City and searches made in July to ensure that all arms and weapons were in safe custody of 'substantial persons', to 'pacify and please the King's grace'.[137] But Henry was not so easily appeased. He was now deeply suspicious of the City (as of so much else besides), and after 1521 determined to end its disobedience by insistently and relentlessly threaten-ing its liberties.

Sedition seemed then a far graver threat than heresy. On 16 June Archbishop Warham sent to Wolsey a priest, Adam Bradshaw, who had composed seditious bills against King and Council from prison, and these were cast around the streets of Maidstone. He was in prison in the first place because of 'his

---

[133] *John Skelton: The Complete English Poems*, ed. J. Scattergood (Harmondsworth, 1983), pp. 245, 262, 271, 278–9, 301.

[134] 'Rede me and be nott wrothe', in *English Reprints*, ed. E. Arber (1871), pp. 30, 52, 59.

[135] Williams, *Medieval London*; R. Bird, *The Turbulent London of Richard II* (1949); C. M. Barron, 'London and the Crown, 1451–61', in J. R. L. Highfield and R. Jeffs (eds.), *The Crown and Local Communities in England and France in the Fifteenth Century* (Gloucester, 1981), pp. 88–109.

[136] CLRO, Repertory 5, fos. 199^v, 204^r; Helen Miller, 'London and Parliament in the reign of Henry VIII', *BIHR* xxxv (1962), p. 140.

[137] CLRO, Repertory 5, fos. 204^{r-v}.

great presumption in pulling down and breaking of such writings and seals as were set up at the abbey of Boxley against the opinions of Martin Luther'. But it was for Bradshaw's sedition, which was 'of more weight', that Warham alerted the Cardinal.[138] But they were wrong to be so complacent. The proclamation of Luther's heresy and the burning of his books had advertised his ideas to a multitude who had never heard of him before, and fostered speculation where silence was intended. Quite contrary to government intention, the demonstration of the error may have done more to spread it than impede it.[139]

Yet when the Bishops proceeded against heretics it was still for the old heresy that they looked. Late in 1521 Bishop Longland of Lincoln conducted a great quest for Lollards. A royal writ issued on 20 October enjoined all lay officers to assist him.[140] From the Cotswolds to Essex three hundred were detected and examined, and twenty-three were named as coming from London.[141] More may have done. Only John 'Father' Hacker, of the Coleman Street conventicle, prime mover of the whole sect, is named by Foxe among the abjured.[142] Thomas Grove, a City butcher, was ordered to perform open penance.[143] All these London Lollards were discovered and tried not by their own Bishop nor his officials but by Longland. As Longland's quest proceeded plague was raging in London, as severe as that of 1518.[144] Infected houses were marked with St Anthony's cross for forty days. No one was allowed to leave London for the fair at Windsor.[145] There were severe food shortages and fears of disorder.[146] This was no time for searches; anyway the plague carried off Fitzjames himself in January 1522.[147]

[138] PRO, SP 1/22, fo. 199^r (*L&P* iii/1. 1353); cited in L. B. Smith, *Tudor Prelates and Politics, 1536–1558* (Princeton, 1953), p. 221.

[139] Cf. the Pope's fears in October 1521: *L&P* iii/2. 1654.

[140] Ibid. iii/2. 1692; Foxe, *Acts and Monuments*, iv, pp. 241–2.

[141] The original examinations and depositions in the *magna abiuratio* of 1521 are now lost, but Foxe had seen and transcribed the records: *Acts and Monuments*, iv, pp. 221–46.

[142] Ibid. iv, p. 242.

[143] Ibid. iv, p. 227.

[144] *L&P* iii/2. 1648, 1650, 1680, 1681.

[145] CLRO, Repertory 4, fo. 89^r; Repertory 5, fo. 217^r; Journal 12, fos. 136^r, 144^r.

[146] *TRP* i. 86; CLRO, Journal 12, fos. 115^v, 119^{r-v}; Repertory 5, fos. 238^r, 254^r.

[147] Hall, *Chronicle*, pp. 628–9.

Fitzjames's successor, Cuthbert Tunstall, Master of the Rolls ('a man of good learning, virtue and sadness [sobriety]'), was a man less inclined to persecution.[148] In October 1522 his *De arte supputandi* was published, dedicated to Sir Thomas More.[149] That Tunstall was a friend of humanists and a humanist himself gave the reformers the greater hopes of him, and the greater disillusion when he disappointed them. William Tyndale came seeking his patronage in 1523, but when turned away, he would write of Tunstall as a 'still Saturn, that so seldom speaketh, but walketh up and down all day, musing, a ducking hypocrite made to dissemble'. George Constantine, too, wrote of Tunstall's 'stillness, soberness and subtlety'.[150] Tunstall's inscrutability made him a valued diplomat, and was to save him more than once in his long career. In Germany, on royal diplomatic service, he had early discovered not only the danger of Luther's ideas, but also the threat of his means of propagating them: Tunstall was among the first to recognize that the rise of printing would threaten the old faith. In January 1521 he wrote urgently to warn Wolsey to forbid printers and booksellers to import Luther's works or to translate them.[151]

Tunstall came to the see of London, the capital of the printing trade, with the certainty that (as Rowland Phillips put it), 'we must root out printing or printing will root out us'.[152] Tunstall's initiative against heresy was directed rather to prevent the dissemination of error and stop the means of conversion than to persecute individuals and make martyrs. The Bishop's first move against the booksellers came in October 1524 when he called two men before him and ordered them not to import books printed in Germany nor to sell books containing Lutheran errors.[153] If it seemed then that the danger came not from England but from Germany, the publication of *The ymage of loue* in October 1525 gave cause for doubt. Wynkyn de Worde, the printer, and John Gough, the translator, were

[148] *L&P* iii/2. 1972; C. Sturge, *Cuthbert Tunstal* (1938).
[149] *RSTC* 24319.
[150] J. F. Mozley, *William Tyndale* (1937), p. 41; Constantine, 'Memorial', p. 63.
[151] Meyer, 'Henry VIII burns Luther's books', p. 180.
[152] Foxe, *Acts and Monuments*, iii, p. 720; iv, pp. 252–3.
[153] GLRO, DL/C/330, fo. 83ᵛ. For Tunstall's regulation of the book trade, see A. W. Reed, *Early Tudor Drama* (1926), pp. 160–75; Sturge, *Tunstal*, pp. 130 ff.

ordered to recover all the copies they had sold.[154] Wyer too was presented for printing a book called *Symbolum Apostolicum*.[155] Worse; there was already popular demand for such forbidden literature. On 27 August 1525 John Cuthbertson preached at Paul's Cross that 'My lord of London's officers had need to take heed, for there is in London that . . . say that if Luther's books were burnt they would be contented to be burnt with them'.[156]

Late in 1525 Wolsey heard the alarming news that the first sheets of Tyndale's New Testament were coming off the press in Cologne.[157] Fearing their imminent arrival in England, he determined upon a vigorous campaign against heretical literature, and a ceremonial denunciation like that of 1521. For Wolsey's purposes it was particularly timely that Robert Barnes had preached an incendiary sermon in Cambridge upon Christmas Eve, 'for his declaration' of his conversion. This was a comprehensive indictment of the superstition of the Church, the 'gorgeous pomp and pride of all exterior ornaments', and the venality of the clergy, particularly of one, the Cardinal. Barnes's recantation, before Wolsey in quasi-papal majesty, was planned as the centre-piece of the book-burning ceremony.[158] But Barnes, encouraged by his London friends, Parnell and Coverdale, proved obdurate in his opinion, and dared to challenge Wolsey, alleging the vanity and mutability of worldly power. Still, Barnes did bear a faggot, and did his penance on Shrove Tuesday, 11 February 1526, before thirty-six mitred Bishops and Abbots, and the Cardinal enthroned. The Church was arrayed in all its splendour, 'to blind the people and to outface me', wrote Barnes.[159] Before the great 'rood of Northern' at St Paul's the offending books were burnt. With Barnes were four Hanse merchants of the Steelyard—Herbert Bellendorpe, Hans Reusall, Hans Ellerdorpe, and

[154] GLRO, DL/C/330, fo. 103[r].

[155] Ibid. DL/C/330, fo. 153[r].

[156] Ibid. DL/C/330, fo. 100[r].

[157] Mozley, *Tyndale*, pp. 57–74.

[158] *L&P* iv/1. 995; A. G. Chester, 'Robert Barnes and the burning of the books', *HLQ* xiv (1950–1), pp. 211–12; J. P. Lusardi, 'The Career of Robert Barnes', *CW* 8/iii, pp. 1371–83.

[159] *A supplicatyon* (1534), sig. Giii[v]; *A supplicatyon* (1531), sigs. Ei–Ei[v]; Lusardi, 'Career of Robert Barnes', pp. 1378–9. Cf. PRO, SP 1/54, fo. 249[v].

Henry Pryknes—all converts to Luther's teachings and importers and propagators of his works, and more dangerously, those of the radical Carlstadt.[160] Perhaps it was with these merchants that William Roper had discussed the works of Luther.[161] After the recantation Barnes became a 'free prisoner' at the Austin Friars, there to read and to sell Tyndale's Testament to City brethren and to visiting Lollards up from the shires.[162]

As copies of Tyndale's New Testament reached London, smuggled by the brethren overseas, all the forces of control were marshalled. On 23 October 1526 Tunstall issued letters to his officials denouncing the maintainers of Luther's sect, and calling in all copies of the New Testament on pain of excommunication and suspicion of heresy.[163] At the same time the Aldermen were ordered to examine everyone in their wards 'concerning certain books of heresy'.[164] Thirty-one of London's booksellers were summoned before Tunstall on 25 October, and warned against importing any books unless approved by Wolsey, Warham, or Tunstall.[165] At Paul's Cross the Bishop denounced Tyndale's Testament as heresy:

> He declared there in his furiousness
> That he found errors more or less
> Above three thousand in the translation.[166]

Humphrey Monmouth, hearing Tunstall's sermon, had destroyed his copies of the Testament, but John Lambert, another reformer who listened, 'lamented greatly', for he could not believe that the translation contained 'hideous errors'.[167]

For all his efforts to outlaw the contraband book trade, Tunstall was powerless to prevent printers producing forbidden

---

160 PRO, SP 1/37, fos. 147ʳ–157ʳ (*L&P* iv/1. 1962). Attempts were made to prevent the Hanse merchants from importing Lutheran books: 'Den wirdigen . . . heren Burgemeysteren der Stat Coelln . . . ex London der derden dach in Martio xvcxxvi' (*RSTC* 16778).

161 Harpsfield, *Life*, p. 86.

162 Strype, *Ecclesiastical Memorials*, i/2, pp. 54–5.

163 Guildhall, MSS 9531/10, fo. 45ʳ; Wilkins, *Concilia*, iii, pp. 706, 727.

164 CLRO, Repertory 7, fo. 140ʳ; Letter Book O, fo. 17ʳ.

165 GLRO, DL/C/330, fo. 123ʳ.

166 'Rede me and be nott wrothe', p. 46; Sturge, *Tunstal*, pp. 133–4.

167 Strype, *Ecclesiastical Memorials*, i/2, pp. 364–7; Foxe, *Acts and Monuments*, v, pp. 213–4.

works, or booksellers selling them to those who did not heed the book burnings or the proclamations. Even though he had Thomas Keyle and John Gough, leaders of the Christian Brethren, before him in 1526, Tunstall, unaware of their activities, failed to charge them.[168] The brethren were absolutely determined. John Parkyns of St Andrew Hubbard avowed in May 1527: 'If I had twenty books of the Holy Scripture translated into English I would not bring none of them in for my lord of London, curse he or bless he, for he doth it because we should have no knowledge, but keeps it all secret to himself.'[169] At the same time William Johnson declared in a London garden that the works of Luther were good and laudable, and were it in his power they should be published throughout England.[170] It was probably the revelation of defiant heresy such as this which prompted the order for a visitation through the diocese of London in the summer of 1527; the *magna abiuratio* which uncovered the continuing activity of the Lollards, now in league with intellectual evangelists down from the universities and bent on propagating the Word.

More heretics were found in London diocese than almost anywhere else in the realm in the 1520s—'no great marvel since unto this diocese there is so great resort and confluence', wrote More.[171] But were the authorities in London persecutors, looking harder for heresy and finding more? Initiatives for the detection of heresy might sometimes come from the lay powers: in September 1520 the City fathers, fearing 'infection' from Cambridge, ordered that 'my Lord Mayor shall cause to be delivered to the doctor of Honey Lane all such books as be specified in the letter as he hath from Cambridge'.[172] But the detection and judgement of heresy lay with the Church: to the secular arm belonged only the ultimate punishment.

Wolsey, as Cardinal Archbishop and Lord Chancellor, combined in his own person spiritual and temporal authority and

---

[168] GLRO, DL/C/330, fo. 123^r.

[169] Ibid. DL/C/330, fo. 137^v.

[170] Ibid. DL/C/330, fo. 138^v. For Johnson, see S. E. Brigden, 'Thomas Cromwell and the Brethren', in *Law and Government under the Tudors*, ed. C. Cross, D. M. Loades, and J. J. Scarisbrick (Cambridge, 1988), p. 43.

[171] More, *Apology*, p. 116.

[172] CLRO, Repertory 5, fo. 65^r.

held the power to judge and to punish heresy. One London heretic, John Gough, judged Wolsey 'a great persecutor'.[173] Thomas More, from the other side of the confessional divide, thought quite the contrary: for him Wolsey was the 'great bell wether' which led its flock astray.[174] Liberal historians have credited the Cardinal with liberal views on toleration, claiming that he used his legatine authority to inhibit episcopal authority in cases of heresy.[175] Certainly, when Wolsey set up, by legatine commission, the episcopal tribunal at Westminster Abbey in November 1527 to try Bilney, Arthur, and Foster, Tunstall complained that he was empowered to try heretics in his own diocese by ordinary authority.[176] The academic reformers called to answer for doubtful views before Wolsey in camera in the legatine Court of Audience were perhaps treated more leniently than if the Bishops had tried them.[177] In 1528 Wolsey vetoed proceedings against alleged heretics in Cambridge (perhaps to assert his superior authority against the Bishop). When Dr Forman protected Lutheran undergraduates in Cambridge the Cardinal turned a blind eye; again, when Forman was discovered running a heretical book trade between London and Oxford he was let off with a 'secret penance', rather than be exposed to the cynosure of an ignorant populace. More found this leniency galling.[178] When the Cardinal examined suspects himself it was particularly heterodoxy regarding the authority of the Church which he sought.[179]

Yet although the Cardinal favoured humanists and intellectuals who were tempted to explore the new ideas, the hundreds of more lowly heretics troubled during his ascendancy would have doubted his leniency, for the persecution continued. Tunstall's own inquisition in 1527 may have begun in response to legatine instructions. Perhaps as many as one hundred and

---

[173] Guildhall, MS 7086/1. Gough scribbled this on the endpapers of the Pewterers' Company account book which he was binding.

[174] Hall, *Chronicle*, p. 764.

[175] See, for example, Pollard, *Wolsey*, pp. 209–15; Clebsch, *England's Earliest Protestants*, p. 277.

[176] Guildhall, MS 9531/10, fo. 130^v; Sturge, *Tunstal*, pp. 136–7.

[177] J. F. Davis, 'Heresy and Reformation in the South-East of England, 1520–1559' (University of Oxford, D.Phil. thesis, 1968), pp. 17–23.

[178] More, *Dialogue concerning Heresies*, i, book 4, ch. xi.

[179] See, for example, PRO, SP 1/47, fos. 81^v–82^r; 1/49, fos. 45^r–47^r (*L&P* iv/2. 4039 (1, 2), 4444).

thirty men and women in London were detected for heresy in Wolsey's time and five were burnt.[180] If there were few martyrs—compared with what was to come—it was not because Wolsey and Tunstall failed to persecute, rather because the heretics, in the way of the Lollards, chose to 'turn rather than burn'. Most abjured, even if only to relapse, and these no Bishop, however relentless, could condemn and send to the stake. On his death-bed, so Cavendish recalled later, the Cardinal sent his desperate last counsel to the King, beseeching

In God's name, that he have a vigilant eye to depress this new perverse sect of the Lutherans that it do not increase within his dominions through his negligence, in such a sort as that he shall be fain at length to put harness upon his back to subdue them, as the King of Bohemia did.[181]

The Cardinal was hated in London, but not for his persecution of heresy (save by heretics). To extirpate heresy was his sacred duty. Heresy was a sin which would call down divine vengeance not only upon the apostate, but upon the community which sheltered him. Heresy was an 'infection' of which society must be purged, which was why the priests of London poured soap ashes on the grave of John Petyt in 1532, 'affirming that God would not suffer grass to grow upon such an heretic's grave'.[182] To be called an heretic was the early sixteenth century's most damaging insult: hence London's outraged protest in 1515 at Bishop Fitzjames's 'perilous and heinous words' after the Hunne affair that all the citizens 'be so maliciously set in favorem hereticae pravitatis' that no impartial jury might be

---

[180] An exact figure is difficult to provide, for Foxe, the main source for the abjurations, does not consistently give the place of origin of the offenders: Foxe, *Acts and Monuments*, iv, pp. 221–40, 582, 585; v, pp. 26–8, 452–3; GLRO, DL/C/330, fos. 111^r, 137^v, 138^v, 175^v; PRO, SP 1/23, fos. 225^r–226^r; 1/37, fos. 147^r–157^v; 1/38, fo. 17^r; 1/47, fos. 79^r, 81^r–82^r; 1/49, fos. 45^r–47^r (*L&P* iii/2. 1922; iv/1. 1962, 2073; iv/2. 4038, 4039 (1, 2), 4444); BL, Harleian MS 421, fos. 7^r, 11^r–15^r (*L&P* iv/2. 3639, 4029); Guildhall, MS 9531/10, fos. 130^r–136^r. For the martyrs, see Foxe, *Acts and Monuments*, iv, pp. 207–16; v, p. 655.

[181] Cavendish, *Life*, p. 179. In May 1530 Cromwell sent Wolsey welcome news: that Luther was dead: R. B. Merriman, *The Life and Letters of Thomas Cromwell* (Oxford, 1902), i, p. 327.

[182] *Narratives of the Reformation*, p. 28; cf. *L&P* xviii/2. 546, p. 300. Those who brought faggots to the burning of heretics were promised forty days' pardon, so Foxe said: *Acts and Monuments*, iv, p. 581.

found among them to try the Bishop's Chancellor.[183] Defamation cases were brought when people called each other 'heretic', or even 'abjured'.[184] Utter social ostracism still awaited the avowedly heterodox. Hunne could find none to trade with him in the City once his apostasy was known; partly through fear of the Bishops and partly because no bargain could be made with a heretic.[185] That he might live in peace one abjured heretic in London tried to hide his heretic's mark of the faggot, and when John Hig complained that no one would employ him because of the faggot he bore, that he would be forced to beg (and that unsuccessfully), the Vicar-General dispensed him.[186] Thomas Grove was prepared to offer a £20 bribe in 1521 to be spared open penance for heresy.[187] When James Bainham came to die a martyr's death in 1532 it was not for himself that he feared, but for his wife, for he knew that men would point her out in the street: 'yonder goeth the heretic's wife'.[188] The London community knew its duty in Christian charity: anyone associated with heresy must be cast out. So it was not for persecution of heresy that the Cardinal and his clergy were hated, at first.

THE CARDINAL AND THE CITY

The disloyalty of the citizenry at the death of the Duke of Buckingham the King vowed he would 'not forget, but intendeth to punish the City with such sharp and grievous punishments which they be not nor shall be able to bear'.[189] He was as good as his word. From the summer of 1521 the implacable Henry moved to threaten the cherished liberties of one of the greatest franchises in his kingdom; seeking to curb the independence of

[183] E. Jeffries Davis, 'The Authorities for the case of Richard Hunne', *EHR* xxx (1915), p. 477; CLRO, Repertory 3, fo. 17ᵛ.

[184] GLRO, DL/C/207, fos. 7ʳ, 10ʳ, 24ʳ; Guildhall, 9064/11, fo. 1ᵛ; Hale, *Precedents*, cxxxiv.

[185] Milsom, 'Richard Hunne's "Praemunire".', p. 82; Foxe, *Acts and Monuments*, iv, pp. 618–19; Strype, *Ecclesiastical Memorials*, i/2, p. 367.

[186] PRO, SP 1/47, fo. 80ʳ (*L&P* iv/2. 4038(3)); Foxe, *Acts and Monuments*, iv, pp. 179, 180; v, p. 29.

[187] Ibid. iv, p. 227.

[188] BL, Harleian MS 422, fo. 90ʳ; Latimer, *Remains*, pp. 223–4. Mrs Bainham was already the widow of another heretic, Simon Fish: Foxe, *Acts and Monuments*, iv, p. 658.

[189] CLRO, Repertory 5, fo. 199ᵛ.

his capital just as his father had done.[190] Henry began to usurp the City's patronage and to treat it as an extension of his own patrimony.[191] In June 1521 he seized the important commercial and mercantile office of the Great Beam for one in his own service, and other great City offices followed.[192] From August 1521 the number of royal requests on behalf of Crown servants for grants of the freedom of the City, the capital's greatest patronage, increased too, and by 1526 the Common Council was driven to try and stop these insistent demands. Neither was the City's ancient and jealously guarded right to elect to its own great offices left untouched.[193] At Michaelmas 1526 one man after another refused to bear the office of Sheriff.[194]

Although at first the City had looked for the Cardinal's 'special good lordship', and hoped for his protection after the troubles of 1517 and 1521, by the mid 1520s the promised amity had turned to mutual and stark hostility.[195] By 1523 the story in Ludgate prison was that Wolsey had said that 'all London were traitors unto his grace'.[196] How had such acrimony arisen? As architect of the ambitious and unpopular foreign policy of the 1520s, Wolsey was blamed for the series of crippling subsidies and loans which such a policy demanded. In turn, Wolsey blamed the City, for London's increasing reluctance to support the King's wars made these European aspirations untenable.[197] Quarrels over money were the foundation of the growing antagonism between the City and the Cardinal, particularly because Wolsey was not only the initiator of the taxation but also the chief commissioner, collecting the City subsidies in person. In 1522, after many 'minatory words', Wolsey succeeded in extracting two loans from a resentful and tax-dodging citizenry, and another assessment

---

[190] Kennedy, 'The City of London and the Crown', chs. 1–3; G. R. Elton, 'Henry VII: a restatement', *Studies*, i, pp. 66–99.

[191] CLRO, Repertory 5, fos. 199ᵛ, 210ʳ–211ʳ; *L&P* iii/1. 1346.

[192] Kennedy, 'The City of London and the Crown', pp. 46–51.

[193] Ibid., pp. 64 ff.

[194] Hall, *Chronicle*, p.718.

[195] Kennedy, 'The City of London and the Crown', pp. 4–9; Miller, 'London and Parliament in the Reign of Henry VIII', pp. 140–1.

[196] PRO, SP 1/23, fo. 320ʳ (*L&P* iii/2. 3076).

[197] Kennedy, 'The City of London and the Crown', ch. 4; Pollard, *Wolsey*, ch. 4; G. W. Bernard, *War, Taxation, and Rebellion in Early Tudor England: Henry VIII, Wolsey and the Amicable Grant of 1525* (Brighton, 1986).

was made in 1523.[198] There was great resentment, as there had been in 1516, over the manner of personal assessment. When yet another demand came in 1525, for the Amicable Grant, 'the commons in every place were so moved that it was like to have grown into a rebellion'.[199] In the capital order was kept, but there was rebellion in the absolute refusal to pay. When the City governors were warned that refusal might 'cost some their heads', they pleaded and were conciliatory, but dared to remind Wolsey that benevolences were illegal according to a statute of Richard III.[200] London's resistance was noted in the shires. So great was the opposition to the Amicable Grant, not least in London, that war plans were perforce abandoned, and peace was made with France in September 1525.[201] Still 'all people cursed the Cardinal and his coadherents as subversors of the laws and liberty of England'.[202]

When Wolsey received his Cardinal's hat in November 1515 John Colet had preached: upon humility: 'whosoever shall exalt himself shall be abased, and he that shall humble himself shall be exalted'.[203] Wolsey took no heed, and his pride was to have consequences for the whole Church. Wolsey's most loyal adherent and most hostile critic alike bear witness to the Cardinal's pride. George Cavendish told of one who, having once humbled Wolsey, now cringingly covered his London house with the Cardinal's arms.[204] Chapuys wrote in 1529 of how nearly every stone of Wolsey's new college in Oxford was emblazoned with his arms.[205] Edward Hall, the common lawyer, from whose pen the most vitriolic portrait of Wolsey comes, said that 'when he was once a perfect Cardinal, he looked then above all estates so that all men almost hated him'.[206] The general opinion was that Wolsey had enchanted

---

[198] CLRO, Repertory 5, fos. 291^r–292^r, 317^v; Hall, *Chronicle*, pp. 630, 645, 656–7; *L&P* iii/2. 2486; J. J. Goring, 'The General Proscription of 1522', *EHR* cccxli (1971), pp. 682–705; Pollard, *Wolsey*, pp. 131–4.

[199] Hall, *Chronicle*, pp. 695–7.

[200] Ibid., pp. 698–9.

[201] *TRP* i. 104–5; CLRO, Journal 12, fo. 305^r.

[202] Hall, *Chronicle*, p. 696.

[203] Pollard, *Wolsey*, p. 57.

[204] Cavendish, *Life*, pp. 5–6.

[205] *CSP Sp.* iv/1, p. 326; 'Rede me and be nott wrothe', p. 54.

[206] Hall, *Chronicle*, p. 583; cf. p. 593. Polydore Vergil hated him too: *Anglica Historia*, pp. 231, 255.

his royal master, for otherwise he could not have gathered so much power.[207] From the mid 1520s the hostility to Wolsey was everywhere apparent, and nowhere more evidently than in London. He, in turn, 'hated sore the City of London and feared it';[208] blaming the City for subversion even where none was intended. At Christmas 1526 a 'goodly disguising' was played at Gray's Inn, a conventional allegory of 'Lady Public Weal', despised and then restored. Wolsey understood this to be a satire against him and imprisoned the author and actors.[209] Even Warham regretted that 'such a matter of play be taken in earnest'.[210] But there was sedition in Gray's Inn of another, more dangerous kind. Simon Fish, a young lawyer there, was of the 'brethren', and one of many City reformers to enjoy a play satirizing the clergy. Foxe has it that Fish took the part of Wolsey in the play.[211] 'Pasquinades' and bills against the Cardinal were nightly circulated, so hostile to Wolsey that there was a general rumour that the King would relieve him of some of his powers. Such was Wolsey's suspicion of Londoners' general evil intent against him that when a bill 'contrary to his honour' was discovered in April 1527 he ordered a great watch to be held throughout the City; much to the consternation of the citizenry, who feared the consequences of his paranoia.[212]

Anger against Wolsey as the King's minister was soon directed against him rather as Cardinal and legate of Rome. Wolsey's ascendancy stirred Londoners' resentment of clerical tyranny, and his personal pride, wealth, and ambition, personifying all clerical vices, led to a general antagonism against the City clergy, especially of the first estate: 'by example of his pride, priests and all spiritual persons waxed so proud, that they wore velvet and silk . . . kept open lechery, and so highly bare themselves by reason of his authorities and faculties that no man durst once reprove anything in them'. So wrote

---

[207] G. L. Kittredge, *Witchcraft in Old and New England* (1929), pp. 109–10.

[208] Hall, *Chronicle*, p. 774.

[209] Ibid., p. 719; R. M. Fisher, 'The Inns of Court and the Reformation, 1530–1580' (University of Cambridge, Ph.D. thesis, 1974), pp. 133–4.

[210] PRO, SP 1/40, fo. 251ʳ (*L&P* iv/2. 2854).

[211] Foxe, *Acts and Monuments*, iv, p. 657.

[212] Hall, *Chronicle*, p. 721. On 16 May John Hubank was charged with setting up seditious bills: CLRO, Repertory 7, fo. 103ᵛ. Cooke, chaplain to the late Duke of Buckingham, also spoke against Wolsey: PRO, SP 1/47, fo. 85ʳ (*L&P* iv/2. 4040).

Edward Hall.[213] The faults of the Cardinal, as well as their
own, would in time be visited upon the London clergy, who
were closest to him. At least six City rectors were also Wolsey's
chaplains.[214] Two of them—Carter and Palsgrave—were at
the centre of the City tithe disputes. It was no coincidence that
the most intense struggles between the City government and the
London curates for a final tithe settlement came at the height of
the capital's resentment of the Church's most spectacularly
wealthy prelate.[215]

Opposition to paying tithe had become part of a wider
opposition to the Church. If the quarrel was not essentially
with the principle of paying tithe, it was with the compulsion to
make payment to undeserving clergy. In 1525 there were
already signs that prevalent criticism of London's clergy might
lead to widespread withdrawal of tithe. At Paul's Cross in
August John Cuthbertson warned: 'there is in London that
hath withdrawn their rights and other their devotions to the
Church'.[216] City tithe controversies went beyond local
disputes. The citizenry were conscious that this was an issue
over which they should unite in corporate resistance to clerical
pretensions. The policy of the City authorities in the 1520s was
to defend any citizen who was prosecuted by the clergy for
tithe, occasionally giving financial aid to those involved in long
and expensive suits, while always attempting to negotiate a
definitive settlement.[217] With time they would become less
conciliatory. One suit became a test case and precipitated a
major dispute. When in 1526 Richard Hearne, a goldsmith of
St Mildred Poultry, was sued by his rector before the Court of
Arches the citizens rallied to contribute to the staggering costs
of £100 for Hearne's defence.[218] Hearne and his partisans

---

[213] Hall, *Chronicle*, p. 593. And others followed Hall: see *A dyalogue betwene one
Clemente a clerke of the conuocacyon and one Bernarde a burges of the parlyament* (*RSTC* 6800.3),
sigs. Diii^v–Div^v.

[214] John Palsgrave, William Cleybroke, William Marshall, William Capon,
Laurence Stubbs, Robert Carter: Pollard, *Wolsey*, pp. 199, 240–1; *BRUO*.

[215] *VCH London*, i, pp. 247–52; S. E. Brigden, 'Tithe Controversy in Reformation
London', *JEH* xxxii (1981), pp. 294–6.

[216] GLRO, DL/C/330, fo. 100^r.

[217] J. A. F. Thomson, 'Tithe Disputes in late medieval London', *EHR* lxxviii
(1963), p. 5.

[218] Lambeth, Carte Miscellane, ix/10, fos. 1–21; B. Walton, *A Treatise concerning the
Payment of Tythes in London*, in S. Brewster, *Collectanea Ecclesiastica* (1752), pp. 85–90.

challenged the clergy in a petition sent to the Court of Common Council in August 1527. They claimed that the citizens' obligations had been laid down in Pope Nicholas's bull of 1453 'purchased' by the clergy, but that the clergy, keeping their parishioners in ignorance, now construed the bull 'after their mind and pleasure'. Any layman daring to question his tithe would be dragged before the partial Church Courts. 'All this was like to grow to a marvellous great grudge in the whole City.'[219] By November 1528 the Common Council had been moved to take an unusually hard line, telling the clergy to abide by the bull and 'if that they will attempt anything to the contrary that then provision shall be made for the defence, &c'.[220] The clergy in turn evolved a counter offensive and organized for their defence.[221] The numbers and proximity of the City clergy gave them greater opportunity for concerted action and organized resistance to lay incursions upon their wealth and privilege than clergy elsewhere in the country.

As the credit of the clergy suffered during the Cardinal's ascendancy, so too did the Pope's, for the hatred earned in England by the legate was transferred from servant to master, or so it seems. During the opposition to the Amicable Grant bills were posted which claimed that the 'Cardinal sent all the money to Rome'.[222] The Sack of Rome in May 1527 was an outrage to all Catholics: or it should have been. Thomas More expressed his horror.[223] But when Clement VII was taken captive in July the 'Communalty little mourned for it, and said that the Pope was a ruffian and was not meet for the room; wherefore they said that he began the mischief and so was well served'. (This, anyway, was Hall's version.)[224] The Cardinal ordered processions in every parish church, and fasting, for the Pope's release, but few fasted, because the clergy who gave the monition would not fast, and if they would not neither would the laity.[225] This was not the first time that the Londoners had

[219] CLRO, Letter Book O, fo. 49[r].

[220] Ibid., Journal 13, fo. 90[v]; Letter Book O, fo. 124[v].

[221] Walton, *A Treatise*, p. 92; CLRO, Journal 13, fos. 122[r]–123[v]; Letter Book O, fos. 140[r]–141[v].

[222] Hall, *Chronicle*, p. 697.

[223] More, *Dialogue concerning Heresies*, i, p. 370; R. C. Marius, 'The Sack of Rome', *CW* 6/ii, appendix C, pp. 773–7.

[224] Hall, *Chronicle*, pp. 727–8, 754.   [225] *L&P* iv/2. 3770, 3764.

shown such contempt for Peter's successor: in 1468 when some City cordwainers obtained a papal bull prohibiting work on the sabbath or the making of shoes with pointed toes ('pykys') 'some men said that they would wear long pykys whether the Pope will or nill, for they said the Pope's curse will not kill a fly'.[226] But to derogate an authority if it seemed immutable posed little danger. And in 1528, as in 1468, papal authority in England was fundamentally unquestioned.

The citizenry's concern in the 1520s was not for the Pope but for their Queen. News of what went on in high places soon reached the City. As early as the summer of 1527 there 'began a fame' that the King, convinced by 'great clerks' that his marriage was 'not good, but damnable', would repudiate Katherine of Aragon for a French princess. The 'rumour sprang so much' that Henry summoned Thomas Seymour, the Mayor, and charged him to 'see that the people should cease of this communication upon pain of the King's high displeasure'.[227] In the City companies a royal letter was read on 15 July, warning against 'persons of light disposition', who sowed 'seditious, untrue and slanderous rumours', which would bring the City to 'wild and insolent demeanour'.[228] The proposed match with the Princess of France was upsetting enough, but as nothing compared with the outrage felt at the King's liaison with Anne Boleyn. It seemed as if Londoners talked of little else. Henry himself wrote to Anne wondering when he should see her next, 'which is better known at London than with any that is about me'.[229] By August 1530 a Venetian observer thought that 'the population here will rebel', should the King marry Anne.[230] Londoners, of anyone in the realm, were most aware of the royal negotiations with the Papacy, and witnessed the arrival of Cardinal Campeggio, and the holding of the legatine court at Blackfriars in the summer of 1529 to arrange an annulment.[231]

---

[226] J. R. Lander, *Government and Community: England, 1450–1509* (1980), p. 119.

[227] Hall, *Chronicle*, p. 728.

[228] Goldsmiths' Company, Court Book D, fos. 237^r–238^r; *Acts of Court of the Mercers' Company*, p. 749.

[229] *L&P* iv/2. 4539.

[230] *CSP Ven*. iv. 701.

[231] Scarisbrick, *Henry VIII*, pp. 224–7.

SUBVERSION

All the political and religious disturbances of the 1520s—the quest for heresy, resentment of a foreign policy which brought crippling taxation, the hatred of the Cardinal—were probably as nothing compared with the catastrophic harvests and recurrent plagues and sweating sickness which the people suffered. War with the Emperor seriously disrupted trade with Spain and Flanders, for the markets were closed to English merchants. So in March 1528, when the country clothiers brought in their cloths for sale at Blackwell Hall none would buy them.[232] In Suffolk the unemployed weavers were mutinous: 'two or three hundred good poor fellows together . . . would have a living', and threatened that 'he that had the most should have least, peradventure'.[233] Wolsey, at Norfolk's behest, called the London merchants before him and berated them for neglecting their social duty: 'you use not yourselves like merchants but like graziers and artificers'.[234] The rich should provide the poor with work: in return the poor owed obedience. Humphrey Monmouth, Tyndale's wealthy patron, had always performed his obligation to the poor, buying cloths in Suffolk and paying for them every week; that is, until his ruinous imprisonment upon suspicion of heresy. Now, before the Bishops in May 1528, he warned that 'if the clothiers fail of their money they cannot set the poor folks to work'.[235]

Wealth and poverty were taken to be part of the natural order. Social distinctions were everywhere apparent in Tudor London; credence and positions of authority were only ever given to citizens who were among the most 'substantial', 'worshipful', or 'honest'; their wealth equalling their power.[236] Few ever challenged the bonds of 'estate' or 'degree', but some did. Philip Smakir vowed in 1513 that 'it were better for the

---

[232] W. G. Hoskins, *The Age of Plunder: The England of Henry VIII, 1500–1547* (1976), p. 184.

[233] *L&P* iv/2. 4012, 4044. There were signs of conspiracy in Kent, and messages passed between Kentish conspirators and unknown Londoners in May: ibid. iv/2. 4299, 4301, 4310, 4331.

[234] Hall, *Chronicle*, pp. 745–6; *L&P* iv/2. 4044.

[235] Strype, *Ecclesiastical Memorials*, i/2, p. 367 (*L&P* iv/2. 4282).

[236] Thrupp, *Merchant Class*, pp. 14–27.

commons of the City to . . . choose a poor man to be Mayor
. . . than to have any rich Mayor or Sheriff'.[237] The governors
of the City always feared for social stability. Economic deprivation might cause some to question the inequalities. The new
fear in London in the 1520s was that with heresy would come
social subversion; the German Peasants' War of 1525 made
that fear seem a reality. Into his commonplace book a Londoner
copied the egalitarian demand of the German peasants: 'And
right shall be done as well to the poor as to the rich without
favour'; 'We be all brethren because we be descended all of
one father'.[238] In London one of the first of the brethren was
indeed stirring social unrest. John Tyndale, merchant taylor,
met a Colchester clothworker in Blackwell Hall in April 1528,
and declined to buy his cloths. Asked what remedy there was
for the slump, he said that he saw none, unless the commons
rose and complained to the King that half of them were out of
work.[239] The actions of reformers on the Continent gave more
reason for alarm at the subversion which heresy brought. In
June 1528 news came from Paris to shock the King: iconoclasts
'worse than Jews' had desecrated a figure of the Virgin with
her Child in her arms.[240] In 1529 Thomas More prophesied
that if ever the new heresy were tolerated God 'would withdraw
his grace and let all run to ruin'. By early 1532 he believed that
his prophecies were coming true:

As folk begin now to delight in feeding their souls of the venemous
carrion of those poisoned heresies . . . Our Lord likewise againward
to revenge it withall, beginneth to withdraw His hand from the fruits
of the earth . . . all this gear hitherto is but a beginning yet.[241]

---

[237] CLRO, Repertory 2, fo. 151[v].

[238] BL, Lansdowne MS 762, fo. 76[v].

[239] *L&P* iv/2. 4129; PRO, SP 1/47, fos. 176[r]–177[r] (*L&P* iv/2. 4145). For Tyndale,
see Brigden, 'Thomas Cromwell and the Brethren', pp. 33, 36, 37.

[240] *L&P* iv/2. 4338, 4409.

[241] More, *Confutation of Tyndale's Answer*, p. 3; A. Fox, *Thomas More: History and
Providence* (Oxford, 1982), ch. 8. More's favourite daughter, Margaret, nearly died
from the sweating sickness, and was saved by his prayers: Roper, *Lyfe*, pp. 28–9.

IV Clergy and Communalty, 1529–1533

ANGER AGAINST the vices and pretensions of the clergy had an old history in London. Wolsey was not the first great clerical Chancellor to be hated by the citizenry: they had murdered another in 1381.[1] At some times antagonism towards the clergy ran deeper than at others. So it was in 1529. Then Chapuys could write from London, 'nearly all the people here hate the priests', and early in 1530 a leading City cleric—Dr Miles of St Bride's—was murdered.[2] The resurgent opposition of Londoners to the priesthood, being more overt and more concentrated than elsewhere in the country, had desperate consequences for the English Church, for it was used by those whose quarrels with the whole clerical estate went deeper by far: by the heretics, who challenged the very nature of sacerdotal power and sought to appeal to the orthodox by insisting on the venality and rapacity of the priesthood, and by those who wished, for political or economic reasons, to see the clergy humbled. From 1529 the anticlerical elements began to make moves from which there could be no return.

Memories of Richard Hunne's death and of the Standish affair still rankled. The suspicion—for some, the certainty—remained long after that Hunne's only heresy had been to challenge the clergy, and that he had been killed. Reminding Londoners of this victim of rapacious priests would have a special resonance in 1529, and some then determined to make political capital of the Hunne legacy. In his Spital sermon in Easter week, before a large audience, Dr Goderidge read a bill announcing that there was money from Hunne's estate to pay towards the repair of the Fleet conduit; 'upon whose soul and

---

[1] R. Bird, *The Turbulent London of Richard II* (1949).
[2] *CSP Sp.* iv/1, p. 367; *Two London Chronicles*, p. 4; Gee and Hardy, *Documents Illustrative*, p. 174; A. G. Dickens, *The English Reformation* (1964), p. 95; C. A. Haigh, 'Anticlericalism and the English Reformation', *History*, lxviii (1983), pp. 391–407.

all Christian souls, Jesus have mercy'.[3] Hunne had always been known as 'a fair dealer among his neighbours' (even More admitted this), 'a singular friend of the poor', and now the citizenry were reminded of the public-spiritedness of this supposed heretic.[4] Goderidge was forbidden to celebrate, ordered to make public recantation at Paul's Cross and to do penance. Roger Whaplode and his accomplice, Thomas Norfolk, were brought before the Bishop for preparing the bill. Whaplode was Hunne's son-in-law; moreover, he was of the brethren.[5]

The 'evangelical brethren' were moving to the offensive. Their cause, they saw, might be forwarded by anticlerical diatribes and political tracts with a wider appeal than serious works of theological exposition. So it was that in 1528 William Tyndale published his *Wicked Mammon* and the *Obedience of a Christian Man*. To a King battling with a Pope and thwarted by a Cardinal, Tyndale's assertion in this latter work that the Church hierarchy had not only nullified God's promises but usurped the magistracy of the prince won favour in a way which Tyndale's other works would not.[6] Henry declared that 'this book is for me and all Kings to read'.[7] Roye and Barlow, apostate Observant friars, published in 1528 their scurrilous *Rede me and be nott wrothe*, which was denounced not only by its target, Wolsey, but also by the more scrupulous Tyndale as a 'railing rhyme'.[8] Of the readership of these works we know little. Another, more effective, blast against the clergy came in 1529: Simon Fish's *Supplicacyon for the Beggers*. This work indicted the minions of Satan: 'Bishops, Abbots, Priors, deacons, Archdeacons, Suffragans, priests, monks, canons, friars, pardoners, and summoners', who had taken all and were as vindictive as they were rapacious. Londoners fighting for a tithe settlement with their clergy would have applauded Fish's jibe: 'Yea, and they look so narrowly upon their profits that the poor wives

[3] Foxe, *Acts and Monuments*, v, pp. 27–8.

[4] More, *Dialogue concerning Heresies*, i, p. 326; *Anglica Historia*, p. 229; Fish, 'Supplicacyon', p. 12.

[5] *L&P* iii/2. 3062(4); Foxe, *Acts and Monuments*, iv, pp. 197–8.

[6] W. A. Clebsch, *England's Earliest Protestants, 1520–1535* (New Haven and London, 1964), pp. 146–55.

[7] Cited in J. F. Mozley, *William Tyndale* (1937), p. 143.

[8] Ibid., p. 122; Clebsch, *England's Earliest Protestants*, p. 147; E. G. Rupp, *Studies in the Making of the English Protestant Tradition* (Cambridge, 1949), pp. 52 ff.

must be countable to them for every tenth egg, or else she getteth not her rights at Easter and shall be taken as an heretic.'[9] The fate of the most celebrated victim of clerical injustice, Hunne, was also here rehearsed.[10] This book certainly fell into the hands of Londoners. George Robinson and George Elyot, leading City merchants, showed copies to the King early in 1529.[11] Robinson was a mercer, rich enough to be a candidate for the shrievalty in 1526, and later (perhaps already) a confidant of Thomas Cromwell.[12] Another mercer possessed this book, and perceived its danger: Thomas More published his refutation of Fish's tract in September 1529; *The Supplycacyon of Soulys*.[13] On Candlemas day 1529 (3 November), the first day of a new and momentous Parliamentary session, copies of the *Supplicacyon for the Beggers* were 'scattered at the procession in Westminster' and cast through the City streets.[14] People in 1529 might have sympathy with the anticlerical views while abhorring the heretical opinions of those who spread them, yet now the door was wide open, said Chapuys, for 'the Lutheran heresy to creep into England'.[15]

THE 'PARLIAMENT FOR THE ENORMITIES OF THE CLERGY'

The summoning of this 'Parliament for the enormities of the clergy' (as Fabyan termed it) followed upon the fall of Wolsey.[16] The failure of the legatine court at Blackfriars to procure the King's desperately desired annulment, and the indignity of Henry's summons before the Rota in Rome proved disastrous for the Cardinal. Through the autumn of 1529 all was uncer-

---

9 Fish, 'Supplicacyon', p. 2.

10 Ibid., pp. 9, 12.

11 Foxe, *Acts and Monuments*, iv, p. 658.

12 Hall, *Chronicle*, p. 718; see below, p. 291.

13 Thomas More, *The supplycacyon of soulys . . . gaynst the supplycacyon of beggars* (London, before 25 Oct. 1529; *RSTC* 18092); Rainer Pineas, *Thomas More and Tudor Polemics* (1968), pp. 158–72; S. W. Haas, 'Simon Fish, William Tyndale and Sir Thomas More's "Lutheran Conspiracy"', *JEH* xxiii (1972), pp. 125–36.

14 Foxe, *Acts and Monuments*, iv, pp. 659, 666–7; *TRP* i. 129. Only one edition of Fish's tract is known to have been printed until it was reissued with another tract in 1546: *RSTC* 10884.

15 *CSP Sp.* iv/1, p. 236.

16 S. E. Lehmberg, *The Reformation Parliament, 1529–1536* (Cambridge, 1970), ch. 1.

tain. Henry was even supporting heretics, and the vultures were gathering.[17] Rowland Philips wrote to Wolsey of one who 'trusts to a great change, and specially the extinction of your authority'.[18] Even Wolsey's own chaplain, John Palsgrave, had composed—albeit secretly—a comprehensive indictment of the Cardinal's misrule: 'we have begun to have the whole power of the Pope'. Palsgrave wrote of Wolsey's 'great enterprises . . . the least of them to our commonwealth much expedient', come to nothing; of his pride 'to have two poleaxes borne after us and to clothe the clerks of England in silk and velvet'; of his 'wasteful expenses' and 'manifest tokens of vainglory'; of his humbling of the nobility ('temporal lords served us oft combe water').[19] Thomas Wolsey was compared to Thomas Becket, and his removal was no less desired by his King.[20] On 18 October 1529 Wolsey, indicted before King's Bench for violation of praemunire, resigned the Great Seal. As he left by barge for Esher on 25 October the Cardinal was awaited by a 'thousand boats of men and women of the City of London . . . expecting my lord's departing . . . to the Tower, whereat they rejoiced'.[21] Nowhere was his fall more welcomed.

The disgrace of Wolsey and the summoning of Parliament gave the expectation of redress and hopes of reform. But those expectations were of diverse kinds. Two Londoners came to this Parliament who would have the greatest influence: Thomas More and Thomas Cromwell.[22] More, now Lord Chancellor, came pledged to combat heresy and determined— all in vain—to stay out of the King's Great Matter. Cromwell, the servant of a fallen master, came to London 'to make or mar'. He was already in touch with reformers, and would in time find a way for Henry to break the deadlock with Rome and to divest himself of his wife. There were other Londoners

---

17 *L&P* iv/3. 5864; *CSP Sp*. iv/1. 160 (pp. 221, 229); J. J. Scarisbrick, *Henry VIII* (1968), p. 228; J. A. Guy, *The Public Career of Sir Thomas More* (Brighton, 1980), pp. 105–10.

18 *L&P* iv/3. 5898.

19 PRO, SP 1/54, fos. 244^r–252^r (*L&P* iv/3. 5750).

20 William Tyndale, *Expositions and Notes . . . on the Holy Scripture and the Practice of Prelates*, ed. H. Walter (Parker Society, Cambridge, 1849), p. 292.

21 Cavendish, *Life*, p. 100.

22 *House of Commons, 1509–1558* (Thomas Cromwell, Thomas More); G. R. Elton, *The Tudor Revolution in Government* (Cambridge, 1953), pp. 77–80; Lehmberg, *Reformation Parliament*, pp. 27–8; Guy, *Public Career of Sir Thomas More*.

in the House of Commons; not only the four burgesses for the City, Paul Withypoll, John Petyt, John Baker, and Thomas Seymour, but also the many who, like Edward Hall and Thomas Cromwell, sat for county boroughs.[23] Moreover, there were all those who had studied in London: the common lawyers.

It was the particular misfortune of the clergy that so many lawyers sat in Parliament, for they had old reasons to question the authority of 'overmighty' clerics. Since the Constitutions of Clarendon the common lawyers had been prepared to attack the Church as an institution while remaining inveterately orthodox and conservative, but their old challenge to spiritual jurisdiction became much more radical during the Reformation Parliament. Disputes between the 'spiritualty' and the 'temporalty' over whether 'men's laws be made only by the princes or the people', and where the Church's power to legislate lay, became more publicly ventilated.[24] A single copy survives of a contemporary pamphlet: *A dyalogue betwene one Clemente a clerke of the conuocacyon and one Bernarde, a burges of the parlyament, disputynge bewene them what auctoryte the clergye have to make lawes. And howe farre and where theyr power doth extende.*[25] Bernard the burgess made a comparison between the Church and the City of London, which could legislate to govern itself, but 'cannot stretch to govern or to correct any other of the King's subjects being no such citizens':[26] so it was for the clergy. During Wolsey's ascendancy any claims that spiritual authority was subject to Parliament and the common law had particular pertinence, and the struggle for precedence more urgency, for it was largely the Cardinal's hatred for the Commons and his

[23] Lehmberg, *Reformation Parliament*, pp. 11, 19–20, 28. For the many Londoners who sat in the Reformation Parliament, see *House of Commons, 1509–1558*.

[24] From 1530 the pamphlets of Christopher St German, circulated in London and refuted by Sir Thomas More, began to provide a juridical framework for the royal supremacy; F. le van Baumer, 'Christopher St German: the Political Philosophy of a Tudor Lawyer', *AHR* 42 (1940), pp. 631–51; J. A. Guy, 'Thomas More and Christopher St German: the Battle of the Books', *Moreana*, 1 (1984), pp. 5–25.

[25] *RSTC* 6800.3. This tract was discovered by H. C. Porter in Selwyn College, Cambridge, and discussed by him in 'Hooker, the Tudor Constitution and the *Via Media*', in *Studies in Richard Hooker: Essays Preliminary to an edition of his Works*, ed. W. Speed Hill (Ohio, 1972), pp. 86 ff.

[26] *A dyalogue betwene . . . Clemente a clerke . . . and Bernarde a burges*, sig. B7ʳ. Cf. More's plea in arrest of judgement: Roper, *Lyfe*, p. 93.

inability to control it which had meant that no Parliament was called between the débâcle of 1523 and his fall in 1529.[27]

Parliament opened with an oration from the new Chancellor explaining that new laws were needed to reform 'divers new enormities . . . sprung amongst the people', abuses furthered by the 'great wether' of the King's flock 'of late fallen as you all know'.[28] Anticlerical feeling was already running high: here was a mandate to curb, 'by law', the abuses of the clergy. The City of London always acted as a powerful lobby in the Commons, and now in 1529 it had a case prepared.[29] In the Mercers' Company a programme for the redress of the City's grievances was ready to present to the House: four articles concerned trade, but the fifth was the one which, according to Hall, 'sore moved' the Commons:

the King's poor subjects, principally of London, been polled and robbed without reason or conscience by the Ordinaries in probating of testaments and taking of mortuaries and also vexed and troubled by citations with cursing one day and absolving the next day, et hec omnia pro pecuniis [and all for money].[30]

This complaint revealed not only lay resentment of clerical exaction but also the fears of the Church's unbounded powers to cite and punish lay men in its courts. In 1529 the Commons was not attacking clerical jurisdiction *a priori*, rather the clerical vices of venality and negligence, but in such a way were the protests voiced that they seemed to be a concerted attack upon the nature of Church authority. When a member of Gray's Inn in the Commons (Edward Hall himself perhaps) could question the customary authority of the Church to take probate fees by putting the case that 'the usage hath ever been of thieves to rob on Shooters Hill [Blackheath], ergo is it lawful?', then the Church's claim to authority from tradition was vulnerable indeed.[31]

---

[27] A. F. Pollard, *Wolsey* (Fontana edn., 1965).

[28] Hall, *Chronicle*, p. 764; *CSP Sp.* iv/1. 211 (pp. 323–5); Guy, *Public Career of Sir Thomas More*, pp. 113–15; Lehmberg, *Reformation Parliament*, pp. 78–9.

[29] H. Miller, 'London and Parliament in the reign of Henry VIII', *BIHR* xxxv (1962), pp. 128–49.

[30] Mercers' Company, Acts of Court, ii, fos. 25ᵛ–26ᵛ; cited in Miller, 'London and Parliament', p. 144. In 1504 the City had complained in Parliament about the testamentary jurisdiction of the Church; ibid., p. 134.

[31] Hall, *Chronicle*, p. 767.

In the past challenges to the clergy had often enough been condemned as heretical; and it was as heresy that the Ordinaries construed the Commons' criticism in 1529. Imputations of heresy threatened soul and body, so when Bishop Fisher compared the Commons with the Hussite 'kingdom of Bohemia', and ascribed their motives to 'lack of faith' the outraged Commons protested its orthodoxy to the King.[32] But Fisher had a point: some of the men most active in curbing clerical abuses in 1529 were already won to reform. When the committee was appointed of men learned in the law to draw up corrective statutes it contained Thomas Cromwell and, probably, John Petyt.[33] Cromwell had old links with the brethren, and Petyt was the patron of Dr Forman and defender of Dr Barnes.[34] In the event, the legislation to reform clerical abuses contained so many loopholes that the clergy found endless ways of evasion.[35] But the laity, in London at least, did not take advantage of the golden opportunity offered them by these statutes to present their defaulting clergy, and to make money from the fines into the bargain. Only seven City clerics were presented before the Barons of the Exchequer in the 1530s.[36] Yet the grievances between the clergy and the laity, voiced concertedly in 1529, were of the greatest consequence. To dare to criticize the clergy, even for evident abuses, was now to incur the suspicion of heresy, and now a campaign against heresy was mounted in the capital, during which suspects were hunted down so implacably that even the orthodox might tremble.

---

[32] Hall, *Chronicle*, p. 766; Lehmberg, *Reformation Parliament*, pp. 87–8, 92. On his death-bed Wolsey compared 'the new perverse sect of the Lutherans' with the Bohemians: Cavendish, *Life*, p. 179.

[33] Lehmberg, *Reformation Parliament*, p. 83.

[34] *A supplicatyon made by Robert Barnes unto kinge henrye the eyght* (1531; *RSTC* 1470), sig. I 2ᵛ; S. E. Brigden, 'Thomas Cromwell and the Brethren', in C. Cross, D. Loades, and J. J. Scarisbrick (eds.), *Law and Government under the Tudors* (Cambridge, 1988), pp. 44–5.

[35] *Statutes of the Realm*, iii, pp. 285–8, 288–9, 292–6. See, for example, P. Heath, *The English Parish Clergy on the Eve of the Reformation* (1969), p. 50.

[36] PRO, E 159/310, m. 53, 56d; 314, m. 2; 315, m. 11, 21, 37. There were only 210 suits in the country between 1530 and 1535, of which only fourteen were pressed to a conclusion: J. J. Scarisbrick, 'The Conservative Episcopate in England, 1529–1535' (University of Cambridge, Ph.D. thesis, 1955), pp. 88–94.

THOMAS MORE AND PERSECUTION

England now had a lay Chancellor, but one who, some said, 'greatly favoured the Bishop and the clergy'.[37] Condemning Wolsey for traducing his trust, More saw his first task in 1529 as the defence of the Church from heresy. The More who wrote *Utopia* had been prepared to speculate upon reform, even to countenance—in the realm of imagination, at least—toleration. But time and experience had turned him into the intransigent champion of Catholic orthodoxy. He had already written his vindication of persecution by the time he was in a position to initiate a campaign.[38] For More, the heretic's death at the stake was of little moment, because he was already destined for the fire eternal. Thus More wrote of his fears for 'young father Frith': 'Christ will kindle a fire of faggots for him, and make him therein sweat the blood out of his body here, and straight from hence send his soul for ever into the fire of Hell.'[39] James Bainham's last words at the stake were: 'The Lord forgive Sir Thomas More', 'my accuser and my judge'.[40] In fact, the Chancellor was empowered to arrest, investigate, and hold suspects for judgement by the Courts Christian, and this More did with a vengeance, but he could not judge or condemn for heresy. Only the Bishops had that power.[41] Yet More did adduce novel powers to himself. The campaign of Catholic repression was codified in 1530. Two proclamations were issued —the first on 22 June, and another later in the year—to 'repress blasphemous books lately made and privately sent into the realm' by Luther's disciples. The June proclamation ordered that offenders should be brought, not before the Bishops, but before the King's Council, sitting either as a board or as a court

---

[37] For example, Hall, *Chronicle*, p. 784.

[38] More, *Confutation*, pp. 3–40; especially pp. 8, 16–17; A. Fox, *Thomas More: History and Providence* (Oxford, 1982), pp. 118–20; L. A. Schuster, 'Thomas More's Polemical Career, 1523–1533', *CW* 8/iii, pp. 1135–1268.

[39] More, *Apology*, p. 122; cf. *Confutation*, p. 1208; *Answer to a Poisoned Book*, p. 197.

[40] Foxe, *Acts and Monuments*, iv, p. 705.

[41] For heresy procedure, see J. F. Davis, *Heresy and Reformation in the South-East of England, 1520–1559* (1983), ch. 2; J. A. Guy, 'The Legal Context of the Controversy: the Law of Heresy', *CW* 10, pp. xlvii–lxvii.

when in Star Chamber.[42] Wolsey had used the prerogative and censorship powers but rarely (for example, against Robert Necton); More did so systematically. Here he could indeed sit as accuser and judge.

The campaign against heresy was conducted from the capital, and from there came most of its victims. More moved fast. Late in 1529, 'about the latter end of the time of Cuthbert Tunstall', a heretic named Stile was burned in Smithfield with his *Apocalypse*.[43] With John Stokesley's promotion to the see of London in March 1530 More found a more willing partner in persecution. Unlike More, Stokesley was a proponent of the Divorce (spending 1530 trailing around the universities of Europe seeking learned opinion). It was this, and his abandonment of his master, the Cardinal, which had won him royal favour.[44] Stokesley had been Wolsey's almoner and arbitrator in the Cardinal's court at Whitehall, but doubts about his judgement had caused him to be 'deposed his office'. He was, according to Hall, 'a man that had more learning than discretion to be a judge'.[45] Stokesley's learning was renowned, and now all directed to the earnest defence of the 'Romish decrees'.[46] At first a champion of Erasmus's translation of the New Testament, he had defended it, as More had done, against the attacks of antediluvian friars; now, like More, he was appalled by the way in which humanism had fostered heresy. All his erudition and all his languages were used henceforth to argue against, rather than to further, the translation of Scripture. He became a renowned persecutor: 'blódy bisshop crysten catte'.[47] Stokesley's lieutenant in his diocese was Richard Foxford, Chancellor and Vicar-General of London: 'a common butcher

[42] *TRP* i. 122, 129. For these proclamations and their dating, see Guy, *Public Career of Sir Thomas More*, pp. 171 ff.; R. W. Heinze, *The Proclamations of the Tudor Kings* (Cambridge, 1976), pp. 279–80.

[43] Foxe, *Acts and Monuments*, v, p. 655.

[44] Hall, *Chronicle*, p. 761; *CSP Sp.* iv/i. 415, 567 (pp. 695, 876). Stokesley had returned to England in September 1530. Scarisbrick, *Henry VIII*, pp. 255–7; Roper, *Lyfe*, p. 38.

[45] Hall, *Chronicle*, p. 585; J. A. Guy, *The Cardinal's Court: the Impact of Thomas Wolsey in Star Chamber* (Hassocks, 1977), pp. 44–5.

[46] For references to Stokesley's learning, see Foxe, *Acts and Monuments*, v, p. 56; Wriothesley, *Chronicle*, i, pp. 105, 107; Strype, *Ecclesiastical Memorials*, i/2, pp. 381–2; John Bale, *Yet a course at the romyshe foxe* (Emden, 1543; *RSTC* 1309), fo. 87ᵛ.

[47] M. Dowling, *Humanism in the Age of Henry VIII* (1986), p. 22; 'The Souper of the Lorde', printed in More, *CW* 11, appendix A, p. 336.

of God's saints', according to Foxe, and duly visited by providential punishment.[48] This triumvirate acted concertedly and
urgently to extirpate heresy during the two-and-a-half years of
More's chancellorship.

This Chancellor's campaign against heresy was desperate
because he knew that time was short. His own animus against
the heretics grew with his awareness of their resolve, and the
converts they were making to their cause.[49] More could write
in his polemic against the brethren of the 'good wife' of the
Bottle at Botolph Wharf who listened to the persuasions of the
learned Dr Barnes only to confute his arguments with homely
good sense, but he feared also that the 'simple' were with the
'wind of every new doctrine blown about like a weathercock'.[50]
He knew, too, that heresy had 'infected' men and women in
high places, who were more than a match for the Bishops in
their cunning and resolve. Tunstall was hopelessly outmatched,
and More himself was outwitted. In Antwerp in the summer
of 1529 Tunstall gladly accepted Augustine Packington's disingenuous offer to buy up for the Bishop a whole edition of
Tyndale's New Testament to burn. But Packington was of a
reforming dynasty, and he was in league with Tyndale: so
although Tunstall had books to burn at Paul's Cross that
winter, Tyndale had the money to print a bigger and revised
edition. Who at home aided the exiles abroad, asked More.
Replied George Constantine gleefully, 'Tunstall'.[51] Thereafter
More took the hunting of book agents upon himself, but he,
too, would be duped. Once More had had misgivings about
writing against heresy at all, lest the people be perversely
intrigued by its novelties.[52] He had been right. In time, his
own works of polemic were read avidly by the brethren, and his
purpose confounded. John Field possessed a copy of More's

[48] Foxe, *Acts and Monuments*, v, p. 64.

[49] For More and the persecution of heresy, see G. R. Elton, 'Sir Thomas More and
the Opposition to Henry VIII', *Studies*, i, pp. 158–61; R. W. Chambers, *Thomas More*
(1935), pp. 284 ff.; H. G. Ganss, 'Sir Thomas More and the Persecution of Heretics',
American Catholic Quarterly, xxv (1900), pp. 531–48; R. Marius, *Thomas More* (1984),
ch. 25; L. Miles, 'Persecution and the *Dialogue of Comfort*; a fresh look at the charges
against Thomas More', *JBS* v (1965), pp. 19–30.

[50] Schuster, 'The Polemical Career of Sir Thomas More', *CW* 8/iii, p. 1266; More,
Correspondence, p. 460.

[51] Hall, *Chronicle*, pp. 762–3; Foxe, *Acts and Monuments*, iv, pp. 670–1.

[52] More, *Apology*, p. 123.

Supplycacyon of soulys, for there was no safer way to read Fish's
Supplicacyon for the Beggers, which was quoted *in extenso* therein.
On More's orders Field was imprisoned, in conditions of priva-
tion unusual even by the appalling standards of Tudor prisons,
from January 1530 until May 1532. (Even after his resignation
as Chancellor, More hunted Field down.)[53] This Chancellor
went a long way in the pursuit and 'persuasion' of his suspects.
For More was sorely tried. The heretics were all around him, in
the places he knew best and held dearest. The Inns of Court
were 'infected'. There was a cell of brethren in his own company, the Mercers, led by George Robinson and Robert
Packington. Even his own family was not safe: his prayers had
saved his son-in-law William Roper, but John Rastell, his
brother-in-law, died in prison in 1536, a heretic.

More acted with such urgency because he knew that his own
days in power were numbered, and that there were heretics in
the highest places of all. While he sat in Star Chamber in
judgement upon book agents in the autumn of 1530 there were
bitter clashes in Council.[54] Since early 1530 (perhaps by
February) Thomas Cromwell was in the royal service, and by
the end of the year a sworn member of the royal Council.[55]
There in the Council, and in the City too, More and Cromwell
struggled for influence. By August 1530 Henry VIII had been
convinced that he was absolute 'as Emperor and Pope in his own
kingdom'; that neither he nor his subjects could be summoned
to Rome, for England was independent of foreign jurisdiction.[56]
Cromwell would show him how the lost privileges might be
reclaimed, but this aim might be thwarted if More remained
Chancellor. On 20 September Chapuys told the Emperor that
More had come within an ace of dismissal for speaking so much
in the Queen's favour.[57] On 19 September, with a grand fanfare
of trumpets, a proclamation was issued which was ominous for
the future, and showed which group now held sway within the
Council. No papal bulls were to be received in England which

[53] PRO, SP 1/78, fos. 246ʳ-247ʳ (*L&P* vi. 1059); Elton, 'Sir Thomas More and
Opposition', *Studies*, i, pp. 159-61.

[54] Guy, *Public Career of Sir Thomas More*, pp. 138-40.

[55] Elton, *Tudor Revolution in Government*, pp. 82-6.

[56] *CSP Sp.* iv/1. 429, 433, 445; Guy, *Public Career of Sir Thomas More*, pp. 131-3;
Scarisbrick, *Henry VIII*, pp. 260 ff.

[57] *CSP Sp.* iv/1. 433 (p. 727).

were prejudicial to the royal prerogative. This proclamation was meant to annoy the Pope and to frighten the Queen's supporters. In London it was much 'mused at'.[58]

While Henry challenged Rome still he saw himself as *defensor fidei*. In October 1530 he boasted to the Papal nuncio: 'I will soon establish a much stricter rule in this my kingdom than has been done in Germany, touching the suppression and punishment of Lutheran errors.'[59] The King's revived fervour had not a little to do with the heretics' new ploy of writing against the matter closest to Henry's heart, and most unpopular with his people—the Divorce. In London More was persecuting men who—ironically—held views close to his own on the King's Great Matter. In 1530 Tyndale published his *Practice of Prelates*, a venomous tirade against Thomas Wulfsee and his venal disciples. (Tyndale thought, wrongly, that the Cardinal was soon to return to favour.) But the tract had a second title: *Whether the King's grace may be separated from his Queen because she was his brother's wife*.[60] Tyndale thought emphatically not. Three thousand copies were circulated through London. Back from his tour of the European universities, urging the King's case for annulment, Stokesley now sought the book-running culprits.[61] Before More and the lords in Star Chamber came 'five or six' London merchants in October, 'for having books against the King's proclamation'. Fines, penance, and imprisonment were enjoined.[62]

Facing their horses' tails, four men—John Tyndale, Thomas Somer, John Purser, and an apprentice from London Bridge—rode from the Tower through the City to Cheapside cross on 19 November to be publicly shamed. Their clothes were festooned with copies of Tyndale's forbidden Testament and other heretical works, and they wore placards proclaiming *Pecasse contra mandata regis*. Into a great fire they tossed the 'infected'

---

[58] *TRP* i. 130; *CSP Sp*. iv/1. 433 (p. 726); Hall, *Chronicle*, pp. 772–3; Guy, *Public Career of Sir Thomas More*, p. 139.

[59] *CSP Sp*. iv/1. 460 (p. 761).

[60] William Tyndale, . . . *The Practice of Prelates*, ed. H. Walter; A. Hume, 'English Protestant Books printed abroad', *CW* 8/ii, p. 1078; Mozley, *William Tyndale*, pp. 163–9.

[61] *CSP Ven*. iii, p. 642.

[62] *CSP Sp*. iv/1. 460 (p. 761); HL, Ellesmere MS 2652, fo. 15ʳ; Guy, *Public Career of Sir Thomas More*, pp. 173–4.

books, and were then set on the pillory.[63] But they were not penitent at all. Thomas Somer rode upon 'a lofty gelding and fierce', lent by one of the brethren, which pranced as the basins rang.[64] They knew that their penance would bring knowledge of Tyndale's writings to the Londoners, hitherto oblivious of them, more immediately than ever their book running could have done; that the persecutors forwarded the books they sought to suppress. Chapuys recognized this too, but the outraged King ordered proclamations to be set up attesting to the decisions of the universities in favour of the Divorce, and condemning *The Practice of Prelates*. Such was the popular speculation about the King's Great Matter which followed, and so eager were the citizens to read the forbidden work, that the proclamations were torn down and burnt (so none survive). The brethren had taken the chance also that December to post defamatory libels on the door of Canterbury cathedral: first against the Archbishop and his Chancellor, then against the King and Council.[65] The penitents of November had influential friends. Latimer wrote to Henry on 1 December, protesting their innocence: 'there is no man . . . that can lay any word or deed against them that should sound to the breaking of any of your grace's laws' (save possession of Tyndale's Testament).[66] Another old friend of theirs kept silent, biding his time: Thomas Cromwell.

Though More might have been losing the struggle for influence over the King, who listened now to the evangelicals about him, still he determined to use his powers as Chancellor to strike down the patrons of reform. He knew well that Londoners of wealth and influence were prime movers in the heretical book trade. John Petyt, who had framed the anticlerical legislation in 1529, with Cromwell, had been patron and protector of evangelical preachers—Barnes, Forman, and Crome—and More 'sore suspected' him of abetting Tyndale.[67] To Petyt's house at Lyon's Quay More went in person, seeking contra-

[63] *CSP Ven.* iii. 642; *CSP Sp.* iv/1. 509 (pp. 820-1); *Two London Chronicles*, p. 5. For these men, see Brigden, 'Thomas Cromwell and the Brethren', pp. 33-8.

[64] Foxe, *Acts and Monuments*, v, p. 453.

[65] *CSP Sp.* iv/1. 509, 522, 539, 547 (pp. 820-1, 834, 847, 852).

[66] Latimer, *Remains*, p. 306.

[67] Lehmberg, *Reformation Parliament*, p. 83; Barnes, *Supplicatyon* (1534), sigs. I 2ᵛ-I 3ᵛ.

band books. Finding nothing, and with no proof against Petyt, More invented some (so it was said later): he had 'gotten a little old priest, that should say he had Tyndale's testament in English and did help him and such other to publish their heretical books'. Is this to be believed? Maybe: desperate times needed desperate remedies. Petyt 'caught his death' in the Tower and was dead by August 1532.[68] That October Thomas Cromwell petitioned the Court of Aldermen on behalf of Petyt's widow: he had been 'a true and loving citizen, and from time to time exceeding painful in the procurement of your common affairs'.[69] Thomas Cromwell was, as More knew, the most powerful of all the City patrons, and the one still too necessary to the King to be struck down.

Heretics were now in favour at Court. Robert Barnes, returned to England under safe conduct, went around the Court in secular dress that December, in conference with Nicholas de Burgo.[70] Simon Fish, too, was in London with a safe conduct.[71] There was talk that Henry wanted Tyndale home also, to convert to friend rather than enemy. Anne Boleyn was jubilant: she sported a new device: 'ainsi sera groingné qui groingne' ('thus it will be: grudge who grudges'). The citizenry were correspondingly downcast, listening to prophecies which foretold that the kingdom would be destroyed by a woman.[72] 'Some worthy English merchants' consulted Chapuys about the possibility of emigrating to Flanders or Spain.[73] 'The Queen's agents in Parliament', Chapuys reported, came to him in December to ask whether any sentence had arrived from Rome 'on which they could ground their opposition'.[74] Well might the opponents of a new order be fearful. In the Council a daring scheme was planned. Cromwell wrote to Wolsey on 21 October to tell him that the earlier decision to indict a few leading clergy, supporters of the Queen, with praemunire was

---

[68] *Narratives of the Reformation*, pp. 26–8, 296 (Petyt's will).

[69] CLRO, Repertory 8, fo. 242ʳ; *House of Commons, 1509–1558* (Petyt).

[70] *CSP Sp.* iv/1. 549. Barnes thought that More sought his life, despite the safe conduct: More, *CW* 8/iii, p. 1394.

[71] Foxe, *Acts and Monuments*, iv, pp. 657–8.

[72] *CSP Sp.* iv/1. 539; R. Demaus, *William Tindale* (1886), ch. 10. E. W. Ives, *Anne Boleyn* (1986), pp. 173–4.

[73] *CSP Sp.* iv/1. 547.

[74] Ibid.; *House of Commons, 1509–1558* (Withypoll).

abandoned: 'the prelates shall not appear in the praemunire: there is another way devised'.[75]

The new plan—no less than to level the threat of praemunire charges against all the clergy in Convocation—was Cromwell's own.[76] Wolsey died before he could witness the clergy's humiliation. He died on 29 November as he journeyed south (fearing that the Londoners would murder him on his arrival).[77] Many rejoiced. The Earl of Wiltshire celebrated by putting on a farce 'of the Cardinal Wolsey going down to Hell' (and had it printed).[78] John Gough wrote at the back of a book he was binding, 'this year died the sumptuous high Cardinal at Leicester: a great persecutor'.[79] But all was still uncertain: Parliament was prorogued, and prorogued again, 'as if they do not know their own mind about the measures to be proposed therein'; or perhaps did not dare to propose them.[80] Convocation met, fearfully, on 12 January 1531. A month later the clergy of the Southern province acquiesced to the momentous claims made by the King. Henry was henceforward, they recognized, 'sole protector and supreme head of the English Church and clergy'; though 'only so far as the law of Christ allows', so perhaps, for many, not at all. For violating the statute of praemunire they pledged to pay a fine of £100,000 (a fine easier to promise than to collect, as Stokesley realized when he was faced by a riot of City priests that September). In return they received the royal pardon.[81] They hoped, too, for his protection against 'many our enemies, especially the Lutherans . . . of late raging against . . . the prelates of the clergy with their famous lies and cursed books . . . everywhere dispersed' (*The Practice of Prelates?*).[82]. But Henry was by now protecting heretics who might write for his cause. Before Con-

[75] R. B. Merriman, *Life and Letters of Thomas Cromwell*, 2 vols. (Oxford, 1902), i, p. 334.

[76] J. J. Scarisbrick, 'The Pardon of the Clergy, 1531', *CHJ* xii (1956), pp. 22–39; J. A. Guy, 'Henry VIII and the Praemunire manœuvres of 1530–1531', *EHR* xcvii (1982), pp. 481–503.

[77] Cavendish, *Life*, pp. 174 ff.; Hall, *Chronicle*, p. 774; *CSP Sp.* iv/1. 522 (p. 833).

[78] Ibid. iv/2. 615 (pp. 40–1).

[79] Guildhall, 7086/1, fo. 220[r].

[80] *CSP Sp.* iv/1. 555.

[81] Elton, *Reform and Reformation*, pp. 142–4; Hall, *Chronicle*, pp. 783–4; PRO, SP 1/67, fos. 8[r]–12[r] (*L&P* v. 387); Sta Cha 2/ii/172.

[82] Lehmberg, *Reformation Parliament*, p. 111.

vocation in February came the early Reformation's legendary
evangelists—Hugh Latimer, Edward Crome, rector of St
Antholin , John Lambert, and Thomas Bilney. They had all
preached erroneously against Purgatory, the veneration of
saints, and the signs and ceremonies of the Church. But
Convocation took no action against them.[83] The nature of
persecution was changing.

The history of persecution is more straightforward than the
history of heresy, for it depends only upon the will of the
authorities to define and investigate error. The error which
the government determined to extirpate changed with political
circumstance. The King's quarrel with the Pope prepared for a
new orthodoxy regarding what constituted authority within the
Church. Before Convocation in January 1531 Dr Crome
appealed to be tried, not before prelates, but by the King, 'the
Archbishop's sovereign'. Acting as Supreme Head, with cure
of souls, Henry in his favourite role as Justinian tried Crome.
He ordered that Crome make public profession of his orthodoxy on such matters as intercession and Purgatory, but as for
his denial of the papal primacy: that was no longer heresy.[84]

THE HERETICS

The nature of heresy was changing also, as a distinctive English
theology developed. William Tyndale, with his vernacular
Scripture, had provided the means of conversion; his fellow in
exile, John Frith, developed a sacramental theology which was
of profound significance for English Protestantism. Frith's
personal example had a lasting influence upon the brethren.[85]
In exile from late in 1528 until his return at the end of July

---

[83] Ibid., pp. 117–18; BL, Harleian MS, 425, fos. 13ʳ–14ʳ; PRO, SP 1/65,
fos. 160ʳ–161ʳ (*L&P* v. 860, 129); Foxe, *Acts and Monuments*, v, pp. 181 ff.; Guy, *Public
Career of Sir Thomas More*, pp. 167–71.

[84] *L&P* v. 148.

[85] For Frith's life and theology, see R. E. Fulop, 'John Frith and his relation to the
Origin of the Reformation in England' (University of Edinburgh, Ph.D. thesis, 1956);
Clebsch, *England's Earliest Protestants*, pp. 78–136; D. B. Knox, *The Doctrine of Faith
in the Reign of Henry VIII* (1961), pp. 43–55. Brinklow urged 'all those which favour
the free passage of the Gospel' to read Frith's works on the sacrament, for he had
written 'invincibly in this matter'; 'Lamentacyon', p. 103. Cranmer believed that
Frith followed 'after the opinion of Oecolampadius'; BL, Harleian MS 6148, fo. 25ʳ
(*L&P* vi. 661).

1532—save for a brief visit to stir his followers in Lent 1531—
and then in the Tower until his martyrdom in July 1533, Frith
was only briefly with the evangelicals.[86] Nevertheless he provided inspiration for the persecuted and writings for their cause.
Simon Fish's *Supplicacyon for the Beggers* had sparked off a major
controversy over the doctrine of Purgatory, and maybe had
won support, but by early 1531 Fish was carried off by the
plague.[87] Frith entered the controversy; dangerous and worthy
opponent to More. In his *A disputacion of purgatorye* of 1531 he
found Purgatory nowhere in Scripture, and denied the doctrine
because it derogated from divine grace, causing redemption to
hang upon man's repentance.[88] It was during his imprisonment
in the Tower (he was arrested at some time between 25 July
and 21 October 1532) that 'young father Frith' wrote his
greatest works, turning at the last to sacramental theology, and
above all to refute the doctrine of transubstantiation, in *A boke
made . . . answeringe vnto M mores lettur.* According to Foxe, the
'first occasion of his writing' was the request of a 'certain old
familiar friend of his, touching the sacrament of the body and
blood of Christ'.[89] For the brethren Frith wrote *A letter, wryten
vnto the faythful folowers of Christes gospel* and *A Mirroure to know
thyselfe* concerning the nature of the true congregation of
Christ.[90] For Frith adversity itself was a sacrament. Even from
the Tower he strengthened and was strengthened by contacts
with the brethren; letters came from William Tyndale, con-
firming him in his purpose, and he made clandestine visits
from the Tower to 'consult with godly men' in the City.[91] One
of these was John Petyt, and it was Thomas Phillips, 'a known
man', long imprisoned at Stokesley's pleasure, who let him
out.[92] Frith wrote that a willing death is a sign of faith and a
triumph over sin: 'they that die with such a courage and such a

[86] Foxe, *Acts and Monuments*, v, pp. 5–6, 801–2; Demaus, *William Tindale*,
pp. 341–2, 409–10.

[87] Foxe, *Acts and Monuments*, iv, pp. 657 ff.

[88] More, *Answer to a Poisoned Book*; *RSTC* 11388; Clebsch, *England's Earliest
Protestants*, pp. 88–94.

[89] *RSTC* 11381; Foxe, *Acts and Monuments*, v, p. 6; Clebsch, *England's Earliest
Protestants*, pp. 117–27; More, *Apology*, pp. 122–4.

[90] *RSTC* 11385.5; 11390.

[91] *L&P* vi. 403, 458; Demaus, *William Tindale*, pp. 411–19; More, *Apology*,
pp. 89–91, 121, 125.

[92] *Narratives of the Reformation*, pp. 25–8.

trust in God, it is a certain sign that they shall be saved'.[93] He went to the stake himself on 4 July 1533.[94]

Men went to the fires of Smithfield in these years—Stile, Richard Bayfield, John Tewkesbury, James Bainham, John Frith, and Andrew Huet[95]—not only nor even because More and Stokesley were persecutors, but because they were prepared to die rather than to abjure. Richard Hilles, an apprentice merchant taylor of London Bridge, was urged persistently by his master to abandon his heresy, 'sometimes calling me opinionative, and sometimes saying we cannot see but that any of ye all will revoke rather than to die'. Hilles dared not say absolutely that 'all articles that are forsaken for fear of death be false, for then I should condemn Saint Peter', but for his part he hoped that God would 'never suffer me to dishonour His blessed truth'.[96] There was a glory in martyrdom. Said Barnes: 'We make nowadays many martyrs. I trust we shall have more shortly. For the verity could never be preached plainly, but persecution did follow.'[97]

Persecution drove the brethren underground. Secrecy was paramount: their operations must be clandestine if they were to be preserved and their cause to succeed. Yet the brethren were, first and foremost, evangelicals. Arthur had prophesied in 1527 that if he should die for preaching the Gospel seven thousand would come in his place.[98] There were hardly seven thousand in 1529, nor even seven hundred. Yet the brethren might have an impact quite out of proportion to their tiny numbers. The Chancellor knew this well: 'as few as they be a man shall always find some',[99] for they were as indefatigable as most of the old faith were complacent.

How many brethren were there in London? More than were ever found. The vigilance of the Chancellor and the Bishop,

[93] *The foundacyon and the summe of the holy scripture* (1535?; *RSTC* 3034); cited in J. Rhodes, 'Private Devotion on the Eve of the Reformation' (University of Durham, Ph.D. thesis, 1974), p. 267.

[94] Frith stood trial at St Paul's on 20 June 1533: Guildhall, MS 9531/11, fo. 126[r-v]; Foxe, *Acts and Monuments*, v, pp. 11 ff.; appendix xxii; PRO, C/85/126/39 (significavit for 3 July 1533).

[95] Foxe, *Acts and Monuments*, iv, pp. 688, 694, 705; v, pp. 15, 18, 655.

[96] PRO, SP 1/74, fos. 107[v]–108[r] (*L&P* vi. 99).

[97] Cited in J. P. Lusardi, 'The Career of Robert Barnes', *CW* 8/iii, p. 1380.

[98] See above, p. 83.

[99] More, *Apology*, p. 160.

and the awesome penalties which awaited those who were discovered, urged circumspection. The good Catholic citizenry, too, were alert to discover the errors of their neighbours—at first. So, many of those who had rejected the teachings and ceremonies of the Church still attended its services and partook of its sacraments, however casuistically. William Lancaster, a tailor, confessed that he would have refused to attend Mass on the feast of the Assumption in 1532 'if it were not for the speech of the people'.[100] When the priests of London poured soap ashes on the grave of a heretic many of the 'Balaamites' gathered to watch approvingly.[101] But More thought that 'good Catholic men' became complacent and tolerant, failing to cast out the apostate as once they had done.[102]

Though it was in London that the most furious quest for heretics was conducted, and though in the capital there were innumerable inquests and methods of detection, still it was the best place to hide. The size and transience of the population in the metropolis allowed its visitors a certain anonymity. The brethren offered shelter and provided safe houses for the like-minded who came for refuge from the persecution. Stacy and Maxwell, 'known men', hid Richard Bayfield, who came down from Essex, and helped him to exile.[103] Thomas Patmore came to dine with his curate at the Bell in New Bridge Street, which had been a Lollard haunt. Andrew Huet was hidden, as others had been, at John Chapman's house in Hosier Lane by Smithfield.[104] An evangelical painter, Edward Freez, fled Yorkshire for Essex and London, even though this was Stokesley's diocese, and dangerous. There, so 'evil willers' reported to the Bishop, he 'kept school in the night in [his] house or in other men's houses and kept in corners and secret places'.[105]

The brethren in adversity protected each other: pledging money, helping each other away to exile, keeping silence, even lying for each other. More excoriated the 'vow breaking brethren' who constantly eluded him.[106] John Purser and

100 Foxe, *Acts and Monuments*, v, p. 39.
101 *Narratives of the Reformation*, p. 28; cf. *L&P* xviii/2. 546.
102 More, *Apology*, p. 158; *Answer to a Poisoned Book*, pp. 4–5.
103 Foxe, *Acts and Monuments*, iv, p. 681.
104 Ibid. v, pp. 36, 16.
105 PRO, SP 1/73, fos. 175^r–176^r (*L&P* v, appendix 34).
106 More, *Apology*, p. 29.

others had stood surety for John Birt, the book binder, yet far from delivering him up, they 'force not to forfeit their bond for brotherhood, but let him slip aside' to exile in the Low Countries.[107] Birt too had urged George Constantine 'to call back his confession again if he might', and 'even as a man armed with faith, go forth in your matter boldly and put them [the Bishops] to their proofs'.[108] Perjury was, for them, justified in a greater cause. George Gower pleaded to Cromwell in 1532 to 'listen to the cries of an oppressed heart', for he was long in durance for his perjury. Sworn against his will by the Bishop and his Chancellor to answer to the activities of John Purser and a man too subversive for him even to name, Gower had given conflicting testimony. More accused him of being 'privy' to their 'evil acts'. Gower denied that he had seen the unnamed heretic until 'I found him in the stocks' (as a penitent for smuggling Tyndale's works in the winter of 1530?); indeed, he was 'always weary of his company and shook him off'. Insisting also that Purser was of 'none acquaintance' with the man, Gower then admitted that he had seen him in Purser's house. But Purser's house was a 'common tavern' which anyone might frequent.[109] More did not believe him, and rightly, for Purser's house was a haunt for heretics, but he could not prove anything against him.

Even those whose heresy had been proven and judged, who had recanted and abjured, on pain of the ultimate penalty if ever they should relapse, dared to continue in their apostasy; some secretly, but others proudly. John Tyndale had been excommunicated in May 1529, and abjured in 1530.[110] With the other book agents he had feigned penitence in November 1530, but nothing would stop his evangelical activities. Throughout the 1530s and 1540s he remained associated with the founder member of the Christian Brethren, John Sheriff, and with John Gough and with Latimer.[111] Men such as Tyndale would deny their faith only to proselytize again: *reculer pour mieux sauter*. Nothing angered the persecutors more than such casuistry. The

---

[107] Ibid., p. 90.
[108] More, *Confutation*, pp. 19–20.
[109] PRO, SP 1/70, fo. 184^r (*L&P* v. 1176).
[110] PRO, C 85/188/28; Foxe, *Acts and Monuments*, v, p. 29.
[111] PRO, E 101/348/38; SP 1/162, fos. 1^r–31^r (*L&P* xiv/2. 255; xv. 936).

Chancellor accused the brethren of bad faith in its double sense. Knowing well that they 'gaily glorifieth' in the examples of godly fortitude under persecution he sought always to deny their courage and undermine their resolve.[112] John Tewkesbury, a leatherseller of St Michael le Querne, was found to have Luther's *Liberty of a Christian Man* (which Tewkesbury had copied in his own hand) and Tyndale's *Wicked Mammon* and *Obedience of a Christian Man*. Once a member of Hacker's sect, a reader of the Lollard Bible and Wycliffe's Wicket, Tewkesbury was converted by these new works to reformed doctrines. He had abjured—at least once; in 1529, and maybe even before— so when discovered by More in 1531 there was no way for him but one. Yet, according to More, if it might have saved his life Tewkesbury would have abjured 'all his heresies again', 'with all his heart', and he urged his fellow prisoner, James Bainham, to recant while he still might. Foxe's story was, of course, different.[113]

More did not doubt that Bainham would falter once he was told of Tewkesbury's recantation. To persuade Bainham to return to the fold the Chancellor employed a dual torture: physical and mental. Foxe told of More standing by while Bainham was racked, waiting for him to name other young gentlemen of the Inns of Court who had been 'infected' with heresy.[114] More admitted to shaking Bainham's resolve by telling him of Tewkesbury's recantation.[115] James Bainham did 'deny God' once, and in 1531 left prison. But his conscience —like Bilney's and Tewkesbury's—could not stand his feigned recantation. He relapsed, publicly and defiantly, asking 'God and all the world forgiveness, before the congregration in those days, in a warehouse in Bow Lane'. He was condemned in April 1532.[116] Latimer feared, at first, that Bainham's convictions were not so extreme as to justify such a sacrifice. Visiting Bainham in prison, he told him that if it was only for denunciation of St Thomas Becket as traitor 'it were madness for a

---

[112] More, *Confutation*, pp. 13–14.
[113] Foxe, *Acts and Monuments*, iv, pp. 668–94; More, *Confutation*, p. 21.
[114] Foxe, *Acts and Monuments*, iv, p. 698.
[115] More, *Confutation*, p. 21. More also claimed that Fish recanted: *Apology*, pp. 75–6.
[116] Foxe, *Acts and Monuments*, iv, pp. 700–4.

man to risk his life', but denial of Purgatory and of satisfactory masses, he agreed, was sufficient cause: 'ye may be sure rather to die in the defence thereof than to recant both against your conscience and the Scriptures'. Latimer warned him to 'beware of vainglory, for the Devil will be ready now to infect you when you shall come into the multitude of the people' to die.[117] Bainham went to the stake on 30 April 1532.

Some of the brethren moved slowly to their final public avowal of a faith which would condemn them. Edward Freez fled Jervaulx Abbey and the life of a monk (to which he had been forced by the Abbot, his 'natural brother').[118] He lived quietly and incognito in Colchester from 1526, save for sorties to the capital, working as a painter, now married and with children. After about five years he was hired to paint cloths to decorate the new inn in Colchester market-place. But Freez was a gospeller and on those cloths he painted Scriptural texts.[119] His letter from the Bishop's prison marks him an evangelical through and through. 'I am Christ's man which gave Himself for our sins to deliver us from this present evil world.' 'Let the earth, air, fire and water take their parts, so is my body consumed which was made for the spirit, and the spirit to seek the praise of God.' 'Christ for his people hath offered once for all not the blood of a sinner, not the blood of goats, calves or bulls, but His own precious blood, whereby we may arise out of our own sins not seven times but seventy times seven times.'[120] This was the letter of a religious fanatic, and a man driven mad in the Bishop's prison.

A wilder fringe of the brethren advertised their heretical views in ways much more likely to convince the orthodox of the outrages of heresy than to convert them. Robert Hudson of St Sepulchre offered the image of St Nicholas the Boy Bishop a dog 'for devotion' on the Bishop's feast day of the Holy

---

117 BL, Harleian MS 422, fo. 90[r] (*L&P* v, appendix 30).

118 *L&P* v. 1203, appendix 34.

119 Foxe, *Acts and Monuments*, iv, pp. 694–5; A. G. Dickens, *Lollards and Protestants in the Diocese of York, 1509–1558* (Oxford, 1959), pp. 30–3. Another brother(?), Angel Freez, was apprenticed to one of the leading London brethren: William Callaway: Goldsmiths' Company, records of members.

120 PRO, SP 1/73, fos. 175[r]–176[r] (*L&P* v, appendix 34). The letter is anonymous, but the life story recounted is surely Freez's.

Innocents in 1531. He thought that the child might like a dog; anyway it was the tenth dog (a tithe).[121] Jasper Wetzell of Cologne, his arms outstretched, blasphemed the crucified Christ on the famous rood at St Margaret Pattens, saying 'he could make as good a knave as he is, for he is made but of wood'.[122] Heretical tracts were broadcast. A broadsheet fixed on the doors of St Paul's was sent to a monk of Evesham in August 1531, who was 'amazed' at the writer's open attacks upon saints, fasting, and pilgrimage to God. 'He does not wish Mary to pray for him, but the day will come when he will be glad of it.'[123]

Despite the dangers in these 'perilous days' the brethren grew ever more daring. The most effective way of evangelizing was also the most dangerous: preaching. A manuscript copy of a sermon which had been preached twice publicly came into the Chancellor's hands. It had a distinctively reforming text: 'He made us by the truth of His word'.[124] Robert Bayfield promised 'pharisee' City priests in 1527: 'I shall make you, and many other more, good and perfect Christian men, ere I depart this City; for I purpose to read a common lecture every day at St Foster's church.' His boldness in preaching against superstitition and idolatry caused him to be arrested. He abjured, but fled into exile to become a book agent.[125] More wrote of Bayfield's duplicity; of his 'suing for remission and pardon of his offence for bringing in those books, and therewith also in selling them here still secretly, and sending over for more'.[126] He went to the stake on 27 November 1531, but not before bringing in hundreds of forbidden books.

Latimer was preaching in London. in these years. Bainham thought 'no man to have preached the Word of God sincerely and purely and after the vein of Scripture', save Crome and Latimer.[127] At St Mary Abchurch in 1531 Latimer preached at

---

121 Foxe, *Acts and Monuments*, v, p. 38.

122 Ibid. v, p. 32.

123 *Letter Book of Robert Joseph*, ed. H. Aveling and W. H. Pantin (Oxford Historical Soc., new series, xix, 1967), p. 155; cited in J. K. McConica, *English Humanism and Reformation Politics* (Oxford, 1963), p. 98.

124 Clebsch, *England's Earliest Protestants*, p. 284.

125 Foxe, *Acts and Monuments*, iv, pp. 681–8; v, pp. 43–4.

126 More, *Confutation*, p. 17.

127 Foxe, *Acts and Monuments*, iv, p. 699.

the request of certain merchants (whose names he conveniently forgot). He defended Bilney, who was then before the Bishops, and he preached with a certain irony a new orthodoxy. 'I once thought the Pope Christ's vicar and lord of all the world . . . Now I might be hired to think otherwise . . . I once thought the Pope's dispensation of pluralities discharged consciences, and that he could spoil Purgatory at his pleasure by a word of his mouth. Now I might be entreated to think otherwise.'[128] But he may have preached more radical views yet—or had his preaching misinterpreted. Asked how 'he fell into' his sacramentarian heresy, John Tyrel, an Irish tailor of Billericay, answered that he had heard Latimer's sermon at Abchurch in midsummer 1531; 'that men should leave going on pilgrimage abroad, and do their pilgrimage to their poor neighbours'; moreover, 'he did set at little the sacrament of the altar'.[129]

As the quest of the Bishop and Chancellor proceeded their prisons gradually filled with recalcitrant brethren. Those who were suspected of heresy, but who would neither confess nor abjure, remained at the Bishop's pleasure: 'sheep ready in the butcher's leisure to slaughter'.[130] So it was for Thomas Phillips. In his own house at Chelsea, More (so he said) 'honestly entreated' Phillips 'one day or twain . . . and laboured about his amendment in as hearty loving manner as I could'.[131] Phillips, a citizen and pointmaker of St Michael le Querne, had been 'chief reader' in the Lollard conventicles, and for his 'forbidden books' came before the Chancellor at Christmas 1529. Failing to move Phillips, More sent him before his Ordinary, and thence—unusually—to the Tower (More thought the Tower desirable, he explained, lest Phillips—like his cousin Holy John, like Hunne, and other desperate heretics—should kill himself).[132] Twelve articles were objected against Phillips; that he denied the existence of Purgatory, the efficacy of pilgrimage, fasting, and prayers to saints, and worst of all, that he thought the blessed sacrament but 'a remembrance of Christ's passion

128 Latimer, Remains, p. 322; L&P v. 607.
129 Foxe, Acts and Monuments, v, p. 39.
130 PRO, SP 2/P, fo. 11ʳ (L&P vii. 155); Strype, Ecclesiastical Memorials, i/2, p. 179.
131 More, Apology, pp. 126–7, 372–3; Elton, 'Sir Thomas More and Opposition', Studies, i, p. 161.
132 BL, Harleian MS 421, fo. 13ʳ (L&P iv/2. 4029); More, Apology, p. 126.

and a signification and token of better things to come'. All these charges he denied, admitting only possession of the Vulgate.[133] The *suspicio* charge could not be proved against him.[134] John Stacy, another 'known man', testified against him, then retracted. Refusing to abjure, Phillips long remained in the Tower, turning it into an evangelical conventicle. There he continued to associate with the brethren, he refused the Lenten fast, and read forbidden books sent in by his friends.[135]

Prison might easily prove another path to martyrdom, so gruesome were the conditions. Freez pleaded that God would open Stokesley's 'spiritual eyes and give grace that you may have pity on the poor men which be in your prisons which be so sore abused with irons that they shall never be able to get their living after'.[136] Thomas Somer, John Petyt, a boy of Colchester, and Christopher, a Dutchman, all died after incarceration.[137] When warning came from Cromwell to Stephen Vaughan in Antwerp at the turn of 1530 that George Constantine had delated him to More along with his fellow book agents, Necton and Bayfield, and had even named the smuggling sailors and told him of the special marks on the fardels of books, Vaughan quite understood Constantine's plight. In 'imminent peril and danger' as prisoner in More's house, bound in irons 'like a beast', remembering his wife 'desperate bewashed with continual tears', such torments would make a 'son forget the father that gat him', so why should Constantine protect Vaughan? Constantine was helped to escape from the Chancellor's clutches and to exile in Antwerp in December 1531.[138] The brethren boasted that a band of two or three hundred would spring an imprisoned brother from the Commissary's house, and evangelical vigilantes did attempt (and fail) to rescue 'a well known open heretic out of the Ordinary's hands'.[139] Only when the government's attitude to heresy changed could the

---

[133] PRO, SP 2/P, fo. 11ʳ (*L&P* vii. 155).

[134] Foxe, *Acts and Monuments*, v, p. 29.

[135] Ibid. v, pp. 29–30; *Narratives of the Reformation*, p. 27.

[136] PRO, SP 1/73, fo. 176ʳ (*L&P* v, appendix 34); see also SP 1/70, fo. 3ʳ (*L&P* v. 982).

[137] *Narratives of the Reformation*, p. 27; PRO, PCC, Prob. 11/24, fo. 113ᵛ; Foxe, *Acts and Monuments*, v, pp. 37, 38, 453.

[138] More, *Confutation*, p. 20; Foxe, *Acts and Monuments*, iv, p. 671; BL, Cotton MS Galba B x. fos. 21ʳ–22ʳ (*L&P* v. 574).

[139] More, *Apology*, p. 157.

prisoners hope for release. Upon More's resignation as Chancellor rumours reached Erasmus that Lutherans were being set free; maybe twenty; maybe forty.[140] Those in Stokesley's prisons remained, including Thomas Phillips and Thomas Patmore, whose plight was not forgotten.

How many were haled before the authorities during this quest for heresy? We cannot be certain. The *acta* and processes against heresy in Stokesley's time are now lost. Foxe had seen the registers; he described the processes against the martyrs and listed the names and errors of 'certain persons' abjured in London diocese under Tunstall and Stokesley. Of this 180 only about a quarter (46) were certainly troubled in the City of London during More and Stokesley's partnership.[141] There were certainly many more; their names to be found in the State Papers, in the chronicles, and elsewhere in Foxe's *Acts and Monuments*: John Chapman, George Gower, Joan Bainham (previously Joan Fish), John Field, John Stanton, Stephen Vaughan, Valentine and Edward Freez, Purser, Parnell's servant, Hilles, and Thomas Alwaye.[142] At the turn of 1531 Londoners witnessed one public punishment for heresy after another. To St Paul's Cross came the abjured to bear faggots and to make public recantation of their error: on 22 October Thomas Patmore, and a glazier (Robert Goldstone?); on 5 November John Parnell's servant and one other; on 27 January 1532 a 'Dutch man', and another penitent on 11 February.[143] The chroniclers record that some were condemned to perpetual imprisonment: Harry Thomson, a tailor, and Patmore on 11 November.[144] For the most resolute there was the 'narrow pathway' of martyrdom: Richard Bayfield was burnt on 4 December 1531, John Tewkesbury on 20 December, and James Bainham on 30 April 1532.[145] Maybe the citizenry looked on with approval as heretics were led to the stake; certainly some stoked the fires; yet even those who thought that heresy brought terrene disorder were genuinely perturbed

---

[140] Elton, 'Thomas More and Opposition', *Studies*, i, p. 159.

[141] Foxe, *Acts and Monuments*, iv, p. 586; v, pp 26–45.

[142] Ibid. v, pp. 16–17; iv, p. 698; M. Dowling, 'Anne Boleyn and Reform', *JEH* xxxv (1984), p. 30; see above, pp. 163, 181–2, 190–1, 193 and below, pp. 206–7.

[143] *Two London Chronicles*, pp. 5–6.

[144] Ibid., p. 5.

[145] Foxe, *Acts and Monuments*, iv, pp. 688, 694, 705; *Two London Chronicles*, pp. 5–6.

about the methods the inquisitors used to discover and judge the error.

THE COMMONS' SUPPLICATION AGAINST
THE ORDINARIES

When Parliament met again on 15 January 1532 the Lower House 'sore complained of the cruelty of the Ordinaries'. Whether this 'infinite clamour of the temporalty . . . against the misusing of the spiritual jurisdiction'[146] arose as a spontaneous protest and demand for redress of grievance, or whether it was orchestrated, the agitation of 1529 revived, as part of a master plan to deprive the clergy of legislative powers, it is clear that the Commons' fear of the unchecked judicial powers which the clergy exercised over them in the Church Courts, especially *ex officio* and in the case of heresy, was deep and genuine.

For the Ordinaries would send for men and lay accusations to them of heresy, and say they were accused, and lay articles to them, but no accuser should be brought forth, which to the commons was very dreadful and grievous; for the party so cited must either abjure or be burned, for purgation he might make none.[147]

On 18 March 1532 the Commons submitted to the King a Supplication against the Ordinaries.[148] 'Much discord, variance and debate' had arisen, and more was likely, not only because of 'new fantastical and erroneous opinions', spread by 'frantic, seditious and overthwartly framed books' of heresy, but also 'by the extreme and uncharitable behaviour and dealing of divers Ordinaries'.[149] Most of the Supplication's nine charges were extremely specific, concerning the real powers of the Church Courts, and the abuses within the system. The Com-

---

[146] Hall, *Chronicle*, p. 784; Lehmberg, *Reformation Parliament*, pp. 138 ff.

[147] Hall, *Chronicle*, p. 784.

[148] The 'Commons' Supplication against the Ordinaries' is printed in Gee and Hardy, *Documents Illustrative*, pp. 145–53, and discussed in G. R. Elton, 'The Commons' Supplication of 1532: Parliamentary Manœuvres in the reign of Henry VIII', *Studies*, i, pp. 107–36; J. P. Cooper, 'The Supplication of the Ordinaries Reconsidered', *EHR* lxxii (1957), pp. 616–41; M. J. Kelly, 'The Submission of the Clergy', *TRHS* 5th series, xv (1965), pp. 97–119; Guy, *Public Career of Sir Thomas More*, pp. 183–8.

[149] Gee and Hardy, *Documents Illustrative*, p. 145.

mons' hostility to the courts might seem well founded, yet more important than whether the courts were inequitable or not was that the laity thought that they were. London, where the clergy could most effectively ally, where lay opposition to the clerical estate was most sharply voiced, and where the Church Courts were most numerous, might provide the best ground to test how far the complaints against the clergy were justified. Unfortunately, the City records for the decade before the Reformation are elusive and fragmentary.[150] The ecclesiastical authorities seem to have been consistently active but whether there was an increase in judicial activity in these years when complaints were mounting is impossible to know. If the laity were choosing to bring cases for the Church to judge (instance) they would seem to have had less claim to attack the system, but Dr Wunderli suggested that the citizenry were looking for justice instead from the secular courts.[151] Yet in the absence of any records of the mayoral and shrieval courts for this period his conclusion must remain unproven. Mrs Bowker has suggested rather, from an incisive analysis of the *acta* of Lincoln diocese, that it was precisely justified fear of the rigorous efficiency and relentless initiative of the courts which provided the impetus behind the Commons' Supplication.[152] Would London have been different?

There was a vital distinction between litigation brought by laymen against laymen—as in marriage and defamation cases—and that brought by the spiritualty against lay offenders—as in cases of failure to pay ecclesiastical dues.[153] The chance survival of one kind of court record rather than another in the

[150] Deposition books exist for cases brought before the Consistory Court between 1520 and 1524 and between 1529 and 1533 (GLRO, DL/C/207–8). Miscellaneous *acta* for the Court of Vicar-General Foxford are recorded (GLRO, DL/C/330), and this book has been edited: C. A. MacLaren, 'An Edition of "Foxford", a Vicar-General's book of the Diocese of London, 1521–1539' (University of London, M.Phil. thesis, 1975). Foxe saw, and transcribed in part, the heresy processes recorded in a register of Bishop Tunstall which has since been lost (BL, Harleian MS 421, fos. 11 ff.). No Commissary Court Act Books for this period survive, but Archdeacon Hale transcribed cases from books which have since disappeared: Hale, *Precedents*, cxci–cccxlii (1496–1529).

[151] Wunderli, *London Church Courts*, esp. pp. 4, 136–9.

[152] M. Bowker, 'Some Archdeacons' Court Books and the Commons' Supplication against the Ordinaries of 1532', in *The Study of Medieval Records: Essays in Honour of Kathleen Major*, ed. D. A. Bullough and R. L. Storey (Oxford, 1971), pp. 282–316.

[153] Ibid., pp. 289–93.

1530s evidently colours the impression of the litigation which occurred. In London the cases which came before the Vicar-General's court differed from the routine instance litigation which came before the more lowly courts of the City, but it is clear that the great majority of cases before Foxford's court were for ecclesiastical offences, brought by the initiative of the court officials.[154] Londoners did ostensibly have cause for grievance.

The Commons reserved their attacks upon the activity of the Church Courts to *ex officio* cases and testamentary proceedings; with reason, for the point about these particular processes was that the laity had no choice but to regard and abide by the decision of the spiritual judges.[155] (If Londoners had indeed abandoned using the Church Courts for their own purposes they had special reason to resent being cited by the clergy and bound by their judgement.) Probate could be granted only by the Church, and the complaint in 1532 was that the process was long and expensive, and rather worse than better since the reforming legislation of 1529.[156] Yet the Commissary Court of London seems to have acted with speed and efficiency. Between 1522 and 1526 30 per cent of a sample of one hundred wills were proved within a fortnight, 35 per cent within a month, 14 per cent within three months, and only 21 per cent took longer.[157] (This also reveals much about how late most testators left it before making their last wills.) It is equally doubtful whether the Commons' complaint in 1532 about executors being cited outside the shire was true for the City. Two cases were brought before the Barons of the Exchequer of defendants being cited outside London diocese in the 1530s; both cases were dismissed.[158] The old resentment against the rival claims of the Prerogative Court of Canterbury for London probate,

[154] The type of cases recorded in 'Foxford' resemble more closely the cases brought before the Bishop's Court of Audience in Lincoln: *An Episcopal Court Book for the Diocese of Lincoln, 1514–20*, ed. M. Bowker (Lincoln Record Society, 61, 1963); 'Some Archdeacons' Court Books', p. 292 n. 1.

[155] For *ex officio* procedure, see B. L. Woodcock, *Medieval Ecclesiastical Courts in the Diocese of Canterbury* (Oxford, 1952); J. A. Guy, 'The Legal Context of the Controversy: The Law of Heresy', *CW* 10, pp. xlvii–lxvii.

[156] Gee and Hardy, *Documents Illustrative*, p. 149 (vi).

[157] Guildhall, MS 9171/10, fos. 1ʳ–75ᵛ. Mrs Bowker found this charge without justification for Lincoln diocese: 'Some Archdeacons' Court Books', pp. 294–6.

[158] PRO, E 159/315, m. 11; 318, m. 12d.

and more recent claims that Wolsey had been arrogating probate cases to himself were clerical, not lay, grudges.[159] Yet
such old antagonisms distorted the perception of how the process actually worked.

For Londoners the fifth charge in the Commons' Supplication
had a special force: that the clergy took money for the 'sacraments and sacramentals of Holy Church', which they were
bound to minister freely 'in charitable and godly wise'. Worse,
that they would even withdraw the sacraments from those who
refused to pay.[160] To exclude the communicant from divine
grace was the most desperate of remedies against the reluctant
tithe-payer, and forbidden, but some London priests were
known to have recourse to it.[161] The citizens' disputes with
their priests over tithe and other dues went back a long way,
but now the King's struggles with the clergy gave their quarrels
a new urgency. The clergy were on the defensive; and the laity
encouraged by the moves which the King was making against
the whole Church to challenge their own priests. The years of
persecution also saw particularly acrimonious tithe disputes in
London.[162] At St Andrew Hubbard the parish united against
its rector in 1531–2 when he cited seven parishioners before the
Bishop's Chancellor for not paying tithes. So incensed were the
parishioners by what they saw as the malicious behaviour of
their clergy that communally they brought a case against the
curate in the Sheriff's court, and won.[163] (The rector meanwhile took the extraordinary step of procuring a writ in
Chancery against Chancellor Foxford, who had tried the
case.)[164] In 1532 the parishioners of All Hallows Honey Lane
petitioned Cromwell for remedy against the rapacity of their
own rector and those of St Benet Gracechurch, St Leonard
Eastcheap, and St Magnus, and finally sent a bill before Star
Chamber. These rectors, showing particular defiance in the
wake of the Pardon of the Clergy, had sued their parishioners

[159] *L&P* iv/1. 1118; Pollard, *Wolsey*, p. 194.
[160] Gee and Hardy, *Documents Illustrative*, pp. 148–9 (v).
[161] Lyndwood, v. ii. 1; C. Hill, *Economic Problems of the Church, From Archbishop Whitgift to the Long Parliament* (Oxford, 1956), p. 90.
[162] S. E. Brigden, 'Tithe Controversy in Reformation London', *JEH* xxxii (1981).
[163] PRO, Court of Requests 2/7/123 b and c; Guildhall, MS 1279/2, fo. 15ᵛ.
[164] PRO, C 1/788/28.

TABLE 1: *Correction Processes in the London Ecclesiastical Courts*

i. *Commissary Court Processes*[a]

	1472	%	1484	%	1493	%	1502	%	1512	%	1513	%	1514	%
Total cases before the Commissary Court	1305		764		1061		547		294		303		328	
Defamation	405	31	171	22	211	20	160	29	129	44	110	36	149	45
Adultery	282	22	236	31	348	33	160	29	44	15	70	23	70	21
Fornication	171	13	89	12	160	15	79	14	58	19	48	16	35	11
Bawdery	248	19	140	18	112	11	65	12	8	3	20	7	24	7
Prostitution	65	5	42	5	61	6	28	5	1	0.3	5	2	14	4
Breach of faith or petty debt	39	3			41	4	5	1	1	0.3				
Tithe					12	1	14	3			14	5	14	4
Probate	9	0.1	12	0.1	26	2	13	3	29	10	13	4	5	
Marriage suits					15	1	14	3			25	8		
Non-observance of Sundays and feast days					18	2	8	1	2	0.7	1		1	

Note:

[a] This information is derived from R. M. Wunderli, *London Church Courts and Society on the Eve of the Reformation* (Cambridge, Mass., 1981), pp. 108, 113, 117, 120, 123, 142–7.

TABLE 1: *Correction Processes in the London Ecclesiastical Courts* (*cont.*)

ii. *Correction Processes before the Vicar-General's Court,*[b]
    1521–Mar. 1532

Disputed testamentary cases	9
Unspecified instance cases	8
Cases arising from failure to pay ecclesiastical dues	2
Morality cases	32
Defamation cases	6
Marriage cases	11
Breaches of ecclesiastical discipline	44
Disorder and contempt of the clergy and court officials	15
Heterodoxy	6

Note:

  [b] GLRO, DL/C/330, fos. 1–214.

before the Court of Arches, demanding payment of tithe according to the old rate, stated in the bull of Nicholas V. The citizens, having learnt a lesson from the royal tactics against the clergy in 1531, borrowed the same legal stratagem, saying that the bull was procured without royal licence, and could not then be obeyed without incurring the charge of praemunire. All this John Colyns noted in his commonplace book. Significantly —and not surprisingly—two of these rectors, John Coke and Edmund Baschurch, were among the most resolute and desperate opponents of the royal supremacy.[165]

The Church had long feared that prevalent criticism of the clergy, and the reception of heretical ideas, would lead to the withdrawal of tithe. Certainly, reformers vehemently denied that the priesthood should be a way to wealth. Adrian Dolevyn had averred in 1523 that 'the shepherds who preach the Word

---

[165] PRO, SP 2/M, fos. 200ʳ–201ʳ (*L&P* v. 1788); B. Walton, *A Treatise concernying the payment of Tythes in London*, in S. Brewster, *Collectanea Ecclesiastica* (1752), pp. 103, 108. Walton copied the proceedings of this suit in Star Chamber, which no longer survive in the Public Record Office. Lambeth, Carte Miscellane, viii/2 a–e. In his commonplace book Colyns recorded that Edmund Grey had incurred praemunire; BL, Harleian MS 2252, fos. 32ᵛ–33ʳ. For Coke and Baschurch, see below, pp. 209, 242–3, 278.

of God should have meat and drink but no money'.[166] Bilney preached at St Magnus in 1527: 'Good people, I exhort you in God that if priests be of evil conversation or will not apply their learning that you help them not, but rather let them starve than give them any penny.'[167] Rather than admit their failings, the clergy might associate the withdrawal of tithe with religious deviance. So, venal, violent, concupiscent Robert Shoter, curate of St Botolph Aldersgate, had accused his tithe-dodging parishioners of heresy.[168] Orthodox Londoners came to fear, with some reason, that non-payment of tithe had become a cause of suspicion of heresy, even a reason for attachment. In both tithe and heresy cases the clergy proceeded *ex officio*, as 'judges and parties' in their 'own causes', and against their clerical accusers the laity had neither defence nor recourse.[169] The association between tithe withdrawal and heresy prosecution, and the mutual suspicion between Londoners and their priests which it engendered, was portentous. In 1532 a bill for London tithe was presented to the Commons; again, in 1534 the close conjunction in time and emotive appeal between the drafting of the Heresy bill and the bill for London tithe was marked.[170]

It was the prevalent and genuine dread of the clergy's ultimate judicial power, in imprisoning and sentencing for heresy, which lay behind the Supplication against the Ordinaries.[171] The twelfth clause expressed the fear. Heresy proceedings initiated by the clergy *ex officio* were, the Commons thought, conducted in an arbitrary, sometimes malicious manner, the scales always weighted against the defendant. The laity had no defence; they were neither allowed to confront their accusers, nor able to justify their views against the sophistical learning of the clergy.[172] Compurgation was not allowed. Persecution had been, until recently, urged by most of the laity; by 1532 many,

---

[166] PRO, SP 1/23, fo. 225ʳ (*L&P* iii/2. 1922); see also SP 1/47, fo. 79ʳ (*L&P* iv/2. 4038).

[167] Guildhall, MS 9531/10, fo. 133ᵛ.

[168] GLRO, DL/C/330, fo. 44ᵛ.

[169] BL, Harleian MS 2252, fos. 34ᵛ–35ʳ; PRO, SP 1/106, fo. 22ʳ; SP 6/2, fo. 44ᵛ; SP 2/M, fos. 200ʳ–201ʳ (*L&P* xi. 325; viii. 453; v. 1788).

[170] Brigden, 'Tithe Controversy', pp. 296–8.

[171] Kelly, 'The Submission of the Clergy'; Bowker, 'Some Archdeacons' Court Books'.

[172] Gee and Hardy, *Documents Illustrative*, pp. 151–2.

although orthodox, distrusted clerical motives. Why so abrupt a change in attitude? Time was when the wealthy and influential had been relatively immune from the penalties imposed in the Church Courts. No longer. Hunne's case and the death of John Petyt in prison had shown that now no quarter was given.

The clergy were thought to condemn men simply for 'gain-saying or displeasing them'.[173] They were neither charitable enough to distinguish between truly erroneous opinion and simple lack of knowledge of 'the high mysteries of our faith', nor prepared to teach and preach the truth. The citizenry had accused their clergy of laxity and arrogance and, precisely because they were orthodox, sought a return to a purer clerical estate. Yet there seemed to be no remedy against clerical offenders, for as Simon Fish had complained to the King:

So captive are your laws unto them, that no man that they list to excommunicate, may be admitted to sue any action in any of your courts. If any man in your sessions dare be so hardy to indict a priest of any such crime, he hath, ere the year go out, such a yoke of heresy laid in his neck, that it maketh him wish that he had not done it.[174]

The Ordinaries, making their abortive attempts at reform, blamed the subversive influence of heresy for causing unjustified attacks upon the life and authority of the clergy. Stephen Gardiner's intemperate reaction to the Supplication against the Ordinaries was that it was the work of 'uncharitable . . . evil and seditious persons', pretending to be moved by the 'zeal of justice', but with motives more sinister.[175] In the Answer of the Ordinaries the response was more conciliatory, but the suspicions of what moved the Commons the same.[176]

As if to spur the Commons in their resolve to check the powers of the clergy, the plight of one victim of a rapacious Bishop was rehearsed to the House in April or May. Thomas Patmore, a City draper, had been condemned to perpetual imprisonment on 11 November 1531. Having twice performed

---

[173] PRO, SP 1/106, fo. 22^r; SP 6/2, fo. 44^v (*L&P* xi. 325; viii. 453); *A dyalogue betwene . . . Clemente a clerke . . . and Bernarde a burges*, sigs. Bi^v, Div^v.

[174] Fish, 'Supplicacyon', p. 8. More was at pains to deny these criticisms throughout the *Apology* and the *Debellation*.

[175] Cited in Lehmberg, *Reformation Parliament*, p. 146.

[176] 'The Answer of the Ordinaries' is printed in Gee and Hardy, *Documents Illustrative*, pp. 154–176.

public penance, in November 1530 and October 1531, he was manifestly unrepentant, and now remained 'wrapt in the Bishop's nets'.[177] Patmore had, with John Tyndale and Thomas Somer, broadcast William Tyndale's forbidden works.[178] His other offences were denial of the intercessory powers of saints and inveighing against images. Pointing at a tapestry of a hunting scene, he had said he had 'as lieve pray to yonder hunter . . . for a piece of flesh as to pray to stocks [images] that stand in walls'. Against the arguments from tradition of a priest of St Peter Cornhill, Patmore had defended the validity of clerical marriage.[179] So, too, did Thomas Patmore, vicar of Much Hadham, Hertfordshire, argue for the right of priests to marry; indeed, he arranged for the marriage of his curate to his Lollard maidservant. Foxe wrote that the two Thomas Patmores, both persecuted by Stokesley, were brothers. Thomas Patmore, the priest, had gone to Wittenberg to sit at the feet of Luther, and returned to become a leader in the Christian Brethren. Stokesley accused him of denying the power of Pope or Bishop to curse, of arguing against clerical celibacy and infant baptism. But because 'he was not publice diffamatus apud bonos et graves, according to the law, he was not by the law, bound to answer'. The brethren appealed to the King and to Anne Boleyn for Patmore, the draper, and Patmore, the priest, alleging the malice of Stokesley.[180] But it is more likely that the two Patmores were one and the same; that the priest lived for a while incognito as a layman, in his father's company, the Drapers, which already contained a cell of the brethren. Stokesley's dislike of Patmore may have gone back a long way, for both were once fellows of Magdalen College, Oxford.[181]

Patmore's case was brought before the Commons by John Stanton, his servant. Thomas More suspected Stanton of being

[177] *Two London Chronicles*, p. 5; Foxe, *Acts and Monuments*, v, pp. 34–5.
[178] BL, Harleian MS 425, fo. 15ʳ.
[179] Foxe, *Acts and Monuments*, v, pp. 34–5.
[180] Ibid. v, pp. 35–7; appendix xiii.
[181] Henry Patmore, citizen and draper of St Peter Cornhill, had only one son Thomas, who was, when his father made his will in 1520, vicar of Much Hadham (preferred by Fitzjames in 1516). But in 1530 a Thomas Patmore became free of the Drapers' Company: PRO, PCC, Prob. 11/19, fo. 222ʳ; Drapers' Company, Repertory 7/1, fos. 2, 5, 13, 25, 49, 124; index of Company members; *BRUO*.

a favourer of heresy and, since we shall meet Stanton later as an adherent of Latimer, he was probably right. The Chancellor intervened, saying that it ill became such a person as Stanton to 'be attorney for his master in the Parliament House', and accused him wildly of being 'at the conveying of certain nuns of St Helen's' (heretics being, of course, by repute both sacrilegious and concupiscent), 'which nuns your said orator never saw, nor had knowledge of them, nor of their departure out of their house'. From the Fleet, Stanton petitioned the King for himself, for his master, and for all Stokesley's other victims imprisoned 'causeless'.[182]

Henry passed the Answer of the Ordinaries to the Speaker on 30 April. He made it clear enough how they should respond: 'We think their answer will smally please you . . .'.[183] On 10 May the King demanded that the Church should renounce all authority to make canons without royal licence.[184] The royal will became more plain and more compelling the following day: to a delegation from the Commons the King expressed his wonder; 'we thought that the clergy of our realm had been out subjects wholly, but now we have well perceived that they be but half our subjects, yea, and scarce our subjects'.[185] On 15 May the Convocation yielded to the royal demands, and the liberties of the English Church were lost. The churchmen, wrote Chapuys, were now of less account than cobblers, for cobblers were free still to assemble, to make their own statutes. The Submission of the Clergy was subscribed on 16 May.[186] Stokesley's name was missing.[187] On that same day Sir Thomas More resigned as Chancellor, determined to keep silent, never more 'to study nor meddle with any matter of this world, but that my whole study should be upon the passion of

---

182 PRO, SP 1/70, fos. 2^r–3^r (*L&P* v. 982). Among Cromwell's remembrances were declarations of Patmore's goods, and supplications by him or on his behalf: *L&P* vii. 923 (xii, xxvi). In 1535 commissioners were appointed for 'determining of the matter of T. Patmer, clerk'; that is, allegations of Stokesley's mistreatment of him: *L&P* viii. 1063. Patmore's case is discussed in Elton, 'Thomas More and Opposition', *Studies*, i, p. 159, and Dowling, 'Anne Boleyn and Reform', p. 40.

183 Hall, *Chronicle*, p. 788; Lehmberg, *Reformation Parliament*, p. 148.

184 Ibid., pp. 149–50.

185 Hall, *Chronicle*, p. 788.

186 *CSP Sp.* iv/2. 951; Kelly, 'The Submission of the Clergy'.

187 Lehmberg, *Reformation Parliament*, p. 152.

Christ and mine own passage out of this world'.[188] So, at least, he claimed.

THE KING'S GREAT MATTER

Late in May 1532 two great fish were caught in the Thames.[189] 'The people consider this a prodigy foreboding future evil'; so they did also the fourteen suicides in the past few days.[190] Another outbreak of the plague threatened. The King's Great Matter was only one among many things to unsettle the Londoners, but when the general state of mind was so disturbed this one issue came to seem the cause of all the rest. In the weeks after the Submission Crisis it was not the King's humiliation of the Pope and his clergy which so distressed the citizenry but his repudiation of his Queen. London's women, especially, rallied in defence of a wife scorned. In the autumn of 1531 a strange riot had been reported. A murderous mob of seven or eight *thousand* women (and men dressed as women) set out to seize Anne Boleyn, who was staying at 'a house of pleasure' by the river. She escaped. No punitive action was taken because this was 'a thing done by women'. Such was the state of agitation in the City that the Venetian ambassador thought this story not incredible.[191] (After all, had not hundreds of priests just rioted outside Stokesley's palace?)[192] Murder was committed at Court in April 1532 when rival retinues of the Dukes of Norfolk and Suffolk contended to avenge the 'opprobrious words' spoken against Anne by the Duchess of Suffolk, Katherine's sister-in-law.[193] A preacher at St Paul's in May 1532, preaching in favour of the Divorce, was interrupted by a woman who told him that he lied, and that 'this example in a King would be the destruction of the laws of matrimony'.[194] The capital was still severely disaffected in the last months of

---

[188] More, *Correspondence*, p. 552. But he withdrew to write with a new urgency: Fox, *Thomas More*, ch. 7.

[189] *Two London Chronicles*, p. 6.

[190] *CSP Ven.* iv. 773.

[191] Ibid. iv. 701.

[192] Hall, *Chronicle*, pp. 783–4.

[193] *CSP Ven.* iv. 761.

[194] Ibid. iv. 768.

1532. Plague raged and took a great toll.[195] When Henry
departed for France, taking Anne with him, trouble was
expected in his absence. On 10 October a watch, on a grand
scale, was kept, and another, secret, watch ordered on 23
October.[196] Ostensibly the City was quieter, but 'the simple
people cannot be ceased of their babbling tales'.[197]

Among Katherine's staunchest supporters were leading City
clergy.[198] Such men recognized early that the King's rejection
of the papal dispensation for his marriage attacked the foundations of papal authority. Some dared to preach openly in
defence of the Queen's cause. In March 1532 a preacher at St
Paul's—most likely Dr Coke, rector of All Hallows Honey
Lane—attacked the Divorce. (Imprisonment with Frith and
the brethren fortified his conservatism.)[199] Before the King
himself on Easter day 1532 at Greenwich the head of the
Observant Franciscans, William Peto, warned the King against
'false counsellors', and told him that by proceeding with the
Divorce the Crown was endangered, for 'great and little were
murmuring'. Should Henry marry Anne the dogs would lick
his blood as they had licked Ahab's.[200] When a royal chaplain
was sent to Greenwich forthwith to refute Peto's arguments,
Henry Elstow, the warden, denounced him and the Divorce.
Peto and Elstow were both sent to prison to await the royal
pleasure, but even there they conspired.[201] Sir George Throckmorton told later of his summons from Peto to Lambeth, of
how he was urged to 'stick' to the Queen's cause in Parliament
'as I would have my soul saved'. From Bishop Fisher,
Katherine's most unswerving champion, Throckmorton had
instructions about how to vote in the Commons on bills for

---

[195] *L&P* v. 1421, 1457, 1466, 1469; *CSP Ven.* iv. 806.

[196] CLRO, Journal 13, fo. 349ʳ; Letter Book O, fo. 248ᵛ; *L&P* v. 1421, 1466, 1469,
1472, 1473.

[197] PRO, SP 1/71, fo. 151ʳ (*L&P* v. 1458).

[198] J. E. Paul, *Catherine of Aragon and her Friends* (1967). For preaching in her
defence, see *CSP Sp.* iv/2. 922, 934; Stow, *Annales*, p. 561.

[199] Thomas Abell, her chaplain, Dr Coke, parson of All Hallows Honey Lane, and
Rowland Phillips, rector of St Michael Cornhill, were in the Tower: *L&P* v. 1256;
PRO, SP 1/71, fos. 119ʳ, 151ʳ–152ʳ, 163ʳ (*L&P* v. 1432, 1458, 1467); BL, Additional
MS, 28,585, fo. 222ʳ (*L&P* vi. 178); *L&P* vi. 842, 1672; PRO, E 36/143, fo. 29ʳ (*L&P*
vi. 1370); *L&P* v. 559(34); vii. 143.

[200] *CSP Sp.* iv/2. 934; Guy, *Public Career of Sir Thomas More*, pp. 192–3.

[201] *CSP Sp.* iv/2. 934 (pp. 427–8); Lehmberg, *Reformation Parliament*, p. 146 n. 1.

Annates, Appeals, and the Supremacy. 'Divers times' he went to St Thomas Apostle to consult the rector Dr Nicholas Wilson, another of the Queen's adherents, who strengthened his resolve. Father Reynolds of Syon confessed him. And in the confessional, the most secret of all places for conspiracy, Reynolds gave uncompromising and terrifying advice: to support the Queen's cause and oppose the anti-papal policies until death, or 'I should stand in a very heavy case at the Day of Judgement'.[202]

By 1533 a loosely organized political opposition to royal policies was forming, at Westminster and in London.[203] At the Queen's Head tavern Throckmorton dined regularly with the like-minded among his fellow members of the Commons while Parliament was in session, and there they planned resistance to all the bills against Rome—Annates, Appeals, and Supremacy. One of the malcontents was surely William Bowyer, elected burgess for the City on Petyt's death in August 1532. Bowyer's name appears among the list of those who opposed the Act in Restraint of Appeals.[204] The Appeals bill, presented on 14 March 1533, met with considerable resistance; in part because of religious qualms, in part because of fears of economic retaliation from the Imperial domains. One of the London burgesses suggested early in April that Parliament offer the King £200,000 (a sum evidently thought sufficient to pay him off, after the Pardon of the Clergy) if he would remit the question of the Divorce to a General Council of the Church. The proposer, according to Chapuys, was 'one who represents this City of London, who was once in Spain, and is one of my most intimate and familiar friends'. This burgess was most likely Paul Withypoll.[205] Chapuys, always hoping for Imperial intervention, insisting to Charles V that the arrival of a few Spanish troops would steel the English people to thwart Henry's designs, gathered those about him in London who shared his hopes and encouraged him to think that his wishes might be

---

[202] PRO, SP 1/125, fos. 247ʳ–256ʳ; printed in Guy, *Public Career of Sir Thomas More*, pp. 207–12.

[203] Elton, 'Sir Thomas More and Opposition'.

[204] *House of Commons, 1509–1558*, appendix vii; (Bowyer).

[205] *CSP Sp.* iv/2. 1058; Lehmberg, *Reformation Parliament*, pp. 174–5; *House of Commons, 1509–1558* (Withypoll).

realized.[206] But the people were far from turning grudge into action, and anyway the cause was lost. Late in January 1533 Henry had married Anne Boleyn. In May his new Archbishop, Thomas Cranmer, pronounced the union lawful.[207]

The more angry the people with the King's marriage, the more incensed the King by their presumption. When half the congregation at the Austin Friars at Easter left in protest as the Prior offered prayers for the new Queen, the King summoned the Mayor and ordered the City's silence, on pain of 'extreme punishment'.[208] Each citizen was to ensure the good demeanour of all his dependants, but such mutterings could not easily be stopped. Elaborate pageants and generous presents were to mark the City's good will and to win favour at Anne's coronation, but trouble was feared.[209] The Duke of Norfolk was asked whether the clergy—Katherine's keenest supporters —should be allowed to attend the coronation.[210] To inhibit resistance to the City's gift of 1,000 marks for Anne the Aldermen went in person to the houses of all those who refused to contribute.[211] On 31 May Anne rode through the City in triumph, to her coronation at Westminster on 1 June.[212] The 'utmost order and tranquillity' in London showed the City loyal, whatever the private feelings of the citizens. Only the absence of Thomas More openly dishonoured the Queen.[213]

Now, for the first time, the most fervent defenders of the Pope and of the old Queen were driven to conspiracy and exile. In the Low Countries Katherine's old supporters were writing and printing books against royal policies and smuggling them back to England. Like the brethren, they did not act without

---

[206] *CSP Sp.* iv/2. 1058; 1072, 1073.

[207] Scarisbrick, *Henry VIII*, pp. 309–13.

[208] *CSP Sp.* iv/2. 1062; *CSP Ven.* iv. 878; Tom Girtin, *The Triple Crowns* (1969), p. 112.

[209] Ives, *Anne Boleyn*, pp. 215–28.

[210] CLRO, Repertory 9, fo. 1ᵛ.

[211] Ibid., Journal 13, fo. 371ᵛ; Repertory 9, fos. 7ʳ–8ʳ. The alien merchant communities were included, save for the Spanish who were tactfully exempted: *CSP Sp.* iv/2. 1073; *L&P* vi. 508. The contribution was levied company by company: for example, the Waxchandlers gave 7s. 6d.: Guildhall, MS 9481/1, fo. 26ʳ.

[212] Hall, *Chronicle*, pp. 798 ff.; *CSP Ven.* iv. 912; *RSTC* 656. The Pewterers' Company spent £2. 4s. 8d. in accompanying the Queen: Guildhall, MS 7086/2, fo. 15ᵛ.

[213] Ives, *Anne Boleyn*, p. 222; Marius, *Thomas More*, pp. 438–40.

help from home. Friars Peto and Elstow were fled from prison to Antwerp: there they received from More Frith's dangerous works upon the sacrament of the altar and his own *Confutation of Tyndale's Answer*, written in the dark days just before and after his resignation.[214] The friars sent back in return copies of Fisher's forbidden Divorce tracts, eighty at a time.[215] 'They be much helpen out of England with money', and were sent so many cloaks and cowls that they could not wear them all, so Vaughan told Cromwell in August 1533, 'but I cannot learn by whom'.[216] He had his suspicions: Stokesley's servant Thomas Docwray, a notary public, was in Antwerp; if Cromwell would question Henry Pepwall, the stationer, another of the Bishop's men, he would discover why.[217] A friar came every week to Peto, and Vaughan learnt that a wealthy London merchant sent him: 'if it is true he is worthy to suffer to make others beware in time'. Surely this merchant was Anthony Bonvisi. He was certainly sending money to Peto.[218] Bonvisi was Thomas More's closest friend, and More himself was implicated in this conspiracy. Vaughan advised Cromwell: 'I can no further learn of More his practices, but if you consider this well, you may perchance espy his craft.'[219] For far from withdrawing from the political fray after his resignation in May 1532 More had involved himself ever more deeply in the polemical cause.[220] In the opening of *The Debellation of Salem and Bizance*, published late in 1533, More boasted that he had thwarted the 'Pacifier' (alias St Germain) by driving away Salem and Bizance, and 'if the Pacifier convey them hither again, and ten such towns with them, embattled in such dialogues, Sir Thomas More hath undertaken to put himself in

---

[214] PRO, SP 1/77, fo. 107ʳ; 1/78, fos. 37ʳ–38ʳ, 39ʳ, 57ʳ–58ʳ, 84ʳ–85ʳ (*L&P* vi. 726, 899–900, 917, 934); Fox, *Thomas More*, p. 191.

[215] PRO, SP 1/78, fo. 85ʳ (*L&P* vi. 934).

[216] Ibid., *L&P* vi. 1324.

[217] *L&P* vi. 934. Docwray's name appears through the London Church Court records. For Pepwall's later allegiance to Stokesley, see below, p. 283. Pepwall's servant, Michael Lobley, turned to reform: Foxe, *Acts and Monuments*, v, p. 38.

[218] PRO, SP 1/78, fo. 84ᵛ; 1/80, fo. 51ʳ (*L&P* vi. 934, 1369); *L&P* vi. 1324.

[219] PRO, SP 1/78, fo. 84ʳ. For More's friendship with Bonvisi, see below, pp. 420–1.

[220] Fox, *Thomas More*, ch. 7; Guy, 'The Political Context of the *Debellation*', *CW* 10, pp. xvii–xxviii.

the adventure alone against them all'.[221] It was a personal crusade from which there could be no return.

Meanwhile, the most dissident of the English religious houses had been infiltrated by Cromwell. In the summer of 1532 he had recruited as informers John Lawrence, a malcontent friar of Greenwich, and also Richard Lyst, an unstable lay brother, once a grocer of London and servant of Wolsey, now a client of Anne Boleyn. They reported assiduously upon the intemperate outbursts of such as Friar Forest.[222] 'I suppose you heard' of Forest's 'last indiscreet sermon at Paul's Cross', wrote Lawrence to Cromwell in November 1532. 'Barking and railing . . . over large of the decay of this realm and the pulling down of churches.'[223] Forest's denunciations were the more disturbing because manifestly true. In February 1532 the priory of Holy Trinity, Aldgate had surrendered itself to the Crown, of financial necessity.[224] No one thought that this would be the last confiscation. At the convent of the Crossed Friars on the eve of St James the Apostle there was treasonable talk. John Driver, the Prior, was exhorting his brethren to remain constant in 'good religion', for 'the King should have a fall shortly'. 'A certain religious man', he foretold, 'should come to him privily who should say that the King's highness was determined to put down certain religious houses'; 'whereas heretofore he was called *Defensor Fidei*, he should be called *Destructor Fidei*'.[225] Three acts of sacrilege shocked the City in that spring and summer. On Good Friday the consecrated host was stolen from St Botolph Aldersgate. In July a young priest was hanged, drawn, and quartered, without first being degraded from his orders, for clipping the King's coin. The Queen had refused to intercede, so Court gossip alleged: there were too many priests already, she said.[226] Blood was shed at All Hallows Bread Street when two priests fought, and

[221] Fox, *Thomas More*, p. 146.

[222] *L&P* v. 1259, 1312(2), 1371, 1525, 1591; Knowles, *Religious Orders*, iii, pp. 208–9.

[223] BL, Cotton MS Cleo. E iv, fo. 29ʳ (*L&P* v. 1525).

[224] J. Youings, *The Dissolution of the Monasteries* (1971), pp. 29–31; *L&P* v. 823.

[225] PRO, SP 1/70, fos. 201ʳ–203ʳ (*L&P* v. 1209).

[226] *Two London Chronicles*, p. 6; *L&P* v. 1165; Ives, *Anne Boleyn*, pp. 194–5; *Grey Friars Chronicle*, pp. 35–6.

no service could be performed there until the priests did penance before a general procession.[227] This was the perversion of Christian society which had been foretold.

Prophecies were circulating, foretelling disaster for the perpetrators of change, and giving hope of the return of the old order.[228] Such prophecies, hallowed by age, were widely believed. Where they told of events which were imminent they were the more emotive and dangerous. Mrs Amadas, the wife of the King's goldsmith, told of the Mouldwarp of ancient prophecy: the sixth king from John, who would be driven from the realm 'as Cadwallider was', and was 'cursed with God's own mouth'. The Mouldwarp was Henry VIII. The year of fulfilment of the prophecy would be 1533, for the destruction of the 'King's part' would come soon, and 'the dragon shall be killed by midsummer'. That Mrs Amadas had been abandoned by her husband gave her especial sympathy for the rejected Queen: it also sufficiently unhinged her to give her visionary powers and the foolishness to utter them. She knew well who to blame for all the change: 'to set up a candle before the Devil she gave four New Year's gifts, one to the King, another to our lady Anne, another to the Duke of Norfolk [who had also treated his wife shamefully] and another to Mr Cromwell'.[229] Other London women were prepared 'to die in the quarrel for Queen Katherine's sake': on 23 August 1533 two women, one pregnant, were stripped to the waist, beaten, nailed by the ears to the Standard, and banished from the City forever; for saying that Queen Katherine was true Queen of England, not Anne.[230] It may have been that they, like others in the City, had been moved by the revelations of the visionary and reputedly saintly Elizabeth Barton, the Nun of Kent.[231]

There was an attempt to turn prophecy into reality. The Nun of Kent had gathered to her cause 'certain spiritual and

[227] *Two London Chronicles*, pp. 6–7; Stow, *Survey*, i, p. 347.

[228] For the power of prophecy in the popular imagination, see Keith Thomas, *Religion and the Decline of Magic: Studies in Popular Beliefs in Sixteenth- and Seventeenth-Century England* (1971), ch. 13; Elton, *Policy and Police*, ch. 2.

[229] BL, Cotton MS Cleo. E iv, fo. 99^{r-v} (*L&P* vi. 923); Elton, *Policy and Police*, pp. 59–61. When her husband died he left a cup worth £100 to the King that he be 'good and gracious' to his widow: PRO, PCC, Prob. 11/25, fo. 39^r.

[230] CLRO, Repertory 9, fos. 21^r, 23^v; *Two London Chronicles*, p. 8.

[231] A. Neame, *The Holy Maid of Kent* (1971).

religious persons . . . appointed as they had been chosen of God to preach', and they were to set forth her revelations to the people that the 'pleasure of God was that they should take him [Henry] no longer for their King'. Her band of clergy were to extol her sanctity and advertise her powers and prophecies to the world, when the time was right. These preachers had gained a great reputation for their holiness and ascetism (not universal virtues among the clergy), 'whereof credence should have been given to their preaching'.[232] The Nun of Kent had found support among a spiritual élite in London and Kent, including Bishop Fisher (More was too circumspect).[233] Henry Gold, rector of St Mary Aldermary, fell under her spell. His scholarship, and his staunch resistance to any reform, had won him the patronage of Warham. In his City parish he preached, it seems, of the outrages of 'men with venom teeth and poison tongues' who slandered the preachers of God, and were unworthy to hear the 'precious and holy Gospels' which they sought; and of vile iconoclasts. He told of some (in Germany?) who burnt a crucifix and hanged an image of the Virgin at the instigation of Carlstadt, 'the destroyer of images'.[234] By 1531 or so Gold's conservatism had turned to treason. Gold had become 'the most busy to sow this mischievous seed' of the Nun's revelations. He had acted as interpreter between the Nun and Pulleo, the papal nuncio. To Katherine he sent messages foretelling that her heirs would succeed to the throne, not Anne's, and had denounced the Boleyn marriage.[235] A further link in this chain of treasonable activity was that Gold had on many occasions visited the Marquis of Exeter, resolute supporter of Katherine and opponent of all religious change.[236] Such men trusted that the new ways could not last long; they expected the imminent return of the old order when the heretics would be overthrown. Against that day they were gathering evidence against the reformers.[237]

---

[232] PRO, SP 1/82, fo. 86ʳ (*L&P* vii. 72).

[233] Elton, *Policy and Police*, p. 408; More, *Correspondence*, 192, 197, 198, 200.

[234] PRO, SP 1/83, fos. 147ʳ–148ʳ (*L&P* vii. 523(4)); *L&P* vii. 523(6). See above, pp. 59–60.

[235] PRO, SP 1/82, fos. 86ʳ, 90ʳ–91ʳ (*L&P* vii. 72).

[236] PRO, SP 1/139, fo. 16ʳ (*L&P* xiii(2). 827(3)).

[237] See below, p. 285.

V The Political Reformation, 1534–1536

THE TIME would soon come when that which it had been heresy even to discuss would be enforced as a new orthodoxy. From 1531 the King determined to change the spiritual allegiance of his subjects from the Pope as sovereign of Catholic Christendom to himself as Supreme Head of the Church of England. The clergy—'scarce our subjects'—had in 1532 relinquished their liberties and become servants of a royal master. The Act in Restraint of Appeals of March 1533, followed by the definitive Act of Supremacy in November 1534, made a revolution in ecclesiastical authority, by statute.[1] Laws are always easier to make than to enforce. The King and his counsellor, Thomas Cromwell, knew well the magnitude of the changes they were introducing, and they feared the political consequences. The nation had still to be persuaded of the royal usurpation of papal sovereignty, and those who could not be persuaded must be punished. London would be the centre of the propaganda campaign; from its printing presses and pulpits the royal will would be declared and on its scaffolds the King's enemies destroyed.[2] The conformity of the capital—'the common country of all England'[3]—must be assured, for if London was not compliant then the Reformation could not succeed. Yet London's loyalty was very far from certain.

Ensuring outward obedience to government command was not at all the same thing as winning hearts and minds. Many still bore 'the crossed keys close within their hearts'.[4] The Break with Rome was anathema to many of the people upon

---

[1] The best introductions and surveys are provided by Bishop Burnet, *The History of the Reformation of the Church of England*, 3 vols. (Oxford, 1865); A. G. Dickens, *The English Reformation* (1964); P. Hughes, *The Reformation in England*, 3 vols. (1951–4); G. R. Elton, *Reform and Reformation: England, 1509–1558* (1977); J. J. Scarisbrick, *Henry VIII* (1968), chs. 8–10.

[2] Elton, *Policy and Police*.

[3] *L&P* xii/2. 91.

[4] PRO, SP 1/162, fo. 127^r (*L&P* xv. 1029(21)).

whom the King and Cromwell relied for its enforcement. The most ardent persecutors and most 'earnest defenders of the Romish decrees' were now bound to punish the Bishop of Rome's staunchest supporters. If this was unfortunate for Henry, it was intolerable to them. William Pavier was Town Clerk of the City, and in 1530 warden of that bastion of conservative opinion, the gild of the Name of Jesus at St Paul's.[5] As Town Clerk, he had officiated at the burning of James Bainham a fortnight before the Submission Crisis. Bainham proclaimed at the stake that 'God's book' should be freely available in English; that the Bishop of Rome was Antichrist; that he knew 'none other keys of Heaven-gates, only the preaching of the law and Gospel'; that there was 'none other Purgatory, than the Purgatory of Christ's blood', but Pavier challenged Bainham's dying words; 'thou heretic'.[6] Soon afterwards, seeing the perverse direction which royal policy was taking, Pavier swore that he would cut his own throat if ever he thought that the King would set forth the Gospel in English. Edward Hall heard the oath and recorded its fulfilment. Before a rood in his chamber, Pavier hanged himself in May 1533.[7] His fellow City governors lived on, to endure and even implement the changes. Thomas More had—ostensibly—left political life after fighting a losing battle against the King and the reformers, but Stokesley remained. With time the Bishop's casuistic submission became more agonizing to him, and his loyalty to the Supreme Head increasingly doubtful. It was upon Stokesley, who hoped always for a 'happy day' when the old ways would return,[8] and upon unwilling City governors that Henry and Cromwell depended for the enforcement of the royal supremacy in London. London's compliance to the new order was won only by a sustained campaign to control, even to break, those in authority there.

The Break with Rome was not meant to augur the end of persecution. Certainly the King never intended that it should, always denying any suggestion that he was following the

[5] Bodleian, Tanner MS 221, fo. 16[r].

[6] Foxe, *Acts and Monuments*, iv, p. 705.

[7] Hall, *Chronicle*, p. 806; Foxe, *Acts and Monuments*, v, p. 63; viii, p. 635; *Two London Chronicles*, p. 7.

[8] *Original Letters*, i, p. 231.

Lutherans.[9] But Pavier had been far-sighted. The burning of John Frith with the London apprentice, Andrew Huet, on 4 July 1533 would be the last for a while.[10] The savagery of More and Stokesley's quest for heretics had turned many in the City against them; if not quite to favour their victims. Yet, though More was out of the way, Stokesley and his clergy were still bent on persecution, and heresy procedure, with its notorious *suspicio* charge, remained unchanged. The Bishop's suspects remained in prison. The case of one of them was taken up by the Commons in February 1534. Thomas Phillips had been long incarcerated, without proof; an earlier appeal to the King as Supreme Head had gone unheard, but now his plea to the Lower House moved it to his defence.[11] Bowyer had been replaced as burgess for London by one 'who talked somewhat against the covetousness and cruelty of the clergy': this was Robert Packington, friend and agent of Cromwell, who may have been the champion of Phillips's cause.[12] The Lords were suspicious: that 'illustrem Senatum' would not help one 'de Heresyi suspectus'.[13]

Phillips remained in durance, but some were determined that a new heresy law must be introduced to prevent any other suspect from being in a similar plight, with no case proven against him but with no hope of release. Once more, as in the 1532 session, bills for heresy and bills for London tithe were prepared concurrently, both directed against the arbitrary powers and the rapacity of the clergy.[14] On 5 March a delegation was sent from the Commons to the King, desiring reformation of the most feared clerical abuses:

In calling many of his subjects to their courts *ex officio* and not knowing their accuser, and to cause them to abjure or else to burn them for pure malice and contrary to justice, and that they were judges and parties in their own causes.[15]

---

[9] *L&P* vii. 232.

[10] Foxe, *Acts and Monuments*, v, pp. 16–18.

[11] S. E. Lehmberg, *The Reformation Parliament, 1529–1536* (Cambridge, 1970), p. 186; J. Guy, 'The Legal Context of the Controversy', *CW* 10, pp. lxii–lxiv.

[12] *The House of Commons, 1509–1558* (Packington).

[13] *LJ* i, pp. 65–6, 71.

[14] *L&P* vi. 120.

[15] BL, Harleian MS 2252, fos. 34ʳ–35ʳ.

The London tithe bill was dropped for a while, but a new Heresy Act was passed on 28 March.[16] The repeal of *De Haeretico Comburendo* allayed some of the laity's worst fears. No longer was it possible to arrest and imprison solely on suspicion: the Church Courts could proceed only on presentment by two witnesses, and suspects were to answer in open court. A writ of wrongful arrest might be sought if the indictment were thought insufficient, or malicious.[17] Speaking against the 'pretended power of the Bishop of Rome' was no longer deemed heretical: this provides the clue to the first reason why heresy persecution was interrupted.

The graver threat to authority in the Church came now not from the evangelicals but from those who denied the royal supremacy. Maybe the gospellers thought the question of authority a sterile issue compared with the propagation of the Word; yet they certainly believed the Roman pontiff not only an usurper, but Antichrist. The King's enemies henceforth would be the most orthodox of Catholics and staunchest defenders of papal primacy; such as he had once been himself. Around him at Court and in his Privy Chamber, Henry gathered those who protected the evangelicals, indeed were evangelicals themselves: Thomas Cranmer, Queen Anne, and Thomas Cromwell.[18] While they held the King's favour the old ways would not return. The conservatives circulated prophecies saying so:

> When A B C is brought down low
> Then shall we begin the Christ cross row[19]

Though Thomas Cromwell had risen high, his contemporaries never forgot how he had begun:

This Cromwell his father was an Irishman born, and a smith by occupation, and after kept a brewhouse in Wandsworth, and there was this Cromwell born, and at the last coming in favour with King Henry VIII, he made him knight, and Lord Privy Seal, and then

---

[16] Lehmberg, *Reformation Parliament*, pp. 186–7; Guy, 'The Legal Context of the Controversy: The Law of Heresy', pp. lxv–lxvi.

[17] 25 Hen. VIII c. 14; *Statutes of the Realm*, iii, pp. 454–5.

[18] M. Dowling, 'Anne Boleyn and Reform', *JEH* xxxv (1984); E. W. Ives, *Anne Boleyn* (1986); D. B. Starkey, *The Reign of Henry VIII: Personalities and Politics* (1985).

[19] *L&P* xii/2. 1212 (iii); cf. xii/1. 318, 841.

Vicegerent, after Lord Chamberlain, and then Earl of Essex, and thus he brought him up of nought.[20]

Having faced the wilderness at Wolsey's fall, Cromwell knew well the mutability of political fortune, especially under such a King.[21] He understood that he served a master who was suspicious, fickle, self-deluding, disloyal, and implacable, and that 'indignatio principis mors est' ('the wrath of the King is death'). As More had told Roper, 'if my head could win him a castle in France it should not fail to go'.[22] Still, Cromwell led Henry toward reform in religion more radical than the essentially conservative King would countenance, and he did so knowing the risk for himself when the King finally reverted. Why? Because Cromwell was a convinced evangelical, and one of the first. The brethren were to be found in the Inns of Court, in the City merchant community, and among English merchants abroad. Cromwell was part of all these worlds and had known some of the brethren of old. His acquaintance with them proves nothing, for London's establishment was small, but his protection of them suggests his support for their cause. Among Cromwell's friends, associates, and agents from the 1520s were men whom it was dangerous for him even to know: Stephen Vaughan, John Coke, Thomas Somer, John Copland, John Petyt, Robert Barnes, Robert Packington, Humphrey Monmouth, perhaps John Purser, and others.[23] In the 1530s the reformers hailed Cromwell as God's special 'instrument' for bringing in the Gospel, and prayed God to 'preserve him long to such good purposes, that the living God may be duly known in his spirit and verity'.[24] On the other hand, Cardinal Pole thought that 'God in his anger at the King' had spared Cromwell's life in 1529 to give it to Satan to use 'as an instru-

---

[20] *Two London Chronicles*, pp. 15–16.

[21] For Cromwell's life, see R. B. Merriman, *The Life and Letters of Thomas Cromwell*, 2 vols. (Oxford, 1902); A. G. Dickens, *Thomas Cromwell and the English Reformation* (1959); B. W. Beckingsale, *Thomas Cromwell: Tudor Minister* (1978).

[22] Roper, *Lyfe*, p. 21.

[23] S. E. Brigden, 'Thomas Cromwell and the Brethren', in *Law and Government under the Tudors*, ed. C. Cross, D. M. Loades, and J. J. Scarisbrick (Cambridge, 1988), pp. 31–50; S. E. Lehmberg, 'The Religious Beliefs of Thomas Cromwell', in R. L. de Molen (ed.), *Leaders of the Reformation* (1984), pp. 134–51.

[24] Latimer, *Sermons*, p. 411; Latimer, *Remains*, pp. 395, 386–7; PRO, SP 1/96, fo. 36[r] (*L&P* ix. 226).

ment' to afflict Henry's soul.[25] Cromwell's enemies and his friends alike were right that Henry's ambitions for caesaropapal authority within his realm would have come to nothing without Cromwell's guiding intelligence and determination to reform.[26]

Closer to the King than Cromwell, and as convinced an evangelical, was the Queen. Seeking reasons for the Break with Rome many chose *chercher la femme*. True, in order for the King to marry Anne he had finally to deny papal authority, but the Queen's interest in reform came long before she conceived such lofty matrimonial ambitions, and her passion, first and last, was the Gospel. In one of her own books of devotion this prayer was inscribed:

Grant us most merciful father . . . the knowledge of thy holy will and glad tidings of our salvation, this great while oppressed with the tyranny of the adversary of Rome . . . and kept close under his Latin letters.[27]

Her courtier, Thomas Wyatt, and her brother, George Rochford, shared her Scriptural piety.[28] In her chamber the Bible in English was left open for her ladies' edification, to turn them from frivolous pursuits.[29] For her chaplains, 'lanterns and light of my court', she chose men won to reform: Matthew Parker, Robert Singleton, William Latimer, Nicholas Shaxton, and others.[30] Even her London silkwomen Anne Vaughan and Joan Wilkinson were fervent gospellers.[31]

In the City Anne used her influence and patronage first to protect, then to advance the gospellers. In 1528 she had interceded with Wolsey to spare the book-running parson of All Hallows Honey Lane. Simon Fish had safe-conduct to return from exile, escaping More's quest through the City and the Inns, because of Anne's sway over Henry.[32] Only the Queen's

[25] *L&P* xiv/1. 200, p. 82.

[26] Elton, *Reform and Reformation*, chs. 4–12; *Reform and Renewal*; 'Thomas Cromwell Redivivus', *Studies*, iii, pp. 373–90.

[27] Cited, Dowling, 'Anne Boleyn and Reform', p. 33.

[28] Constantine, 'Memorial', p. 65; H. A. Mason, *Humanism and Poetry in the Early Tudor Period* (1959).

[29] Bodleian, MS C Don. 42, fos. 31ᵛ–32ʳ; Dowling, 'Anne Boleyn and Reform', p. 33.

[30] Ibid., pp. 37–41.

[31] *Lisle Letters*, 4, pp. 49, 859; 5, pp. 1102, 1117, 1125; see below, pp. 418–19.

[32] See above, p. 128.

intervention finally secured the release of Thomas Patmore, long 'wrapt in the Bishop's nets' (that and his payment of £100 fine to Stokesley: where did he get that money from unless from powerful friends?).[33] An academic generation of reformers from Gonville Hall found favour with Anne: one of these was Dr Crome, and she persuaded Cranmer in 1534 to prefer him to St Mary Aldermary.[34] That benefice was vacant only because Henry Gold, one of her most inveterate enemies among London priests, was in the Tower. In 1534, at the height of her power, she sought to evict another who had preached against her: a letter came to the Drapers' Company from the Queen asking for the voidance of St Michael Cornhill, Rowland Phillips's benefice. But she was thwarted.[35] All over the country Anne was hated as a promoter of reform, and she was never forgiven for ousting Queen Katherine from the King's affections. In the London Black Friars Dr Maydland knew by 'his science which is necromancy' that a day would come when those of the New Learning would die, and the 'people of the old learning shall be advanced and set up'. He trusted to 'see the King to die a violent and shameful death, and also to see that mischievous whore the Queen to be burnt'.[36]

The first test of the capital's allegiance to the royal supremacy—political revolution and religious schism—came in the spring of 1534 when the clergy and the citizenry were called universally to swear an oath, according to the Succession Act.[37] They were adjured

to be true to Queen Anne, and to believe and take her for the lawful wife of the King and rightful Queen of England, and utterly to think the Lady Mary daughter to the King by Queen Katherine, but as a bastard, and thus to do without any scrupulosity of conscience.

[33] Foxe, *Acts and Monuments*, v, p. 35.

[34] BL, Harleian MS 6148, fo. 79ᵛ (*L&P* vii. 693). St Mary Aldermary was a Canterbury peculiar.

[35] CLRO, Repertory 8, fo. 264ʳ; A. H. Johnson, *The History of the Drapers of London*, 5 vols. (Oxford, 1914–22), ii, pp. 61–2.

[36] PRO, SP 1/99, fo. 67ʳ (*L&P* ix. 846). For numerous other outbursts against her, see Elton, *Policy and Police*, pp. 11, 24, 58, 60, 107, 137, 277–9, 282, 300, 331, 347, 354, 357, 384.

[37] Gee and Hardy, *Documents Illustrative*, pp. 232–43; Elton, *Policy and Police*, pp. 222–6. Only the Spanish merchant community, with special reason to be loyal to Queen Katherine, was exempt: *CSP Sp.* v, p. 58.

Never before had a spiritual instrument of commitment been used as a political test. 'Dame Perjury' led her followers to Hell; almost everyone hated the Boleyn marriage, yet the penalty for refusing this oath was misprision of treason.[38] The Londoners were the first to be sworn, and Thomas More, never allowed to remain *oblitus oblivescens* as he claimed to wish, was first to be called. The oath he was offered at Lambeth on 13 April (one specially contrived for him) he could not swear without, he said, 'jeoparding my soul to perpetual damnation'.[39] From a Lambeth garden, More watched Latimer 'very merry' with Cranmer's chaplains, and looked on as the London clergy 'played their pageant'. Rowland Phillips, who was expected to refuse, swore then went to drink in the buttery 'valde familiariter', in order to vaunt his compliance before the Bishops.[40] 'The remnant of the priests of London that were sent for, were sworn and . . . had such favour at the Council's hands, that they were not lingered nor made to dance any long attendance . . . but were sped apace to their great comfort.'[41] Alone of the London clergy, Dr Nicholas Wilson, an old defender of Queen Katherine, refused to swear, and he joined More and Fisher in the Tower.[42]

The royal command for the citizens to swear came on 18 April, and the City's officers swore that day. Twenty-three commissioners were appointed to take the oath of every citizen in his gild on 20 April. In the Guildhall the Mayor and Recorder 'sat . . . and tooketh the oath of every person there being going out'. (Once the compliance of the masters was assured their apprentices and servants, aged over twenty, were sworn on 5 June.)[43] No one refused. Even More's fool swore.[44] But

---

[38] Wriothesley, *Chronicle*, i, p. 24. For perjury and its penalties, see S. E. Brigden, 'Religion and Social Obligation in early sixteenth-century London', *P&P* 103 (1984), pp. 86–92.

[39] More, *Correspondence*, p. 502; Elton, *Policy and Police*, pp. 223–4.

[40] More, *Correspondence*, pp. 503–4. Phillips had been a leader of opposition in the lower house of Convocation, and exempted from the Pardon of the Clergy; Lehmberg, *Reformation Parliament*, pp. 71, 127.

[41] More, *Correspondence*, p. 504.

[42] Ibid., pp. 503, 532–8.

[43] CLRO, Repertory 9, fos. 55ʳ, 56ᵛ; Journal 13, fo. 407ʳ; Letter Book P, fo. 38ʳ; *L&P* vii. 490 (Chapuys gives the date as 16 Apr.); Mercers' Company, Acts of Court, ii, fos. 66ᵛ–68ᵛ. The Coopers' Company purchased a 'copy of the oath that the King sent that every man was sworn by'; Guildhall, MS 5601/1 (no folio numbers). *Two London Chronicles*, p. 9.   [44] More, *Correspondence*, p. 529.

perhaps some contrived to abstain. A London scrivener, James Holywell, fled to Rome where he boasted that 'when every man was sworn to the King he was not, nor would be'.[45] The execution of the Nun of Kent, Henry Gold, and her other adherents on the very day that the oath was demanded had shown Londoners compellingly enough what would happen to those who denied the King, or even spoke disparagingly of his purposes.[46] Ignorance of the oath's implications might also have explained the universal compliance. Yet there were already some in London who found an advantage in renouncing allegiance to Rome and knew how to use anti-papal legislation for their purposes. In 1534 James Bacon, who was involved in a bitter and protracted matrimonial suit, brought information to the King against his judge, Dr Dakyns, the Commissary of the Dean and Chapter of St Paul's, who had, contrary to statute, sued an appeal to the 'Court of the Bishop of Rome calling himself Pope'.[47] Everyone who had sworn the Succession Oath had implicitly renounced the papal primacy, as More, Fisher, and Wilson had clearly seen. The swearing of the Oath of Supremacy a few months later could leave no possible doubt, for therein papal jurisdiction and papal supremacy were specifically denied.

The clergy was the one group in society whose explicit compliance to the royal supremacy was demanded. Every member of the ecclesiastical hierarchy had been sworn to the Pope as head of the Church, and looked to him as the fount of authority, but after 1534 their allegiance must be transferred to the King. Their influence over the laity, for whom they had 'cura animarum', made their conformity politically imperative.[48] For the swearing of the Oath of Supremacy, the clergy belonging to corporate institutions were required to take a comprehensive vow denying papal power and recognizing Henry as Supreme Head, second only to Christ.[49] The secular and parish clergy were ordered simply to subscribe to the state-

[45] *L&P* viii. 763.   [46] PRO, SP 3/5, fo. 59^r (*L&P* vii. 522).

[47] PRO, SP Q, fo. 28^{r-v} (*L&P* vii. 1605). For Dakyns, see C. A. Haigh, *Reformation and Resistance in Tudor Lancashire* (Cambridge, 1975), pp. 43, 82, 109–10. For Docwray, Dakyns's fellow judge, see above, p. 212.

[48] M. Bowker, 'The Supremacy and the Episcopate: the Struggle for Control, 1534–40', *HJ* xviii (1975), pp. 227–44; Bowker, 'The Henrician Reformation and the Parish Clergy', *BIHR* l (1977), pp. 30–47.

[49] Elton, *Policy and Police*, pp. 227–9.

ment—not an oath—that 'Romanus episcopus non habet maiorem aliquem jurisdiccionem a deo sibi collatem in hoc regno anglie quam alius externus episcopus'. The royal supremacy was here only implicit, not expressly commanded, and the parish clergy might claim that they accepted the repudiation of the Pope with no acquiescence to the royal supremacy; maybe.

The potential challenge which this oath posed to the ecclesiastical hierarchy—both in compliance to its claims, and in its practical administration—was pre-empted by the Crown's assumption of one of the Church's powers, the visitation.[50] The original intention of the King may have been to respect ecclesiastical autonomy and trust the traditional hierarchy of Bishops, Archdeacons, and Deans to enforce the oath, but their loyalty was still highly uncertain in 1534, and the Supreme Head evidently doubted his officers. The Crown seized upon the Church's established procedure of metropolitan visitation, with its suspension of episcopal authority, in order to frighten the Bishops into compliance. On 10 May 1534 Stokesley was informed of Cranmer's impending visitation in June.[51] Cranmer's visitation aroused a storm of opposition from the Bishops, and not least from Stokesley. Stokesley's protest largely concerned the Archbishop's title of 'Apostolicae Sedis legatus' and the respective authorities of legate and primate within the Church. Yet Stokesley's resistance was apparently provoked by the prospect that Cranmer would extend his metropolitan visitation indefinitely rather than by the royal supremacy, for just as those with a grievance had once appealed from the provincial to the Pope, now Stokesley and his chapter were appealing to the King.[52] But the Supreme Head was moved not at all and the visitation went ahead.

[50] Bowker, 'Supremacy and Episcopate', pp. 228–34.

[51] Guildhall, MS 9531/11, fo. 59ʳ.

[52] Ibid., fos. 61ʳ–62ʳ; Strype, *Memorials of Cranmer*, ii, appendix xv, pp. 704–8; *L&P* vii. 1683; Bowker, 'The Supremacy and the Episcopate', pp. 231 ff. An anonymous tract written at the end of 1535 repeated the arguments against the visitation; that it was in derogation of the royal supremacy, and that the Archbishop's Court of Audience infringed the authority of another Bishop in his diocese (that is, the Bishop of London): *L&P* viii. 705. It is likely that this tract was written by Bishop Stokesley. Professor Lehmberg doubts that the tract was written by Gardiner (to whom the editors of *Letters and Papers* ascribe authorship), but suggests no alternative author: 'Supremacy and Vice-Gerency: a reexamination', *EHR* cccxix (1966), p. 231 n. 2.

At some time during the visitation the swearing of the Oath of Supremacy by the religious corporations of London and the denial of papal primacy by the secular clergy of the City was effected. The austere London Carthusians refused ever to swear, and were sent to prison, there in chains to reflect upon the consequences of disobeying their sovereign.[53] But Stokesley swore.[54] For the most part, the heads of the City's religious houses and colleges set their seals, and their members swore in the early summer of 1534, before the London parish clergy were summoned in late June and early July.[55] Three hundred and fifty-six London priests certainly subscribed; more may have done so.[56] No resistance was reported, but Edward Field, the Master of Whittington College, subscribed with the proviso that he blamed Cranmer for this treachery.[57] Many must have subscribed uneasily, casuistically. George Rowland of the Crossed Friars later dismissed his acquiescence, for 'an oath loosely made may be loosely broken', and for Friar Forest an oath taken by the 'outer man' could not bind or compromise the loyalty of the 'inward man'.[58] Some had expected wider refusal. In October 1534 a letter came to James Beck in London from a friar imprisoned in Stamford (with the admonition to read and burn it; 'for it is not unknown to you what hurt hath chanced by letters . . .'). Asking for news of his fellow Grey Friars in London and of the Greenwich Observants, he marvelled to hear that some had sworn to the supremacy.[59] Yet to deny the King meant certain imprisonment and likely death. The passage of the Treason Act in November 1534 (to come

[53] Knowles, *Religious Orders*, iii, pp. 222–40.

[54] Foxe, *Acts and Monuments*, v, pp. 71–3; *L&P* viii. 121. Cranmer had taken a further move to intimidate Stokesley. He cited him for admitting Edmund Close as rector of St George Botolph Lane, on 26 June 1535, during the visitation and in derogation of it: Guildhall, MS 9531/11, fo. 65ᵛ; *VCH London*, i, p. 262.

[55] *L&P* vii. 665, 921; Rymer, *Foedera*, xiv, pp. 492–527.

[56] PRO, E 36/63, fos. 65ʳ, 94ʳ, 95ʳ, 102ʳ; E 36/64, fos. 1ʳ–5ʳ, 71ʳ. The documents are undated, save for the subscription of the clergy of the deanery of Barking, who signed on 1 July 1535. The signatures of approximately half the London parish clergy who should have sworn are extant. It is more likely that the documents of some parishes are lost altogether than that the rest evaded signing.

[57] PRO, E 36/63, fo. 102ʳ.

[58] PRO, SP 1/102, fo. 73ᵛ (*L&P* x. 346). The Crossed Friars were sworn in Apr. 1534; *L&P* vii. 665. BL, Harleian MS 530, fo. 120ʳ (*L&P* xiii/1. 1043). See also *L&P* xiii/2. 829.

[59] PRO, SP 1/86, fo. 137ʳ (*L&P* vii. 1307).

into force the following February) had extended the definition of that crime to include treason by word. Anyone who spoke 'maliciously' or in derogation of any of the 'dignities or titles' of the King could be accounted a traitor. Now to deny the royal supremacy, even to fail to acknowledge it, was high treason.[60] Bishop Stokesley, who had contrived not to subscribe during the Submission Crisis, swore a comprehensive oath in February 1535 acquiescing to the royal supremacy. Later, he is said to have repented: 'Oh! that I had holden still with my brother Fisher and not left him when time was.'[61] But the conscience and courage of Fisher, More, and the Carthusians were given to few.

Withdrawn from the world though living on the edge of the City, the London Carthusians had provided the Londoners with an inspiring example of the austere spiritual life.[62] The citizens' devotions to these monks had continued while they despised the worldly lives of their parish priests. To the Charterhouse Sir Thomas More had withdrawn in his youth to contemplate taking orders; now others, who like More found the events of recent days intolerable, yearned for the peace of the cloister. Sebastian Newdigate, gentleman of the Privy Chamber, who had earlier discovered the alarming activities of the Christian Brethren, left the Court for the London Charterhouse.[63] So too the Vice-Chamberlain, Sir John Gage, renounced his office and his wife in January 1534, intent upon becoming a Carthusian.[64] The new Prior of the London Charterhouse, John Houghton, was esteemed as saintly before ever a martyr's death seemed a possibility, and he praised the monks under his rule as 'angels'.[65] So austere was the Carthusian life that one monk there, Andrew Boord, could not bear its 'rigorosity', and he turned agent to Cromwell, to the disgust of all the citizens

---

[60] 26 Hen. VIII, c. 13; *Statutes of the Realm*, iii, pp. 508 ff.; Gee and Hardy, *Documents Illustrative*, pp. 247–51; Elton, *Policy and Police*, pp. 282–92.

[61] *The Earliest English Life of St John Fisher*, ed. Hughes (1935), p. 160; cited in Scarisbrick, *Henry VIII*, p. 327 n. 1.

[62] E. M. Thompson, *The Carthusian Order in England* (1930), pp. 371–485; L. Henricks, *The London Charterhouse, its Monks and its Martyrs* (1889); Knowles, *Religious Orders*, iii, pp. 222–40.

[63] Roper, *Lyfe*, p. 6; Knowles, *Religious Orders*, iii, p. 227; Foxe, *Acts and Monuments*, v, appendix xiii.

[64] *L&P* vii. 14.

[65] Knowles, *Religious Orders*, iii, pp. 225–6.

who encountered him.[66] The monks sought to live in peace, in prayer and contemplation, but the King demanded otherwise. The Succession Oath did not violate their faith, so they swore it;[67] but the Supremacy Oath was anathema. They refused absolutely to swear, for they could not deny Christ's trust to St Peter.[68] At Syon too, Richard Reynolds vowed to spend his blood in the Pope's quarrel: 'he doth this as a thousand thousand that be dead'.[69] On 26 April an agonized, and surely an intimidated jury brought a guilty verdict, under the new Treason Act, against Houghton, Reynolds, and their fellows.[70] On 4 May the first Carthusians left the Tower for Tyburn to suffer the ghastly death of traitors. More watched them depart, rejoicing that they came as 'bridegrooms' to Christ.[71](Their quartered remains were impaled on the City's gates. Houghton's arm was stolen away as the sacred relic of a martyr for the faith.)[72] An appalled Chapuys reported that the whole Court attended, including Henry, disguised as a borderer.[73]

In the Tower, Thomas More thought on last things and waited for death.[74] 'I would', he told Cromwell, 'never meddle with the world again, to have the world given to me.'[75] His reasons for refusing the oath in April 1534, and thereafter, he determined to keep 'secret in my conscience'. His adamant silence could never be taken for consent, but neither could it be proven treasonable dissent, for the Treason Act could punish only expressed denial of the King's titles and purposes. Silence seemed the best defence against the 'two-edged sword' which was the Act of Supremacy: for 'if I should answer one way I

---

[66] *L&P* xi. 297. For Boord's writings, see *RSTC* 3373–3387.

[67] *L&P* vii. 728; Knowles, *Religious Orders*, iii, p. 230.

[68] Ibid. iii, pp. 230–1; appendix iii.

[69] *L&P* viii. 566; J. A. Guy, *The Public Career of Sir Thomas More* (Brighton, 1980), p. 211.

[70] Husee reported on 27 Apr. that Cromwell 'hath had much ado with the judges and sergeants about certain of the Charterhouse'; *Lisle Letters*, 2. 375; Thompson, *Carthusian Order*, p. 397.

[71] Guildhall, MS 1231, fos. 8ᵛ–23ᵛ (an eye-witness account of the martyrdom of the Carthusians); *L&P* viii. 726; Wriothesley, *Chronicle*, i, p. 27; Roper, *Lyfe*, pp. 80–1.

[72] See below, p. 427.

[73] *L&P* viii. 666.

[74] For More's last months, see More, *Correspondence*, 200–18; A. Fox, *Thomas More: History and Providence* (Oxford, 1982), part iii; R. Marius, *Thomas More* (1985), chs. 28–9.

[75] More, *Correspondence*, p. 552

should offend my conscience, and if I should answer the other, I put my life in jeopardy'.[76] Willingly to court death was a sin: to die at the block, if that death might be avoided, would be suicide.[77] By the end of May 1535 Henry determined to be rid of both More and Fisher, for—to the King's fury and humiliation—the Pope had made Fisher Cardinal.[78] The reluctant Cromwell, whom once More had acknowledged as his 'special tender friend', who 'tenderly favoureth' him in his plight, was bound to do 'his Grace's pleasure'.[79] Cromwell, cynically enough, made last attempts to persuade More to a pragmatic compliance. Had he not himself as Chancellor compelled heretics to say yes or no to questions about the Pope's primacy?

and why should not then the King, since it is a law made here that his grace is head of the Church here, compel men to answer precisely to the law here as they did then concerning the Pope?

But the Pope's laws were universal, binding on all Christendom, answered More. Well, said Cromwell,

they were as well burned for the denying of that, as they be beheaded for denying of this, and therefore as good reason to compel them to make precise answer to the one as to the other.[80]

During one of his interrogations, putting cases with Richard Rich, More had offered more than silence concerning the supremacy; enough to condemn him.[81] Cromwell had never sought More's death but now, bound upon his allegiance to a master who did, he found ways to secure a guilty verdict. One of the jurors sworn for the trial on 1 July had an ancient grudge against More, and thought himself doubly the Chancellor's victim. John Parnell was the defeated defendant in a long and costly suit, and he was also of the brethren; a defender of Robert Barnes and married to Lucy, the widow of the persecuted John Petyt. He had tried once, and failed, to charge

[76] Ibid., 216. More's contest with his King from prison is discussed, definitively, in Elton, *Policy and Police*, pp. 400–19.

[77] More, *Correspondence*, p. 559; Marius, *Thomas More*, pp. 129, 473–4.

[78] Elton, *Policy and Police*, p. 407.

[79] More, *Correspondence*, p. 506.

[80] Ibid., pp. 557–8.

[81] Roper, *Lyfe*, pp. 84–6; Guildhall, MS 1231, fos. 2ᵛ–8ᵛ; J. D. M. Derrett, 'The Trial of Sir Thomas More', *EHR* lxxix (1964), pp. 450–77; Elton, *Policy and Police*, pp. 409–15.

More with corruption; now he had his revenge.[82] More was condemned as a traitor.

The summer of 1535 saw a series of executions of the Pope's defenders. Twenty years on, Cardinal Pole reminded the citizens that none had ever been vouchsafed such examples of steadfastness in the faith as they.[83] On 19 June three monks of the London Charterhouse were hanged, drawn, and quartered; though with less show than their Prior in the previous month, 'for fear of the displeasure of the people which was shown at the death of the others'.[84] Henry thought it necessary to 'appease the murmurings of the world' by ordering sermons to be preached against the traitors who would deny his supremacy. But these new preachers, like Robert Barnes, were denounced as 'false prophets', and their 'abhominable sermons' disturbed rather than placated the citizens.[85] Fisher's execution on 22 June was 'of very many lamented', for his learning and piety were revered. The Pope 'sent the Cardinal's hat as far as Calais, but the head it should have stood on, was as high as London bridge or ever the hat could come to Bishop Fisher'.[86] More went to the block on 6 July. 'I cannot tell', wrote Hall, who knew him, 'whether I should call him a foolish wise man, or a wise foolish man', for he could not even go to his death without jesting with his executioners. The authorities did not take his death so lightly, and two days later a watch was ordered through the City.[87] The reports of the Catholic ambassadors of the people's sorrow at these bloody executions was probably little exaggerated. Although some would watch plays showing the King cutting off the heads of his clergy, and may have watched the reality with equanimity, many were appalled. 'They begin to murmur, for ever since these executions began it has rained continuously, and they say it is the vengeance of God.'[88] Late in May a man was committed to the Tower for

[82] Roper, *Lyfe*, pp. 61–3; Harpsfield, *Life*, pp. 343–4, 349–50; Guy, *Public Career of Sir Thomas More*, pp. 75–7. Parnell's lasting rancour about the loss of the case is revealed further in his will: PRO, PCC, Prob. 11/27, fos. 103^r–104^r.

[83] Strype, *Ecclesiastical Memorials*, iii/2, pp. 490–7.

[84] *L&P* viii. 846; Wriothesley, *Chronicle*, i, p. 29.

[85] *L&P* viii. 876.

[86] Hall, *Chronicle*, p. 817; *L&P* viii. 948, 1104, 1105.

[87] Hall, *Chronicle*, pp. 817–18; CLRO, Journal 13, fo. 446^v; Letter Book P, fo. 63^v.

[88] *L&P* viii. 949.

foretelling, 'this month rainy and full wet, the next month death and the third month wars'. He must stay there 'until experience shows the truth of his prophecies'.[89] After the worst summer weather that anyone could remember dearth and plague would follow.[90]

With the swearing of the Oaths of Succession and Supremacy the clergy and citizenry of London had ostensibly renounced the Roman obedience and bowed to the King as Supreme Head. Despite almost universal acquiescence, neither Henry nor Cromwell were under any illusion that the schism was welcome, or that they had the people's support. They knew that many looked to the time when the old learning would return, and counselled steadfastness until that day. The parish clergy, now confronted by changes which outraged their conservative consciences and threatened their monopoly of spiritual knowledge, might associate the changes in authority within the Church with the heresy which sought to abolish any real distinction between priest and layman. Very many of them were disaffected, but they were bound upon their allegiance to the Supreme Head to propound the changes.[91] Moreover, they were now supervised by the laity: a circular letter came to the Sheriffs in June 1535 ordering them to make 'diligent search' to ensure that the Bishop was promulgating the royal supremacy without secret resistance. This order was publicly posted, humiliating Stokesley before his flock.[92] The Londoners, who had thought their priests 'overmighty', might perhaps now rally to their defence when they saw them so humbled. The King and Cromwell feared resistance, and with reason, but the days of collective attacks upon the clergy had passed. Now they moved against the leaders of the conservative hierarchy and individual dissidents, while conducting a campaign to persuade and convert the more amenable.

---

[89] Ibid. viii. 771.

[90] See below, pp. 240–1.

[91] From June 1535 parish clergy were ordered to preach the royal supremacy; *TRP* i. 158. For the reaction of the clergy to the changes, see Haigh, *Reformation and Resistance*, ch. 8; A. G. Dickens, *Lollards and Protestants in the Diocese of York, 1509–1558* (Oxford, 1959), pp. 138–53; M. Bowker, 'The Henrician Reformation and the Parish Clergy'; Elton, *Policy and Police*, pp. 230 ff.

[92] Ibid., pp. 238–40.

I.

All the Bishops of the old order—Fisher, Standish, Nixe,
Tunstall, Gardiner, Longland—suffered intimidation in the
1530s in an attempt to crush their opposition to the King's
designs.[93] But no Bishop was so persistently humiliated,
taunted, and tested as Stokesley. He was, as a reformer recognized, 'much harassed by Cromwell, and others', and 'died
miserably, almost worn out with grief'.[94] Cromwell, always
hating the 'snuffing pride of prelates', liked nothing better than
seeing the haughty and sanctimonious Bishops discomfited,[95]
but there was a particular rancour, even vindictiveness, in his
exchanges with Stokesley. Cromwell conducted a sustained
campaign against Stokesley throughout the 1530s. He knew
well that the Reformation in London would succeed in spite of
its Bishop rather than through his efforts, but other, older
reasons lay behind his animus. The opening round of hostilities
may have come in 1534 when the Abbess of Wherwell was
charged with incontinence, for her 'suspect familiarity' was—
allegedly—with none other than Stokesley.[96] No charges were
preferred against him, but he must have lived in dread.

The battle lines were drawn between Cromwell and Stokesley
over who should be licensed to preach in London, and who
should be silenced. Control of the pulpits would mean that the
promulgation of official measures would be assured—even if
hearing were not believing. The only pulpit of national importance was at the cross outside St Paul's cathedral. Cromwell,
Vice-Gerent in spirituals, determined to order the Paul's Cross
sermons and to use them as a vital element in his propaganda
campaign to establish the Henrician Church.[97] Preachers must
be found to denounce the usurped power of the Bishop of

[93] J. J. Scarisbrick, 'The Conservative Episcopate in England, 1529–1535'
(University of Cambridge, Ph.D. thesis, 1955), pp. 270–318; L. B. Smith, *Tudor
Prelates and Politics, 1536–1558* (Princeton, 1953), p. 184; Bowker, 'Supremacy and
Episcopate', p. 242; *L&P* viii. 592; xvi. 449.

[94] *Original Letters*, i, p. 231. But see below, p. 308.

[95] See, for example, *L&P* xii/2. 241. Elton, 'Thomas Cromwell *Redivivus*', *Studies*,
iii, p. 377.

[96] PRO, SP 1/85, fos. 10^r–13^r (*L&P* vii. 907). For earlier charges against him, see
A. F. Pollard, *Wolsey* (Fontana edn., 1965), p. 12.

[97] M. Maclure, *The Paul's Cross Sermons, 1534–1642* (1958); Elton, *Policy and Police*,
pp. 213–16.

Rome, and the King's critics must be debarred. Yet Paul's Cross was the pulpit of Stokesley's own cathedral, and his was the right to appoint the preachers. In December 1533 Stokesley was instructed to order carefully what was said at the Cross, but his censorship, lacking at the height of the Divorce crisis, was soon found wanting again.[98] In March 1534 John Rudd, a chantry priest at All Hallows Barking, had ruined the effect of his mandatory sermon against the Nun of Kent by a gratuitous attack (or so the authorities perceived it) upon those who had falsely accused her of violating penance and confession.[99] Edward Leighton also preached less than dutifully that summer.[100] Because 'the Word of God had lately suffered much obloquy' Cranmer invited the reforming luminary Latimer to preach the Lenten sermons at the Court in 1534, even though Latimer had twice been banned from London diocese the previous year.[101] By May 1534 Cranmer was assuring loyal preachers for the Cross by appointing them himself (using a City gospeller, John Blage, as his messenger).[102] Stokesley's authority was again usurped when Paul's Cross preachers were appointed over his head for the six weeks of the crucial Parliamentary session of November to December 1534.[103] The Bishop was not trusted.

But more than silence was required of the conservative Bishops. Their avowed approval of the schism might have a greater propaganda effect than all the invective of the reformers against the usurped powers of the Pope. So, one by one, they came to the Cross to preach the validity of doctrines they once had found, probably still found, anathema.[104] One of Henry's causes which Stokesley could countenance supporting was the Divorce; after all, it was his services in this matter which had brought him to his (now how unwelcome) eminence. Accordingly, Stokesley preached on 11 July 1535 of the invalidity of

98 L&P vi. 1487.
99 Ibid. vii. 303.
100 PRO, SP 1/85, fo. 61ʳ (L&P vii. 981).
101 BL, Harleian MS 6148, fo. 41ʳ (L&P vii. 29–30); PRO, SP 6/9, fos. 10ʳ, 84ʳ (L&P vi. 1214 (i,ii)). Cromwell possessed a copy of Stokesley's ban of Latimer (L&P vii. 923 (xxxix)).
102 BL, Harleian MS 6148, fo. 46ᵛ (L&P vii. 616).
103 Foxe, Acts and Monuments, v, p. 68.
104 Maclure, Paul's Cross Sermons, pp. 185–6; Elton, Policy and Police, pp. 214–15.

the Aragon marriage. Cromwell gleefully told Chapuys that he would have given a thousand marks for the Emperor to have heard this sermon and that he looked forward to sending him a copy.[105] But Stokesley had disingenuously declined Cromwell's invitation to provide the text; he preached without notes, and could not, so he said, now recall the exact words.[106] More likely, he distrusted Cromwell's motives and feared alterations which might even touch upon doctrine. Stokesley preached the royal supremacy, zealously enough: one thought that 'sugar would scarcely have melted in his mouth'.[107]

Stokesley could not countenance being present, and thereby conferring his approval, at sermons at the Cross which outraged his own faith. So, rather than hear Friar Hilsey 'rail' 'in mine own church' 'in mea contumelia' and against 'remembrance of souls departed', he appointed instead Simon Matthew, a conservative prebendary of St Paul's. (This in the same week that he had defied Cromwell by not providing the sermon for publication.)[108] Stokesley had summoned Hilsey and required him to conform to the Catholic doctrine of Purgatory, as Latimer and Crome had done, but unavailingly.[109] No doctrine did Stokesley defend more staunchly than this one: he had preached himself in the Shrouds of St Paul's in April 1535 on the virtue of satisfactory masses for the dead.[110] 'When I am departed hence', Stokesley told Cromwell, 'I shall suffer for your pleasure the friar to rail at the Cross at his pleasure, but I doubt not but he will set forth more fervently some pernicious doctrine.'[111] Cromwell acquiesced to the Bishop's plea, this time, but had his revenge: on 20 October 1535 Cromwell gave

---

[105] *L&P* viii. 1019, 1105.

[106] PRO, SP 1/94, fos. 50ʳ, 98ʳ (*L&P* viii. 1043, 1054); see also BL, Cotton MS Titus B i, fo. 425ʳ (*L&P* viii. 527).

[107] PRO, SP 1/155, fo. 65ʳ⁻ᵛ (*L&P* xiv/2. 613). Elton, *Policy and Police*, p. 161.

[108] PRO, SP 1/94, fos. 98ʳ–99ʳ (*L&P* viii. 1054); Elton, *Policy and Police*, p. 214 n. 5. Cromwell's remembrances for June 1535 noted 'the sermons made in London on Sunday last, and how well Symonds behaved himself': BL, Cotton MS Titus B i, fo. 475ʳ (*L&P* viii. 892). The government printed a sermon by Matthew: *A sermon made in the cathedrall churche of saynte Paule* (30 July 1535; *RSTC* 17656).

[109] PRO, SP 1/94, fo. 98ᵛ (*L&P* viii. 1054); *L&P* vii. 1643 (misdated to 1534).

[110] Foxe, *Acts and Monuments*, v, p. 601. Foxe gave the year as 1534, but this was certainly the sermon given in 1535. Stokesley preached in the Shrouds. This only happened when it was wet, and May 1535 was the wettest month that anyone could remember.

[111] PRO, SP 1/94, fo. 98ᵛ (*L&P* viii. 1054).

Hilsey a licence to appoint preachers to the Cross.[112] Thereafter, while Cromwell's influence lasted, Hilsey occupied the Cross more than any other Bishop, and 'when any abuse should be showed to the people either of idolatry or of the Bishop of Rome, he had the doing thereof by the lord Vice-Gerent's commandment'.[113]

The authority to license or silence preachers lay with each Bishop in his diocese. Preaching in the capital throughout the 1530s were Stokesley's conservative partisans. Stokesley needed to find men who were not touched by reforming doctrines, but who would still denounce the Pope; but often he found men whose theology he approved, but whose consciences were more scrupulous than his own, for their licences to preach they used as licence to inveigh against the new order. The names of one hundred clerics admitted to preach by Stokesley are known, and of these at least twelve were to find themselves in trouble for open resistance to royal policy in the 1530s.[114] Such was the concern about seditious preaching, particularly against the Divorce, that at Easter 1533 Cranmer instructed the Bishops of London, Winchester, and Lincoln to revoke all preaching licences and to issue new ones.[115] Again, in November 1534 he revoked all licences to preach in London, and did so once more early in 1536.[116] All these directives suggest that the City preachers were not fulfilling their duty to advertise the supremacy; that, worse, some were denouncing it, and their Bishop was powerless, or unwilling, to control them. If Stokesley was not to be relied upon, his reformist opponents would use their superior authority to infiltrate his diocese with men whose views he abominated: so the Archbishop kept Richard Champion, Thomas Lawney, and Thomas Garrett as his preaching chaplains, and the Vice-Gerent gave preaching licences to Thomas Rose and John Cardmaker.[117] These men were

---

[112] BL., Additional MS 48022, fos. 87^r–88^r.

[113] Wriothesley, *Chronicle*, i, p. 104.

[114] GLRO, DL/C/330, fos. 261^r, 261^v, 266^v, 267^v. For their resistance to the changes, see below, pp. 257–61.

[115] BL., Harleian MS 6148, fo. 20^r (*L&P* vii. 463).

[116] GLRO, DL/C/330, fo. 242^v; Elton, *Policy and Police*, pp. 244–5.

[117] See below, pp. 238, 243, 263, 265–6, 311–12, 316–17, 401; BL, Additional MS 48022, fo. 88^r. From Jan. 1536 the radical John Cardmaker had a Vice-gerential licence to preach.

leading evangelicals. Conservative City priests baulked at handing over their pulpits to heretical firebrands such as these: at St Margaret Lothbury John Forde said that he would accept none except those with the Bishop's licence, whose conservative views could, presumably, be relied upon.[118] But Stokesley's authority was already usurped.

Once Cromwell became the lay Supreme Head's lay lieutenant in spirituals, Vice-Gerent, he was empowered to inhibit or take over episcopal jurisdiction.[119] The trial of Anabaptists in London in May 1535 had been conducted by Vice-gerential commission, rather than by the Church Courts of the Ordinary.[120] Even so was the persecuting Bishop thwarted. In September 1535 letters came to the Bishops inhibiting them from exercising powers which had been delegated to the Vice-Gerent. A general visitation under Cromwell's direction was to follow.[121] On 2 October Cranmer wrote to Stokesley, inhibiting his episcopal visitation.[122] On 1 October St Paul's and other 'religious places' within the City were visited by Thomas Legh. At St Paul's a relic of Our Lady's milk was removed, for it was used for 'covetousness in deceiving the people'.[123] Like other Bishops, Stokesley secured a partial withdrawal of the inhibition on 19 October, which restored his powers to ordain, to institute to benefices, and to grant probate.[124] But little by little Cromwell ensured that Stokesley's control within his own diocese was undermined.

Cromwell used his patronage and influence to place compliant men in London benefices, increasing, in time, the power of the reforming party and reducing the influence of the conservatives. He would appeal directly to the patrons of benefices, and such bequests could hardly be refused.[125] Cromwell made certain that Stokesley lost control even over dignities and

---

[118] PRO, E 36/120, fos. 178ʳ–180ʳ (L&P xiii/1. 1492).

[119] Lehmberg, 'Supremacy and Vicegerency', p. 227; Bowker, 'Supremacy and Episcopate', pp. 234 ff.   [120] L&P viii. 771, 846.

[121] Wilkins, Concilia, iii, p. 797.

[122] Guildhall, MS 9531/11, fos. 67ʳ–68ʳ.

[123] Wriothesley, Chronicle, i, p. 31.

[124] Guildhall, MS 9531/11, fo. 68ʳ. No ordinations were carried out between 22 May and 18 Dec. 1535 in the diocese of London. C. J. Kitching, 'The Probate Jurisdiction of Thomas Cromwell as Vicegerent', BIHR xlvi (1973), pp. 102–6.

[125] L&P iv/3. 5410; v. 1227; PRO, SP 1/99, fo. 143ʳ (L&P ix. 992); x. 213; xii/1. 874; xiii/1. 669; VCH London, i, p 272.

canonries within his own cathedral for, given the chance, the
Bishop presented conservatives.[126] In 1537 Stokesley wrote to
Cromwell granting him, at the King's command, the collation
to the prebend of Islington, and lamented that this surrender of
promotions left him 'destituted of learned men' (that is, of
conservative caste).[127] Stokesley's authority was diminished in
every area almost. When early in 1536 there were stirrings at
the Crossed Friars, Hilsey, as their Visitor, entered the house
to quell the trouble, incurring the wrath of the Bishop's apparitor, who saw this as another breach of episcopal authority.
This clash of jurisdictions was debated for a whole afternoon in
Convocation on 10 March, but no conclusion was reached.[128]
The party of reform, led by Cromwell and Hilsey, had usurped
the Bishop's powers in order to enforce the Reformation in
London. But the acrimony between Stokesley and Cromwell
was not only about authority; not this sort of authority at least.

In January 1536 Stokesley wrote to Cromwell telling him
that he would have sent 'my books of the canon law and the
schoolmen favouring the Bishop of Rome', as well as Fisher's
tracts against the Divorce which had recently been proscribed
(in a proclamation issued in London on 15 December, a month
earlier than anywhere else in the realm). He would never wish
to 'keep unaware any that maintain that intolerable and
exorbitant primacy'.[129] The insolence and sarcasm in this offer
could hardly have been missed. There was no more learned
champion of the 'Romish decrees' than Stokesley; he would
with his 'devilish divinity' and 'old rusty sophistry' allege the
centrality of patristic and conciliar authority, and defend
'unwritten verities' against all those who asserted the sole rule
of Scripture. 'Ye are far deceived if ye think there is none other
Word of God but that which every souter and cobbler do read
in his mother tongue': so Stokesley challenged.[130] Once he had

---

126 Cranmer complained to Cromwell in Feb. 1539 that in a disputed presentment
in Essex Stokesley had instituted Hugh Payne, an erstwhile Observant and 'a seditious
person'; *L&P* xiv/1. 244.

127 PRO, SP 1/124, fo. 254ʳ (*L&P* xii/2. 729).

128 BL, Cotton MS Cleo. E iv, fo. 131ʳ (L&P x. 462).

129 PRO, SP 1/89, fo. 32ʳ (*L&P* viii. 55); *TRP* i. 161; CLRO, Repertory 9, fo.
145ʳ.

130 Foxe, *Acts and Monuments*, v, pp. 379, 383; Alexander Alesius, *Of the auctorite of the
word of God agaynst the bisshop of london* (1544; *RSTC* 292).

defended Erasmus's translation of the New Testament, but now he was appalled that the liberty to read the Bible in English would 'infect' the people with heresy.[131] It was Stokesley's agents who had hunted William Tyndale in Antwerp, and maybe they who found him. Tyndale was betrayed and martyred in Antwerp in October 1536.[132] But Cromwell, too, had read Erasmus's New Testament when it first appeared, and had learnt it by heart. Like his evangelical friends, he was passionately convinced that the Word should go forth. In 1534 Cranmer commissioned the Bishops to prepare an English translation of the New Testament. Even Gardiner had translated his two Gospels by June 1535, but nothing was forthcoming from the Bishop of London: 'I have bestowed never an hour upon my portion, nor never will . . . for I will never be guilty to bring the simple people into error.' His portion was the Acts of the Apostles, and these, jested Thomas Lawney, Cranmer's chaplain, 'were simple poor fellows and therefore my lord of London disdained to have to do with any of their acts'.[133]

Cromwell's campaign against Stokesley in the 1530s was first and last political, but it was personal too. For betrayal earlier in their careers Cromwell sought revenge. Both had served Wolsey, but Stokesley had fallen, rather plummeted, from the Cardinal's favour: 'in the Star Chamber openly put to rebuke and awarded to the Fleet'. In promoting the Divorce Stokesley found 'good occasion . . . to revenge his quarrel' against Wolsey, and 'busily travailed to invent some colourable device'; revealing it to the King, he disgraced the Cardinal and advanced himself.[134] Cromwell never forgave those who had brought about his master's fall; neither Stokesley nor Lord Darcy.[135] But if Cromwell had enemies to prosecute in the 1530s, he also had friends to advance.

Without the compliance of the capital the reforms would never succeed: Cromwell could ensure its conformity if anyone could. He was a citizen and— except for ventures abroad in his 'ruffian' youth—he had lived there always. He understood its

---

[131] See above, p. 71.
[132] J. F. Mozley, *William Tyndale* (1937), ch. 13 (esp. p. 300).
[133] *Narratives of the Reformation*, pp. 277–8.   [134] Roper, *Lyfe*, pp. 38–9.
[135] For his animus against Darcy, see G. R. Elton, 'Politics and the Pilgrimage of Grace', *Studies*, iii, p. 209.

institutions and how they worked; he knew where power lay; whom he should trust and whom not. He had friends among the leading merchants and City governors. Men like Alderman Monoux addressed him as 'my faithful beloved friend, Thomas Cromwell', and Cromwell himself wrote to the Merchant Taylors as his 'Right well beloved friends'.[136] These were more than conventional courtesies. Such old connections had a special significance in the 1530s, for upon men like Richard Gresham and Ralph Warren he relied for information, and he called upon them when he needed sound men to serve. Then as now, there were close links between those who governed the City and those who governed the country, but in the same way, there is little evidence of the table talk. Yet sometimes there are glimpses of those old friendships in the 1530s: writing to Cromwell in 1537 of the preaching of a 'hot gospeller' of more enthusiasm than latinity, Gresham told him wryly that the man ended his sermon, 'In nomine Patris et filius et Spiritus Sanctus', as if this might amuse the harassed minister.[137] Cromwell turned to Londoners to help him in his giant task; in turn, some had hopes that he would bring a change in the nature of government: Sir John Aleyn wrote to him in 1535 'for the commonwealth of all subjects, the which I know well hath ever been in your mind'.[138]

It was essential that conformist men serve in the City to police and enforce the changes. But disapproval of the royal supremacy and disinclination to prosecute those who resisted it may have provided a new and compelling reason for refusing to serve the City. By 1535 many were discharging themselves of their obligation to take office, thereby breaking their oath to the City and perjuring themselves.[139] Maybe the time and expense involved dissuaded most, but for some a conservative conscience intervened. In 1535 Richard Farmer, a wealthy grocer, extricated himself from serving as Alderman, and perhaps as Sheriff, by paying a £100 fine. He was an old opponent of the Divorce, and still in 1540 treasonably harboured a popish

---

136 *L&P* iv/2. 2387, 2400; Merriman, *Life and Letters of Thomas Cromwell*, i, p. 356.
137 PRO, SP 1/124, fo. 155ʳ (*L&P* xii/2. 624).
138 *State Papers*, i, p. 443; Elton, *Reform and Renewal*, p. 36.
139 *L&P* viii. 208. J. E. Sherwood, 'Religion, Politics and the Twelve Great Livery Companies of London, 1509–1547' (Pennsylvania State University, Ph.D. thesis, 1972), pp. 24 ff.

chaplain.[140] At St Andrew Holborn 'there was never so many fines or amerciaments paid for refusing the office of warden in the parish in any one year' as in 1536–7.[141] The attitudes of the leading citizens towards the new order were usually inscrutable, but Cromwell knew which men favoured reform. These men would be of use to him. Opponents of the supremacy must be excluded from office, even if it was their turn to serve.

The Communalty had always fought tenaciously to control appointment to its offices, and fiercely resisted any encroachment upon its autonomy, but in the 1530s, breaching his own City's cherished liberties, Cromwell intervened year after year in the City elections to ensure that the right men served. The Sheriff elected for Middlesex for 1535–6 (nominated by the Mayor) was none other than Humphrey Monmouth, once ostracized as Tyndale's patron.[142] Cromwell challenged Audley for the nomination of the Undersheriff for Middlesex that autumn, and won: Monmouth and Cotes promised that not only the nomination, but 'they and all they have shall be at the King's disposal'.[143] Cromwell feared some 'stir' in London, and with reason. The wettest of summers was followed by dearth and by plague. 'I fear that these great humidities will engender pestilence', warned Vaughan at the beginning of August.[144] As 'the common sickness waxeth very busy' in the capital the Court withdrew to the country and banished visitors.[145] On 22 August Sir John Aleyn reported widespread unemployment, and urged that 'if the King would make a present [of £10,000] to the City he would do a great deal of charity and cause many men, women, and children to be set to work, who will beg or steal if remedy be not found'.[146] There was 'great penury of wholesome bread'; 'what was sold for a half penny when you were here is now a penny', wrote Broke to Cromwell on 4 September.[147] Early in October

---

[140] CLRO, Journal 13, fo. 461ᵛ; see below, p. 313.

[141] Edward Griffith, *Cases of Supposed Exemptions, from Poor Rates . . . with a preliminary sketch of the Ancient History of St Andrew Holborn* (1831), p. xiii.

[142] *L&P* ix. 273–4.

[143] *L&P* ix. 273–4, 342, 370; CLRO, Journal 14, fo. 107ʳ.

[144] *L&P* ix. 47.

[145] *L&P* ix. 29, 41, 47; *TRP* i. 157, 160. For reports of the plague, see *L&P* ix. 85, 90, 99, 106, 119, 132, 137, 172, 259, 358, 413, 484, 549, 690, 700.

[146] *L&P* ix. 152.

[147] *L&P* ix. 274.

Cromwell rode from the City, risking the plague, to ensure that a conformable Mayor be elected, whether the citizens liked him or not, for the City was so unsettled that to have a man of influence and experience was vital.[148] Sir John Aleyn, of the royal Council and close to Cromwell, was their man. At the Guildhall on 12 October the King's Sergeant read a letter from the King commanding that Aleyn be chosen. So he was, but at no wish of his own, for reputedly he spent £100 in bribes trying to free himself.[149] For the following three years the mayoral elections were fixed, and Cromwell's candidates elected. In 1536 Sir William Hollis, whose rightful turn it was to be Mayor, was ousted by Ralph Warren. There was controversy among the citizens over the election, with partisans for the rival candidates taking sides for reasons which were clearly religious.[150] For by 1536 a different kind of Reformation was being promulgated.

II.

The King's commandment to all his clergy in June 1535 had been that they should 'publish and show to the people how the Pope hath usurped and taken upon him contrary to Christ's faith, and that his authority and pardons should be extinct and put down . . . for ever more, and his name to be blotted or put out of the mass book forever'.[151] Most London priests had faint-heartedly subscribed the oath which denied papal primacy, but few could have signed away their allegiance with a clear conscience. The martyrdom of the Carthusians, of Fisher and More, in May, June, and July 1535; the revulsion at hearing the martyrs denounced as traitors by 'false prophets' like Barnes; the horror of having to promulgate the royal supremacy themselves had caused some to defy the King's command. It was often the presence of restless and gleeful radicals in their congregations which caused these conservative clergy to resist, and certainly those of the New Learning who denounced them

---

[148] *L&P* ix. 594.

[149] CLRO, Repertory 9, fo. 130ᵛ. Extraordinarily, and doubtless significantly, the reference in the Repertory to the mayoral election breaks off abruptly. Wriothesley, *Chronicle*, i, pp. 31–2.

[150] See below, pp. 250, 294, 321.

[151] Wriothesley, *Chronicle*, i, p. 30.

before the authorities.[152] Seeing the monks of Syon brought from the Tower to Westminster for trial for their 'obstinacy' at the end of April, goaded Sir Roger to avow that 'God hath ordained here on earth a Vicar-General, which is the Bishop of Rome, who hath power over all the world, as well Christian as heathen'. Challenged by a servant of Henry's, who was present in the tavern, Sir Roger avowed the Gelasian doctrine of the two swords of temporal and spiritual justice; but this was not good enough for a King who would wield both swords.[153] As Robert Augustine, a Carmelite friar, preached at St Bride Fleet Street on 10 June, he was so incensed by his audience and by the slanders of the heretics who had preached before him, that he insisted that anyone who wrote 'otherwise than charity would require against the Bishop of Rome, he would regard him . . . no other ways than he would a schismatic, paynim or jew'. 'Also he said we should see a new time of the Bishop of Rome if we did live.' He did not expect to live.[154]

Sir Philip, priest at St Mary Woolchurch, hearing 'that heretic' Barnes preach, was sufficiently outraged to risk the Tower (though 'he cared not a fart' for it) for holding still to the papal primacy. 'Saint Peter being with God knoweth what a man requireth of him and heareth him in earth out from Heaven, therefore they were false heretics that would not believe in such like authorities.'[155] Other London priests were urging resistance to 'this enacted matters' and obedience to the Pope in the most secret of places, and where their advice was binding: the confessional. One priest claimed, not incorrectly, that treason uttered in confession need not be divulged.[156] At All Hallows Honey Lane Dr Coke used the confessional to 'corrupt' his parishioners with treason, as his predecessor had used it to inculcate reform. He counselled them to 'stand strongly in the old faith', for 'all these new preachings and New

---

[152] See below, pp. 277–9, 281–2.

[153] PRO, SP 1/92, fos. 72ʳ–73ʳ (*L&P* viii. 595).

[154] PRO, SP 1/92, fo. 127ʳ (*L&P* viii. 624): Elton, *Policy and Police*, pp. 19–20. In Aug. 1538 Augustine was granted a dispensation to hold a benefice: conservatives would be needed when the reaction came: *BRUO*.

[155] PRO, SP 1/94, fos. 1ʳ–2ʳ (*L&P* viii. 1000).

[156] *L&P* viii. 406–7; Keith Thomas, *Religion and the Decline of Magic: Studies in Popular Beliefs in Sixteenth- and Seventeenth-Century England* (1971), p. 154; Elton, *Policy and Police*, pp. 27, 346.

Learning were contrary to God's laws'.[157] In confession also, at the Crossed Friars in April 1536, George Rowland alleged his adherence still to the Pope as 'head of the universal Church of Christ', condemning the usurped supremacy of the King, who seized all for money. Convinced that almost none were yet converted to the New Learning, he counselled John Stanton to be 'steadfast in your faith, and not wavering'. But he spoke to the wrong person: Stanton was of the brethren. Swiftly, Hilsey prohibited confessions at the Crossed Friars and sent in the evangelical Cardmaker.[158] 'Under umber of confession', so it was reported on 24 August 1536, certain London curates and religious commanded their 'ghostly children' to 'hold honour due to the Bishop of Rome', and promised that the King and Queen and all the gospellers 'shall shamefully be put to death within two years'.[159] If this was the reaction while only distant matters of authority were altered, what would they do if doctrine were questioned and ancient ceremonies put down?

In the summer of 1536 new orders came which caused men to believe that the Catholic faith itself was threatened; and they were right. The worst fears of the prescient More and Fisher were realized. Now not only were papal pardons put down but, so it seemed, Purgatory also. In July 1536 Thomas Starkey warned Henry against the preachers 'who, under the colour of driving away man's tradition, have almost driven away virtue and holiness. With the despising of Purgatory the people begin to disregard Hell and Heaven.'[160] The Ten Articles of religion produced by Convocation in August 1536 mentioned only three of the seven sacraments.[161] In the North, Hall reported, they warned: 'see, friends, now is taken from us four of the seven sacraments, and shortly ye shall lose the other three also and thus the faith of Holy Church shall utterly be suppressed'.[162] Images were to be allowed, to 'stir' men to lament their sins,

---

[157] PRO, SP 1/99, fo. 97ᵛ (*L&P* ix. 846: this passage omitted). Coke's accuser was a reformer: John Maydwell.

[158] PRO, SP 1/102, fos. 73ʳ–74ʳ (*L&P* x. 346); BL, Cotton MS Cleo. E iv, fo. 131ᵛ (*L&P* x. 462); *L&P* x. 494. Elton, *Policy and Police*, pp. 27–30.

[159] BL, Cotton MS Cleo. E iv, fo. 127ʳ (*L&P* xi. 355); see also PRO, SP 1/106, fo. 3ʳ (*L&P* xi. 302); *L&P* xiv/1. 1074.

[160] *L&P* xi. 156.

[161] *Formularies of Faith*, pp. 1–20.

[162] Hall, *Chronicle*, p. 820.

but the people must beware of idolatry; saints might be honoured as 'elect servants of Christ', but 'all grace and remission of sin must proceed only by the mediation of Christ and no other'. Praying for souls departed was allowed, since it was mentioned in Maccabees, but 'there is no certain place named, nor kind of pains expressed in Scripture' as Purgatory, and 'such abuses' must be clearly 'put away which under the name of Purgatory have been advanced; as to make men believe that by the Bishop of Rome's pardons, or by masses said to "Scala Coeli" . . . or before any image, souls might clearly be delivered of Purgatory . . . straight to Heaven'.[163] The Injunctions of August 1536 which followed the Act, promulgated by Thomas Cromwell, Vice-Gerent and Lord Privy Seal, could not be dismissed, for they provided that the Ten Articles, with their profound implications for faith and worship, be declared.[164] 'To the intent that all superstition and hypocrisy . . . vanish away', the clergy were forbidden to 'extol . . . any images, relics or miracles, for any superstition or lucre, nor allure the people . . . to the pilgrimage of any saints'; rather, commanded to tell them that they would 'profit more their soul's health if they do bestow that on the poor and needy which they would have bestowed on the said images and relics' (so the Lollards, Colet, and Erasmus had persistently urged). The clergy were now bound to teach children the Pater Noster, the articles of faith, and the ten commandments in English, reciting one clause a day 'till the whole be taught or learned little by little'.

The doctrine of Purgatory came under attack while almost everyone still lived in dread of its pains. The legislation of the spring of 1536 for the suppression of the smaller religious houses further undermined the existence of Purgatory, for monks' lives were, supposedly, dedicated to praying for souls.[165] 'The great and fat Abbots' in the Lords had consented, said Hall, in hopes that by sacrificing the smaller houses their

---

[163] *Formularies of Faith*, p. 17. A. Kreider, *English Chantries: The Road to Dissolution* (Harvard Historical Studies, xcvii, Cambridge, Mass., 1979), pp. 122–4.

[164] Gee and Hardy, *Documents Illustrative*, pp. 269–74. For their enforcement, see Elton, *Policy and Police*, pp. 247–51. The order limiting the number of holy days maybe had less impact in London where all dedication days had been held on a single day since 1523: see above, p. 72.

[165] Kreider, *English Chantries*, ch. 4.

greater ones would be saved. 'But the great Abbots were putrified old oaks and they must needs follow: and so will other do in Christendom, quod Doctor Stokesley . . . or many years be passed' (whether with approval or fatalistic pragmatism).[166] Thomas Dorset, a priest at St Margaret Lothbury, claimed that Henry VIII himself had appeared before the Lower House on 11 March and 'delivered them a bill and bade them look upon it and weigh it in conscience'. The bill was probably for the Dissolution, and the fate of the monasteries must have been common talk in the capital for this priest to have known about it. He had gone 'as an idler, to Lambeth, to learn the news'. At Paul's Cross on 12 March Latimer inveighed against Bishops, Abbots, Priors, priests as 'strong thieves'.[167] Londoners knew well that there were covetous, worldly, concupiscent religious in London: there had been scandals enough:[168] but they had witnessed too the austerity of the Carthusians and of the reformed religious houses in and around the City. There is no reason to suppose that the citizens had been anxious to see the wholesale destruction of their religious houses. Monasticism had not been repudiated even by those who would soon benefit from its destruction. Only one religious house was suppressed in the City in 1536: Elsing Spital.[169] Wriothesley, writing from London, said of the Dissolution in general 'it was pity the great lamentation that the poor people made for them'.[170] 'Two men of law' found themselves accused of disaffection for failing to report a treasonable speech against the Dissolution.[171] Many must have felt uneasy and resentful, but would they act in defiance of the changes?

The events in high politics in the spring and summer of 1536 made it seem impossible that the reform would ever be

---

[166] Hall, *Chronicle*, pp. 818–19; cited in Lehmberg, *Reformation Parliament*, p. 226.

[167] BL, Cotton MS Cleo. E iv, fos. 131^r–132^r (*L&P* x. 462); Lehmberg, *Reformation Parliament*, pp. 227–8.

[168] See, for example, PRO, SP 1/105, fos. 177^r–180^v (*L&P* xi. 168); *Two London Chronicles*, p. 6; *Three Chapters of Letters relating to the Suppression of the Monasteries*, ed. T. Wright (Camden Society, xxvi, 1843), xvii, xviii, xxv. Such evidence, hardly ever impartial, was never less so than at the time of the Dissolution.

[169] *L&P* x. 1087(10), 1238; xii/2. 796(14), 1311(25), 1338.

[170] Wriothesley, *Chronicle*, i, p. 43.

[171] R. M. Fisher, 'The Inns of Court and the Reformation, 1530–1580', (University of Cambridge, Ph.D. thesis, 1974), p. 121.

reversed.[172] On 7 January Queen Katherine, in whom so many conservative hopes were still vested, died. Queen Anne 'wore yellow for mourning', but her jubilation was short-lived.[173] Henry was soon admitting 'as it were in confession, that he had made this marriage, seduced by witchcraft', which was proved because God would not give him a son.[174] On the very day of Katherine's funeral Anne miscarried.[175] The demure charms of Jane Seymour had already captivated the King. She had been 'well taught . . . by those intimate with the King, who hate the concubine'. By April she was, calculatingly, returning the gifts of her royal suitor; 'she had no greater riches in the world than her honour, which she would not injure for a thousand deaths'.[176] When Henry left the May day jousts precipitately 'many men mused, but most chiefly the Queen'.[177] By the following day she was arrested, charged with incest with Rochford, and multiple adultery with half the King's Privy Chamber. In the Tower Anne plaintively protested her innocence: 'I think that the most part of England prays for me . . . And then she said, shall I be in Heaven, for I have done many good deeds in my days?'[178] All was uncertain. 'Here are so many tales that I cannot tell what to write', confessed John Husee, Lord Lisle's London agent, on 13 May. 'If all books and chronicles were totally revolved . . . which against women hath been penned . . . since Adam and Eve . . . they were nothing compared to the sins committed by Anne'; though even those would not be as many as were imputed to her in London.[179] On 17 May Rochford, Norris, Weston, Brereton, and Smeaton died 'very charitably'; Anne went to the block 'boldly' two days later.[180] Sir Richard Page and Sir Thomas Wyatt remained in the Tower ('what shall become of them God best knoweth'), where Wyatt watched from the Bell Tower the

---

[172] The story of the savage politics of this spring and summer is told by E. W. Ives, *Anne Boleyn*, chs. 15–16; D. B. Starkey, *The Reign of Henry VIII*, ch. 6; R. M. Warnicke, 'The Fall of Anne Boleyn: A reassessment', *History*, 70 (1985), pp. 1–15.

[173] Hall, *Chronicle*, p. 818. So did the King: *L&P* x. 141.

[174] *L&P* x. 199.

[175] *L&P* x. 282.

[176] *L&P* x. 601.

[177] Hall, *Chronicle*, p. 819.

[178] *L&P* x. 797. For Anne's charitable works, see Bodleian, MS C Don. 42, fo. 25[r].

[179] *Lisle Letters*, 3. 695; *L&P* x. 866.

[180] *Lisle Letters*, 3. 698, 697.

execution of his friends. 'These bloody days have broken my heart . . .'[181]

'I promise you, there was much muttering of Queen Anne's death', said Constantine.[182] The reports of the last words of the Queen's innocent suitors were diverse—in significant ways. Chapuys told the Emperor that Rochford had died saying that he had 'deserved death for having been so much contaminated and contaminating others with these new sects, and prayed everyone to abandon such heresies'.[183] But the story one gospeller told another was quite contrary. Rochford had said:

I desire you that no man will be discouraged from the Gospel for my fall. For if I had lived according to the Gospel as I loved it, and spoke of it, I had never come to this. Wherefore said he, Sirs, for God's love, leave not the Gospel, but speak less and live better.[184]

The reformers feared that the disgrace of their great patron, Anne, would injure their cause. Cranmer had written sorrowfully to Henry, hoping for her innocence, but condemning her, if guilty: 'God sent her this punishment for that she feignedly hath professed His Gospel in her mouth and not in heart and deed.'[185] (Barnes hoped to console himself with the mastership of Bethlem Hospital, which was vacant by the death of Rochford.)[186] All those who loved the old ways had rejoiced at the death of the 'concubine', and looked for the restoration of Princess Mary. London women, who had been devoted to Queen Katherine, were her daughter's particular champions. At the end of August the Princess was proclaimed heir apparent at one City church.[187] But they were soon to be bitterly disappointed. Cromwell, with his rivals savagely dispatched, was higher in favour than ever, and the King, married to Jane Seymour, 'hath come out of Hell into Heaven'.[188] With all

---

181 *The Collected Poems of Sir Thomas Wyatt*, ed. K. Muir and P. Thomson (Liverpool, 1969), clxxvi; see also *Sir Thomas Wyatt and his Circle: Unpublished Poems*, ed. K. Muir (Liverpool, 1961), xxvii.

182 Constantine, 'Memorial', p. 64.

183 *L&P* x. 908, p. 379.

184 Constantine, 'Memorial', p. 65. For an almost identical account of Rochford's confession, see *Chronicle of Calais*, p. 46.

185 BL, Cotton MS Otho C x, fo. 226ᵛ (*L&P* x. 792).

186 *L&P* x. 880.

187 *L&P* x. 908; xi. 359.

188 *L&P* x. 908 (Henry was immediately 'banqueting with ladies'); *Lisle Letters*, 3. 713.

influence at Court now lost, the opponents of the heretics in high places might turn to raise, or to join, loyal rebellion in the country.[189]

On 1 October the commons of the North rose. Within a week 'ten thousand are up between Louth and Horncastle', all sworn to the rebel oath: 'Ye shall be true to Almighty God, to Christ's Catholic Church, to our sovereign lord the King, and unto the Commons of this realm; so help you God and Holydam, and by this book'.[190] The Pilgrims of Grace soon gathered forty thousand to the defence of the threatened religious houses and their 'old ancient customs'; 'the country rises wholly as they go before them'.[191] 'First beginner' of the Lincolnshire rebellion was Thomas Kendall, vicar of Louth. Favoured by Stokesley for his learning and for his resolute conservatism, and rewarded with promotions in London and Essex, Kendall had been the Bishop's chosen emissary to Essex in 1532 to combat the opinions of heretics.[192] Still in 1536, inciting rebellion against Cromwell's heretic agents, 'he had himself no other desire than to establish the faith and put down schismatic English books which deceive the unlearned'.[193] Others in London shared his sympathies, as soon became apparent. But would they rally to a rebel cause? The City's governors feared so. They knew well that there was a mass of uncoordinated resentment to the changes in religion; what they could not tell was where the City's loyalties would lie in a rebellion against the new order and the heretics around the King. But now this would be tested. The avowed aim of the rebels, or some of them, was to march south. Robert Aske, the leader of the Pilgrims, a Percy client and a lawyer of Gray's Inn, heard from a 'secret friend' in London that the 'South parts long for our coming'.[194]

'Special regard' must be had for London.[195] Strict pre-

---

[189] This is the contention of Professor Elton: 'Politics and the Pilgrimage of Grace', *Studies*, iii, pp. 183–215.

[190] *L&P* xi. 552.

[191] For the best account of the rising, see M. H. and R. Dodds, *The Pilgrimage of Grace, 1536–7, and the Exeter Conspiracy*, 2 vols. (1915).

[192] *BRUO*.

[193] *L&P* xi. 970; see also 828(1), 843, 854(ii), 968, 1062, 1224.

[194] *L&P* xi. 1128.

[195] *L&P* xi. 580(3).

cautions were taken immediately to prevent the unstable and dispossessed in the City from joining the rebels. Sanctuary men were, by 7 October, locked in their churches, so Chapuys reported. Rich Bishops and City merchants were urged to buy up stocks of unsold cloth, lest the clothworkers made their servants redundant, and therefore more likely to go over to the Pilgrims.[196] Orders came on 8 October for nightly watches through the City. Particular vigilance was to be had for 'valiant beggars' and masterless men. Each householder must answer for his servants. For the time all games were forbidden on Sundays and holy days.[197] The City governors were alert for any signs of disaffection during this state of emergency, and they found many: though some were only tangentially connected to the Pilgrimage of Grace, all were dangerous. 'This is a very hard world with poor men for to live', so an unemployed fuller complained at the height of the rebellion. 'Rich men that had work to put forth would not put it forth because of this busy world.' His friend promised 'on this side Christmas . . . you shall see another manner a change', for 'he did know a hundred good fellows that would see another remedy'.[198] Christopher Morece reported to Cromwell that he found 'few persons in London to serve the King in this business': Cromwell, for his part, had sent sixty or eighty masons who were at work on this house.[199] With each request for armed men to send north against the Pilgrims the dutiful, and frightened, City companies complied with alacrity.[200] London was not only threatened by the invasion of wild borderers, but perhaps also threatened from within by all those who sympathized with the Pilgrims' cause. Moreover, the City had its own internal dissensions in these days.

[196] *L&P* xi. 576.
[197] CLRO, Repertory 9, fos. 198ᵛ, 200ᵛ; Letter Book O, fo. 21ʳ.
[198] CLRO, Repertory 9, fo. 222ᵛ.
[199] *L&P* xi. 640, 576.
[200] Each company was ordered to provide a certain number of armed men: CLRO, Repertory 9, fos. 199ᵛ, 209ᵛ. They complied readily: the Mercers' Company prepared 408 men; the Armourers 35; the Carpenters 50: Mercers' Company, Acts of Court, ii, fo. 95ᵛ; Guildhall, MSS 12071/1, fo. 7ʳ (Armourers' Company); 4329/1, fo. 9ʳ (Carpenters' Company); 11571/1, fo. 8ʳ (Grocers' Company); 9481/1, fo. 51ʳ (Waxchandlers' Company); 7086/2, fo. 30ʳ (Pewterers' Company); 15333/1, fos. 132ʳ–134ʳ (Vintners' Company); Goldsmiths' Company, Court Book F, fo. 39ʳ; Johnson, *History of the Drapers*, ii, p. 62.

A letter came again from Henry expressing the royal will that his candidate for the mayoralty be elected, and so he was: Ralph Warren was chosen instead of Alderman Hollis, whose rightful turn it was. The royal intervention was resented, and an ineffectual protest made:

Item. That Mr Recorder . . . shall ride to the King humbly beseeching his grace to be contented that the commons shall have their free election as they have used afore this time.[201]

This election caused controversy among Londoners, already so unsettled by the news from the north. The exclusion of Alderman Hollis was as welcome to some as it was resented by others, and the partisanship was all to do with religion. In the watch at St Benet Fink in the first days of October there was a revealing quarrel between Robert Ferme and Thomas Multon:[202]

FERME. Shall not Mr William Hollis, our Alderman, be Mayor?
MULTON. No.
FERME. Why?
MULTON. Marry, a good cause why. I was afore him a little since, and Mr Alderman called me heretic knave, and trusted to God to see a day among us (of the New Learning).
FERME. What mean you by that?
MULTON. Marry, there is now a great rising in Lincolnshire.
FERME. What then?
MULTON. Yea, marry. Mr Hollis hath a brother in Lincolnshire which hath a great stock lying there of Mr Hollis to maintain these fellows withall, and this is the day that I mean . . . I had brought him before the King's Council afore this if I had not been a poor man.

Multon insisted that Hollis was a 'maintainer' of the Lincolnshire rebels, and 'because of this he is not Mayor of London for this year' (Multon may even have been right, for it was certainly because of his conservative views that Hollis was excluded.)[203] Multon was indeed a heretic, and a quarrelsome one, and because of this out of sympathy with his fellows in the watch. He challenged another of them: 'What say you to the putting down of the holy days?' Woodhouse answered circum-

[201] CLRO, Repertory 9, fo. 201^r.
[202] CLRO, Repertory 9, fos. 210^r–211^r.
[203] For Hollis, see below, pp. 320–1, 421.

spectly: 'Every man as he list, and his mind serveth him.' Said Multon: 'They that do not hold with the same be traitors', and he called Woodhouse 'traitor', and Woodhouse responded, 'heretic'. 'By the Mass', said Multon, 'half London be traitors', and for so 'slandering the City' he was sent to prison.[204]

Yet there were traitors in the City, and London's governors knew where some might be found. On 10 October the Aldermen were commanded to summon all the priests in their wards, and to demand that they deliver up all their arms. No priest or friar between the age of sixteen and sixty was permitted any weapon, save his meat knife.[205] In London, as in the North, the clergy were the chief agents to advertise and advocate the Pilgrims' cause. Sir George Throckmorton heard that 'divers men' and priests of the capital were making copies of the Pilgrims' demands and of Aske's letters, and that these were 'universal at London'.[206] The parish clerk of St Leonard Eastcheap dutifully handed in letters given to him by a chaplain, Sir Richard Scarlet: one was a copy of the rebel oath; the other an exhortation for the oath to be sworn and the cause followed.[207] William Gibson, a priest of Whittington College, declared that 'the Northern men rose in a good quarrel, and he trusted to see a good day'.[208] At St Mary at Hill the rector was Alan Percy, son of the Earl of Northumberland. His brothers, Ingram and Thomas, were leaders of the Pilgrimage, 'the lock, key and wards of the matter'; but he was circumspect (or apathetic).[209] Where the clergy counselled support for the rebels their influence may have been profound, but it was not compelling.

The Pilgrimage of Grace never came south of the Trent.

---

[204] CLRO, Repertory 9, fo. 212ʳ.

[205] CLRO, Repertory 9, fo. 200ʳ. The Northern clergy were inciting support for the rebel cause: M. Bowker, 'Lincolnshire 1536: Heresy, Schism or Religious Discontent?', *Schism, Heresy and Religious Protest, Studies in Church History*, 9, ed. D. Baker (1972), pp. 198 ff.; S. Brigden, 'The Northern Clergy in the Pilgrimage of Grace' (University of Manchester, B.A. thesis, 1973). See also Brinklow's allegations; 'Complaynt', p. 53.

[206] *L&P* xi. 1405–6.

[207] CLRO, Repertory 9, fo. 223ᵛ.

[208] *L&P* xiii/2. 1202.

[209] Percy was rector at St Mary at Hill through four reigns: 1521–60; Hennessy, *Repertorium*, p. 305.

Even if it had, was there ever any real chance that London would join the rebels? Marmaduke Neville's conclusion that 'Ye Southern men thought as much as we . . . but if it came to battle you would have fought faintly' contains much truth.[210] Prosperity and the social order must always be defended, and the citizens had too much to lose. Had Londoners heard that Thomas Kendall clapped his Louth parishioners 'of the back, bidding them to go forward justly, for they should have goods and riches plenty at London' they would have trembled.[211] Yet when the Pilgrimage of Grace failed many may have sorrowed. A Smithfield cobbler gave a Pilgrim down from Holderness some shoes for only 6*d.*: 'for ye have done very well there of late; and would to God ye had come to an end, for we were in the same mind as ye were'.[212] The only Londoners to suffer after the rising were John Dingley of the order of St John of Jerusalem, and Robert Branceter, a London merchant, who 'stirred divers outward princes to levy and make war' against the King.[213]

The Pilgrimage of Grace encouraged Londoners who shared the rebels' convictions to speak out, or worse, against the heretics around them. Rivalries between the partisans of the old faith and the new were very near the surface in these weeks, and ancient grudges were revealed, and revenged. The Northern rebels were threatening to come south to take Cromwell's head.[214] Many Londoners wished them well. That November Margaret Williamson vowed that 'or this time twelve month his head should be cut off, or else to take her neck'. She was sent to Maidstone prison, out of the way.[215] In St Magnus parish, as the conservative parishioners 'began to have some hope of a change of affairs', they came to Richard Hilles, a known evangelical, 'with menaces', when he would not pay for candles before the altar.[216] On a misty November morning

[210] PRO, SP 1/112, fo. 245ʳ (*L&P* xi. 1319); Elton, *Policy and Police*, p. 65.

[211] *L&P* xii/1. 69.

[212] *L&P* xii/1. 201, p. 95.

[213] S. E. Lehmberg, 'Parliamentary Attainder in the reign of Henry VIII', *HJ* xviii (1975), p. 686.

[214] See, for example, *L&P* xii/1. 199. There were rumours in Lincolnshire—wishful thinking—that Richard Cromwell had been murdered in Chancery Lane: *L&P* xii/1. 201, p. 89.

[215] CLRO, Repertory 9, fo. 222ʳ.

[216] *Original Letters* i, p. 231.

Robert Packington was shot in Cheapside, on his way from his house to his daily Mass at St Thomas Acon.[217] Many heard the shot, 'but the deed doer was never espied or known'. Some suspected vigilante clerics, resorting to illegal measures now that their arbitrary powers to charge with heresy were gone, for Packington had spoken often against the 'covetousness and cruelty of the clergy', and he was one of the brethren. To Hall the parallels with the murder of Hunne were striking. But Robert Singleton, the evangelical chaplain of the late Queen, was accused of the murder: the conservatives were attempting to shift the blame to a reformer.[218] Some gospellers were terrified. Rose Hickman's mother used to read evangelical English books to her children, but after Packington's murder, 'she charged us to say nothing of her reading to us for fear of trouble'.[219] Outraged gospellers had to be saved from themselves. Robert Barnes preached an impassioned sermon at Packington's funeral on 15 November.[220] The following day Marshall, Field, and Goodale, and 'another of that sort of learning' (John Bale?) were imprisoned by their patron, Cromwell. The next day Barnes went to prison also, the principal scapegoat.[221] Cromwell had moved against these leading reformers lest they stir a further conservative backlash, and to calm the distracted City. (Moreover, he still had need of them, and wanted them safe.)

The atmosphere in the capital was as tense as could be. John Husee was sending empty letters to Lord Lisle that December: he dared not write while 'there is diverse here that hath been punished for reading and copying with publishing abroad of news'.[222] At the Guildhall chapel Robert Winke was notorious

[217] Hall, *Chronicle*, p. 824; Wriothesley, *Chronicle*, i, p. 59; *Narratives of the Reformation*, pp. 296–7; Foxe, *Acts and Monuments*, v, p. 250 (misdated to 1538); Stow, *Survey*, i, p. 261; Lehmberg, *Reformation Parliament*, p. 21; *House of Commons, 1509–1558* (Packington). Packington's will reveals his reformed beliefs: PRO, PCC, Prob. 11/27, fo. 32ᵛ. For his association with Cromwell and Vaughan, see PRO, SP 1/90, fo. 212ʳ (*L&P* ix. 346); *L&P* viii. 303.

[218] Foxe, *Acts and Monuments*, v, p. 600.

[219] 'Recollections', p. 97.

[220] PRO, SP 3/4, fo. 4ʳ (*L&P* xi. 1097). Packington willed that Barnes preach his funeral sermon: PRO, PCC Prob. 11/27, fo. 32ᵛ.

[221] PRO, SP 1/111, fo. 187ʳ; SP 1/114, fo. 54ʳ (*L&P* xi. 1097; xii/1. 40); *L&P* xi. 1111, 1164.

[222] PRO, SP 1/240, fos. 200ᵛ–201ʳ (*L&P Addenda*, i. 1148); Elton, *Policy and Police*, p. 371.

as 'a great disturber of divine service', and in his tavern adjoin-
ing he 'made merry' during the Mass with people of 'suspect
living'. Early in December the Court of Aldermen banned him
from their chapel lest his blasphemy further disturb the citizens.[223] A secret watch was ordered on 17 December. The
King planned a procession through his capital on 22 December,
but the City's governors were unsure how to receive him.
Emissaries were sent to Cromwell to discover whether the
clergy 'shall receive the King's grace in copes or not, with
ringing of bells and censing or not':[224] such was the uncertainty
about the definition of orthodoxy now; such was the concern,
too, to please the King and avert interference; such, above all,
was the anxiety not to disturb the fragile status quo between the
warring religious elements in the City .

[223] CLRO, Repertory 9, fos. 229^r, 230^v; PRO, C 1/810/42.
[224] CLRO, Repertory 9, fo. 232^{r-v}.

VI   The Reformation in Religion, 1534–1539

THE BREAK with Rome had fostered the passage of reform. This was a consequence which Henry had never desired, but never guarded against. Early in 1534 he disclaimed 'any intention to follow the Lutheran sect, for he says he does not intend to touch the Sacraments, but only the vices and abuses of the Church'.[1] But others knew that he was inconstant, led by those with ideas far more radical than his own, and that the transformation of authority within the Church would allow the spread of reforming ideas. Needing men to propound the royal supremacy, Henry, or rather Cromwell, found men who believed that the Pope was not only a usurper but Antichrist also, and who held other reforming tenets which they were determined to evangelize. Since the challenge to the Supreme Head came not from the gospellers but from the Pope's Catholic supporters, so during 'all the seditious time' of the 1530s it was not heretics who were persecuted, but rather conservative traitors to the Supreme Head who were hunted down.[2] In November 1536 Thomas Multon was imprisoned for 'slandering the City' for his allegations that there were traitors abroad in the capital during the Pilgrimage of Grace. Yet for all his avowal of a comprehensive heretical credo he was, it seems, never punished. He had said before the watch at St Benet Fink:

Holy bread and holy water was nothing worth, for he can take bread and dip it in water and make as good holy bread as the priest. And also the houseling at Easter was nothing worth, for he could at London bridge foot buy a half penny worth of cakes, and bless them and make them as good as the priest doth, for the priest mocketh you houseling that he doth housel you withall. He hath no power, no more than the said Multon had himself, to consecrate the body of Our Lord. And ye believe ye cannot tell what. Confession is nought for a

[1] *L&P* vii. 232.
[2] Wriothesley, *Chronicle*, i, p. 104. Elton, *Policy and Police*.

man may as well confess himself to the yew tree in the churchyard as well to the priest. . . . Our Blessed Lady was no better than his wife.

Multon sounded for all the world like a Lollard, but he claimed to be of the 'new fashion'.[3] Multon's immunity from prosecution for uttering such heresies shows both the dangers of those days and how far reforming ideas were spreading, unchallenged.

A DIVERSITY OF PREACHING

Some of the English Reformation's leading evangelists were preaching in the capital in the 1530s. 'The Gospel was never more sincerely preached in the time of the apostles than it hath been of late in London'; so claimed Henry Brinklow in 1545, who was himself one of those who 'laboured in the vineyard of the Lord' for the Gospel's sake.[4] Through hearing the Word would men's minds be illumined, so the 'preachers of novelties' expounded Scripture urgently and often.[5] The silent Catholic clergy, they claimed, had been responsible for a thousand years of darkness, seeking to keep spiritual knowledge secret to themselves, 'angered to lose their glory and give it to Christ'.[6] The reformers had a mandate to assert the royal supremacy, but to them authority within the Church was a sterile issue compared with Scriptural truth, and they passed on quickly to expound the matters of faith which inspired them. In a sermon preached on Good Friday 1535 on the text for the day, *Christus passus est pro nobis*, the preacher supported the King's supremacy, but asserting Scripture as the rule of faith, allowed men's traditions as valid only when in accord with Scripture.[7] Some of these evangelists used their freedom to denounce 'the Bishop of Rome and all his cloisters'[8] as licence to deny also Purgatory, the intercessory powers of saints, even the Real Presence in the Mass. And their impassioned preaching could not fail to have an impact—one way or another—on those who came to listen. Thomas Corthorp, curate of Harwich, complained in 1535 that

---

[3] CLRO, Repertory 9, fo. 211ʳ.
[4] Brinklow, 'Lamentacyon', p. 96; PRO, PCC, Prob. 11/31, fo. 158ʳ.
[5] PRO, SP 1/99, fo. 201ᵛ (*L&P* ix. 1059).
[6] PRO, SP 1/105, fo. 104ʳ (*L&P* xi. 138).
[7] *L&P* viii. 451.
[8] PRO, SP 1/117, fo. 160ʳ (*L&P* xii/1. 757).

'the people now a days would not . . . believe the sayings of the captains of the Church, but when a new-fangled fellow doth come and show them a new story him they do believe'.[9] Catholic ambassadors reported, appalled, the outrages preached in London, and in October 1535 Dr Ortiz in Rome heard that

It is publicly said that the Mass is a great abuse; that Our Lord is not in the sacrament of the eucharist, and only was so when He consecrated it; that saying the Ave Maria is folly; and that Our Lady cannot help those who pray for her and invoke her aid, for she is only a woman like others. Blasphemous words are said of images.[10]

But such utterances had confused and disturbed the citizens: 'the King's loving subjects be in great doubts musing whom they should believe'. At Bethlem without Bishopsgate on 15 August 1535 Corthorp warned his audience that the new preachers had introduced 'divisions and seditions among us', never seen before, for 'the Devil reigneth over us now'.[11]

Yet preaching in London also were some of Bishop Stokesley's 'popish' disciples.[12] The sermons of divines who had been persuaded to accept the royal supremacy themselves, while otherwise remaining conspicuously orthodox, would be of the greatest use to the Henrician Church, for they might successfully win over the resistant. John Whalley suggested to Cromwell in April 1535 that even the Carthusians might be induced to conform if Rowland Phillips, Dr Buckmaster, Simon Matthew, and 'other of the popish sort' were called upon to preach against the papacy.[13] But such men might equally well be goaded to defend the old ways against heretic fanatics. Stokesley had given preaching licences to many who fell foul of the new order. Friar William Storme had been sent to the Fleet in 1533 for preaching too vehemently in favour of fasting, pilgrimages, and prayers to saints.[14] In March 1534 John Rudd had defended the Nun of Kent at Paul's Cross.[15]

9 PRO, SP 1/99, fo. 203[r] (*L&P* ix. 1059).
10 *L&P* ix. 681.
11 PRO, SP 1/99, fo. 201[v] (*L&P* ix. 1059).
12 See above. pp. 233. 235.
13 PRO, SP 1/92, fo. 77[r] (*L&P* viii. 600).
14 *L&P* vi. 1690; GLRO, DL/C/330, fo. 266[r].
15 *L&P* vii. 303. Since Rudd signed the Oath of Supremacy he must have been free by the summer of 1535; PRO. E 36/64, fo. 1[r].

Dr Nicholas Wilson, whom Stokesley had appointed late in 1533 to preach and to acquire heretical works in order to refute them, was silenced by imprisonment after his refusal to subscribe to the Oath of Succession.[16] Rowland Phillips had surprised many by conforming in 1534, but he was soon in trouble again.[17] William Clyff and Richard Hilyard were conservative northern clerics.[18] Robert Augustine and Robert Cronkherne had defied the new orthodoxy.[19] Battle was waged in the City pulpits between the evangelicals, free at last to expound the Gospel, and the 'captains of the Church'. This diversity of preaching caused dissension and confusion among the Londoners. In October 1533 Anthony Waite told Lady Lisle, 'many preachers we have here, but they come not from one master; Latimer many blameth, and as many doth allow'.[20]

The new politic orthodoxy which sought to purify the Church of superstitious abuse while protecting the essentials of Catholic doctrine was doubly threatened: reformers challenged the faith itself, and conservatives determined that doctrine and worship should never be changed. In 1536 the Convocation had adumbrated an orthodoxy which was hard to defend against the rival confessions: they allowed that Purgatory existed, but denied that the soul's passage thence to Heaven might be purchased; they allowed that the saints were God's elect, but denied their sanctifying powers for those who prayed to them.[21] Stokesley had in May 1535 defended the efficacy of prayers for souls departed, and thereafter counselled his disciples to defend Purgatory, while it was still permitted by statute.[22] But at St Paul's Cross through the spring and summer of 1536 this central doctrine was consistently undermined by official preachers. Cranmer preached there on 6 February, ostensibly against the usurped power of the Pope, but what most disturbed the citizens, Chapuys reported, was

---

[16] GLRO, DL/C/330, fos. 234[r], 267[v]; *L&P* vi. 1467.

[17] GLRO, DL/C/330, fo. 267[v]; PRO, SP 3/14, fo. 14[r] (*L&P* ix. 583).

[18] GLRO, DL/C/330, fo. 267[v]; A. G. Dickens, *Lollards and Protestants in the Diocese of York, 1509–1558* (Oxford, 1959), pp. 20, 24, 28, 47, 50; *L&P* xiv/2. 684, 723–4, 748–9, 750.

[19] See below, pp. 259, 279.

[20] *Lisle Letters*, 1. 58.

[21] *Formularies of Faith*, pp. 14–17.

[22] Foxe, *Acts and Monuments*, v, p. 601; BL, Cotton MS Cleo. E iv, fo. 131[r] (*L&P* x. 462).

his attack—'newest and most strange to the people'—upon Purgatory. His broadside against the superfluity of unbeneficed priests, praying for souls, and his promise that 'they shall away all that, saving they that can preach', was evidently preparatory to the seizure of endowments.[23] Latimer followed Cranmer to the Cross, and on 12 March, perhaps urged by Queen Anne, he gave a radical oration upon the theme of injustice, denouncing wealthy clergy ('strong thieves') who feasted while the poor suffered.[24] Latimer's attack was reinforced by a sermon by one of Anne's chaplains on 2 April. Robert Singleton denied the existence of Purgatory: if prayers for the souls of the dead were useless then chantry priests were redundant, he argued, and monastic wealth acquired by deception.[25]

One who had been preaching the contrary, bound by a vow made to the Blessed Virgin, came before three evangelical Bishops, Cranmer, Latimer, and Shaxton, early in March 1536. Robert Cronkherne, one of Stokesley's 'popish' coterie, confessed a celestial vision.[26] He had been 'rapt into Heaven', where he saw the Trinity in a blue mantle. 'From the middle upwards they were three bodies'; downwards only one, with two legs and two feet. Cronkherne had spoken with Our Lady: 'she took him by the hand, and bade him serve her as he had done in times passed. And bade him to preach abroad that she would be honoured at Ipswich and Willesdon as she hath been in old times.' Cronkherne vowed, but failed, to prove the existence of Purgatory to these sceptical Bishops from a certain verse of the psalter: Stokesley, he said, had shown him. Though the Bishops were infuriated by this priest's claims to mystical revelations, they were more gravely embarrassed by the challenge to their new orthodoxy from a fellow evangelical, who demanded further, more radical, reform. The following week John Lambert came before them, charged with heresy: he

---

[23] *L&P* x. 282.

[24] BL, Cotton MS Cleo. E iv, fos. 131ᵛ–132ʳ (*L&P* x. 462); S. E. Lehmberg, *The Reformation Parliament, 1529–1536* (Cambridge, 1970), p. 227.

[25] Robert Singleton, *A sermon preached at Poules Crosse . . . 1535* (*RSTC* 22575).

[26] GLRO, DL/C/330, fo. 266ᵛ; BL, Cotton MS Cleo. E iv, fo 131ʳ (*L&P* x. 462). John Bale, *Yet a course at the romyshe foxe* (1543; *RSTC* 1309), sig. Eiiiᵛ. For Cronkherne and confession, see Elton, *Policy and Police*, p. 346. Such admonitions were anyway dangerous in London, where loyalty to shrines remained strong: *L&P* viii. 1069; ix. 681.

had said 'it was sin to pray to saints'.[27] Lambert, like Latimer, had been converted by Bilney, and from Cambridge he went to Antwerp, to Tyndale and Frith. Forty-five articles of heresy were brought against him in 1532 by Warham and More, and his answers were a reformer's credo.[28] Upon Cranmer's elevation to Canterbury Lambert was freed, and from 1533 he lived in London by the Stocks, teaching Latin and Greek, and contemplating abandoning the priesthood for marriage and membership of the Grocers' Company.[29] The evangelicals who tried him in 1536 could not say, nor yet prove from Scripture, that prayers to saints were 'necessary or useful', but they were insistent that 'he might not make sin of it'. Latimer was 'most extreme against him'.[30] Uncompromising reformers like Lambert would undermine the campaign to extirpate superstition, and allow the conservatives to challenge official policy while claiming to protect the truth from extremism.

In the way of revolutions, the moderate Reformation in religion was being taken over by those with ideas more advanced than those of the original innovators. Cromwell and the evangelical Bishops needed to prevent the zealots from threatening the reforms they had made. The authorities moved to check preaching. In January 1536 a royal letter was sent to the Bishops regretting the preaching of divisive and extreme views; in April preachers were warned to avoid expressing all reformed opinions, save the denial of papal primacy; and all licences were revoked in July 1536.[31] It was at this time that Cromwell perforce sought Stokesley's suggestions for preachers at Paul's Cross until Michaelmas, knowing that he would name conservatives. From these Cromwell would appoint 'a convenient mixture' with his own men.[32] Stokesley's chaplains did preach at the Cross that summer, to the disgust of the reformers. William Marshall had learnt from his study of Italian writers of the courtier's duty to tell the truth: on 18 August he, unlike the flatterers, dared to tell Cromwell of how reform was in danger. Symonds, who had preached on 11 August, and Dr Buck-

[27] BL, Cotton MS Cleo. E iv, fo. 131^r (*L&P* x. 462).
[28] Foxe, *Acts and Monuments*, v, pp. 181–225.
[29] Ibid., pp. 225–6.
[30] BL, Cotton MS Cleo. E iv, fo. 131^r (*L&P* x. 462).
[31] *L&P* x. 45; Wilkins, *Concilia*, iii, p. 807; Elton, *Policy and Police*, pp. 243–6.
[32] PRO, SP 1/105, fo. 198^r (*L&P* xi. 186).

master, 'trusty disciples and chaplains' of Stokesley, should
'give a reckoning'. 'For God's sake', implored Marshall, 'help
to punish those that be proud, stubborn and obstinate against
the Word of God': especially the Bishop of London. 'If a mean
man and favourer of the Word of God, out of ignorance' had
preached as Stokesley had just done twice, he would have
burned or abjured. Marshall could 'rehearse a rabblement of
popish preachers' who had gone unpunished. Men of wealth
and influence in London 'but yet ignorant of the Word of God
and great enemies to the same', were saying, 'God save the
Bishop of London, and the vicar of Croydon, and such other
old preachers, for let men say what they will they preach well,
and if they preached not well they should not be suffered to
preach.'[33] In July 1536 the vicar of Croydon, Rowland Phillips,
was imprisoned, once again, for speaking out against royal
policy.[34] Provoked by the excesses of the radicals, he could
conform no longer. In a sermon the previous summer he had
endeavoured to win over the Carthusians (so Stokesley reminded Cromwell), but now he joined them.[35] From Rowland
Phillips's examination, and from the protest made, at the same
time, by the Lower House of the Southern Convocation, alleging sixty-eight articles of heresy against the reformers,[36] we
may discover the nature of the ideas which were spreading, and
test whether the allegations were correct.

THE EVANGELISTS

Rowland Phillips appeared before the Archbishop's court at
the end of July 1536.[37] His opposition to the royal supremacy
was long standing and well known, yet only the ninth question
tested his allegiance to the Henrician Church:

[33] PRO, SP 1/106, fo. 22^r (*L&P* xi. 325). For Marshall, see Elton, *Reform and Renewal*, pp. 18, 25–6, 62, 76.

[34] That Phillips was arrested is suggested by Stokesley's note in his own hand asking Bedyll to intercede with Cromwell for him: PRO, SP 1/105, fo. 198^r (*L&P* xi. 186).

[35] PRO, SP 1/92, fos. 77^r, 82^r; 1/105, fo. 198^r (*L&P* viii. 600, 602; xi. 186).

[36] Strype, *Ecclesiastical Memorials*, i/2, pp. 260–6.

[37] What follows is drawn from Rowland Phillips's examination. The editor of *Letters and Papers* dates the examination to 1537, but there is no supporting evidence for his arrest in that year: PRO, SP 1/123, fos. 125^r–128^r (*L&P* xii/2. 361).

Whom he meant by the Catholic Church when he said that the Catholic Church shall never err in things that be necessary for salvation?

He meant the universal multitude of Christian people, as well laymen as clergy, subjects as rulers.

The questions which were put to him concerned not his own disloyalty, rather the errors which were inculcated by his confessional opponents. Asked first whom he knew that 'trusted to be saved by faith and baptism and have left all good works', he named 'no special person', but thought 'it is about two years ago since the people came into that error', abandoning 'prayers, fasting and alms deeds'. The next question suggested the arrival of an alarming heresy: 'whether he knoweth any persons that doth evil works and leave all good works, which thinketh they do well therein and that they may do so without peril of damnation?' This was antinomianism. Questions upon the centrality of Scripture followed; whether the apostles, evangelists, and the Church preached it 'wholly, sincerely, dilucide et praecise'; 'whom he hath heard say that they would not have the Old Testament meddled withall, for it was but figures and shadows?' They lied, Phillips averred, who 'said the truth hath been kept from the people, and they have been mistaught this five or six hundred years'; indeed, 'the people had not been led in any darkness or error as concerning the faith, but have been taught as came from the feet of the apostles'. They lied, too, when they said that the people worshipped stocks and stones, and they 'slandered other men' when they accused them of preaching for 'promotions and vain glory' (Phillips was himself a famous preacher, and an archpluralist).[38] The interrogators asked questions which revealed their own, still moderate, convictions: whom had Phillips heard 'say of men's traditions, which come originally of Scripture and of the revelations of the Father or the doctrine of the Son or by instruct of the Holy Ghost, that they be but men's traditions?' He answered: 'the most part of them that have preached at London this two years'.

Phillips confirmed what was known already: that the most influential and charismatic preachers of reform were Robert

---

[38] See above, p. 72.

Barnes, Hugh Latimer, Edward Crome, rector of St Mary
Aldermary, and Richard Champion, rector of St Vedast
Foster. They expounded the cardinal tenet of Lutheranism:
'that faith which justifieth, of necessity bringeth forth good
works'. Phillips was not in agreement. Who denounced 'all
bodily observance as frivol and vain, all ceremonies of religion,
and all vocal prayer, calling it lip labour?' Latimer and Crome.
It followed from their injunction that God must be worshipped
in spirit. Some had said 'that we should not pray because God
knoweth our thoughts already', but Phillips did not know
whom. Who 'in Mass do use to clap their finger on their lips
and say never a word?' Phillips named no one, but knew that
'some great men in Court did so'; and even priests at St Paul's.
The proto-Puritan Scottish friar John Maydwell damned all
singing and organ playing. The churches, so lavishly adorned
by the devotions of the people, were scorned as useful 'only to
keep men from the rain or else to buy and sell in'. Some who
prayed for souls departed did so in sophistical ways: 'craftily'
intending not the dead, but 'souls departed from God by sin'.
Why did Cranmer seek from a conservative opponent corrob-
oration for what he knew well already: that reckless, eccentric,
anticlerical sermons were being preached by the evangelicals
whom official policy had loosed? Were not Richard Champion
and Thomas Lawney his own chaplains and Thomas Garrett
his favoured preacher?[39] The Archbishop needed to distance
himself from the zealots.

In the Lower House of Convocation a protestation of 'abuses
. . . worthy special reformation', was being prepared. Even as
Rowland Phillips, their erstwhile prolocutor, lay imprisoned,
they condemned sixty-eight offending doctrines, many of them
the same as those Phillips contested with his interrogators.[40]
To the 'slander' of the realm, 'disquietness of the people, damage
of Christian souls', 'light and lewd persons' challenged the
most cherished and profound convictions and practices of the
Catholic community. Worst of all, they said that 'the sacrament of the altar is not to be esteemed'. 'Is it anything else but

---

[39] *Lisle Letters*, 3, p. 453; 4, pp. 213–14; 5, pp. 111, 182; *Narratives of the Reformation*,
p. 276; *L&P* xi. 1424.

[40] The protestation is undated, but Strype places it in 1536: Strype, *Ecclesiastical
Memorials*, i/1, p. 380; i/2, pp. 260–6.

a piece of bread, or a little pretty piece round Robin?' Those who would deny communion in both kinds to the laity they called Antichrists, but this was the utraquist heresy. Anyone who was present at the Mass but did not receive the sacrament with the priests did not partake, they said.[41] This was to attack the still almost universal belief that to witness the Mass was a benediction, and brought the sinner closer to God.

The heretical views which the Convocation alleged were 'preached, thought, and spoken' were often radical, even militant, sometimes crass; yet when the allegations are compared with the accounts of evangelical sermons and ideas current in London it appears that they were not unfounded. Convocation claimed that the new preachers derided holy water as 'conjured', 'juggled', fit for making sauces and medicines, no different from river or ditch water.[42] And so they did. Robert Barnes asserted in June 1535 that there was no distinction between the water in the Thames and that hallowed by priests.[43] In Essex the previous March another apostate friar, Robert Ward, decried one who promised that holy water could wash away venial sins: not so, this was superstition, and worse, for it was 'robbery of the honour of Christ's blood'.[44] Barnes, too, in his militant *Supplicatyon* of 1531 (which Vaughan had thought he 'shall seal . . . with his blood') had held that the papists were Antichrists for giving some of the honour of redemption to human works instead of to Christ.[45] The reformers, so Convocation alleged, allowed no distinction between mortal and venial sins anyway; 'all sins, after the sinner be once converted, are made, by the merit of Christ's passion, venial sins . . . sins clean forgiven'.[46]

'Alonely faith justifieth before God'; this was the first message of the 'new sect'; the first 'common place' in Barnes's *Supplicatyon* of 1531, and *Sentenciae ex doctoribus collectae* (1530).[47]

---

[41] Strype, *Ecclesiastical Memorials*, i/2, p. 260, i; p. 261, vi–vii.

[42] Ibid. i/2, p. 262, xviii; p. 264, li; p. 265, lxii–lxiv.

[43] PRO, SP 1/94, fo. 1ʳ (*L&P* viii. 1000).

[44] *L&P* viii. 625.

[45] PRO, SP 1/68, fo. 56ᵛ (*L&P* v. 533); Barnes, *Supplicatyon* (*RSTC* 1470); W. A. Clebsch, *England's Earliest Protestants, 1520–1535* (New Haven and London, 1964), pp. 65–8.

[46] Strype, *Ecclesiastical Memorials*, i/2, p. 264, xlviii–xlix.

[47] Cited in J. P. Lusardi, 'The Career of Robert Barnes', *CW* 8/iii, p. 1388.

Indeed, 'it is damnable heresy to affirm that faith only doth *not* justify, seeing that Holy Scripture so teacheth', claimed Miles Coverdale later, in defence of his maligned mentor, Barnes.[48] But Barnes, back in England by the summer of 1534 from his mission to the Lutherans, had returned with more moderate opinions, and revised his *Supplicatyon*.[49] Sir William Shelley, judge in the Common Pleas, thought even the new *Supplicatyon* (1534) 'the most detestable and most unshameful book of heresy that ever he heard'; so extreme that Barnes 'would make a man that were wavering to return in his old manners'.[50] In 1531 Barnes had repudiated the instrumentality of good works in the justification of sinners: by 1534 he had come to a more moderate view of the relation between faith and works. Faith alone could justify a man before God, but works as a necessary fruit of justification made an outward declaration of the 'children of justification before the world'. Only faith might receive Christ's promise, but good works, following faith, would bring a reward. The reasoning of those who asked 'If faith only justifieth, what need we do any good works &c?', he condemned as 'fleshly and damnable'.[51] But there were some who preached the solifidian doctrine of Luther in more uncompromising ways than Barnes, and others who, hearing, believed. The wills of some London citizens were already expressing Protestant tenets of Christ's promise of redemption.[52] 'The captains of the Church' in Convocation chose to cite the more radical views expressed: 'that it is sufficient enough to believe though a man do no good works at all', 'we need not to do anything at all, but to believe and repent'.[53] How many gospellers concentrated in their sermons upon the justificatory tenets of reform is not clear, yet Crome, Garrett, Barnes, Wisdom, Lambert, Lawney, and Champion were certainly preaching

---

[48] Coverdale, *Remains*, p. 339.

[49] Lusardi, 'The Career of Robert Barnes', *CW* 8/iii, pp. 1396–7; *L&P* vii. 874.

[50] PRO, SP 1/239, fo. 83ʳ (*L&P Addenda*, i/1. 953); Elton, *Policy and Police*, pp. 32–3.

[51] Coverdale, *Remains*, p. 341; Clebsch, *England's Earliest Protestants*, pp. 65–8.

[52] See, for example, the wills of Humphrey Monmouth and Robert Packington; Strype, *Ecclesiastical Memorials*, i/2, pp. 368–74; PRO, PCC Prob. 11/27, fo. 32ᵛ; and ch. 9(i) below.

[53] Strype, *Ecclesiastical Memorials*, i/2, p. 264, lv; p. 265, lviii.

impassioned sermons, proclaiming 'only belief, only, only, nothing else'.[54]

Since salvation depended upon the sinner's belief in the sole power of Christ to redeem him, then mediation was worthless, whether in this world or beyond. 'Saints are not to be invocated or honoured, and that they understand not, nor knowing nothing of our petitions, nor can be mediators betwixt us and God.' Prayers to the saints were useless; as to 'hurl a stone against the wind'.[55] This is what Convocation alleged, and so the reformers had preached. Latimer, Crome, Garrett, and Barnes had denied, even in the time of persecution, the intercessory powers of the Virgin and the saints, and all had been in trouble before, once at least.[56] Now they might preach freely, and so they did, against the superstitious veneration of saints for their magical powers, and against oblations to their lifeless images, as though they heard their votaries. Thomas Garrett inveighed still against offerings to images.[57] But when Lambert said that praying to saints was a sin Latimer opposed him: their sanctity was not to be impugned, even though Scripture could not show prayers to them to be 'necessary or useful'.[58] John Bale thought it 'no robbery from God' 'to honour Our Lady and the saints', but thought also that 'neither she nor they were content with the superstitions used by many'. He denounced the 'spiritual whoredom of those who leave Christ for idolatry'.[59] The radical John Swinnerton, who fled to London when life became too hot for him elsewhere, had said that 'Our Lady was not of such honour as the people gave unto her, and preached against oblations done to saints in the Church of God'.[60] 'I have heard say in sermons', John Stanton told his popish confessor in the Crossed Friars in February 1536, 'that the images in the churches be stocks and stones and can not do us no good.' 'If I should give to the image of such a saint a candle, a man of

---

[54] PRO, SP 1/123, fos. 125^r-8^r (*L&P* xii/2. 361); *Letters of Gardiner*, p. 169.

[55] Strype, *Ecclesiastical Memorials*, i/2, p. 261, xv; p. 262, xxii; p. 263, xxxix.

[56] See pp. 115, 158-9, 187 above.

[57] PRO, SP 1/113, fo. 109^v (*L&P* xi. 1424).

[58] BL, Cotton MS Cleo. E iv, fo. 131^r (*L&P* x. 462).

[59] *L&P* xi. 1111; J. W. Harris, *John Bale: A Study in the Minor Literature of the Reformation* (Urbana, Ill., 1940), pp. 68-70.

[60] PRO, SP 1/113, fo. 109^v (*L&P* x. 1424); *L&P* x. 804, 850; Elton, *Policy and Police*, pp. 212-13.

wax or a penny I think it should nothing profit.'[61] Our Lady
was a 'maintainer of bawdery', so a 'false prophet' preached at
St Bride Fleet Street in June 1535 (perhaps he meant in encour-
aging pilgrimages).[62] Convocation alleged that some said worse:
'that Our Lady was no better than another woman, and like a
bag of saffron or pepper, when the spice is out, and that she can
do no more with Christ than another sinful woman'.[63] So said
the Lollards and the Anabaptists. So, some alleged, preached
Robert Barnes. At Barking church in 1537, or so Standish
had heard, Barnes 'in declaring the canticle Magnificat' spoke
slightingly of the Blessed Virgin. Coverdale had heard quite
the contrary from merchants in the congregation. Yet Barnes
always preached with irony.[64]

If the holy saints in Heaven could not intervene for forgiveness of sinners, could the Church on earth? Salvation
rested upon faith and upon repentance: it followed, for the
reformers, that 'confession auricular, absolution, and penance
are nothing necessary or profitable in the Church of God'; 'that
it is sufficient to make . . . confession to God only'; that the
clergy had no power to absolve; that it was lawful to confess to
a layman as well as to a priest.[65] John Stanton relayed the
message of Latimer's sermon of 22 February 1536 to his outraged confessor: 'that no man of himself had no authority to
forgive sins and . . . if a man came to confession and be not
sorry for his sins the priest had no power to forgive him'.[66]
Friar Ward in his offending sermon in March 1535 had desired
priests to reform themselves in ministering the sacrament of
penance; no longer to delude the people 'that they will have
forgiveness by their absolution'.[67] Though he never denied the
necessity of auricular confession, said John Bale, 'no priest can
assoil [absolve]' those who would not reconcile themselves to
those whom they had offended.[68] For another apostate friar,

---

[61] PRO, SP 1/102, fo. 74^r (*L&P* x. 346).
[62] PRO, SP 1/121, fos. 83^r–84^r (*L&P* xii/2. 65). The editors of *Letters and Papers*
misdate this document.
[63] Strype, *Ecclesiastical Memorials*, i/2, p. 263, xl.
[64] Coverdale, *Remains*, pp. 347, 350–1; see also below, pp. 317–18.
[65] Strype, *Ecclesiastical Memorials*, i/2, pp. 262–3, xxvi–xxxii; p. 265, lix–lx.
[66] PRO, SP 1/102, fo. 73^r (*L&P* x. 346).
[67] *L&P* viii. 625.
[68] *L&P* xi. 1111.

Frances Eliot, 'confessions was but a counsel and no command-ment'.[69] Thomas Multon had avowed in October 1536, in the way of the Lollards, that 'a man may as well confess himself to the yew tree in the churchyard as well as to the priest'.[70] This was to express contempt for the priesthood as well as confession, and of course the two were related. Why confess to an 'unlearned buzzard'; how could a priest absolve 'so long as priests may keep whores without danger of death?'[71]

One of the effects of Luther's abolition of the absolute distinction between priest and layman was the denial of priestly celibacy. The vow of chastity might be against God's law. Barnes in his *Supplicatyon* (1534) argued, 'By God's word it is lawfull to priests, that hath not the gift of chastity, to marry wives', and John Rastell petitioned Cromwell in 1534 that priests might marry.[72] Some were already marrying, including Friar Ward.[73] In November 1535 Anthony Waite wrote that 'it is preached here that priests *must* have wives'. Lord Lisle thanked him for the news, 'which I would a been gladder of twenty years agone that I might have made some priest cuckold'.[74] In Convocation they had associated the claim that priests could marry with another charge more extreme and dangerous: 'that all things ought to be in common' (and this only a year after the anarchic commune of the Anabaptists was set up in Munster). But here the clergy in Convocation were imputing to the gospellers views more radical than all but a very few would countenance, in order to discredit the 'new sect'. It was said, they claimed, that 'God never gave grace nor knowledge of the Holy Scripture to any great estate or rich man'.[75]

The constant fear of sixteenth-century governments was that religious radicalism would become confused with a general antagonism to the ruling orders; that with the new contention between the faiths an extra dimension would be added to the

[69] PRO, SP 1/113, fo. 109ʳ (*L&P* xi. 1424).

[70] CLRO, Repertory 9, fo. 211ʳ. Anne Hudson, *The Premature Reformation: Wycliffite Texts and Lollard History* (Oxford, 1988), pp. 294–301.

[71] Brinklow, 'Lamentacyon', p. 111.

[72] Lusardi, 'The Career of Robert Barnes', *CW* 8/iii, p. 1397; Lehmberg, *Reformation Parliament*, p. 215.

[73] Wriothesley, *Chronicle*, i, p. 83; *L&P* xiii/2. 571.

[74] *Lisle Letters*, 2. 477, 479.

[75] Strype, *Ecclesiastical Memorials*, i/2, p. 261, xi, xiii; i/1, p. 380.

tension between the rich and the poor; that some might use the Gospel to preach a socially egalitarian message. The Gospel was being adduced by the reformers to warn the rich of the cardinal sin of covetousness, to exhort them to deal mercifully with the poor, and to console the oppressed. This was nothing new. Preachers had for centuries challenged the ethics of the wealthy, and alleged the righteousness of poverty.[76] After all, Christ had pointed out the difficulties for the rich of entering the kingdom of Heaven. Yet the message was expounded with new urgency and might easily become seditious. From the pulpit of St Bride's in June 1535 an evangelical had preached the terrible parable of the rich man and Lazarus (Luke 16: 19–26). The papist who followed him remarked sarcastically that 'although the rich man had no name, yet right well he might have named his steward, called Nemo [No one]'.[77] But some soon did give names to the rich men.[78] Such preaching by 'apostates, abjured persons' would arouse social unrest, it was feared, even unleash a wider chaos. 'The Devil reigneth over us now.'[79]

The gospellers expected rejection. Bale did not care when he was taunted as apostate friar, 'no more than Paul when he reported him once a Pharisee and a persecutor'. They were proud of being called 'new learned fellows and teachers of new doctrine', for so the 'Pharisees called Our Saviour Christ'. They in turn reported stalwart conservatives to prevent the trouble which would 'upsurge . . . if that such papists be not sequestered from such seditious preaching'.[80] And they were strong in their certainty of the truth they expounded. Barnes wrote:

When I am dead, the sun and moon, the stars and the element, water and fire, yea, and also the stones, shall defend this cause against them

---

[76] G. R. Owst, *Preaching in Medieval England: An Introduction to Sermon Manuscripts of the Period c.1350–1450* (Cambridge, 1926), pp. 17–21, 182–4, 295–6; *Literature and Pulpit in Medieval England* (Oxford, 1966), pp. 237–375; N. Cohn, *The Pursuit of the Millennium* (1962 edn.); H. C. White, *Social Criticism in Popular Religious Literature of the Sixteenth Century* (New York, 1944).

[77] PRO, SP 1/121, fo. 83ʳ (*L&P* xii/2. 65). Thomas Rose had preached the same parable in Hadleigh in 1534, to the dismay of the inhabitants: BL, Harleian MS 6148, fo. 1ʳ (*L&P* vii. 355).

[78] See below, pp. 319–20.

[79] PRO, SP 1/99, fo. 201ᵛ (*L&P* ix. 1059).

[80] *L&P* xi. 1111.

(meaning the cause of God's Word against the spiritualty) sooner than the verity should perish.[81]

If the ideas described by Convocation were so freely divulgated, unpunished, who then would be so outrageous, and what heresies so appalling, that they would be persecuted? During 'all the seditious time' only those who were guilty of disturbing the peace as well as of apostasy, convicted of social as well as religious radicalism, were hunted down. Or so it seems, for we surely cannot know all who were detected. The cessation of persecution upon the resignation of More had been none of the clergy's choosing, and now their powers to accuse and judge were curtailed by the Heresy Act of 1534.[82] The City authorities were not eager to search the consciences of citizens. But some heresies were so extreme, so repellent, that they could not be tolerated, even for a time.

Anabaptism alarmed the authorities more than any other heresy. From the Continent Cromwell was receiving disturbing reports in 1534 from his agents: 'divers places are infested with this new sect of rebaptizement', numbering 6,000 in East Holland already.[83] By 1535 it was feared that Anabaptists and their ideas were entering England. As sacramentaries, and worse, the Anabaptists were guilty of the gravest error, but their apostasy was tainted by sedition. Every government in Europe was haunted by the spectre of Munster. Anabaptism might be adopted as easily as any other Continental heresy: not least because Anabaptists and Lollards held certain doctrines in common.[84] In 1532 a group of radical sectaries had been discovered in St Michael Bassishaw, meeting to read Scripture and disseminate heretical books. Led by Bastiane, a Fleming, their 'Bishop and reader', they were clearly Anabaptists, holding 'strange opinions' concerning the humanity of Christ. John Gough, the ubiquitous gospeller, printed for them the 'Confession of Geneva'.[85] The spread of Anabaptism was not confined

---

[81] Coverdale, *Remains*, p. 346.   [82] See above, pp. 218–19.

[83] *L&P* vii. 317, 394, 397; viii. 198.

[84] G. Williams, *The Radical Reformation* (Philadelphia, 1962); Dickens, *Lollards and Protestants in the Diocese of York*, pp. 10–11, 21; I. B. Horst, *The Radical Brethren: Anabaptism and the English Reformation to 1558* (The Hague, 1972).

[85] PRO, SP 1/237, fo. 290^r (*L&P Addenda*, i/1. 809). Horst, *Radical Brethren*, pp. 50–3.

to London, but in the capital foreigners abounded, and it was here that the first Anabaptists were discovered, and their error publicly displayed.

In March 1535 a proclamation was issued, ordering the 'great number' of foreign Anabaptists to depart the realm within twelve days, upon pain of death.[86] By displaying himself in his role as *defensor fidei*, Henry demonstrated to conservatives at home and abroad his determination to protect orthodoxy from the spread of strange heresies. The burning of foreigners, whose heresy had not yet found support even among radical spirits in England, was both effective propaganda to persuade his detractors, and uncontroversial. The action against the Anabaptists in 1535 showed clearly that the impetus and authority had passed from Bishop to Vice-Gerent. Towards the end of May over twenty Dutch Anabaptists were captured and sent for trial at St Paul's.[87] Their 'no less strange than damnable' opinions were quintessentially Anabaptist:

That Christ hath not the nature of God and man; that Christ born of the Virgin Mary took no part of the substance of her body; that the bread consecrated by the priest is not the incarnate body of Christ; that baptism given in the state of innocence to children does not profit; that if a man sins deadly after being baptized he shall never be forgiven.[88]

In the way of sectaries, they claimed the inspiration of the Holy Spirit.[89] Before a Vice-gerential commission, upon which sat Cromwell, Barnes, and Stokesley, about half these Anabaptists were convicted of doctrinal heresy and sentenced to be burnt. A man and a woman suffered on 4 June, and others were sent elsewhere to share their fate and to provide a dreadful warning.[90] Londoners saved their compassion for the Catholic martyrs who went to the block in the same month.

---

[86] *TRP* i. 155; R. W. Heinze, *The Proclamations of the Tudor Kings* (Cambridge, 1976), p. 135. The proclamation allowed no possibility of the existence of English Anabaptists, and appealed to English xenophobia.

[87] *L&P* viii. 771, 846; Foxe, *Acts and Monuments*, v, p. 44. These heretics were tried not by the Bishop in his Consistory, but before a special commission. This was the first test of the new Vice-Gerent in spirituals and his Court of High Commission: S. E. Lehmberg, 'Supremacy and Vicegerency: a reexamination', *EHR* cccxix (1966), p. 227.

[88] *L&P* viii. 771; Stow, *Annales*, p. 571.

[89] *L&P* viii. 846.

[90] Wriothesley, *Chronicle*, i, p. 28; *L&P* viii. 826.

Thomas Merial, a tiler of St Sepulchre, was the first Londoner to be indicted under the new Heresy Act. Provoked by Stokesley's sermon in the Shrouds commending masses for souls departed, Merial avowed strange heresies at the May watch in 1535 (and again at the end of July at the Mount at St Giles Cripplegate): 'Christ never died nor shed His blood for us, but only for them who were in limbo patrum.' More prosaically, he claimed that his wife was as good as Our Lady.[91] To deny that Christ's passion could redeem those who came after Him, as Merial and a few other eccentrics did, seemed to contradict the essential Christian message.[92] Merial might never have been haled before the Bishop, for all this, had he not spoken before John Twyford: not only was Twyford a 'furious papist', the builder of the fires of Smithfield for Tewkesbury, Frith, Bayfield, and Bainham, but he also had an old grudge against Merial for 'striking his boy'. Foxe heard that Twyford 'allured' ten men to his tavern, 'craftily overcame them with wine', and persuaded them to depose against Merial. Maybe: but Foxe's informants were Gregory Newman, Merial's wife, William Tomson, and others, all known gospellers.[93] Indicted before Middlesex assizes, Merial appealed to Cromwell. He claimed that Twyford, 'a great blasphemer and railer upon all the true preachers of the Word of God', had presented him maliciously for his defence of the reformers; moreover, that the jury was packed with the Bishop's servants, who would not allow him to speak in his own defence.[94] For his heresy Merial did open penance, bearing a faggot at Paul's Cross on 19 November 1535.[95]

Another heretic of the lunatic fringe made this appeal: 'Good Lord Cromwell, help me out of prison'. 'The raging of my tongue against the Church hath been for lack of discretion, which in time past hath lacked.'[96] An understatement, this.

[91] PRO, SP 1/94, fo. 227^r (L&P viii. 1129); Foxe, Acts and Monuments, v, p. 601. Foxe gives 1534 as the year of Merial's presentment, but the outburst was provoked by Stokesley's sermon in the Shrouds of St Paul's in May 1535.

[92] For the same heresy, see Elton, Policy and Police, p. 377.

[93] Foxe, Acts and Monuments, v, pp. 601–2.

[94] PRO, SP 1/162, fos. 154^r–155^r (L&P xv. 1029(47)). Significantly, Merial's appeal was to Cromwell.

[95] Two London Chronicles, p. 13. Miles Hogarde remembered this heretic later; The displaying of the protestantes (RSTC 13557; 1556); p. 18.

[96] PRO, SP 1/144, fos. 154^r–155^r (L&P xiv/1. 647).

The appellant was William ('frantic') Collins.[97] In April 1536 he came before the Common Council as a 'common brawler, fighter and common scold'. His actions were as provocative as his convictions:

Shooting dispitefully at the picture of Christ at St Margaret Pattens, and also for neglecting and despising the sacrament of the altar, and all other sacraments of holy Church, and for railing against the said sacraments commonly.[98]

He had fired an arrow at the foot of this famous rood—an object of pilgrimage, especially for Spanish sailors—challenging it to defend itself and punish him, if it were able.[99] In a London church, distracted, so Foxe said, by the infidelity of his wife, Collins had mocked a priest during the most sacred part of the Mass: as the priest elevated the host Collins lifted a dog over his head.[100] Some said it was not even for this that he was imprisoned, but for using Scripture to denounce 'the nobility and great men of the kingdom', for saying that in Scripture, 'especially the prophets . . . there was any [no] mention made of unrighteous judgments, or the cruel treatment of neighbours and dependants'. Those whose reckless evangelism turned to violence or to a crusade against the social order threatened the peace of the City. Collins's pleas went unheard: in prison he remained, collecting verses from Scripture to console the faithful, and promising destruction for 'bloody and deceitful men'.[101]

'Spiritus ubi vult spirat.' From a window in his mansion house in Whitechapel, from a tree in his garden, the 'hot gospeller' John Harrydance, another heretical bricklayer, expounded the Gospel, day and night at 'intempestive hours'. He preached twenty times, two-hour sermons, between June and August 1538, to anyone who would listen (and to some who would not: a neighbour playing bowls threatened to throw a bowl at his head unless he stopped). The curate of Whitechapel cursed Harrydance from the pulpit—as Hunne had been cursed there twenty years before—and 'cried out blood

---

97 For Collins's earlier troubles, see above, p. 117.
98 CLRO, Journal 13, fo. 476ʳ.
99 *Original Letters*, i, pp. 200–1.
100 Foxe, *Acts and Monuments*, v, p. 251.
101 *Original Letters*, i, p. 201.

and fire' upon him. But Harrydance scorned adversity, thinking it 'no marvel if the world doth persecute holy men and setters forth of light', for Christ would reward the faithful. He challenged a baker in his parish: 'it is as fit for me to be burned as for thee to bake a loaf'.[102] The authorities tried repeatedly, and failed, to silence him. Summoned before the City fathers in August 1537 for unlicensed preaching, Harrydance promptly repeated his sermon (Gresham did not think he would bother to relay his message to Cromwell).[103] He was of old a Bible man: 'he all these thirty years hath endeavoured himself to learn the Scripture, but he can not write or read'.[104] This illiterate nevertheless 'declared Scripture as well as he had studied at the universities', and gathered a following of laymen and clerics: sometimes only a handful, but Edward Underhill remembered seeing a thousand at these sermons. 'By reason the people followed him', Cranmer examined Harrydance in 1538, but could prove nothing against him, save his unlicensed preaching.[105] Once Harrydance had been a Lollard, and perhaps was a Lollard still. In the way of the Lollards, he turned rather than burned. At Paul's Cross on 8 December 1538 he bore a faggot.[106]

Robert Gore, a City draper, appeared before the Court of Aldermen in April 1538 for utterances as seditious as they were heretical. No one, he said, should give any credence to them 'that wear crowns' (the tonsured clergy), 'for they were false harlots all in general, as well the Bishops being of the King's Council, as the other sort'. There was 'no more virtue in the holy unction' than in grease or butter, and the King had been deceived by his anointment at his coronation. Let men accuse him of heresy if they liked: 'he is contented to be burnt in this cause, for then he should not die for cold'.[107] The papists had

[102] PRO, E 36/120, fos. 133^r–135^r (*L&P* xiv/2. 42); Elton, *Policy and Police*, pp. 162–4. For other heretical bricklayers, see BL, Cotton MS Cleo. E v, fo. 398^r (*L&P* ix. 230).

[103] PRO, SP 1/124, fos. 118^r, 155^r (*L&P* xii/2. 594, 624); CLRO, Repertory 9, fos. 261^v, 262^r.

[104] PRO, SP 1/124, fo. 155^r (*L&P* xii/2. 624).

[105] *Narratives of the Reformation*, p. 171; Wriothesley, *Chronicle*, i, pp. 82–3.

[106] *Two London Chronicles*, p. 15.

[107] CLRO, Repertory 10, fo. 28^r; cf. Strype, *Ecclesiastical Memorials*, i/2, p. 262, xix; A. Duke, 'The Face of Popular Religious Dissent in the Low Countries', *JEH* xxvi (1975), p. 53.

promised such a fire for Latimer, for maligning the Virgin
Mary in a ballad which Gore may have heard:

Pulchra es, amica mea, et macula non est in te.
This text Christ said by her, as in Scripture is told.
Wherefore it is pity thou shouldest die for cold.[108]

Gore went to prison, but lived on, a heretic.[109]

Heresies concerning the eucharist were more feared than
any other. By 1538 the authorities were in no doubt that 'the
error of the sacrament of the altar was . . . greatly spread
abroad in this realm, and daily increasing more and more'.[110]
In the 1530s only one man was indicted before the City fathers
specifically charged with heresy: Adrian Bukstyll appeared on
5 June 1537, and was promptly sent before Stokesley.[111] The
arrival of alien sacramentaries (and Bukstyll probably was one)
was specially feared. The plea of the Duke of Saxony's Chancellor to Cranmer in June 1538 that Atkinson should be allowed
to perform his penance at his parish church rather than Paul's
Cross was firmly refused, diplomatic privilege notwithstanding.
It was necessary that he should recant at the Cross so that as
many people as possible should know his perilous error and
learn from his example.[112] In August 1538 five men reported
Richard Reynolds, a stationer, before the Court of Aldermen
for saying that 'he might better be confessed of a temporal man
than of a spiritual man, and likewise to receive his housel of a
temporal man than of a spiritual man'. Worse, he thought 'the
Mass was nought, and the memento was bawdery, and after
the consecration it was idolatry'.[113] To despise the mementos,
the two intercessory prayers before and towards the end of
the canon, was to impugn Purgatory and to deny the Mass
as a propitiatory sacrifice, benefiting the living and the dead.

---

[108] Strype, *Ecclesiastical Memorials*, i/2, p. 180.
[109] GLRO, DL/C/614, fo. 7ᵛ.
[110] PRO, SP 1/133, fo. 174ʳ (*L&P* xiii/1. 1237).
[111] CLRO, Repertory 9, fo. 253ʳ. The Court's precise reference to the statute
25 Henry VIII c. 14 in this indictment indicates the concern of the Aldermen to be
seen to be acting strictly within the law in this novel area of authority.
[112] PRO, SP 1/133, fo. 174ʳ (*L&P* xiii/1. 1237).
[113] CLRO, Repertory 10, fo. 42ʳ. There is no record of a Richard Reynolds
printing in London at this time, but Thomas Reynolds was a well-known printer of
reformist literature: E. G. Duff, *A Century of the English Book Trade* (1905), pp. 130–1;
E. G. Duff et al. (eds.), *Handlists of Books printed by London Printers* (1913), p. 111.

Maybe Reynolds had been touched by the eucharistic doctrines of the Swiss Reformation, but from such brief, if evidently heretical, utterances it is impossible to discover and define exactly the genesis of the heresies. The Lollard legacy of sacramentarianism did much to incline English reformers away from the Lutheran doctrines of communion to the more radical commemorative beliefs of the Swiss reformers. The views of the Lollards, who were learning from the reformers, underwent a sea change too. In his will, written in 1538, Nicholas Durdant of Staines, a member of a Lollard dynasty, bequeathed his soul to God, 'trusting undoubtedly that by the merits of his only Son Jesus Christ to be one of the elect . . . to whom it shall be said Venite Benedicite'.[114] This was no longer the native heresy.

Heretical views were spreading in the capital. The preaching of reformers and zealots, private speculation upon Scripture, sometimes beyond its confines, and reading reforming works had led to a variety of opinion, held with varying degrees of conviction. The heretical attacks upon the Church were aimed primarily against the excessive and magical practices of late medieval popular piety. How many went beyond denial of confession and penance to be converted to the justificatory tenets of the reformed faith, and how many rejected the Catholic sacramental system along with the Virgin Mary and her cult can never be established. But those who were won to the new faith and those who were 'wavering' were surely very few compared to those who had never questioned the old faith at all. Could it be that they would allow the sacred mystery of the Mass to be so reviled without defending it? Ellis, a servant of the Marquis of Exeter, railed against Mrs Cowper, one of the City's new Bible-reading Protestants, and told her that 'the day will come there shall be no more wood spent upon you heretics, but you will be tied together, sacked, and thrown into Thames'. Asked who would do this, Ellis replied confidently: the Bishop of London.[115]

---

[114] PRO, PCC, Prob. 11/27, fo. 172^r; Foxe, *Acts and Monuments*, iv, pp. 177–8, 226, 228.

[115] PRO, SP 1/139, fo. 1^r (*L&P* xiii/2. 820(1), 817).

WAITING FOR 'THE WORLD TO TURN'

The year 1538 saw the reformers at the height of their power—but most imminently threatened. In their triumph their downfall was prefigured, for reform had spread further and faster than the people could bear. Very many looked for a 'good day' when the old ways would return. Reaction would not come until the King reverted, but there were already signs that he could not countenance such heresy spreading in his name.[116] But at the same time the reforming party, knowing that its time was short, made moves which were meant to bring the Word to the people so that they would never lose it, and to silence the opposition.

As long as the reformers at Court remained ascendant, reformers in the City were able to charge with disloyalty all those who challenged their own views. Anyone who dared to question the new order could be branded as papist or traitor and brought to answer. Resistance to the royal supremacy among the laity and clergy of London seemed to be growing rather than receding after 1534. Not until the autumn of 1537 was any Londoner brought before the Aldermen for open allegiance to the papacy. Elizabeth Tyse declared that 'the Pope should bear as great authority as ever he did', and when told that she deserved to be burnt, she replied that 'she trusted God to see them that held against the Pope to be hanged or burnt'.[117] In January 1538 Mitchell was put on the pillory for 'railing against the King's Council'. He had sent a bill to the King defaming all the City's officers—Justices, Sergeants, Common Councillors, the Mayor and Aldermen, Sheriffs, and members of the wardmote enquests—as 'high heretics, otherwise high rank traitors' for their complicity in the Reformation.[118] The following January Harry Totehill was reported for his insistent loyalty to the Pope and St Thomas Becket.[119] It was not that they were alone, or even exceptional, in holding against the new order. Mutterings against the royal policies may have been everywhere, but

---

[116] G. Redworth, 'A Study in the Formulation of Policy: the Genesis and Evolution of the Act of Six Articles', *JEH* xxxvii (1986), pp. 45–50.

[117] CLRO, Repertory 9, fo. 277^r.

[118] CLRO, Repertory 10, fo. 16^v; Letter Book P, fo. 138^v.

[119] *L&P* xiv/1. 47.

few would be reported if most people still felt the same way. Disaffection was only reported when someone was angry, conscious of his duty, or expectant of some reward. Once the gospellers began to make converts to their cause it became more and more dangerous to inveigh against the new ways, unless in private and among the like-minded.[120] As reform advanced, so the more zealous radicals lay in wait for known papists to betray themselves. The supporters of the Pope and 'good religion' would have been wise to remember those languishing in prison for transgressing the Treasons Act, and most were circumspect. All those clerical opponents of the royal supremacy who were detected in the City in these years were discovered for one of three reasons: either because they determined to take a stand against the changes; or because they happened to be indiscreet or defiant in the presence of keen reformers; or because they were so unpopular that their conservatism was used by their neighbours as a way to be rid of them.

On St Thomas's day 1535 Thomas Baschurch threatened that as soon as high Mass was over he would proclaim the King a traitor. He had fallen into despair.[121] Baschurch, the absentee rector of St Leonard Eastcheap and erstwhile secretary to Warham, abhorred the changes in religion, but in his City parish the citizens used praemunire to thwart his tithe claims, and a reforming zealot, Stephen Caston, was preaching.[122] Yet Baschurch's despair was of an even deeper kind. Convinced that he, like Esau, was 'created unto damnation', he tried repeatedly to kill himself. Then he saw a way in which he might bring upon himself the penalties for speaking or writing treason (although, as More had known, to compass death in this way was suicide, and would certainly damn him): 'If I can not be rid this way, I shall be rid another way.' Baschurch wrote a book: *Rex tanquam tyrannus opprimit populum.* By 1538 he was dead, of cause unknown.

The reformers attended the sermons of popish priests, wait-

---

[120] Elton, *Policy and Police*, ch. 1.

[121] *L&P* x. 113.

[122] Caston had preached so zealously that Cromwell sent him away to Ireland, out of harm's way: PRO, SP 1/118, fo. 204ʳ (*L&P* xii/1. 960). Caston had been in the deanery of the Arches in the summer of 1535 to swear the Supremacy Oath: PRO, E 36/63, fo. 102ʳ.

ing and hoping for them to defy the new order and castigate the heretics. Often the sport was to provoke them to treasonable outbursts. Robert Augustine was reported in June 1535 because he rose to the challenge of a reformer in his audience: 'When I alleged the mischievous and proud usurping of the Bishop of Rome that so used King John', said William Lilly, Augustine 'laid the chronicles of this realm were false and how he was accused maliciously . . . of false heretics'.[123] When Thomas Corthorp, the curate of Harwich, inveighed against 'new learned fellows' and their heresies at Bethlem without Bishopsgate in August 1535, he had in his congregation some of the City brethren—George Elyot (who had shown the King Fish's *Supplicacyon for the Beggers*), William Carkke, William Lacy, John Merifield, and others—as well as a royal servant. Hearing Thomas Bacon, a London salter, defend 'the order of preaching appointed by . . . the King', Corthorp replied, 'Well, this preaching will last but a while, for I trust to see a day.'[124] The rector of St Michael Wood Street, Thomas Jennings, spoke 'contemptuous and abused words' against the 1536 Injunctions before his parishioners who promptly reported him. Four of them were certainly of reformist sympathies: the rest may well have harboured old grudges after a tithe dispute in the parish in 1529.[125]

Where a priest had alienated his whole parish the parishioners were likely to be less forgiving if he also spoke out against the new order. So it was for William Barton, chantry priest at St Olave Jewry. He insisted that papal power was God-given, and that Henry was a traitor to God to usurp it.[126] Yet a personal animus lay behind Barton's vehement denial of the King's right to own or grant anything belonging to the Church, for his own chantry had been taken away from him because of his negligence and unsatisfactory behaviour, and was granted to

---

[123] PRO, SP 1/92, fo. 127ʳ (*L&P* viii. 624).

[124] PRO, SP 1/99, fos. 201ᵛ–202ʳ (*L&P* ix. 1059).

[125] PRO, SP 1/113, fo. 110ʳ (*L&P* xi. 1425). In 1540 four of his accusers—Robert Daniel, Robert Andrew, Thomas Gilbert, and William Hickson—were 'suspected to be sacramentaries and rank heretics, and procurers of heretics to preach'; Foxe, *Acts and Monuments*, v, p. 445; CLRO, Letter Book O, fo. 141ᵛ.

[126] PRO, SP 1/162, fo. 127ʳ (*L&P* xv. 1029(21)). A petition was sent to Audley by Laurence Elveden, a scrivener, charging Barton and his neighbour, Richard Smith, with treason.

another by royal letter patent in 1537.[127] Barton had refused to leave, brought an action of trespass against his successor and the parish, and scorned the King's seal; 'made under a bush, and not worthy to be regarded or obeyed unto'. 'The vengeance of God will shortly strike this realm', he promised, and had a book of prophecy to show that Henry would be torn apart by his own mule. 'The King was a cuckold and should worthily die'. He looked, too, for the day when the heretic heads of Audley and Cromwell would roll.[128] Not surprisingly, he was imprisoned 'for a long season'. In Barton were personified most of the vices of which the laity could complain in their clergy. He was one of only seven London priests to be presented before the Barons of the Exchequer in the 1530s: this was for firing his crossbow.[129] There were suspicions, too, that he had been clipping the King's coin.[130] By 1540 Barton was back in St Olave Jewry, looking for trouble and for those who had reported him. He was as negligent as ever: 'he very seldom would say Mass, scant once in a week', and spent festival and holy days in the alehouse.[131] When he failed to come to church on St Stephen's day 1540 'the most and chief men' of the parish, with the Alderman's deputy, went to arrest him and sent him to the Counter. Barton complained before Star Chamber that certain of his parish maliciously 'conspired his death', and sought 'utterly to impoverish him by infinite suits as well in the spiritual as in the temporal laws'. The following Easter the parish clerk took the chalice from Barton's hand as he was about to serve the Easter Mass, and there in the church Barton threatened him: 'I will strike thee with my dagger'. Barton was to be stopped from serving not because he was 'out of charity' (though he was), but because he was singing Mass at five o'clock in the morning, before his parishioners arrived, so that he could breakfast early with his friend, Richard Smith,

---

[127] GLRO, DL/C/208, fos. 116^r, 118^r, 134^v, 147^v, 148^r; PRO, C 1/831/ 26–7. 1537 is the *terminus ante quem* of the treason case.

[128] PRO, SP 1/162, fo. 127^r (*L&P* xv. 1029(21)).

[129] PRO, E 159/315 m. 21.

[130] PRO, Sta Cha 2/xix, fo. 219.

[131] Numerous appearances before the Consistory Court could not make Barton amend: GLRO, DL/C/3, fos. 3^v, 47^v, 48^v, 51^v, 52^v, 53^r, 70^v, 75^r, 90^r, 91^r, 98^r, 126^r, 130^r, 168^v.

a 'great inquieter and disturber of his neighbours'.[132] No priest in London was so troublesome and unworthy as Barton. If he had served his parishioners, while refusing to obey his King would he still have been delated as an adherent of Rome? Perhaps not.

In May 1538 three parishioners at St Mildred Bread Street reported William Bell, their rector, for reading at the Easter Mass the forbidden name, 'St Gregory Pope'. Relations in this parish were hardly harmonious, for Bell brought a case in Consistory against his curate and was himself ousted from his parsonage.[133] But Bell's conservatism also contrasted unfavourably with the charismatic preaching of Latimer in the same weeks. Londoners had come to realize what they had been missing. Reforming parishioners, reading and speculating upon Scripture themselves, now looked to their clergy to preach the Gospel, 'our ghostly food and edifying', and to share their spiritual knowledge. Again and again they were disappointed, especially in London where the contrast between the evangelism of Latimer and his coterie and the recalcitrance of their own parish priests was marked. In St Benet Gracechurch reforming parishioners tried to oust their curate, Master Edward Laborne, and to replace him with a man whose convictions matched their own. They found a good pretext to remove Laborne when in May 1538 he preached a sermon, 'sounding both of sedition and unchristianly'.[134] He had exhorted the parish to pray for souls departed, and when 'he lacked Scripture', resorted to an unfortunate metaphor: 'for an old dog . . . is worth the whistling to, and when he is dead his skin is worth a penny'. To this certain of the parish took exception: they had heard Latimer preach. Laborne inveighed against the 'young wits' of the parish, reminding them that the missionary St Augustine had landed in the Isle of Thanet with a wooden cross and a picture of Christ. True, they answered

---

[132] PRO, Sta Cha 2/iii, fos. 199ʳ–200ʳ; xix, fos. 211ʳ, 219ʳ; xxiii, fo. 294ʳ; xxv, fo. 73ʳ. The outcome of this trouble is unknown, but in Sept. 1545 Barton was discharged from prison by the Court of Aldermen, whence he had been sent by the Mayor for his 'misdemeanour'; CLRO, Repertory 11, fo. 220ᵛ.

[133] CLRO, Repertory 10, fos. 31ᵛ, 241ᵛ, 242ʳ; GLRO, DL/C/3, fos. 3ᵛ, 4ᵛ, 7ʳ, 9ᵛ, 11ᵛ. He was also involved in a tithe dispute later; CLRO, Repertory 11, fo. 64ʳ.

[134] PRO, SP 1/132, fo. 218ʳ; E 36/120, fos. 107ʳ–108ʳ (*L&P* xiii/1. 1111 (1,2));

knowingly, but he was 'the legate of a reprobate master, the Pope'. Laborne sought to show how St Paul failed to prove predestination. This 'erroneous preaching' offended his audience, not only because of Laborne's presumption, but also because if 'he rebuked Paul he would . . . avert men from reading him'. Laborne's sophistical learning threatened the 'edifying' of his parish when 'he spoke . . . to the advancement of philosophy that he would to be regarded so much more than Scripture'. Laborne was one of those learned City priests who determined not to cast pearls before swine by sharing their knowledge with their flocks. There was no doubt of his learning: he had dedicated his exposition of Psalm XX to Henry VIII.[135] Once he had been a schoolmaster and client of the Countess of Salisbury and the Poles.[136] Any association with that family was more than unfortunate in 1538, for they were found to be conspiring against the King and the heretics around him at Court.[137]

Cromwell still looked for a chance to bring down London's leading conservative—the Bishop—and in 1538 it looked as though he had found it. Opponents of the old ways thought of Stokesley as 'founder of the faith of Christ'. Sheriff Bell of Gloucester and his City friends looked to Stokesley and the Duke of Norfolk to use their influence to restrain Latimer and his followers.[138] Though hardly a man of the people, Stokesley might become a dangerous focus for resistance. In 1538 Cromwell chose to use the Bishop's old association with Syon Abbey (so fiercely resistant to the supremacy until its capitulation in January 1536) to punish him. On 29 May 1538 a writ was brought against the Bishop in the King's Bench, charging him with praemunire. The specific charge was that Stokesley had taken the profession of several monks in 1537, thereby acknowledging a papal bull, and had performed various 'papistical rites and superstitions'. Associated with the Bishop were several of the Syon community, and three unnamed secular clergy of the City. Stokesley's response was to confess his guilt and to throw himself on the King's mercy. He knew well

---

[135] BL, Royal MS, 7F xiv.

[136] *L&P* iii/1. 411.

[137] For the Exeter conspiracy, see M. H. and R. Dodds, *The Pilgrimage of Grace, 1536–7, and the Exeter Conspiracy, 1538*, 2 vols. (Cambridge, 1915).

[138] *L&P* xii/2. 534; PRO, SP 1/115, fos. 66ʳ–67ʳ (*L&P* xii/1. 308); Elton, *Policy and Police*, pp. 121–2.

enough that Cromwell was after him, and wrote reminding Cromwell of his own 'pains to persuade them of Syon to renounce the Bishop of Rome'.[139] The court bailed the Bishop in swingeing sureties, and judgement was postponed, awaiting the King's decision.[140] In the event, the 'King was better to him than he deserved', and Stokesley received a general pardon on 3 July.[141] On 17 August a release, a quitclaim, and a free gift of lands in Surrey were enrolled in Chancery: from Stokesley to Cromwell, his friends and relatives.[142] Why? Payment for favours? Stokesley's friends had feared that he would lose all. Sir Geoffrey Pole was 'very sorry', and had defended the Bishop against his detractors when he was 'in trouble for the praemunire'. Stokesley had been his 'good lord, and given him the keeping of a park, and lent him money in his necessity'.[143] For Stokesley to have such friends was dangerous.

The King feared the Courtneys and the Poles: he knew well that they might present a rival claimant to his throne; that they hated his proceedings in religion; that they regretted not joining the Pilgrimage of Grace. 'Nothing grieved her husband so much in all his life as the putting out of the Privy Chamber' in 1536, said the Marchioness of Exeter.[144] By August 1538 Henry had evidence that the Marquis and his brothers were conspiring against him—they spoke of insurrection, saying 'Beware of the third'—and he determined to be rid of them.[145] Sir Geoffrey Pole was broken by interrogation: Husee heard at the end of October that he was 'in such despair' that he had tried to take his life.[146] Pole confessed that he thought 'the world in England waxeth all crooked, God's law is turned upso down'.[147] He had talked with More's daughters, Alice Clement

[139] PRO, KB 9/539/7; KB 27/1107, Rex rot. 20; SP 1/132, fo. 204ʳ (*L&P* xiii/1. 1096); Elton, *Policy and Police*, p. 161; Knowles, *Religious Orders*, iii, pp. 220–1.

[140] *L&P* xiii/1. 1095. The Court bailed Stokesley on surety of 10,000 marks of his own, and 500 marks from each of six others. His sureties included Henry Pepwall, the stationer, who had been Stokesley's agent sent to Antwerp to aid the friars and hunt heretics: PRO, SP 1/78, fo. 85ʳ (*L&P* vi. 934). For Pepwall's printing, see E. G. Duff, *Westminster and London Printers, 1476–1535* (1906), p. 149.

[141] *L&P* xiii/1. 1519(3); Wriothesley, *Chronicle*, i, p. 81.

[142] *L&P* xiii/2. 119.

[143] PRO, SP 1/139, fo. 25ᵛ; SP 1/138, fo. 211ᵛ (*L&P* xiii/2. 803).

[144] *L&P* xiii/2. 802.

[145] *L&P* xiii/2. 957; Bodleian, MS Ashmole 861, fos. 335ʳ–336ʳ.

[146] *Lisle Letters*, 5. 1259.

[147] PRO, SP 1/138, fo. 192ʳ (*L&P* xiii/2. 797 (p. 309); 986(10)).

and Margaret Roper, and they said, as he did, that 'they liked
not this plucking down of abbeys, images and pilgrimages and
prayed God to send a change'.[148] The Poles spoke more darkly.
Geoffrey had promised: 'we should do more, and here when
the time should come, what with power and friendship, nor it is
not the plucking down of these knaves [at Court] that will help
the matter; we must pluck down the head'.[149] In his house at
the Red Rose in St Laurence Pountney the Marquis spoke
treasonably (just as Buckingham had once done there): 'I trust
to see a change of this world'; 'to have a fair day upon those
knaves which rule about the King'; and, clenching his fist, 'to
give them a buffet one day'. He had dreamt that the King was
dead.[150] Henry, Lord Montague and Sir Edward Neville were
also implicated in these treasonable conversations (though,
thought Montague, 'it would be a strange world as words were
made treason').[151]

Stokesley knew these enemies to the King well, but once
he had been circumspect. At the end of April 1536 Chapuys
reported that 'the brother of Lord Montague' told him at
dinner that, when asked whether Henry would abandon Anne,
Stokesley refused to give his opinion to anyone except the
King.[152] By 1538 he was far less discreet. Hugh Holland told
Cardinal Pole that Geoffrey went to dine with the Bishop
'often'.[153] Stokesley had told Sir Geoffrey of his deep resentment of the usurpation of his authority: 'he was but a cipher,
for the Lord Privy Seal first, and then the Bishop of Rochester
have appointed heretics to preach at Paul's Cross'.[154] There
was a whole bundle of letters between the Bishop and Pole, but
these letters Pole had burnt, 'for the keeping of letters might
turn a man's friends to hurt'.[155] The letters contained treason.
The interrogators suspected Stokesley of intriguing with the
conspirators: Pole was asked what 'conference' he had had

    [148] *L&P* xiii/2. 695(2), 830(v).
    [149] *L&P* xiii/2. 800.
    [150] *L&P* xiii/2. 771, 804(5), 954, 956, 979(7, 15).
    [151] *L&P* xiii/2. 829(2), 875 (ii).
    [152] *L&P* x. 752.
    [153] PRO, SP 1/138, fo. 193ʳ (*L&P* xiii/2. 797).
    [154] PRO, SP 1/139, fo. 38ʳ (*L&P* xiii/2. 695(2), 829, 830(vii)).
    [155] *L&P* xiii/2. 875; PRO, SP 1/139, fo. 31ʳ (*L&P* xiii/2. 829); SP 1/139, fos. 45ᵛ,
72ᵛ; *L&P* xiii/2. 957.

with Stokesley upon the 'changes'; what heresies Stokesley had
claimed had been preached at the Cross; what Stokesley thought
would be the outcome of the 'proceedings within the realm';
what 'likelihoods, causes . . . he hath opened at any such
conference that this world should not continue'; whether he
was in communication with Cardinal Pole, and whether 'he had
commended his doings'.[156] Sir Geoffrey had confessed all: it is
hard to believe that the conspirators did not have enough evidence to convict Stokesley. Nevertheless he was not arraigned,
nor condemned with the noble conspirators that December.
But a writer railing against popish traitors saw the fact that
Stokesley died a natural death as sure proof of the King's
infinite mercy.[157] Pole had been saving the manuscripts of the
martyred More for posterity, and collecting evidence against
the reformers.[158] Among Pole's letters to be burnt was one
revealing that Latimer, before he was Bishop (that is, before
1535) had sent a crown overseas: surely to aid the brethren in
exile. Pole was dismayed: 'What have ye burnt that also?
Those letters were shown before the Council, and my lord of
Norfolk told me I might keep those letters well enough.'[159]
Hopes that the conservative nobles and Bishops might thwart
the plans of the heretics now faded. In 1538 the gospellers were
firmly in power, and might realize their old ambition to bring
the Gospel into England.[160]

WINNING, AND LOSING, THE GOSPEL

Cranmer had once despaired of seeing an official translation of
Scripture: in 1534, thwarted by Stokesley's adamant refusal to
translate Acts, he told Cromwell despondently that they should

---

[156] *L&P* xiii/2. 695, 828.
[157] PRO, SP 1/155, fo. 65 (*L&P* xiv/2. 613).
[158] PRO, SP 1/139, fo. 21ʳ (*L&P* xiii/2. 828); More, *CW* 12, pp. xxiii, xxvi.
[159] *L&P* xiii/2. 829.
[160] For the making of the English Bible, see A. W. Pollard, *Records of the English Bible* (1911); C. Anderson, *Annals of the English Bible*, 2 vols. (1845–55); J. F. Mozley, *Coverdale and his Bibles* (1953); *Incidents in the Lives of Thomas Poyntz and Richard Grafton, two citizens and grocers of London, who suffered loss and incurred danger in common with Tyndal, Coverdale and Rogers in bringing out the Bible in the vulgar tongue*, collected and confused by J. A. Kingdon (1895); D. Wilson, *The People and the Book: the Revolutionary Impact of the English Bible, 1380–1611* (1976).

wait until the 'day after doomsday'. But early in 1537 the Archbishop 'thought it now . . . a meet time to restore the old doctrine of Christ, according to the Word of God and the old primitive Church'.[161] A commission was appointed, to be presided over by the Vice-Gerent, to settle doctrine within the Henrician Church, and to 'establish a godly and perfect unity and concord in Scripture'.[162] Seemingly by chance, Cromwell encountered the rabid Scot, Alexander Alesius, returned from Germany, and invited him to propound the reformist argument.[163] Gardiner, 'the Pope's chief champion', and Stokesley, led the defence 'with all subtle sophistry'.[164] Stokesley had done all he could to evade this encounter with the reformers, even, so it was said, feigning illness,[165] but came then to argue fiercely for the existence of all seven sacraments, even if not proved in Scripture, and for the centrality of patristic and conciliar authority. So Alesius showed in his tract: *Of the auctorite of the word of god agaynst the bisshop of london*. Said Stokesley: 'And if ye think that nothing pertaineth unto the Christian faith, but that only that is written in the Bible; then err ye plainly with the Lutherans; for St John saith, "that Jesus did many things which be not written".'[166] When Cranmer, Cromwell, and the Bishops 'which did defend the pure doctrine of the Gospel', saw 'him flee . . . unto his old rusty sophistries and unwritten verities', 'they smiled a little one upon the other'.[167] They had won the day, 'and set their hands to a godly book of religion called *The Bishops' Book*' (otherwise known as *The Institution of a Christian Man*).[168]

By the summer of 1537 Cromwell had gained Henry's approval to license Miles Coverdale's translation of Scripture: 'If there be no heresies then, in God's name, let it go forth among

[161] *Narratives of the Reformation*, p. 223; Wilson, *People and the Book*, p. 60.

[162] Foxe, *Acts and Monuments*, v, pp. 378–80; J. Guy, 'Scripture as Authority: Problems of Interpretation in the 1530s', in A. Fox and J. Guy, *Reassessing the Henrician Age: Humanism, Politics and Reform, 1500–1550* (Oxford, 1986), pp. 199–220.

[163] Foxe, *Acts and Monuments*, v, p. 378. For Alesius's theological position, see G. Wiedermann, 'Alexander Alesius' Lectures on the Psalms at Cambridge, 1536', *JEH* xxxvii (1986), pp. 15–41.

[164] *Narratives of the Reformation*, p. 223.

[165] *L&P* xi. 1355.

[166] Foxe, *Acts and Monuments*, v, p. 383.

[167] Ibid.

[168] *Narratives of the Reformation*, pp. 223–4; *Formularies of Faith*, pp. 21–211.

our people.'[169] On 13 August Cranmer wrote expressing undying gratitude to Cromwell: 'you shall have a perpetual laud and memory of all God's faithful people and favourers of His Word'; Cromwell's work for the Gospel would not be forgotten at the Day of Judgement.[170] By the end of 1534, John Rogers had left his City cure, Holy Trinity the Less, to become chaplain to the Merchant Adventurers in Antwerp, and there to join Tyndale in his great work of translation and exposition of the Gospel. This he continued, protected and helped by his evangelical friends Richard Grafton and Thomas Poyntz, even after Tyndale's martyrdom in October 1536.[171] In Antwerp, Grafton, a London grocer and a merchant adventurer, printed Rogers's (rather his pseudonym Matthews's) text, and by the summer of 1537 was sending copies to Cromwell in England. Grafton wrote to Cromwell sometime in 1537 asking that they, who had set this work forth 'with great study and labours', might have sole licence to print, for pirate editions which falsified the text were being published. 'Look how many sentences are in the Bible, even so many faults and errors' these pirate printers made, 'for their seeking is not to set out God's glory, and to the edifying of Christ's congregation (but for covetousness)'. Grafton could soon sell out his first edition of 1,500 copies if Cromwell would insist that every curate have a Bible 'that they may learn to know God', especially if he compelled the 'papistical sort' to buy one. There were enough of these in London alone to 'spend away a great part of them'.[172] The following summer Grafton, with Edward Whitchurch, a London stationer, was in Paris, at Cromwell's behest, printing a new English Bible, using Matthews's text revised by Coverdale.[173] Soon every curate in every parish, whether he wished it or not, was bound to provide a copy of this Bible for his parishioners, and to expound it for them.

The campaign to reform the commonwealth; to bring the

---

169 Pollard, *Records of the English Bible*, pp. 214–22.

170 Strype, *Memorials of Cranmer*, i, pp. 82–3.

171 Kingdon, *Incidents in the Lives of Thomas Poyntz and Richard Grafton*.

172 BL, Cotton MS Cleo. E v, fo. 325ʳ; Strype, *Memorials of Cranmer*, ii. xx; *L&P* xii/2, appendix 35.

173 Foxe, *Acts and Monuments*, v, pp. 410–11, 825–7; Mozley, *Coverdale and his Bibles*, chs. 6–7.

Gospel to the people; to make war upon idolatry and superstition, culminated in Cromwell's Injunctions of September
1538.[174] Curates were to buy a Bible 'of the largest volume in
English' and 'expressly [to] provoke, stir and exhort' their flock
to read this 'very lively Word of God'. In their quarterly (at
least) sermons they must persuade their hearers to 'works of
charity, mercy and faith . . . and not to trust . . . in any other
works devised by men's phantasies besides Scripture'. No
longer were the people to devote their worship to processions
singing *Ora pro nobis*, 'to candles, or tapers to images or relics,
or kissing or licking the same'; nor to set lights before pictures.
Only the lights before the sacrament, the rood loft, and the
sepulchre were to remain. For Cromwell, who had known the
New Testament by heart since his youth and had long wanted
others to know it, this was the fulfilment of his evangelical
ambition. But in this proclamation too was his own fall prefigured.

As reforming ideas spread, so did reckless image-breaking. By
October 1533 images were already being taken down and cast
away 'as stocks and stones of no value', so John Rose, a priest
at St Antholin, reported in disgust. Some Londoners even
pricked the fallen images mockingly to see whether they would
bleed.[175] Until 1534, such iconoclasm was always condemned
as heresy and vandalism, and lay and Church authorities alike
protected the images, but in that year an order came to halt
discussion of the veneration of images from the pulpits.[176] This
moratorium was a sure sign of the disturbance this subject
raised, but it presaged also a change in official religious policy.
Soon images and relics became the object of official as well as
popular attack. During the visitation of the London religious
houses in October 1535 the girdles of Our Lady and St Elizabeth
(used even by Katherine of Aragon to ease childbirth) were

---

[174] Gee and Hardy, *Documents Illustrative*, pp. 275–81. For their enforcement, see
Elton, *Policy and Police*, pp. 254–5, 259–61.

[175] PRO, SP 1/79, fo. 222^r; BL, Cotton MS Titus B i, fo. 454^v (*L&P* vi. 1311,
1381).

[176] J. F. Davis, 'Heresy and Reformation in the South-East of England,
1520–1559' (University of Oxford, D.Phil. thesis, 1968), p. 302; *L&P* ix. 357, 358.
In 1534 John Rastell sought Cromwell's support for legislation prohibiting offering to
images; Lehmberg, *Reformation Parliament*, p. 215.

removed.[177] In September 1535 Audley sent Cromwell 'a book, lately printed'—*A treatise declaring and showing that images are not to be suffered in churches*—and warned him that it would 'make much business if it goes forth'.[178] Yet Thomas Godfray was 'compelled again to print this little treatise' within the year.[179] Its translator was William Marshall, who was soon taken into that circle of reformers and thinkers advanced by Cromwell.[180] His translation of the relief ordinance of Ypres, *The forme and maner of subuention or helpying for pore people* (1535), inspired Cromwell's attempt to establish a reformed poor law in the following year.[181] In the capital the reformers preached of the 'spiritual whoredom' of idolatry. Before Convocation in 1536 Latimer complained, as so often before, of the abuse in venerating relics, shrines, and images as 'juggling deceits', which led the people to confound false miracles for true ones. For the reformers—Latimer, Colet, Gough, and Cromwell— the wealth of the shrines was an affront, because the poor, 'Christ's image', languished; wretched, naked, and starving.[182]

The year 1538 saw a concerted campaign against abused images. The vanity of those images which had been worshipped as idols was to be publicly exposed before the people they had deceived. To that end, the most spectacular were brought to London, 'a jolly muster' to be dishonoured and destroyed.[183] At Paul's Cross on 24 February Hilsey revealed the revered Rood of Grace of Boxley Abbey as a fraud: it was believed to be 'moved by the power of God' and to do miracles, but he showed the strings and pulleys which worked it. The monks of Boxley 'time out of mind . . . had gotten great riches in deceiving the people': now the people's devotions must be given to the poor. The Bishop broke the image, and left it for

---

[177] Wriothesley, *Chronicle*, i, p. 31; Guildhall, MS 9531/11, fo. 68ᵛ.

[178] *L&P* ix. 358; *A treatise declarynge . . . that pyctures and other ymages . . . are in no wise to be suffred in the temples or churches* (*RSTC* 24238).

[179] Ibid., sig. Giii.

[180] Elton, *Reform and Renewal*, p. 62 n. 79.

[181] Elton, 'An Early Tudor Poor Law', *Studies*, ii, pp. 137–54.

[182] *L&P* xi. 1111; Latimer, *Remains*, pp. 233, 333; Latimer, *Sermons*, pp. 23, 55; cited in J. B. Phillips, *The Reformation of Images: The Destruction of Art in England, 1535–1660* (Berkeley, Los Angeles and London, 1973), p. 53. John G . . . *The myrrour or lokynge glasse of lyfe* (1532?; *RSTC* 11499), sig. Iiii.

[183] Latimer, *Remains*, p. 395; Wriothesley, *Chronicle*, i, pp. 74–7, 80, 83–4; Bodleian, MS Ashmole 861, fo. 335.

the 'rude people and boys' to smash.[184] Husee wrote to Lisle despairingly, resignedly, in March: 'Pilgrimage saints go down apace, as Our Lady of Southwark, the blood of Hailes, St Saviour and others.' To Lady Lisle he confessed, 'I doubt not the resurrection will after': 'I can no more, but God be lauded in all his works.'[185] In May 1538 Latimer wrote to Cromwell about a plan: 'And sir, if it be your pleasure, as it is, that I shall play the fool after my customable manner when Forest doth suffer . . .'. Cromwell, who 'loved antiquities' and who knew how to expound ancient prophecy, had devised a grim fulfilment to the ancient prophecy that the image of Darvel Gaderen, 'the great god of Wales', should set a forest on fire.[186] Instead of being hanged, drawn, and quartered for his treason of denying the royal supremacy, 'Forest the friar, that obstinate liar' burned alive with the image, while Latimer preached.[187] Some of the ten thousand at the Cross were moved by this savage spectacle to further reformation.

Late that same night a group of radical vigilantes determined to take the drive against images into their own hands. They pulled down the famous rood at St Margaret Pattens, to which pilgrims came from afar to venerate.[188] Ten of the offenders were caught and brought before the Court of Aldermen: twenty more were never found. In defence of their sacrilege, they claimed that they were told by Dr Crome, who was told by Latimer, that Cromwell had ordered the removal of the rood. Maybe so. Leader of the iconoclasts was the radical John Gough, and among his followers were Flemings and parishioners of St Margaret Pattens: two pewterers, a cutler, an ironmonger, a cordwainer, a basket maker, even a constable.[189] One of them—John Pet, a painter—may have gone to the stake in 1539, for Foxe listes 'John, a painter' among the martyrs.[190]

[184] Wriothesley, *Chronicle*, i, pp. 74–6.

[185] *Lisle Letters*, 5. 1129, 1131. The resurrection to which Hussee refers is probably the pageant of the resurrection in St Nicholas, Calais.

[186] Latimer, *Remains*, pp. 391–2; 375; Hall, *Chronicle*, p. 826; Keith Thomas, *Religion and the Decline of Magic: Studies in Popular Beliefs in Sixteenth- and Seventeenth-Century England* (1971), p. 27; *L&P* xi. 790.

[187] Wriothesley, *Chronicle*, i, pp. 78–81; BL, Harleian MS 530, fo. 120[r]; Bodleian, MS Ashmole 861, fo. 335[r]; Foxe, *Acts and Monuments*, v, p. 408.

[188] *Original Letters*, i, p. 200; Stow, *Survey*, i, p. 209.

[189] CLRO, Repertory 10, fo. 34[v]; Wriothesley, *Chronicle*, i, p. 81.

[190] Foxe, *Acts and Monuments*, v, p. 654.

Such iconoclasm was long associated with the Lollards, but it was also typical of more radical Continental Protestantism, of Zwinglianism and Anabaptism, and the part played by Gough and the Flemings suggests that such advanced views were spreading in the capital among a few zealots. Ever since June 1528 when the news came that iconoclasts in Paris had broken a statue of the Virgin, it had been feared that such desecration would spread to London.[191] The Sunday after the rood was pulled down there was 'a great fire' at St Margaret Pattens— God's plague upon the iconoclasts, it was said.[192]

The authorities feared opposition to their campaign against superstitious image worship; from conservatives who lamented the sacrilege, and from radicals who demanded the instant removal of all images, abused or not. In July George Robinson, a City mercer, reported in disgust seeing St Uncumber in St Paul's 'in her old place . . . with her grey gown and silver shoes, and a woman kneeling before her . . . to God's dishonour'. 'I beseech God to put in the King's grace mind to take them all away, and so shall he have the blessing that King Josias had.'[193] On 23 August Gresham received orders, via Barnes, for the destruction of the rood of Grace ('ungrace' to the reformers) and the figure of St Uncumber in the cathedral. The Dean of St Paul's had the venerated rood and images removed urgently and by night, lest there be trouble either from those who would desecrate or those who would defend them.[194] By the morning of 24 August they were gone.

In the late summer of 1538 one of London's most famous sons was dishonoured. The cult of St Thomas Becket was put down. In September his shrine at Canterbury, for so long a place of pilgrimage, was destroyed; his festival day was abrogated.[195] Becket had been, so Stow boasted, clerk to the Sheriff, parson of St Mary at Hill, prebend of London:[196] would the citizens not resent his disgrace? Friar Ward had warned Cromwell in 1535 that pardoners were glorifying St Thomas as a

[191] *L&P* iv/2. 4338, 4409.

[192] Wriothesley, *Chronicle*, i, p. 81; *L&P* xiii/2. 596.

[193] PRO, SP 1/134, fo. 183ʳ (*L&P* xiii/1. 1393). He wrote to Cromwell.

[194] PRO, E 36/120, fo. 107ʳ; SP 1/135, fo. 247ʳ (*L&P* xiii/1. 1111; xiii/2. 209); Wriothesley, *Chronicle*, i, pp. 83–4.

[195] *TRP* i. 186; Wriothesley, *Chronicle*, i, pp. 86–7.

[196] Stow, *Survey*, i, p. 105.

saviour of the poor, telling the people of 'the articles of St Thomas and the liberties of the Church of England'.[197] The parallels between Henry VIII and Henry II were dangerously close, and the saint's legend might easily become a focus for resistance to the new order. Friar Forest had claimed, treasonably, that 'that blessed man St Thomas of Canterbury suffered death for the rights of the Church'.[198] But Lollards and reformers vilified Becket: Robert Man and John Hewes impugned his sanctity, and Bainham had called him a 'thief, and a murderer and a devil in Hell'.[199] In London all signs of him were now to be removed. The windows depicting the untimely spectacle of 'the King [Henry II] kneeling naked before the monks as he should be beaten at the shrine of St Thomas' were promptly taken out of St Thomas Acon on 3 October.[200] In St Dunstan in the East and St Mary Magdalen Milk Street, too, the images of St Thomas were taken away.[201] The Mercers' Company, so long associated with Becket, quickly and pointedly stopped calling him saint in their records.[202] Yet it was not until a year later that the arms of the City replaced the image of St Thomas on the common seal of the City.[203] Hussee, observing the desecrations and the desecrators, reported their activities to Lady Lisle with deep irony:

Mr Pollard . . . hath been so busied, both night and day, in prayer, with offering unto St Thomas' shrine . . . with other dead relics, that he could have no idle worldly time to peruse your ladyship's book . . . Howbeit, when his spiritual devotion is past I doubt not he will . . . apply his worldly causes accustomed.[204]

---

[197] PRO, SP 1/92, fo. 128ʳ (*L&P* viii. 626). See also SP 1/71, fos. 3ʳ–4ʳ (*L&P* v. 1271).

[198] PRO, SP 1/132, fo. 155ʳ (*L&P* xiii/1. 1043(3)).

[199] Foxe, *Acts and Monuments*, iv, p. 702; v, p. 34; J. F. Davis, 'Lollards, Reformers, and St Thomas of Canterbury', *University of Birmingham Historical Journal*, ix (1963), pp. 1–15.

[200] PRO, SP 1/92, fo. 128ʳ (*L&P* viii. 626); *Lisle Letters*, 5. 1238; Mercers' Company, Acts of Court, ii, fo. 271ʳ.

[201] Guildhall, MSS 2968/1, fo. 94ʳ; 2596/1, fo. 75ʳ.

[202] Mercers' Company, Renter Wardens' Accounts, fo. 10ᵛ; see also Guildhall, MS 5445/1 (Brewers' Company).

[203] CLRO, Journal 14, fo. 158ᵛ; Letter Book P, fo. 197ʳ; *VCH London*, i, p. 269. It was not until even later that the seal of the Bridgehouse was changed: Letter Book Q, fo. 3ʳ.

[204] *Lisle Letters*, 5. 1217.

In August 1538 Sir Richard Gresham wrote to Cromwell of a providential accident. At St Mary Spital the roof and the rood loft had fallen down: surely 'it was the will of God that the house should be converted to some better use'.[205] He shared with Cromwell the aspiration for 'charitable reformation'; that Christian devotions should be spent upon the poor and the sick. To that end he had ordered weekly collections to be taken for the City's poor at the Paul's Cross sermons from Lent 1538.[206] Above all, Gresham hoped that the wealth of the Church might be used to relieve the poor (though he hoped for some of it for himself too, and bought Fountains Abbey and its lands from the Crown).[207] He had written to the King in August (and now asked Cromwell to make sure that Henry read the letter), urging that London's dissolved religious houses might be used for the 'relief, comfort and helping of the poor and impotent people . . . and not to the maintenance of canons, priests and monks, to live in pleasure; nothing regarding the miserable people lying in every street'.[208]

Soon London was transformed, traumatically and irrevocably. The dissolution of the religious houses changed the face of the City, forever. The houses were soon seized by the nobility of England and the wealthiest of the citizens, and the hundreds of religious who had found their homes there were cast out, homeless and purposeless.[209] On 12 October the houses of the Crossed Friars, the Austin, Black, and Grey Friars were surrendered, and the following month, with the London Charterhouse, St Helen's priory, and St Thomas Acon, they were suppressed.[210] All the religious of London changed their habits

---

[205] *L&P* xiii/2. 13.

[206] Wriothesley, *Chronicle*, i, pp. 77–8.

[207] *L&P* xiii/2. 671.

[208] BL, Cotton MS Cleo. E iv, fo. 222ʳ; Strype, *Ecclesiastical Memorials*, i/1, pp. 409–11.

[209] This large and important subject has not yet found its historian, but see: M. Honeybourne, 'The extent and value of property in London and Southwark occupied by the Religious Houses, Inns of Bishops and Abbots and churches and churchyards before the Dissolution' (University of London, M.A. thesis, 1948); E. Jeffries Davis, 'The Transformation of London', in *Tudor Studies*, ed. R. W. Seton-Watson (1924), pp. 287–314. The innumerable grants and sales of monastic land to and by London citizens and the nobility are calendared in *Letters and Papers*.

[210] *L&P* xiii/2. 648, 652, 806, 807, 808, 809, 908, 1024; Wriothesley, *Chronicle*, i, p. 88.

for those of secular priests.[211] Meeting Friar Bartley (Barclay?) by Paul's churchyard, Cromwell threatened: 'If I hear by one o'clock that this apparel be not changed, thou shalt be hanged immediately for example to all others.'[212] (But Barclay's offences were more serious than failing to 'leave his hypocrite's coat': for he had been preaching dangerous subversion in the West Country; describing the providential fire in St Margaret Pattens following the iconoclasm.)[213] Yet the citizens made no stand to defend the religious, for all their earlier reverence for them. In June 1537 (a month after the remaining London Carthusians had finally sworn the Oath of Succession) the Court of Aldermen debated whether 'the citizens should make labour to the King's highness for the continuance and standing' of the Charterhouse, which 'for divers of their great offences shall be suppressed'.[214] No move was ever made. Yet the sympathies of most of the citizens were probably with the monks, however much they pitied the poor. It was more likely that they wished the dissolved houses were still homes for the religious praying for souls than shelters for the indigent. Gresham's aspirations for reform were, like Cromwell's, ahead of their time. There were in London many who had resented Gresham's election to the mayoralty the previous October (once again, by royal command), and once more it was division in religion which occasioned the contention. They wanted a conservative: 'Mr Hollis which had been put off three years by the King's letter'.[215] And there was trouble again at the mayoral election in 1538. All was confused. Husee told Lisle on 14 October: 'The Lords sit daily in Council . . . Mr Forman is Mayor elect and is now absent, whether it be with his will or against his will, God knoweth.'[216] Once more the conservative Hollis was excluded, but his time would come soon.

There were signs that reform could not continue much longer. An embassy of Lutheran divines sent from Saxony and Hesse to negotiate rapprochement with England disputed daily at Lambeth through September, but on the 27th Husee wrote,

---

211 Wriothesley, *Chronicle*, i, p. 87.
212 Foxe, *Acts and Monuments*, v, p. 396; Wriothesley, *Chronicle*, i, p. 82.
213 *L&P* xiii/2. 596.
214 CLRO, Repertory 9, fo. 255ʳ.
215 Wriothesley, *Chronicle*, i, p. 67.
216 *Lisle Letters*, 5. 1251A.

'the Lubeckers are this day dispatched'.[217] Henry sent his
regrets diplomatically enough: 'as the matter of our negotiations concerns the glory of Christ and the discipline of religion,
it requires mature deliberation'.[218] But over the doctrines of
communion in one kind, private masses, and priestly celibacy
there could be no rapprochement.[219] Henry was fearful that the
worst heresies were spreading within his Church, even in his
name. Thinking 'no office more friendly among those that
govern than to admonish each other of dangers', the Duke of
Saxony and the Landgrave of Hesse sent Henry on 25 September an alarming letter discovered from an Anabaptist. Its
writer rejoiced that 'In England the truth silently but widely is
propagated and powerfully increases; God knows for how long!
The brethren have issued a printed book *De incarnacione Christi*.'
The German princes warned the King not to 'slay good men'
along with Anabaptists in his fury to repress the sect, but
Henry was not to be pacified.[220] A commission was established
on 1 October (on which sat, among others, Stokesley and
Gwent, Barnes and Crome) to search out and examine Anabaptists.[221] Soon they found some.

The proclamation issued on 16 November 1538 expressed all
the uncertainty in religious policy, all the confusion and rivalry
at Court and in the Council. For nearly three years Gardiner
had been away in France. On 28 September he returned to
Court—'the Bishop of Winchester came this night', reported
Husee—and his arrival was ominous for Cromwell.[222] In the
November proclamation the Vice-Gerent's hand and his pre-
occupations were apparent in only the last two of ten sections.[223]
The ninth ordered that the clergy instruct their parishioners in
the 'true meaning and understanding of Holy Scripture, sacramentals, rites and ceremonies' so that all 'superstitions, abuses

---

217 *Original Letters*, i, p. 106; *Lisle Letters*, 5. 1233; Strype, *Ecclesiastical Memorials*, i/1,
pp. 384–5; Wriothesley, *Chronicle*, i, pp. 81–2.
218 *L&P* xiii/2. 497.
219 Burnet, *History of the Reformation*, i, pp. 401–2; Scarisbrick, *Henry VIII*,
pp. 403 ff.
220 *L&P* xiii/2. 427, 265.
221 Wilkins, *Concilia*, iii, p. 836 (*L&P* xiii/2. 498).
222 *Lisle Letters*, 5. 1233, 1235; *L&P* xiii/2. 442.
223 *TRP* i. 186. The drafting and inwardness of the proclamation is discussed in
Elton, *Policy and Police*, pp. 255–7, and Redworth, 'A Study in the Formulation of
Policy'.

and idolatries' associated with the rites be removed. The last section attacked Thomas Becket; rewriting his history (Sadler had been doing some historical research)[224] and putting down his festival. For the rest, the proclamation demonstrated the Supreme Head's will for religion in his Church. The draft of the first part was corrected in the royal hand, and the choice of penalties revealed something of Henry's implacability. Books were being imported which reflected the beliefs of 'Anabaptists and sacramentarians', the proclamation declared, and the expression of 'fanatical opinions' led to disputes on the sacraments 'in open places, taverns and alehouses'. Henceforward no books were to be brought into England without royal licence; printing in England must be licensed by the Privy Council; the sale of Bibles was restricted to approved versions. Anabaptists were to be executed, if already found; the rest should leave, fast. Only 'learned men in Holy Scriptures' from the universities might 'reason . . . upon the . . . Holy and Blessed Sacrament' and its mysteries—others would do so 'upon pain or loss of their lives and forfeiture of their goods without any favour or pardon'. Clergy who had married, in defiance of their vows, were to be deprived. How such disparate orders ever came together remains mysterious: maybe the conservatives in Council were prepared to compromise with the reformers, but why should the King? On the very day that the proclamation was set forth Henry proved his determination that justice should be executed against the heretics and purity in religion preserved.

Clad all in white ('as covering, by that colour and dissembling, severity of all bloody judgment'), the King came to Westminster with his nobles in train on 16 November to dispute with John Lambert.[225] Lambert had had warning enough that his zeal in advancing his views threatened the campaign against superstition and to bring in the Gospel. In 1538, after hearing the reformer Dr Taylor preach at St Peter Cornhill, Lambert had challenged his teaching on the eucharist. Taylor had conferred with Barnes who, fearing 'that it would breed among the people some let or hinderance to the preaching of the

---

[224] *L&P* xiii/2. 685.
[225] Foxe, *Acts and Monuments*, v, pp. 229–34; *Lisle Letters*, 5. 1273; Dugmore, *The Mass and the English Reformers*, pp. 177–82.

Gospel (which was now in good forwardness) if such sacramentaries should be suffered', reported Lambert to Cranmer.[226]
On 14 November Hussee heard that Lambert would dispute
'concerning the Sacrament against eight doctors. . . . It
is verily thought that Lambert will to the fire, which if he
deserve, God send him.'[227] In prison William Collins prepared Lambert for his ordeal, gathering consoling verses from
Scripture: 'Blessed are they which are persecuted for righteousness' sake, &c'.[228] The chronicler Wriothesley heard that to
Lambert were imputed the worst heresies of the Anabaptists—
that infants should not be baptized, that Christ was not corporally present in the sacrament of the altar, and incarnate of
the Virgin[229]—but the dispute turned upon his views of the
Real Presence. The King led the questioning fiercely:

THE KING. Answer as touching the sacrament of the altar, whether
dost thou say, that it is the body of Christ, or wilt deny it?
LAMBERT. I answer, with St Augustine, that it is the body of Christ,
after a certain manner.
THE KING. . . . tell me plainly, whether thou sayest it is the body of
Christ, or no.
LAMBERT. Then I deny it to be the body of Christ.
THE KING. Mark well! for now thou shalt be condemned even by
Christ's own words, 'Hoc est corpus meum'.[230]

Stokesley, using the example of water boiling so long upon a
fire that it evaporated, insisted that 'it is nothing dissonant
from nature, the substances of like things . . . be oftentimes
changed one into another; so that nevertheless the accidents do
remain; albeit the substance itself, and the matter subject, be
changed'.[231] Lambert could not be persuaded. His *Treatise upon
the Sacrament* (addressed to the King from custody in Lambeth)
showed Lambert to be a follower of the early English Reformation's most influential theologian, John Frith. As Christ 'is no
more corporally in the world, so can I not see how he can be
corporally the sacrament, or His holy supper'. Lambert followed St Augustine's teaching that in the sacrament Christ is

226 Foxe, *Acts and Monuments*, v, pp. 227–8.
227 *Lisle Letters*, 5. 1268.
228 *Original Letters*, i, p. 201.
229 Wriothesley, *Chronicle*, i, p. 89.
230 Foxe, *Acts and Monuments*, v, p. 230.
231 Ibid. v, pp. 232–3.

sacrificed mystically daily, because His offering, once for ever made, is thereby signified and represented.[232]

Thus, O most gracious and godly prince! do I confess . . . that the bread of the sacrament is truly Christ's body, and the wine to be truly His blood . . . but in a certain wise, that is to wit, figuratively, sacramentally, or significatively . . . But . . . can I not find the natural body of Our Saviour to be there naturally, but rather absent both from the sacrament, and from all the world, collocate and remaining in Heaven, where He, by promise, must abide corporally, unto the end of the world.[233]

Three or four Anabaptists went to the stake on 22 November: 'it is thought more of that sect shall to the fire'.[234] On the same day Lambert was burnt for his 'great heresy'.[235] To Wyatt in Spain on the 28th Cromwell wrote of Henry triumphing in disputation over this 'miserable heretic sacramentary'.[236] Yet Foxe related how on his way to Smithfield Lambert stopped at Cromwell's house, 'and so carried into his inward chamber, where Cromwell desired him of forgiveness'.[237] Cromwell had never favoured sacramentaries; rather he had feared that their radicalism would prejudice the evangelical cause and threaten the free passage of the Word. Yet perhaps he had known Lambert among the City brethren, and regretted that his beliefs had brought him to martyrdom.[238]

It seemed then as if the days of the reformers in power at Court were numbered. They had always known how fickle was the King they served and how fearful the risks they ran. But they were sure too that the Gospel, once it had reached the people, could never again be hidden. The evangelicals' purpose was clear: as one of the Christian Brethren had said, they had put books into the people's hands; once they were in the people's heads, 'they should have no further care'.

---

[232] Foxe, *Acts and Monuments*, v, pp. 237–50. Lambert's theology is discussed in Dugmore, *The Mass and the English Reformers*, pp. 127–30, 177–82.

[233] Foxe, *Acts and Monuments*, v, p. 249.

[234] *Lisle and Letters*, 5. 1285; Stow, *Annales*, p. 575.

[235] Wriothesley, *Chronicle*, i, p. 89; *Lisle Letters*, 5. 1285.

[236] *The Works of Henry Howard, Earl of Surrey, and of Sir Thomas Wyatt, the elder*, ed. G. F. Nott, 2 vols. (1815–16), ii, pp. 326–7.

[237] Foxe, *Acts and Monuments*, v, p. 236.

[238] S. E. Brigden, 'Thomas Cromwell and the Brethren', in *Law and Government under the Tudors*, ed. C. Cross, D. M. Loades, and J. J. Scarisbrick (Cambridge, 1988), pp. 31–50.

VII The Act of Six Articles and its Victims

FACTION STRUGGLES in Court and Council produced shifting policies towards reform: first advancing it, then repressing it. Yet while political intrigue might determine the incidence of persecution and decide the victims, events would show that the new religion was spreading independently of it, so preventing the full restoration of Catholic orthodoxy, whatever the policy of King and Council. The religious history of London was intimately linked with the feuds of contending factions at Court in these confused years. The tensions witnessed in the capital between the 'papists' and 'the new sect', and the popular disturbance there, underlay all the machinations in high politics and influenced the outcome.

'ONE PERFECT UNITY IN RELIGION'

The savage proscription of speculation upon the Mass did not silence the evangelicals, who became ever bolder. As the Easter of 1539 approached there came new orders to ensure orthodoxy. A proclamation in February demanded the observation of 'the ceremonies of holy bread, holy water, procession, kneeling and creeping on Good Friday to the cross'.[1] But the gospellers desired a Church thoroughly reformed. They sent news of religious affairs in London to their brethren on the Continent on 8 March:

The ceremonies are still tolerated, but explanations of them are added; so that now the holy water . . . is for no other purpose than to refresh our minds with the remembrance of the sprinkling of the blood of Christ; the bread signifies the breaking of the body of the same; the *pax* . . . sets before our eyes the reconciliation of God and man . . . The Mass is not asserted to be a sacrifice for the living and the dead, but only a representation of Christ's passion.

[1] *TRP* i. 188.

Even 'these things are retained for the sake of preventing any disturbances'.[2] For there was still trouble from reckless radicals. Images were being removed, whether they were abused or not. George Crane trampled on the palms strewn to hallow the streets for the Palm Sunday procession. On Palm Sunday (30 March), while in every parish hung the great Lenten veil painted with the Passion story, hiding the chancel and rood from the people, John Stokes and Richard Gorton taunted that at Easter, when the veil was drawn, 'before the crucifix ye shall see a little jackanapes'; 'it were alms [meritorious] to beat it down with clubs and burn it'.[3]

The Lenten fast was being broken this year.[4] At St Vedast Foster (Dr Champion's parish) a woman ate bacon on Maundy Thursday, for 'savouring of her mouth'. Shriven by Sir Thomas Bodnam, she confessed this fault and was told that if her conscience 'thought it good it were none offence'. She was 'houselled at Easter, but only to God'.[5] Thomas Freebarn's wife ate pork to satisfy the craving of pregnancy and was betrayed with her husband. Stokesley accused Freebarn of speaking against pilgrimages, holy bread, and holy water, of not going on Palm Sunday procession: 'thou art not a Christian man'. Sent before the Court of Aldermen, Freebarn was ordered to stand on the pillory. The brethren rallied to his defence. The wife of Michael Lobley, the evangelical book binder, went to her friends in Cromwell's household, asking them to intercede with their master. Barnes and Barlow spoke for Freebarn, and Cromwell summoned the Mayor and asked for leniency. Sheriff Wilkinson comforted the distraught Mrs Freebarn, and her husband was released.[6] But Cromwell's protection of his friends among the brethren made him vulnerable: not for much longer could he save them. In Holy Week John Worth heard from Ball, the Recorder of London's servant, that 'there was one

---

[2] *Original Letters*, ii, p. 624.
[3] CLRO, Repertory 10, fo. 89[v].
[4] *L&P* xiv/2. 71.
[5] CLRO, Repertory 10, fo. 90[v].
[6] Foxe, *Acts and Monuments*, v, pp. 385–7. For Lobley's troubles under Stokesley: ibid. v, pp. 38, 412. Freebarn was named, with other reformers, among those 'gentlemen not to be allowed in my lord's [Cromwell's] household . . . but when they have commandment or cause necessary to repair thither': *L&P* xiii/2. 1184.

hanged for eating flesh on a Friday against the King's commandment'.[7]

The signs in London were diverse: thoroughgoing reaction seemed imminent, yet the Gospel prospered still. The City parishes dutifully bought their copies of the Great Bible in English, as the November proclamation had ordered,[8] but the consequences of this Gospel freedom were ominous. Soon the reformers were reading the Bible 'in high and loud voices, and specially during the saying of Mass', to drown the idolatry.[9] The gospellers made a notable and public convert. Sir Nicholas Carew, Master of the Horse and another victim of the savage politics of the Court, went to the block on 3 March repenting his erstwhile 'folly and superstitious faith' which had brought him to 'so shameful a death'. He thanked God for his imprisonment in the Tower, for there 'he first savoured the life and sweetness of God's most holy Word, meaning the Bible in English'. Thomas Phillips, More and Stokesley's prisoner, turned gaolkeeper, had been his evangelist.[10] Some were writing verses against the Mass; others performed interludes.[11] John Bale had been released from prison in 1537 by Cromwell 'to write comedies'.[12] In January 1539 his *King John* was performed by Tower Hill, which provoked one in the audience, in his cups, to defend the Pope and Becket.[13] Priests were marrying, breaking their vow of chastity, and reformers dared to preach the validity of clerical marriage, even before the King, who most opposed it.[14] But a ballad circulating in London in 1539 suggests how many of the people also lamented this clerical

7 *Lisle Letters*, 5. 1415.
8 Guildhall, MSS 2895/1, fo. 101^r (St Michael le Querne); 4956/1, fo. 154^r (All Hallows Staining); 1432/1, fo. 62^r (St Alphage); 1454/53 (St Botolph Aldersgate); 2968/1, fo. 94^r (St Dunstan in the West); 3907/1 (St Laurence Pountney); 1002/1a, fo. 80^v (St Mary Woolnoth); 1279/2, fo. 34^v (St Andrew Hubbard).
9 Foxe, *Acts and Monuments*, v, pp. 443, 452.
10 *Original Letters*, ii, p. 625; Hall, *Chronicle*, p. 827.
11 *Lisle Letters*, 5. 1423. Husee had found an interlude called *Rex Diabole* for Lady Lisle in Oct. 1538, and was trying to find others concerning 'these new Scripture matters'; ibid. 5. 1242.
12 PRO, SP 1/114, fo. 54^r (*L&P* xii/1. 40); J. W. Harris, *John Bale: A Study in the Minor Literature of the Reformation* (Urbana, Ill., 1940), pp. 67, 100.
13 *L&P* xiv/1. 47(2).
14 Wriothesley, *Chronicle*, i, p. 83; *Original Letters*, ii, p. 624.

disobedience, and how the violation of oaths was thought to be turning Nature in her course:

> When cats unto mice
> Do swear obedience
> Then put in priests' wives
> Your trust and confidence.[15]

'The saying is', so John Worth told Lisle on 15 May, 'that there is an act passed that if any priest or married man be taken with any man's wife, to suffer death. God save the King.'[16]

When all was so uncertain men watched for signs of the politic direction in religion. Now the King's own orthodoxy was at last unambiguous. Henry went on procession about the Court at Westminster on Holy Thursday; and 'the high altar was garnished with all the apostles upon the altar, and Mass by note, and the organs playing'. On Good Friday the King crept to the cross, 'and so served the priest to Mass the same day, his own person, kneeling on his Grace his knees'. Every Sunday the King received holy bread and holy water and 'doth daily use all laudable ceremonies, and in London no man upon pain of death to speak against them'.[17] Henry was seriously alarmed: by the spread of heresy and sacrilege he witnessed at home; by the divisions which the rise of the 'new sect' had generated; and, in the spring of 1539, by the real threat of invasion from a Catholic League against a heretic and excommunicate England.[18] In London on 8 May a resplendent muster of 16,000, with all the City's military might arrayed, was held to terrify the King's enemies (though there had been fears that this would risk further disturbance in the capital).[19] The King had more reason to be fearful once he heard of the dissensions in his bridgehead on the Continent: Calais.

This garrison town had become an enclave for gospellers, where radical heresies were preached along with Scripture. So

---

[15] PRO, SP 1/156, fo. 187ʳ (*L&P* xiv/2, appendix 48).

[16] *Lisle Letters*, 5. 1415.

[17] Ibid.

[18] J. J. Scarisbrick, *Henry VIII* (1964), chs. 12–13; G. Redworth, 'A Study in the Formulation of Policy: the Genesis and Evolution of the Act of Six Articles', *JEH* xxxvii (1986), pp. 42–67.

[19] Stow, *Survey*, i, pp. 91, 103–4; Wriothesley, *Chronicle*, i, pp. 95–7; *Lisle Letters*, 5. 1415, 1386, 1391, 1406; Bodleian, MS Ashmole 861, fo. 337ʳ; S. E. Lehmberg, *The Later Parliaments of Henry VIII, 1536–1547* (Cambridge, 1977), p. 60.

scared was Lord Lisle, the governor of Calais, that one sect
might rise against the other that he slept in armour.[20] From as
early as 1537 Lisle had written insistently and often to Cromwell
warning him that sacramentarian heresy was spreading.[21]
Cromwell ignored him. He promoted his own men, and hoped
for the furtherance of the Word.[22] In February 1530 Clement
Philpot, Cromwell's body servant, sent word to the gospellers
in Calais: 'my lord doth know them all in Calais . . . that doth
favour the Word of God and them that do not favour it'.
'There is', he added, 'both sorts [that is, orthodox and re-
formed] in London as well as there is in Calais; but I trust the
most part good [reformed], or else I would they were good.'[23]
All through 1538 Cromwell ignored Lisle's pleas and protected
the reformers. Part of the correspondence is lost; however,
notes taken by a seventeenth-century historian from a letter
among the manuscripts of Sir Robert Cotton's library show
how emphatic was his protection:

A sharp letter of Cromwell to Lord Lisle taxing him for persecuting
those who favour and set forth God's word and favouring those who
impugn it. Also for suffering bruits to be scattered that the Bishop of
London is Vicar-General of England, and all English books to be
called in, &c.[24]

But Cromwell could protect the reformers no longer once the
King discovered how far heresy had permeated his garrison,
even among the 'saddest sort', and how far the unity of Calais
was threatened. Because of the invasion threat the Earl of
Hertford was sent in February 'to view the strength of Calais'.[25]
Cromwell's secret would out. In March Cromwell jotted a
remembrance for 'a device in the Parliament for the unity in
religion'.[26] Throughout April he was ill with tertian fever,
away from Court.[27] His conservative opponents, long excluded, now had their chance to return and could use the

[20] *Lisle Letters*, 5, p. 514.
[21] Ibid. 5. 1160, 1166, 1178, 1189, 1190, 1498, 1498A. Three letters written by
Lisle around Easter 1538 are missing.
[22] See *Lisle Letters*, 5, pp. 391, 489, 491, 675.
[23] PRO, SP 1/143, fo. 72^r; *Lisle Letters*, 5, p. 392.
[24] Bodleian, Jesus College MS 74, fo. 198^v.
[25] *Lisle Letters*, 5. 1362; Bodleian, Jesus College MS 74, fo. 275^v.
[26] *L&P* xiv/1. 655.
[27] *Lisle Letters*, 5. 1389, 1394, p. 450, 1396.

disturbing discoveries in Calais for their purposes.[28] The King
needed no convincing.

In April 1539 a draft proclamation was prepared, attacking
both those who sought to restore 'the usurped power of the
Bishop of Rome, the hypocrite religion', and their opponents
who used the Bible 'to subvert and overturn as well the Sacraments of Holy Church as the power and authority of princes
and magistrates'. The fervent partisans of either sect, who
disputed 'so earnestly . . . in churches, alehouses, taverns',
slandering each other as 'papist' or 'heretic', were denounced.[29]
The proclamation was never issued. Another way was found.
Cromwell sought to recover himself: he wrote to Lisle, feigning
astonishment: 'I cannot a little marvel' that Lisle, knowing
Cromwell's desire to repress error and establish 'one perfect
unity in religion', should never have vouchsafed the news that
Calais was infiltrated by sacramentaries.[30] Lisle was terrified,
fearing himself 'half undone'.[31] In his London garden Hertford
interviewed the circumspect Husee about 'what business there
was ado concerning the Sacrament': 'my Lord Privy Seal',
said 'things should be surmised and scant believed'.[32] Though
Cromwell's own position began to look shaky, still at the end of
May he was commanding Lisle to use 'all things with charity',
to handle 'matters mildly', not to initiate a general quest.[33]

In London, Convocation opened on 2 May, with Stokesley
singing the Mass of the Holy Ghost and with a sermon from
London's future Bishop, Nicholas Ridley; 'Timor Domini
initium sapientiae'.[34] A committee of Bishops sat daily, often
with the King in attendance, disputing the doctrine of transubstantiation, to establish doctrine and a 'thorough uniformity
in religion'.[35] By 4 May Husee (with inside knowledge) wrote
to Lisle that the forthcoming Parliament would settle, once and
for all, 'the Christian religion of this realm'. A way would be

[28] For Cromwell's exclusion of his enemies, and their return, see G. R. Elton,
'Thomas Cromwell's Decline and Fall', *Studies*, i, pp. 192–200, 205.
[29] *TRP* i. 191; R. W. Heinze, *The Proclamations of the Tudor Kings* (Cambridge,
1976), pp. 139–40.
[30] *Lisle Letters*, 5. 1403, 1443.
[31] Ibid. 5. 1435.
[32] Ibid. 5. 1428.
[33] Ibid. 5. 1429.
[34] Wriothesley, *Chronicle*, i, pp. 94–5.
[35] *Lisle Letters*, 5. 1422.

found to silence all those who dared to 'frame so excellent a thing' as the Mass 'after their carnal judgement' 'ere this Parliament be prorogued', so he wrote on 20 May. 'It will not be long', he told Lady Lisle on the 26th, 'ere your ladyship shall hear the best news that may be had'.[36] From Calais at the end of the month came the sacramentaries to Lambeth for trial, and pleas from the Calais Council that Cranmer's Commissary, who had protected them, be removed.[37] Still on 1 July the word was that 'my Lord of Canterbury handleth them very gently', even though three days earlier the most savage penal act against heresy had been passed: the Act of Six Articles.[38]

In this act the doctrinal orthodoxy of the Church of England was defined, and the King cleared himself, he hoped, of any taint that he was following the evangelicals.[39] Parliament and Convocation had 'accorded' in a statement of orthodoxy on central matters:

1. . . . in the most blessed Sacrament of the altar . . . is present really, under the form of bread and wine, the natural body and blood of our Saviour Jesus Christ . . . and that after the consecration there remaineth no substance of bread or wine, nor any other substance, but the substance of Christ, God and man.

2. Communion in both kinds is not necessary, *ad salutem*, by the law of God, to all persons . . .

3. Priests . . . may not marry, by the law of God.

4. . . . vows of chastity . . . by man or woman made to God . . . ought to be observed . . .

5. . . . it is meet and necessary that private masses be continued . . .

6. . . . auricular confession is expedient and necessary . . .

This was a penal act: it was meant to be. The intransmutable penalty for denying transubstantiation was death by burning: no longer was the heretic given the chance to recant or abjure. Offences against the other articles were variously subject to the

---

36 Ibid. 5. 1400, 1421, 1423, 1427.
37 Ibid. 5. 1410, 1424, 1434, 1437, 1456, 1457.
38 Ibid. 5. 1468. For the making of the Act, see Elton, 'Thomas Cromwell's Decline and Fall', *Studies*, i, pp. 205–10; Redworth, 'Genesis of the Act of Six Articles'; Lehmberg, *Later Parliaments*, pp. 65–6, 68–74, 82–3.
39 The Act is printed in Gee and Hardy, *Documents Illustrative*, pp. 303–19, and discussed in Elton, *Reform and Reformation*, pp. 287–8.

penalties of felony and praemunire. New commissions with powers of inquisition were added to the ordinary process of the spiritual law. Cranmer wrote a refutation of the act, for which they would have 'laid thousands of pounds to hundreds in London' that he should have been sent to the Tower. When Cranmer's refutation fell into the wrong hands his secretary would have gone to the stake had not two City brethren, Blage and Cromwell, intervened to save him.[40] Savage though the act was, very many, like Husee, rejoiced at its passage. The French ambassador, Marillac, wrote on 9 June (the day of the bill's second reading in the Lords): 'the people show great joy at the King's declaration touching the Sacrament, being much more inclined to the old religion than to the new opinions'.[41]

With time, the 'new sect' had grown intolerable to orthodox Londoners—most Londoners—who had longed for reaction. Adherents of the old faith were deeply moved that summer when Dr Wattes, a man of 'great and high learning', preached and read lectures daily in the City. He so 'clerkly confuted' the ideas which Latimer had 'sown among the Londoners', and so convincing were his arguments, that 'many which before that time were mighty of the new opinions were turned by his reading'.[42] His sermons in the capital and the crowds they drew gave cause for alarm.[43] The indignant Wattes was examined by Cranmer and Barnes, 'grieving because the one of them was abjured [Barnes] and the other suspected of heresy [Cranmer]'—and sent to prison.[44] A delegation of citizens, led by an Alderman of Gracechurch Street, went straight to the Archbishop and offered to bind themselves in £1,000 for Wattes's surety if he were committed for 'heresy' (*sic*), for 'ten thousand of London' were anxious for his release, so it was said, and would have come to plead for him themselves had they not been restrained. Wattes was sent to Kent, away from possible vigilantes.[45] When one Catholic preacher arrived with

---

[40] *Narratives of the Reformation*, pp. 248–9; Foxe, *Acts and Monuments*, v, pp. 388–91.
[41] *L&P* xiv/1. 1092.
[42] PRO, SP 1/155, fo. 96ʳ; 1/157, fo. 25ʳ (*L&P* xiv/2. 750; xv. 31).
[43] BL, Cotton MS Titus B i, fo. 477ʳ (*L&P* xv. 438). He had caused similar contention in Salisbury earlier: Elton, *Policy and Police*, p. 102.
[44] PRO, SP 1/157, fo. 23ʳ (*L&P* xv. 31).
[45] PRO, SP 1/155, fo. 196ʳ (*L&P* xiv/2. 750); *L&P* xv. 259; *Original Letters*, ii, p. 627; Lambeth, Cranmer's register, fo. 69ᵛ.

the charisma of a Latimer the wavering might return to the fold and the orthodox might be stirred to as great a fervour as the reformed.

As for Latimer, he resigned his bishopric, and Lisle's correspondents delighted to see him 'in a priest's gown and a sarcenet tippet . . . and never a man after him'. Long before he had foretold that one day he would put the ocean sea between himself and Stokesley; now he made for exile, but was captured at Gravesend. Shaxton resigned rather than subscribe to the act, and the rumour was that Crome would also. 'God send them all as they deserve!'[46] Yet, far from resigning, Crome went on to preach with more than usual zeal. On Relic Sunday (13 July) he protested the innocence and integrity of the maligned Shaxton and Latimer, goaded by the common talk in the City which slandered them as 'false knaves and whoresons', and inveighed against the 'lie mongers' everywhere in London: in vintners' and barbers' shops and at Bishops' tables.[47] William Jerome, vicar of Cromwell's own parish of Stepney, openly denounced the burgesses of Parliament as 'butterflies, dissemblers and knaves' for traducing the Reformation.[48]

All London celebrated the customary observances on Relic Sunday, save Dr Crome, who would have no special service at Aldermary. On 16 July he preached in his parish of requiem Mass: he had read in the Gospels of a burial with Jesus, Mary, and Martha among the mourners, but there was nothing of vestments, tapers, dirges, or masses. Masses were 'a good institution of man, and . . . he would not dispraise it, but as well away as there, and nothing in the Mass of Christ's institution, but only the holy consecration'. At the burial, Martha spoke only to Christ, though the apostles were there, so 'did not as we have done calling upon this saint and that saint that they should be helpers'.[49] At St Bride Fleet Street, on the very day that the act came into force, 13 July, Christopher Dray, a plumber, said that the sacrament of the altar 'was not offered up for remission of sins, and that the body of God was not there,

[46] *Lisle Letters*, 5. 1475, 1476, 1478; Wriothesley, *Chronicle*, i, p. 101; *Original Letters*, i, p. 215.
[47] PRO, SP 1/153, fo. 27ʳ (*L&P* xiv/2. 41); Foxe, *Acts and Monuments*, v, appendix xvi.
[48] PRO, SP 1/158, fo. 124ʳ (*L&P* xv. 414).
[49] PRO, SP 1/153, fo. 25ʳ (*L&P* xiv/2. 41).

but was a representation and signification of the thing'.[50] A
heresy which warranted the death penalty. Some of the evangelicals would never be silenced, but others lived in terror.
'Since Parliament' Hilsey could find no one willing to preach at
Paul's Cross. Stokesley lay in wait for the unguarded, and
summoned one of Hilsey's chaplains, who had preached at the
Cross on 20 July. The wheel had come full circle: let Stokesley
provide the preachers now, Hilsey besought Cromwell. 'For
lack of another' he would preach himself on the following
Sunday [27th], but 'with more fear than ever he did in [his]
life'.[51]

Stokesley longed to make a reality of the reaction. But no
time was left. He died on 8 September; his dying words a boast
'that he had sent thirty one heretics unto the infernal fire' (or
so Foxe wrote).[52] Stokesley's funeral on Holy Rood day was
splendid: St Paul's was draped in black from the high altar
to the west door; there was a procession, with the parish
crosses of the City borne by eighty London priests, tapers carried by thirty poor men, the Mayor and Aldermen in attendance. Dr Hodgkin, the suffragan, preached the funeral sermon,
of Stokesley's 'two talents of science and authority', and of 'the
great study and steadfastness he had taken . . . in all the schism
and division time'. As he told how the late Bishop 'received the
sacrament of the altar with so great reverence' all present had
'weeping eyes'.[53] Here was a Bishop who might have kept
London Catholic.

THE FALL OF THOMAS CROMWELL

The passage of the 'bloody act' seemed to augur a period of
savage repression, yet there was no swift return to orthodoxy,
no immediate inquisition. In November 1539 at St Martin
Vintry Alexander Seton dared to preach that private masses for
souls departed were otiose, and to avow the justificatory tenet
'that faith only did justify, and that works were not . . . profit-

---

50 PRO, SP 1/243, fo. 75ʳ (*L&P Addenda*, i/2. 1463).
51 *L&P* xiv/1. 1297.
52 Foxe, *Acts and Monuments*, iii, p. 104.
53 Wriothesley, *Chronicle*, i, pp. 105–7.

able . . . but only to . . . testify our faith'.[54] The reverse for Cromwell and the reforming party had been temporary, and there would be no quests under the Six Articles while they were ascendant in the Council and the counsels of the King. Henry was found a new wife. On 6 January 1540 he married Anne of Cleves, but instantly he sought to be rid of her: 'I like her not'.[55] The events of the spring and summer of 1540 confused even those who lived through them, so shifting were political fortunes, and so uncertain was the prospect of reform or reaction.[56]

In the early months of 1540 the favour of the reformers in high places and the exclusion of conservatives from power[57] allowed the progress of reform among the people. In February it was reported that 'the Word is powerfully preached by . . . Barnes, and his fellow ministers. Books of every kind may safely be exposed to sale.' At the end of the month still 'good pastors' were 'freely preaching the truth' without incurring the penalties of the Six Articles.[58] But there were signs which boded ill for the reformers. Gardiner and Dr Nicholas Wilson were chosen to preach at Paul's Cross in Lent. Barnes, too, was appointed, but he failed to recognize how ominous was the choice of his rivals, admitting later: 'I have been deceived myself.'[59] Gardiner made a double offensive against Barnes on 13 and 15 February, vituperatively attacking the reformers in league with the Devil, particularly his cohort of apostate friars who had 'left retailing of Heaven in pardons' to peddle heretical ideas with 'abuse of Scripture'. The abuses of the Church had gone, but evil doctrines had arisen to replace them, and now the 'Devil hath excogitate to offer Heaven without works', so that to enter Heaven 'needs no works at all, but only belief, only, only, nothing else'.[60] Barnes replied a fortnight later at the Cross and, reviling Gardiner for the 'evil herbs' he had 'set

[54] PRO, SP 1/243, fo. 78ʳ (*L&P Addenda*, i/2. 1463).

[55] Scarisbrick, *Henry VIII*, p. 370.

[56] Part of the following account appears already in my 'Popular Disturbance and the Fall of Thomas Cromwell and the Reformers, 1539–40', *HJ* xxiv (1981), pp. 257–78. I am grateful to the editors for permission to re-publish that part.

[57] Gardiner and Sampson had been excluded: *L&P* xiv/2. 423, 750(1, 2).

[58] *Original Letters*, ii, pp. 627, 614.

[59] Wriothesley, *Chronicle*, i, p. 113; *Original Letters*, ii, p. 616.

[60] *Letters of Gardiner*, pp. 168–9.

in the garden of Scripture', he threw down the gauntlet. This
sermon met (albeit according to a reformer) with 'almost
universal applause'.[61] An incensed Gardiner complained to the
King—judge of heretics—who tried the errant Barnes before a
secret council.[62] Gardiner 'prepared the Tower for his school
house', and there Barnes and another reformer came to be
'taught'.[63] Of Barnes's fellow nothing is known, but on 12 April
1540, like Richard Hunne, a man was found hanged in Gardiner's prison.[64] The main points of controversy between
Gardiner and Barnes were evidently the sufficiency of Christ's
passion for salvation, and the nature of grace, free will, and
penance.[65] For Barnes the doctrines of Purgatory and soli-
fidianism were the touchstone of their contention. He declared
that he had stood alone in his avowal of justification by faith,
for although many approved none would declare this tenet,
save Latimer.[66] But he had underestimated at least two of his
fellow reformers.

On 7 March William Jerome gave a sermon at Paul's Cross
as seditious and subversive as any by Barnes. He preached
firstly that 'no magistrates had power to make that thing which
of itself is indifferent to be not indifferent', and then averred
the cardinal Protestant doctrine: 'the promise of justification is
without condition, for he that putteth a condition unto it doth
exclude grace'.[67] Gardiner personally annotated the extant
report of this sermon. By the first article he wrote: 'Doctor
Barnes's book where he teacheth that men's constitutions bind
not conscience', thus making obedience to princes 'an outward
behaviour only', and by the second, it 'engendreth such an
assured presumption and wantonness that we care not greatly

---

[61] *Letters of Gardiner*, pp. 169–70; Hall, *Chronicle*, pp. 837–8; *L&P* xv. 306.

[62] *Letters of Gardiner*, p. 172.

[63] Hall, *Chronicle*, p. 838.

[64] Wriothesley, *Chronicle*, i, p. 115; Foxe, *Acts and Monuments*, v, p. 430; Brinklow, 'Complaynt', p. 29; John Bale, *Yet a course at the romyshe foxe* (1543; *RSTC* 1309), p. 32.

[65] BL, Cotton MS Cleo. E v, fos. 98ʳ, 99ʳ (*L&P* xv. 312); Foxe, *Acts and Monuments*, v, pp. 432–3. See Davis, *Heresy and Reformation*, pp. 73–4; J. P. Lusardi, 'The Career of Robert Barnes', *CW* 8/iii, pp. 1408–11; C. L. Hughes, 'Coverdale's *Alter Ego*', *BJRL* 65 (1982).

[66] *Original Letters*, ii, pp. 616–17.

[67] PRO, SP 1/158, fos. 50ʳ, 120ʳ (*L&P* xv. 345, 411(2)); Foxe, *Acts and Monuments*, v, appendix viii.

whether we obey God or no'.[68] By 12 March Barnes, and William Jerome and Thomas Garrett with him, had been ordered to recant in Easter week.[69] Jerome preached at St Mary Spital on Monday 29, Barnes on 30, and Garrett on 31 March before the Mayor and Communalty—and Gardiner.[70] Significantly, the King ordered that report be made to him.[71] Jerome recanted his offending articles. He stressed the binding power of law over men's consciences, affirmed that 'without works of penance no man may after his lapse attain to his justice again. Nor that Christ's passion is anything available to men that do not repent', and finally retracted his slander of Parliament of the previous year. Moreover, Jerome averred the necessity of all the sacraments, and of penance, contrition, attrition, and renovation as a means to salvation.[72] Yet Jerome's recantation was preceded by a sermon which attested his belief in the very articles he was bound to retract. Taking the Gospel for the day, he determined to 'use the story of Christ's resurrection to declare . . . our election'. He preached the Old Testament promises of Christ's advent and man's election, and rehearsed the Pauline teaching, 'like as he died for our sins, so he is risen again for our justification'. Jerome urged his followers to 'talk of Christ, and read His word, and hear it preached'.[73] After such a declaration of reformed faith Jerome's recantation was a mockery.

Barnes's recantation the following day was as unconvincing. He prefaced his articles by a declaration of his belief in the sufficiency of the passion of Christ, 'only Redeemer and justifier', thus contradicting his subsequent admission that penance was necessary for justification after the Christian had come into faith and baptism. But Barnes did deny absolute predestination, and upheld the efficacy of good works. He asserted that the commands of rulers should be obeyed, and not only through fear of punishment but through conscience.[74] After his recantation

[68] PRO, SP 1/158, fo. 50^r (*L&P* xv. 345).

[69] *Correspondance de Marillac*, p. 169; *L&P* xv. 334.

[70] *Grey Friars Chronicle*, p. 43.

[71] PRO, SP 1/158, fo. 124^r (*L&P* xv. 414).

[72] PRO, SP 1/158, fos. 120^r, 124^r–125^r (*L&P* xv. 411(2), 414); Foxe, *Acts and Monuments*, v, appendix xxi.

[73] Foxe, *Acts and Monuments*, v, appendix xxi.

[74] Guildhall, MS 9531/12, fo. 37^r; Foxe, *Acts and Monuments*, v, appendix vii. The extant text may be a digest of the recantations of the three reformers, for although

Barnes called for Gardiner's forgiveness: this eventually grudgingly given, Barnes 'plainly and directly preacheth the contrary of that he had recanted, so evidently as the Mayor . . . asked whether he should from the pulpit send him to ward'.[75] Garrett's recantation was as patently insincere as that of his fellows. To one of their friends at Court it was reported 'how gaily they had all handled the matter, both to satisfy the recantation and also . . . to utter out the truth, that it might spread without let of the world'.[76] On 3 April Barnes, Garrett, and Jerome were imprisoned in the Tower 'by the King's own commandment'.[77]

'Even for the least' the 'new sort' had 'their feathers plucked' that Easter.[78] Any sign of subversion or unorthodoxy was dealt with severely. It was wise to apply for a licence before eating meat that Lent, because those who dared to break the Lenten fast were haled before the Court of Aldermen.[79] Two women, one a priest's wife, were troubled for that offence on 22 March.[80] Ten or twelve 'burgesses' of London joined the preachers in the Tower by 10 April; also fifteen or twenty Flemings (all Anabaptists, according to Marillac, but there is nothing to identify them or their heresy).[81] Serving men dared to make a 'fray' at a sermon of Gardiner's on 11 April.[82] By 24 April some of the dissidents had been released without punishment, while the fate of the others remained undecided, 'on account of the contention which has arisen among the Bishops about their doctrine'.[83] Maundevild, a Frenchman, an Italian painter, and an unnamed Englishman, all of 'very low condition', were burnt on 3 May for 'speaking irreverently of the Holy Sacra-

claimed to be by Barnes, it is signed by all three: J. F. Davis, 'Heresy and Reformation in the South-East of England, 1520–1559' (University of Oxford, D.Phil. thesis, 1968), p. 319.

[75] Hall, *Chronicle*, p. 838; *L&P* xv. 425; *Letters of Gardiner*, p. 174.

[76] Wriothesley reported that Garrett 'recanted nothing': *Chronicle*, i, p. 114. Foxe, *Acts and Monuments*, v, p. 433; *Letters of Gardiner*, p. 174.

[77] Wriothesley, *Chronicle*, i, p. 114; *Correspondance de Marillac*, p. 175; *L&P* xv. 485.

[78] *L&P* xv. 697.

[79] Guildhall, MS 9531/12, fo. 19^r.

[80] CLRO, Repertory 10, fo. 169^v.

[81] *L&P* xv. 566; *Correspondance de Marillac*, p. 175.

[82] Wriothesley, *Chronicle*, i, p. 115.

[83] *L&P* xv. 566.

ment' and for refusing communion that Easter.[84] All these executions were at Gardiner's instance.

There had been a change of direction in high politics. The arrest of the preachers and the keen interest that Henry was taking in their heresy was ominous for the reforming party. It was particularly dangerous for Cromwell, for he had long favoured Barnes, and Gardiner might well use Barnes's disgrace to blacken his rival.[85] On 10 April Marillac wrote that Cromwell and Cranmer 'do not know where they are' and that Cromwell was 'tottering'.[86] But Cromwell was still masterminding Parliamentary business; higher in favour than ever, or so it seemed. On 18 April he was created Earl of Essex and Great Chamberlain. The wheel of fortune was at its height for him. Now far from overtly supporting reform Cromwell was circumspect, giving no support to Barnes, and on 20 April disassociating himself from Alexander Seton.[87] However, by May political fortunes had changed again, and the conservatives seemed imminently threatened by the reforming party. In the City, Richard Farmer, a leading merchant, was condemned in the King's Bench to perpetual imprisonment and forfeiture of his goods for supporting a popish chaplain, and two other wealthy citizens of the same opinion left the country secretly, taking their property with them.[88] On 26 or 27 May Dr Wilson and Bishop Sampson went to the Tower, and others of their faction feared the same fate.[89] Cromwell's coup of late May seemed completely successful and secure—Cranmer had taken over the pulpit at St Paul's in place of Sampson and was preaching contrary to Gardiner's Lenten sermons, and there were rumours that Barnes would be released and Latimer reinstated[90]—yet his triumph was short-lived. On 10 June Cromwell was suddenly arrested. 'Many lamented, but more rejoiced',

[84] Stow, *Annales*, p. 579; *L&P* xv. 651; *Original Letters*, i, pp. 200–1; Foxe, *Acts and Monuments*, v, p. 654.

[85] Elton, 'Thomas Cromwell's Decline and Fall', *Studies*, i, pp. 214–15; PRO, SP 1/120, fo. 172ʳ (*L&P* xii/1. 1260); Lusardi, 'The Career of Robert Barnes', p. 1403.

[86] *L&P* xv. 486.

[87] Lambeth, Cranmer's Register, fo. 69ʳ.

[88] *L&P* xv. 697, 598, 615, 650, 730, 939, 1005; *PPC* vii, pp. 11, 22; Hall, *Chronicle*, p. 838; Wriothesley, *Chronicle*, i, p. 119; Stow, *Annales*, p. 580.

[89] Elton, 'Thomas Cromwell's Decline and Fall', pp. 219–20.

[90] *L&P* xv. 736–7.

especially the clergy and those who opposed the changes in religion.[91]

Cromwell's opponents proceeded immediately to prepare an attainder bill against him, denying him a trial and declaring his guilt.[92] The charges against him were long and specific, accusing him of overweening power, abuse of office, but overwhelmingly of heresy. He had caused 'damnable errors and heresies to be inculcated', spread heretical literature, licensed known heretics to preach, set them at liberty and refused to brook complaints against them. Allegedly, on 31 March 1539 at St Peter le Poor in London Cromwell had vigorously defended Barnes and other preachers, avowing that he would never turn away from reformed doctrine, even if the King and all his subjects did so; rather 'I would fight in the field in my own person with my sword in my hand against him and all other'.[93] The decisive charges in the attainder were the accusations that Cromwell had said that it was lawful for every Christian to minister the sacrament, and possessed sacramentarian books in English. Henry was convinced that Cromwell was a sacramentary, and it was for this that he allowed him to fall.[94] The common talk was that Cromwell would be burned at the stake as a heretic, but he went to the scaffold on 28 July.[95]

At the block Cromwell protested his guilt in offending God and his King: 'I have been the causer of my fall.' To confess at the last was conventional: it was also binding upon the Christian who would die 'charitably'. Cromwell's last speech, as it was reported, has the air of truth. He had 'travelled the world' when he was young and 'came of low stock'. The King had raised him high, but Cromwell had brought his own downfall. Now if there were 'any man or woman . . . in London' whom he had offended he besought forgiveness, and 'of charity I forgive all the world'. To the last he protested his innocence of heresy: 'I believe in the laws ordained by the Catholic Church,

---

[91] Hall, *Chronicle*, p. 838; *Original Letters*, i, p. 202; *L&P* xv. 766.

[92] *L&P* xv. 498(60); S. E. Lehmberg, 'Parliamentary Attainder in the reign of Henry VIII', *HJ* xviii (1975), p. 693.

[93] *L&P* xv. 498(60); Foxe, *Acts and Monuments*, v, p. 399.

[94] *L&P* xv. 766, 824; Foxe, *Acts and Monuments*, v, p. 402; Scarisbrick, *Henry VIII*, pp. 379–80; Elton, 'Thomas Cromwell's Decline and Fall', pp. 224–7.

[95] *Original Letters*, i, pp. 202–3.

and in the holy sacrament without any grudge.'[96] Cromwell had been an evangelical. If he had ever moved to a more radical view he left no sure sign, unless it were of the friends and servants he promoted. His death left his evangelical friends bereft and grieving. It was perhaps of Cromwell that Wyatt wrote:

> The pillar perished is whereto I lent
> The strongest stay of mine unquiet mind.[97]

In Fulham churchyard Thomas Stokes said that Cromwell had 'at the hour of his death departed very sorry and penitent, and died like a Christian man'. For this he was sent to Newgate on 7 August.[98] These were troubled days.

If Cromwell was removed ostensibly for his dangerous sympathy to heretics there could be no better way to illustrate his complicity in a heretical movement than to purge three radicals. Barnes, Garrett, and Jerome were exempted accordingly from the general pardon and condemned by attainder, without trial.[99] That their deaths proceeded from such an enactment rather than by the Act of Six Articles, which—had they been guilty—would have acted equally swiftly and inexorably, again points to the political motive behind their fall. It also suggests that the extreme charge of sacramentarianism could not have been proven against them. The three reformers went to the stake on 30 July, two days after Cromwell's death. The execution on the same day of three conservative priests—Thomas Abell, Richard Featherstone, and Edward Powell—also confirmed the political nature of the purge. For these three men, old supporters of Katherine of Aragon and the Pope, had been long imprisoned.[100] Rather than an exhibition of Henry's scrupulous equity, their hanging, drawing, and quartering as traitors alongside the burning heretics was part of the coup against Cromwell; a cynical sacrifice, Winchester's device 'to

---

[96] Bodleian, Fol. Δ 624, p. 462. I am most grateful to Mr Gary Hill for showing me his transcript of this manuscript, the chronicle of Anthony Anthony, a London evangelical.

[97] *Collected Poems of Sir Thomas Wyatt*, ed. K. Muir and P. Thomson (Liverpool, 1969), ccxxxvi; *Chronicle of King Henry VIII of England*, trans. M. A. S. Hume (1889), p. 104.

[98] PRO, C 1/1063/75. I am indebted to Dr Steven Gunn for this reference.

[99] *L&P* xv. 498(ii cap.49); *Original Letters*, i, p. 207.

[100] J. E. Paul, *Catherine of Aragon and her Friends* (1966).

colour his own tyranny', to confuse the issue and silence the murmurings of the people.[101] In the last they manifestly failed.

THE MARTYRDOM OF BARNES, GARRETT, AND JEROME

The vilest heresies had been imputed to Barnes, Garrett, and Jerome. But although supposedly 'their heresies were so many' yet there was 'not one there alleged as special cause of their death'.[102] Protestant observers were bewildered, for these preachers 'never once opened their mouths expressly against that statute' either in their public preaching, or private conversation, except when they 'were with honest and godly men'.[103] Luther complained that the 'cause why Barnes was martyred is still hidden, because Henry must be ashamed of it'.[104] The victims, too, sought to know the truth: Barnes found himself condemned to die, 'but wherefore, I cannot tell'.[105] Why did they die?

Barnes had been 'slandered to preach that Our Lady was but a saffron bag', an imputation that he denied Christ's incarnation, and thus of Anabaptism. But Barnes's intention, he protested, had always been 'utterly to confound and confute all men of that doctrine as are the Anabaptists'.[106] In 1532 Tyndale had warned Frith that to question the presence of Christ's body in the sacrament would bring fierce opposition from Barnes, and in 1538 Barnes, fearful that such teachings would prejudice the progress of the Word, had been one of the instigators of charges against John Lambert. Barnes was no sacramentary.[107] Of William Jerome's faith or guilt less may be said, but he, too, was probably accused of Anabaptism, for he insisted in his recantation that Christ was 'very man in deed', and had not a 'phantastical body, as some heretics hath

---

[101] Elton, *Reform and Reformation*, pp. 292–4; Foxe, *Acts and Monuments*, v, p. 420.

[102] Hall, *Chronicle*, p. 840; William Turner, *The huntyng of the Romysh Vuolfe* (Emden, 1554; *RSTC* 24356), sig. Eiiii^v.

[103] *Original Letters*, i, p. 210.

[104] *L&P* xvi. 106.

[105] Foxe, *Acts and Monuments*, v, p. 435.

[106] Ibid. v, p. 434. Barnes had sat on the commission to try Anabaptists in 1535.

[107] *L&P* vi. 403; Foxe, *Acts and Monuments*, v, p. 228. W. A. Clebsch, *England's Earliest Protestants* (New Haven and London, 1964), chs. 4–5.

said'.[108] Thomas Garrett was one of the first and most effective evangelical Lutherans in England, but there was no sign that he held more extreme convictions. Maybe these men had moved gradually to a more radical view of the eucharist; maybe popular opinion in London had so changed that they now preached publicly what they believed privately; however, there is no evidence that any of the three reformers was guilty of convictions as radical as those of which they were accused, of the sacramentarianism which alone carried the death penalty. Yet Barnes and Jerome had in the past been guilty of sedition.

The death of Barnes and the others was politically expedient, but not simply to bring down Cromwell. These men were not innocents, caught in the toils of high politics, but were dangerous in their own right, for they had a popular following. Barnes was always a disruptive and impassioned preacher, and London conservatives blamed him particularly for all the confusion in the realm.[109] In July 1540, with nothing left to lose, Barnes preached a sermon to fulfil all the authorities' worst fears. The reformers' protestations at the stake were so provocative, and their deaths 'so moved the people that if they had had a leader there might have been a great tumult'.[110] At the stake, Barnes took his last chance to preach and play to the crowd, creating a posthumous legend which the conservatives would find it impossible to suppress. He appealed to his audience whether any there knew 'wherefore I die', or who 'of my preaching hath taken any error', and pointedly forgave Gardiner and his faction if they had 'sought or wrought' his death 'through malice or ignorance'.[111] He began by asserting his unequivocally Protestant conviction in the sufficiency of Christ's passion for salvation, and went on openly to mock Catholic teachings. Now when Barnes declared Our Lady 'the most pure Virgin that ever God created and a vessel elect of God' the sarcasm was clear, and at this point he was stopped.[112] 'If saints do pray for us', Barnes told the Sheriff waiting to carry out the death

---

[108] Foxe, *Acts and Monuments*, v, appendix xxi.

[109] Lusardi, 'The Career of Robert Barnes', p. 1369 n. 1.

[110] *Correspondance de Marillac*, p. 209; *L&P* xv. 953.

[111] Barnes's sermon is printed in Foxe, *Acts and Monuments*, v, pp. 434–6.

[112] Barnes had impugned the sanctity of the Virgin before. Standish stressed Barnes's use of irony; Coverdale, *Remains*, p. 333.

sentence, 'I trust to pray for you within this half hour.' Barnes
denied that he was 'a preacher of sedition and disobedience',
and even argued that resistance was never justified, but he
intended, to the last, to urge reform. He followed his avowal of
loyalty by four requests to the King so heavy with irony that
they could only be seen as disruptive. The fourth: 'that matrimony be had in more reverence than it is; and that men, for
every light cause invented, cast not off their wives', was
clearly, mockingly, directed to the faithless King. Two days
before, on the very day that Cromwell went to the block,
Henry had married his fifth wife, Catherine Howard.[113] But
the first request was most disturbing of all: 'whereas his grace
hath received into his hands all the goods and substance of the
abbeys'—here the disquieted Sheriff tried vainly once again to
silence him—but Barnes continued, insisting that the King
should bestow his goods 'to the comfort of his poor subjects,
who surely have great need of them'.

Barnes's following derived in part from his preaching of social
ideas as emotive and powerful as Protestant doctrine itself. The
reformers had looked to the advent of a Christian commonwealth with the dissolution of the religious houses; not only
would monastic abuses be removed, they hoped, but monastic
wealth would be distributed to the poor. They were to be sadly
disappointed, for the new wealth was only creating a greater
schism between the rich, who became richer, and the poor, who
were even more neglected than formerly, 'which thing above
all other infidelities shall be our damnation', as the reformers
lamented.[114] The London pamphleteer Henry Brinklow thought
that Barnes and the others had died for advocating the cause of
the poor: 'Well, the poor well feeleth the burning of Doctor
Barnes and his fellows, which laboured in the vineyard of the
Lord. For according to their office, they barked upon you to
look upon the poor.'[115] These reformers who had lamented the
'covetousness of the possessioners' were now attacked by the
very order they criticized, who determined to 'procure the ruin

---

[113] Scarisbrick, *Henry VIII*, pp. 429–30.
[114] Brinklow, 'Lamentacyon', p. 80.
[115] Ibid., p. 91.

of the said doctors'.[116] Richard Hilles certainly thought that it was for financial reasons that the King chose Barnes, Garrett, and Jerome as victims to placate and win over the clergy, 'the ignorant and rude mob', and the 'obstinate part of the nobility and citizens', that they 'might pay more readily the money granted . . . by Parliament'.[117]

The gospellers had adduced Scripture to warn the wealthy of the cardinal sin of covetousness. Chilling accounts of the fate of the selfish rich, in parables, had been used in medieval sermons and *exempla* to exhort the upper orders to deal mercifully with the poor and to console the oppressed. The purpose was moral, but it might leave the rich justifiably fearful of the vengeance of the lower orders, and the distressed looking for salvation upon earth as well as in the world to come.[118] The biblical parables were being preached by the gospellers to new effect at the Reformation, to the alarm of the authorities. In the crowded City the extremes of wealth and poverty were most marked. 'In every parish in London', so John Gough lamented, 'the poor image of God' perished while the rich looked on.[119] On 7 July 1540 the 'frantic' evangelist, William Collins, who had used the Gospel to preach a message of social equality, went to the stake in Southwark.[120]

Reformers were preaching the terrifying parable of the rich man and Lazarus.[121] Some in London began to take the general warning literally and to name the damned rich men. Eleanor Carkke came before the Court of Aldermen in October 1540 for slandering the late and fabulously wealthy Alderman Kitson.[122] Mrs Carkke had heard in a sermon of the Holy

---

[116] *L&P* xv. 486.

[117] *Original Letters*, i, p. 210.

[118] See, for example, G. R. Owst, *Literature and Pulpit in Medieval England* (Oxford, 1966), pp. 237–375.

[119] John G . . . *The myrrour or lokynge glasse of lyfe* (1532?; *RSTC* 11499), fo. 13ʳ.

[120] Wriothesley, *Chronicle*, i, p. 119. Dr Horst suggests that Collins may have been the first English Anabaptist: *The Radical Brethren: Anabaptism and the English Reformation to 1558* (The Hague, 1972), pp. 90–2.

[121] See above, p. 269.

[122] Eleanor Carkke was widow of Cromwell's cousin, John Whalley, and had remarried, to William Carkke, one of the brethren: *Lisle Letters*, 3. 771a–772, p. 493; 4. 922–922a, p. 255; 5. 1236, 1572. Thomas Kitson was wealthy enough to build Hengrave Hall, and for his widow to be courted by a Gentleman of the Privy Chamber: D. B. Starkey, 'The Age of the Household', in S. Medcalf (ed.), *The Later Middle Ages* (1981), p. 234.

Ghost's dread warning to a rich man: 'this night the Devil shall fetch thy soul from thee', and then ventured the opinion that Kitson had 'used himself like unto the rich man that Gospel maketh mention of'.[123] The Gospel preachers had led the way. Latimer himself preached this parable in 1550; probably not for the first time.[124] When Crome preached in July 1539 of the integrity of the Bishops after their resignation, he made the analogy that 'the honesty of a man lieth not in his substance or riches', and the natural corollary was surely evident.[125] Later, in 1541 William Tolwin, rector of St Antholin, was found with a work of the Anabaptist Balthasar Hubmaier, and forced to recant the anarchic doctrine of the Anabaptists, that all things should be 'common among the people'.[126] With good cause were the gospellers feared as subversive.

In July 1540, just as the faction struggle reached its climax and the coup was accomplished, a major inquisition for heresy began in the City, the first quest under the Act of Six Articles.[127] Persecution was the conservatives' first objective: personal ambitions apart, this was the reason they had sought power, and this was their first move when the reformers fell. 'In four-teen days' space', five hundred Londoners were arrested, and no preacher or layman 'which had spoken against the suprem-acy of the Bishop of Rome, but he was wrapped in the Six Articles'.[128] This was the version of events in Edward Hall's chronicle, but this part was written not by Hall but by Richard Grafton. Hall had first opposed the bill for the Six Articles in

[123] CLRO, Repertory 10, fos. 178^v, 194^r. Significantly, it was the wife of London's most conservative Alderman, William Hollis, who reported her.

[124] Latimer, *Sermons*, p. 280.

[125] PRO, SP 1/153, fos. 27^r–28^r (*L&P* xiv/2. 41).

[126] Bale, *Yet a course at the romysh foxe*, sig. Giiii^v.

[127] Hall maintained that the inquisition took place 'in short time after the passage of the act', while Foxe dates the first quest to 1541 (*Chronicle*, p. 828; *Acts and Monu-ments*, v, pp. 443–51). The purge could come only with the conservative triumph in 1540. Hilles reported that just before the grant of the general pardon on 15 July 1540 many were imprisoned in London for holding reformist opinions (*Original Letters*, i, p. 208). Original indictments for twenty suspects were endorsed 'billa vera' by Hollis, the Mayor, on 17 July 1540 (PRO, SP 1/243, fos. 60^r–80^r; *L&P Addenda*, i/2. 1463). Thomas Lancaster, a priest accused of disseminating heretical books, was in the hands of the City authorities by 1 July 1540 (CLRO, Journal 14, fo. 210^r). *VCH London*, i, p. 277.

[128] Hall, *Chronicle*, p. 828; J. Gairdner, *Lollardy and the Reformation in England*, 4 vols. (1908–13), ii, pp. 200–2; *House of Commons, 1509–1558* (Hall).

the Commons, but prudently he had voted for it later. Grafton was himself a victim of the quest, and hardly its most impartial observer, but he may have been right. We know the names of at least two hundred Londoners who were arrested.[129] (Their place in City society and their heresies we shall examine in a later chapter.)

The task of extirpating heresy fell to the City rather than the ecclesiastical authorities, and now London had a Mayor who was eager to persecute. In October 1539 William Hollis, excluded from the mayoralty since 1536, was at last elected. His friendship with Bonvisi, and the reformers' suspicions of him, suggest the extremism of his views.[130] The orthodox among the citizenry were now given their chance to report their evangelical neighbours, whose heresies they had been forced to tolerate throughout the 1530s. There were fears that the Act of Six Articles would encourage 'false witnesses'. Awareness that by this draconian act the automatic and immutable penalty for sacramentaries was death may have made some draw back. Hilles explained that his bitterest enemies were unwilling to inform against him, and thus 'to be regarded in the sight of all as guilty of treachery against their neighbours'.[131] Yet reports were made to the London juries of all manner of religious dissidence, and these juries were, apparently, composed of men so zealous that they inquired

not only who denied the Sacrament to be Christ's very natural body, but who also held not up their hands at sacring time, and knocked not on their breasts: And they not only inquired who offended in the Six Articles but also who came seldom to the church, or in communication contemned priests or images.[132]

If the party of reaction had ever thought that a return to orthodoxy would be easily accomplished, this first drive against heresy under the new act proved that it would not. Such was the zeal of the juries that the City's prisons filled with suspect heretics. A year later the Recorder of London accused them of

---

[129] Foxe, *Acts and Monuments*, v, pp. 443–51.
[130] *Memorials of the Holles Family, 1493–1656*, ed. G. Holles (Camden Society, 3rd series, lv, 1937), p. 19.
[131] *Original Letters*, i, pp. 232–3.
[132] Hall, *Chronicle*, p. 828.

doing as 'much as in them lay to make an uproar among the King's people'.[133] No one remained long imprisoned. Henry ordered that the persecution stop when he heard from Crome of the 'cruel treatment of some preachers and citizens of London'. This was Hilles's story.[134] According to Hall (or Grafton), it was the Lord Chancellor who interceded for the prisoners, believing that they had been charged through malice. All were released, bound to appear in Star Chamber on All Souls' day.[135] None were called. The heresy quest was halted on 1 August—the day following the deaths of the papists and reformers—by royal command.[136] The reaction had proceeded too far, too fast, widening the religious divisions and setting citizen against citizen.

No one could have regarded the executions of July 1540 with indifference or equanimity: Cromwell and his reforming friends were either revered or reviled, and they were not soon forgotten. Fifteen years on, when witnesses tried to remember when a particular event happened, it was the death of Barnes they used as a landmark.[137] The guilt and fall of Cromwell, Barnes, and their fellows soon provoked a ballad controversy in London, 'one envying against another, one reviling and reproving another, one rejoicing at another's fall and adversity'.[138] Thomas Smith ('a great and notable papist called Trolling Smith') fired the first shot with his scurrilous *Ballad on Thomas Cromwell*, which accused the fallen minister of corruption ('both plate and chalice came to thy fist'), falseness, and heresy:

> Thou did not remember, false heretic
> One God, one faith, and one King Catholic
> For thou has been so long a schismatic.
> Sing troll on away, troll on away &c.[139]

---

[133] Foxe, *Acts and Monuments*, v, p. 442.

[134] *Original Letters*, i, p. 208.

[135] Hall, *Chronicle*, p. 828; Foxe, *Acts and Monuments*, v, pp. 451, 441; Strype, *Ecclesiastical Memorials*, i/1, pp. 565–6.

[136] CLRO, Letter Book P, fo. 219ᵛ.

[137] GLRO, DL/C/209 (no folio numbers).

[138] Coverdale, *Remains*, p. 325 (preface to his *Confutation of the Treatise of John Standish*). The ballads are printed in E. W. Dormer, *Gray of Reading* (Reading, 1923). *The Book of Broadsides* (Society of Antiquaries, 1866), pp. 2–5 gives the titles and provenance.

[139] Dormer, *Gray of Reading*, pp. 76–8.

William Gray, who had written *The Fantasy of Idolatry* for Cromwell, came to the defence of his disgraced master, writing
A Ballad against Malicious Slanderers, which John Gough printed.
Gray could not deny that Cromwell died justly, as a traitor, but
he could challenge the other charges:

> For he was a traitor, and thou art the same
> Troll away, papist, God give thee shame.

> The sacrament of the altar, that is most highest
> Cromwell believed it to be the very body of Christ
> Wherefore in thy writing, on him thou liest.[140]

Smith countered with *A Little Treatise against Seditious Persons*,
and *A Treatise declaring the despite of a secret seditious person that
dareth not show himself.*[141] Though Smith urged unity between
'true trollers', the exchange continued.[142] Were these ballads
being sung in the streets of London? Certainly the Council
moved swiftly to stop their printing, so sensitive and controversial was the issue. Smith and Gray came before the Council
in December 1540, and Grafton admitted printing the ballads
under Banks's colophon, instead of his own.[143]

In his *Little Treatise against Seditious Persons* Smith referred to a
martyr's legacy, treasured by the reformers:

> The confession of an heretic that lately did offend
> And amongst others suffered for his deserving
> Secretly they embrace as a most pleasant thing.

This was probably the confession, perhaps a last letter, from
Barnes which was smuggled away and printed in Germany. This
confession may well have called forth John Standish's *A lytle
treatise . . . against the protestacion of R. Barnes at the tyme of his death*
(October 1540), which was answered in turn by Coverdale,
from exile.[144] The attacks upon Barnes were more vituperative, more retributive and jubilant, than those upon Cromwell.
In 1540 a set of verses deriding Barnes, 'the vicar of Hell', were

---

[140] Ibid., pp. 79–82.

[141] Ibid., pp. 83–6; *Book of Broadsides*, p. 3.

[142] There were nine ballads in all: *Book of Broadsides*, pp. 2–5.

[143] *PPC* vii, pp. 103, 105.

[144] Hughes, 'Coverdale's Alter Ego', pp. 110–15; Lusardi, 'The Career of Robert
Barnes', pp. 1414–15.

printed by Redman for Banks, 'cum privilegio Regali': *This lytle treatyse declareth the study and fruits of Barnes borned.* Each verse treated one aspect of Barnes's error; the refrain gleefully celebrated his end.

> Of repentance
> O how like a Christian man he died
> Stiffly holding his hands by his side
> Saying, if ever were any saint, that died
> I will be one, that must needs to be tried.
> Without repentance, the Devil was his guide.
> All this was said like a false friar
> Yet all could not save him from the fair fire.[145]

But, as Coverdale knew, to glory in the deaths of Barnes and his fellows was not only uncharitable, but impolitic, and the rest of the world might condemn Henry for allowing such 'jesting and railing ballads' to go forth with his imprimatur. The executions of July 1540 made martyrs for the Protestant cause. Barnes, Garrett, and Jerome were mourned by the brethren, and their martyrs' deaths were celebrated long after. In 1548 was published *The metynge of doctor Barons and doctor Powell at paradise gate.*[146]

---

[145] *RSTC* 1473.5. I am grateful to the Librarian of Shrewsbury School for sending me a copy of their unique copy of this broadside.
[146] *RSTC* 1473.

VIII Politics and Persecution, 1540–1546

THE PERSECUTION of heresy in the last years of Henry VIII's reign was sporadic and uncertain, for these were years without a leader, save for the inconsistent and bellicose monarch, and of no certain purpose or directive in religion. For all the feuding at Court and in the Council, no faction was able to win lasting favour or ascendancy.[1] A profound change had occurred in the nature of politics: now the factions struggled not for power alone, but for a cause. The avowed aim of the conservatives, led by Gardiner and Norfolk, was to restore the country to orthodoxy. In order to achieve this end they must overthrow their rivals at Court, and permanently, for they knew from their own experience that exiled opponents were dangerous, and might return. They knew also, after the quest through London in the summer of 1540, how disturbing was any attempt to extirpate heresy unilaterally among the people, and how futile. The leaders of the reaction perforce changed their tactics.

Thereafter the drive against heresy was directed selectively against the leaders of reform in the Church and nation. The aim was to uncover the networks of patronage and persuasion among the nobility and gentry, leading citizens, the common people, the reformist writers and printers of London, and the most determined of evangelical clergy. The vulnerability of the reform-minded in high places to charges of association with advanced religious views was increased by the moves of reforming writers to appeal specifically to a noble readership. Now the reformers dedicated tracts to the nobility; not only because they wished to win the most influential in society to the reformed faith, but also to solve the problem of obedience to authority

---

[1] For the history of Henry's last years, see J. J. Scarisbrick, *Henry VIII* (1968), chs. 13–14; D. Wilson, *A Tudor Tapestry* (1972); J. Gairdner, *Lollardy and the Reformation in England*, 4 vols. (1908–13); D. Starkey, *The Reign of Henry VIII: Personalities and Politics* (1985), ch. 7; Elton, *Reform and Reformation*, chs. 13–15.

that such opposition raised.[2] There were high hopes of that 'foolish, proud boy', Henry Howard, Earl of Surrey, who, in the way of things, had rebelled against the religious conservatism of his father, Norfolk. In the summer of 1539 George Constantine had urged Cromwell that Surrey be sent in the mission to Cleves, 'for that he should there be fully instructed in God's Word. For if the Duke of Norfolk were as fully persuaded in it as he is in the contrary, he should do much good.'[3]

The downfall of Cromwell had shown that there was no one so high, so intimate with the King, that he could be invulnerable to attack if his faith were once suspect. Yet in his last years Henry kept about him men, and women, who hated the restoration of orthodoxy, and privately would not conform to it. These courtiers had religious sympathies close to the fallen minister's and very far from Henry's own: hardly surprisingly, for one of Cromwell's last schemes had been the reform of the royal household, and there he had placed some of his own clients.[4] Others migrated to the royal service when their master fell, so that in daily attendance upon the King in his last years were Cromwell's men. Among the newly formed band of Gentleman Pensioners were reformers.[5] Sir Anthony Denny, Chief Gentleman of the Privy Chamber and later Groom of the Stool, was one of Henry's closest friends and confidants: he was also thoroughly committed to the ideal of a reformed commonwealth, as his will reveals. The lessons of his humanist education at St Paul's school had stayed with him.[6] To be in the King's Privy Chamber was to be his constant companion, and some of the Gentlemen were zealous Protestants: Thomas Cawarden, Richard Morison, Christopher Askew, Philip Hoby (Gentleman Usher), Sir George Blage, squire of the body, and

[2] For example, William Turner's *The huntyng of the Romyshe Vuolfe* (*RSTC* 24356) was dedicated to the young lords and gentlemen of the counties; see J. Loach, 'Pamphlets and Politics, 1553–1558', *BIHR* xlviii (1975), pp. 36–8.

[3] Constantine, 'Memorial', p. 62.

[4] G. R. Elton, *The Tudor Revolution in Government* (Cambridge, 1953), pp. 385–97.

[5] For Cromwell's men, and their commitment to reform, see M. L. Robinson, 'Thomas Cromwell's Servants: the Ministerial Household in Early Tudor Government and Society' (University of California at Los Angeles, Ph.D. thesis, 1975), pp. 373, 438 ff. For a list of the Gentlemen Pensioners in 1539, see *L&P* xiv/2. 783.

[6] Starkey, *Reign of Henry VIII*, pp. 133–6; *House of Commons, 1509–1558* (Denny). PRO, PCC, Prob. 11/32, fos. 285ᵛ–286ᵛ.

a favourite of the King ('my pig'), Sir Thomas Sternhold, groom of the Robes, George Ferrers, a page, Sir Thomas Weldon, first master of the household, and John Lascells, a sewer, would all be troubled for their religion. Blage expected 'this life to leave through fire at Smithfield'.[7] As Henry's health declined, he grew daily more dependent upon his favourite physicians, Dr Buttes and Dr Huick.[8] Constantine 'durst adventure' his life that 'the diet of Gwaiacum' would heal the King, and meant to tell Buttes about it. Constantine knew Buttes not only because they were both doctors but also because both were of the new sect, as was Dr Huick. Huick had had a dramatic conversion to the 'Gospel truth' as early as 1532, when (so he told Morison) 'the veil of Moses was raised from his eyes'.[9] Some of Henry's leading councillors kept as their personal chaplains men with extreme views. John Assheton, who was later tried for beliefs which verged on Socinianism, had been Audley's chaplain. The Duke of Suffolk retained as chaplains the reformers Thomas Lawney, Alexander Seton, John Willock, and John Parkhurst, who were doubtless chosen not by the Duke, but by his redoubtably reformist Duchess Katherine Willoughby d'Eresby.[10] In England, as in France, leading ladies at Court had a powerful influence upon the spread of evangelical doctrine. Henry could hardly be unaware that reform had spread at his Court: sometimes he chose to ignore it, sometimes not. He had lived to regret abandoning Cromwell, the most faithful servant he had ever had, 'on light pretexts',[11] and now, from time to time, protected his friends troubled for religion.

Returning to Court in September 1540, John Lascells asked 'what news there were pertaining to God's holy Word, seeing we have lost so noble a man which did love and favour it?' When he heard that the conservatives still prevailed, he counselled William Smethwick, sewer in the Queen's Chamber, 'not to be too rash or quick in maintaining the Scriptures, for if he would

---

[7] For their biographies, see *House of Commons, 1509–1558.*

[8] For payments to them, see *L&P* xiv/2. 781. Starkey, *Reign of Henry VIII*, p. 117.

[9] Constantine, 'Memorial', p. 76; *L&P* xii/1. 212.

[10] J. F. Davis, 'Heresy and Reformation in the South-East of England, 1520–1559' (University of Oxford, D.Phil. thesis, 1968), pp. 406–7.

[11] Scarisbrick, *Henry VIII*, p. 383.

let them alone or suffer a little time, they would overthrow themselves'.[12] If the gospellers could wait the world would turn again: meanwhile they must profess the Word in secret. Well aware how vulnerable their commitment to reform made them, and conscious of the risks they ran, still these courtiers determined to profess the new faith, and they maintained their contacts in the evangelical underworld which grew in London under persecution. The contradictory official policy of setting forth the Bible in English, while forbidding the discussion of central issues of faith, had itself fostered the creation of gospelling conventicles. Under persecution, a Protestant underground grew up, shadowy to all but the new sect. By chance the Council uncovered in 1541 something of the hidden world of London conventicles. From Salisbury John Erley, a disciple of Latimer's, had written to Mr Mathew, an upholsterer of Cornhill, or to Mr Downes, asking them for news 'for the consolation of good willers': 'You be not only associate and in company daily with such manner of men as be favourable to the Word of God, but also with such as can partly delate whether that it be of any likelihood that God's Word shall have free passage or no.' Mathew had promised 'some godly matter in metre or rhyme to give courage unto little children to read the Word of God'. Let Richard Downes, he besought, 'give some counsel to Pittes for the avoiding of the worst'.[13] Richard Downes had long been convinced that the Bible must go forth in English. An old friend of Stephen Vaughan's, he had carried messages, and the works of Tyndale, from Vaughan in the Low Countries to Cromwell in 1531 and 1532.[14] In December 1542 Downes made his will, remembering his reforming friends and expressing a reformer's creed:[15]

I trust to be saved by no man's works . . . but only by the merits of Jesus Christ, and trust in Him to be my mediator unto God . . . and that Christ's death and blood shedding on the cross may be between me and my offences, for after that the rulers of the high priests had

---

[12] *L&P* xvi. 101.

[13] PRO, Sta Cha 2/34/28II. Elton, *Policy and Police*, p. 36. Pittes, a clothworker like Downes, lived to offend against religious orthodoxy again, in Mary's reign: GLRO, DL/C/614, fo. 49[v].

[14] *L&P* v. 247, 753.

[15] PRO, PCC, Prob. 11/29, fo. 113[r].

caused Christ to be crucified and was dead and buried yet He rose again and overcame death and all to save sinners, whereof I am one.

To his partner John Merifield, another of the brethren, he left a ring inscribed 'Christ is risen for our justification'. His apprentice, Alexander Ettis, he freed from two years of his apprenticeship, if he would be guided by his parents, who were more friends of Downes in the new sect. Like Merifield, William Ettis and his wife had been troubled during the quest in the summer of 1540 for harbouring reformist preachers at St Matthew Friday Street.[16] Because 'St Paul saith that he that laboureth is worthy to eat' Downes followed his exhortation and left 4s. to the poor water bearers in Cornhill 'who labour truely for their living'. Though a gospeller, Downes was no sacramentary, for he bequeathed a wax taper to burn before the blessed sacrament. Even under persecution, Downes and his friends, and other groups like theirs, continued to meet in each other's houses to expound Scripture, as they had done before.

Yet many of the gospellers despaired now of the Bible ever being freely read or known. 'The tenth man in London, neither the hundredth man in the whole realm', knew the Gospel.[17] The people would sink into 'more bondage and blindness than before', so Constantine lamented in 1539.

Who is there almost that will have a Bible but he must be compelled thereto? How loath be our priests to teach the commandments, the articles of the faith, and the Pater noster in English! Again, how unwilling be the people to learn it. Yea, they jest at it, calling it the new Pater noster and New Learning.[18]

Many, indeed most, even in London, evidently shared Norfolk's view that 'it was merry in England before this New Learning came up'[19] (how many we shall explore in the next chapter). If the reaction was generally welcomed, the authorities could wait for the gospellers' energies to wane, and for a calm and unity in religion to return. Only the most reckless and extreme evangelists, those who preached doctrines most embarrassing to the authorities, were hunted down.

---

16 Foxe, *Acts and Monuments*, v, p. 446.
17 BL, Royal MS 17 B, xxxv, part 2, fo. 10[v].
18 Constantine, 'Memorial', p. 59.
19 *L&P* xvi. 101.

A POLITIC ORTHODOXY

Religious controversies continued to divide the capital through the autumn of 1540. The more daring evangelicals preached defiantly, 'with more zeal than ordinary'. In October Bonner was moved to inhibit all unlicensed preaching after the sermons of John Willock, curate at St Katherine Coleman and one of the rabid Scottish reformers exiled to London, had disturbed the 'Christian flock'.[20] Willock had denied the existence of Purgatory, and consequently the efficacy of prayers for souls or the intercession of saints. He preached against confession, and just as priests were renouncing their wives as they were commanded, he argued for clerical marriage.[21] The ballad war waged between Smith and Gray that autumn was probably less disruptive than the contest of pulpit oratory between Dr Crome and Dr Wilson, which stirred a 'variety of opinions and contention' among the citizens.[22] Erley reported the essence of their dispute: 'The variance between Crome and Wilson was not whether private masses be profitable or necessary or not, but whether they profited the souls departed or not . . . for they did both agree that masses were profitable to the now living.'[23] Of Wilson's arguments we have no report, but he defended 'human traditions'. (As he had been exempted from the general pardon that July for denying the royal supremacy, the sword of Damocles still hung over him.)[24] Crome had 'many enemies by report', who gathered almost thirty offending passages from his sermons.[25]

Crome had preached that 'no works can justify in the same manner as Christ does, nor do they so satisfy as He satisfied by suffering for us'. No truth, Crome averred, is necessary to be believed unless expressly revealed by Scripture; to offer masses for the dead is superstition, introduced 'by means of a vision, yea, rather a delusion of Satan, in the time of Pope Gregory'. To the King's embarrassment, Crome had questioned, even

---

[20] Guildhall, MS 9531/12, fo. 21^v.

[21] Foxe, *Acts and Monuments*, v, pp. 446, 448; iv, p. 586.

[22] Guildhall, MS 9531/12, fo. 26^v.

[23] PRO, Sta Cha 2/34/28 I.

[24] *Original Letters*, i, p. 211; *L&P* xv. 498.

[25] *Original Letters*, i, pp. 212–13.

derided, the contradictory nature of a royal policy which had, in *The Institution of a Christian Man* of 1537, maintained that masses *scala coeli* could not avail the dead, only to insist in 1539 that private masses be continued. Moreover, if masses were profitable to the dead, why then were monasteries dissolved? To those who called the gospellers 'seditious preachers', Crome replied, 'you are seditious parties, who defend superstition', ignoring the voice of Christ. 'The Church of Christ is suffering, and ever will suffer, persecution.' 'The world tried to persuade' those who had suffered that summer; it might try to persuade Crome, but he trusted that he would not falter. The King's judgement on 18 January 1541 was that Crome must recant.[26] As Crome reflected, a commission was established on 19 January to seek out heresy.[27]

The articles which Crome was ordered to recant on 13 February enshrined the precise and politic orthodoxy of the moment.[28] What Kings gave, they could take away (God 'turneth' their hearts 'as He lusteth'). First, Crome must declare that Scripture, so freely given in 1538, might be 'restrained from the lay people'. Prayers, fasting, and alms deeds, he must declare, were profitable to the dead. Only God knew 'the measures and times of His own judgements and mercies', but men might know that 'masses public and private be a sacrifice both for the quick and the dead', if they could not know for whom, or when. Prayers to the saints in Heaven were 'good and laudable'. Kings and princes might ordain what their people should observe in religion. Apostolic and patristic teachings not expressly revealed in Scripture should nevertheless be believed. The authority of the Church, Crome was to assert, was not above Scripture, but to the Church was given authority to expound and explicate it. The King was Supreme Head, under Christ, of the Church, both of the clergy and the laity. Two imputations which derogated from the politic Reformation Crome was specifically to recant:

---

[26] Ibid. i, p. 214.

[27] Guildhall, MS 9531/12, fo. 18ᵛ; Foxe, *Acts and Monuments*, v, pp. 440–1.

[28] For the process against Crome, and his recantation, see Guildhall, MS 9531/12, fo. 26ʳ⁻ᵛ; Foxe, *Acts and Monuments*, v, appendix xvi; R. S. Brodie, 'The King's Judgement against Dr Crome', *TRHS* new series, xix (1905), pp. 295–304.

such preachers as say that the Mass is available for souls departed doth not thereby, as farforth as I know, go about to deprive the King of his supremity nor yet to bring in again the Bishop of Rome.

Although masses and other suffrages be profitable for souls departed yet the King . . . and his high court of Parliament have lawfully and justly suppressed the abbeys and monasteries.

Crome's strictures upon an inconstant policy must be withdrawn: not least because the chantries still stood, and more were still being endowed. Crome, the past master of the ambiguous recantation, preached as commanded on 13 February. Yet, as he prefaced the required articles with a reforming sermon and announced that he read them only because he was thus ordered, no one could have failed to see that he recanted not at all in spirit. Hilles feared that 'the clergy will not let him off thus'.[29]

Bonner, whom the reformers had once trusted to further their cause, as a 'fast friend of the Gospel of Christ', now turned persecutor, as the times demanded.[30] By March 1541 Wyatt was reporting upon the Bishop's changed religious attitude.[31] In Paris, Bonner had promised Grafton and Whitchurch that he would further the freedom of the Gospel as much as Stokesley had hindered it, and assured them that six copies of the English Bible would soon be placed in St Paul's. (This was done before Bonner's departure for France late in February 1540.)[32] Yet in London he soon witnessed the excesses which the liberty to read Scripture had fostered, and began to denounce the common people as too irreverent and irresponsible to be entrusted with its sacred mysteries.[33] With the Bible in St Paul's came an admonition that it be read reverently, without interpretation, but soon certain bold spirits were 'brabbling' Gospel truth in the cathedral and other City churches. A new order was sent in May 1541 that if the 'insolent and indiscreet' behaviour of the gospellers did not stop the Bible would be taken from them.[34]

[29] *Original Letters*, i, p. 215.

[30] Foxe, *Acts and Monuments*, v, pp. 151, 160, 412; G. M. V. Alexander, 'The Life and Career of Edmund Bonner, Bishop of London, until his deprivation in 1549' (University of London, Ph.D. thesis, 1960).

[31] *L&P* xvi. 640.

[32] Foxe, *Acts and Monuments*, v, p. 412.

[33] Strype, *Ecclesiastical Memorials*, i/2, pp. 473–4.

[34] Guildhall, MS 9531/12, fo. 26ᵛ; Foxe, *Acts and Monuments*, v, appendix xiv.

John Bale wrote in 1541 that Bonner's intent had always been malicious; that 'he set up Bibles in St Paul's not purposing any Christian erudition to the people, but as snares to catch them by . . . to know which were the busy Bible men of London that he might speak with them at leisure'.[35]

So, in 1541 Bonner discovered John Porter. 'Great multitudes' came to hear Porter reading from the Bible in St Paul's. Through 'diligent reading of the Scripture', and by listening to sermons Porter had become, so Foxe heard, 'very expert' in this 'godly exercise'.[36] But Foxe did not admit how Porter had used the freedom to read the Bible as licence to expound forbidden doctrine. It was in Calais that Porter had been won to the new faith. Now he believed that Christ 'shed His precious blood for the redemption of all sinners that trust unto Him, and hath left the eternal life by His Word here amongst us'. But he had also forsaken the Mass: it 'was made but by man. Why should we set by it?' Even in Lord Lisle's house, he had said 'that he would make the bread which he did eat as good as the sacrament'.[37] Dismissed by Lisle for his 'evil fashions', Porter fled home to London to hide. (He had a brother in Watling Street.) Husee thought it would be easy enough to find Porter: 'on the holy days tailors walk abroad'. By 15 August the Mayor had Porter, weeping and 'very repentent', in the Counter.[38] Bonner was right to fear that the most radical heresies might be dilated by the 'Bible men'.

Through the spring of 1541 the quest against heresy proceeded, but the vindictive zeal which had animated some of the citizens in the past summer to report their reformist neighbours had gone. Now Bonner doubted, as Fitzjames had done, whether there was a jury in London which would bring a guilty verdict against a heretic: 'In London they ever find nothing'.[39] On 5 April the Bishop instigated his own quest, ordering his clergy to mark and report any parishioner who failed to confess that Lent or to receive communion at Easter.[40] The Court of Aldermen was moved to imprison John Starkey on 7 April.

---

[35] John Bale, *Yet a course at the romyshe foxe* (*RSTC* 1309), sig. Mᵛ.
[36] Foxe, *Acts and Monuments*, v, pp. 451–2.
[37] *Lisle Letters*, 5. 1515, 1515A.
[38] Ibid. 5. 1509, 1514, 1518, 1521.
[39] Foxe, *Acts and Monuments*, v, p. 441.
[40] Guildhall, MS 9531/12, fo. 20ʳ.

Starkey had been indicted in 1540, with others in St Magnus, for harbouring preachers of the New Learning, but nothing could suppress Starkey's conventicle: more 'naughty and evil disposed persons' resorted to his house than ever.[41] Bonner was aghast and angry that the juries would let the guilty go free. But they had their reasons. The issue of general pardons in 1540 and 1541 made the Act of Six Articles seem arbitrary; while the order to the City authorities on 12 July that heretics might be burned without the issue of the writ *de haeretico comburendo* only emphasized the absolute and draconian nature of this law.[42] On the same day, as if to silence protest, orders came to the Sheriffs that nightly watches be held.[43] The jury which sat in judgement upon young Richard Mekins that July could find no verdict. Having received no guidance, it doubted its competence to judge his heresy; not least because the witnesses gave conflicting evidence. Mekins may have been the boy who slandered the City in August 1540, and he may have been the reforming son of a reforming father, Thomas Mekins,[44] but even if he were guilty the jury drew back. Knowledge that the immutable penalty for the guilty was death, with no chance of abjuration, and that Mekins was only about fifteen, made the jurors hesitate.

Mekins had averred that 'Barnes died holy', and impugned the Mass. One witness had claimed that Mekins thought the sacrament 'nothing but a ceremony; and the other nothing but a signification'.[45] Hilles had heard that Mekins held 'the Lutheran opinion touching the eucharist. He did not altogether deny a corporal presence, but asserted, as our Wycliffe did, that the accident of bread did not remain there without the substance.'[46] On 30 July Mekins went to the stake as a sacramentary, lamenting that ever he had known Barnes or followed his heresy. (Though Barnes had never been a sacramentary.) The authorities needed to blacken Barnes's memory, and as Grafton said, Mekins was so terrified that he would have

[41] CLRO, Repertory 10, fo. 204ᵛ; Foxe, *Acts and Monuments*, v, p. 444.
[42] CLRO, Repertory 10, fo. 214ᵛ. Because this writ became redundant there are no significavits for heresy in these years.
[43] CLRO, Repertory 10, fo. 214ʳ.
[44] Ibid., fo. 170*. For Thomas Mekins, see above, pp. 121–2.
[45] Foxe, *Acts and Monuments*, v, pp. 441–2, 653.
[46] *Original Letters*, i, p. 221.

confessed that the twelve apostles taught him.[47] Mekins
became a martyr, celebrated as an innocent victim of a persecuting regime. John Bale suggested that the jury which condemned him at last had been rigged: 'if the wardmote enquest
will not condemn your accused, bring out a false of your own,
of Rome runners, pardoners, parish clerks and bell ringers, as
ye did now of late for Richard Mekins, a poor simple lad of
17 years of age'.[48] Some pitied Mekins, as an innocent led
astray; some thought that Bonner should have 'laboured' to
save him; but this evidence comes all from the reformers.
Bonner acquired then his reputation as a persecutor: 'what a
plague is this, that in no man's time alive, was ever any Christian Bishop reigning over the City of London, but every one
worse than other'.[49]

The most resolute champion of orthodoxy, Dr Richard
Smith, preached at St Antholin on 13 November upon the
theme of reconciliation (2 Corinthians 5: 20). Alexander Seton
was goaded to reply. Smith's insistence that 'man, by his
works, may merit' Seton denounced. The nature of free will
and grace lay at the heart of the contention between Smith and
Seton. Seton denied the freedom of the will: 'Paul saith, Of
ourselves we can do nothing; I pray thee then where is thy
will?' 'There is nothing in Heaven or earth . . . that can be
mean towards our justification; nor yet can nor may any man
satisfy God the Father for our sin, save only Christ, and the
shedding of His blood.' It followed, claimed Seton, that the
man who, like Smith, preached that works might merit
'preacheth a doctrine of the Devil'. For his City audience,
Seton alleged an analogy of a man making a payment to prove
his assertion that 'if thy works do merit one little jot or tittle
towards thy justification, then is Christ false of His promise,
who said He would do all together'. Persecution would never
deter him: 'If you ask me when we will leave preaching only
Christ: even when they do leave to preach that works do
merit.'[50]

---

[47] Foxe, *Acts and Monuments*, v, p. 442; Wriothesley, *Chronicle*, i, p. 126; Hall,
Chronicle, p. 841.

[48] Bale, *Yet a course*, sig. Fi[v], fo. 24[v].

[49] Brinklow, 'Complaynt', p. 93.

[50] Foxe, *Acts and Monuments*, v, pp. 449–51. Smith had been given a special licence
to preach by Bonner: Guildhall, MS 9531/12, fo. 13[v].

It was William Tolwin, a fellow reformer, who had allowed
Seton to preach at St Antholin, without episcopal licence. Both
men made their submissions and bore faggots at Paul's Cross
on 18 December.[51] John Bale reprinted Tolwin's shameful
recantation in his *Yet a Course at the Romyshe Foxe*, together with
an explication and refutation of each of Tolwin's retractions as
a warning to the gospellers to stand firm against Bonner,
'satellite of Satan'.[52] Tolwin had admitted firstly that he had
been a 'great favourer and receptor of heretics', and that he
possessed heretical books. The discovery of eighteen books in
Tolwin's possession proved that new and dangerous heresies
were spreading, for the works covered the spectrum of heretical
opinion. Some were works which not even John Bale knew,
yet. There were old Lollard works, adopted by the reformers,
like *A boke called thorpe and Oldcastell*, and *The dore of holy scripture*
(which had been 'perused' by Barnes, Turner, and Seton).[53]
There were Lutheran works: the *Catechisms* of Melanchthon,[54]
Urbanus Rhegius[55] and of Sarcerius,[56] and the *Postylles* of
Corvinus.[57] By Luther himself there was a book of prayer, and,
apparently, *The boke of the Counterfayt bysshop* (which does not
seem to have been a translation of any of his known works).[58]
Tolwin also possessed a copy of Watt's work, translated by
William Turner as *The Old God and the Newe*, and of his contem-
porary, Lancelot Ridley's, *Commentaryes upon Saynte Paule to the
Ephesyanes*.[59] *The Postylles upon the Epystles as Gospels* Bale could

---

[51] Wriothesley, *Chronicle*, i, p. 132.   [52] Bale, *Yet a course*, sigs. Ci, Ciiii.

[53] *The examinacion of master William Thorpe preste accused of heresye. The examinacion of syr
J. Oldcastell* (W. Tyndale? G. Constantine? 1530; *RSTC* 24045); *The dore of holy scripture*
(J. Gough, 1540; *RSTC* 3033). (The prologue to the Wycliffite Bible.)

[54] This work was noted particularly by Bale: *Yet a course*, p. 7. Perhaps the work was
The confessyon of the fayth of the Germaynes (1536; *RSTC* 908), or else an edition of another
of Melanchthon's works not yet translated into English, or not known to have been.

[55] This work may have been *A comparison betwene the olde learnynge & the newe*,
translated by William Turner, of which three editions were printed by John Nicholson
at Southwark, 1537–8, for John Gough: *RSTC* 20840, 20840.5, 20841.5.

[56] The work may have been the *Common Places of scripture ordrely set forth*, translated
by Taverner (John Byddell, 1538; *RSTC* 21752.5).

[57] The first known English edition of this work was not published until 1550: *RSTC*
5806.

[58] *The images of a verye chrysten bysshop, and of a counterfayte bysshop* (Wyer, for William
Marshall, 1536?; *RSTC* 16983.5).

[59] *A worke entytled of the olde god & the newe* (J. Byddell for W. Marshall, 1534; *RSTC*
25127). (Bale thought this by 'Hermanus a Germane': *Yet a course*, fo. 48ᵛ.) *A
commentarye in Englyshe vpon sayncte Paules epystle to the Ephesyans* (1540; *RSTC* 21038).

not identify, unless by Taverner. There were more radical works also: two unnamed books by John Frith (one of which Bale could not identify, but the other he thought to be *A disputacion of purgatorye*), and a work of Zwingli, *Swinglius of Gods provydence*.[60] Most extreme of all was the *Catechysme of Pacimontanus* (the alias of the Anabaptist leader, Balthasar Hubmaier). None of Hubmaier's works were definitely known to have reached England by then. Even Bale knew little of him (or was he disingenuous?): 'But I conceive here the better opinion of him, for that my lord hath condemned him among these men, whose doctrine I know to be pure and perfect.'[61] Thus might extreme radicalism spread by association with more moderate reform.

Tolwin denied all the heresies contained in these works, one by one: the teachings against the sacraments, against free will (thereby 'making God the author of sin'), against good works, confession, clerical celibacy, against communion at Easter and in one kind, against the invocation of saints. Lastly, Tolwin denied two arguments advanced in these works which set those who held them against society: refusal to 'give an oath before a judge', and the proposal 'to have all things in common amongst the people'.[62] These were the anarchic teachings of the Anabaptists. Well might the authorities have feared a threat to the social order as well as to orthodoxy. On 20 December, two days after these recantations, another London curate was committed to the Tower. Robert Wells of St Mary Colchurch had dared to imagine the prophet Daniel's vision of the four kingdoms 'as written of the King's person', and to expound it 'lewdly and traitorously, therein depraving his majesty's godly proceedings'.[63] Comparisons between Henry VIII and Belshazzar were not far to seek. Yet there were signs that Henry was moving again to free England from a Babylonian captivity.

In the autumn of 1541 Henry turned increasingly to Cranmer, and 'began a little to set his foot again in the cause of religion'. On his progress to Yorkshire, the King saw the old faith yet unreformed, the worship of 'feigned miracles' continuing. He

---

[60] Bale, *Yet a course*, fo. 47ᵛ.

[61] Ibid., sig. Giiiiᵛ; I. B. Horst, *The Radical Brethren: Anabaptism and the English Reformation to 1558* (The Hague, 1972), p. 94.

[62] Bale, *Yet a course*, fo. 88ᵛ.

[63] *PPC* vii, p. 285.

sent orders to Cranmer on 4 October for the injunctions against superstitious worship at shrines, against setting up lights before images, to be enforced.[64] Bonner, in turn, ordered an inquisition on 14 October throughout his diocese to discover where idolatry continued.[65] But the reformers complained that although 'divers idols be taken away', Bonner 'shamed not . . . to set up other in their places', replacing the lost image of the Virgin at St Paul's by one of St John the Baptist.[66] Soon, for reasons not of religion, the Howards were disgraced. That autumn all the Court knew, save the King himself, that his adored new Queen was unfaithful to him. Who would dare tell him? At Mass on All Souls' day Cranmer delivered a damning account of Catherine's treasonous adultery to an incredulous Henry. Norfolk made a politic withdrawal to the country. On 13 February 1542 the Queen, with Lady Rochford, went to the block.[67] But Catherine's fall and the Howards' disgrace signalled no lasting reversal of fortunes for the reformers.

Parliament met on 16 January. London sent two new burgesses: John Sturgeon and Nicholas Wilford.[68] There is a suspicion that religion played some part in their election, as it may have done in many others during these years. Sturgeon was of the brethren, a friend of Latimer from earlier days. He had dared to protect preachers of the New Learning even in 1540, despite his prominence in City society.[69] Nicholas Wilford had a brother, Robert, whose continuing allegiance to Rome could be used to prejudice Nicholas's election. That Robert Wilford's father-in-law, Richard Farmer, had been attainted in 1540 for harbouring a popish chaplain made him the more vulnerable.[70] On Christmas day 1541, as he attended the Mayor to Mass at St Paul's, Robert Wilford was challenged in this most public forum for being a 'maintainer' of the Pope. With his blessing, so he was accused, the arms of the Bishop of Rome remained still in the window of his parish church, St

[64] Foxe, *Acts and Monuments*, v, pp. 462, 463, 832.
[65] Guildhall, MS 9531/12, fo. 33^{r-v}; Foxe, *Acts and Monuments*, v, p. 695.
[66] Brinklow, 'Complaynt' and 'Lamentacyon', pp. 61, 87.
[67] Scarisbrick, *Henry VIII*, pp. 430–3.
[68] *House of Commons, 1509–1558* (John Sturgeon and Nicholas Wilford).
[69] *L&P* x. 1201; Foxe, *Acts and Monuments*, v, p. 444.
[70] *House of Commons, 1509–1558* (Richard Fermour).

Bartholomew the Little.[71] Nicholas Wilford was elected any-way. If the new sect hoped for reform to be advanced in this Parliament they were to be disappointed.

By royal command Convocation debated whether the Great Bible might be retained 'without scandal, error, and manifold offence to Christ's faithful people'. Gardiner listed a hundred words which he judged should only go forth in Latin or Greek ('to teach the laity their distance'). Cranmer announced on 10 March the King's will to have the Great Bible examined in the universities.[72] From that time on the printing of the English Bible was halted. The loss of the Bible was a disaster for the reformers, a victory for all those who feared the consequences of the freedom of the Word. In the time taken for the Bible's revision, the conservatives hoped (so the reformers claimed) that the King would be dead, they would be immune from his displeasure, and the Bible forgotten. 'How many read it' even now?[73] In Convocation's lower house they protested against the plays and interludes performed in London.[74] If the Bibles gathered dust in the City churches the gospellers would take the Word to the people in ballads or plays.

A wave of repression began in the spring of 1542. Bonner sent out an index of prohibited books to his clergy. He was the only Bishop to do so, but only Bonner had purveyors of forbidden works outside his own cathedral, in Paul's churchyard, and only in London was there a large market for such heady works. Most of the banned books were by English reformers: Frith, Roye, Joye, Tyndale, Barnes, Fish, Rastell, and Marshall. All the works of Luther and Calvin were outlawed, but none by Zwingli or the sectaries (despite the revelation that Tolwin owned some).[75] On 22 March a new commission was estab-lished to seek out offenders against the Six Articles.[76] John Porter, the gospeller and sacramentary, died a martyr's death, in Newgate prison, despite the efforts of his City friends to

---

[71] CLRO, Repertory 10, fos. 236ᵛ, 238ʳ.
[72] S. E. Lehmberg, *The Later Parliaments of Henry VIII, 1536–1547* (Cambridge, 1977), pp. 163–5.
[73] J. F. Mozley, *Coverdale and his Bibles* (1953), p. 264.
[74] Lehmberg, *Later Parliaments*, p. 165.
[75] Guildhall, MS 9531/12, fo. 40ᵛ.
[76] Ibid., fo. 38ʳ.

release him. To the last he proselytized among his fellow
prisoners.[77]

GARDINER 'BENDS HIS BOW AGAINST
THE HEAD DEER'

Around Candlemas 1543 the Earl of Surrey with a band of
noble companions ran wild through London at dead of night,
breaking windows in churches and grand City houses. From
the river they cast stones at the 'queens' (whores) in the Bank-
side Stews, in traditional misrule. With Surrey were his boon
companions: Sir Thomas Wyatt, the reckless son of the late,
and by Surrey lamented, poet; Sir William Pickering; and
Thomas Clere, who would die to save Surrey at Montreuil.[78]
The pride of the Earl was renowned: he boasted himself a
prince, and it was said already that 'if ought other than good
should become of the King, he is like to be King'.[79] When Sir
George Blage rebuked him for the escapade, Surrey was regret-
ful: he 'had liever [rather] than all the good in the world it were
undone . . . but we shall have a madding time in our youth'.
He feared, rightly, that the King and Council would hear of
it.[80] But he found also a moral intent behind his attack on
London. In the Fleet he wrote a 'satire against the citizens of
London':

> London! hast thou accused me
> Of breach of laws, the root of strife?

In Wyatt Surrey had found religious as well as poetic and
personal inspiration, and in Surrey's poetry, as in Wyatt's, are
found aspirations animated by evangelical beliefs.[81] When
Surrey wrote of the judgement which would fall upon London,

---

[77] Bale, *Yet a course*, sigs. Fi^r–v, Ii; Brinklow, 'Complaynt', p. 54; Mozley, *Coverdale and his Bibles*, pp. 265–9.

[78] PRO, SP 1/176, fo. 156^r (*L&P* xviii/1. 327); *APC* i, p. 104. For Surrey's life, see *The Works of Henry Howard, Earl of Surrey, and of Sir Thomas Wyatt the Elder*, ed. G. F. Nott, 2 vols. (1815–16), i, pp. i–ccxxxv. For Surrey's elegies upon Wyatt and Clere, see *Henry Howard, Earl of Surrey, Poems*, ed. E. Jones (Oxford, 1964), pp. 27–8, 32.

[79] PRO, SP 1/175, fo. 85^r; SP 1/176, fo. 151^r (*L&P* xviii/1. 73, 315(1)); Constantine, 'Memorial', p. 62.

[80] PRO, SP 1/176, fo. 156^r (*L&P* xviii/1. 327).

[81] For what follows, see H. A. Mason, *Humanism and Poetry in the Early Tudor Period* (1959), pp. 241–8.

as on Babylon, for its vice and pride, he did no more than prefigure the beliefs he would express in Chapter III of his *Ecclesiastes*:

> Then thought I thus: one day the Lord shall sit in doom
> To view his flock and choose the pure: the spotted have no room.
> Yet be such scourges sent in that each aggrieved mind,
> Like the brute beasts that swell in rage and fury by their kind,
> His error may confess . . . [82]

Again and again, Surrey's religious poems reveal his coruscating hatred and contempt of 'greedy gain', 'this failing wealth'; of men who with 'wicked works' sucked 'the flesh of Thy elect' and tempted 'the living God'.

> Another righteous doom I saw of greedy gain . . .
> The plenteous houses sack'd. [83]

To Blage he dedicated his version of Psalm 73, which promised:

> So shall their glory fade; thy sword of vengeance shall
> Unto their drunken eyes in blood disclose their errors all. [84]

So it would be for the citizens of London, within their 'proud towers' and 'wicked walls':

> Thy dreadful doom draws fast upon.
> Thy martyrs' blood, by sword and fire,
> In Heaven and earth for justice call.

And it was Surrey himself who would appear, 'in secret silence of the night', as the instrument of divine wrath.

> A figure of the Lord's behest,
> Whose scourge for sin the Scriptures show.
> That, as the fearful thunder clap
> By sudden flame at hand we know,
> Of pebble stones the soundless rap
> The dreadful plague might make thee see
> Of God's wrath . . .

---

[82] *Surrey, Poems*, p. 93. Professor Jones finds convincing evidence that the paraphrases of Ecclesiastes and the Psalms were written in Surrey's last days, during his final imprisonment: p. 153.

[83] Eccles. 4; Ps. 73; *Surrey, Poems*, pp. 95, 100. R. Zim, *English Metrical Psalms: Poetry as Praise and Prayer, 1535–1601* (Cambridge, 1987), pp. 88–98, 106–9, 144–9, 193–5, 197–8.

[84] *Surrey, Poems*, p. 100 (Ps. 73).

. . .
And greedy lucre live in dread
To see what hate ill-got goods win;
The lechers, ye that lust do feed,
Perceive what secrecy is in sin;
And gluttons' hearts for sorrow bleed,
Awaked, when their fault they find.
In loathsome vice each drunken wight
To stir to God, this was my mind.
Thy windows had done me no spite;
But proud people that dread no fall,
Clothed with falsehood and unright
Bred in the closures of thy wall.[85]

Surrey was not alone in denouncing the citizens and their rulers. Others who 'with God's Word do fight against Antichrist and his members' lamented their ungodliness.[86] Gardiner complained that 'the Aldermen of London, who should be held in reverence were, in *The lamentacion of a christian against the citie of London* of Roderyck Mors, 'noted for the lewdest men' in London.[87] London's gospellers early accused their magistrates of conspiring with the Bishops, as agents of persecution, repressing the Word.[88] And for 1543 Sir William Bowyer was elected Mayor, a man so conservative that he had opposed the bill in Restraint of Appeals in Parliament a decade earlier.[89] Yet it was not Sir Richard Gresham nor William Birch, Surrey's victims, behind their shattered windows, who had most imminent reason to be fearful that spring.

At Easter the gospellers 'quailed', looking 'every hour to be clapped in the neck', for the word had gone round that Gardiner, 'the Fox', 'had bent his bow to shoot at some of the head deer'. The new alliance with the Emperor made the King eager to display his orthodoxy, and no one 'bare so great a swing about the King' as Gardiner.[90] In Convocation Cranmer won what looked to be the last concession for the advancement of the

---

85 *Surrey, Poems*, pp. 30–1.
86 PRO, PCC, Prob. 11/31, fo. 158ʳ.
87 *Letters of Gardiner*, p. 160.
88 For example, Brinklow, 'Lamentacyon', p. 80.
89 *House of Commons, 1509–1558* (William Bowyer).
90 Foxe, *Acts and Monuments*, v, p. 486.

Gospel—in every parish the curate was to read the English Bible, chapter by chapter; first the New Testament, then the Old—but Gardiner had him in his sights.[91] At Canterbury the conservative prebends were gathering evidence against their Archbishop, 'the greatest heretic in Kent'. As his chaplain, Dr Champion, rector of St Vedast, was buried hot coals were poured into his grave, 'as though he had been an heretic worthy burning'.[92] On 16 March Simon Heynes, Dean of Exeter and an adherent of Cranmer, was accused of 'evil opinions' and 'seditious preaching', and was sent to the Fleet.[93] Soon gentlemen of the Court joined him: Thomas Weldon, a master of the household, was found guilty of maintaining a sacramentarian priest. 'For like causes' Thomas Sternhold (composer of the metrical psalms) went to the Fleet on 18 March, and on the 19th Sir Philip Hoby followed. Sir Thomas Cawarden and Edward Harman were also implicated.[94] Heresy had spread among a group at Windsor. 'There is a marvellous sect of them', said Gardiner, 'for the Devil cannot make one of them to bewray the other.'[95]

Catholic proprieties were being flouted with bravado in the City early in 1543. Henry had graciously conceded that his subjects might eat white meats that Lent, but many flagrantly defied the Lenten fast altogether.[96] Fifteen men of high birth were summoned before the Privy Council early in April for such a misdemeanour, and eight committed to the Fleet, among them Surrey and his companions.[97] The Council had waited until now to teach these noblemen a lesson. The meat for Surrey's household (only the best, 'for peers of the realm should thereof eat, and besides that a prince') came from a butcher, Andrew Castle, whose zeal as well as business instinct impelled him to break the Lenten fast. His wife was a noted

---

[91] Wilkins, *Concilia*, iii, p. 863.

[92] *L&P* xviii/2. 500, 546 (p. 300). For the plots against Cranmer, see Foxe, *Acts and Monuments*, viii, pp. 23–34; Strype, *Memorials of Cranmer*, i, chs. 26–7; Gairdner, *Lollardy and the Reformation*, ii, pp. 359–76, 394–421.

[93] *APC* i, p. 97.

[94] Ibid. *House of Commons, 1509–1558* (Cawarden, Hoby, Sternhold, Weldon).

[95] Foxe, *Acts and Monuments*, v, pp. 464 ff.; Hall, *Chronicle*, pp. 858–9; Gairdner, *Lollardy and the Reformation*, ii, pp. 384–91.

[96] Foxe, *Acts and Monuments*, v, pp. 463–4; Guildhall, MS 9531/12, fo. 52[v].

[97] *APC* i, pp. 103–4, 106, 114; Bodleian, Jesus College MS 74, fo. 258[r].

'brabbler' of the New Testament.[98] Orders came to the Mayor and Aldermen on 30 March to seek out 'abuses customably used within the City, touching eating flesh in the Lent, and breaking of windows in the night . . . and the licentious manner of players'. Butchers who had sold meat were summoned, but they revealed a politic forgetfulness about the names of their customers.[99] In many parishes the citizens withdrew their tithe that Easter, whether through pious dissatisfaction with their clergy, or through more impious motives. Both the City clergy and laity sent bills to Parliament in that session, expressing their grievances and demands; the clergy no doubt strengthened in their resolve by the political fortunes of the conservatives.[100]

Groups of players staged interludes in the capital that spring. These plays often took for their theme satire upon the Catholic establishment, or worse.[101] Three years earlier, in 1540, the keeper of Carpenters' hall had staged a play in which priests were mocked and called knaves, and at St Giles Cripplegate Henry Patinson and Anthony Barber urged their 'boys' to sing a sacramentarian ballad.[102] Bonner's injunctions of 1542 forbade 'common plays, games or interludes' to be acted in the City churches, but to little avail.[103] On 2 April bills advertising performances of such interludes were torn down, and special inquisition was made for those who had dared to post on the Mayor's gate slanders against the most redoubtably conservative rectors in London, Dr Wilson and Dr Weston.[104] Yet the players were sheltered by leading citizens and courtiers. George Tadlowe, William Clycheman, and Thomas Hancock staged interludes in their houses, and the Lord Warden kept a company of players.[105] George Tadlowe was the 'great persuader' who urged Ralph Robinson to translate More's *Utopia*, and

[98] PRO, SP 1/175, fos. 85ʳ, 86ʳ (*L&P* xviii/1. 73); Foxe, *Acts and Monuments*, v, p. 444.

[99] *APC* i, pp. 103–4, 108, 112; CLRO, Repertory 10, fo. 324ᵛ.

[100] S. E. Brigden, 'Tithe Controversy in Reformation London', *JEH* xxxii (1981), pp. 299–300; H. Miller, 'London and Parliament in the reign of Henry VIII', *BIHR* xxxv (1962), p. 148.

[101] T. W. Craik, *The Tudor Interlude* (Leicester, 1967); J. N. King, *English Reformation Literature: the Tudor Origins of the Protestant Tradition* (Princeton, 1982).

[102] Foxe, *Acts and Monuments*, v, pp. 445, 446.

[103] Guildhall, MS 9531/12, fo. 39ʳ; *L&P* xvii. 282(17).

[104] CLRO, Repertory 10, fo. 322ᵛ.

[105] Ibid., fos. 322ᵛ–323ʳ; *L&P* xviii/1. 392, 401.

maybe the interludes in Tadlowe's house in Lent 1543 had a humanist theme.[106] But not so surely the 'unlawful disguising' of a band of twenty artisans on Sunday, 8 April.[107]

The London booksellers, too, were called to answer. On 8 April eight men were imprisoned by the Privy Council for printing unlawful books: Whitchurch, Bedyll, Grafton, Lant, Keyle, Mayler, Petit, and Middleton.[108] Protestant zeal, as well as the profit motive, had led each of these men to dare to print forbidden works during the reaction. Richard Grafton, with Edward Whitchurch, his partner in the printing of Matthews and the Great Bible, had been indicted during the quest for heresy in the summer of 1540. Soon afterwards, Grafton was in trouble again, for printing Gray's ballads and disseminating Melanchthon's letter against the Six Articles. Associated with him then were others of the City brethren, with whom he had long been in league. John Mayler had been troubled also in 1540, for speaking contemptuously of the Mass: he called it 'the baken god', and said they called it 'the miss' abroad, 'for all is amiss in it'.[109] Mayler was the printer of all Thomas Becon's works, with the ubiquitous John Gough as their publisher.[110] Richard Lant was Robert Singleton's 'scholar'.[111] In October 1540 John Petit had been before the Consistory Court for breaking the sabbath by selling books at Paul's Cross.[112] Both Petit and Keyle may have had fathers who were Christian Brethren in the first generation of reform.[113] Five of these printers were released on 22 April, bound to declare how many books and ballads they had sold within the past three years, and which merchants had imported forbidden works. Three days later twenty-five booksellers were before the Privy Council, each bound in recognizances of £100 to answer these same questions.[114] The authorities grew desperate to stem the flood of reformist literature.

[106] King, *English Reformation Literature*, p. 111; *House of Commons, 1509–1558* (Tadlowe).

[107] *APC* i, p. 109.   [108] *APC* i, p. 107.

[109] See above, pp. 287, 323, and below, pp. 405–6, 412.

[110] *RSTC* 1713, 1714, 1717, 1723, 1724, 1731, 1734, 1738–40, 1742–3, 1749–50, 1756, 1775–6.

[111] Foxe, *Acts and Monuments*, v, p. 601.

[112] GLRO, DL/C/3, fo. 3ʳ.   [113] See above, pp. 125, 184–5.

[114] *APC* i, pp. 107, 117, 120; R. W. Heinze, *The Proclamations of the Tudor Kings* (Cambridge, 1976), p. 290.

The Council did uncover an underworld of printing and patronage in London; of printers, curates, schoolmasters in league. 'A certain erroneous book named a Postilla upon the Gospels' was delivered to Richard Turke for printing.[115] This work was translated by Stephen Cobbe, a schoolmaster who had been harboured by John Gough. Cobbe was associated with William Reed, reforming curate of All Hallows Honey Lane. On 4 May a commission of clerics was set up to examine Cobbe and Reed together:[116] Cobbe, at least, was released, to offend again. He became schoolmaster in Stephen Vaughan's household, and there he was 'cherished as a jewel', imparting the reformed faith.[117] For all their vigilance, the authorities could never keep pace with the zealous brethren. Even as the inquisition into the heretical book trade was conducted in the spring of 1543 Henry Brinklow was writing his bitter *The complaynt of Roderyck Mors*. The prelates, Mors lamented, have 'bewitched the Parliament house in making such viperous acts as the beast of Rome never made himself'. 'How shamefully have they . . . driven men from reading the Bible.'[118]

The reaction went further when the conservatives, ascendant in the counsels of the King and dominant in Convocation, determined that orthodox theology should be firmly enshrined in statute. In April the long-awaited revision of the now embarrassingly reformist *Bishops' Book* of 1537 was prepared in Convocation, by 'learned men of divers judgements', said the King, and by himself: *The King's Book* or *Necessary Doctrine and Erudition of a Christian Man*. Joye thought that 'it savoureth everywhere' of Gardiner's 'damnable doctrine'.[119] The prince's right to decide who might read Scripture, and who might not, was soon made statutory. The Act for the Advancement of the True Religion, passed in May 1543, allowed that 'the highest and most honest sort of men' profited from studying

---

[115] *APC* i, p. 120. This work was probably *A Postill . . . or collection of most godly doctrine upon every gospell throughout the year* (*RSTC* 5806) of Corvinus. Tolwin had possessed a copy in 1541; see above, p. 336.

[116] *APC* i, pp. 115, 126, 128; GLRO, DL/C/3, fos. 220ʳ, 221ᵛ.

[117] PRO, SP 1/208, fo. 39ᵛ (*L&P* xx/2. 416). For Vaughan's household, see below, p. 418.

[118] Brinklow, 'Complaynt', pp. 54, 57.

[119] *Formularies of Faith*, pp. 212–377; L. B. Smith, *Tudor Prelates and Politics, 1536–1558* (Princeton, 1953), pp. 244–9.

Scripture: not so the 'lower sort', who in their freedom to read the Bible in English had been misled, and fallen into error and division. Upon pain of a month's imprisonment, 'no women, nor artificers, prentices, journeymen, serving men of the degree of yeomen or under, husbandmen nor labourers' could now read the Bible 'privately or openly'.[120] The loss was profound. The reformers complained that the measure proceeded from the same fear and shame which had first led the Church to 'lock up the Scriptures in a strange language . . . lest the people by reading of them should perceive their works to be nought'.[121] But there were, they said, darker motives, and more sinister consequences.

Gardiner had long feared that free reading of Scripture might 'beguile the people into the refusal of obedience'.[122] And had not Collins, whom he had burnt, 'exclaimed against the nobility and great men of the kingdom', and adduced against them 'many passages of Holy Scripture?'.[123] The clergy thought that 'vice, uncharitableness, lack of mercy' had reigned since the people had had the Scriptures in English, and this theme became a commonplace in their sermons.[124] In 1544 William Scott preached at Paul's Cross, lamenting the decline of the clergy in popular esteem, and the prevailing lack of reverence for holiness and degree. Apprentices, he said, had found in the Bible the excuse to be more unruly than ever:

They will talk much of the Scripture, God's Word, and yet will not learn thereof to be obedient and gentle unto their masters: they will talk much of Paul, and yet it doth nothing move them that Paul in so many places doth beat and inculcate in that servants should be obedient and faithful unto their masters.[125]

Such complaints were not without justification.

---

[120] 34 & 35 Hen. VIII. c. 1; Lehmberg, *Later Parliaments*, pp. 186–8, 190. In July 1543 Bullinger was sent a copy of the proclamation 'fixed up in public': *Original Letters*, i, p. 356.

[121] Bale, *Yet a course*, sig. Cii^v.

[122] Quoted in Smith, *Tudor Prelates and Politics*, p. 245.

[123] See above, p. 273.

[124] 'A Supplication of the Poore Commons' (EETS, extra series, xiii, 1871), p. 64.

[125] William Scott, *Two notable sermones lately preached at Paules Crosse*, sig. Hiv^v; cited in J. W. Blench, *Preaching in England in the late fifteenth and sixteenth centuries* (Oxford, 1964), pp. 240–4.

The reformers, for their part, saw the loss of the Word as a conspiracy: 'the rich of the City of London take part, and be fully bent with the false prophets, the Bishops'.[126] Roderyck Mors (Brinklow) knew that wealthy 'citizens will not have in their houses that lively Word of our souls'.[127] Perhaps the English Bible was forbidden in many citizens' houses as once it had been in the Marquis of Exeter's.[128] Tracy, too, had lamented how 'this wicked generation' would, in their socially segregated churches, put the Bible 'in some pew, where poor men durst not presume to come'.[129] How dared they take the Word away, asked William Turner in *The huntyng & fyndyng out of the romishe fox*, his diatribe against Gardiner published in September 1543: 'Died not Christ as well for craftsmen and poor men as for gentlemen and rich men, and would not Christ that the poor labouring men should have wherewith they might comfort their souls as well as rich men and gentlemen?'[130] When the Gospel was outlawed in 1543 its most charismatic exponents had to be silenced also. Robert Wisdom was so passionate a gospeller that at Easter 1541 he had exhorted his Essex parishioners to 'take the Scripture in their hands, when they meet together on the Sundays and holydays at the ale-house, and to talk, and commune, and reason of it'. 'This is', he avowed, 'no heresy.'[131] Three leading gospellers, foremost in 'building the Temple', were the last victims of the conservative purge in 1543: Thomas Becon (hailed by Wisdom as 'the man of God'), Robert Singleton, and Wisdom himself. The Bishops 'laid them by the heels' through envy of their success, alleged Wisdom: they said, 'Lo, all the world goeth after him. What shall we do? This fellow hath an exceeding audience. If we let him alone thus, all will believe him.'[132]

All three preachers had been troubled before. In 1536 Wisdom had outraged the citizens of Oxford because 'he preached

---

[126] Brinklow, 'Lamentacyon', p. 80.
[127] Ibid., p. 79.
[128] PRO, SP 1/139, fo. 1ʳ (*L&P* xiii/2. 820).
[129] 'A Supplication of the Poore Commons', p. 67.
[130] Cited in D. Wilson, *The People and the Book* (1976), p. 88.
[131] Strype, *Ecclesiastical Memorials*, i/2, pp. 473–4.
[132] Ibid. i/2, p. 478. D. S. Bailey, 'Robert Wisdom under Persecution, 1541–1543', *JEH* ii (1951); J. F. Davis, *Heresy and Reformation in the South-East of England, 1520–1559* (1983), pp. 75–8.

Christ and spake against the ungodliness and false doctrine of the papacy'.[133] Hunted from Oxford, he went to Essex, and later to London, where he became curate at St Margaret Lothbury and a leading gospeller of the conventicles.[134] Conservative priests began to collect evidence against him, so that when he went to hear Dr Wilson preaching in London in July 1543 he was arrested. The Bishop made a pact with Wisdom, promising 'by God and by his faith and baptism' that he could go free if he confessed. Persuaded by Whitchurch and his uncle that he would not survive a 'seven night' in the Bishop's prison, and trusting Bonner's promise, Wisdom wrote a confession, 'which bill (for all his baptism) the Bishop of London laid up in store'.[135] In a Lenten sermon in 1543 at St Mary Aldermary, Dr Crome's parish, Wisdom again offended orthodoxy, and was committed once more to the Lollards' Tower. His reforming brethren rallied: William Carkke and Henry Brinklow stood surety for him on 14 May.[136]

Robert Singleton had been a 'lantern and light' of Anne Boleyn's household, one of her reforming chaplains.[137] In 1536 he had preached at Paul's Cross against Purgatory, and the conservatives had nearly had their revenge.[138] Thomas Becon had recanted already in 1541 for his preaching in his native Suffolk and Norfolk. Thereafter he turned to proselytizing in Kent, where he established a network of support and patronage among the gentry, and to writing devotional treatises and works of controversy. It was for his links with the most radical of the London book trade that Becon was arrested in this purge.[139]

On Relic Sunday in July 1543 Wisdom, 'in the midst of these two penitents' Singleton and Becon, professed himself

---

[133] PRO, SP 1/105, fo. 104ʳ (*L&P* xi. 138).

[134] Wisdom was probably witnessing wills in St Margaret Lothbury by 1538: Guildhall, MS 9171/10, fos. 311ʳ, 328ʳ. Foxe, *Acts and Monuments*, v, p. 444.

[135] 'A revocation of that shameful bill that Winchester devised and Wisdom read'; Foxe, *Acts and Monuments*, v, appendix xxii*.

[136] Guildhall, MS 9531/12, fo. 45ᵛ.

[137] M. Dowling, 'Anne Boleyn and Reform', *JEH* xxxv (1984), p. 40. Singleton had been one of Cromwell's informants. He sent him 'a little tittling upon a sinister and seditious sermon preached at the Charterhouse called the Sheen': BL, Cotton MS Cleo. E v, fos. 407ʳ–408ʳ.

[138] See above, pp. 253, 259.

[139] D. S. Bailey, *Thomas Becon and the Reformation in England* (Edinburgh, 1952).

sorry for having 'slandered the doctrine of our religion and defamed the charity of the public ministers of justice'. Wisdom's recantation was composed by Gardiner: its purpose was to retract publicly those themes in Wisdom's sermons which the Bishop wished to be disavowed. With time the doctrines which particularly embarrassed the authorities changed: so it was that in 1543 Wisdom must declare that 'this is a realm of justice and no persecution of them that be good'. Though Wisdom had preached that 'men cannot live well in Christ but they be persecuted', he now acknowledged that 'I have known no man particularly to have been persecuted for the truth'; and that Frith and Lambert had died justly as sacramentaries; Barnes, Garrett, and Jerome for their 'false and untrue doctrine'. His 'companion' Becon's assertion in *Of David's Harp* that 'persecution is a token of the true Gospel' Wisdom now condemned. All sectaries—Anabaptists, Sacramentaries, Adamites, Arians, and Sabellianists—must be purged. All those 'of a lower sort, as we be of here' must never think themselves vindicated by persecution. Moreover, Wisdom recanted, as a 'very detestable . . . opinion derogating the grace of God', his denial of man's free will, and retracted his attacks upon the veneration of saints.[140] Wisdom would soon reproach himself bitterly for his recantation, and in prison wrote 'A Vindication of himself against certain articles charged upon him',[141] and 'A revocation of that shameful bill that Winchester devised and Wisdom read'.[142]

In his 'Vindication' Wisdom inveighed against the 'Nembroths, hunters and persecutors . . . and mighty . . . builders of Babel, now begun to fall', and rehearsed the beliefs that he had preached and had, so miserably, recanted. Accused of denying man's free will, he allowed that 'God of His rich mercy would, for Christ and in Christ' accept men's works 'as perfect and well pleasing, and reward them with the crown of immortal glory'. And yet he insisted that 'a man by the power and strength of his own free will only is not able to do good or

---

[140] Guildhall, MS 9531/12, fos. 43ᵛ–44ʳ; printed in Foxe, *Acts and Monuments*, v, appendix xii.

[141] BL, Harleian MS 425, fos. 4ʳ–6ᵛ; printed in Strype, *Ecclesiastical Memorials*, i/2, pp. 463–79.

[142] Foxe, *Acts and Monuments*, v, appendix xxii*.

think good': 'the Holy Ghost must come and create a new
heart and govern him into all truth'. 'All men's traditions shall
be plucked up by the root', Wisdom wrote; indeed, some of
them had been already: 'The abolishing of the Roman Bishop,
the throwing down of abbeys, the destruction of sects, the
putting away of pelting pardons, and the rooting out of famous
idols, teach us plainly that Heaven and earth may pass, but the
Word of God shall not pass.' He urged more radical reformation: 'The day will come, when the very root of all Popery,
even your masses, will be plucked up by the root; and all the
world shall know how shamefully ye abuse the Holy Supper of
the Lord.' This Christ was not to be worshipped in idolatry:
'we have a living Christ', only mediator, 'not a Christ of
clowts'. Purgatory, for Wisdom, was an invention of 'dead
holiness', for there are no 'dead ghosts', no 'puling silly
ghosts'. 'Souls departed do not come and walk and play bo-
peep with us.'[143] Wisdom recanted once, twice, three times in
his life. He saw little choice, for otherwise he would stand in the
Bishop's grace, which is as 'the grace of the butchers of Eastcheap to the poor lambs brought to their market'.[144] Wisdom
left London for Yorkshire.[145]

Thomas Becon recanted first the pride that had led him to
adopt a kingly pseudonym, Theodore Basile, and his spiritual
pride. 'Did ever man say of his own book that it containeth as
much of Christ in a few lines as the Bible and doctors teach of
Christ in many?' Secondly, he retracted some 'specialities' of
his writings and sermons: his attacks upon prayers to saints, his
allegations of priestly incontinence, his denial of the efficacy of
prayers for the dead, his derogation of the sacrament of the
altar, his derision of the sacraments of confirmation and
extreme unction. Then, one by one, he ripped his books in
pieces.[146] Yet Becon seems to have appended to his recantation
a denial. His retraction, unlike Wisdom's, was feigned and
recognized as such by all who heard it.

[143] Strype, *Ecclesiastical Memorials*, i/2, pp. 463–79.

[144] Foxe, *Acts and Monuments*, v, appendix xxii*.

[145] A. G. Dickens, *Lollards and Protestants in the Diocese of York, 1509–1558* (Oxford,
1959), pp. 194–7.

[146] Guildhall, MS 9531/12, fos. 44^r–45^r; printed in Foxe, *Acts and Monuments*, v,
appendix xii.

Singleton's recantation was utter and abject. 'I am an un-learned, fantastical fool. Such hath been my preaching and such hath been my writing, which I here before you all tear in pieces.'[147] Soon after, Singleton was charged with complicity in a plot, and in March 1544 he suffered: not for heresy, but for treason.[148] A puzzled Foxe wrote how Singleton was charged as a 'stirrer up of sedition and commotion', although 'his purpose was ever to preach the Gospel to the people'.[149] What Singleton plotted is unknown, but in December 1543 the Recorder and all the learned counsel of the City were to debate certain seditious words spoken by an unnamed preacher, and to peruse recent statutes to see whether these constituted treason.[150] Later, Gardiner wrote to Somerset, 'Your Grace, I doubt not, remembereth Singleton's conspiracy.'[151] Was Singleton really a traitor, or did his enemies charge him falsely, as before in 1536? Certainly, treason was always a more dangerous charge than heresy: never more so than in 1543 when the conservatives at Court had failed in their attempt to 'shoot down the head deer'.

POPISH PLOTS

Allegiance to the Pope continued long after the Break with Rome. Many looked for the time when England would return to the papal obedience; how many the authorities could not discover. Yet they feared, and knew, that there were some desperate conservatives who not only continued to oppose the royal supremacy but were prepared to plot for the restoration of the Roman primacy. Cardinal Pole, exiled in Italy, was thought to be at the centre of any conspiracy, giving leadership and sanction. The memory and example of Sir Thomas More were an inspiration. In London one group in particular was united by its old association with More, and it kept faith. Some

---

[147] Guildhall, MS 9531/12, fo. 45[r].

[148] *L&P* xviii/2. 546, pp. 334, 359; Gairdner, *Lollardy and the Reformation*, ii, p. 382; C. A. Haigh, 'The Reformation in Lancashire to 1558' (University of Manchester, Ph.D. thesis, 1969), pp. 252–6.

[149] Foxe, *Acts and Monuments*, v, pp. 600–1, 696.

[150] CLRO, Repertory 11, fo. 18[r].

[151] Foxe, *Acts and Monuments*, vi, p. 52.

of its members, like William Rastell, John Clement, and John Heywood, were of More's family. Their influence was the greater because of their association with printing, for they might more easily spread their ideas. William Rastell had printed More's polemical works—until 1534, when it became too dangerous. Rastell then turned to law, but continued to collect and edit More's manuscripts to print when the time was right. Heywood was a printer still.[152] More's adherents were perceived as disaffected and dangerous. In 1540 one of his sons-in-law, Giles Heron, had suffered the penalty for treason.[153] When the reformers gathered strength these conservative coteries were vulnerable.

At Court, early in 1544, the leaders of reform waited their chance to strike against the conservatives who had so nearly brought them down in the previous year. They looked for pretexts to destroy their rivals; for associations which would threaten them. Soon they found them. Two men had given hostages to fortune by being involved in the secret enquiries about Cranmer surrounding the Prebendaries' Plot: John Heywood and Germain Gardiner.[154] Better still that Germain Gardiner was the Bishop of Winchester's nephew and secretary, more conservative even and far less circumspect than his uncle, for association with a traitor would threaten the leader of the conservatives. Author of *A letter of a yonge gentylman, wherin men may se the demeanour & heresy of J. Fryth late burned* (printed by William Rastell in 1534), Germain Gardiner had early pinned his colours to the mast in the confessional dispute. He blamed Frith for spreading heresy, 'tending to nothing else but to the division and rending asunder of Christ's mystical body, His Church, the pulling down of all power and utter subversion of all commonwealths'.[155] John Heywood, More's brother-in-

---

[152] For the More circle, see A. W. Reed, *Early Tudor Drama* (1926) and his introduction to Thomas More, *The English Works*, ed. W. E. Campbell, A. W. Reed, R. W. Chambers, and W. A. G. Doyle-Davidson (1931), i, pp. 1–23. For Rastell as More's editor, see *CW* 2, pp. xxix–xxxii.

[153] Lehmberg, *Later Parliaments*, pp. 126–7; *House of Commons, 1509–1558* (Heron).

[154] *L&P* xviii/2. 546, pp. 325–6; Strype, *Memorials of Cranmer*, i, pp. 163–8; Gairdner, *Lollardy and the Reformation*, ii, pp. 405–13, 416.

[155] *RSTC* 11594. This passage is cited in B. Camm, *The Lives of the English Martyrs*, 2 vols. (1904), i, p. 544.

law, of known conservative views, was suspect.[156] Also implicated in the supposed conspiracy was John Larke, a City rector specially favoured by More. The Chancellor had nominated him vicar of his own parish of Chelsea in 1530.[157] When George Joye chose a priest to vilify in his *The subversion of Moris false foundacion*, he deliberately named one who was close to More: two nuns departed their convent, he wrote, 'lest they should have been made harlots by a vicious priest called Sir John Larke'.[158] Of the other men indicted little is known: John Eldrington, John Beckingsale, John Ireland, a priest of Kent, Roger Ireland, John More, and William Daunce. All were charged with conspiring against the royal supremacy.[159] The leading conspirators were found guilty at Westminster on 15 February. All were exempted from the general pardon.[160] On 7 March they went to their deaths: save for John More, who was pardoned, and John Heywood, who escaped a traitor's death by recanting on the way to the scaffold. In July Heywood did public penance.[161] Maybe these men were guilty of treasonable intent, but they were also victims of the continuing faction struggles at Court.

None longed more ardently for the overthrow of the royal supremacy than the exiles. Abroad they might plot, perhaps to little effect, but with less fear of retribution. An act of attainder in 1542 had condemned Richard Pate and Seth Holland, his chaplain, for their allegiance to the Pope and Cardinal Pole. Rome was their sanctuary until Mary's reign.[162] In April 1544 Henry Cole, Bonner's 'traitorous' popish Chancellor, one of Pole's right scholars', was pardoned for his treasonable journeys to Rome, for his communication with Pole, and also for

[156] Reed, *Early Tudor Drama*, pp. 62-3.

[157] Hennessy, *Repertorium*, p. 120; Cresacre More, *Life of More* (1726), p. 278. Larke was a pluralist. In Oct. 1540 the Chelsea churchwardens had presented him for neglecting the church and holding the benefice vacant. In 1541 Larke had negotiated for a pension upon his resignation of St Ethelburga Bishopsgate: GLRO, DL/C/3, fo. 3ᵛ; Guildhall, MS 9531/12, fo. 32ᵛ.

[158] *RSTC* 14829; More, *Apology*, p. 365.

[159] *L&P* xix/1. 444(6).

[160] PRO, KB/9/557, m. 1-10 (the indictment is printed in Gairdner, *Lollardy and the Reformation*, ii, p. 411); 35 Hen VIII c. 18.

[161] *Grey Friars Chronicle*, p. 46; Foxe, *Acts and Monuments*, v, p. 528; *L&P* xix/1. 444(6); Guildhall, MS 9531/12, fos. 44ʳ, 61ʳ; Wriothesley, *Chronicle*, i, p. 148.

[162] Lehmberg, *Later Parliaments*, pp. 147-8.

his suspicious dealings with Michael Throckmorton, 'a false traitor', in Paris.[163] The grant of pardon, itself remarkable, did not chasten Cole. In January 1545 news came that Cole was on his way to Rome, for an unknown but subversive purpose.[164] In May 1545 Friar Elstow was still active, receiving and forwarding letters for Pole.[165] In the same month Selby, quondam monk of the Sheen Charterhouse, and 'distract of his wit', was arrested for possession of a 'lewd writing subscribed with his own hand against the Primacy'.[166] More important perhaps than the reality of these shadowy conspiracies was the conviction of many that there were conservative clergy plotting subversion; all these fears amplified by the allegations of the reformers. During the summer of 1545 when dearth, tempests, and a threatened invasion from France deeply unsettled the City there were persuasive rumours that the clergy were in league with Catholic Europe.[167]

In London there were cells of ultra-conservatives, keeping the faith and conspiring. Many, many more, though remote from the plotting, secretly shared some of their hopes. The reformers were not sanguine about the prospects of mass conversion in a capital sunk still in hypocrisy and 'popish leavings': and if it was thus in London, 'the head city of the whole realm, and so nigh the King's presence, it must needs be worse farther off'.[168]

'SINGULARITY SO RULETH'

During 1544 the persecution abated. In Parliament legislation was introduced early in the year to prevent all those 'secret and untrue accusations and presentments' which zealots had 'maliciously conspired' against offenders against the Act of Six Articles. Now no one might be arraigned by the Act unless

---

[163] *L&P* xix/1. 444(11). Brinklow, 'Complaynt', p. 61. For Cole, see E. L. C. Mullins, 'The Effects of the Marian and Elizabethan Religious Settlements upon the Clergy of London' (University of London, MA thesis, 1948), p. 254.

[164] PRO, SP 1/197, fo. 53ʳ (*L&P* xx/1. 40).

[165] PRO, SP 1/200, fo. 225ʳ (*L&P* xx/1. 696(2)).

[166] *APC* i, p. 161.

[167] Wriothesley, *Chronicle*, i, p. 159; CLRO, Repertory 11, fo. 215ʳ; *APC* i, p. 234. Wilson, *Tudor Tapestry*, p. 179.

[168] Bale, *Yet a course*, sig. Biiʳ.

presented by a jury of twelve, or by special commissioners, or by Justices of the Peace. Gardiner and Norfolk were powerless to prevent its passage.[169] At Court the reforming party again prevailed. The King ordered an English litany, which seemed to presage a change in ceremony.[170] Henry had married his sixth wife, Katherine Parr, in July 1543. His new Queen was a humanist and inclined to reform, and around her she gathered those of like mind.[171] The Earl of Hertford, another of the reforming party, had delighted the King by his victorious raid in the Scottish Lowlands in May.[172]

For Henry now thought of nothing but war. By early 1544 England was at war with both her 'ancient enemies', France and Scotland, and Henry devised mighty campaigns against them.[173] These wars would cost his subjects dear, for Henry financed his military ambitions not only by raising one tax after another—fifteenths, tenths, prests, loans, subsidies—and by selling off monastic lands in a precipitate and profligate way, but also by systematically manipulating the currency for fiscal purposes. The year 1544 saw the beginning of the 'Great Debasement': so called because of its scale and the revenue generated.[174] The impact upon the standard of living in London was immediate and considerable. For seven years from 1544 the average rate of inflation was very nearly 10 per cent a year; an increase alarming and almost unprecedented.[175] Writing forty years on, Stow claimed that because of the debasements, 'rents of lands and tenements, with the prices of victuals, were raised far beyond 'their former rates, hardly since to be brought down'.[176] In 1544, apart from the general subsidy levied the year before, Londoners were forced to make

---

[169] 37 Hen VIII c. 4; Gairdner, *Lollardy and the Reformation*, ii, p. 410; Lehmberg, *Later Parliaments*, p. 198.

[170] Wriothesley, *Chronicle*, i, p. 148; F. E. Brightman, *The English Rite*, 2 vols. (1915), i, pp. 174–90.

[171] M. Dowling, *Humanism in the Age of Henry VIII* (1986), pp. 235–7; Scarisbrick, *Henry VIII*, pp. 456–7.

[172] Ibid., p. 444.

[173] For Henry's foreign policy, see R. B. Wernham, *Before the Armada: The Growth of English Foreign Policy, 1485–1588* (1966), pp. 150–3; Scarisbrick, *Henry VIII*, ch. 13.

[174] C. E. Challis, *The Tudor Coinage* (1978), pp. 248–55.

[175] S. R. Rappaport, 'Social Structure and Mobility in Sixteenth-Century London' (University of Columbia, Ph.D. thesis, 1983).

[176] Stow, *Survey*, i, p. 57.

extra contributions: a £1,000 prest for wheat, a levy of five hundred men for France, a loan of £20,000 from the livery companies, and a further loan of £3,000 from the citizenry.[177] In January 1545 there came a new demand for a benevolence. Not even all the Aldermen complied: Alderman Reed was conscripted into the army for Scotland (a grim prospect), and Alderman Roach was imprisoned for protesting the illegality of the tax.[178] When the harvest failed, as in 1545, the suffering was extreme. A new urgency entered the social complaints of the reformers.

Battle continued to be waged between rival preachers in the pulpits of the capital. The order given to the Sheriffs on 29 November 1543 that two of their servants should be in attendance at every Paul's Cross sermon suggests that the authorities were expecting trouble or had encountered it.[179] The conservatives yearned for the old consent in doctrine. In November 1544 William Chedsey, preaching at Paul's Cross, lamented:

the time hath been, when that those which have occupied this place, have laboured . . . to have pacified and set at quiet the weak and feeble consciences of their audience. And now . . . singularity so ruleth: that he that can best dispute and reason a new matter in the pulpit, he is the best preacher.[180]

In a capital distressed by war and dearth the evangelicals grew ever bolder. At Paul's Cross on 21 December William Scott bewailed the prevailing lack of reverence for holiness and degree. 'Singularity doth now destroy and setteth men at variance.'[181] Bills and tracts were circulating wildly. An anonymous bill against the Lord Mayor was sent to the Lord Chancellor that autumn, and late in the year the Privy Council received a *Supplication touching the Church*, which urged further reform, excoriated the ignorance of the clergy, and complained of the withdrawal of the Bible from the people.[182] Such tracts not only expressed forbidden doctrine, but also took the form of

---

177 Wriothesley, *Chronicle*, i, pp. 147–8.
178 Ibid. i, pp. 151–2; *L&P* xx/1. 98.
179 CLRO, Repertory 11, fo. 13ᵛ; Letter Book Q, fo. 94ᵛ.
180 Chedsey, *Two notable sermones lately preached at Pauls Crosse* (*RSTC* 5106.5), sig. Ciiiᵛ; cited in Blench, *Preaching in England*, p. 256.
181 Ibid., pp. 240–4.
182 CLRO, Repertory 11, fo. 100ʳ; *L&P* xix/2. 797.

political tirades against the forces of reaction. On 15 December sacramentarian tracts were broadcast in the City streets impugning the Mass, Gardiner, and the Bishops.[183] After such outrages it seemed likely that reaction would follow. A fourth commission under the Act of Six Articles was established on 5 January 1545.[184]

A nest of radicals was soon discovered. William Hertwell, a pewterer, had confessed by 15 January that he had lately scattered about London 'seditious bills against God's laws and the King's'.[185] Implicated with him were an apprentice to Richard Keyle, the reforming printer, Thomas Chilton, servant to Lady Lambert, William Mitchell, a smith, Thomas Thompkins, a weaver, Garrard Morris, a chandler, and Peter Shaklady, servant to Banks, a skinner. Though forbidden, Hertwell and his friends in the heretical underground were still 'meddling with Holy Scripture', and reading aloud together other outlawed books: 'Frith's book' (perhaps *A boke made by John Frith . . . answeringe vnto M mores lettur*), George Joye's *A present consolacion for the sufferers of persecucion*, and the 'Lord Cobham's book'. The activities of this conventicle again demonstrated the potent influence of the old heresy upon the new, for the last work was a Lollard hagiography resurrected by the reformers.[186] Thomas Thompkins avowed that he 'liketh' these books 'so well that he can be contented to shed his blood in that quarrel; that the matter in them contained is the very truth': a brave boast, and one that he would fulfil, but not until Mary's reign.[187]

---

[183] *Grey Friars Chronicle*, p. 48.

[184] Guildhall, MS 9531/12, fo. 67^r.

[185] CLRO, Repertory 11, fos. 158^r–160^r. (Perhaps these were the sacramentarian tracts against Gardiner disseminated on 15 Dec.)

[186] None of these men were charged under the Act of Six Articles. Chilton was indicted for reading books prohibited by royal proclamation; surely that following the Act for the Advancement of the True Religion. 'The Lord Cobham's book' was either *A boke of Thorpe or John Oldecastell*, a Lollard tract early adopted by the reformers, or more likely, John Bale's recent edition of the trial of Sir John Oldcastle, Lord Cobham. The *Boke of Thorpe* was already rare by the 1540s, for when Tolwin was charged with possessing a copy in 1541 even Bale had not heard of it (M. Aston, 'Lollardy and the Reformation', *History*, xlix (1964), p. 150). This group probably had a copy of Bale's *A brefe chronycle concernynge the examinacyon . . . of syr J. Oldecastell* published (probably in Antwerp) in 1544 (*RSTC* 1276). That another edition was published in the following year is an indication of its popularity (*RSTC* 1277).

[187] Foxe, *Acts and Monuments*, vi, pp. 717–22.

Now Thompkins and his sacramentarian friends went free.[188] As before, it was heretics with friends in high places whom the Bishops sought.

Again the Lenten fast was broken by nobles who scorned the Church's prohibitions. City butchers confessed to having the Earl of Arundel, 'my lady Richmond', and the Earl of Sussex among their customers that Lent.[189] Only three were certainly troubled under the new quest. Anne Askew, Joan Sawtry, and Robert Lukine, servant to Sir Humphrey Browne, Justice of the King's Bench, were arraigned as sacramentaries before the commission at the Guildhall on 13 June. No witnesses appeared to accuse Askew or Sawtry, and Lukine's accuser was found to be acting maliciously. The jury acquitted all three.[190] That summer also Thomas Daye, imprisoned in 1542 for denying auricular confession, was discharged by royal pardon.[191] The Queen's intervention had saved Stephen Cobbe, the reforming schoolmaster, in October 1544. Catherine sent her servant Warner (another of reforming views) to plead for him before the Court of Aldermen.[192] But in September 1545 Cobbe was in trouble again, summoned before Bonner's Chancellor for 'matters of religion'.[193]

As disaster followed disaster in the summer of 1545—dearth, tempests, a threatened invasion from France, and interminable demands for taxes and men[194]—the citizens turned against their traditional scapegoats and denounced foreigners and priests as traitors. In May the Privy Council intervened to prevent London innkeepers from charging foreigners extortionately.[195] Close watch was kept for the French spies active in the capital.[196] Rumours circulated wildly. Bills addressed to the Mayor were broadcast in the City streets foretelling the

[188] Chilton, Hertewell, and Morris were whipped through the City; CLRO, Repertory 11, fos. 158^r–160^r, 179^r.

[189] Ibid., Repertory 11, fos. 174^v, 176^r.

[190] Askew had been imprisoned 'for certain matters concerning the vi articles' since March, at least. Ibid., Repertory 11, fo. 174^v. Wriothesley, *Chronicle*, i. p. 155.

[191] Ibid. i, p. 156.

[192] CLRO, Repertory 11, fo. 117^r; *House of Commons, 1509–1558* (Edward Warner; John Marshe).

[193] PRO, SP 1/208, fo. 39^v (*L&P* xx/2. 416).

[194] Wriothesley, *Chronicle*, i, pp. 156–9; *APC* i, pp. 172–3.

[195] Ibid. i, p. 163; *L&P* xx/1. 755.

[196] PRO, SP 1/201, fos. 157^r, 164^r (*L&P* xx/1. 831; cf. 930, 1014).

treason of 'certain priests and strangers that would fire the City'. These were shown at once to the Privy Council on 20 July, and a nightly watch to keep surveillance over aliens was ordered.[197] Two 'Dutch men' with handguns were attached within the week after and, as if to prove the truth of the bills, another two were arrested in September, suspected of burning down St Giles Cripplegate.[198] That year the traditional Mid-summer watch was abandoned, by royal command.[199] By early November the dearth in London was severe enough to disturb the Privy Council: 'since the harvest there hath not come 200 quarters of wheat to London . . . they have a great lack'. Certain merchants were hoarding 'to their singular benefits'. Deprivation presaged instability: if the shortage 'be not provided before Parliament [it] will breed a greater dearth than were expedient'.[200]

As Parliament met on 23 November the divisions in the King's Council and in his Privy Chamber were translated for a time to Westminster. Rival factions, each bound by distinct family loyalties and by opposed beliefs, determined in this arena to influence policy, and the struggles were transferred from Court to the Commons, and to the Lords. Anxious and uncertain, Wriothesley sent to Paget on 11 November a bill 'which was let fall yesterday, as I was going to Mass, in my dining chamber', and urged Paget to show it to the King. 'You know that when these naughty books were brought unto me, I could do no less than send them to his Highness, and also travail, as much as I could, to find out the author, wherein though I have not much prevailed, yet some be angry with my doing.'[201] The 'some' to whom Wriothesley referred were perhaps his reformist enemies around the King; men who specially hated Wriothesley, 'picture of pride, of papistry the plat'. Gardiner also had been sent a scurrilous book: *The lamentacyon of a christian agaynst the cytye of London* of Roderyck Mors. The disseminators of this tract hoodwinked even Gardiner: allegedly it was printed 'in Jericho in the Land of Promise by

---

197 Wriothesley, *Chronicle*, i, p. 159; CLRO, Repertory 11, fo. 215^{r-v}.
198 Ibid., fos. 217^r, 223^r, 226^r.
199 Wriothesley, *Chronicle*, i, p. 156.
200 *State Papers*, i/2, p. 835.
201 Ibid. i/2, p. 840.

Tom Truth'. Could it be by George Joye, Gardiner wondered, for when the author wrote of joy 'he writeth that word with a great letter'?[202] Gardiner would have had more reason to lament so 'many books and scrolls cast abroad in London this year and the offender never found out' had he known that the author of the *Lamentacyon* was not 'a knave lurking in a corner as Joye doth at Antwerp', but Henry Brinklow, the City mercer,[203] who almost certainly had connections with the Protestant zealots at Court.

'A great hurly-burly' followed 'the examination of certain books covertly thrown abroad', and a member of the group of advanced Protestants in the Commons was implicated. Sir Peter Carew was found to have had one of the forbidden books, and was imprisoned for a while. Octavian (Otterden?), a priest, 'the great setter forth of the said books', remained in the Tower.[204] In the Commons was a Protestant coterie deeply opposed to the conservative leaders in the Lords. Seven of this faction in the Commons were of the King's Privy Chamber (George Blage, George Carew, Thomas Cawarden, Thomas Darcy, John Gates, William Sharington, and Thomas Sternhold), and three others, including Peter Carew, were attached by family connections.[205] Matters of religious policy were at issue in this Parliament.

Roderyck Mors (Brinklow) had urged the King that for reform in the commonwealth it must be 'down with all your vain chantries, all your proud colleges of canons, and specially your forked wolves the Bishops'. Henry took some of this advice: the chantries would be dissolved, but to pay for the war and to fortify England's coast against invasion, not, as Brinklow urged, to provide for the poor.[206] There had been rumours

---

[202] Brinklow, 'Lamentacyon', p. 120; *Letters of Gardiner*, pp. 159–60; *L&P* xx/2. 732.

[203] Brinklow wrote that he was 'banished my native country . . . for speaking God's truth', but he was certainly in England in 1543, and at the time of writing his will was still engaged in business in the City; Guildhall, MS 9531/12, fo. 45ᵛ; PRO, PCC, Prob. 11/31, fo. 158ʳ.

[204] PRO, SP 1/212, fos. 45ʳ–48ʳ (*L&P* xx/2. 995); Lehmberg, *Later Parliaments*, p. 222.

[205] David Starkey's review of Lehmberg, *Later Parliaments*, in *JEH* xxix (1978), pp. 227–8; Lehmberg, *op cit.*, pp. 211–15.

[206] Brinklow, 'Complaynt', pp. 47–8; A. Kreider, *English Chantries: The Road to Dissolution* (Harvard Historical Studies, xcvii, Cambridge, Mass., 1979), p. 164.

that the chantries would go. Vaughan wrote hopefully from Antwerp in May: 'If any colleges be put down, I would I had a piece of some good thing for my money.'[207] But it was not until 15 December, when a bill for dissolution was introduced in the Lords, that Henry's intentions were clear. There was great opposition from the Commons.[208] Petre wrote to Paget: 'the book of colleges &c escaped narrowly, and was driven even to the last hour, and yet then, passed only by division of this House'.[209] Another bill caused greater dissension: the bill 'for abolition of heresy and against books containing false opinions'. In the Lords it passed only at the fourth reading; in the Commons too there was trouble. There might be adverse diplomatic consequences also if it passed. So 'the bill of books, albeit it was at the beginning set earnestly forward, was finally dashed'.[210] Henry had encouraged Cranmer to think how 'somewhat to further the reformation of the corrupt religion' by ending such popish ceremonial as ringing bells on All Hallows night, veiling images in Lent, and creeping to the cross on Good Friday. With the removal of the ceremonies, Cranmer knew, the people would think 'the honour of Christ is taken away' unless they were instructed.[211] No reform came, for Henry, fearing that the moves would threaten the league with the Empire and France, changed his mind: 'I am otherwise resolved.'[212] The advance for the reforming party in 1545 had been more apparent than real.

HERESY IN HIGH PLACES

In January 1546 Hooper sent to Bullinger the 'news from England'. It was dire. 'As far as true religion is concerned, idolatry is nowhere in greater vigour. Our King has destroyed the Pope but not popery.' 'The impious Mass', invocation of

---

[207] PRO, SP 1/200, fo. 231ʳ (L&P xx/1. 700).
[208] Kreider, English Chantries, ch. 7; Lehmberg, Later Parliaments, pp. 220–2.
[209] PRO, SP 1/212, fo. 110ʳ (L&P xx/2. 1030).
[210] PRO, SP 1/212, fos. 45ʳ–48ʳ (L&P xx/2. 995); Lehmberg, Later Parliaments, pp. 222–3; L. B. Smith, 'Henry VIII and the Protestant Triumph', AHR 71 (1966), p. 1257.
[211] Foxe, Acts and Monuments, v, pp. 561–2; Thomas Cranmer, Miscellaneous Writings, ed. J. E. Cox (Parker Society, Cambridge, 1844), pp. 415–16.
[212] Smith, 'Henry VIII and the Protestant Triumph', pp. 1257–8.

saints, Lenten fasts, confession, clerical celibacy, and Purgatory
'were never before held by the people in greater esteem than at
the present moment'. 'The chief supporters of the Gospel'—
Audley, the Duke of Suffolk, Dr Butts, Baynton, Poynings,
Wyatt—'are dying every hour'.[213] Gardiner's return from the
Imperial Court in March 1546 presaged worse to come. With
such men in authority in England, wrote Richard Hilles in
April, 'it is not probable that the Gospel will be purely and
sincerely received there'.[214] The struggle between the parties of
reform and reaction gained a new urgency in the early months
of 1546 as it became evident that the King was not long for this
world, and that the party ascendant at his death would hold
power in the new reign. The contest would be fiercest at Court,
but the association of the Protestant groups around the King
with reforming groups in the City would often make London
the arena for the disputes.

The persecution of 1546 revealed the strength and cohesion of
the party of reform in London as well as the power and determination of the conservatives. On Passion Sunday (3 April)
Dr Crome dared to preach a sermon at the Mercers' Chapel
which reinforced his old attacks on orthodoxy. As before, he
challenged the policy which insisted upon the validity of the
doctrine of Purgatory while abolishing by statute the chantries
which existed to pray for souls.[215] Taking the text 'Christ is our
high shepherd' (Hebrews 9), he told of how Christ 'entering
into the holy place once for all, not with strange blood but with
His own precious blood, hath found plentiful and eternal
redemption'. The Bishop of Rome 'I said and say again . . .
hath wrongly applied the sacrifice of the Mass making it a
satisfaction for sins of the quick and the dead'. The Mass was
a sacrifice certainly, but of 'thanksgiving to our only shepherd
for His once offered offering which hath made a satisfaction of
all the sins of them which believe and cleave to Him by faith'.
'Yea, it is eucharistia, . . . sacrificiam laudis. Yea, and it is to
us a commemoration of Christ's death and passion'.[216] It was

---

213 *Original Letters*, i, pp. 36–7.
214 Ibid. i, p. 153.
215 Foxe, *Acts and Monuments*, v, p. 537.
216 BL, Harleian MS 425, fos. 65ʳ–66ʳ (*L&P* xvi. 814: the document is here dated
to 1541, wrongly).

not that popular opinion had so altered since Crome's recantation five years earlier that he might now, without penalty, preach publicly what he believed privately; but that he took what might be his last chance desperately to avow his central Lutheran convictions. Crome was one of the City's most fashionable preachers, attracting large crowds, and for him to preach against the Catholic doctrine of Mass as well as attacking Purgatory was a turn of events alarming and dangerous to the conservatives. The five principal sermons that Easter were devoted to refuting that supreme heresy, the denial of transubstantiation.[217]

Crome was ordered to recant on 9 May. Preach he did, but he withdrew none of his opinions, rather declaring himself more convinced of them than ever.[218] Taking for his text John 10: 11, 'I am the Good Shepherd', he avowed once again that Christ had given His life for His sheep, 'which sacrifice once offered hath satisfied for the sins of all that are, were or shall be saved until the end of the world'. Crome then turned to attack 'my brethren the priests', who were 'wonderfully offended with him', for challenging their livings, and for speaking against the Bishop of Rome. For his part, he 'would not have them to abuse the blessed sacrament for a living'. Although the Pope, 'the idle beggar', had been put down, yet his 'staff' remained with which 'many poor men are daily beaten'. Again, Crome urged his audience to 'judge what they would, if they durst', of the doctrinal implications of the King's dissolution of the Pope's 'fond foundations', the chantries and religious houses. Crome's recantation did not fail by ambiguity alone this time, but by clear defiance. He would not recant, and attacked those who were circulating rumours that he would, to 'the greater slander of God's Word and of me'.[219]

Crome's sermons were the talk of London, even of the countryside. To Oundle on 11 May came a report of outrageous sermons in the capital. A preacher had openly asked

'Where is God? Is he in Paul's?', saying, 'No, No, your creed shewith plainly that God is in Heaven and sitteth of His Father's right hand and shall come again at the Day of Judgement, but He is not come yet.'

[217] *Grey Friars Chronicle*, p. 50.   [218] *CSP Sp.* viii, p. 266; *L&P* xxi/1. 938.
[219] BL, Harleian MS 425, fos. 65^r–66^r.

There were interludes played too, reviling the sacrament of the altar. In one a player asked, 'how should *hoc est corpus meum* serve to certify a hundred persons?'[220] A 'lewd bill set up lately upon a church door in London, against one of them which deposed against Crome' was sent to the Privy Council by the Mayor by 13 May, and other 'lewd books and writings' were circulating among 'light persons, which meddle further in these matters than their capacities be able to comprehend'.[221] Speculation about whether Crome would recant or not divided the citizens.

'What news at the Court?' they asked a royal guard who came into a shop in Bow Lane: that the heretic Crome had been broken by the Council and would now recant, he jubilantly replied. But Richard Wilmot, who was one of the ardent Bible-reading Protestants among the City apprentices, protested: 'then it is contrary to the truth of God's Word, and contrary to his own conscience, which shall before God accuse him'. Wilmot's master was outraged, threatening that he 'would be hanged or burned, swearing that he would take away all his books and burn them'. Wilmot's godfather would have been incensed to hear of his spiritual welfare, for he was the redoubtably conservative Sir Richard Rich. Wilmot and his evangelical friend Thomas Fairfax were sent before the Mayor for questioning. Wilmot had known Crome 'but a while', two years. Crome's sermon on Passion Sunday he thought 'no heresy'; for if it were St Paul's Epistle to the Hebrews was heresy. Wilmot had read in Scripture that 'Christ said to His disciples "the time shall come . . . that whosoever killeth you, shall think that he shall do God high service"', and now he expected to suffer himself at the hands of Bonner and Gardiner. He might indeed have done so had not his company, the Drapers, intervened to save him.[222] Walking in St Paul's after Crome's sermon on 9 May, Sir George Blage betrayed himself likewise in conversation with two fellow members of the Commons. 'I heard say', said Edward Littleton, that Crome

---

[220] A. G. Dickens, 'Early Protestantism and the Church in Northamptonshire', *Northamptonshire Past and Present*, viii (1983–4), p. 34.

[221] *State Papers*, i/2, p. 846. The Earl of Bath's players were arrested for performing lewd plays on 2 May: *APC* i, p. 407.

[222] Foxe, *Acts and Monuments*, viii, pp. 517–22. Foxe heard this account from Wilmot himself.

preached 'that the Mass profiteth neither for the quick, nor for
the dead'. Blage concurred with Crome. 'Wherefore then?'
asked Littleton. Replied Blage, 'Belike for a gentleman, when
he rideth a hunting to keep his horse from stumbling.' Littleton
and Calverley remembered this conversation, embroidered
upon it, and would use it as evidence to accuse Blage later.[223]

The Council recognized the consequences of Crome's failure
to recant: his steadfastness was 'to the confirmation of them,
that be of his faction, and to the danger of many others, which
heard him'. Worse, they knew that others 'were as much to be
blamed, or more, than himself'.[224] Crome was called to answer
for his 'rashness and indirect proceedings'.[225] The Council
now wanted more than his retraction: it wanted the names of
his followers, 'sundry persons of divers qualities', in the Court
and the City. The King's will 'how we shall use the calling and
ordering of them, as a matter wherein we would be loath to
offend in doing too much, or too little' was first sought, and the
royal sanction was given for a thoroughgoing purge which was
to have far-reaching repercussions.[226] Under persistent interrogation Crome 'accused divers persons as well of the Court as
of the City'.[227] In London the Council sought and found some
who 'specially comforted Crome in his folly', for in the capital
Crome had an old following. Crome had ignored Dr Heynes's
warning to 'beware of the brethren of London and not to yield
to their fantasies',[228] and their exhortations had strengthened
him in his resolve.

Within a week many of Crome's bold counsellors had been
rounded up. John Lascells was arrested not because Crome
named him but 'because himself boasted abroad that he was
desirous to be called to the Council, and would answer to the
prick'.[229] Brave words. The King's own physician, Dr Huick,
'to hand' because of some marital controversy, was soon examined for religion also. When Crome had shown him the recantation articles, Huick 'showed himself to mislike the same, and

[223] Foxe, *Acts and Monuments*, v, p. 564; *House of Commons, 1509–1558* (Littleton and
Calverley).

[224] *State Papers*, i/2, pp. 848, 846.   [225] *APC* i, p. 414.

[226] *State Papers*, i/2, pp. 843–4.

[227] Wriothesley, *Chronicle*, i, pp. 166–7.

[228] *State Papers*, i/2, pp. 846, 843.

[229] Ibid. i/2, p. 844. For Lascells, see Wilson, *Tudor Tapestry*.

thought they could not be maintained with good conscience'. By 13 May the Council informed the King that 'we look for' Latimer, now returned to London, John Cardmaker, the sacramentarian vicar of St Bride, and others.[230] Two days later William Morres and Nicholas Shaxton were called to answer regarding 'matters of Crome'. To the Tower on 17 May went Huick, Lascells, 'the Scottish priest' (John Willock), Worley, a page of the Pallet Chamber, and William Plaine, a City skinner and gospeller; all for their erroneous opinions and for 'dissuading of Crome'. Likewise Crome's 'man' was imprisoned for being less than 'frank' in his declaration, and Pawley, servant of the reformist Earl of Arundel, was forbidden to depart London without licence, because he 'had been with Crome before the time of his last sermon'.[231]

First among Crome's confederates was Latimer, his old ally in preaching the Gospel, now returned to London, the first centre of their mission. Before the Council on 13 May Latimer confessed that he had 'indeed, been sundry times in his [Crome's] company . . . and that he had said somewhat touching his recanting, or not recanting', but 'couching his words so' that he did not clearly implicate himself with Crome's defiance or his beliefs. Latimer had, seemingly, always been circumspect in his eucharistic preaching, but once before he had 'set at little the sacrament of the altar', and now there were suspicions that he had done so again. He wished to speak with the King before he answered the Council; he had been 'deceived that way' before. Again, Heath and others tried to 'fish out the bottom of his stomach'.[232] Of Dr Huick, the Council, for all its 'travail' had only 'long writings and small matter'. 'He trusteth so well', wrote Gardiner, that he insisted on having two or three Gentlemen of the Privy Chamber to report upon all the proceedings to the King. As for Lascells, he too was circumspect, despite his earlier boasts. He refused to answer 'to that part of his conference with Crome, that toucheth Scripture matter', without the King's express commandment and protection:

[230] *State Papers*, i/2, pp. 844, 846.

[231] *APC* i, pp. 417, 418–19, 420, 423.

[232] *State Papers*, i/2, pp. 848–9; Foxe, *Acts and Monuments*, v, p. 39; vii, p. 463; H. S. Darby, *Hugh Latimer* (1953), pp. 163–5.

'for he saith, it is neither wisdom nor equity, that he should kill himself'. Cardmaker was found to be 'of the same sort, but yet not so bold as the rest'. The Scottish priest was dismissed as 'very ignorant'; 'he hath framed his sayings after his audience'; 'more meet for Dunbar than for London'.[233] Worley, the page from the Court, had been arraigned already on 5 May for his 'unseemly reasoning . . . upon Scripture matters and for his sinister opinions'.[234] Still messages came to Crome urging steadfastness. William Gray, a London plumber, sought to prevent him from fulfilling his promise to the King for a 'plainness at Paul's Cross'.[235] While Crome and his confederates waited in prison the Council discovered a shadowy conspiracy afoot in London, in which this group was implicated.

On 7 June 'Weston the luteplayer' was arrested for conspiring with Barker, Laynam, Lascells, and others 'upon prophecies and other things stirring to commotion against the King's Majesty'.[236] A nightly watch was ordered through the capital on the following day.[237] The link between prophecy and reform might be compelling, for the persecuted might find in prophecies some consolation, interpreting in opposite ways the same prophecies in which the conservatives had placed their hopes. It was no wonder that John Lascells at Court and William Weston in the City were in contact, for both had been in Thomas Cromwell's service. Barker too served a master whose convictions were reformed. He had been with the Earl of Surrey at Boulogne. That the dissemination of prophecies was a profession for Laynam emphasizes rather than diminishes their significance, for people were keen enough to pay to hear them. Cromwell had dismissed Laynam as a 'mad prophet', but Cromwell's adherents listened to Laynam now, and he had been proved right before.[238]

Laynam foretold obscurely of future 'fields' in England, of the advent of the 'dead man' and the return of the Pope, the

[233] State Papers, i/2, p. 850.
[234] APC i, p. 402.
[235] Ibid., p. 466.
[236] Ibid., p. 449.
[237] CLRO, Repertory 11, fo. 289ᵛ.
[238] For Laynam's long career of prophecy, see Keith Thomas, Religion and the Decline of Magic: Studies in Popular Beliefs in Sixteenth- and Seventeenth-Century England (1971), p. 401.

decline of the clergy, the fall of Charing Cross, of the King's imminent exile from his realm, of his fight with priests, and the impending demise of Queen Catherine. Now Laynam once more identified Henry with the Mouldwarp of Merlin's prophecy, the sixth King from John, who would be driven forth from the land to return to 'do wonders'. The King would 'curse both priest and clerk; for wrong wise works looketh after wrecks with clerks unwisely wrought' (a reference to the divisions of opinion amongst the clergy). It was the clergy who would expel the King, and 'causa ruina populi sunt sacerdotis [*sic*] mali'. Leaning on a rail by St Nicholas Shambles, Barker and Laynam spoke of the exile of the King: 'it must be this year, for this is the last year of all tribulation'. In the spring of 1546 this group had some expectation of their reforming hopes coming to pass. Weston and Barker had talked secretly as they walked in the fields between Cripplegate and Moorgate, Barker bound to silence: 'if you should bewray me; there were but your Yea and my Nay'. Weston said:

What news? Do ye not hear of the going down of these colleges and chantries? Marry! I trust to see the day that every priest shall be glad to say Mass in chalices of wood, and once within this twelvemonth ye shall see that every boy in the street shall spit in the priests' faces and hurl stones at them.[239]

His prophecy was fulfilled,[240] but his confederates and he went to prison for a while. Barker was to be released at the intercession of his mother, the landlady of the Rose tavern in Newgate. The prophecies had spread among reforming circles. In Arundel's household they were known: John Geffrey, the Earl's servant, had a work of John Bale's 'with erroneous words by him uttered upon the same, with certain prophecies'.[241]

At Paul's Cross on 27 June Crome at last recanted. His submission had, so it was reported, 'a very good effect upon the common people, who are greatly affected'.[242] Crome utterly denied the sacramentarian heresy. He confirmed the

[239] PRO, SP 1/220, fos. 60ʳ–69ʳ (*L&P* xxi/1. 1027).
[240] *TRP* i. 292.
[241] *APC* i, p. 509.
[242] *CSP Sp.* viii, p. 291; Wriothesley, *Chronicle*, i, p. 167; Foxe, *Acts and Monuments*, v, p. 836.

sacrifice propitiatory of the Mass, the doctrine of transubstantiation. While he affirmed the efficacy of prayers for the dead,
the irony of preaching of Purgatory even as the commissioners
for the dissolution were active in the City can hardly have been
missed. He blamed his deviance from the right path upon
'lewd and ungodly books and writings', warning his listeners to
beware of them, 'for under the fair appearance of them was
hidden a dangerous accombrance of Christian consciences'.
(On 8 July a new proclamation was issued in which forbidden
works were named and banned.) Now Crome admitted that he
had been led astray by the 'persuasions of certain perverse
minded persons', and exhorted 'all men to embrace ancient-
ness of Catholic doctrine and forsake newfangledness'. Thus
was orthodoxy at last proclaimed by London's most influential
preacher, to the satisfaction of conservatives in the Council and
City, and the dismay of the brethren.[243]

Before the persecution of the Crome circle had ended, the
purge of another group, associated with the first but with even
more powerful connections, had begun. The old political ploy
of bringing down the reforming party by associating it with
arrant heresy was used with a vengeance that summer by the
conservatives, who found in Anne Askew their ideal instrument.[244] On 24 May the Privy Council had given orders for the
arrest of Robert Wisdom and Anne Askew: Wisdom escaped to
exile, but not so Askew, who was brought back to London for
questioning.[245] Anne Askew's story was told only by partial
witnesses: for her, Bale and Foxe; against her, later, Robert
Parsons the Jesuit.[246] To Bale she was 'a young wench of

[243] Guildhall, MS 9531/12, fo. 109ᵛ; Ellis, *Original Letters*, ii, p. 177; *TRP* i. 272;
Wriothesley, *Chronicle*, i, pp. 168–9. The chronicler of the Grey Friars, jubilant at
Crome's concession, appended the recantation articles to his account, but they do not
survive there: *Grey Friars Chronicle*, p. 51.

[244] The story of Anne Askew and her friends is told in Wilson, *Tudor Tapestry*. For
her posthumous portrait, see R. Strong, *The English Icon: Elizabethan and Jacobean
Portraiture* (1969), p. 93.

[245] *APC* i, p. 424; *L&P* xxi/1. 898, 1491.

[246] *The First Examinacyon of A. Askewe, lately martyred in Smythfelde: with the Elucydacyon
of Iohn Bale* was printed in Marburg in 1546, and was followed by *The Lattre
Examynacyon of A. Askewe* in the next year (*RSTC* 848–53); reprinted in the *Select Works*
of John Bale, ed. H. Christmas (Parker Society, Cambridge, 1849), pp. 135–246.
Foxe followed Bale's account: *Acts and Monuments*, v, pp. 537–50. Robert Parsons,
A Treatise of Three Conversions of England (1604) is printed in part in *Narratives of the
Reformation*, pp. 307–11.

worshipful house, and of elegant beauty and rare wit &c'; to Parsons she was a wanton, abandoning her husband 'to gad up and down the country a gospelling and gossiping, where she might and ought not. . . . but specially she delighted to be in London near the Court.'[247] Fleeing Lincolnshire, 'three score priests bent against her', and a husband she hated, she had come to London late in 1544 to seek a legal separation in Chancery and to find like-minded brethren. Family and friends from among the East Midlands gentry had congregated in the Court and capital before her, many of them convinced reformers. Her brother Edward had left Cranmer's household for the King's, and was Gentleman Pensioner and Cup Bearer at the Court (there, too, their brother Christopher had been Gentleman of the Privy Chamber until his death in 1543); her sister Jane had married George St Paul, the Duke and Duchess of Suffolk's steward; Christopher Brittayn, her cousin, was a lawyer of the Middle Temple; and her old neighbour John Lascells of Sturton, a lawyer of Furnivall's Inn, was also at Court.[248] Anne was distantly related to Sir John Aleyn, twice Lord Mayor of London.[249] Among such elevated friends she evangelized, but *spiritus ubi vult spirat*, and her passion to 'follow the liberty of the new Gospel' and to make 'new gospellers and proselites' led her among company such a gentlewoman would never otherwise have sought. Her 'dear sister, disciple and handmaid' was the peripatetic Lollard and Anabaptist-to-be, Joan Bocher, who smuggled books of heresy to her at Court.[250] Anne Askew also had gospelling friends among London artisans and apprentices.

All the while she was watched. From his window opposite Anne's lodgings by the Temple, Wadlow, then 'a great papist', spied upon her. 'At midnight she beginneth to pray, and ceaseth not in many hours after, when I and others apply our sleep or do worse.' But she was not always alone, and her contacts were reported to Wriothesley, Wadlow's master in Chancery.[251] By March 1546 there was evidence enough

---

[247] Ibid., pp. 308–9.

[248] Wilson, *Tudor Tapestry*, pp. 94–6, 115–16, 160, 162, 191–2.

[249] *L&P* xiv/1. 210; Wilson, *Tudor Tapestry*, p. 153.

[250] *Narratives of the Reformation*, pp. 307, 309; J. F. Davis, 'Joan of Kent, Lollardy, and the English Reformation', *JEH* xxxiii (1982), p. 231.

[251] *Narratives of the Reformation*, p. 40.

against her: she was arrested and brought before the heresy commissioners at Saddlers' Hall. Accused of saying that 'God was not in temples made with hands', she confessed 'that she said no less'. Had she said that if an evil priest ministered the sacrament it was 'the Devil and not God'? She answered that 'his ill conditions could not hurt my faith, but in spirit I received, nevertheless, the body and blood of Christ.' Had she the 'spirit of God' in her? 'If I had not, I was but a reprobate or castaway.' Asked whether 'private masses did help souls departed', she replied that 'it was great idolatry to believe more in them, than in the death that Christ died for us'. On the central sacrament, of the Mass, she equivocated. From the quest she was sent before the Mayor, Martin Bowes. He asked her, as it was said she had asked others, 'whether a mouse, eating the host, received God or no'? She simply smiled. In the Counter she was left to contemplate, with no visitors, save a priest sent to 'tempt' her to fall. Her cousin Brittayn came to plead for her release. On 24 March she was examined before Bonner, who offered to summon City clerics and men she esteemed as 'learned and of a godly judgment': Crome, David Whitehead, and John Huntingdon. Bonner counselled her to unburden her conscience, but she parried all his questions: she would tell him whether she believed the sacrament in the pyx was only bread if he would tell her 'wherefore Stephen was stoned to death'. 'I believe', so she maintained throughout, 'as the Scripture doth teach me.'[252] At last Bonner determined, as politic, to construe her answers as a confession, and to publish her submission.[253] This favour towards her, he said, was because she had 'good friends' and came of 'worshipful stock'. Among her 'good friends' at the examinations were Edward Hall, the chronicler, and Alexander Brett, whose later stand against papistry would be as fatal as hers.[254] But Christopher, a servant to Sir Anthony Denny, protested against Bonner's worldly reasoning: 'Rather ought you, my lord, to have done it . . . for God's sake than for man's.'[255] Anne's submission was

---

[252] Foxe, *Acts and Monuments*, v, pp. 540–1; *Select Works of John Bale*, pp. 148 ff.; *Narratives of the Reformation*, p. 41.

[253] Foxe, *Acts and Monuments*, v, pp. 541–2; Guildhall, MS 9531/12, fo. 109ʳ.

[254] For Brett, see below, pp. 537, 541–2.

[255] Foxe, *Acts and Monuments*, v, p. 542.

of the closest concern to Denny and her reforming friends at
Court; not through friendship alone, but because a confession
from her could endanger them. Anne left prison and—persuaded
that it was too perilous—London, for home. There she was
betrayed.

On 19 June Anne and her husband appeared before the
Council at Greenwich. Gardiner, Wriothesley, and Paget challenged her in turn upon her views on the Mass. How could she,
asked Paget, avoid the very words of Christ, 'Take, eat, this is
my body'? The bread, argued Anne, was a symbol; as the vine,
a door, a rock, the Lamb of God: 'All these do signify Christ,
like as the bread doth signify His body.' Christ did not 'bid them
hang up the bread in a box and make it a god, or bow to it'.[256]
This was, it seems, the sacramentarian heresy of the Swiss
Reformers, and of many of the Lollards before them; that the
Lord's Supper was primarily a memorial rite, and that the
elements remained unchanged. Anne was sent to Newgate 'to
remain there to answer to the law'.[257] The conservatives were
determined not only to prove Anne's heresy, but to force her to
retract and to deny it. If she remained obdurate there could be
no way for her but the stake: an appalling prospect because she
was 'of worshipful stock', and because her fortitude as a martyr
would only strengthen the cause they sought to destroy. Neither
could they allow her to rot in prison, for she had influential
friends, and was too well known. But if she could be broken, she
might bring men and women of influence down with her. Even
before the Council, when fellow reformers Lisle and Parr counselled her to 'confess the sacrament to be flesh, blood and bone',
she reproached them: 'it was a great shame for them to counsel
contrary to their knowledge'.[258] Even Crome had succumbed,
and now when Anne, 'sore sick, thinking no less than to die',
sought consolation, it was Latimer she asked to see.

Anne's purpose under interrogation had been at first to avoid
avowing heretical doctrine while not being forced to confess
orthodoxy; to live but not to sully her conscience. But at last in
Newgate she wrote a confession of belief: 'by me Anne Askew,

256 Ibid. v, pp. 544–5. Paget's examination is printed by Bale, but omitted by
Foxe: *Select Works of John Bale*, pp. 203, 205.

257 *APC* i, p. 462.

258 Foxe, *Acts and Monuments*, v, p. 544.

that neither wish death, nor yet fear his might; and as merry as one that is bound toward Heaven'. Therein she declared again a commemorative view of the Mass: 'the bread is but a remembrance of His death, or a sacrament of thanksgiving for it, whereby we are knit unto Him by a communion of Christian love'.[259] The Mass in her time was 'the most abominable idol in the world: for my God will not be eaten with teeth, neither yet dieth He again'.[260] Anne was arraigned at the Guildhall on 28 June, the day following Crome's 'canting, recanting, decanting or rather double decanting'. With her were Christopher White, a City merchant who had made an 'erroneous book' and was a sacramentary, a Colchester tailor, and Nicholas Shaxton.[261] Before the quest Anne declared again that the Son of God was 'glorious in Heaven', never in the sacrament: 'as for that you call your God, it is a piece of bread . . . Let it but lie in the box three months, and it will be mouldy.' 'God is a spirit, and will be worshipped in spirit and truth.' Anne was condemned for heresy, though with dubious legality, for no quest of twelve men was empanelled according to the act.[262] Shaxton and White abjured, 'and the talk goeth that they shall chance to escape the fire for this voyage'.[263] Even after Anne's condemnation the conservatives sought to break her. Her most implacable adversaries were not the Bishops, even Gardiner, but the lay leaders of the conservative faction, whose positions would be most imperilled upon Henry's death: Wriothesley and Rich. They determined to use Anne, while she still lived, to bring down their rivals at Court, for they knew with a compelling urgency that if the reformers once gained ascendancy their cause was lost; not only would personal ambitions be thwarted, but all hopes of defending the old faith would be dashed.

In the Tower Anne was charged upon her obedience to reveal others of her sect, especially her associations with the great ladies of the Court: the Duchess of Suffolk, the Countesses of Hertford and Sussex, Lady Denny and Lady Fitzwilliam. Who

[259] Foxe, *Acts and Monuments*, v, p. 545.
[260] Ibid. v, p. 549.
[261] Ellis, *Original Letters*, i, pp. 177–8; *APC* i, p. 462.
[262] Foxe, *Acts and Monuments*, v, pp. 546, 837.
[263] *L&P* xxi/1. 1180.

helped to maintain her in prison? Her maid pleaded her cause in the City streets to the apprentices, 'and they, by her, did send me money; but who they were, I never knew'. The Countess of Hertford, she admitted, sent a man in a blue coat with 10*s*.; a man in a violet coat brought 8*s*. from Lady Denny. Through the indiscretions of the wives their husbands might be betrayed. 'Then they said, there were of the Council that did maintain me: and I said, No.' In desperation, Wriothesley and Rich put her to torture to force her to name names.[264] 'She hath been racked since her condemnation (as they say)', so Otwell Johnson reported: 'a strange thing in my understanding. The Lord be merciful to us all.'[265] In prison, 'weary and painful as . . . patient Job', Anne wrote to her 'instructor' Lascells: she heard of the councillors' alarm that the King would learn the common talk of the shocking and unusual torture and condemn them for it.[266] So, it seems, he did. In early July George Blage's enemies brought evidence that he had reviled the Mass by saying: 'What if a mouse should eat the bread?': well, elevate the mouse in the pyx too. Blage was arrested and thought to die. But the King, hating the heresy, hated too the secret interventions against his courtiers, 'even into his Privy Chamber'. He saved Blage, his 'pig', from roasting, but he would not save Lascells, and Lascells would not save himself by recanting.[267]

A great stage was built at Smithfield on 15 July for the spectacle of the burning.[268] On the following day, while Shaxton preached his recantation, his fellow reformers went to the stake, Anne Askew so broken by the rack that she could not stand. Anne Askew, John Lascells, John Hadlam (an Essex tailor), John Hemsley (an apostate Observant friar) were martyred before a great crowd.[269] From nearby St Bartholomew's

---

[264] Foxe, *Acts and Monuments*, v, p. 547; *Narratives of the Reformation*, pp. 43, 303–7.

[265] *L&P* xxi/1. 1180.

[266] Foxe, *Acts and Monuments*, v, p. 548.

[267] Ibid. v, pp. 564, 551–2; *House of Commons, 1509–1558* (Blage); *Narratives of the Reformation*, pp. 41–3.

[268] CLRO, Repertory 11, fo. 298[r].

[269] Guildhall, MS 9531/12, fo. 108[r–v]; Foxe, *Acts and Monuments*, v, appendix xvii; *Grey Friars Chronicle*, p. 51. There is some confusion in the sources over the name of the priest who suffered: Foxe names Nicholas Belenian of Shropshire; the Grey Friars chronicler, Hemsley; Stow names Nicholas Otterden (is this the 'Octavian' who was the disseminator of heretical books with Carew in Nov. 1545?): Foxe, *Acts and Monuments*, v, pp. 550–1; Stow, *Annales*, p. 592.

members of the Council watched, with who knew which mixed emotions. In their desperation to force Anne Askew to implicate their rivals at Court in her heretical affinity, Rich and Wriothesley had turned the rack themselves. But though Anne had died, they had lost. For within six months of Anne's martyrdom the 'Romanist party' in the Council was overthrown. In the King's last days the reformers returned to favour, and could defeat the conservative rivals who had so nearly condemned them.[270] But the reformers could not remember Anne's suffering with clear consciences. Paget had interrogated Anne, challenging the beliefs which he would soon play a part in advancing.[271] For Cranmer the crisis of conscience was most acute. Soon, maybe even within the year, he was converted to a eucharistic position which, though still far from the one for which Anne had died, was contrary to the Catholic orthodoxy that he was still, as Archbishop, bound to uphold.[272]

But in July no one could have foretold another revolution in Court politics nor another reversal in religion. Continued persecution seemed certain. 'For fear of death' about sixty Protestants (including some of Anne's closest supporters) fled to Flanders or to the north; away, anyway, from London.[273] Though Bale claimed that Anne's fortitude had strengthened the cause and converted many, Robert Crowley feared that Shaxton's shameful recantation had convinced many others to revert to the old faith.[274] The Londoners who watched the burnings were deeply divided in their reactions. No human sympathy for the suffering could diminish Catholic certainty that the punishment was deserved. Admiration and outrage animated their reformed neighbours. As the martyrs died, the

---

[270] For the political reversal, see Starkey, *The Reign of Henry VIII*, ch. 8.

[271] In some editions of Bale's *Examinacyons* of Anne Askew the pages were glued in order to conceal Paget's association with the persecution. King, *English Reformation Literature*, p. 79.

[272] P. Brooks, *Thomas Cranmer's Doctrine of the Eucharist* (1965); C. W. Dugmore, *The Mass and the English Reformers* (1958), pp. 182 ff.

[273] PRO, SP 1/227, fo. 170ʳ (*L&P* xxi/2. 596); *L&P* xxi/1. 1491; Dickens, *Lollards and Protestants in the Diocese of York*, pp. 34–5. Among the fugitives were Wisdom and Whitehead. Vaughan dared keep Cobbe, his children's schoolmaster, no longer, lest men suspect his own orthodoxy: *L&P* xxi/1. 1494; xxi/2. 52.

[274] See King, *English Reformation Literature*, p. 75; Robert Crowley, *The confutation of .xiii. articles, wherunto N. Shaxton, late byshop subscribed and caused to be set forth in print M.C. xlvi. when he recanted* (*RSTC* 6083).

martyrologists remembered, rain fell and thunder cracked, as
'God's own voice'. 'For fear of damnation' John Louthe could
not stand by in silence, and he called down upon the councillors
present 'a vengeance of you all that thus doth burn Christ's
member'. At that a Catholic carter struck him.[275] It is to discover the extent of the divisions between Catholic and reformed
in London, and how and among whom the divisions were
created, that we now turn.

[275] *Narratives of the Reformation*, pp. 44–5.

IX The Old Faith and the New

AT CHRISTMAS 1545 Henry VIII had made a moving last oration to Parliament: 'I hear . . . and alas the while, that the special foundation of our religion, being charity between man and man is so refrigerate . . . as there was never more dissension.' The cause of this 'lack of love' was contention over what was religious truth. In taverns and alehouses men discussed the mysteries of faith, and they depraved ceremonies, and each other. 'Papist' and 'Lutherian' were 'names devised by the Devil . . . for the severing of one man's heart by conceit of opinion from the others'. And the King—'charitable man to mine even-christened'—wept, as though these divisions were none of his making. And as they listened the members of Parliament, too, were moved to 'water their plants'. There was, Henry declared, but 'one Truth and Verity'.[1] On this, at least, no one disagreed. Everyone, of whatever confession, claimed that there was one true faith, and one universal, catholic Church, with a monopoly of spiritual truth. But as to which Church this should be there was division. 'Merry in the Lord' after his condemnation, John Lascells boasted: 'My Lord Bishop would have me confess the Roman church to be the catholic church, but that I can not, for it is not true.'[2] At the Reformation, because of the Reformation, division in religion was inevitable, because everyone agreed that anyone not of their Church was against it, heretic and schismatic.

Religion was a public duty as well as a matter of private conscience. Belief was hard to hide. Yet many Londoners, prudently acquiescent to the changes, concealed their private beliefs, at least in public, behind conventional forms. Beyond the conformity demanded by the monarch, they believed as they wished, and real religious beliefs changed by means, and for reasons, more subtle than royal command. The evidence is

---

[1] Foxe, *Acts and Monuments*, v, pp. 534–6; Hall, *Chronicle*, p. 865; PRO, SP 1/212, fos. 108ʳ–110ʳ; 111ʳ–112ʳ (*L&P* xx/2. 1030); S. E. Lehmberg, *The Later Parliaments of Henry VIII, 1536–1547* (Cambridge, 1977), pp. 229–31.

[2] *Narratives of the Reformation*, p. 43.

difficult to interpret. The most uncompromising Catholics, of course, looked upon the changes in public religion and private mores with mounting horror. On 16 January 1547, as a new age of reform was beginning, John Feckenham preached a despairing sermon before the Londoners. 'We have lost Christ.' The youth of England, he proclaimed, were brought up in a lamentable contemporary progression: 'from pride to lechery, from lechery to theft, and from theft to heresy'. The clergy were scorned and left impoverished by the withdrawal of first fruits and tenths. 'The sanctimony of life was abandoned.' Time was when all had been loyal to virtue and ceremony, and the converts to the New Learning few and circumspect; now, on the contrary, so many were disaffected from the Church that good Catholics dared not 'use fasting on Wednesdays and Saturdays, and beads . . . for fear they should be laughed to scorn'. 'What a world it shall be' when this heretical generation shall have the rule . . . it will be treason shortly to worship God'.[3] Maybe this was the sort of sermon that an ultra-Catholic, on the eve of his exile, would make. Surely reform had not progressed so far, so universally, even in London.

Sometimes seemingly insignificant or isolated incidents reveal much more of popular religious allegiance than propaganda or overt opposition. Though many hated the changes, the great majority silently repined. Very few were so aggressive in the defence of the old ways as the two parishioners of St Bartholomew the Little who threatened to hang an Alderman if he dared to remove the papal arms from the church windows.[4] But in other parish churches where, if there was protest and defiance, we know nothing of it, the papal arms simply remained.[5] Londoners had done nothing to prevent the Dissolution of the monasteries, but this did not mean that they did not lament it. In 1540, soon after the execution of the three Carthusian Abbots for the treason of loyalty to the Pope, a beggar in London found a sure way of winning the sympathy and support of the citizens. He 'asked alms, saying that "they put my master to death", declaring that the Abbot of Glastonbury was his master'.[6] How

[3] PRO, SP 1/228, fos. 55^r–56^r (*L&P* xxi/2. 710).
[4] CLRO, Repertory 10, fo. 238^r.
[5] See below, pp. 430.
[6] CLRO, Repertory 10, fo. 156^r.

are the convictions of the mass of Londoners to be discovered? Perhaps hardly at all, but there are signs.

TESTAMENTS OF FAITH

Personal hopes for salvation and confidence in the means of attaining it were expressed at the last: in wills. While the old faith was still unchallenged almost every testator had entrusted his soul to the Virgin and saints in Heaven as intercessors to the Almighty, and ensured that as many prayers as possible should be said for the wealth of his soul and for his passage from Purgatory. How would the soul be bequeathed if Purgatory was believed to be imaginary, and once men were assured of the absolute sufficiency of Christ's saving passion for sinful man's redemption? If popular religious beliefs were changing, then contemporary wills would testify to a shift in allegiance of the faithful away from a devotion to the Virgin and her heavenly followers to a renewed confidence in Christ's perfection as 'saviour' and 'redeemer of all the world'.[7] Certainly no absolute distinction could ever be subtle enough, for every Catholic of course insisted upon the saving power of Christ as the essence of faith, and in the 1520s, and long, long before, Catholic men and women who were untouched by Luther's theology of grace made wills which indicated that not everyone was convinced by the doctrine of salvation by works, nor confident that man of himself could deserve redemption. Thomas Mirfyn, Lord Mayor in 1518, trusted through the merits of Jesus to be a saved soul. To his creator, 'firmiter sperans per ipsum salvari . . . quamvis indignus', Roger Pritchard bequeathed his soul. Gerard Danet trusted 'by the merits of Christ's passion to come to the endless bliss which He bought me unto and all mankind upon the mount of Calvary with His most precious blood'. Peter Potkins besought 'His maker to show that tender and infinite charity of His passion that He suffered for mankind . . . that my ghostly enemy may have no power of me'.

---

[7] Professor Dickens pioneered the study of wills to examine belief: *Lollards and Protestants in the Diocese of York, 1509–1558* (Oxford, 1959), pp. 171–2; 'The Early Expansion of Protestantism in England, 1520–1558', *ARG* lxxviii (1987), pp. 213–17. On methodology, see M. L. Zell, 'The use of religious preambles as a measure of religious belief in the sixteenth century', *BIHR* l (1977), pp. 46–9.

All of these testators, making their wills in the early 1520s, hoped too for the Virgin and her 'heavenly Court' to be mediators to God for them, and asked for prayers for their departed souls.[8] Faith and works might be inseparable: Agnes West bequeathed her good deeds to God; she also left her faith to Him.[9] Yet, for all the increasing concentration in late medieval piety upon Christ's historic death and passion, upon *tristitia Christi*,[10] as the evangelists preached many became convinced that the essential spiritual application of it to man's salvation was only rediscovered by Luther and his followers.

In their wills the converts to the new faith avowed emphatically their absolute reliance upon the mercy of Christ and the saving power of His passion. Edward Hall, the chronicler, writing his will in 1546, expressed this certainty: 'First, I yield and give my soul to her maker and redeemer by whose passion, and not by my deserts, I trust *only* to be saved, for He hath washed away my sins, I doubt not, by His precious blood.'[11] While the new faith was persecuted overt declarations of evangelical belief might be dangerous; if not for the dead testator, then for those he left behind, or at the least for the certain execution of the will. William Tracy, member of the Reformation Parliament, in his testament trusted to be saved by the merits of Christ alone, spurning intercessory masses and the mediation of saints. This will condemned him as heretic, and in 1531 Tracy's body was exhumed and burnt by the clergy.[12] There were real reasons for circumspection. Even some of the bravest of the early brethren—including Purser, Somer, Petyt, and Lome—made wills which expressed nothing of their known evangelical belief. Bequeathing their souls to God alone, or even to the Virgin and saints, they looked for all the world like their conventional Catholic fellow citizens.[13] True, Somer and

---

[8] PRO, PCC, Prob. 11/20, fos. 99ʳ, 7ʳ, 8ᵛ.

[9] J. A. F. Thomson, 'Clergy and Laity in London, 1379–1529' (University of Oxford, D.Phil. thesis, 1960), pp. 65–6.

[10] More, *De Tristitia Christi*; *CW* 14 (bibliography); J. Rhodes, 'Private Devotion on the eve of the Reformation' (University of Durham, Ph.D. thesis, 1974).

[11] A. F. Pollard, 'Edward Hall's Will and Chronicle', *BIHR* ix (1932), pp. 171–7.

[12] Hall, *Chronicle*, p. 796. The will was printed in 1535: *RSTC* 24167. Only the lay authorities could burn heretics: the errant cleric who carried out the sentence against Tracy was himself castigated: Guildhall, MS 9531/11, fo. 127ʳ.

[13] PRO, PCC, Prob. 11/24, fos. 113ᵛ, 167ᵛ (Somer and Petyt); 11/25, fo. 49ʳ (Purser); 11/27, fo. 110ʳ (Lome).

Petyt were in the Tower, and Purser dying, when their wills were written, but similar, if considerably lesser constraints, may have bound other evangelicals. Perhaps the first in London to avow evangelical tenets in a will was Joan Birrell of St Botolph Billingsgate, who in January 1530 bequeathed her soul to God 'in whom I have perfect faith, hope and trust by the merits of His most blessed and bitter passion my soul shall be saved'. (Did she too face opposition? Her will, without probate granted, was copied into the register by a different hand, perhaps by a later scribe.)[14] Others followed her; though, at first, a tiny minority. Of 638 London wills written between January 1530 and August 1539 (just after the passage of the Act of Six Articles) 41 (6 per cent) expressed solifidian tenets.[15]

But at the Reformation, as before, thinking upon the saving power of Christ's passion did not always detract from belief in the power of intercessory prayers to hasten the soul's passage to God. The preaching of the evangelicals may have led the citizens to look with more hope and faith to Christ as redeemer without altogether lessening the devotion to the Virgin or the fears of Purgatory. So Thomas Crull, a grocer, committed 'my soul, my faith, my death to my Lord God, my . . . redeemer Christ Jesu, to Our Lady Virgin Saint Mary and all the holy company of Heaven, with all the works and deeds that ever I did by grace to my Lord God, the worker of them, and the rewarder of them, by His charitable mercy and without any desert in my behalf'.[16] Some may have wanted to believe in justification by faith alone, but could not quite, yet. In 1536 John Goche, a mercer, bequeathed his soul, 'hoping to be saved by the merits of the passion of Jesu Christ . . . beseeching that blessed Virgin Our Lady Saint Mary with all the holy company of Heaven to pray for me that it may be so'.[17] In the wills of many Londoners apparently solifidian preambles are followed by bequests for Catholic practices: for trentals, obits, dirges, and continued prayers for souls.[18] Easy distinctions

---

[14] Guildhall, MS 9171/10, fo. 150^r.

[15] Ibid., MS 9171/10, fos. 142^r–338^r; PRO, PCC, Prob. 11/24, 27; *LCCW* 106–20.

[16] PRO, PCC, Prob. 11/28, fo. 65^r.

[17] Ibid. 11/28, fo. 293^r.

[18] See, for example, ibid. 11/27, fos. 46^v–47^v, 77^v, 97^v, 215^r; 11/28, fos. 43^r, 78^r, 84^r, 89^v, 288^r; 11/31, fo. 103^r; Guildhall, MSS 9171/11, fos. 57^r, 58^v, 105^r, 118^r, 128^r, 134^v, 153^v; 9052/1a, fo. 33^r.

between orthodox and evangelical may be delusive. In 1543
Robert Cavarden, painter stainer, made his will, 'faithfully
believing through the merits of Christ's most precious passion
to be clearly forgiven'. Cavarden was no evangelical: he was
one of the 'conversi' of the London Charterhouse, surely
opposed to reform, and he asked for sung mass and dirge and
prayers for his soul and all Christian souls.[19]

Old ways of worship did not change easily or quickly, and
men were not likely to jeopardize the chance of remission for
their souls in Purgatory, or prejudice hopes of eternal life if
they could atone in some way in their wills. The wavering,
tempted by the new ideas in their lives, may often have reverted
to the Catholic fold on their death-beds, and the presence of
priests as witnesses doubtless helped to convince them. Their
sovereign when he died thought still upon Purgatory, and
entrusted his soul to the Virgin and saints, and so too did most
of the citizens.[20] 'The holy company of Heaven' was still called
upon by the great majority of testators in the 1530s (85 per cent
of 638 testators making wills between January 1530 and
August 1539).[21] Maybe those about to die, making wills, were
more resistant to change than the 'youth of England' brought
up in heresy; moreover, it is likely that conservative priests had
a say in the expressions of faith; but the conclusion that the
great majority of citizens were as yet unswayed by the new
ideas is, even so, irresistible. The wills of the 1540s provide
evidence of the conversion of some, and the confusion of
many.[22] A majority—but now a declining majority—still
remembered the Virgin and saints in Heaven and sought their
intercession (477 of 702 testators—67 per cent—writing wills
between late 1539 and January 1547). A fifth of Londoners
now bequeathed their souls to God or to Christ alone, neither
expressing a belief in the sufficiency of Christ's passion, nor of
assurance in the power of prayer to release the sinful soul from
Purgatory. For all their brevity, it is more likely that they
trusted in the old ways than in the new. Had the reformers read

[19] *LCCW* 155; *L&P* vii. 728–9.

[20] L. B. Smith, 'Henry VIII and the Protestant Triumph', *AHR* lxxi (1966),
p. 1264.

[21] See n. 15 above.

[22] Guildhall, MSS 9171/11; 9052/1a; PRO, PCC, Prob. 11/28, 31.

the preambles of Londoners' wills in the 1530s and 1540s they might well have thought that they had preached in vain.

Yet by the 1540s, in the last years of Henry's reign, the number of testators avowing the cardinal justificatory tenets of Protestantism had doubled: 13 per cent made unequivocally evangelical wills. True, evangelical wills were often written and witnessed by evangelical clerics who had a powerful influence. So parishioners at St Mary at Hill, whose wills were composed by their curate William Erith, committed themselves uniformly 'body and soul to my maker and redeemer Jesu Christ, beseeching Him as He with His blood bought me to have mercy on my soul'.[23] But it was naturally in the parishes where the pioneer evangelists served that the first conversions were made: at All Hallows Honey Lane, St Mary Aldermary, St Bride, St Margaret Lothbury, St Martin Ironmonger Lane, St Vedast Foster, St Antholin.[24] Just as conservative parishioners chose conservative priests to witness their wills, so evangelical parishioners chose the Gospel preachers, and so, too, they found evangelical scriveners like Henry Bright and William Carkke.[25]

The evangelicals' hope for salvation lay all in Christ's promise. Sir Anthony Denny, one of the Greek gifts from the Trojan horse of St Paul's school, made his will in August 1549, appealing as a plaintiff to 'Christ, my advocate', crying with the prophet, 'Lord, I suffer violence, answer for me, aid me, calling the matter to Thy court of mercy.' Denny could not 'commit the pleas of my cause to any works', for 'certes, my sheet anchor is to trust in Thy promises'.[26] Roger Townshend trusted by the 'only merits' of Christ's passion, 'excluding all my deserving, merits or good deeds', his 'poor soul to be redeemed from all sins, death and Hell none of them all never to have any power over it'.[27] These evangelicals trusted to 'arise with the elect and have the fruition of His godhead'.[28]

[23] Guildhall, MS 9171/11, fos. 162[r], 171[r].

[24] See above, pp. 113–14, 128, 187, 222, 263 and below, pp. 398–404.

[25] See, for example, PRO, PCC, Prob. 11/27, fo. 269[v]; 11/31, fo. 17[v]; Strype, *Ecclesiastical Memorials*, i/2, p. 374. For Carkke's own will, written in 1548, see PRO, PCC, Prob. 11/32, fo. 217[r].

[26] Ibid. 11/32, fos. 285[v]–286[r].   [27] Ibid. 11/27, fos. 169[v]–170[r].

[28] See, for example, Guildhall, MS 9171/11, fos. 145[v], 156[v]; PRO, PCC, Prob. 11/27, fo. 172[v].

And no durance in Purgatory awaited them, for they believed
in no such place. It followed that the ringing of knells, the singing of funeral dirges, and psalms, and contingents of mourners
in black, and elaborate burials 'with pomp and pride of this
world' were a delusion and vanity which they spurned.[29]
Denny lamented the 'abused use . . . of mourning that hath of
long time crept into executors and other dissembled friends',
compelling the dying to 'call upon their executors' unmindful-
ness with the unkind cry of Miseremini mei vos saltem vos amici
mei'.[30] Some forbade the 'bestowing of black garments'.[31]
Henry Brinklow threatened to disinherit his wife should she
wear at his funeral 'a worldly, fantastical, dissembling black
gown for me'.[32] But these were cries in a Catholic wilderness.
Londoners continued to order funerals as elaborate as they
could afford, and not only for worldly show. For their souls'
health they required the multiplied prayers of all who would
attend.

Only belief in the existence of Purgatory and the conviction
that the prayers of those in this life could avail the dead in the
next, that there was communion between the living and the
dead, can explain the continued bequests for propitiatory
masses, the desire for elaborate rites of Catholic funerals, trentals, and obits. Attacks upon the Church left many fearful for
the traditional ways, sometimes questioning of them, but still
they left money to endow them. The citizens' preoccupation
with life after death and concern for atonement were traditionally manifested by the endowment of chantries for souls.
Though desire for immortality for the founder's name might
also have lain behind the foundation of perpetual chantries,
any endowment for masses for souls always signified a belief in
the intercessory power of prayers to release souls from Purgatory. Few permanent chantries were endowed on the eve of the
Reformation, but even until the dissolution of the chantries
Londoners left bequests to endow priests to pray for them, their
families, friends, 'all Christian souls', for as long as they could

[29] See, for example, Guildhall, MS 9171/11, fos. 101^r, 164^r; PRO, PCC, Prob.
11/27, fos. 149^r, 189^r.
[30] Ibid. 11/32, fo. 285^v.
[31] Ibid. 11/27, fo. 189^r.
[32] Ibid. 11/31, fo. 158^r.

afford. By 1537 11 per cent of testators whose wills were registered in the Prerogative Court of Canterbury bequeathed
money for chantry priests to sing for a total of 46 years; of the
wills proved between 1539 and 1541 twenty-two contained
bequests for twenty-six priests to sing for 63 years.[33] As attacks
mounted against the Church the endowment of chantries was a
particularly conscious and optimistic act of orthodox piety.
After the Dissolution of the monasteries, no religious foundation
could be thought secure from sequestration, for the doctrine of
Purgatory was implicitly undermined (for all that the authorities
might deny it), and the future of the chantries was uncertain.
Still, in the wake of the monasteries' dissolution, the citizens
had faith, or wished to keep faith, in the future of chantries and
felt the need for prayers. In the Commissary Court registers
9 per cent of testators left money for chantry priests between
1522 and 1530; 2 per cent between 1530 and 1539, and only
seven of the 530 who made wills in the last years of Henry's
reign.[34] This was a significant decline, yet many who would
not leave money for chantries did ask for prayers for souls, so it
was surely not doubt about the mediatory power of prayers for
the dead which deterred them. But after 1545 chantries could
no longer legally secure gifts and endowments, whatever the
people's wishes.

There had never been any assurance that a citizen's provision
for his soul in the after-life would be fulfilled by his friends and
executors, nor that masses endowed in perpetuity would last
forever, but attachment to chantries at the Reformation may
be traced by examining the performance of bequests. In
London, unlike some other towns, perpetual chantries had
been established by individual citizens rather than by the Communalty. In the sixteenth century the religious and economic
functions of the gilds of London combined still in the companies' trusteeship of religious foundations. The companies
were invested with large sums: to maintain chantries, obits,
and lights for their departed members. Sixty-one of the City
chantry priests at the Dissolution were appointed and paid by
the companies. There was usually a handsome margin of profit

---

[33] PRO, PCC, Prob. 11/27 and 28.
[34] Guildhall, MSS 9171/10 and 11.

between the money spent on the services and the income from the endowment; sometimes the companies provided only the minimum services and profited thereby from their religious trust.[35] So in 1540 a bill was brought in the Star Chamber against the master and wardens of the Goldsmiths' Company because masses were no longer sung in St John Zachary for Robert Butler's soul. The Goldsmiths claimed that the income had been stolen and priests could not be paid.[36]

In a few parishes chantries were falling into abeyance before the Dissolution, and their endowments had been unofficially expropriated. In 1548 seven parishes failed to provide the chantries they were endowed to maintain.[37] Where the endowment was no longer used for its original purpose of providing masses for souls, religious attitudes may have been changing: in St James Garlickhithe the annual income of £8, once used to provide a lamp, had since 1546 been given to the poor.[38] In parishes which were reformist strongholds it was reformed conviction which lay behind the expropriation of chantries and the expulsion of chantry priests. At St Matthew Friday Street in 1540 the churchwardens tried unlawfully to evict a chantry priest from his house. One of these churchwardens, William Ettis, was indicted in the same year for harbouring Gospel preachers.[39] All Hallows Lombard Street had an income of £5. 16s. 8d. to provide a chantry priest, but none had been found after 1540; St Magnus parish had by 1548 failed for twenty years to find a priest; in All Hallows Honey Lane there was an income for an obit and a chaplain, but neither was provided: all three were parishes with early reforming

---

[35] PRO, E/301/34. They also maintained eleven priests outside the City. For the companies' deliberations about whether the endowment income would cover the expenditure, see G. Unwin, *The Gilds and Companies of London* (4th edn., 1963), p. 206; A. H. Johnson, *The History of the Worshipful Company of Drapers of London*, 4 vols. (Oxford, 1914–22), ii, pp. 35, 84; *Acts of Court of the Mercers' Company, 1453–1527*, ed. L. Lyell and F. D. Watney (1938), pp. 529–30, 566, 667–8, 747, 772; Mercers' Company, Acts of Court, ii, fos. 2ᵛ, 5ʳ, 11ᵛ, 14ʳ, 38ʳ; Goldsmiths' Company, Court Book D, fos. 137ʳ, 209ʳ. The profit made by the companies is evident from the Chantries certificates: *LMCC*.

[36] Goldsmiths' Company, Court Book D, fos. 340ʳ–351ʳ. Cf. Johnson, *History of the Drapers*, ii, p. 84.

[37] *LMCC* 6, 20, 23B, 24, 52, 56, 76.

[38] Ibid. 20, 16.

[39] PRO, C1/981/85–7; Foxe, *Acts and Monuments*, v, p. 446.

traditions.[40] But legislation halted the endowment of chantries before—perhaps long before—most people had been convinced by the reformers who branded the chantries as superstitious and urged that the money be given to the poor instead.

Obits, the anniversary masses for souls, were still endowed and performed in the early days of the Reformation. Between 1537 and 1541 7 per cent of testators with wills proved in the Canterbury Prerogative Court asked for obits.[41] One of the more rigorously enforced duties of the gildsmen of early Tudor London was to attend memorial masses for dead members of their fraternity, a lasting indication of the religious as well as economic foundation of the companies. As with chantries, not all obits bequeathed were ever established, nor those established diligently observed. In 1531 Robert Atkinson left £50 to the Skinners for a perpetual obit, but none was ever kept.[42] After the dissolution of the religious houses obits traditionally kept in them could no longer be performed, and the Crown seized the endowments. The companies objected, for their own profit from the trusteeship was lost—and the donor's soul imperilled. In certain cases the obits were transferred to the parish churches and continued there.[43] Obits had often enough been neglected in the past because people could not be bothered to attend: in 1519 the Drapers acknowledged the difficulty by allowing that only one-half of the fellowship need attend at one time. Even the priests were castigated for their absence by the Goldsmiths in 1539.[44] But at the Reformation there were new reasons for refusing to come: in February 1546 William Callaway and John Coke were fined for staying away from an obit. Callaway was of the brethren.[45] In 1548 all obligation to attend

[40] *LMCC* 23B, 24, 52.

[41] PRO, PCC, Prob. 11/24, 27, 28.

[42] PRO, PCC, Prob. 11/24, fo. 84[r]; J. J. Lambert, *The Records of the Skinners' Company of London, Edward I to James I* (1933), pp. 171–2.

[43] Goldsmiths' Company, Court Book F, fos. 131[r], 157[r]; *L&P* xiii/2. 809; Johnson, *History of the Drapers*, ii, p. 64.

[44] Ibid. ii, p. 33; Goldsmiths' Company, Court Book F, fo. 143[r]. In the last years of Henry's reign the Grocers were attending twelve obits a year; the Merchant Taylors twenty-seven; and the Armourers attended two in Aug. 1541: Guildhall, MSS 11571/5, fos. 129[v], 152[v]; 298/2. 4; 127071/1, fos. 107[r], 132[r].

[45] Goldsmiths' Company, Court Book H, fo. 47[r]; Foxe, *Acts and Monuments*, v, pp. 446, 525–6; Guildhall, MS 9531/12, fo. 32[r–v].

obits was removed when they, with chantries, were declared superstitious. The chantry survey of 1548 discovered over 360 obits in London still performed, dissolved them, and expropriated their considerable income for the Crown.[46] The costs of recovery for the City and its companies of all the lost revenues from the endowments for obits and chantries of which they had been trustees proved very heavy (see Table 2).

The persistence of loyalty to the religious gilds, brotherhoods founded to pray for brothers and sisters languishing in Purgatory as well as to give alms to suffering living brethren, is further evidence that popular piety was slow to change. Yet though the fraternities had continued to play so vital a part in the devotional and social lives of the London laity until the very eve of the Reformation, there were soon signs that allegiance was failing. While the need to succour living brethren remained, indeed grew more urgent, the spread of reforming ideas and the gradual abandonment of the doctrine of Purgatory challenged the gilds' first, spiritual purpose. The falling number of bequests to the fraternities was already marked by the 1540s: almost a quarter of testators whose wills were proved in the London Commissary Court between 1522 and 1539 left bequests to the gilds; between 1539 and 1547 only 9 per cent did so.[47] Hesitation to endow institutions which were clearly ripe for the royal plucking doubtless gave some pause, like the scrivener who left 40s. to the gild of the Name of Jesus only '*if* the said fraternity do continue'.[48] But most fraternities were funded by subscription, not endowment. It was the constant propaganda of the reformers that there was no Purgatory, that the 'making of brotherhoods' was yet another clerical money-spinner, which began to undermine them. George Marshall, of the Christian Brethren, had informed his Essex congregation in 1527 that he 'would occupy his chamber as a free chamber and not as a fraternity or gild, to pray for Hob, Gib and Piers'.[49] A few still did make bequests to gilds while avowing evangelical tenets —like Thomas Crossley who left a torch to his brotherhood at St Margaret Bridge Street—for loyalty to the fraternities

[46] *LMCC*. 212 were kept by the parishes, 156 by the companies.
[47] Guildhall, MSS 9171/10 and 11.
[48] PRO, PCC, Prob. 11/27, fo. 177ᵛ.
[49] GLRO, DL/C/330, fo. 139ʳ.

TABLE 2: *Payments made to the Crown by the City and Companies for the Recovery of Chantry Assets, 1550*[a]

	£	s.	d.
Mercers	3,935	3	4
Grocers	1,718	0	0
Drapers	1,082	6	0
Goldsmiths	2,125	6	8
Fishmongers	1,901	6	8
Merchant taylors	2,006	2	6
Vintners	361	10	0
Skinners	923	11	8
Salters	972	12	6
Ironmongers	242	10	0
Saddlers	645	3	4
Haberdashers	810	3	4
Ale brewers	362	6	8
Clothworkers	405	3	4
Tallow chandlers	162	13	4
Wax chandlers	64	0	0
Armourers	211	0	0
Bakers	49	0	0
Leathersellers	189	6	8
Cutlers	13	6	8
Curriers	36	13	4
Cordwainers	23	6	8
Carpenters	8	0	0
Dyers	126	13	4
Joiners	20	0	0
Founders	20	0	0
Cooks	29	5	0
Fletchers	4	0	0
Grey tawyers	9	6	8
Coopers	19	0	0
Pewterers	27	0	0
Barbers surgeons	2	13	4
Mayor and Communalty	210	13	4

Note:
[a] Mercers' Company, Register of Writings, ii. fo. 114ᵛ; Acts of Court, ii, fo. 238ᵛ.

died hard.[50] Yet if the doctrine of Purgatory was once undermined and the greater part of the brotherhood (the dead) was beyond the power of prayer, the gilds would lose most of their *raison d'être*. In 1548 their fate was sealed with the Chantries Act, which outlawed them as superstitious, and made their demise absolute, seizing the endowments of eighteen brotherhoods within the City and dissolving those whose existence depended upon quarterages.[51] But surely Purgatory had been officially denounced long before most people relinquished their fears that souls lingered there, or their hopes that the living could intercede for them. Few yet shared the certainty of Bartholomew Gibbs that no mortal supplications could determine his fate. He willed in 1549 that 'you bury me as quietly as you can without any ceremony, for when the day is ended the night is come with me'.[52]

Yet now, instead of wishing for prayers for their departed souls, some citizens were determined to bring the Word of God to the living. Bequests for thirty sermons instead of the traditional thirty offices for the dead—a trental—mark a clear evangelical impulse. Humphrey Monmouth asked for sermons from Barnes, Latimer, Crome, and Taylor, and his wishes were followed.[53] He was the first of many Londoners to ask the reformers to preach. Alice Wethers left 8*s.* each for thirty sermons at All Hallows Bread Street 'of such men as can and will preach God's Word truely and sincerely', naming Barnes, Crome, Heath, and John Thixstyll.[54] Others asked for John Hardyman and Thomas Garrett, reformers both, to preach.[55] In March 1543 Stephen Hawkyn requested sermons 'to the setting forth of God's Word and His glory by such learned persons' as his executors chose. But, as a new wave of reaction began, he foresaw the dangers, and asked that if his curate

[50] Guildhall, MS 9171/10, fo. 315[r].

[51] *LMCC* 7, 12, 14, 18, 20, 22, 25, 43, 47, 48, 72, 75, 92, 93, 94, 100, 105, 106, 107. Amongst these were some gilds founded before 1389: St Giles at St Giles Cripplegate, Salve Regina at St Magnus, St Katherine at St Katherine Colchurch, Our Lady at St Dunstan in the West, St Fabian and St Sebastian at St Botolph Aldersgate: H. F. Westlake, *Parish Gilds of Medieval England* (1919), pp. 180–8, 236–8.

[52] PRO, PCC, Prob. 11/34, fo. 73[r].

[53] PRO, PCC, Prob. 11/27, fos. 93 ff.; SP 1/139, fo. 136[r] (*L&P* xiii/2. 856); Wriothesley, *Chronicle*, i, p. 72.

[54] PRO, PCC, Prob. 11/27, fos. 232[v]–233[v].

[55] Ibid., fos. 149[r], 189[r], 227[r], 245[r].

refused to allow such preaching in his church the money be given to the poor instead.[56] The citizens were divided in faith, but the divisions among the clergy were deeper and of greater consequence.

Time was, only a little before, when a priestly vocation or entry to the religious life had been serious prospects for many London men and women. In their wills citizens remembered sisters who were nuns, cousins who were priests, or made provision for their children if they entered the cloister.[57] In most families there were somewhere members who had taken vows. In the diocese of London 840 men had offered themselves for ordination during the episcopate of Fitzjames, on the very eve of the Reformation. But when the Reformation came the number of candidates dramatically declined. Bishop Tunstall ordained 306 men (1522–30), Stokesley 97 (1530–9), and Bonner 46 (1540–7).[58] Certainly, no religious could be presented for ordination from their houses after the Dissolution, and neither were there now monastic titles to guarantee ordinands; certainly the examination of ordinands had been particularly stringent under Stokesley. But the real reasons for the decline lay far deeper.[59] No one of the old faith could be anxious to serve as priest in a Church so changed, in which further change was presaged. In a new world where the Word was, at least for some, more central than the sacraments, the status and function of the clergy had been profoundly challenged and transformed.

THE CLERGY

The Conservative

The reformers, as well as conservatives, despaired of London. 'To be a citizen of London is no more harm before God than it is to be a dweller of any other quarter of the world', wrote John

[56] PRO, PCC, Prob. 11/31, fo. 10ʳ.

[57] See, for example, ibid. 11/24, fo. 90ʳ; 11/27, fo. 259ᵛ; CLRO, Repertory 2, fo. 135ᵛ.

[58] Guildhall, MSS 9531/10, fos. 152ʳ–163ʳ; 9531/11, fos. 128ʳ–136ᵛ; 9535/1, fos. 14 ff.

[59] The problem is stated and analysed in convincing detail in M. Bowker, 'The Henrician Reformation and the Parish Clergy', *BIHR* l (1977), pp. 30–47.

Bale, 'but to be of the diocese is to be a member of Antichrist's kingdom.'[60] For in London the forces of reaction were united— unregenerate City curates, and their Bishop, in league with repressive Aldermen.[61] How could the people's faith change if their clergy, the spiritual guides, remained staunch in the old faith? For the Catholic clergy the challenge of the reformed faith to their monopoly of spiritual knowledge, its attack upon their instrumental role in the sacraments, the new demands for scriptural knowledge and preaching had been anathema. Worst of all was the heresy taught. Rather than hear heretics preach at Paul's Cross, a priest living in Eastcheap said that he would rather hang, and so he did, by his own hand.[62] After the passage of the Act of Six Articles the clergy were given the chance to win back the wavering and to attack the New Learning. If they were challenged they could adduce a double authority: the teachings of the Church and the royal command. Asked by Laurence Maxwell in 1540 by what authority priests could hear confession, John Morris, curate of St Mary Aldermanbury, replied: 'It is so determined by the King and his Council, therefore it is past argument of good Christian people.'[63]

While most London clergy were loyal to the politic orthodoxy of Henry VIII's last years, others were not. Two of Bonner's own chaplains—Covert and Feckenham—preached sedition.[64] The authorities could never be certain that the conservative clergy, while teaching Catholic doctrine, were not also conspiring to persuade their parishioners to remain loyal to the Pope and to the worst abuses of unreformed Catholicism. The reformers played constantly upon this uncertainty, alleging that the priest would 'secretly in confession or by some other crafty means, poison his flock with man's traditions and popish doctrine'.[65] There had been evidence enough of their doing so in the 1530s. Some doubtless did appeal to the superstitious elements of popular Catholicism, though surely few as desperately as the priest who feigned the miracle of transubstantiation by

---

[60] John Bale, *Yet a course at the romysh foxe* (Emden, 1543; *RSTC* 1309), fo. 15ᵛ.
[61] See Brinklow, 'Complaynt' and 'Lamentacyon'.
[62] *Narratives of the Reformation*, pp. 23–5.
[63] PRO, SP 1/243, fo. 74ʳ (*L&P Addenda*, i/2. 1463).
[64] *PPC* vii, p. 182; see above, p. 379.
[65] 'A Supplycacyon to our moste Soueraigne Lorde Kynge Henry the Eyght' (EETS, extra series, xiii, 1871), p. 46.

secretly pricking his finger at Mass and allowing his blood to flow upon the altar.[66] Where the convictions of the priest were shared by his parishioners there would be no resistance and no report; moreover, the citizens were again, as in the 1520s, fearful of malicious accusations by their clergy. So if superstition was practised 'under title of devotion, or else of commendable rites of Holy Church, the King shall never know of it. Let him command what he list, yet shall it be as they will have it.'[67] Even when there were moves in Henry's last years to alter ceremonies and to make services more available to the people by having them in English, the clergy could ignore the orders. Allegedly, the clergy refused to use the new primer in English and Latin, and on the King's return from Boulogne in 1544 did their best to have it recalled, causing 'all such parishes as they might command to use their old Kyrie Eleison again'.[68]

Certain clergy had particular and compelling reasons to cling to orthodox Catholic doctrine. After the Dissolution of the monasteries the 'great multitude of chantry priests, soul priests', still owed their livelihoods to singing for souls, and their survival depended absolutely upon the continuing acceptance of the doctrine of Purgatory. If Purgatory were no longer a place their livings were lost. Chantry priests were particularly despised by the reformers as upholders of superstitious piety, but they were also urgently needed by the citizens for the celebration of divine service. Without them there would be too few priests to minister Mass at Easter and on holy days. Where the incumbent of a parish was customarily absent, 'a strawberry priest', their presence was the more necessary. The citizens had recognized this need, requiring the priests who sang for their souls to attend divine service in the church in which they prayed.[69] Because of the 'great number of houseling people' in St Sepulchre and St Botolph Aldersgate it was recommended in 1546 that a chantry priest be retained as assistant in these churches, and the rector of St Michael Bassishaw petitioned

---

[66] Wriothesley, *Chronicle*, i, p. 152. Cf. C. W. Dugmore, *The Mass and the English Reformers* (1958), p. 77.

[67] Bale, *Yet a course*, sig. Av.

[68] 'A Supplication of the Poore Commons', (EETS, extra series, xiii, 1871), p. 69.

[69] See, for example, PRO, PCC, Prob. 11/20, fos. 24[r], 64[r], 150[v], 171[r], 183[v]; Prob. 11/28, fos. 95[v], 199[v], 257[r].

that a priest be retained there since the parish was so under-served.[70] Chantry priests were responsible for serving the cure in thirteen parishes in 1548.[71]

Many of the priests singing for souls in the City churches in the 1540s had once prayed for them in priories and abbeys, for many religious who were dispossessed at the Dissolution stayed in London or, as priests had always done, came there seeking preferment. Though the citizens had defended their religious houses only feebly many still venerated the monastic ideal, and perhaps the presence of the erstwhile religious strengthened their conservatism. (Though, of course, there were notable apostates among the religious.) In February 1541 Richard Billingsley of St Sepulchre bequeathed his best jacket to Thurstan Hickman, a former Carthusian; £7 to Mr Trafford to sing for his soul; and 5s. each to twenty-three London Carthusians who still lived.[72] Between 1536 and 1546 eighty-two religious from the City houses were given dispensations to hold benefices.[73] At least eleven of the City's beneficed clergy in the 1540s had once been in religious orders.[74] Of the 317 London chantry priests recorded in the 1546 survey twenty-seven were found to be pensioned; and there were probably more ex-religious.[75] Dr Egleby, quondam Prior of the Black Friars of Derby, was admitted by the Goldsmiths' Company to Lichfield's chantry in St John Zachary in January 1540, and Richard Tregoose, a former London Carthusian, stayed in the City until his death in 1546.[76] Two priests at St Dunstan in the East came from Reading Abbey; two other London chantry priests came from St Helen Bishopsgate and St Mary Spital, and another from St Neot's.[77]

---

[70] PRO, E 301/88, m. 1, 4d; E 301/89, fo. 3ᵛ.

[71] PRO, E 301/34, 2, 5, 34, 35, 53, 68, 69, 75, 82, 95, 107, 115.

[72] Guildhall, MS 9171/11, fo. 55ʳ.

[73] *Faculty Office Registers, 1534–1549*, ed. D. S. Chambers (1966).

[74] William Atherwolde, John Senock or Denman, Richard Benese, Thomas Paynell, John Foxe, William Jennings, John Dey, John Joseph, Edward Kirkby, Thomas Brigotte, Thomas Munday. See S. E. Brigden, 'The Early Reformation in London, 1521–1547' (University of Cambridge, Ph.D. thesis, 1979), p. 318; E. L. C. Mullins, 'The Effects of the Marian and Elizabethan Religious Settlements on the Clergy of London' (University of London, M.A. thesis, 1948), appendix.

[75] PRO, E 301/88.

[76] Goldsmiths' Company, Court Book F, fo. 129ʳ; Guildhall, MS 9531/12, fo. 205ᵛ.

[77] PRO, E 301/88.

Some of the most 'papistical' priests in London were erstwhile religious. No surprise. Thurstan Hickman and William Peryn, chantry priests at St Paul's, were both ex-monastics, still fiercely opposed to royal supremacy. Peryn had gone into exile in the 1530s, came home with the reaction, and returned again in 1547. Hickman, too, left for exile on Edward's accession.[78] To St Leonard Foster Lane came Thomas Munday, like Hickman a survivor from the London Charterhouse. But it seems that his parishioners did not share his resolute conservatism. When Munday seized burial tapers as his due, the parish sued a writ of trespass against him in the Sheriff's court, and Munday retaliated by bringing a case in Chancery. At 'their great cost and charges' the parishioners found another priest to serve the cure, a man less resistant to change certainly, but perhaps also a reformer.[79] In July 1543 Munday brought this curate, Miles Northby, before the Consistory Court and Northby was forbidden to serve.[80] The 'godly men' of the parish of Stepney 'wearied' too of the 'strong, stout, popish prelate' who came to serve there. He was Henry More, quondam Abbot of St Mary Graces by the Tower. So hating the reforms, he determined to silence the Gospel preachers who came to his parish by singing in the choir during their sermons, or ringing the bells, and challenging them in the pulpit.[81]

But most of the City clergy had silently acquiesced to the new ways, while probably hating them even more than their parishioners did, 'wishing that all things was now as it was twenty years since'. To these clergy was given the task, after the reaction, of reviving confidence in the old ways of worship among Londoners of wavering conscience and even of winning back the converted. A study of the wills of the clergy themselves reveals something of their attitudes. Of forty-five clergy whose wills were written in the last years of Henry's reign, and recorded in Bonner's register, thirty-six expressed confidence in the intercessory powers of the Virgin and saints. None expressed a reformist credo. Eighteen left books, mostly the tools

---

[78] BRUO (Peryn); Stow, Annales, p. 594; Guildhall, 298/2.4; Merchant Taylors' accounts, 1545–6; Wriothesley, Chronicle, i, pp. 184–5.

[79] Hennessy, Repertorium, p. 127; PRO, C 1/1029/41–2.

[80] GLRO, DL/C/3, fos. 249ʳ–250ʳ.

[81] Narratives of the Reformation, p. 157. More had followed a Protestant martyr, William Jerome, to that cure.

of their priestly craft: Mass books, breviaries, hymnals, antiphoners. Two bequeathed copies of the conservative *King's Book*
of 1543.[82] Three left Bibles: 'the whole Bible in small volume',
a Latin New Testament. Others owned Fabian's *Chronicle*,
the *Golden Legend*, the *Destructionum Viciorum* of Alexander
Carpenter. To Robert Marten were left copies of Thomas
More's works.[83] There is little enough here to suggest reforming influence. Though most of the City clergy probably longed
for religious unity and a return to obedience, the obstacles to
achieving it were formidable.

Bonner sought to establish a Catholic preaching clergy in
order to refute the errors of the reformers and to assert the
authority of the Catholic Church and its teachings. Leading
Catholic clergy, like Richard Smith and William Clyff, were
granted episcopal licences to preach.[84] The Bishop's concern
for the standard and tenor of preaching in the capital appears
from his Injunctions of 1542 which stressed the clergy's duty to
preach.[85] Now they were ordered not to rehearse sermons
made within the last two or three hundred years—but to take
the Gospel of the day and rehearse it according to the interpretation of some Catholic Father. The new and forbidden
doctrines must never be mentioned. Instead, their sermons
must calmly and rationally reassert the righteousness of such
disputed tenets as the efficacy of good works, prayers, and ceremonies. They were to explain the significance of the sacraments, especially why the Mass was to be held so holy. Bale
was unjust when he wrote that Bonner 'thinketh three times in
the year . . . is enough for preaching, and therein only . . .
concerning ceremonies and tithes'.[86] The fervour with which
the reformers mocked the 'wise false prophets', like Standish,
Peryn, and Smith, who preached the miracle of transubstantiation suggests that their sermons found eager audiences.

The first concern of the Catholic clergy was to confirm faith
in those doctrines which the reformers rejected, alleging that

---

[82] 'A Supplication of the Poore Commons', p. 63; Guildhall, MS 9531/12,
fos. 179^v ff.
[83] Ibid., fos. 192^v, 199^r, 200^r; MS 9171/11, fo. 66^v; PRO, PCC, Prob. 11/28,
fos. 162^v, 243^r, 265^v.
[84] Guildhall, MS 9531/12, fos. 13^v, 18^r.
[85] Ibid., fo. 38^v; *L&P* xvii. 282.
[86] Bale, *Yet a course*, sig. H.

they had no foundation in Scripture. So, in the spring of 1540 Gardiner had disputed Barnes's teaching of *sola fides*, and in the autumn Wilson challenged Crome's denial of the efficacy of good works and prayers for the dead. Smith preached in November 1541 of man's reconciliation to God, urging the merits of good works for salvation.[87] The greatest contention surrounded the sacrament of the altar. Crome's preaching upon the Mass had so deeply disturbed the City in 1546 that the five principal sermons that Easter reasserted the Catholic doctrine.[88] Peryn preached, perhaps in that Lent of 1546, in defence of transubstantiation. It was a miracle, revealed through faith, not reason:

This miracle in this sacrament is not wrought . . . to persuade faith . . . For in this sacrament natural experience contendeth openly against faith, and not only reason, but also all our senses are led captive, against all natural experience, into the sole and only Word of God.[89]

Though the Catholic sermons might seem now, in cold print, to lack the charisma and energy of the sermons of the evangelicals, there were probably many who welcomed that restraint. Certainly crowds had flocked to hear the reformers in the 1530s, but surely very many had stayed away in disgust. Henry Gold had claimed that 'people would rather go without their matins and Mass on Sunday than have to hear the Word of God preached'.[90] The silencing of the gospellers brought a welcome calm to the City, for the while. It did not stop their campaign to convert.

The Reformed

After the preaching campaign of the early Reformation was halted a second wave of reform began as the clergy, with cure of the citizens' souls, used their particular influence to evangelize. Wherever a reformer held a cure there were conversions in his parish. Patrons of benefices knew this, and wherever they

[87] See above, pp. 309–10, 330, 335.
[88] *Grey Friars Chronicle*, p. 50.
[89] *Thre godlye and notable sermons, of the sacrament of thaulter* (1546; *RSTC* 19785.5–6), fos. 6ᵛ, 9ᵛ; cited in J. W. Blench, *Preaching in England in the late fifteenth and sixteenth centuries* (Oxford, 1964), p. 349.
[90] *L&P* xii/2. 530; xiii/1. 715; vii. 523(4).

could they appointed clergy who shared their own convictions, conservative or reformed. So the conservative Sir Richard Southwell appointed the quondam Carthusian, Munday, to St Leonard Foster Lane. Where Bonner had powers of appointment he instituted leading conservatives: Hugh Weston to St Botolph Bishopsgate and John Standish to St Andrew Undershaft. Yet sometimes they made mistakes. Sir Richard Rich, surely unknowingly, placed the radical John Hardyman in St Martin Ironmonger Lane. On the other hand, Cranmer used his patronage in the City to further reform: to the Canterbury peculiars he instituted known reformers: John Ponet to St Michael Crooked Lane, John Joseph to St Mary le Bow, and Richard Marsh to St Pancras.[91]

It is possible to trace the activities of a band of more than fifty reforming clergy in London in Henry's last years. Some were summoned before the Council, the Court of Aldermen, the commissions established under the Act of Six Articles, or the Consistory Court, for religious offences.[92] The influence of others, sometimes more quiescent, is found in the wills of citizens.[93] Whenever a reforming coterie is found within a parish there was also an evangelical curate there. Clergy were marrying. As early as 1525 Rowland Taylor had married *non in facie ecclesie*, and others followed him. Friar Ward had married in 1538.[94] In February 1546 the marriage of Richard Marsh and Anne Norton, widow, was entered in the register of his parish, St Pancras, and in the following year the marriages of his fellow priests, Adrian Arnold and Thomas Kirkham appear.[95] Whenever a priest married he declared implicitly his

[91] Hennessy, *Repertorium*, pp. 93, 111, 127, 277, 284, 308, 311.

[92] Sixteen London clerics are named by Foxe as indicted during the troubles of 1540 (though some were, in fact, indicted later): *Acts and Monuments*, v, pp. 443–51. The Consistory Court Act Books record processes *ex officio* against eighteen London clergy for breaches of ecclesiastical discipline: GLRO, DL/C/3 & 4. No cause papers survive to elucidate the brief entries.

[93] For example, William Erith, curate of St Mary at Hill: *LCCW* 193; PRO, PCC, Prob. 11/31, fo. 155ᵛ; Guildhall, MS 9171/11, fos. 162ʳ, 171ʳ; William Tolwin, rector of St Antholin; ibid., MSS 9171/10, fo. 322ʳ; 9171/11, fo. 58ᵛ; Philip Bale, curate of St Michael Queenhithe; PRO, PCC, Prob. 11/27, fos. 31ᵛ, 193ᵛ; Guildhall, MS 9171/11, fo. 174ʳ; Richard Bostock at St Botolph Aldersgate; ibid., fos. 3ʳ, 14ʳ.

[94] See above, p. 111; Wriothesley, *Chronicle*, i, p. 83.

[95] *HSR* xliv–v (St Pancras, Soper Lane), pp. 440–1.

allegiance to reform. Every conversion made among the clergy
themselves greatly strengthened the cause. Even the redoubtably orthodox could be won over. John Huntingdon, ordained
by Stokesley in 1538, had reported Seton for heresy in 1541,
and wrote a tract condemning the new faith, *A Genealogy of
Heresy*. But between the writing of Bale's reply to it, *A mysterye of
inyquyte*, and its publication in 1545, Huntingdon's views had
changed, and he adopted reform with all the zeal of a convert.[96]

As the liberty to read Scripture and preach reform was
removed, some counselled their parishioners to conform: 'the
commandment of the King must appear in your outer manners'.[97] But they were less circumspect themselves. John
Hardyman openly declared in February 1540 that 'confession
is confusion and deformation', that it was 'mischief . . . to
esteem the sacraments to be of such virtue', for in so doing
'they take the glory of God from Him', and that faith in Christ
was sufficient to justify without any other sacrament.[98] His
curate, John Coyte, could not bear to listen to the confessions
of his parishioners, he said, especially any 'standing to bawdery
or filthiness'. But his scruples, like Hardyman's, were surely
also reforming ones. Coyte refused to go in general
procession.[99] At St Vedast Thomas Bodnam allowed a parishioner to be houselled at Easter 1539 'without confession, but
only to God'.[100] In March 1540 Richard Bostock had preached
in St Botolph Aldersgate against auricular confession, alleging
that in Henry's reign it had 'killed more souls than all the bills,
clubs and halters had done'. 'Thou may not come to me to spit
thy venom in my ear, for if thou do I will spit it in thy bosom
again.' In Kent in Easter week 1546 Bostock asserted that there
was 'heresy' in hallowing holy bread and water.[101] John
Willock came from Scotland to serve at St Katherine Coleman.
He had gravely disturbed the 'Christian flock' by denouncing
holy water, confession, prayers to saints, and by denying the

96 Guildhall, MS 9531/11, fo. 136ᵛ; *A mysterye of inyquyte contayned within the heretycall
genealogye of P. Pantolabus* (Geneva, 1554; *RSTC* 1303), sig. A2; Foxe, *Acts and
Monuments*, v, pp. 449, 539.
97 PRO, SP 1/243, fo. 61ʳ (*L&P Addenda*, i/2. 1463).
98 Ibid.
99 GLRO, DL/C/3, fos. 291ᵛ, 293ʳ.
100 CLRO, Repertory 10, fo. 90ᵛ.
101 PRO, SP 1/243, fo. 80ʳ (*L&P Addenda*, i/2. 1463); *APC* i, pp. 418, 421, 492.

efficacy of prayers for souls, saying that there was no Purgatory. In 1546 he was preaching in the City still, aligned with a group of radicals, and proselytizing to the converted: 'he hath framed his sayings after his audience'.[102]

The spirit of the Christian Brethren lived on in London during the persecution, and perhaps part of its old organization. Thomas Lancaster, a priest of Calais, was imprisoned sometime before July 1540, accused of smuggling heretical books. His confederate was the radical publisher, John Gough. The work may have been Lancaster's own *The ryght and trew understandynge of the supper of the Lorde*.[103] George Parker had disseminated sacramentarian tracts in the 1520s: in 1540, as curate of All Hallows the Less, he was discovered with copies of the Lutheran *Unio Dissidentium*.[104] Another of the founding Christian Brethren may have kept the cause alive in London. Incriminating letters to the Council were discovered in the house of Marshall, a former priest now married, who had spread books against the King in Danzig.[105]

Radical doctrines concerning the Mass were being taught. John Cardmaker, vicar of St Bride Fleet Street, asserted in 1540 that it was 'as profitable to a man to hear Mass as to kiss Judas's mouth who kissed Christ our Saviour, &c'. For Cardmarker, the Mass was idolatry and profanation, and in Edward's reign he would preach the crudest denials of Christ's presence in the sacrament. Yet in 1546 he was found to be 'not so bold as the rest' of the heretical circle in which he moved.[106] Thomas Cappes, chaplain in St Mary Magdalen Old Fish Street, asserted that the sacrament of the altar 'is but a memory and a remembrance of the Lord's death'. Two years later he was still

---

[102] Guildhall, MS 9531/12, fo. 21^v; Foxe, *Acts and Monuments*, v, pp. 446, 448. Summoned before the Vicar-General on 23 Oct. 1540, he failed to appear: GLRO, DL/C/3, fo. 10^v. *State Papers*, i/2, p. 850.

[103] CLRO, Journal 14, fo. 210^r; Foxe, *Acts and Monuments*, v, pp. 448, 498; *RSTC* 15188. The tract was printed by Whitchurch for John Gough. The editors of *RSTC* give the tract's date as 1550, but this is too late, for Gough was dead by then.

[104] Foxe, *Acts and Monuments*, v, pp. 448–9. He was summoned to appear before the Vicar-General in Nov. 1541, and assigned to read a commentary upon St Matthew. He failed to show his letters of ordination: GLRO, DL/C/3, fos. 93^v, 95^r.

[105] *APC* i, p. 419.

[106] Foxe, *Acts and Monuments*, v, p. 448; *State Papers*, i/2, p. 850. In 1542 Cardmaker was informed against for holding lands to farm in Devon: PRO, E 159/321, m. 20. For Cardmaker's later career, see below, pp. 435, 451, 498, 573, 608.

celebrating, but now in St Michael Wood Street.[107] Stevens, another reformer, had followed Willock to St Katherine Coleman. He was accused of 'seducing' Cage, a servant of the Duke of Norfolk, 'touching . . . the most blessed sacrament'. Though Stevens protested that his parish would affirm his innocence, he was soon inciting them to iconoclasm.[108] Of the City clergy who came before the Council in 1546 for holding the sacramentarian heresy—Edward Crome, John Olde, chaplain to Lord Ferrers, Cardmaker, Stevens, Robert Crome, rector of St Leonard Foster Lane, and Willock—none suffered, then.[109]

The ceremonies which each church conducted, and the forms of worship followed, were determined by the incumbent, and even under persecution some City clergy still defied the authorities. William Tolwin of St Antholin and Dr Crome of St Mary Aldermary had abandoned orthodox ceremonies by 1540, especially the hallowing of holy water.[110] In 1539 Crome's fellow priest at Aldermary complained that Relic Sunday was celebrated with special services everywhere in London, save there.[111] William Bull, curate of Colchurch, ministered the sacraments in English at Easter 1542.[112] At All Hallows Honey Lane William Reed, the curate, had sung the *Te Deum* in English, and on the festival of the Purification 1543 neglected to sing lauds. By May he was in the Tower for his part in disseminating a forbidden work.[113] Reforming curates were abandoning their popish vestments. Edmund Frevell, curate of St Antholin, John Tovye, and John Smith, a celebrant in St Botolph Aldgate, were all summoned by the Vicar-General to appear in clerical dress.[114] John Birch, a priest of St George Botolph Lane, was noted by Foxe as a 'busy reasoner in certain opinions not agreeing with the Pope's Church'. Summoned before the Vicar-General in November 1541, he was assigned

[107] PRO, SP 1/243, fo. 62^r (*L&P Addenda*, i/2. 1463); GLRO, DL/C/3, fos. 175^v, 176^v, 186^r, 205^r, 213^v. He said that he had been ordained in Lincoln diocese.

[108] *APC* i, p. 394; see below, pp. 430–1.

[109] *APC* i, pp. 394, 414, 418–19, 440, 466, 479.

[110] Foxe, *Acts and Monuments*, v, pp. 446, 448.

[111] PRO, SP 1/153, fo. 27^r (*L&P* xiv/2. 41).

[112] GLRO, DL/C/3. fos. 142^{r-v}, 148^r. *PPC* vii, p. 285.

[113] GLRO, DL/C/3, fos. 220^r, 221^v. Perhaps he was the William Rede who had been preaching heterodoxy in Lancashire earlier: C. A. Haigh, *Reformation and Resistance in Tudor Lancashire* (Cambridge, 1974), pp. 42, 43, 82–4, 99.

[114] GLRO, DL/C/3, fos. 137^r, 233^r, 235^r, 259^r.

to read chapters from Scripture with a Catholic gloss.[115] Adrian Arnold, curate at St Clement Eastcheap, had preached against royal injunctions. By the summer of 1543 he had moved to the reforming parish of All Hallows Honey Lane and was in trouble again.[116] Robert Feron, curate at St Bartholomew the Little, was cited for some offence unknown, as were Thomas South, parish priest of All Hallows Lombard Street, and a priest named Some.[117] The reforming influence of such curates is to be found in their parishioners' wills. In their own wills too the clergy expressed their convictions. William Caterick, BD, celebrant in St Alban Wood Street, was called before the Vicar-General in March 1544. That he had been evangelizing seems clear from the creed revealed in his will of September 1548:

the Father, the Son and Holy Ghost redeemed me from Hell and sin by the second person Jesus Christ suffering His most painful passion for the redemption of my sins and rising again for to make me righteous before God . . . I appeal to Him, here utterly forsaking mine own merits and righteousness.[118]

In the early Reformation very few parishioners, even those moving toward reform, were yet used to trusting their individual conscience or to finding faith only in Scripture; rather they were accustomed to authority and looked, as always before, to their parish clergy for guidance. Yet in the London parishes the messages might be diverse as reforming curates contended with reactionary rectors. So it was in St Leonard Foster Lane and St Botolph Aldgate. At St Sepulchre the curate was determined to resist any innovation, but the vicar had been instituted by Thomas Cromwell, and was followed by John Rogers in 1550. Thomas Artes, curate at Aldermary, clearly disapproved of Crome, his rector. At St Mary at Hill the time-serving Alan Percy allowed first one priest, William Wegen, and then another, William Erith, to spread reform.[119] The radical Cardmaker was assisted at St Bride Fleet Street by

[115] Foxe, *Acts and Monuments*, v, pp. 448–9; GLRO, DL/C/3, fos. 93ᵛ, 100ᵛ, 102ʳ, 104ʳ.

[116] Ibid., fos. 211ʳ, 235ʳ, 237ʳ, 239ʳ.

[117] Foxe, *Acts and Monuments*, v, pp. 448, 451; PRO, PCC, Prob. 11/27, fos. 95ʳ, 226ᵛ.

[118] GLRO, DL/C/3, fos. 294ᵛ, 297ʳ; Guildhall, MS 9531/12, fo. 215ᵛ.

[119] Foxe, *Acts and Monuments*, v, p. 28; Hennessy, *Repertorium* , p. 383; see above, pp. 384, 396, 402 and below, pp. 431–2.

a group of chantry priests who, like many of his parishioners, were surely appalled by their vicar's views. But once a preacher like Cardmaker arrived, conversions—and trouble—followed. In January 1544 Hugh Eton paraded through St Bride 'in fond fashion disguised' during the service 'before the most blessed sacrament', disturbing the priest at Mass and 'bringing all the people there in a great tumult'. His penance was to sit in the cage in Fleet Street in his disguise. Christopher Dray, a fellow parishioner, stood surety for him.[120] Dray too was won to reform. In July 1539 he had declared that the 'sacrament of the altar was not offered up for sins, and that the body of God was not there, but was a representation and signification of the thing'. He denied the whole sacramental system of the Church.[121] In 1545 Cardmaker and Dray were witnesses to the will of another reformer, William Arden.[122] The conclusion that Cardmaker had won them to reform seems inescapable.

THE PASSAGE OF REFORM

The sacred mystery of the Mass was being 'jangled' in the ale-houses of London. A group of neighbours were out drinking in the Bell tavern, Aldgate at the end of May 1539 when one said, 'Masters, let us make our reckoning that we may go to church and hear our High Mass.' 'Tarry', said Giles Harrison, and taking bread he lifted it over his head, and bowing over a cup of wine made the sign of the cross, and raising the cup, asked, 'Have ye not heard Mass now?' He was no radical from the ranks of the dispossessed, but the King's ale brewer and one of the richest men in the City.[123] When such as he dared to 'contemn, deprave or despise the blessed sacrament' there was more reason to be fearful of the spread of reform. The first quest under the Act of Six Articles discovered sixty who rejected Catholic teaching upon the Mass, but the heresies were diverse.[124] Thomas Plummer of St Matthew Friday Street allowed still that faith might overcome reason and the senses in

[120] *LMCC* 107; CLRO, Repertory 11, fo. 35ᵛ; Letter Book Q, fo. 102ʳ.

[121] PRO, SP 1/243, fo. 75ʳ (*L&P Addenda*, i/2. 1463).

[122] PRO, PCC, Prob. 11/31, fo. 50ᵛ.

[123] Ibid. SP 1/243, fo. 72ʳ (*L&P Addenda*, i/2. 1463); E 179/144/123(50); CLRO, Repertory 12, fo. 123ʳ; *Diary of Machyn*, p. 10.

[124] Foxe, *Acts and Monuments*, v, pp. 443–51.

this matter, for some: 'the blessed sacrament was to him that doth take it, so; and to him that doth not it was not so'.[125] But John Athee's view of transubstantiation was that 'God might . . . if he would, turn it into a chicken's leg'.[126]

Varying views upon the nature of the sacrament of the altar were spreading, their provenance difficult to discern. When Richard Mekins was indicted in 1541 there was confusion over the nature of his heresy. The witnesses disagreed: 'one affirmeth that he should say that the sacrament was nothing but a ceremony: and the other nothing but a signification'. But Mekins had been converted by Barnes and it is unlikely that he had gone so far. Hilles said that Mekins was of the 'Lutheran opinion', and 'did not altogether deny a corporal presence, but asserted as our Wycliffe did, that the accident of bread did not remain there without the substance'.[127] Thomas Trentham, a pinner of St Giles Cripplegate, declared the Mass to be 'a good thing', but 'not as men took it, very God'.[128] Theological sophistication may have been beyond some of those discovered in 1540, but there is no doubting the subtlety of some alehouse arguments or that laymen felt free to discuss high matters of doctrine.[129]

The indictments of 1540 reveal that some had gone so far as to hold a commemorative view of the sacrament, denying even a spiritual presence. Lollards and reformers alike might exclude the possibility that Christ could be in the sacrament, believing that after the Ascension He is in Heaven. William Wyders of Holy Trinity the Less denied the sacrament to be Christ's body, saying that it was only a sign.[130] For Richard White, haberdasher of St Olave Jewry, Christ was not in the sacrament, but 'in Heaven above'.[131] John Mayler, a grocer and printer of St Botolph Aldersgate, called the eucharist the 'baken god' of

---

[125] Ibid. v, p. 446.

[126] Guildhall, MS 9531/12, fo. 254ᵛ; Foxe, *Acts and Monuments*, v, p. 528; appendix xv.

[127] Ibid. v, pp. 441–2; *Original Letters*, i, p. 221.

[128] Foxe, *Acts and Monuments*, v, p. 445.

[129] Cf. M. Bowker, *The Henrician Reformation: the diocese of Lincoln under Bishop Longland, 1521–1547* (Cambridge, 1981), pp. 166–7.

[130] Foxe, *Acts and Monuments*, v, p. 443. His will expressed solifidian beliefs: Guildhall, MS 9171/11, fo. 131ʳ.

[131] PRO, SP 1/243, fo. 69ʳ (*L&P Addenda*, i/2. 1463); Foxe, *Acts and Monuments*, v, p. 447.

bread, and said that 'there is more abomination in the Mass than in all other things that is spoken of yet'.[132] For four parishioners of St Giles Cripplegate the Mass was a thing of 'pieces and patches'.[133] It was not to be worshipped. At St Martin Outwich a group of parishioners ostentatiously walked around the church at sacring time, leaving their caps on, averting their gaze from the elevated host, and in St Thomas Apostle thirteen were seen to give 'small reverence at the sacring of the Mass'.[134] Some showed their contempt more overtly. At St Giles Henry Patinson and Anthony Barber urged their 'boys to sing a song against the sacrament', and Thomas Grangier and John Dictier were 'common singers against the sacraments and ceremonies'.[135] Most emphatic of all was Henry Bird of All Hallows the Great who threatened a defender of the Mass: 'Dost thou call it thy maker? Call it no more so, for if thou do I will kill thee, for it is no better than a piece of bread.'[136]

'He would not believe in that thing that the knave priest made', said John Athee of Middlesex in 1543, 'but in God that is in Heaven.'[137] Disavowal of the Mass was often clearly associated with a rejection of priestly authority, with the heresy that the unworthiness of the priest vitiated the sacrament. Increasingly, the clergy was being mocked. Seeing a priest prepare for Mass, William Clinch of St John Walbrook said, 'Ye shall see a priest now go to masking.' In the Carpenters' Hall an interlude was played in which 'priests were railed on and called knaves'. An apprentice in St Mildred Bread Street said that he would prefer to hear dogs howling than a priest singing matins and evensong.[138] Consequent upon the rejection of priestly authority was a widespread refusal to confess. In 1540 29 Londoners were charged with offending against the sixth of the Six Articles which insisted upon auricular confession. Laurence Maxwell denied that the power of absolution was given to any but the Apostles; confession was, then, 'an invention of the Bishop of

---

[132] Mayler was the printer of Becon's works. PRO, SP 1/243, fo. 79ʳ (L&P Addenda, i/2. 1463).

[133] Ibid., fo. 76ʳ.

[134] Foxe, Acts and Monuments, v, pp. 444–5.

[135] Ibid., p. 445.

[136] PRO, SP 1/243, fo. 71ʳ (L&P Addenda, i/2. 1463).

[137] Guildhall, MS 9531/12, fo. 254ᵛ.

[138] Foxe, Acts and Monuments, v, pp. 443, 444, 446.

Rome'. John Sutton and his wife in St Giles asserted that 'no knave priest should know their minds, but only God that forgiveth all sins', and Robert Norman refused to have himself or any of his servants 'shriven of a knave priest'.[139]

'A sermon preached is better than the sacrament of the altar', so these reformers thought. In their houses they harboured the outlawed preachers—Barnes, Garrett, Jerome, Willock, Ward, Rose, Wisdom, Smith—and held 'disputation of heresy'. These conventicles were found in the parishes of St Magnus, Aldermanbury, St Michael Wood Street, St Matthew Friday Street, and elsewhere.[140] Where learned clerical evangelists attended conventicles the purity of reformed doctrine was likely to have been maintained. But when the reforming movement went underground the consequent fragmentation of the faithful led to an even greater variety of opinion, deriving from individual and esoteric interpretation of Scripture and the heretical works which were circulating. Eclectic and speculative Lollardy was open to, and influenced by, more radical ideas from the Continent. Robert Plat and his wife of St Benet Fink were 'great reasoners in Scripture'. Yet they could not read: 'they had it of the Spirit'.[141]

John Sampye and John Goffe (Gough) of St Mary at Hill scorned an anthem of Our Lady, saying that there was heresy in it.[142] The Reformation had brought a revolution in what was considered heresy and what was not. The quest of 1540 discovered heresies of various provenance. Among those indicted were members of old Lollard circles in the City, some of whom may have turned to reform. Such were William Ettis of St Matthew Friday Street, Laurence Maxwell, and, almost certainly, John Sempe. Sempe had been in trouble before for his refusal to swear oaths in 1521 or to attend processions in St Mary at Hill in 1529. But by 1543 his views had altered. In his will he commended his soul to Christ 'by merits of whose blessed passion is all my whole trust'. Though Sempe was unorthodox, he was no Lollard sacramentary, for he bequeathed a taper to

---

[139] PRO, SP 1/243, fos. 74ʳ, 77ʳ, 67ʳ, 70ʳ (*L&P Addenda*, i/2. 1463); Foxe, *Acts and Monuments*, v, p. 445.

[140] Foxe, *Acts and Monuments*, v, pp. 448, 444, 445, 446; *Original Letters*, i, pp. 210, 232–3.

[141] Foxe, *Acts and Monuments*, v, p. 446.

[142] Ibid. v, p. 447.

burn 'at Mass times . . . in the honour of the blessed sacrament'.[143] Within months Gough also made his will. From the
beginning he had been one of London's most zealous reformers:
translator of *The ymage of loue* in 1525, publisher of Colet's
English paternoster, author of the *Myrrour or lokynge glasse of lyfe*,
iconoclast. His will expressed his 'faithful hope', found in
Scripture, that despite a 'life in the corruptible flesh of my fore-
father Adam', being penitent, like Job, and committing his soul
'only to His glorious custody in a strong faith and hope and in
none other', he would have the 'reward of Christ's redemption
at my latter coming'. Gough had long laboured in the 'vineyard of the Lord', and still relied upon his old evangelical
friends. John Tyndale was witness to his will.[144]

Networks of friendship, old association, and family ties
bound the first reformers. But their conversion may have been
determined by other, deeper causes than such personal or acci-
dental bonds as neighbourhood or common upbringing. Sometimes the reformers themselves suggested that the divisions
between adherents of the old faith and the new mirrored other
divisions in society; that the liberating evangelical doctrines
brought hope to the poor, thereby alienating their oppressors.
The cause of Protestantism and the cause of the poor were soon
associated; at least in the propaganda of the rival partisans.
The claim was that the wealthy citizens of London tried to keep
the Word, the truth, hidden from their poor dependants, and
that they were responsible both for religious persecution and
social oppression. Henry Brinklow accused the 'rich of the City
of London' of being 'fully bent with the false prophets . . . the
Bishops . . . to persecute and put to death all and every godly
person'.[145] 'Died not Christ as well for craftsmen and poor
men as for gentlemen and rich men?' the gospellers asked.
Why then was the Word taken away from the people?[146]
Contention over the freedom to read the Bible in English
exposed, so they said, a rift between the social orders: 'the
great part of these inordinate, rich, stiff-necked citizens will not
have in their houses that lively Word of our souls, nor suffer

---

[143] BL, Harleian MS 425, fo. 13^r, (*L&P* iv/2. 4029); Guildhall, MS 9171/11, fo. 101^{r-v}. His will was witnessed by his reformed curate, William Erith.

[144] Guildhall, MS 9171/11, fos. 132^v–133^r.

[145] Brinklow, 'Lamentacyon', p. 80.   [146] See above, pp. 346–8.

their servants to have it'. And it was true that in some house-holds the Gospel was outlawed even before statute ordained that women, and the poor, were too socially irresponsible to read it for themselves.[147] Historians, too, have asked of earlier heretical movements, and of the Reformation, whether doctrinal differences, often of great theological subtlety, could really have engendered such powerful and enduring movements, and whether the real and underlying causes should be sought elsewhere.[148] The Reformation was not only a religious revolution: it was also a social one, perhaps created by one. The religious transformation came at a time of profound social and economic instability, and in such a crisis the poor must have suffered most acutely. Did they, in their distress, find a special salvation in the new faith? Was reform the province of the oppressed?

Conversely, the new faith with its doctrine of election might have a particular attraction for those who, through their industry and consequent worldly success, could already believe themselves to be marked by God's special favour. Economic achievement might come to be identified with spiritual salvation. The Protestant Reformation has been seen as an episode in the bourgeois revolution which reorganized society for the benefit of commerce and capital formation. These theories, propounded by one great sociologist, Weber, and by the followers of another, Marx, have had a profound influence upon historians of the English Reformation and English 'Revolution'.[149] Both the doctrine of providence and the coming of capitalism properly belong to a later generation of reform, but maybe the theories may be pertinently applied to London in its first years of Reformation.

---

[147] Brinklow, 'Lamentacyon', p. 79. See above, p. 347; *L&P* xvi. 101.

[148] See, for example, A. H. M. Jones, 'Were Ancient Heresies national or social movements in disguise?', *Journal of Theological Studies*, new series, x (1959), pp. 280–98; H. Hauser, 'La Réforme et les classes populaires', *Revue d'histoire moderne et contemporaine*, 1 (1899–1900); N. Z. Davis, 'Strikes and Salvation in Lyon', in *Society and Culture in Early Modern France* (1975); P. Benedict, *Rouen during the Wars of Religion* (Cambridge, 1981).

[149] M. Weber, *The Sociology of Religion*, trans. M. Fischoff (1965); R. H. Tawney, *Religion and the Rise of Capitalism* (1926); Christopher Hill, *The Century of Revolution, 1603–1714* (Edinburgh, 1961). For a remarkable investigation of a relation between wealth and conscience in another society, see Simon Schama, *The Embarrassment of Riches: An Interpretation of Dutch Culture in the Golden Age* (1987).

Was it true that religious choices were made, consciously or not, for reasons of social and economic interest as much as through doctrinal conviction? If hopes of treasure in this world as well as in Heaven helped to win allegiance to the new faith it was not because any such promises were given. The workings of God are mysterious, and it was not for men to speculate upon the divine plan.[150] It was the purpose of the preachers to tell of Christ's purchase of men's salvation, rather than to forecast worldy fortune. Yet the evangelical preachers, using Gospel parables to address contemporary ills in society, did call upon the rich with a new urgency to succour the poor, and did warn with special vehemence of the sin of covetousness.[151] And their audiences may, contrary to the preachers' intent, have interpreted the message as socially egalitarian.

The Lollards were the first to renounce the sacraments and ritual of the Catholic Church. What alternative consolations did they find among their bands of 'known men'? It has been suggested that the 'spirit of sturdy self help' found in their sect reflected that of their practical occupations as artisans: textile workers or carpenters.[152] Perhaps so, but in London on the eve of the Reformation the composition of that sect had been changing. Some of the wealthiest citizens were touched by Lollardy, and their association with that sect brought them into contact with London's dispossessed; even with water carriers and paupers. (Not that all of them would 'have to do with poor men'.)[153] Did the new faith, like Lollardy, command a wide social following? A detailed study of the Londoners indicted for heresy during the quest of 1540 might discover whether there was an association between social status and faith. Of the 190 laymen and women charged then, 57 may be traced in the lay subsidy assessments for London of 1536 and 1544. For the rest of the 190, they may have been too young or too poor to be counted; anyway, assessors only ever looked to find the subsidy demanded, not to search out every citizen liable to pay.[154]

---

[150] Keith Thomas, *Religion and the Decline of Magic: Studies in Popular Beliefs in Sixteenth- and Seventeenth-Century England* (1971), pp. 88–9.

[151] See above, pp. 269, 318–20, 361 and below, pp. 472–5.

[152] Thomas, *Religion and the Decline of Magic*, p. 663.

[153] See above, pp. 96–8.

[154] PRO, E 179/144/92–3, 96–114, 122–3. S. Rappaport, 'Social Structure and Mobility in sixteenth-century London', II, *London Journal*, 10 (1984), p. 108.

Although this sample is small, and although the quests discovered some of the heretical community rather than others, the conclusion to be drawn from it is clear. Among those indicted were some of London's wealthier citizens, members of the Inns of Court and of the premier companies. But there were also servants and shoemakers. (See Table 3.) Apprentices can hardly be categorized, for they might become rich in time. The new faith drew its adherents from across the social spectrum in the City, transcending the usual barriers of rank and wealth. Yet the indictments of 1540 may provide evidence for the history of persecution rather than of heresy. If citizens can be discovered openly declaring their own personal beliefs, at a time when they were beyond fears of prosecution or hopes of advancement, will the conclusion about the social composition of the reformed community be any different? An analysis of the first London wills expressing reformed convictions might reveal an association between religious allegiance and trade and status, though the doubts regarding the true expression of belief in wills remain (see Table 4). Again, it seems clear that the new faith was, in terms of its social composition, a very broad Church.

The reformed community was diverse, linked by faith before common commercial interest or social rank. Certainly some trades had powerful reasons for desiring the preservation of Catholicism: for example, fishmongers made money from the Lenten fast and goldsmiths from selling Catholic liturgical vessels. Yet there were evangelical goldsmiths and fishmongers, and the only organmaker whose will is preserved was emphatically Protestant. Here certainly was religious conviction at odds with economic interest. Surely those members of the City companies who adopted the reformed faith were moved rather by the influence of fellow members than by the practice or status of being a clothworker, a draper, or a mercer. So, when John Lambert contemplated becoming free of a City company he thought of the Grocers, because there, he knew, he would find the like-minded.[155] And in the Mercers' Company, where an influential cell of the brethren were gathered—Locke, Packington, Keyle, Brinklow, Robinson, Cawarden, among

---

[155] Foxe, *Acts and Monuments*, v, pp. 225–6. In this Company leading evangelicals were to be found: John Petyt, Geoffrey Lome, John Blage, and others.

TABLE 3: *Wealth and Occupational Status of those indicted during the first quest under the Six Articles, 1540*

£500 +	£100–£499	£50–£99	£20–£49	£10–£19	£5–£9	Less than £5
John Browne	Master Blage, grocer	William Stokesley	Richard Manerd	Mrs Brisley	Mrs Marshall	Anne Bedicke
Elizabeth Statham	Ralph Symond	Ralph Clervis, grocer	Richard Phillips	William Aston, mason	William Wyders, butcher	Roger Butcher
	Thomas Bele		Mrs Castle, butcher's wife	Nicholas Newell	William Thomas	John Grene
	John Sturgeon, haberdasher		Thomas Langham	John Gough, stationer	John Williamson	Agnes Palmer
	Nicholas Barker, armourer		William Beckes		Richard Bilby, draper	William Selly
	William Ettis, girdler		James Banaster, cordwainer		John Mayler, grocer	Alexander Frere
	Giles Harrison, beer brewer		John Starkey, fishmonger		Henry Patenson	Thomas Gilbert
			Christopher Smith		John Cockes	John Grene
			Richard Grafton, printer & grocer			Thomas Grangier
			John Bush			John Richmond
			Robert Daniel			Robert Causy
			William Petingale, clothworker			John Palmer
			William May, clothworker			John Curteys, mercer
			John Benglosse, clothworker			John Merifield, clothworker
			Henry Foster			Thomas Plummer
			William Callaway, goldsmith			Herman Peterson
			John Gardiner, fishmonger			
			Christopher Dray, plumber			

others—they chose a known Lutheran, Sebastian Harris, as their chaplain, so that he might minister to them and convert others.[156]

The compelling reasons for conversion in the first generation of reform were not prevailingly economic. The choices made in faith were as likely to have brought divisions within as between the social orders.[157] Deeper, more atavistic, reasons than material self-interest determined religious allegiances. Catholic preachers and writers claimed that a desire for 'parasite liberty' lay at the heart of conversions, and it was true that the new faith brought many freedoms. We have seen already, and shall see again, that the younger generation were in the vanguard of reform, finding in Protestantism excitement, liberation, consolation.[158] What the Reformation also provided was the occasion for more intense religious experience and emotional engagement, because both the old faith and the new were so urgently in need of defence. Women were often observed to be most fervent in the practice of their faith. It was the wives of London who on the eve of the Reformation were found praying piously upon their beads and reading Our Lady's primer daily in their churches; they who, in turn, became earnest 'brabblers of Scripture', so earnest that that right was denied them by statute.[159] Women had been full members of the Catholic confraternities, in their own right, and later in the Protestant underground created under persecution in the reigns of Henry, Mary, and Elizabeth they formed about half of the godly congregations.[160] Women were not silent in these congregations and were not only, nor even, following their husbands. Indeed, the authorities grew alarmed by the ardour with which London wives supported causes; their championship of Queen Katherine and their fury against Anne Boleyn, their consolation of the

[156] Mercers' Company, Acts of Court, ii, fo. 58ᵛ; BL, Harleian MS 421, fo. 15ʳ (L&P iv/2. 4029(3)). John Whalley, of the brethren, left Harris the advowson to a benefice: PRO, PCC, Prob. 11/27, fos. 218ᵛ–219ʳ.

[157] See P. Collinson, The Religion of Protestants: The Church in English Society, 1559–1625 (Oxford, 1982), p. 241.

[158] S. E. Brigden, 'Youth and the English Reformation', P&P 95 (1982), pp. 37–67.

[159] See above, pp. 15–16, 343–4, 347.

[160] J. J. Scarisbrick, The Reformation and the English People (Oxford, 1984), p. 25; Foxe, Acts and Monuments, v, pp. 443–51; GLRO, DL/C/614; C. Burrage, Early English Dissenters, 2 vols. (Cambridge, 1912), ii, pp. 9–11.

TABLE 4: *Analysis of the Preambles of London Wills written between 1529 and 1546, according to Status, Company Membership, or Trade*

	Catholic	Ambivalent	Evangelical
Armourer	3	1	
Attorney	1		1
Auditor of the Duchy of Lancaster	2		
Baker	4		2
Barber surgeon	13	2	2
Basket maker	1		
Blacksmith	4		
Bookseller	1		
Bowstring maker	1		
Bowyer	3		
Brewer	24	3	1
Bricklayer	2		
Brickmaker	1		
Broderer	2		
Butcher	4	1	2
Carman	2		
Carpenter	11	1	
Chandler	2		
Cheesemonger	1		
Clerk to the Counter	1		
Clothdresser	1		
Clothworker	29	8	3
Cobbler	3	1	
Cook	3	1	2
Cooper	1	2	
Cordwainer	5	1	1
Currier	3	1	
Cutler	7	3	
Doctor of Law	1		
Doctor of Physic	1		
Draper	33	5	5
Dyer	8		
Fishmonger	24	4	4
Fletcher	3	1	
Founder	2	2	1
Freemason	2		
Fruiterer	3	1	
Furrier	1		
Gaoler			1
Gardener	1		
Gentleman (including lawyers of the Inns)	20	4	3
Girdler	6	1	1
Goldsmith	24	1	2
Grey tawyer	1		
Grocer	21	6	7
Haberdasher	20	8	3
Hatmaker	2		

Table 4, cont.

	Catholic	Ambivalent	Evangelical
Horner	1		
Hosier	1		
Innholder	9	1	1
Ironmonger	8	2	
Joiner	2		
Leatherseller	12		1
Linenweaver	1		
Mariner	2		1
Marshall	1		
Mercer	19	5	9
Merchant taylor	46	8	7
Minstrel		1	
Organ maker			1
Painter stainer	6		
Pasteler	3		1
Pewterer	8	1	1
Plasterer	3		
Plumber	2		
Porter	2		
Poulterer	7		
Poursuivant	2		
Proctor of the Arches	1		
Saddler	3		2
Salter	8		1
Scrivener	6	1	
Servant	5		
Sherman	1		
Singing man			1
Skinner	24	5	1
Smith	1		
Spurrier	1		
Stationer	6		2
Stockfishmonger	3		1
Tailor	15	4	1
Tallowchandler	13	2	
Tappissary	1	1	
Tiler	5		
Upholsterer	1		
Vintner	8		
Waterman	1		
Waxchandler	8		
Weaver	2		
Woodmonger	1		
Woolman	1	1	1
Woolpacker	2	1	
Yeoman	1	1	
	77.3%	12.7%	10%

Marian prisoners for the faith.[161] Women marched through the City streets in their hundreds, in the Catholic processions, but also to greet Bishop Bonner upon his release from prison ('as many women as might kissed him'), and to fête the ousted ministers of Elizabeth's reign, and to conduct their heroes to their exile.[162] This female religious enthusiasm is usually to be glimpsed rather than counted. The wills of London widows show their allegiances to the rival faiths to have fallen in roughly the same proportions as their male neighbours', but wills for the most part record affiliation rather than intensity of religious experience. (See Table 5.) We cannot know how many women converted others to an evangelical vocation and spurred them to action; how often the courage and zeal of women strengthened their husbands' faltering resolve. But we can guess. There were women prophets, like Anne Bokkas, 'the Light of the Faith'. Not only in England, but in the whole of Reformation Europe women manifested a remarkable zeal and endurance in faith which sustained the persecuted minorities, whether Protestant or Catholic.[163]

TABLE 5: *Analysis of the Preambles of London Wills written by Women between 1529 and 1546*

Catholic	Ambivalent	Evangelical
138 (70.7%)	38 (19.5%)	19 (9.8%)

For all the attempts to find universal explanations for conversion, it is clear that the influences which won converts were individual and personal: the charisma of a preacher, the proselytizing of a curate in a parish, the evangelical ethos within a household or company. Though London was 'no grange',[164] and neither in Henry's reign, nor later, could parish boundaries

---

[161] See above, pp. 208–9, 214, 247 and below, pp. 604, 625.

[162] *Grey Friars Chronicle*, p. 82; P. Collinson, *The Elizabethan Puritan Movement* (1967), p. 93.

[163] *APC* iv, p. 403. See Benedict, *Rouen and the Wars of Religion*, pp. 86, 92; Davies, *Society and Culture in Early Modern France*, ch. 3; R. Bainton, *Women of the Reformation in Germany and Italy* (Boston, 1971).

[164] Collinson, *Elizabethan Puritan Movement*, p. 84.

prevent earnest gospellers, or Catholics, from traversing the City to hear sermons by the most charismatic preachers of either confession, still the little worlds of the London parishes could preserve their own distinctive religious character. In one parish the process of Protestant evangelism and indoctrination could continue undisturbed, if certain proprieties were observed; meanwhile, in a neighbouring parish only streets away the old faith could be preserved unchanged by those who wished it so. Though close in distance St Nicholas Cole Abbey and All Hallows Honey Lane were worlds apart in attitude. The City was not a place of 'anonymity and freedom',[165] but there were many secret places and the citizens could band together within the City in little cells of the like-minded.

REFORMATION ALLEGIANCES AND DIVISIONS

At the Reformation men and women were caught up in a revolution not at first of their own making, but in time made by them. They faced, for the first time, fundamental choices in faith, and religion which once had bound them now divided them. The confusion which many felt, and the anger, reaches us from their letters and their actions. For opponents of the New Learning it seemed as though they were on the brink of chaos: 'the Devil reigneth over us now'. For converts to the new faith trauma often came with emancipation as 'the veil of Moses was lifted'. Richard Hilles wrote to Cromwell of how his master 'wept unto me and exhorted me to revoke', hoping that he would 'return again from Christ'.[166] Families were divided as children chose different confessional paths from their parents. Hooper's father was, he wrote, 'so opposed to me on account of Christ's religion' that he would become 'not a father, but a cruel tyrant'.[167] But in John Stow's family the divisions were otherwise: he was devoted to the old faith, his mother not;

[165] J. P. Boulton, 'The Limits of Formal Religion: the Administration of the Holy Communion in late Elizabethan and early Stuart London', *London Journal*, 10 (1984), pp. 135–54.

[166] PRO, SP 1/74, fos. 107ᵛ–108ʳ (*L&P* vi. 99). But see also below, p. 572.

[167] *Original Letters*, i, p. 34; see also, for example, Foxe, *Acts and Monuments*, viii, p. 209; PRO, SP 6/6, fo. 94ʳ (*L&P* vii. 146); *L&P* vii. 667; Haigh, *Reformation and Resistance*, p. 167.

and his brother claimed that for twenty years John had not asked their mother's blessing.[168] Even Thomas More's family was rent by religious conflict: William Roper was won back to the true faith, in part by his father-in-law's prayers, but John Rastell was lost, and died in prison, a heretic.[169]

The Reformation made friendships which were tested. The networks of the evangelical brethren were close, and like other family relationships extremely complex. Henry Brinklow remembered in his will the men who 'laboured in the vineyard of the Lord to bring the people . . . to the knowledge of Christ's Gospel'. One of them, Stephen Vaughan, Cromwell's old friend, married Brinklow's widow. 'I have long known the woman', he told Paget.[170] In Vaughan's household the schoolmaster, Stephen Cobbe, was 'cherished as a jewel'. Cobbe was associated with Gough in the printing of radical works, and when he was in trouble again later Warner, a servant of Katherine Parr, interceded for him. Anne Vaughan married Henry Locke, of another evangelical family, and she became John Knox's especial friend.[171] There were many other evangelical marriages. Widows of the brethren, like Margery Brinklow, took others of the fraternity as their husbands. (They could not compromise their faith, and perhaps no one else would have them.) Richard Downes's widow married Edward Underhill; John Whalley's widow married William Carkke; Simon Fish's married Bainham; Lucy Petyt married John Parnell.[172] Richard Grafton married into the family of the luminary preacher, Edward Crome.[173]

Maybe godly parents arranged ideologically sound marriages for their children, though few as assertively as Joan Wilkinson who threatened to deprive her daughter of her inheritance if

---

[168] Stow, *Survey*, i, pp. xiii, lv.

[169] See above, pp. 110, 182; PRO, SP 6/9, fos. 120ʳ–130ʳ (*L&P* x. 248); *L&P* xi. 1487.

[170] PRO, PCC, Prob. 11/31, fo. 158ʳ; SP 1/214, fo. 168ʳ; *L&P* xx/1. 95, 105, 106; W. C. Richardson, *Stephen Vaughan: Financial Agent of Henry VIII* (Baton Rouge, 1953), pp. 21–2.

[171] PRO, SP 1/208, fo. 39ᵛ (*L&P* xx/2. 416); *APC* i, pp. 120, 115, 126; CLRO, Repertory 11, fo. 117ʳ; *House of Commons, 1509–1558* (Edward Warner); P. Collinson, 'The Role of Women in the English Reformation illustrated by the life and friendships of Anne Locke', *Godly People* (1983), ch. 10.

[172] *House of Commons, 1509–1558* (Petyt, Underhill); see above, pp. 163, 319–20.

[173] PRO, PCC, Prob. 11/45, fo. 139ʳ⁻ᵛ. I owe this reference to the kindness of Miss Susan Wabuda.

she did not 'by the rule of God' agree to marry the husband found by her executors, a man 'utterly abhorring papistry'.[174] At the wedding of Mary Blage to Richard Goodrich, Rowland Taylor was present, for this was the evangelical wedding of the year in the City.[175] Mary's father, John Blage, a grocer in Cheapside, was Cranmer's agent and a sustainer of the brethren. It was to him that Cranmer's secretary appealed when the Archbishop's incriminating account of the Six Articles fell into the hands of his enemies.[176] Evangelical masters often chose evangelical apprentices, or converted them, and Blage's apprentice was Richard Grafton. Both men were prime movers in setting forth the Bible in English: at his death in 1552 Blage bequeathed a copy of Matthew's Bible, the Bible which Grafton had printed.[177] When the reaction came in 1540 Grafton and Blage were confederate in opposition to it, and found themselves in deep trouble. Not only were they both indicted during the first quest for their failure to conform, but also at the turn of the year they were found to be disseminating copies of Melanchthon's tract against the Act.[178] Mary's marriage to Goodrich failed, and Goodrich, divorcing Mary, married the widow of his reforming friend (and Mary's cousin?) Sir George Blage. Upon Edward's death Goodrich and his new wife thought upon exile for religion's sake.[179]

The 'sustainers' who would aid the exiles of Mary's reign had been long associated. Wills of reformed citizens in Edward's reign already named them in conjunction, for a common commitment to reform created an allegiance. As the brethren contemplated last things and made their wills they naturally enough called upon their evangelical friends to be witnesses and executors, and left them tokens.[180] To his friend Merifield, a patron of

[174] PRO, PCC, Prob. 11/42B, fos. 233ʳ–234ʳ.

[175] GLRO, DL/C/208 (without folio numbers).

[176] Foxe, *Acts and Monuments*, v, pp. 389–90.

[177] For evangelical masters with evangelical apprentices, see above, pp. 97–8, 193 n. 119, 329; J. A. Kingdon, *Richard Grafton, Citizen and Grocer of London* (1901); PRO, PCC, Prob. 11/35, fo. 111ʳ.

[178] Foxe, *Acts and Monuments*, v, pp. 350–8, 443–4; *PPC* vii, pp. 97–8, 100–1, 103, 105–6.

[179] PRO, C 1/1354/66–8; GLRO, DL/C/331, fo. 182ʳ; DL/C/208; *House of Commons, 1509–1558* (Richard Goodrich, George Blage). It is surely likely that John and George Blage were related.

[180] See, for example, PRO, PCC, Prob. 11/27, fos. 218ᵛ–219ʳ; 11/28, fos. 253ʳ–255ʳ; 11/34, fos. 140ʳ, 228ʳ.

preachers under persecution, Richard Downes bequeathed a
ring inscribed 'Christ is risen for our justification'.[181] Thomas
Sternhold, a member of Henry's Privy Chamber, and composer of the English metrical psalms, called Huick, Taylor, and
Whitchurch to witness his will.[182] As he lay dying, John Purser
called Barnes's 'scholar' Thomas Parnell to his bedside to witness his will, and in time Purser's son Dick was taken into
Thomas Cromwell's keeping.[183] The confederacy especially
needed to rally when their friends were in trouble, as they so
often were. They stood surety for each other; 'vow breaking
brethren', wrote More. When William Callaway of the St
Matthew Friday Street conventicle was in Bonner's custody
John Gardiner stood surety for him; so too Carkke and Brinklow
came forward to aid Robert Wisdom; and Alexander Brett
dared to help Anne Askew.[184] But these men could also be
called upon to testify against the conservative enemies of their
faith: so Cromwell co-opted John Parnell to the jury to try his
old enemy Sir Thomas More, and on the London jury which
tried the Marquis of Exeter and his associates for treason sat
Richard Downes.[185] But in time the Catholic authorities would
find Catholic jurors to try the reformed.[186] The Reformation
had created deep divisions within London society, which could
be exploited for political purposes.

The same ties of loyalty and friendship united the Catholic
opponents of reform. For twenty years Sir Thomas More was a
'continual nurseling in the house of Bonvisi', and in his 'ship-
wreck' in the Tower More counted this constant friendship as
one of the 'brittle gifts of fortune'.[187] Until he was forbidden,
Bonvisi sent More meat, wine, and messages in prison.[188] In
his last days More wrote with a coal to his friend, looking to
their future companionship in a place 'where no wall shall dis-
sever us, where no porter shall keep us from talking together';

---

[181] PRO, PCC, Prob. 11/29, fo. 113ʳ.
[182] PRO, PCC, Prob. 11/32, fo. 282ᵛ.
[183] PRO, PCC, Prob. 11/25, fo. 49ʳ; *L&P* xiv/2. 782, pp. 336, 338–41.
[184] Guildhall, MS 9531/12, fos. 32ʳ⁻ᵛ, 45ᵛ, 109ʳ.
[185] See above, pp. 229–30; *L&P* xiii/2. 979(6); 986(23).
[186] See below, p. 554.
[187] More, *Correspondence*, 34, 217.
[188] *L&P* viii. 856 (38, 39, 43, 45).

that is, Heaven.[189] And still in 1547 Bonvisi supported More's family, and in 1549 left with them for exile.[190] Once he had been a friend of both More and of Cromwell and humanists like Starkey, but in time he had to choose. He became the 'patron and second father' to London's ultra-Catholics: both William Hollis and John Story made him executor to their wills. Story trusted Bonvisi to find a suitable—a Catholic—husband for his daughter, and to persuade his widow against returning to a schismatic England.[191] When William Bowyer made his will he chose Clement Smith, most committed of Catholics, to be his overseer. Both men had steadfastly resisted the New Learning, standing firm in the old faith though their conformity was required.[192] The wills of conservatives, like the wills of their reformist rivals, reveal their old associations. When John Twyford, a 'furious papist', made his will in 1549 he called his 'ghostly father', Henry George, as witness, and he remembered Roger Cholmley's chaplain. George was one of the most resolutely conservative of City curates and Cholmley was, as Recorder of London, persecutor of heretics.[193] The allegiances of the most committed of conservatives and of reformers are plain; harder to discover, but surely as significant, are similar friendships between like-minded but less well-known families and neighbours throughout the City.

If there were friendships there were also feuds. In the parish of St Sepulchre, Twyford, who brought faggots to the burnings, and Merial, who might have been burnt himself for his strange and emphatic heresies, hated each other. Merial had, moreover, beaten Twyford's boy.[194] Such quarrels were commonplace, and the religious divisions exacerbated and justified other, older antagonisms. As it was at Court, so it was in the City. The contention between the faiths was revealed not least in the writing of its history. Between John Stow and Richard Grafton, chroniclers of the Reformation, there was lasting

---

[189] More, *Correspondence*, 217.
[190] *CPR Edw. VI* i, p. 203; see below, pp. 453-4.
[191] *CSP Ven.* vi/i, p. 111; Harpsfield, *Lyfe*, p. 138; *L&P* ix. 867; x. 273; Inner Temple, Petyt MS 538/47, fos. 66ʳ–68ʳ.
[192] *House of Commons, 1509–1559* (William Bowyer and Clement Smith).
[193] PRO, PCC, Prob. 11/33, fo. 39ᵛ; see p. 272 above, and below, p. 440.
[194] Foxe, *Acts and Monuments*, v, p. 601.

antagonism.[195] They impugned each other's scholarly veracity, for they hated each other's faith. Grafton was troubled for reform; Stow, later, for popery.[196] Grafton scorned Stow's chronicles as 'memories of superstitious foundations, fables and lies foolishly *stowed* together'. Stow, in turn, accused Grafton of plagiarism—'as his own conscience (if he had any) can well testify'—and worse, of mendacity.[197] For example, 'Thomas Cooper saith that thirty Germans [of 1,166] taught the abrogation of the sacrament of the altar', but all Grafton said of this heresy was that 'they taught a reformation'. Moreover, Grafton neglected to mention in his chronicle the usurpation of Lady Jane Grey whose proclamation he had himself printed.[198] Stow's *Survey of London* can be read as a lament for a lost world. And he admitted and justified the omissions he made in the name of faith and justice: he did not mention the new monuments set up in the City churches by iconoclasts, for these were men 'worthy to be deprived of that memory whereof they have injuriously robbed others'.[199]

[195] The story of their quarrel is told in Stow, *Survey*, i, pp. xlviii–liii.
[196] Ibid. i, pp. xvi–xvii.
[197] Ibid. i, pp. ix–xii.
[198] Ibid. i, p. l.
[199] Ibid. i, p. xxxi.

X   The Religion of Edwardian London

ALREADY AT the accession of Edward VI the great religious houses of the City were fallen into ruin, 'profane and desolate'. The choirs and naves, once gilded and adorned with image upon image, stood empty, open to the sky.[1] The priory church of Christ Church Aldgate, its stone 'proffered to whomsoever would take it down' by its new owner, was torn down stone by stone. At the Austin Friars Sir William Paulet stored coal and corn in the steeple; his son sold 'the monuments of the noble men there buried' to the highest bidder.[2] Soon the alabaster and marble tombs of the illustrious dead at the Grey Friars were scattered.[3] Worse desecration and, some said, providential punishment were to follow. In the precinct of the Black Friars the inhabitants lost both parish church and friary: Sir Thomas Cawarden, Master of the Revels, stowed tents and pavilions and stabled horses in St Anne's, threatening the dismayed parishioners that if they did not take down the sacrament hanging over the altar, then he would.[4] Soon after he became Lord Protector, the Duke of Somerset began to build a palace worthy of so vainglorious a ruler, plundering stone from the churches and priories of the City, profaning tombs, blowing up the bell tower at St John of Jerusalem at Clerkenwell, which long ago— so wrote John Stow, elegiacally—had been 'gilt and enamelled, to the great beautifying of the City, and passing all other that I have seen'.[5] Stow looked on in horror at the iconoclasm and desecration, condemning so 'many that of a preposterous zeal or of a greedy mind spare not to satisfy themselves by so wicked

---

[1] E. Jeffries Davis, 'The Transformation of London', in *Tudor Studies*, ed. R. W. Seton-Watson (1924), pp. 287–314.

[2] Stow, *Survey*, i, pp. 142, 177; *CPR Edw. VI* i, p. 136.

[3] Ibid. i, pp. 317–22; *Grey Friars Chronicle*, pp. xviii–xxii, 54.

[4] PRO, C 1/1330/39; 1405/39–41. For Cawarden's commitment to reform, see pp. 326, 343 above, and pp. 590, 619 below.

[5] W. K. Jordan, *Edward VI: The Young King* (1968), pp. 498–9; Stow, *Survey*, ii, pp. 84–5.

a means'. But soon Miles Partridge, who had won Jesus bell tower at St Paul's from Henry VIII at dice, would suffer with Somerset for such impiety.[6] Many Londoners shared Stow's distress at the despoliation, and his pleasure at the fall of Edward's councillors and courtiers.

But other Londoners had high and urgent hopes of reform. As the courtiers were swift to despoil the religious houses, so the citizens might hasten to alter the churches, edified by the devotions of centuries. Within days of Edward's accession, one City church was transformed. Gone were the rood, the images and pictures of saints; in their place, whitewashed walls, the arms of the new King, and Scriptural messages.[7] 'Thou shalt make no graven images, lest thou worship them', so Gardiner heard, 'is newly written in the new church, I know not the name, but not far from the Old Jewry.'[8] The church was St Martin Ironmonger Lane, and there the reforming rector, John Hardyman, and the churchwardens had put an end to idolatry, in joyous expectation of the reformation to come under a new Josiah who would destroy the Temple of Baal. They acted too soon: it was not for the people of their 'arrogancy and proud hastiness' to 'run before they be sent, to go before the rulers, to alter and change things in religion without authority'.[9] But the godly thought that they had waited too long already for reformation, and looked for their new rulers to rebuild the Temple. Edward, raised in the true faith, was heralded as the 'godly imp'.[10] The councillors ascendant at his father's death were known to be won to reform. Had not the Earl of Hertford (now Duke of Somerset, Lord Protector) welcomed Latimer to his house after the Bishop's resignation in 1539; had not his players performed interludes with (surely) reformist messages in the City, and his Countess been suspected of protecting the martyred Anne Askew?[11] The King's other uncle, Thomas

---

[6] Stow, *Survey*, ii, p.75; i, p. 330.

[7] *APC* ii, pp. 25–6; *CSP Sp.* ix, p. 45.

[8] Foxe, *Acts and Monuments*, vi, p. 61.

[9] *Documentary Annals*, i, pp. 53–4.

[10] See, for example, Cromwell's scaffold speech: Bodleian, Fol. Δ 624, facing p. 462.

[11] M. L. Bush, *The Government Policy of Protector Somerset* (1975), p. 102; CLRO, Repertory 11, fo. 157ʳ; J. N. King, *English Reformation Literature: The Tudor Origins of the Protestant Tradition* (Princeton, 1982), pp. 274–5; see above, pp. 374–5.

Seymour, had swiftly married the evangelical Katherine Parr.[12]
For the rulers of Edward's reign, minded to reform, the central
problem would be how to teach 'the weak . . . to flee all old
erroneous superstitions . . . the Bishop of Rome's traditions',
while persuading the extremists to 'tarry the time which God
hath ordained for the revealing of all truth'.[13]

A world-weary Paget told Somerset in the summer of 1549
that 'what countenance so ever men make outwardly [in
religion] was to please them in whom they see the power
resteth'; that, even in matters of faith, the people conformed,
sheep-like, to the rulers' commands.[14] Maybe so, but in
London the truth was more complicated. In Edward's reign,
with a King so young, with councillors vying desperately for
power, their position always unstable, the religious inclinations
of Londoners may, conversely, have had an impact upon the
direction taken in national religious policy. The course of
politics, and the vulnerability of the governors, will be the
subject of another chapter, but it is important to remember
that all the machinations in high politics lay behind the course
of the religious change which will be the subject of this one. For
their part, the governors of Edwardian England always
remembered that London was 'a mighty arm and instrument
to bring any great desire to effect, if it may be won to a man's
devotion', and that Londoners might have to be placated.[15]
The Duke of Northumberland, outwardly a prime mover
behind reform, claimed at his fall in August 1553 that he would
have restored the 'true religion'—that is, the old religion—
later:

If it had pleased God to have granted him life and had remained in
authority, he would have put it [reform] down himself before ever
one year had come to an end, and added that for to ruin the hearts of
the citizens of London because they loved new things he would not
before do it.[16]

---

[12] Jordan, *Edward VI: The Young King*, pp. 368–71.
[13] *Documentary Annals*, i, pp. 53–4.
[14] PRO, SP 10/8, no. 4.
[15] Stow, 'An Apology of the City' (*c*.1575); *Survey*, ii, p. 206.
[16] BL, Harleian MS 353, fo. 142ʳ.

The citizens of London who 'loved new things' influenced their governors; their pent-up zeal was at some times a powerful impetus to reform, and at others a deterrent.

I. TRUE AND FALSE IMAGES

On 8 February 1547 every City church held a solemn dirge for Henry VIII, tolled a knell and celebrated a requiem in Latin.[17] This would be almost the last time in Edward's reign that every parish would follow the old rite or pray for a soul in Purgatory. Gardiner had intended to offer a dirge for the dead King in Southwark on 6 February, but at the same time the Earl of Oxford's players announced a rival, mock celebration; 'a solemn play to try who shall have the most resort, they in game, or I in earnest'. Surely, Gardiner wrote to Paget, this was 'a marvellous contention, wherein some shall profess in the name of the commonwealth, mirth, and some sorrow at the one time', and he asked despairingly for 'uniformity in the commonwealth'.[18] But there was none to be had. Soon the penitential Lenten fasts were being disregarded. Lent was, so Tonge and Joseph declared from the City pulpits, 'one of Christ's miracles, which God ordained not man to imitate and follow'.[19] At Paul's Cross in April Dr Glasier preached that the Lenten fast was an economic provision rather than a divine one.[20] Rhymes were broadcast and interludes played to 'deprave the Lent'. The reforming brethren made rhymes of Bonner prophetically lamenting the fall of Gardiner ('Stephen Stockfish'), and soon life imitated art. Gardiner protested. Somerset replied philosophically: 'The people buy these foolish ballads of Jack a Lent. So bought they in times past pardons, and carols, and Robin Hood's tales.' 'Lent remaineth still, my lord, and shall, God willing.'[21] But Somerset's reply was disingenuous, for his sympathies were already with reform. Soon after Edward's accession, Thomas Dobbe came up to London. He had been ejected from St John's College, Cambridge for 'addicting his mind to the Christian state of matrimony'. In St Paul's at the

[17] Wriothesley, *Chronicle*, i, p. 181.
[18] *Letters of Gardiner*, pp. 253–4.
[19] Foxe, *Acts and Monuments*, vi, p. 32.
[20] Stow, *Annales*, p. 594.
[21] Foxe, *Acts and Monuments*, vi, p. 35; CLRO, Repertory 11, fo. 339ᵛ.

most sacred point in the Mass, as the priest elevated the host, Dobbe 'with godly zeal' exhorted the congregation not to honour the visible bread as God. He was at once arrested, accused before Cranmer, and cast into the Counter, where he died. But he was not meant to be a martyr: he died before Somerset's pardon could reach him.[22] Somerset was already inclined to look leniently even upon such heresy, but Cranmer, who had but recently condemned Anne Askew for perhaps the same offence, could not, yet.

Men of deep, but opposing, conviction sensed immediately the extremity of the religious reforms to follow. Dr Richard Langrysh, having resisted joining his friends in exile for the old faith during the reforming 1530s, took a more desperate course in 1547. In May he leapt from St Magnus steeple into the Thames and 'wilfully drowned'.[23] In despair too, the quondam monks of the London Charterhouse chose exile. On 3 April John Foxe, now rector of St Mary Magdalen Old Fish Street, fled to Louvain with the aid of Thomas Munday, rector of St Leonard Foster Lane, and Thurston Hickman, a fellow Carthusian, conveying with him treasured relics, including the arm of John Houghton, their martyred Prior. Other 'late religious persons of their confederacy' planned to follow. William Peryn, returned to England from exile during the conservative reaction of 1543, went back to Louvain in 1547.[24] For these extreme conservatives, and others who would join them later, exile was the only course to remain undefiled. Reading the signs, they had fled even before any official changes in religion were introduced. Official reform did soon follow, and in its wake unofficial reform with disturbing implications.

Injunctions were issued on 31 July 1547 (the day of publication of the *Homilies*)—the first measure to effect reform. Like the Injunctions of 1536 and 1538 before them, these were intended for 'the advancement of the true honour of God, the suppression of idolatry and superstition'.[25] All curates were to

[22] Foxe, *Acts and Monuments*, v, pp. 704–5; J. Ridley, *Thomas Cranmer* (Oxford, 1962), p. 264.

[23] *L&P* xi. 1350; *Two London Chronicles*, p. 44.

[24] Wriothesley, *Chronicle*, i, pp. 184–5; *BRUO* (Peryn). Hickman and Munday were pardoned in Feb. 1548: *CPR Edw. VI* ii, p. 1.

[25] *Documentary Annals*, i, pp. 4–20. R. Strong, 'Edward VI and the Pope', *Journal of the Warburg and Courtauld Institutes*, xxiii (1960), pp. 311–13.

declare the Word of God, and to exhort parishioners 'to faith, mercy and charity', warning them that 'works devised by men's fantasies as wandering to pilgrimages', venerating images, 'or kissing or licking of the same, praying upon beads', had not only no promise of reward in Scripture, but 'contrari-wise, great threats to God, for they be things tending to idolatry and superstition'. There was to be no praying upon beads. Abused images were to be utterly cast away, and no candles to be lit before pictures of saints. Now lights were allowed only before the sacrament on the high altar 'for the signification that Christ is the very true light of the world'. The only innovations ordered in the conduct of divine service were the orders to read the Epistle or Gospel in English, not Latin, and the prohibition of procession. Yet in the Injunctions were signs of doctrinal change to come. Where the form of bidding the common prayers had always besought God to lessen the temporal pains of the departed soul and to ensure its present felicity, now the new form which was appended to the Injunctions referred only to the future destiny of the dead: 'that they with us, and we with them at the day of judgment, may rest both body and soul with Abraham, Isaac and Jacob in the kingdom of Heaven'.[26] Still masses for souls continued, but for how much longer? By September the Imperial ambassador was convinced that in future masses were 'not to be allowed on behalf of the dead'.[27] Already chantry priests were enjoined in the penultimate Injunction to devote themselves now not to prayer but to the godly education of youth.

By 23 July the Mercers' Company had already ordered that their almsmen should wear upon their gowns the badge of a maiden's head instead of beads. At St Paul's school the picture of Jesus was removed, along with the tablet over the tomb of Colet, its founder.[28] The order in the Injunctions to take away and destroy 'all shrines . . . pictures, paintings, and all other monuments of feigned miracles' was carried out with a will in the City churches. Soon, 'by the assent of the most part of the parishioners there' (so they claimed later), the churchwardens

---

[26] Cited in A. Kreider, *English Chantries: The Road to Dissolution* (Cambridge, Mass., 1979), p. 186.

[27] *CSP Sp.* ix, p. 148.

[28] Mercers' Company, Acts of Court, ii, fos. 215^r, 286^v.

began to sell wholesale the ornaments and treasures of their churches, given and honoured over decades, even centuries. Silver crosses, monstrances, tabernacles, wooden images from the rood lofts, candlesticks, chalices, altar cloths, vestments, copes of the 'boy bishop', cloths to hang before the saints in Lent and other 'old trumpery', Latin service books, paxes, pyxes, censers, sanctus bells, holy water stocks, churchyard crosses were sold within a year of Edward's accession in half of London's churches.[29] In early September the Imperial ambassador reported that images were being removed all over the City: 'they will not even leave room for them in the glass'.[30] In their place, Scriptural messages. At the dissolved Grey Friars all the altars were pulled down, the tombs of the nobility in alabaster and marble carried away, the walls and stalls of the choir demolished. Soon after, the neighbouring churches of St Ewen and St Nicholas Shambles were pulled down; their parishioners, together with some from St Sepulchre, transferring to the newly created parish of Christ Church in the old Grey Friars.[31] One by one, the churches were transformed, reformed from idolatry: the brightly painted walls now whitelimed, pulpits replaced tabernacles, Scriptural messages replaced wall paintings, plain glass stood in the windows.[32] Some parishes claimed that the precipitate sales of goods were necessary to finance the required reform. At St Bride there was in 1547, so they said, 'no money in the church box wherewith to reform . . . the idolatrous images . . . and to garnish the church with Scripture'.[33] But maybe they were disingenuous. It was clear that for some zealots in the parishes reform could not come soon enough.

---

[29] In 1552 the commission established to prepare inventories of church goods preliminary to their seizure required the churchwardens to give a list of all the goods sold, year by year, since the first year of Edward's reign. Their answers are printed in Walters, *London Churches*. For the earliest desecrations, see ibid., pp. 96, 104, 108, 111–12, 122, 134, 149, 153, 165, 170, 223, 229, 246, 257–8, 265, 280, 289, 293, 308, 314, 322, 329, 337, 340, 351, 355, 369–70, 376, 383, 393, 400, 405, 415, 422, 427, 439, 443, 460, 463, 476, 511, 516, 521, 557, 609, 617.

[30] *CSP Sp.* ix, p. 148.

[31] *Grey Friars Chronicle*, pp. xviii–xxii, 54; Stow, *Survey*, i, pp. 317–22.

[32] For the expenditure, see Walters, *London Churches*.

[33] Ibid., p. 223. At St Leonard Foster Lane, St Dunstan in the East, St Olave Silver Street, and St Swithun also they claimed that plate was sold to pay for repairs and refurbishment: ibid., pp. 337, 246, 556, 377, 608.

The City iconoclasts were so enthusiastic, so indiscriminate that by mid-September the Council took measures to restrain them. 'All images and pictures in every church to the which no offering nor yet prayer is made', the Council resolved, must remain. Any already removed must be set up again, and the Mayor and Aldermen were enjoined to punish churchwardens and clerics who had gone beyond their authority. Windows portraying St Thomas Becket (even now; though his image had been denounced in 1538) must be altered as carefully as possible; where the Bishop of Rome was still portrayed in the glass his crown should be painted over. 'If any doubt rise', the City governors should consult their 'friend' Russell, or another councillor (during Somerset's absence in Scotland).[34] Clearly they feared trouble. After great debate about how to proceed, the Court of Aldermen decided on 22 September that every Alderman or his deputy 'in the most secret, discreet and quiet manner' should visit the City churches, and shutting the doors behind them to prevent any 'gathering of the people together', take notes of which images remained, which had 'any offerings and were prayed unto, and which not, and who took them down, and what is done with them, and . . . what misdemeanour was done in taking down of them'.[35] At the end of September the Council still sought the punishment of the iconoclasts, but dared not order the restoration of images wrongfully torn down, lest 'it might engender contention among the people upon the point whether they were abused or no'.[36]

In every parish of London the battle for and against reform, for idolatry or iconoclasm, would be fought. So in the spring of 1547 William Peryn had preached at St Andrew Undershaft the spiritual benefits of worshipping the pictures of God and of saints, but by June orthodoxy had so changed that he recanted, and went, once more, into exile.[37] Doubtless he left behind many parishioners who shared his beliefs, but soon there followed a 'hot gospeller' to that parish. John Stow saw Sir Stephen there, often 'forsaking the pulpit . . . preach out of a

---

[34] CLRO, Journal 15, fo. 322^r; Letter Book Q, fo. 210^v.
[35] Ibid., Repertory 11, fo. 373^r.
[36] APC ii, p. 518.
[37] Stow, Annales, p. 594.

high elm tree in the midst of the church yard', saying that the
names of the churches might be altered; the names of the days
of the week changed; fish days kept on any day except Friday
or Saturday; Lent kept at any time except between Shrovetide
and Easter. When he named the great maypole at St Andrew
Undershaft an idol, the people, 'after they had dined to make
themselves strong', marched to tear it down, 'mangled it and
after burned it'.[38] Though the authorities feared, rightly, the
reckless crusaders against idolatry, the iconoclasts themselves
sometimes took their lives in their hands, for their conservative
neighbours, who treasured the images, vowed vengeance.
When Edward Underhill, who was 'all of the spirit', took it
upon himself to remove the pyx from the altar at Stratford le
Bow, the wife of Justice Tawe and other women of Stepney
'conspired to have murdered me'. 'The Lord preserved'
Underhill, who went into hiding 'in a secret corner, at the
nether end of Wood Street'.[39]

At St Botolph Aldgate a running battle developed between the
conservative farmer of the parish, William Green, and his unregenerate curate, William Rufford, and a reforming contingent
led by the churchwardens, newly elected.[40] The churchwardens
had signalled their religious intentions by selling the parish's
Latin service books 'which the people did not understand', and
other tokens of idolatry, immediately upon Edward's accession.
They bought books of psalms in English 'to the end that the
people should understand to praise God the better', but Rufford
refused to use them. The quarrel began, ostensibly, over tithe,
but the real contention was of deeply opposed religious convictions; how could the parishioners pay tithe to so unworthy a
minister? The Mayor had settled the tithe dispute in Green's
favour in September 1547, but in April 1548 the parish prepared a supplication to Somerset, whom they hoped would be
sympathetic, for remedy against Green, and his curate, who
'could not preach or teach well, nor no curate for him'.[41]

---

[38] Stow, *Survey*, i, pp. 143–4.

[39] *Narratives of the Reformation*, pp. 159–61.

[40] The story is revealed in the—partial—churchwardens' accounts of St Botolph
Aldgate: Guildhall, MS 9235/1, fos. 1ʳ–8ᵛ.

[41] CLRO, Journal 15, fo. 327ᵛ; Repertory 11, fo. 455ᵛ; Guildhall, MS 9235/1,
fo. 4ʳ; *CPR Edw. VI* i, p. 95.

Robert Owen was, however, one of those zealous church-wardens of whom the City authorities had taken note, and in May 1548 he was ordered to return those symbols of Catholic practice—the pyx, the chalice, and the clapper of the church bell—which he had taken away.[42] Seeing that the parish determined to have services in English, Green and Rufford contrived to have Owen arrested and sent to prison late in August 1548. On 4 October the parish paid a 'God's penny' (an earnest) to William Dabbs, the minister 'elected . . . by the [reformist] faithful of the parish', and on the following day seven parishioners went before the Lord Protector to 'have his good will' that Dabbs should replace Rufford. 'More of his lordliness than of wisdom' the Mayor banished Robert Owen from the City, but conceded nevertheless on 6 October that Dabbs should replace Rufford that Christmas.[43] In the reign of Mary the conservative parishioners at St Botolph would have their revenge upon the radicals by making them pay for the restoration of the Catholic ornaments which they had destroyed,[44] but meanwhile the despoliation continued and reformed services were held.

In other parishes throughout the City those of the old faith tried to save what relics and images they could: only Catholic sentiment or a traditional piety could explain why little, broken, wooden images were bought for a few pence.[45] In the houses of Catholic citizens the images remained, despite the Injunctions. In John Clement's house in Bucklersbury there were no signs of Reformation: images hung still on the walls, and in his study the portrait of a new saint, Sir Thomas More, his father-in-law.[46] Even at St Paul's they hid an image of Our Lady, to save it from the visitors.[47]

While reformist elements in his diocese sought to hasten the pace of reform and could find no distinction between true and false images, Bishop Bonner, who opposed even the moderate reform, was powerless to restrain them. He became the more impotent during the royal visitation of the autumn of 1547 to

[42] CLRO, Repertory 11, fo. 459ʳ.
[43] Guildhall, MS 9235/1, fos. 6ʳ–8ʳ.
[44] See p. 592 below.
[45] See, for example, Walters, *London Churches*, pp. 88, 196.
[46] PRO, C 1/1337/18.
[47] Wriothesley, *Chronicle*, ii, p. 1.

ensure obedience to the Injunctions and royal proclamation, for then episcopal authority was inhibited. Bonner, like Gardiner, proved resistant. When the Visitors, led by Sir Anthony Coke, presented themselves at St Paul's early in September, Bonner insolently demanded to see their commission, and protested that he would observe the Injunctions and *Homilies* only 'if they be not contrary and repugnant to God's law and the statutes and ordinances of the Church'.[48] 'Whether for fear, or for conscience' Bonner later recanted this registered protest; meanwhile he was sent to the Fleet, as an example to others of his persuasion.[49] 'The City of London is very glad to hear of the Bishop, their pastor, being in the Fleet', wrote Thomas Seymour confidently to his brother on 15 September,[50] yet he spoke only for a part, a minority of the citizenry. Many more repined as reforms in religion were introduced while their Bishop lay in prison. That September, according to the Injunctions, the litany was, for the first time, sung in English in St Paul's, 'between the choir and the high altar, the singers kneeling half on one side, and half on the other side', and the Epistle and the Gospel read in English during the celebration of high Mass.[51] On 17 November the great rood at St Paul's, and all the images in the cathedral, were pulled down; with such urgency that two of the workmen were killed and others injured: a divine judgement, surely, against the iconoclasts.[52] Although images were being removed, and masses for the dead might not long continue, still the Mass was celebrated, 'the Holy Sacrament is held in all reverence, and confession is observed'.[53]

FAREWELL TO 'MISTRESS MISSA'

But outrages against the Mass followed the iconoclasm. For some of the radicals the idolatry of worshipping the images of

[48] Foxe, *Acts and Monuments*, v, pp. 742–4; Jordan, *Edward VI: The Young King*, pp. 163–5.
[49] *Grey Friars Chronicle*, p. 54.
[50] Cited in D. E. Hoak, *The King's Council in the reign of Edward VI* (Cambridge, 1976), p. 215.
[51] Stow, *Annales*, p. 594.
[52] *Grey Friars Chronicle*, p. 55.
[53] *CSP Sp.* ix, p. 148.

saints in churches was as nothing compared with the idolatry of worshipping a visible God—'a Jack in the box'—in the form of bread at every elevation of the host. In the Guildhall chapel in the summer of 1547 Giles Strother stood 'contemptuously at the levation time of the sacrament of the altar, with his cap upon his head, not giving any manner of honour or reverence thereunto'. He was seen to rip a mass book into tiny pieces. A sacramentarian tract was found in his handwriting. His visitors in prison were watched.[54] Although the Lord Protector had, reportedly, abandoned the Mass in his own household by December 1547,[55] it was not thought timely for subjects to follow. The first act of Edward's first Parliament was the Act against Revilers of the Sacrament and for Communion in both kinds (1 Edw. VI c.1).[56] The sacrament of the altar 'hath been of late marvellously abused', and the revilers were to be punished by fine and imprisonment. A proclamation was issued on 27 December against the 'irreverent talkers of the sacrament', those who sought blasphemously to determine whether 'the body and blood . . . is there really or figuratively . . . whether His blessed body be there, head, legs, arms, toes, and nails . . . naked or clothed; whether he is broken and chewed, or he is always whole'. It was not given to 'human and corrupt curiosity' to 'search out such mysteries as lieth hid in the infinite and bottomless depth of the wisdom and glory of God'. Rather they must accept the teachings of St Paul, that the sacrament is 'the partaking of the blood of Christ', 'and that the body and blood of Jesu Christ is there: which is our comfort, thanksgiving, love token of Christ's love towards us, and of ours as His members within ourself'.[57] The precise manner of the Real Presence was not defined, but real partaking of the body and blood of Christ was affirmed.

The first part of the 'Act against Revilers of the Sacrament and for Communion in Both Kinds' failed signally. 'Both the preachers and other spake against it [the proclamation] and so

---

[54] CLRO, Repertory 11, fo. 362[r].

[55] *CSP Sp.* ix, p. 221.

[56] The Act is printed in Gee and Hardy, *Documents Illustrative*, pp. 322–8. Its significance is discussed in C. W. Dugmore, *The Mass and the English Reformers* (1958), pp. 127–37.

[57] *TRP* i. 296; Dugmore, *The Mass and the English Reformers*, pp. 116–17.

continued.'[58] The Mayor and Aldermen attended upon the Lord Great Master on 11 January 1548, requiring his help 'for the obtaining of the proclamation for the staying of the common eating of the flesh upon the fish days and the irreverent talking and railing against the blessed sacrament of the altar'.[59] There was wanton disregard for Lenten fasting and widespread abuse of the Mass. On the second Sunday in Lent (25 February) preachers at St Martin Orgar and St Dunstan in the East had both spoken dubious 'words concerning the Mass', and were reported.[60] Yet it was harder to prevent discussion regarding the sacrament of the altar when the definition of its nature had been profoundly altered by the second part of the same act: 'for Communion in both kinds'. The Bishops had debated central questions regarding the Mass in January 1548, and their conclusions were included in the *Order of Communion*, completed on 8 March, to be used from that Easter onwards. The communion was to be given to the people in both kinds, immediately after the communion of the priest, otherwise without change of any rite or ceremony of the Latin Mass. Still the priest would not 'minister the communion' at every Mass, but only at Easter and from time to time. So it was that in May 'Paul's choir with divers other parishes in London sang all the service in English, both matins, Mass, and evensong; and kept no Mass, without some received the communion with the priest'.[61] The anniversary mass of Henry VII at Westminster was sung in English, leaving out all the canon after the Creed save the Pater Noster.[62] That Whitsun the three Spital sermons were given by Dr Tonge, Dr Rowland Taylor of Hadleigh, and John Cardmaker of St Bride Fleet Street, reformers all. The significance of their appointment was not missed: even at the Spital sermons, the 'preachers inveighed against the Mass'.[63]

The Grey Friars chronicler reported despairingly the dismal events of 1548: preachers everywhere denounced the blessed sacrament of the altar. Cardmaker and a preacher who read

[58] *Grey Friars Chronicle*, p. 55.
[59] CLRO, Repertory 11, fo. 403ᵛ.
[60] Ibid., fo. 423ʳ.
[61] Wriothesley, *Chronicle*, ii, p. 2; *Grey Friars Chronicle*, p. 55; Dugmore, *The Mass and the English Reformers*, pp. 123–6.
[62] Wriothesley, *Chronicle*, ii, p. 2.
[63] Ibid. ii, p. 3; *Two London Chronicles*, p. 45.

twice weekly at St Paul's—albeit in Latin—said that the sacrament was but bread and wine. Alone, William Leighton dared to defend the Mass: 'he preached in every place that he preached against them all'.[64] William Leighton was prebend of Kentish Town and a canon of St Paul's, whose orthodoxy had been reinforced at Louvain in the late 1530s.[65] His sermons at St Paul's every Sunday occasioned 'much controversy and business . . . and sitting in the church, and of none that were honest persons', wrote the Grey Friars chronicler.[66] After the recantations of Smith and Peryn in the summer of 1547 few of the Catholic establishment, save Gardiner, would stand to the defence of the Mass and of 'unwritten verities', in public at least. Bonner, after his contention with the commissioners, had turned to sullen acquiescence, for the time. Still, the radicals knew that the enemies of the Gospel were very many, and sought other ways once 'a way had been taken' against preaching.

A flood of tracts and ballads appeared against 'Mistress Mass'. 'Railing bills' calling the sacrament 'Jack of the box' and 'Round Robin' were posted on church doors.[67] Thirty-one tracts against the Mass and Corpus Christi were issued in 1548 alone.[68] The cause was urgent. In *A new dialogue called the endightment agaynste mother Messe* (17 December 1548) Brother Verity and Brother Knowledge pondered whether Christ's Gospel could flourish when so many opposed it.

BROTHER VERITY. . . . There is an abhomination that sitteth sore in men's consciences, that we must see a remedy for the Gospel's sake, which is the mother of all mischief and the abhomination of desolation.

BROTHER KNOWLEDGE. Truly, her name is . . . Mother Mass, a woman that hath brought the people into a devilish trade . . . she vaunts herself to be a God of Gods.

Whereas Brother Verity 'with the doctrine of Jesus Christ' had brought 'a few to the true knowledge', she with 'filthy traditions

---

[64] *Grey Friars Chronicle*, pp. 56–7.

[65] *BRUO.*

[66] *Grey Friars Chronicle*, p. 56.

[67] Foxe, *Acts and Monuments*, vii, p. 523.

[68] These tracts are discussed in King, *English Reformation Literature*, pp. 284–9, and P. M. Took, 'The Government and the Printing Trade, 1540–1560' (University of London, Ph.D. thesis, 1978), pp. 135 ff.

and wicked laws', by threats and persecution, had the many in
thrall. She promised that 'she can deliver the silly souls that
hath been long piteously punished in the pains of Purgatory';
she said that 'she can make fair weather and rain, and heal all
sickness and bring departed souls out of Hell'. Worse, this
'blasphemous monster' 'saith she can purchase remissicn of
sins by the offering up again of Christ's body and blood, and
can with five words make both God and man'. She threatened
the people that 'except they do hear and see her play in juggling
garments' every morning 'they should not prosper and go
forward in her business'.[69] Gospelling poems and satires gleefully told of Mistress Missa wandering desolate, bewailing
her fate:

> I was as proud as Lucifer that fell
> Which did presume to the almighty throne
> Whose portion therefore was the dark pit of Hell.
> So on earth compared to me was none
> I thought myself with God to be all one
> Privy of his mysteries and divine purveyance
> And now, alas, unhappy is my chance.[70]

In mock dirge the loss of Mother Mass was lamented. In the
tract *Vpcheringe of the messe* 'mistress missa', the harlot from the
Southwark Stews, 'Winchester gosling', returned whence she
came—the whore of Babylon back to her father in Rome. The
poet bade her an obscene farewell:

> A good mistress missa
> Shall ye go from us thissa?
> Well yet I must ye kissa
> Alack, from pain I pissa.

In a mockery of the Requiem, the poet sang to the departing
Mass:

69 William Punt, *A new dialoge called the endightment agaynste mother Messe* (*RSTC* 20499). For Punt, see King, *English Reformation Literature*, p. 288; C. H. Garrett, *The Marian Exiles: A Study in the Origins of Elizabethan Puritanism* (Cambridge, 1938), pp. 263–4.
70 Bodleian, Tanner 47(2), sig. Aii[v].

> Requiem eternam
> Lest penam sempiternam
> For vitam supernam
> Ad umbram infernam
> For veram lucernam
>
> . . .
>
> She may no longer stare
> Nor here with you regnare
> But trudge ad ultra mare
> And after habitare
> In regno plutonico
> Et Eve acronyco
> Cum cetu babilonico
> Et cantu diabolico.[71]

Yet this was not the poet's last encounter with Mother Mass, for Luke Shepherd may have composed at least nine satires against her, 'proper books against the papists'.[72] Edward Underhill credited 'Mr Luke, my very friend of Coleman Street, physician' with the dialogue between a simple ploughman and a priest who encountered each other on the eve of Corpus Christi, and 'reasoned together of the natural presence in the sacrament': *John Bon and Mast Parson* (1547). 'The papists were sore grieved' with this book, so Underhill remembered, and the Mayor, Sir John Gresham, was alarmed. But Underhill protested that it was 'a good book . . . there is many of them in the Court', and Gresham, perusing it, agreed that 'it was both pithy and merry'.[73]

For the printing trade of London 1548 was a year of great freedom and prosperity. Radical young printers like Scoloker, Daye, Seres, Powell, and Stoughton set up presses, and were joined by refugees from the Continent like Walter Lynne and Stephen Mierdman. In 1546 to 1547 there had been twenty-five known printers operating in the City; by 1548 there were thirty-nine. In 1548 225 works were printed in London, compared with a hundred or so in the previous year. The vast

---

[71] Luke Shepherd, *The vpcheringe of the messe* (*RSTC* 17630).

[72] For Shepherd and the nine anonymous works attributed to him, see King, *English Reformation Literature*, pp. 252–70, 502.

[73] *Narratives of the Reformation*, pp. 171–2. In the Repertory of the Court of Aldermen for 10 Jan. 1548 is a recognizance for John Day, the printer of the tract: King, *English Reformation Literature*, p. 95.

increase in production was accounted for almost entirely (80 per cent) by the flood of Protestant polemic.[74] The authors and printers of the Mass tracts dared to defy the proclamation of December 1547 because they believed in their message,[75] and because there was a booming market of those who wished to read it. Since surely only a small proportion of the City community was yet committed to the new faith, how is the voracious demand to be explained? The Londoners 'loved new things', and could not but be excited by the controversy. The crudeness of many of the tracts surely attracted the salacious. The writers often lampooned men in authority, and institutions known especially to them, and topical allusions in the tracts grounded the debate squarely in London, at the moment of Reformation.

Mother Mass was known to be found in London, and to have many followers who wished her to stay.

> And all the silly souls
> That heareth Mass in Paul's
> And in places beside
> In London that is wide
> Where Mass is sung or said
> And be nothing afraid
> That she shall go away
> But tarry while she may
> For she must long continue
> She hath such great retinue.[76]

Where is, asked Sergeant Wisdom, Mother Mass's 'most abiding and common trade where I may lay wait for her?' Easy, replied Brother Knowledge:

Her most abiding is at Paul's or at St Faith's parish under Paul's, or at Ludgate in St Martin's parish, and if you chance not of her there, walk to St Sepulchre's church, and if you look well you shall find her in one of those places without fail.[77]

---

[74] Took, 'Government and the Printing Trade', pp. 144–8; King, *English Reformation Literature*, pp. 88–9.

[75] For the religious views of individual printers, see Took, 'Government and the Printing Trade', p. 184.

[76] *The vpcheringe of the messe* (*RSTC* 17630).

[77] Punt, *A new dialoge . . . agaynste mother Messe*, sig. Avi[r].

There was 'a curate, not far without Newgate, of a parish large' whom Luke Shepherd lampooned as the drunkard *Doctour Doubble Ale*. This unregenerate curate determined to

> keep his old conditions
> for all the new commissions
> And use his superstitions
> And also mens traditions
> And sing for dead folk's souls
> And read his bead rolls.

He hoped still to defend the old faith by sheltering his parishioners, even now, even in London, from the defiling new ways:

> And let these heretics preach
> And teach what they can teach
> My parish I know well
> Against them will rebel
> If I once them tell
> Or give them any warning
> That they were of the New Learning
> For with a word or twain
> I can them call again.[78]

Doctor Double Ale was Henry George, curate of St Sepulchre.[79] Brother Knowledge was right to think that Mother Mass was to be found in this parish. If the wills which George witnessed at St Sepulchre in Edward's reign are examined, they are found to be prevailingly Catholic, never reformed.[80] The writers of the tracts knew well the popish strongholds in London because so many of them were Londoners themselves—William Punt, Randall Hurlestone, Phillip Nicolls, Luke Shepherd, John Champneys, Edward Underhill, Thomas Mountain, John Mardeley. The tracts, written by Londoners, printed and sold by the London book trade, were bought and read by the citizenry.

Matters of doctrine were being laid before the people, and explained, in the most basic terms: the novelty of being consulted was appealing. But there was always the danger that

---

[78] Luke Shepherd, *Doctour Doubble Ale* (1548; *RSTC* 7071); sigs. Aii[v], Av[v].

[79] The identification is explicit: ibid., sig. Avi[r].

[80] Guildhall, MS 9171/12, fos. 8[v], 9[r], 55[r], 67[v], 108[v]; PRO, PCC, Prob. 11/34, fo. 35[v].

the great 'numbers of books . . . abroad . . . very greedily devoured of a great sort' might be understood otherwise than their authors intended. Philip Nicolls in his *Here begynneth a godly newe storye of .xii. men that moyses sent to spye owt the land of Canaan* asked:

Do they read with such judgement that they receive the good and reject the bad? No. No. The weightier matter the sooner passed over and the less thanked the author. But trifling matters finely handled are esteemed.

But better that the reforming message was spread, even if misinterpreted, than for the gospellers to remain silent.[81] So it was, almost certainly, that Londoners would read works with a serious reforming intent, for entertainment. Some, of course, bought the pamphlets precisely in order to be shocked: like the conservative vicar of Skipton, Yorkshire, in London upon the Earl of Cumberland's business in January 1549, who purchased 'a testament of heresy, the confession of the maker thereof'.[82] Yet surely only those who had begun to doubt the power and efficacy of the Mass could enjoy these ballads and lampoons.

Most vocal among Mother Mass's opponents were the forward youth of London. The hero of *Doctour Doubble Ale* was a cobbler's boy, fired by the Gospel. 'The boy doth love no papistry', and his 'reading and sobriety, and judgement in the verity' made him worthier by far to have cure of souls than the drunken curate.

> Yet could a cobbler's boy him tell
> That he read a wrong Gospel.

The outraged Doctor Double Ale called upon his congregation:

> Is there no constable among you all
> To take this knave that doth me trouble?
>
> . . .
>
> And so the poor lad
> To the Counter they had.[83]

---

[81] P. Nicolls, *Here begynneth a godly newe story of .xii. men that moyses sent to spye owt the land of canaan* (*RSTC* 18576–7), sigs. Aiii^v, Aiiii^r; cited in Took, 'Government and the Printing Trade', p. 150.

[82] *Clifford Letters*, ed. A. G. Dickens (Surtees Society, clxxii, 1962), pp. 101–4; cited in Took, 'Government and the Printing Trade', p. 24.

[83] Shepherd, *Doctour Doubble Ale*, sig. Aiiii^r–v.

In Edwardian London art imitated life, and life art. At Christmas 1552 Thomas Cheeseborough's son insulted the curate at St Peter Paul's Wharf by 'washing' in mockery of the Mass 'with handkerchief in his hands'. When the curate, Mr John Holland, expelled the child from church for 'profaning' and 'playing the boy in the choir', the boy's mother challenged him: 'My boy shall be here, and . . . he is more worthy to be here than thou, for thou canst not read, and thou art a very drunken and a knave priest.'[84] At Corpus Christi 1548 two boys and a girl had dared to profane the blessed sacrament at St Margaret New Fish Street, even on this sacred festival, and in September Charles Tilby, a lad of thirteen, was whipped naked at St Mary Woolnoth for throwing his cap at the blessed sacrament at Mass as it was elevated.[85]

A threat more alarming by far than that posed by young reforming hotheads came from the heretical fringe. London became a haven for Anabaptists and Libertines, sectaries who could be tolerated by no government in Europe; indeed many of those in London were exiles. The reforms made by the 'lawful authority of the magistrates' were endangered—now more seriously than ever—by these extremists. The reformers had removed persecution, taken the 'sword from the adversaries' hand', with the repeal of the heresy laws, and had given the Gospel instead; only to see the freedom of the Gospel turned to licence and inspiring the most desperate beliefs. John Champneys of Stratford le Bow, knowing that the 'elect of God and regenerate in Christ' preferred 'the truth' however 'rudely written' to 'clerkly eloquence', set forth his millennial manifesto in *The Haruest is at hand wherein the tares shall be bound, and cast into the fyre & brent* (1548). He denounced the clergy: 'they themselves know not what the regeneration of the Spirit of Christ is, but what the Devil and the subtlety of man's wit by outward learning knoweth'.[86] These were not antichristian Roman clergy whom he condemned: rather the licensed preachers of the Edwardian commonwealth. For Champneys the true Gospel

---

[84] GLRO, DL/C/209, fos. 50r–52r.

[85] CLRO, Repertory 11, fos. 464v, 495v.

[86] John Champneys, *The haruest is at hand wherein the tares shall be bound, and cast into the fyre & brent* (RSTC 4956), sig. Aiiiiᵛ. King, *English Reformation Literature*, pp. 91–2.

had been so persecuted since the apostles' time that the godly—
the 'marked men'—might not profess it. Champneys appeared
before a commission, headed by Cranmer, on 27 April 1549 at
St Paul's and abjured extreme heresies. He held the antinomian heresy of the Dutch Libertines, believing that there
were two 'homines', the inner man of the spirit, and the outer
man of the flesh: the outer man might sin, but the inner man,
regenerate in Christ, never could. To His elect, Champneys
claimed, Christ allows 'their bodily necessities of all earthly
things'.[87] On 28 April 1549 Champneys bore a faggot at Paul's
Cross for his heresies.[88] In May 1549 Michael Thombe, a
butcher of St Mary Magdalen Old Fish Street, called before
Cranmer at Lambeth, confessed his Anabaptist beliefs: 'that
Christ took no flesh of Our Lady; and that the baptism of
infants is not profitable, because it goeth before faith'. In May
he bore a faggot at Paul's Cross, twice; 'because he made a
mock at the first time'.[89]

'Alas! not only are those heresies reviving among us which
were formerly dead and buried, but new ones are springing up
every day', wrote an appalled John Hooper to Bullinger in
June 1549. Anabaptists were flocking to his daily lectures.
They denied that Christ was incarnate of the Virgin; they
contended that 'a man who is reconciled to God is without sin,
and free from all stain of concupiscence, and that nothing of the
old Adam remains in his nature. A man who is thus regenerate
cannot sin.' (The antinomian heresy, this.) Some maintained
heresies worse still in their conventicles, Hooper feared: 'denying
that Christ is Saviour of the world', even calling 'that Blessed
Seed a mischievous fellow and deceiver of the world'. Hooper
feared trouble, and rightly.[90] A fragmentary text survives of an
interlude called *Love Feigned and Unfeigned* (1547–9?) which
articulates incendiary Anabaptist arguments for the abolition

[87] Lambeth, Register Cranmer, fo. 78ʳ; J. F. Davis, *Heresy and Reformation in the
South-East of England, 1520–1559* (1983), pp. 103–4; Strype, *Memorials of Cranmer*, i,
pp. 254–5.

[88] Wriothesley, *Chronicle*, ii, p. 10; *Grey Friars Chronicle*, pp. 58–9. Laurence Clerk of
Whitechapel was one of the sureties for Champneys, and Clerk himself would be
bound to keep the peace on 20 July 1549, at the height of the rebellions: CLRO,
Repertory 12, fo. 113ᵛ.

[89] Strype, *Memorials of Cranmer*, i, p. 257; *Grey Friars Chronicle*, pp. 58–9.

[90] *Original Letters*, i, pp. 65–6.

of private property and the levelling of social ranks.[91] Whether it was ever performed, and before whom, is unknown, but Joan Bocher would maintain at her burning in May 1550 that there were a thousand Anabaptists in London and Essex.[92] The threat was real enough.

By the autumn of 1548 tolerance of dissent had given way before the urgent necessity of enforcing the social order. To prevent dissension all sermons were halted by proclamation in September.[93] The Mercers' Company ordered each householder to 'look to his apprentices or other servants that they do not run to Paul's a gazing or gaping as they have been wont to do heretofore, for because there shall be no sermons there yet a while'.[94] Just as preaching was inhibited to prevent the dissemination of divergent views which engendered chaos, so the Bishops sought to end doctrinal anarchy by establishing a uniform doctrine upon the central question of the eucharist. In the House of Lords they disputed between 14 and 18 December. At extreme poles of the debate were Ridley and Bonner; Bonner soon to be quondam Bishop of London, and Ridley his successor. Cranmer had been converted by Ridley, his chaplain, to Ratramnian doctrine on the eucharist, in 1546, or so it seems, and it was this doctrine which Cranmer had enshrined in the new Prayer Book and on which the debate turned.[95] 'As Christ', Ridley held, 'took upon him manhood, and remaineth God; so is bread made by the Holy Ghost holy and remaineth bread still.' 'Concerning the outward thing it is very bread. But according to the power of God is ministered the very body.' 'The manhood is ever in Heaven; His divinity is everywhere present . . . Christ sits in Heaven. And is present in the Sacrament by His working.' For this doctrine Lambert had died; so Bonner objected. But for all Bonner's protests, it was

[91] King, *English Reformation Literature*, p. 311.

[92] I. B. Horst, *The Radical Brethren: Anabaptism and the English Reformation to 1558* (The Hague, 1972), pp. 109–11. For Joan Bocher's career, see J. F. Davis, 'Joan of Kent, Lollardy and the English Reformation', *JEH* xxxiii (1982), pp. 225–33.

[93] *TRP* i. 313; Wriothesley, *Chronicle*, ii, p. 6.

[94] Mercers' Company, Acts of Court, fo. 224ʳ (cited in Hoak, *The King's Council*, p. 215); Clothworkers' Company, Court Book I, fo. 203ᵛ.

[95] The debate is printed in F. A. Gasquet and E. Bishop, *Edward VI and the Book of Common Prayer* (1890), pp. 395–443, and discussed in pp. 157–81. H. C. G. Moule, *Bishop Ridley on the Lord's Supper* (1895); P. Brooks, *Thomas Cranmer's Doctrine of the Eucharist* (1965); Dugmore, *The Mass and the English Reformers*, pp. 127–31.

this doctrine which would be maintained in the first Book of Common Prayer, which according to the first Act of Uniformity must be introduced on Whit Sunday 1549.[96]

It was not in the structure and order of the service that *The Supper of the Lord, and the Holy Communion, commonly called the Mass* in the 1549 Prayer Book differed from the Latin Mass, but there was a fundamental transformation in the rite.[97] No longer was the Mass clearly a placatory sacrifice of Christ 'here really present'; rather a thanksgiving: 'to celebrate the commemoration of the most glorious death of Thy son'. Yet to stress 'His one oblation once offered', and to acknowledge Christ's saving passion—'a full, perfect and sufficient sacrifice, oblation and satisfaction for the sins of the whole world'—was never to deny belief in the Real Presence. The Prayer Book laid emphasis upon the presence of Christ in the sacrament, spiritually eaten and drunk:

He hath left in those holy mysteries, as a pledge of His love, and a continual remembrance of the same, His own blessed body, and precious blood, for us to feed upon spiritually, to our endless comfort and consolation.

Yet, for all the attempt at reconciliation, the loss of ceremonies, of the elevation of the host, the use of English instead of Latin, the clear reforming intent which lay behind the new rite, made the Prayer Book an abomination to all those of conservative mind. Whereas all those who wished for further reformation condemned the Book as but half reformed, or worse. Martin Bucer expressed his reservations in his *Censura*, and Peter Martyr also presented criticisms to Cranmer.[98] Cranmer had defended the loss of ceremonies: 'Christ's Gospel is not a ceremonial law . . . but it is a religion to serve God, not in bondage of the figure or shadow, but in the freedom of the spirit.' But he knew well that many 'think it a great matter of conscience to depart from a piece of the least of their ceremonies'; conversely now, many be 'so new fangle that they would

---

[96] F. Procter and W. H. Frere, *A New History of the Book of Common Prayer* (1902), p. 49.

[97] For the discussion which follows, see Dugmore, *The Mass and the English Reformers*, pp. 131–7; F. E. Brightman, *The English Rite*, 2 vols. (1915), i, pp. lxxviii–cxxix.

[98] Ibid. i, pp. cxlii–cxliii; Strype, *Memorials of Cranmer*, ii, appendix lxi.

innovate all thing'.[99] How would the people in their parish churches interpret the new service and receive the new order of celebration?

If purchasing the new service books is evidence of the approval of them, or of using them, then the London churches seem to have been compliant.[100] But the possession of Prayer Books tells nothing of how or whether they were used, in which spirit. The rebellion which began in the West Country against the new Prayer Book found few resonances in London, yet the City governors could hardly be complacent. In the weeks of insurrection trouble came from the radicals, from gospellers like the cordwainer who expounded Scripture at St Paul's in July.[101] At St Botolph Aldersgate the windows were broken 'in the commotion time with shooting of guns',[102] but whether this was an expression of religious conviction, and of which persuasion, is unclear. The confusion of worship and belief was manifest in London through the summer of 1549. Corpus Christi, when the host was carried ascendant, was a sacred Catholic festival; to the reformers it was idolatry. John Bon had asked Master Parson in the 1547 tract:

> But tell me master Parson one thing, and you can;
> What saint is Copsi Cursty, a man, or a woman?[103]

How would Corpus Christi be celebrated in the City in that troubled June? The Grey Friars chronicler recorded the confusion: 'in diverse places in London [Corpus Christi] was kept holy day; and many kept none, but did work openly, and in some churches service and some none, such was the division'.[104]

---

[99] *The Two Liturgies . . . set forth by authority in the reign of King Edward VI*, ed. J. Ketley (Parker Society, Cambridge, 1844), pp. 155–6.

[100] Guildhall, MSS 1016/1, fos. 3r–4v; 593/2, fo. 5r; 4956/1, fos. 175v, 177v; 1432/1, fo. 99v; 1279/2, fo. 52r–v; 1568/1, pp. 19–20; 4570/1, fos. 98r, 101r; 2596/1, fos. 97v, 98v, 99r; Walters, *London Churches*, pp. 213, 319, 324, 333, 368, 387, 546, 548, 560, 564. (It is not always clear whether these references are to the 1549 or 1552 prayer books.)

[101] CLRO, Repertory 12, fo. 117v.

[102] Walters, *London Churches*, p. 203.

[103] *Tudor Tracts*, p. 161.

[104] *Grey Friars Chronicle*, p. 59.

CONSERVATIVE RESISTANCE

Following his protest and imprisonment in 1547, Bonner had complied throughout 1548 in advancing the reformation, 'although not most forward'; sending mandates for abolishing images, the use of candles, ashes, palms, and such ceremonies, for abrogating the Mass, setting up Bibles and ministering in both kinds.[105] But from the spring of 1549 his opposition became plain. Seeing the political instability—the execution of Thomas Seymour in March and the risings in the summer—and hating each alteration in religion, he 'began somewhat, as he durst, to draw back and slack his pastoral diligence', giving licence to those of conservative views to worship as they pleased.[106] Even in St Paul's privy masses were celebrated in private chapels and 'remote places'. The Council, aware of this silent resistance, determined to end it, for, as ever, if the religious changes were seen to be thwarted even in London then enforcement would be harder farther away. So it had proved in the West.

At the end of June 1549 a commandment came to Bonner to stop the private masses in St Paul's; 'the Apostles' Mass and Our Lady's Mass, now falsely named Our Lady's Communion and the Apostles' Communion . . . for the fondness of the name, a scorn to the reverence of the communion of the Lord's body and blood'. No communion was to be held, except at the high altar.[107] Another order came on 23 July, rebuking Bonner for his negligence in introducing the Prayer Book in his diocese.[108] On both occasions Bonner sent admonitions to the Dean and Chapter to comply,[109] but even on 27 and 29 July some were found to be hearing Mass at the French ambassador's residence at Creechurch, and were 'greatly rebuked' by the Council.[110] The Bishop was accused of condoning the

---

[105] Guildhall, MS 9531/12, fos. 111[r], 111[v], 268[r]; Foxe, *Acts and Monuments*, v, pp. 716–18.

[106] Ibid. v, p. 745.

[107] *Documentary Annals*, i, pp. 65–6; Foxe, *Acts and Monuments*, v, p. 723; *Grey Friars Chronicle*, p. 59.

[108] *Documentary Annals*, i, pp. 66–8.

[109] Guildhall, MS 9531/12, fo. 219[v]; Foxe, *Acts and Monuments*, v, pp. 723–4, 727.

[110] *Grey Friars Chronicle*, p. 61.

secret Mass. 'Through the evil example and slackness of you in preaching and instructing', some in London were 'forgetful of their duty to God', coming ever more rarely to 'prayer and to the holy communion'. Worse, 'diverse . . . do frequent and haunt foreign rites of masses'. Bonner was warned to look to reformation, and ordered to preach at Paul's Cross three weeks later, and every quarter thereafter; to celebrate at St Paul's himself at every major feast; to summoh all those who failed to attend communion and who went instead to forbidden masses; to ensure that the derelict churches of London be repaired and tithes be paid; to punish adulterers, and to reside in London for the time being.[111] To test Bonner's allegiance the Council required him to denounce the rebels, promising them 'what masses or holy water so ever they pretend' they would be 'swallowed down alive into Hell'. He must insist that 'extern rites and ceremonies be but exercises of our religion and appointable by superior powers'. Lastly, he must declare that which Gardiner had denied: that the regal authority of the King was not impaired by his minority.[112] On 18 August Bonner pointedly 'did the office at Paul's both at the procession and the communion discreetly and sadly',[113] but would he comply in his sermon? The sermon of 1 September was to be Bonner's Rubicon: so the Council intended. Somerset, struggling desperately to maintain his power, even his life, saw that to sacrifice Bonner might be a way to save himself, for to remove the Bishop might win him the allegiance of a vocal and powerful (if not representative) section of City society.[114]

Bonner preached nothing at all of Edward's authority, and little enough concerning the rebels. He came to chastise the Londoners, and his audience was restive, hostile. Urging them to listen to him in silence, he besought them to pray with him and for him, for he was not 'so uncharitable but he would pray for them which hated him'. He denounced the corruption of cities: 'Marry, sir . . . into a city (as Pythagoras doth say) four things may enter and creep . . . first pleasure, after security,

---

[111] PRO, SP 10/8, nos. 36, 37, 57; Guildhall, MS 9531/12, fo. 220ᵛ; Foxe, *Acts and Monuments*, v, pp. 729–30, 763.

[112] Ibid. v, pp. 745–6.

[113] *Grey Friars Chronicle*, p. 62.

[114] The suggestion is made by Professor Jordan: *Edward VI: The Young King*, p. 217.

then violence, and at the last ruin, destruction and desolation.'
He warned of Christ's promise that 'in the latter days faith
shall decay so sore, and charity wax so cold that in comparison
it was not to be counted as any faith at all'. So it was now in
London. The good works of 'men of old time'—'praying,
fasting, paying of tithes in alms deeds'—were lost. 'For now
charity is waxen so cold that very few men come to church at
all, and if they do come to church, they come to walk, to hear
some news, and to meet with their friends and not to pray.'
Bonner urged his hearers to pray in humility, innocence, and
purity of life and love to neighbours, and he exhorted them to
come more often to holy communion. Here it was that his
sermon diverged from dutiful obedience to Council precept to
resistance to government order. The reason, he averred, why
so few came, why devotion to the communion was 'very cold',
was because many now held false beliefs concerning the blessed
sacrament of the altar.[115] Here he declared his own conviction,
as a reformer recorded (his own hostile comments in paren-
thesis): 'that after the consecration the very body of Christ that
was born of the Virgin Mary (if that were a body), and the self
same body that did hang on the cross (if that were a body)'
were really present in the sacrament.[116] Thus avowing his own
'true and Catholic belief', Bonner prayed, 'even from the
bottom of all my very heart, that they should share it'.[117]

'For the which good sermon', the reformer observed, 'men
doth much marvel that he is not committed to the Tower.'[118]
John Hooper and William Latimer, rector of St Laurence
Pountney, reported Bonner's disobedience to the Council, and
that same afternoon Hooper, 'having a great rabblement with
him of his damnable sect', inveighed against Bonner's sermon
and denied the Real Presence.[119] Within a week Cranmer,
Smith, Petre, Ridley, and May were commissioned to hear

[115] An account of Bonner's sermon was sent to Somerset: BL, M 485/52, vol. 198,
fos. 34ʳ–46ʳ (*HMC* Salisbury MSS, i. 333). Bonner had himself composed the Homily
upon charity of 1547, and would compose that of 1554: J. N. Wall, 'The Book of
Homilies of 1547 and the continuity of English Humanism in the sixteenth century',
Anglican Theological Review, lviii (1976), p. 77.

[116] NLS, Adv. MS 34.2.14, fo. 35ʳ.

[117] BL, M 485/52, vol. 198, fos. 45ᵛ–46ᵛ.

[118] NLS, Adv. MS 34.2.14, fo. 35ʳ; *Original Letters*, ii, p. 557.

[119] Guildhall, MS 9531/12, fo. 221ᵛ; Foxe, *Acts and Monuments*, v, pp. 747–8, 752.

evidence against Bonner. There followed seven hearings between 10 and 23 September. All the while, Somerset's rivals in the Council manœuvred against him, and preachers in the capital stirred the people with radical and demagogic sermons against the Mass and Bonner, its champion. The Council sought his deprivation now with ruthless determination, but Bonner fought hard. He denied both the legality of the commission and the integrity of his accusers.[120] 'Merchant Latimer' and 'Merchant Hooper' (denying their ordination) he despised as 'manifest and notable heretics and seducers of the people'; 'familiarly haunting . . . with sacramentaries'.[121] The matter, he said, turned not upon his disobedience in failing to uphold the King's authority during his minority; instead upon his views on the nature of the Mass. To Cranmer's question, 'What presence?', Bonner answered him, mocking the Archbishop's own vacillation regarding this central doctrine: 'What believe *you*, and how do *you* believe, my lord?'[122]

The commissioners examined also his compliance to the previous Injunctions. Did the Bishop know that 'certain persons' had attended Mass in Latin, after the old rite?[123] He knew of none, 'except it be in the house of my lady Mary's grace, or in the houses of the ambassadors, nor yet there, nor in any of them, but by flying and not assured report'.[124] Further witnesses were brought to testify against Bonner on 16 September—Sir John Cheke, Henry Markham, John Joseph, John Douglas. Against these also Bonner objected: let detailed questions be put to them; whether any of them had been 'reputed for a sacramentary'; whether they had 'wished me . . . to be deprived or put in prison . . . rejoicing thereof.'; 'whether any of them have been in times past a friar preacher . . . monk . . . professing solemnly poverty, chastity and obedience', and hereafter married? The answers to these questions—being positive—impugned the good faith of the witnesses.[125] John Joseph, rector of St Mary le Bow (by Cranmer's presentment),

---

[120] Guildhall, MS 9531/12, fo. 222r; PRO, SP 10/8, no. 57; *Documentary Annals*, i, pp. 71–3; *CPR Edw. VI* i, pp. 165–6; Foxe, *Acts and Monuments*, v, p. 751.

[121] Ibid. v, pp. 752, 755.

[122] Ibid. v, pp. 752, 759, 765.

[123] PRO, SP 10/8, no. 58; Foxe, *Acts and Monuments*, v, p. 763.

[124] Ibid. v, p. 768.

[125] Ibid. v, pp. 770–2, 780–1.

was an erstwhile friar preacher, now married, who preached most vehemently against the old faith, and, wrote Foxe, 'converted not a few to sincere religion'. He, with Chambers and Cheke, and doubtless the others, held reformed views upon the Mass.[126] With each hearing Bonner's intransigence and contempt for his judges became more apparent. When Sir Thomas Smith 'rebuked the Bishop greatly' Bonner 'gave him many shocking words openly'.[127] Micronius told Bullinger that Bonner behaved with such effrontery 'that you would rather call him a buffoon than a Bishop'.[128] On 20 September Bonner comforted his despondent chaplains: 'Why . . . so sad and heavy in mind? . . . I am . . . joyful of this my trouble, which is for God's cause.'[129]

At last Bonner gave clear leadership to the opponents of Reformation. He charged his chaplains to entreat the Mayor and Aldermen never to listen to the sacramentaries preach, lest they be infected themselves by 'poisoned doctrine', and also 'give a visage to the encouragement of others'.[130] Rather than listen to 'vile beasts and heretics' rail against the blessed sacrament, Bonner had 'departed yesterday [15 September] from the heretic prater's uncharitable charity'. Still on 20 September at St Paul's, Cardmaker preached that 'if God were a man he was six or seven feet of length [tall] . . . and if it be so, how can it be that he should be in a piece of bread in a round cake on the altar?' That night Bonner was committed to the Marshalsea. The Council used the radical preachers to justify the trial of the Bishop to his adherents and to delight his enemies. Hooper preached, at Cranmer's behest, at Paul's Cross on 22 September, and 'there he spake much against the Bishop of London'. Cardmaker, playing to his audience, regretted on the 25th that he would not be able to read at Paul's two days later, 'for because he must needs be at the session . . . at Lambeth for the Bishop of London'. 'But it was not so', protested the Grey Friars chronicler, 'for the Bishop came not there.'[131] On the 29th the preacher in the Shrouds 'spoke much against the

---

126 For Joseph, Cheke, and Chambers, see pp. 426, 562, 571, 628 above and below.
127 *Grey Friars Chronicle*, p. 63. (My interpretation here differs from that of the editor of the chronicle.)
128 *Original Letters*, ii, pp. 557–8.   129 Foxe, *Acts and Monuments*, v, p. 785.
130 Ibid.   131 *Grey Friars Chronicle*, p. 63.

Bishop'. Bonner was summoned to Lambeth on 1 October, and there was deposed by Cranmer: 'as much as lay in his [Cranmer's] power', judged the chronicler—'but mark what followeth'.[132]

On 8 October the Lord Protector was proclaimed a traitor and sent to the Tower. Bonner's deprivation, so ruthlessly sought, had been part of a political campaign by Somerset to prove his reforming credentials and to save himself. Some were convinced of his favour to the Gospel, others not; and popular opinion on this matter was crucial and might be manipulated. So, at Hampton Court on 6 October Somerset had heard that his rivals 'are not ashamed to send posts abroad to tell that we are already committed to the Tower and that we would deliver the Bishops of London and Winchester out of the prisons and bring in again the old Mass'.[133] On 7 November 1549 Hooper told Bullinger of Bonner's deprivation, and wrote presciently of his own 'sharp and dangerous contest with that Bishop . . . Should he be again restored . . . I shall, I doubt not, be restored to my country and my Father which is in Heaven.'[134] Meanwhile, Scudamore reported to Hoby in February 1550: 'It is openly spoken that there shall be more quondam Bishops in England shortly.'[135]

In the Marshalsea Bonner took his imprisonment with a typical impatience. Stories circulated through London of his discomfiture. The Knight Marshal, whose reforming instincts were compounded by the Bishop's refusal to pay him, withdrew all comforts from Bonner, taunting him: 'Cod's head' and 'grossum caput', 'swearing that he had no more learning than a dog, unless it were in the Pope's degrees'. The 'patient prelate boldly answered . . . that if the Marshal had spoken thus in Cheapside he would have cut away a piece of his face'.[136] So with 'great gentleness' (wrote the Grey Friars chronicler ironically) Bonner's bed was taken away.[137] Soon

---

[132] Wriothesley, *Chronicle*, ii, p. 24; *Grey Friars Chronicle*, p. 64.

[133] PRO, SP 10/9, fo. 6^r. Bonner called Somerset and Smith 'my deadly enemies': Foxe, *Acts and Monuments*, v, p. 797.

[134] *Original Letters*, i, pp. 69–70.

[135] NLS, Adv. MS 34.2.14, fos. 5^v, 29^r.

[136] Ibid., fo. 9^r.

[137] *Grey Friars Chronicle*, p. 65.

enough Bonner would be discovered at the centre of a ring organizing Catholic propaganda.

With their Bishop now powerless to shelter them, and with rulers more committed than ever to reform, there seemed no way for London's most resolute Catholics but exile. The year 1549 saw an exodus of those who 'to see a Mass freely in Flanders, are content to forsake, like slaves, their country'. So Ascham scorned them, shunning his compatriots on a visit to Louvain in 1551.[138] The exiles, for their part, had resolved 'in this perilous time of trial of the corn from the moveable chaff' no longer to stay among heretics. On 21 December William Rastell followed his father-in-law, John Clement, John Story, Anthony Bonvisi, and others to Louvain. Story, writing his will in exile in May 1552, would thank God for His great mercy, 'leading me, a wretched sinner, out of my native country, the which (being swarmed out of the sure ship of our salvation) I beseech Almighty God of His infinite mercy to restore again to the unity of the same vessel being our mother the holy Catholic Church'.[139] Story's exile followed his imprisonment in November 1548 by an angry House of Commons when he denounced the Statute of Uniformity and its initiators: 'Woe unto thee, O land, where the King is a child.'[140] With every day in exile he hated more the regime and the heresy which had compelled him to leave. His wife he bound by a 'promise made to God and me . . . that she at no time until the land of England be restored to the unity of Christ's Church will return thither', and he secured her vow by giving Anthony Bonvisi charge of her allowance.[141] Story would have his revenge upon the heretics. Clement, Rastell, Balthasar, and Bonvisi could not have returned to England during Edward's reign, even had their resolve weakened, because their houses, goods, and books were seized on 7 February 1550, the day after Somerset's release from the Tower, and the very day of

---

[138] Cited in J. K. McConica, *English Humanists and Reformation Politics under Henry VIII and Edward VI* (Oxford, 1965), p. 271.

[139] Inner Temple, Petyt MS 538/47, fos. 66^r–68^r.

[140] *House of Commons, 1509–1558* (Story).

[141] Inner Temple, Petyt MS 538/47, fos. 66^r–68^r. Story's daughter and grandson became recusants: V. H. H. Green, *The Commonwealth of Lincoln College, 1427–1977* (Oxford, 1976), pp. 136–7.

Bonner's formal deprivation and condemnation to perpetual imprisonment.[142]

The Catholic community in exile studied to provide polemic to oppose England's rulers, and looked always to the time when the old faith would be restored. In Liège since 1541, Sir Geoffrey Pole treasured the manuscripts of More's writings, even censoring passages where he thought More advanced perilously close to a reformed position.[143] William Rastell, surely in contact with Pole, prepared his edition of More's complete works.[144] Bonvisi was More's greatest friend, Clement his son-in-law. At the university of Louvain such as Smith, Story, and Peryn made 'forts . . . against the open truth of Christ's Gospel'.[145] The exiles conspired with friends at home to bring works of Catholic polemic and instruction into England. In March 1551 the Council discovered a plot to smuggle from Paris 'Dr Smith's most false and detestable books', refuting the works of Cranmer and Peter Martyr.[146] This enterprise was not directed to a whole nation suffering in heretical bondage, rather to certain known sympathizers. Books, often with accompanying letters, were sent to the Catholic establishment: to Bonner in the Marshalsea, to Walter Prince, servant to Sir Edmund Peckham (who received twenty books in Latin, eighty in English), to John White, Warden of Winchester College, to Reynolds, a priest (probably Robert Reynolds, Gardiner's Commissary at Winchester), to Bonner's chaplain Dr Gilbert Bourne. Catholics in London and Southwark eagerly sought the books: John Cawood, the printer 'dwelling at the sign of the Holy Ghost', Baldwin Wotton, Beard, a tailor in Fleet Street (later the Marian promoter), Anne Alford, Seton, Mr Bowle, a priest of Trinity Lane, White, Royar, a printer, and the King's crossbow maker whose son was apprentice in Paris.[147]

Robert Caly set up a Catholic press in exile at Rouen. There

---

[142] Wriothesley, *Chronicle*, ii, pp. 33–4; PRO, C 1/1418/23–6; *CPR Edw. VI* iv, p. 171; A. W. Reed, 'John Clement and his books', *The Library*, vi (1926), pp. 329–39.

[143] More, *Dialogue of Comfort*, pp. xxvi–xxviii.

[144] A. W. Reed, *Early Tudor Drama* (1926), pp. 86–8.

[145] Foxe, *Acts and Monuments*, viii, p. 649.

[146] *APC* iii, p. 232; W. K. Jordan, *Edward VI: The Threshold of Power* (1968), pp. 249–50.

[147] BL, M 485/53/200, fos. 86^r–87^r (*HMC, Salisbury MSS*, i. 346).

he printed Thomas Martin's *Traictise declaryng . . . that the pretensed marriage of priestes is no marriage*, and almost certainly Smith's works.[148] William Seth was the carrier. Here Bonner's part in the enterprise appears, for Seth had been the Bishop's long-suffering servant until All Hallowtide 1550 when, Seth said, Bonner 'fell out with me and did beat me out of his chamber at the Marshalsea'. In prison Seth read to his master: a French chronicle and an English book containing 'matters of religion'. His own reading was of French books, the New Testament, Marcus Aurelius, and the 'Bishop of Winchester's book of the sacrament' (doubtless Gardiner's *An Explicacion and assertion of the true catholyke fayth, touching the sacrament*, which Caly printed at Rouen). Letters came to Bonner from Dr Baines in Paris urging resignation, 'for he was neither the first that had suffered persecution, nor should be the last'; but unavailingly. It was Bonner who sent Seth to Paris with letters of introduction to the Catholic exiles. That the Council was right to fear the continuing resistance of the Bishop and his adherents from prison and exile is suggested by Baines's gift to Bonner of a little book in French 'concerning the answer which the Commons of Devonshire made to the King of England in the time of the late commotion'.[149]

Edward VI, it was said, knew the name and religion of each of his magistrates:[150] he must have known, too, that many of them, though far from urging resistance, were unsound, even in London. The Catholic opponents of the new orthodoxy had in London 'justices abroad fit for their hands', reluctant to enforce the changes in faith. Bonner had written to the Aldermen urging them to shun reformist sermons, fully confident that they shared his views. London's gospellers castigated the City rulers as 'veteran papists'.[151] In May 1548 the King's own uncle, Sir Clement Smith, and the Recorder of London, Robert Broke, attended Mass at St Gregory by St Paul's. The curate 'reciting the common prayers at the choir door in the high Mass time prayed . . . that Almighty God might send

---

148 *RSTC* 17517. Took, 'Government and the Printing Trade', p. 176.
149 BL, M 485/53/200, fos. 88ʳ–93ʳ (*HMC, Salisbury MSS*, i. 347–8).
150 Foxe, *Acts and Monuments*, v, p. 700.
151 John Bale, *Yet a course at the romysh foxe* (1543; *RSTC* 1309); *Original Letters*, ii, p. 636.

the King's Council grace and bring them out of the erroneous opinions that they were now in.' Would the magistrates arrest him, according to their duty? No. 'Hearing the same his prayer', Smith and Broke 'laughed thereat'.[152] In March 1551 when Sir Anthony Brown and Sergeant Morgan were sent to the Fleet for hearing Mass, Smith 'which a year before heard Mass' was also 'chided', so his nephew recorded.[153] If their magistrates and their clergy did not lead them to reform, how were the people to be converted? 'That many-headed monster is still wincing', reported Hooper in February 1550; 'partly through ignorance, and partly fascinated by the inveiglements of the Bishops and the malice and impiety of the Mass priests.'[154] Though the Catholic activists were few, the silent Catholics were very many.

Many—perhaps even most—Londoners lingered still in a religious half-world. Under cover of the new rite all the old ceremonial uses not expressly forbidden might be preserved. At Christmas 1549 Hooper wrote despairingly to Bullinger of a City not yet half reformed. 'The public celebration of the Lord's supper is very far from the order and institution of Our Lord.' Sometimes the Lord's supper, though administered in both kinds, was being celebrated three times a day as the idolatrous Mass had been. 'Where they used heretofore to celebrate in the morning the *Mass* of the apostles; they now have the *communion* of the apostles; where they had the *Mass* of the Blessed Virgin; they now have . . . the *communion* of the Virgin.' Instead of the high Mass, the high communion. Vestments and candles before altars remained; they chanted the hours. 'That popery may not be lost, the mass priests' chanted in English in the same tone and manner they had always used in Latin. 'God knows to what perils and anxieties we are exposed by reason of men of this kind.'[155] Still in the autumn of 1550 the communion service at St Paul's, perhaps in private chapels, was so conducted as to seem 'a very Mass'.[156] The fall of

[152] CLRO, Repertory 11, fo. 465ʳ. *House of Commons, 1509–1558* (Clement Smith and Robert Broke).
[153] *Edward VI Chronicle*, p. 56.
[154] *Original Letters*, i, p. 76.
[155] Ibid. i, p. 72.
[156] *APC* iii, p. 138.

Somerset had raised hopes, spurred by the rumours of 'divers unquiet and evil disposed persons', that 'they should have their old Latin services' again, as though the promulgation of the Prayer Book 'had been the only act of the . . . Duke'. So on Christmas day 1549 the order of the first Act of Uniformity was again proclaimed, and all old Latin service books were ordered to be defaced and removed.[157] But the new faith still waited to be evangelized.

[157] *TRP* i. 353.

XI   The Creation of the Commonwealth

THE REFORMERS had a grand vision of the Christian commonwealth which they would create. Reformation in society must follow reform in religion. The accession of a godly King promised the chance to turn hopes to reality. As Henry VIII lay dying, he summoned to him those who would rule the realm in the new reign: Hertford, Dudley, Denny, and Paget. Men of consuming personal ambition, they determined in the King's last days to manipulate the royal will to secure their own wealth and political advantage. But they also did something surprising. On 13 January 1547, a fortnight before the King's death, the City hospitals of St Bartholomew and Bethlem were granted to the Communalty, out of the Church's control. These grants, like those of the spoils they took for themselves, were signed with the dry stamp which they had seized for their purposes.[1] This ambivalence—of self-seeking and cynical politicians with fleetingly altruistic ambitions—marked the reign. Could a commonwealth be created?

THE ADVANCE OF REFORM

The evangelical exiles of Henry's reign hastened home upon the accession of the new King to rebuild the temple. 'Hotlings' like John Hooper, Thomas Becon, and William Turner returned to places of influence around the Court, and condemned those who had borne with the persecution. Thomas Smith complained to the Duchess of Somerset in 1550 that such men 'kneel upon your grace's carpets and devise commonwealths as they like, and are angry that other men be not so hasty to run straight as their brains crow'.[2] Their evangelism could not be

---

[1] For the machinations at Henry's death-bed, see below, p. 488. *Memoranda relating to the Royal Hospitals of the City of London* (1836), pp. 7–11; appendices iv–v.

[2] BL, Harleian MS 6989, fo. 146ʳ, cited in M. L. Bush, *The Government Policy of Protector Somerset* (1975), p. 67.

tempered by prudence. Hooper returned from exile in May 1549 and to safety 'in London and in the family of the Lord Protector'.[3] Though he found that 'the Gospel of Christ . . . is daily striking root more deeply' and 'great, great . . . is the harvest', still the labourers in the harvest were few. A few godly men expounded the Holy Scripture; two read at St Paul's four times a week, and Hooper hastened to join them. 'Having compassion upon the ignorance of my brethren', he began in June 1549 to preach twice daily 'to so numerous an audience that the church cannot contain them'.[4] Dr Smith wrote sarcastically of his impact: after Hooper began to preach in London the people held him as 'a prophet, nay they looked upon him as some deity'.[5]

Refugees from European persecution found a haven in England. The 'antichrists' on the Continent were 'athirst, athirst' for the blood of the godly, wrote Peter Martyr to Bucer in January 1549, urging him to flee.[6] Leading Continental divines came at the anxious invitations of Cranmer and Somerset—Bucer, Peter Martyr, Fagius, Ochino, Dryander, Utenhove, à Lasco, Tremellius, de la Rivière, Poullain, Vauville, Micronius, ab Ulmis, Veron, Alexander.[7] Some of the exiles found positions at the universities. Many others came to London, some to benefices there.[8] Arriving in England, they found 'fallow ground here, such as the Antichrist is wont to leave', for in England, as in France and Italy, preaching was neglected, and the clergy 'neither very learned, nor zealous in matters appertaining to Christ's kingdom'.[9] These men were to give a lead to England's infant reformed Church, but also to guide their own countrymen who had fled, like them, to London.

There had long been fraternities of foreigners worshipping together in the capital. The Germans who honoured the Holy

[3] *Original Letters*, i, pp. 48–69.
[4] Ibid. i, pp. 55, 65.
[5] Strype, *Ecclesiastical Memorials*, ii/1, p. 66.
[6] *Original Letters*, ii, p. 475.
[7] See, *inter alia*, W. K. Jordan, *Edward VI: The Young King* (1968), pp. 189–205; C. Hopf, *Martin Bucer and the English Reformation* (Oxford, 1946).
[8] Peter Alexander became rector of All Hallows Lombard Street in Aug. 1552; John Veron became rector of St Alphage in Jan. 1553: Hennessy, *Repertorium*, pp. 78, 86, 125, 293.
[9] *Original Letters*, ii, pp. 536, 539; Gorham, *Gleanings*, p. 78.

Blood of Wilsnak met in the Crossed Friars in 1459 and then in 1491 at the Austin Friars: in this gild 'none shall be received but if he be born beyond the sea'. A gild of Dutchmen honoured St Katherine in the Austin Friars in 1495.[10] In the 1540s London's foreign community still wished to worship together, and now with a new urgency, for many were exiled for their advanced beliefs. In England they found a refuge from persecution, but a Church insufficiently reformed, with few who could preach, and none in their language. Already in December 1548 Ochino wrote to Musculus in Strasburg, urging him to come to minister to the 'Germans' in London: a community, he claimed, five thousand strong.[11] In Canterbury Utenhove had gathered that 'through the efforts of Master John à Lasco a church has been granted . . . in London to the Germans'. 'The doctrine of eternal life may be not a little advanced, if the Lord should breathe benignantly . . . on these undertakings.'[12] To Hardenburg came a plea from Bucer that August for 'a faithful preacher in the language of Brabant': in London there were 'six or eight hundred Germans, all godly men most anxious for the Word of God'.[13] Unofficially, strangers' churches had been established in the capital by the summer of 1549; churches, however, without a discipline or ceremony, and threatened. 'Marvellous is the subtlety of Antichrist in weakening the churches of Christ.'[14] During this crisis of Reformation the strangers' churches would stand as a powerful example of a Church fully reformed for an English Church which was struggling to reform itself.

Hooper—'the future Zwingli of England'—and the returned exiles would tolerate no compromise of the 'religion of Christ', and fought hard that 'the idol of the Mass may be thrown out'.[15] During the coup of the autumn and winter of 1549 the godly had feared that their cause was lost and that the papists

---

[10] H. C. Coote, 'The Ordinances of some secular gilds of London, 1354–1496', *TLMAS* iv (1871), pp. 44–55.

[11] *Original Letters*, i, p. 336. For the establishment of the stranger churches, see A. Pettegree, *Foreign Protestant Communities in Sixteenth-Century London* (Oxford, 1986).

[12] Gorham, *Gleanings*, pp. 74, 78.

[13] *Original Letters*, i, p. 352; ii, p. 540.

[14] Ibid. ii, p. 561; Gorham, *Gleanings*, p. 79.

[15] M. Morris West, 'John Hooper and the Origins of Puritanism', *Baptist Quarterly*, xvi (1955–6), p. 24; *Original Letters*, i, p. 79.

would regain 'their kingdom'. In the event Hooper could report by Christmas 1549 their providential delivery.[16] Yet they now found themselves thwarted, not just by the papists, but by that reformed—for Hooper and the Zwinglians half-reformed—religious and political establishment, the pharisee Bishops. The form of worship enshrined in the Book of Common Prayer was 'very defective . . . and in some respects indeed manifestly impious', judged Hooper. The Ordinal of March 1550 was so objectionable that Hooper sent it that Bullinger might know the 'fraud and artifices' whereby the Bishops 'promote the kingdom of Antichrist'.[17] Here was the crisis of reformed Protestantism: theological and structural. What was to be the nature of the sacraments in the reformed Church?

In the City Hooper preached once or twice a day throughout Lent before great crowds. 'God was with them; for he opened their hearts to understand.' His subject: the epistle of Titus, seven chapters of St John's gospel, and the prophet Daniel, especially the third beast in the seventh chapter, 'a subject well suited to our times'. At Court he preached upon the Psalms. His purpose: to refute transubstantiation. Hooper chose the prophet Jonah as the subject for his 1550 Lenten sermons before the Court, a subject which would allow him to 'touch upon the duties of individuals': to show that the 'doctrine we preach is one and the same with the prophets and apostles, and not new', and to 'declare which way the sinful world may be reconciled to God'.[18] Jonah had been sent by God to Nineveh to urge 'amendment of life', and to warn, if the citizens failed, of imminent destruction. So, 'God hath used, from the beginnings of commonwealths' to send the 'most preachers of truth', in His mercy, to the greatest cities, 'as it is to be seen in these days what God had shewed upon London'. That God had sent a godly King, magistrates, and preachers to England was 'a very token that the sins of England is ascended up into His sight'. And which were the sins? 'Miserable and cursed is our time of God's own mouth, that there be such dumb Bishops,

---

[16] Ibid. i, p. 71. For the crisis, see below, pp. 497–501.

[17] *Original Letters*, i, pp. 78–85.

[18] Ibid. i, p. 75; *Early Writings of John Hooper*, ed. S. Carr (Parker Society, Cambridge, 1843), p. 445.

unpreaching prelates, and such ass-headed monsters in the Church of God', and such uncharitable gentlemen. The judgement was at hand for London and England as it had been for Nineveh. The Ninevites, who had listened, were infidels: 'but in this our miserable time of God for sin', supposed Christians were led by 'fraud, guile, treason, heresy, superstition, papistry, ignorancy, arrogancy, malice' to call into question 'whether carnally, corporally, really the precious body of Christ be present'. The ignorance of believing in transubstantiation brought idolatry; idolatry, eternal damnation. 'I am appointed', claimed Hooper, 'to remove ignorancy.' And he proceeded to advance seven reasons why Christ could not be corporeally present in the sacrament.[19]

There were signs that Hooper and his coterie did not 'water and plant in vain' in the Lord's vineyard. In London 'the truth is especially flourishing'. Even the City's Aldermen 'who were veteran papists', sensing which way the theological winds blew, 'have embraced Christ'.[20] Spontaneous conversions were made. Roger Holland, an apprentice, abandoned both papistry and riotous living when he began to attend the daily lectures at All Hallows and the weekly sermons at St Paul's.[21] Every Monday there was an improving 'lecture for servants' at St Benet Gracechurch.[22] Already 'many altars have been destroyed in this City since I arrived here', reported Hooper at the end of March 1550.[23] Certainly, at St Matthew Friday Street, St Benet Gracechurch, St Botolph Aldgate, St Laurence Pountney, St Mary Magdalen Milk Street, St Martin Ludgate, St Martin Outwich (and maybe others, for the chronology is not always clear from the churchwardens' accounts) altars had been pulled down by zealous parishioners 'long afore' the orders came.[24] 'There should be among Christians no altars',

[19] The sermons are printed in *Early Writings of John Hooper*, pp. 431–558.

[20] *Original Letters*, i, p. 79; ii, p. 636.

[21] Foxe, *Acts and Monuments*, viii, p. 473. All Hallows Bread Street was Robert Horne's parish; All Hallows London Wall John Rogers's. The daily lectures could have been at either church.

[22] Guildhall, MS 5681/1, p. 34.

[23] *Original Letters*, i, p. 79. On 8 Mar. the Mercers' Company ordered that the altars in their chapel be removed and sold: Mercers' Company, Acts of Court, ii, fo. 241ᵛ.

[24] *Two London Chronicles*, p. 22. Guildhall, MSS 1016/1, fo. 8ʳ; 1568/1, fos. 31ᵛ, 33ʳ; 9235/1 (unfoliated); 3709/1 (unfoliated); 4570/1, fo. 97ʳ; 2596/1, fo. 96ʳ. Walters,

Hooper had preached that Lent, for to sacrifice was idolatry.[25] In Rochester also Nicholas Ridley independently ordered that altars be replaced by communion tables, for 'when we come unto the Lord's board, what do we come for? To sacrifice Christ again, and to crucify Him again, or to feed upon Him that was once only crucified and offered up for us?'[26] On 1 April Ridley was translated to the see of London; 'a pious and learned man', judged Hooper, adding presciently, 'if only his new dignity do not change his conduct'.[27]

On 19 April Ridley 'came into the choir [of St Paul's] at the communion time, and . . . he and the Dean received and Master Barnes, and the two took the host of the priest in their two hands. And . . . the Bishop commanded the light of the altar to be put out or he came into the choir': so the Grey Friars chronicler recorded the beginning of a new religious regime, for Ridley's celebration was further advanced than the Prayer Book.[28] The new Bishop instituted forthwith a visitation to reform all the abuses which remained in his diocese; those which Bonner had condoned.[29] Belief in Purgatory, invocation of saints, veneration of images and relics, justification by good works, and the traditional uses of rosaries, crucifixes, sepulchres, holy bread, palms, ashes, and candles were all condemned in his injunctions. Ridley sat in visitation at St Paul's and other City churches, hearing a sermon everywhere, sometimes preaching himself. He summoned before him the parson, curate, and six parishioners of every parish, and personally

---

London Churches, pp. 377, 379. R. Hutton, 'The local impact of the Tudor Reformations', in The English Reformation Revised, ed. C. A. Haigh (Cambridge, 1987), p. 125.

[25] Early Writings of John Hooper, pp. 488, 492.

[26] The Works of Bishop Ridley, ed. H. Christmas (Parker Society, Cambridge, 1851), pp. 321–4.

[27] Original Letters, i, p. 79.

[28] Grey Friars Chronicle, p. 66. It was not until the revised service of 1552 that the host was delivered into the hand rather than the mouth: E. Brightman, The English Rite, 2 vols. (1915), i, p. cliii.

[29] Guildhall, MS 9531/12, fos. 288ᵛ–306ʳ. The visitation articles and injunctions are printed in Visitation Articles, ii, pp. 230–45, and Works of Bishop Ridley, pp. 319–21. On 12 May 1550 Scudamore sent copies to Hoby: NLS, Adv. MS 34.2.14, fo. 53ʳ. The visitation records are not extant, but some of the presentations are recorded in GLRO, DL/C/3. Ridley was frequently commended by Bullinger's correspondents for his zealous stewardship of his diocese: Original Letters, i, pp. 79, 185, 187–8, 323.

examined every cleric 'privately of their learning'.[30] Hales reported at the end of May that Ridley 'threatens to eject those who shall not have come to their senses before the next visitation, and if I know the man, he will be as good as his word'.[31] There were no purges of the London clergy, but whenever there was a vacancy a reformer was appointed. To City parishes came some of the most committed of evangelicals; very many of whom would never compromise their reformed beliefs or conform to political circumstances.[32]

Ridley's first concern was to satisfy himself that none of his clergy still clung to popish superstitions, while assuring also that they did not lead their parishioners to reform more advanced than that enshrined in the Book of Common Prayer. He knew well the popish evasions which Bonner had sanctioned, and now specifically forbade the 'counterfeiting of the popish Mass'; the 'kissing of the Lord's board', washing of hands after the Gospel or communion, licking the chalice, crossing, saying the *Agnus* before the communion, making any elevation of the sacrament, or setting any light 'upon the Lord's board'. Ridley exhorted the curates and churchwardens to set up 'an honest table' 'so that the ministers with the communicants shall have their place separated from the rest of the people', and to take down all other altars.[33] Hooper had urged that the chancel be shut up 'that separateth the congregation of Christ one from another'; that preacher, minister, and people be all together to 'hear and see plainly what is done, as it was used in the primitive church'.[34] Ridley would not go so far. But this was not the only cause of their contention.

Ridley's determination to establish doctrine and authority within the English Church and to bring a 'godly unity' to his divided diocese was imminently threatened by the zeal of Hooper and à Lasco. The stranger churches now founded in London, with their mission to provide an example of purified

---

[30] Wriothesley, *Chronicle*, ii, pp. 38, 41. W. K. Jordan, *Edward VI: The Threshold of Power* (1968), pp. 270–5.   [31] *Original Letters*, i, pp. 187–8.

[32] E. L. C. Mullins, 'The Effects of the Marian and Elizabethan Religious Settlements upon the Clergy of London, 1553–1564' (University of London, M.A. thesis, 1948); B. L. Beer, 'London Parish Clergy and the Protestant Reformation, 1547–1559', *Albion*, xviii (1986), pp. 375–93. See below, pp. 570, 575–6.

[33] *Visitation Articles*, ii, pp. 241–4.

[34] *Early Writings of John Hooper*, pp. 491–2.

liturgy, and Hooper's campaign to purge the Church of all idolatry became thorns in Ridley's side.[35] Writing later, of the troubles of the exiles in Frankfurt, Ridley insisted that, in respect of *adiaphora*, the godly man should 'forbear the custom of his own country' and conform to local practice; so he had tried, and failed, to convince à Lasco 'when he was with us'.[36] At à Lasco's behest, 'for avoiding all sects of Anabaptists and such like'—the Arians, Marcionists, Danists, and Libertines sent by the Devil to impede the Gospel's passage—and 'in view of the duty of a Christian prince to relieve fugitives from papal tyranny', the King granted the Austin Friars to the strangers.[37] This was to be a 'temple . . . wherein the meetings of Germans and other strangers may be celebrated, administered by themselves according to the Word of God and the Apostolic observance': it was also to be a 'body politic' (*corpus corporatum et politicum*), governed independently by à Lasco, as superintendent, and four ministers—Deloenis, Micronius, de la Rivière, and Vauville.[38] Not only did they have, they rejoiced, 'the pure ministry of the Word and sacraments', but also 'we are altogether exempted . . . from the jurisdiction of the Bishops'.[39] Their church was to provide a model of 'a Church rightly reformed', on a Zwinglian pattern, for an English Church yet half reformed.[40] Hooper and à Lasco found common cause in creating a reformed Church and provided mutual support against a hostile Bishop.[41]

[35] For the stranger churches and their foundation, see Pettegree, *Foreign Protestant Communities*; P. Collinson, 'Calvinism with an Anglican Face: the Stranger Churches of Elizabethan London and their Superintendent' and 'The Elizabethan Puritans and their Foreign Reformed Churches in London', in *Godly People: Essays on English Protestantism and Puritanism* (1983), chs. 8, 9; Baron F. de Schickler, *Les Églises du refuge en Angleterre*, 3 vols. (Paris, 1892); *Actes du Consistoire de l'église française de Threadneedle Street, Londres*, i (1560–65), ed. E. Johnston, *HSL*, xxxviii (1937); J. Lindeboom, *Austin Friars: History of the Dutch Reformed Church in London, 1550–1950* (The Hague, 1950); B. Hall, *John à Lasco, 1499–1560: A Pole in Reformation England* (London, 1971).

[36] *Works of Bishop Ridley*, pp. 534–5; cited in Collinson, *Godly People*, p. 218.

[37] *Original Letters*, ii, p. 560; Gorham, *Gleanings*, p. 79; *Edward VI Chronicle*, p. 37.

[38] *CPR Edw. VI* iii, p. 317. À Lasco had returned to England on 13 May 1550, and was given a patent of denization on 27 June: ibid. iii, p. 316; *Original Letters*, ii, p. 560.

[39] *Original Letters*, ii, pp. 567–8; *CPR Edw. VI* iii, p. 317.

[40] À Lasco later defined their hopes: *Actes du Consistoire*, i (1560–1565), p. xiii, cited in Collinson, *Godly People*, p. 248 n.

[41] For Hooper's relations with the ministers of the stranger churches, see Gorham, *Gleanings*, p. 222; *Original Letters*, ii, pp. 565, 572.

There followed a controversy between Ridley and Hooper, long and bitter, which split English Protestantism.[42] Hooper, offered the see of Gloucester in April 1550 (perhaps at the behest of the returned Somerset), declined it on grounds of his opposition to the *Ordinal*, 'both by reason of the shameful and impious form of the oath' (he would be required to swear by saints), and 'those Aaronic habits' he would be forced to wear. He refused to be called 'Rabbi' or to become a 'pie' vested in black and white.[43] Hooper had already delivered before the Court, in his Lenten sermons on Jonah, his manifesto against ordination.[44] Ridley maintained that Hooper's protest was otiose, since the matters to which he objected were indifferent, yet to Hooper they lay at the heart of his campaign to extirpate all idolatrous remnants of an antichristian past. His stand threatened the reforms already established; his insistence upon unassailable conscience, carried to its limits, would destroy the institution of the Church. Bucer warned Hooper in November 1550: 'this controversy afflicts me exceedingly, since it places such an impediment in the way of the ministry of yourself and others'.[45] Hooper's contention that the Bishops were 'children of the world, superstitious and blind papists', was countered by accusations of Anabaptism from Ridley.[46] 'The Bishop of London was most violent against him before the Council.'[47] By February 1551 Hooper had submitted, but great damage had been done to the cause of reform. Hooper's principal, some said his sole, supporter in this dangerous controversy was à Lasco,[48] who was conducting all the while

---

[42] The reformers wrote to each other constantly of the controversy: *Original Letters*, i, pp. 86–90; ii, pp. 558–60, 410, 466, 665, 416, 566–8, 573, 672–3, 426, 486–7, 584–6. For the vestiarian controversy, see J. H. Primus, *The Vestments Controversy: An Historical Study of the Earliest Tensions within the Church of England in the Reigns of Edward VI and Elizabeth* (Kampen, 1960); B. J. Verkamp, *The Indifferent Mean: Adiaphorism in the English Reformation to 1554* (Athens, Ohio, 1977); Morris West, 'John Hooper and the Origins of English Puritanism'; Jordan, *Edward VI: The Threshold of Power*, pp. 293–8. Foxe was with Hooper in the controversy: *Acts and Monuments*, vi, pp. 642–3.

[43] *Original Letters*, i, pp. 87, 187.

[44] *Early Writings of John Hooper*, p. 479.

[45] Gorham, *Gleanings*, pp. 200, 208.

[46] C. Hopf, 'Bishop Hooper's notes to the King's Council', *Journal of Theological Studies*, xliv (1943), pp. 194–9.

[47] *Original Letters*, ii, p. 573.

[48] Ibid. ii, pp. 675, 583.

his own campaign to purify the Church and was thereby contesting the Bishop of London.

Though the King had given the Austin Friars to the strangers in July 1550 their opponents immediately found ways of barring them. À Lasco could not open the church for weekly sermons until, claimed the Lord Treasurer, the Austin Friars was decorated befitting a royal gift. But the delay was politic. Paulet was, Micronius suspected, the agent of the Bishops. Why else did he insist that 'we foreigners must either adapt English ceremonies or disprove them by the Word of God'?[49] For the while the governors of St Bartholomew's allowed the 'preachers of the French nation' to use the Grey Friars.[50] When Micronius preached on 21 September the audience was greater than the church could hold. Soon a reformed discipline was instituted. On 5 October the congregation appointed four elders 'according to apostolic ordinance, to assist the minister, not indeed in the ministry of the Word, but in the conservation of doctrine and morals in the church'. A week later they elected four deacons to take charge of the 'poor and exiled for Christ's sake'. So far, so good. But on 20 October Micronius wrote in great distress to Bullinger that the 'privileges of our German church are in the greatest danger . . . we must be fettered by the English ceremonies, which are intolerable to all godly persons'.[51]

Still, the foreign reformers hoped not only to introduce a pure form of worship in their own church, but to inspire the English to follow: a forlorn hope perhaps in the spring of 1551. The xenophobia of Londoners was at its height, occasioned, or so the Spanish ambassador thought, by the sight of a thousand foreigners gathering to worship at the Austin Friars.[52] In truth, the social distress and political crisis of this month were more compelling reasons for alarm. The pure administration of the sacraments was still forbidden, yet in April Utenhove could report that the Word 'is proclaimed in all its purity . . . by our friend Martin Micronius, who preaches in a popular manner,

[49] Ibid. ii, p. 569; *CPR Edw. VI* iv, p. 15.
[50] St Bartholomew's Hospital, Ha 1/1, fo. 9ᵛ.
[51] *Original Letters*, ii, pp. 570–3.
[52] *CSP Sp.* x, pp. 278–9; A. Pettegree, 'The Foreign Population of London in 1549', *HSL* xxiv (1984), pp. 141–9.

like the clergy at Zurich'. 'Prophesyings' or 'collation of the Scriptures' were to be held in Latin every Monday, after à Lasco's lectures upon St John, and on Wednesdays after Gualter's Latin sermons on Genesis.[53] The influence of the strangers upon the English Church would be more powerful yet if their views upon the sacraments were adopted. In January 1551 à Lasco had written to Bullinger to congratulate him upon the concordat regarding the doctrine of the eucharist made between him and Calvin in the *Consensus Tigurinus* (1549), and at more or less the same time he published his own *Forma ac ratio tota ecclesiastica ministerii, in peregrinorum, potissimum vero Germanorum ecclesia* in which a Zwinglian liturgy was enshrined. This liturgy had an influence upon Cranmer and those now preparing a reformed Prayer Book, yet they would never go so far.[54]

Whatever the people's wishes, by the end of 1550 every City church had a communion table. Alone, St Nicholas Cole Abbey resolutely kept its altar, despite the jeremiads of the reformed preachers.[55] Yet the rite enjoined in the 1549 Prayer Book was still ambivalent regarding the central doctrine of the Mass, and the removal of altars did not remove the belief in the Mass as miracle and sacrifice. Nor were the tables set up without protest. In June 1550 when the high altar at St Paul's was taken down there was a 'fray' and a man killed. Fights continued in the cathedral, 'and nothing said unto them'.[56] The contention was surely about the sacrament, for even while Kirkham preached at Paul's Cross in June that the communion was only a memorial of Christ's death there were still priests in Ridley's own cathedral who dared to 'counterfeit' the Mass for the faithful. In October the Council sent 'three or four honest gentlemen' to observe the 'usage of the communion in Paul's', for the word was that 'it was used as the very Mass'.[57] Corpus Christi was not kept as a holy day that year, but at the feasts of

[53] *Original Letters*, ii, p. 587.
[54] *RSTC* 16571. This work and its influence upon the making of the second Edwardian Book of Common Prayer is discussed in Brightman, *The English Rite*, i, pp. cxlvi–cxlix, clvi ff.; C. W. Dugmore, *The Mass and the English Reformers* (1958), pp. 162–8.
[55] *Two London Chronicles*, pp. 22, 24.
[56] *Grey Friars Chronicle*, pp. 67, 68.
[57] Ibid., p. 67; *APC* iii, p. 138.

the Assumption and Nativity of Our Lady (15 August and 8 September) 'was such division through London that some kept holy day and some none. Almighty God help it when His will is!' pleaded the Grey Friars chronicler.[58] 'Against Easter' 1551 Ridley made more uncompromising alterations at St Paul's: the table was moved to the midst of the choir, the veil drawn so that none should see the sacrament but those who received it. The iron grates of the choir were walled up, 'because many people come thither daily and worship the sacrament'.[59] There were very many left puzzled, perplexed, bereft, without the traditional ways of worship and propitiation. The loss was never felt more keenly than in times of disaster. The divine hand was seen to punish a faithless people that spring, but the faithlessness was perceived in quite contrary ways by papists and reformers.

COVETOUSNESS AND CHARITY

The spring of 1551 was a time of grave political instability and dire economic distress, of portents and prodigies. As so often and so disturbingly before, the savage rivalries of Edward's councillors were played out in London, and the religious affiliations of the citizens manipulated for political ends. In mid-March Princess Mary rode with a great retinue through the capital, each rider carrying a forbidden rosary, to mark her determination never to bow to the demands of her brother and the Duke of Northumberland that she abandon her religion.[60] The manipulation of the coinage continued. At St George's day a shadowy conspiracy to murder the Duke and his adherents, led by the ousted Somerset, was discovered. Ghostly soldiers were seen floating in the air; three suns shone together in the sky; the earth shook:[61] all these prodigies, real and

---

[58] *Grey Friars Chronicle*, p. 67.

[59] Wriothesley, *Chronicle*, ii, p. 47; *Two London Chronicles*, p. 22; *Grey Friars Chronicle*, p. 69.

[60] *Diary of Machyn*, pp. 4–5. For the campaign against Mary, see Jordan, *Edward VI: The Threshold of Power*, pp. 256–64.

[61] *Grey Friars Chronicle*, p. 69; Cooper, *Chronicle*, p. 350. For the conspiracy, see below, pp. 507–11.

imagined, were taken as marks of divine wrath. Soon worse
would follow.

In July 1551 'came the sweat into London'. It was called
'Stop Gallant', for there were some dancing in the Court at
nine o'clock who were dead by eleven o'clock.[62] Letters from
London reported that between 7 and 20 July 938 people had
died there. Eight hundred died within the first week. Of seven
householders who supped together, six were dead by morning.[63]
'Faithful neighbours' who witnessed the wills of friends 'visited
by the hand of God with sickness' were smitten themselves.[64]
The very suddenness, wrote Hooper, who himself escaped, 'is
a most remarkable token of divine vengeance'.[65] The 'warning
to England' was the more plain because 'it seemed that God
had appointed the sickness only for the plague of Englishmen':
'it followed the Englishmen' even abroad, cutting them down
though 'none other nation infected therewith'. Men should
draw lessons: 'Pray the Lord', besought Hooper, 'that we may
always be waiting in the fear of God for the day of death'. Some
were repentant, for a time: the 'sweat' 'being hot and terrible
enforced the people greatly to call upon God, and to do many
deeds of charity'.[66]

It was charity that was lacking, many said, and for lack of
charity that God sent a plague to 'chastise the poor realm of
England'. It was precisely of want of charity that each faith
accused the other. So, the Catholics, hating the reformers as
heretics and iconoclasts, bewailed with the Edwardian innovations and desecration the accompanying loss of spiritual solace.
After matins at St Martin Ludgate on 11 July, as the 'sweat'
raged through the City, Margaret Harbottle denounced her
curate:

he and such as he is was the occasion that God did plague the people
so sore, because that they would not suffer them to pray upon their
beads. And thereupon, in despite of the King's injunction, shook her

---

[62] *Narratives of the Reformation*, p. 82; *Edward VI Chronicle*, p. 71.

[63] *CSP Ven.* v, pp. 541–2; BL, Harleian MS 353, fo. 107^r; Stow, *Annales*, p. 605.

[64] PRO, PCC, Prob. 11/34, fos. 131^v, 133^r, 163^v, 275^v. Will Register Bucke is full
of the hasty wills drawn up in the first days of July 1551.

[65] *Original Letters*, i, p. 94.

[66] Grafton, *Chronicle*, ii, p. 525; Cooper, *Chronicle*, p. 351; *Original Letters*, i, p. 94.

beads at the said Sir Nicholas, saying contemptuously that she would that he should know that she did, and would, pray upon her beads. And that men did die like dogs because they cannot see their maker borne about the streets as they have seen it in time past.[67]

No longer was the blessed sacrament reserved in City churches nor carried in visitation of the sick: would the faithful then die unabsolved? The reformers likewise sought the reasons—'best known unto His high and secret judgement'—why God sent 'this sharp rod' of 'sudden death', and they found the answer in men's 'wicked living', their 'uncharitableness'; above all, in 'that greedy and devouring serpent of covetousness'.[68] These vices the reformers sought ardently to amend.

With a reformation in religion should come a reformation in morals. Upon taking up the cross some gospellers self-consciously forswore their former sinful lives. Dicers and 'whore hunters' had been Edward Underhill's companions until he 'fell to reading the Scriptures and following the preachers'. Converted, he wrote a ballad denouncing such 'falsehood and knavery', reserving special odium for the dicers. Now despised among 'lords and ladies, gentlemen, merchants, knaves, whores, bawds and thieves', Underhill lived in London 'as dangerously as Daniel amongst the lions'.[69] Of course, the Devil is known specially to tempt the sanctimonious, and critics waited for those who pretended reform to fall. The rulers of the commonwealth were warned: 'Take heed . . . for a little sin in you is much noted of the commons. Thus shall all those back biters have always an occasion to slander Christ's Gospel, and say "these gospellers be such whore mongers, such swearers, such proud men, such covetous persons".'[70] So the divorce of the Marquis of Northampton came as a grave embarrassment,

---

[67] GLRO, DL/C/3, fo. 110ʳ, cited in J. G. Davis, *Heresy and Reformation in the South-East of England, 1520–1559* (1983), p. 99. St Martin Ludgate was known to be a conservative parish, see above, p. 439.

[68] *HMC Salisbury MSS*, i, p. 90. Ridley sent orders to the preachers of his diocese on 25 July to 'tell unto men their sins': *Works of Bishop Ridley*, pp. 334–5.

[69] *Narratives of the Reformation*, pp. 158, 175. For another diatribe against gambling, see *Original Letters*, i, pp. 281–8.

[70] Philip Nicolls, *Here begynneth a godly newe story of .xii. men that moyses sent to spye owt the land of canaan* (1548; *RSTC* 18575), sig. Ciiᵛ; see also *Original Letters*, ii, pp. 672, 647, 416, 441.

and the Council was forced to issue a proclamation restating the virtues of Christian matrimony.[71]

In no other diocese was there such corruption of morals, so high a rate of marital breakdown, as in London. 'Adultery and fornication are maintained and kept openly and commonly in the said City of London . . . whereby the wrath of God is provoked against our people': thus was Bonner charged in 1549. Yet the Bishop and his officials, charged to search out the capital's adulterers and fornicators, faced an uphill task. The Consistory Court in mid-sixteenth-century London heard more cases of unsuccessful marriages than were heard then, or earlier, in other dioceses.[72] Correction of the morals of the citizenry was not left to the Church Courts alone: it never had been, but now the City fathers—in alliance with Bishop Ridley, in this as in other matters—began to act against sexual delicts. In April 1550 Scudamore reported approvingly: 'other occurrences I have not worthy of advertisement, but that they begin to punish vice very earnestly in London, the Lord be praised'.[73] Another chronicler found nothing to say of Rowland Hill's mayoralty, save that 'this Mayor was a good minister of justice and a great punisher of adultery'.[74] In April 1550 Hill charged all the wardmote enquests of the City 'to sit and inquire of all misrule done . . . since Candlemas', and to present fresh indictments, 'upon which indictments the Lord Mayor sat *many* times'. Upon the convicted bawds, whores, and scolds 'he caused execution to be done immediately by riding in carts with ray hoods . . . so that he spared none'.[75]

The reformers had called for a godly commonwealth; for Christians to leave worshipping Christ in images of wood and stone devised of 'man's fantastical imagination', and to give their devotions instead to Christ's poor image, the suffering poor: 'it is Christ Jesus himself that in the needy doth suffer

[71] Jordan, *Edward VI: The Young King*, pp. 365–7; R. Houlbrooke, *Church Courts and the People during the English Reformation, 1520–1570* (Oxford, 1979), pp. 70–1; *TRP* i. 303.

[72] Houlbrooke, *Church Courts and the People*, pp. 85, 87; R. M. Wunderli, *London Church Courts and Society on the eve of the Reformation* (Cambridge, Mass., 1981), pp. 84–7, 144; Foxe, *Acts and Monuments*, v, pp. 729–30.

[73] NLS, Adv. MS 34.2.14, fo. 39^r.

[74] *Two London Chronicles*, p. 45.

[75] Wriothesley, *Chronicle*, ii, p. 36.

hunger, thirst and cold'.[76] Surely this commonwealth might be
achieved now, as images were removed and chantries abandoned, and as the wealthy could read in Scripture Christ's
commandments? Yet, to the horror of the reformers, 'carnal
gospellers', although knowing the Gospel and speaking of it,
ignored its precepts; thereby they 'do crucify the son of God
making a mock of Him'.[77] The hypocrisy of seizing the wealth
of the Church, ostensibly for godly purposes, while using it to
reward the rulers was spectacular. 'Covetousness, which is the
root of all evil', and hypocrisy were vices specially condemned
by the Edwardian reformers. The warnings from the reformers
of divine vengeance became more insistent as they witnessed,
especially in London and Westminster, the gadarene rush
of courtiers and citizens rewarding themselves with chantry
lands[78] while the intended beneficiaries—the poor—suffered
worse than ever.

> Ye bring the curse of God daily on your head
> For poverty with you hath the street for his bed.
> Repent ye citizens of London.[79]

The rulers had brought in the Gospel, but ignored its commandments. None had been more guilty than Somerset. Great
noblemen and councillors had acquired the dissolved City
houses of the ousted monks and friars—Holy Trinity Aldgate
went to Audley, the Charterhouse to Lord North, St Bartholomew's to Rich, the lazar hospital of St Giles to Dudley, the
Black Friars to Cobham—but Somerset's acquisitiveness was
greater than any of theirs.[80] Late in 1547 he began to build a
palace worthy of his pride and wealth. The church of St Mary
le Strand, Bishops' palaces, and tenements were razed to make
way for Somerset House. Religious houses and St Paul's cloister
were plundered for stone. Worse, the charnel house of the
cathedral was destroyed, and the bones of citizens long-dead

---

[76] Thomas Lever, *Sermons*, in E. Arber (ed.), *English Reprints* (1870), p. 78.

[77] John Mardeley, *Here beginneth a necessarie instruction for all couetous ryche men* (*RSTC* 17319), sigs. Aii^v, Biii^r.

[78] Grant after grant is recorded in *CPR Edw. VI.*

[79] William Samuel, *A warnyng for the cittie of London. That the dwellers there in may repent their euyll lyues for feare of Goddes plages* (*RSTC* 21690.8).

[80] E. J. Davis, 'The transformation of London', in *Tudor Studies*, ed. R. W. Seton-Watson (1924), pp. 87–314.

scattered, without a resting place. Even at the Reformation no one was accustomed to quite such conspicuous consumption nor such desecration. 'Many well disposed minds conceived a hard opinion' then of Somerset. 'These actions were in a high degree impious, so did they draw with them both open dislike from men and much secret revenge from God.'[81] Every Catholic was, of course, shocked, but so too were the Protector's reforming allies. They saw him become 'cold in hearing God's Word', neglecting sermons while he supervised his masons.[82] In August 1549 John Mardeley (once Somerset's own servant) was before the Council for writing *Here beginneth a necessarie instruction for all covetous ryche men to behold and learne what peril and danger they be brought unto, if they haue their consolation . . . in Mammon* with its clear rebukes of the Lord Protector: 'I made gorgeous fair works, builded my houses, orchards and gardens of pleasure, I gathered silver and gold plenty, &c. And I see all is vanity under the sun.'[83]

Yet Somerset's fall could not bring moral regeneration at Court or in Council. The lands of the quondam Bishops were added to the rest of the prizes to be gained by acquisitive citizens and courtiers.[84] The names of the London residences of Edward's councillors betray their origins as Bishops' houses: John Russell's Carlisle Inn, Edward Seymour's Chester Inn, Paget's Exeter Inn, and William Parr's Winchester House.[85] In his Lenten sermon at Court Stephen Corthop preached of the greed of rulers who 'spared neither abbey lands nor chantry lands, Bishops' lands nor yet the lands of the ancient Crown'. And, 'belikening the rulers to a sponge', he urged the King to 'take these greedy sponges and to wrest out all the moisture again'.[86] In the City, too, the reformers called upon the

---

[81] Jordan, *Edward VI: The Young King*, pp. 498–9; Stow, *Survey*, ii, pp. 84–5, 92–3; BL, Harleian MS 2194, fo. 21ᵛ.

[82] John Knox, *An admonition or warning that the faithful Christians in London, Newcastel, Barwycke & others may auoide Gods vengeaunce* (*RSTC* 15059), sigs. Biiiᵛ–ivʳ.

[83] *APC* ii. 311. This was the first instance of censorship under Somerset's regime: P. M. Took, 'The Government and the Printing Trade, 1540–1560' (University of London, Ph.D. thesis, 1978), p. 165.

[84] NLS, Adv. MS 34.2.14, fos. 5ᵛ, 57ᵛ. F. M. Heal, *Of Prelates and Princes: a study in the economic and social position of the Tudor episcopate* (Cambridge, 1980).

[85] C. L. Kingsford, 'Historical Notes on Medieval London Houses', *London Topographical Record*, x (1916), pp. 56–8, 83–5, 87–8, 117–19; xii (1920), pp. 57–60.

[86] NLS, Adv. MS 34.2.14, fo. 57ʳ⁻ᵛ.

citizens to repent. 'O merciful Lord', preached Lever in Lent 1550, 'what a number of poor, feeble, halt, blind, lame, sickly, yea with idle vagabonds and dissembling caitiffs mixed among them, lie and creep, begging in the miry streets of London and Westminster.' Their presence and their suffering, he warned the magistrates and the rich, was 'to thy great shame before the world and to thy utter damnation before God'.[87] As the poor remained unrelieved, Lever preached at St Mary Woolchurch in 1551 upon the fearful twelfth psalm:

Surely even now in great scarcity, lack and dearth of meat, drink, clothing . . . the sorrowful sighs of the poor and needy must either sink into the rich men's hearts and from them procure . . . comfort unto the needy members of Christ's body; or else directly ascend unto the throne of God's justice, and from thence bring down indignation and vengeance, to destroy wicked worldlings, which being themselves in wealth and abundance, feel no grief at their brothers' distress and misery.[88]

In this year of the 'sweat', Londoners, heeding the warnings, were moved 'to do many deeds of charity: But as the disease ceased, so the devotions quickly decayed'.[89] The reformers lamented 'the charity of rich men . . . grown thorough cold'. To be in London was to witness the sight of the merchants profiting by speculating in the fall of money while the poor bore the brunt, to see them hastening to acquire the lands of the dissolved gilds; to be in the capital was to see the poor suffering in the streets and to know the compelling need to provide for the destitute. Yet the number of paupers grew so quickly that the citizens could never relieve them all: the citizens had not abandoned their Christian duty of charity, but 'London cannot relieve England'.[90]

'The number of poor did so increase of all sorts, that the churches, streets and lanes were filled daily with a number of loathsome lazars, botches and sores.'[91] These were the vagrant poor of England who swelled importunate to the capital,

[87] Thomas Lever, *Sermons* in *English Reprints*, pp. 29, 77–8.

[88] Id., *A meditacion vpon the lordes praier* (*RSTC* 15544), sig. Aii^r.

[89] Grafton, *Chronicle*, ii, p. 525.

[90] John Howes, *Contemporaneous Account in Dialogue-form of the Foundation and Early History of Christ's Hospital and of Bridewell and St Thomas' Hospitals* (1889), p. 57.

[91] Ibid., p. 8.

seeking work or alms. It had always been so, but as the population rose, London drew ever more of the dispossessed. Londoners themselves suffered unusual and desperate hardship in these years, as prices rose meteorically. Between 1544 and 1551, as Steven Rappaport has shown, prices in London rose by 89 per cent. The rate of inflation for 1549 to 1551 was 21 per cent per year. The harvests of 1547 and 1548 were abundant, but those of 1550 and 1551 failed disastrously. The price of flour doubled, and the size of the halfpenny loaf, the staple diet of the poor, shrank dramatically.[92] In February 1551 the governors of St Bartholomew's Hospital, perceiving that 'the half penny loaf (now being very small) is too little for two men at a meal', increased the ration by half.[93] But later that year the 'fall of money', and the attempts to restore the coinage, hit the poor hardest, as the City merchants speculated on the changing rates.[94] The destitution was so dire that in October 1547 the City governors had introduced, for the first time, a compulsory weekly poor rate.[95] Collectors sought relief for London's notorious poverty in the richer shires. The l'Estranges of Hunstanton gave a shilling in June 1550 to 'the men that gather for the poor in London'.[96] But no measures sufficed. By January 1553 the City governors acknowledged that the compulsory poor rate was failing to alleviate the distress, and urged the Aldermen to discover defaulters and to cause the parish collectors for the poor to 'excite' the parishioners to 'enlarge their weekly devotions towards the poor'. Would not 'the Lord of glory, whose work this is', give a heavenly reward?[97]

The reformers' opponents accused them of denying the merit of good works, of failing in charity. So Gardiner, restored to power in 1553 and now vindictive, would summon Thomas Mountain, rector of St Michael Paternoster Royal, and challenge him: 'This is one of your new broached brethren that speaketh against all good works.' Mountain denied the charge: 'in good works every Christian man ought to exercise himself

---

[92] S. R. Rappaport, 'Social Structure and Mobility in Sixteenth-Century London' (Columbia University, Ph.D. thesis, 1983), ch. 7.

[93] St Bartholomew's Hospital, Ha 1/1, fo. 12^v.

[94] See below, pp. 505–6.

[95] CLRO, Journal 15, fo. 325^v.

[96] C. Pendrill, *Old Parish Life in London* (1937), p. 178.

[97] CLRO, Journal 16, fo. 256^v.

all the days of his life, and yet not to think himself justified thereby, but rather to count himself an unprofitable servant'.[98] As 'an unprofitable servant' Richard Canesby renounced all his good works in his bequest of his soul to God, trusting only in the merits of Christ's passion.[99] Some of the new faith disavowed in their wills any possibility that their salvation might be promoted by their charity; saying perhaps, as John Mayor did when he left 20s. to the poor in 1551, that he did it 'not to the intent to pray for me, but to be thankful unto God and to pray for the King', or like Edward Fison, 'I do not suppose that my merit be by the good bestowing of them [his goods], but my merit is in the faith of Christ only'.[100] But to deny that good works adduce to salvation was never to abandon charity. Asked by Gardiner what good works had been done by his 'unprofitable' fraternity in Henry and Edward's days, Mountain could boast of 'many notable things worthy of perpetual memory'; not least

they did erect many fair hospitals; one for orphans and father-less children, wherein they might be taught to know their duty and obedience both to God and man . . . In the other houses, my lord, is the blind, the lame, the dumb, the deaf and all kind of sick, sore and diseased people; they have always with them an honest learned minister to comfort them . . . that they might take patiently . . . God's visitation . . . Are not all these good works, my lord?[101]

Charity in Edward's reign was expressed through one magnificent, beneficent project which sought to solve the problems of poverty and covetousness at a stroke: the creation of the five hospitals of St Bartholomew's, Christ's, St Thomas's, Bethlem, and Bridewell, 'the perfect spectacle of true charity and godliness unto all Christendom'.[102] The setting up of 'Christ's holy hospitals and *truly* religious houses' in the houses where monks

[98] *Narratives of the Reformation*, p. 181.

[99] PRO, PCC, Prob. 11/34, fo. 149ʳ.

[100] Guildhall, MS 9171/12, fos. 80ʳ, 128ʳ; see also PRO, PCC, Prob. 11/34, fos. 138ʳ, 198ʳ; Prob. 11/40, fo. 276ʳ.

[101] *Narratives of the Reformation*, p. 182.

[102] Foxe, *Acts and Monuments*, vii, p. 559. For the foundation of the hospitals, see Howes, *Foundation of Christ's Hospital*; N. Moore, *The History of St Bartholomew's Hospital*, 2 vols. (1918); E. G. O'Donoghue, *Bridewell Hospital*, 2 vols. (1923–9); P. Slack, 'Social Policy and the Constraints of Government, 1547–1558', in J. Loach and R. Tittler (eds.), *The Mid-Tudor Polity, c.1540–1560* (1980), pp. 94–115.

and friars had lived 'in pleasure, nothing regarding the miserable people',[103] was a great work of reformed piety. The godly zeal which animated their foundation is evident in the daily services established for the inmates of St Bartholomew's. This was the anthem:

Being made the servants of God by faith in the merits and blood shedding of His most dear son, our saviour Jesu Christ, we are certain and sure to be saved, and that no damnation can happen unto us, so that we walk not in the wicked desires of the flesh, but in the heavenly and virtuous life praised and commended by God.[104]

The 'poor in their extremes and sickness' were visited by the hospiteller who was charged to minister to them 'the most wholesome and necessary doctrine of God's comfortable Word'.[105] Also he was to comfort the prisoners of Newgate, exhorting them with 'wholesome lessons of the Scriptures to call them from desperation to the life everlasting'. Such was the concern of the governors for the godly education of the hospitaller that William Payne was sent to Oxford to learn Greek.[106] The hospitals stood as a reproach to Catholics; there were, as Gardiner remarked sourly, instead of 'godly learned and devout men that served God day and night . . . a sort of scurvy and lousy boys'. When the world changed, under Mary, Gardiner and Bonner would mount a campaign against them.[107]

The creation of the hospitals, this 'most blessed work of God', was the work of the City fathers and the Bishop in a novel partnership. St Bartholomew's had begun to receive the sick poor in 1547, but one hospital could never alleviate all the City's poor, especially when its revenue was 'consumed in fees and wages to stipendiary priests and other superfluous officers'.[108] In December 1548, and March and August 1549, supplications came to the Lord Protector from the Mayor and Communalty for revision of the scheme for the hospitals, but

[103] Foxe, *Acts and Monuments*, vii, p. 559; T. Bowen, *Extracts from the Records or Court Books of Bridewell Hospital* (1798), appendix i.

[104] Thomas Vicary, *The Anatomie of the Bodie of Man*, ed. F. J. and P. Furnivall (EETS, extra series, liii, 1888), p. 321.

[105] Ibid., p. 307.

[106] St Bartholomew's Hospital, Ha 1/1, fos. 76^r, 2^r.

[107] *Narratives of the Reformation*, p. 183. See below, pp. 622–4.

[108] CLRO, Journal 16, fo. 26^v; 'The Ordre of the Hospital of S. Bartholomewes', in Vicary, *Anatomie*, pp. 291–2.

although Somerset may have been sympathetic, he was soon to fall from power.[109] In 1550 London had a Bishop who burned to be 'advocate . . . in the poor men's cause'. Lever had high hopes of Ridley, as a 'goodly overseer and godly Bishop', to ensure relief for the poor of London and Westminster.[110] Even at the stake, Ridley's worry was for the poor men to whom he had granted generous leases, now left to the mercy of Bonner.[111] Some of the citizens who founded the hospitals were moved, as was Martin Bowes, by pragmatism more evidently than by faith, but others, like Richard Grafton, John Calthorp, and John Marsh, were 'indued with godliness' (as they would prove again later).[112] When Ridley came to bid farewell to the City of London it was Sir Richard Dobbes whom he 'esteemed above other', remembering how Dobbes's 'heart . . . was moved with pity, and as Christ's high honourable officer' he pleaded the cause of the poor: the model of the godly magistrate. It was Dobbes who, as Mayor in 1551–2, broke tradition by inviting the Bishop into the 'very Council Chamber' to intercede for the poor.[113]

At Court in 1552 Ridley preached upon charity. The King was moved: 'I am in the highest place and therefore am the first that must make answer unto God for my negligence.' Edward ordered the Bishop to take counsel with the City fathers for ways of relieving the poor. Together, Ridley and the citizens devised a suit to the King for the royal palace of Bridewell to be used to house the able-bodied poor and to teach them to be useful members of the commonwealth.[114] To Cecil Ridley wrote movingly:

Good Master Cecil. I must be suitor unto you in our good Master Christ's cause, I beseech you to be good to him. The matter is, Sir, alas! he hath lain too long abroad . . . without lodging in the streets of London both hungry, naked and cold.[115]

[109] CLRO, Repertory 12, fos. 18ᵛ, 57ᵛ; Journal 16, fo. 26ᵛ.

[110] Lever, *Sermons*, in *English Reprints*, p. 78.

[111] Foxe, *Acts and Monuments*, vii, p. 550.

[112] Howes, *Foundation of Christ's Hospital*, pp. 13–14, 17–18; Slack, 'Social Policy and the Constraints of Government', pp. 110–15; see below, pp. 552–4.

[113] Foxe, *Acts and Monuments*, vii, p. 559.

[114] Grafton, *Chronicle*, ii, p. 529; CLRO, Repertory 12, fos. 498ᵛ, 510ᵛ; Howes, *Foundation of Christ's Hospital*, pp. 8, 17–19.

[115] *TED* iii, p. 306.

The citizens, too, appealed to Edward for Bridewell in the name of the poor: 'hear us speaking in Christ's name, and . . . have compassion upon us that we may lie no longer in the streets for lack of harbour'. They promised that for this good work Christ, 'which already hath crowned your majesty with an earthly crown, shall . . . crown your grace with an everlasting diadem'.[116] As he lay dying, Edward made a last gift of the Bridewell and Savoy palaces to the City.[117]

The citizens too 'gave frankly, the work was so well liked'. Preachers and ministers, at Paul's Cross and in the parishes, urged Londoners to make their weekly benevolences. Innholders gathered money for the hospitals in their taverns; wardens in their companies.[118] There were poor boxes in every City church for the collection of alms, and immediately after the offertory the minister was charged to admonish the communicants: 'Now is the time, if it please you, to remember the poor men's chest with your charitable alms.'[119] Forms of covenant were drawn up in which householders would pledge: 'T. W. do frankly give . . . forthwith . . . towards the erection of the houses of the poor —— and weekly towards the maintenance . . . of the same poor ——'.[120] Gifts came to Christ's Hospital from men 'nameless' for clothing the children; from 'divers persons' for Christmas pies; collections were taken at cock fights.[121] All these were voluntary gifts, in addition to the compulsory poor rate. As they made their wills, too, the citizens remembered the new hospitals: between 1550 and 1551, when only St Bartholomew's was established, 15 per cent of testators whose wills were proved in the Prerogative Court of Canterbury left it money; between 1554 and 1556 twice as many left legacies to all the new hospitals.[122] 'God in secret brought great things to pass in the advancement of this foundation': many citizens

---

[116] Howes, *Foundation of Christ's Hospital*, p. 19.
[117] Grafton, *Chronicle*, ii, p. 531; *Memoranda relating to the Royal Hospitals*, pp. 57–77.
[118] Howes, *Foundation of Christ's Hospital*, p. 12; Guildhall, 12819/1, fo. 7ʳ.
[119] *Visitation Articles*, ii, p. 244. (Ridley's injunctions).
[120] Howes, *Foundation of Christ's Hospital*, p. 12.
[121] Guildhall, MS 12819/1, fos. 3ʳ, 3ᵛ, 4ʳ, 5ʳ.
[122] PRO, PCC, Prob. 11/34, 36. Not all bequests were honoured by executors. In 1557 the governors of Christ's Hospital ordered searches to be made through City wills for bequests made: Guildhall, MS 12806/1, fo. 5ᵛ.

made anonymous gifts, without hope of reward or recognition, in this world at least.[123]

Though the poor might lie hungry, cold, and shelterless in the City streets, their plight was not unpitied, nor was it a mark of the citizens' uncharitableness.[124] Though social critics alleged that charity was grown cold, gifts to the poor may even have increased. Londoners' legacies to the poor of their City multiplied. On the principle that the volition behind the gift is of more significance than its size (if not to the recipient)—'The Lord accept this of my poverty, as the mite offered by the poor widow'—this increase is discovered by counting the number of benefactors rather than the size of the benefaction.[125] Three times as many testators whose wills were proved in the London Commissary Court in Edward's reign remembered the poor than in the last years of Henry's reign.[126] Of Londoners whose wills were proved in the Prerogative Court of Canterbury between 1520 and 1522 and 1537 and 1539 a third left money to relieve the poor, while by Edward's reign this had risen to a half (1550-1) and remained a half still in Mary's reign (1554-6).[127] How is this growing impulse to relieve the poor to be explained?

The increase in charitable giving coincided—exactly—with the advance of Protestantism. The abandonment of Purgatory, and with it the need to care for the welfare of dead souls, left money available for the succour of the living, whose need was the greater. The dissolution of the monasteries, chantries, and fraternities had the incidental consequence of releasing income for works of charity, though it did not follow that this money would be devoted to the poor. The reformers called constantly upon the rich to relieve their poor brethren, but we cannot know how many consciences their exhortations stirred. What is

---

[123] Howes, *Foundation of Christ's Hospital*, p. 17.

[124] For what follows, see S. E. Brigden, 'Religion and Social Obligation in early sixteenth-century London', *P&P* 103 (1984), pp. 102-8.

[125] Guildhall, MS 9171/12, fo. 152ʳ. This is the method adopted by J. A. F. Thomson, 'Piety and Charity in Late Medieval London', *JEH* xvi (1965), pp. 178-95.

[126] Guildhall, MSS 9171/11, 12 (12 per cent between 1539 and 1547; 32.4 per cent between 1547 and 1553).

[127] PRO, PCC, Prob. 11/20, 27, 34, 36 (between 1520 and 1522 34 per cent; between 1537 and 1539 31 per cent; between 1550 and 1551 44 per cent; and between 1554 and 1556 52 per cent).

clear is that the increase in charity came just at the time when the plight of the poor became both more desperate and more manifest. Perhaps, and most likely, it was the enormous rise of pauperism in the 1540s and the sight of people suffering in the streets which brought home (literally) to Londoners the compelling social necessity, as well as Christian duty, to provide for the destitute.

Some have said that the new religion inspired a new attitude to charity: that while Catholic almsgiving was indiscriminate, aimed at achieving the blessings of the poor, Protestant donations implied no notion of reward.[128] Certainly some of the new faith did emphatically deny any possibility that their salvation proceeded from anything but faith in Christ's redemptive passion, but the distinction between the charity animated by the old faith and the new was not so marked, nor the inspiration so different. In truth, they often seemed the same. Many of the reformers, as well as those of the old faith, held as certain that charity brought a heavenly reward, despite all the reformist sermons denying that works were necessary for salvation. Henry Brinklow told of the 'reward of everlasting life to them which, to their power, have provided to do for the widow and fatherless', and Robert Crowley concluded that 'though the beggars be wicked, thou shalt have thy reward'.[129] The Mercers were promised that God, 'which seeth in darkness', would recompense all those who freely gave to the new hospital of St Bartholomew's 'not only in prospering them . . . in the fruition of mercy during this mortal life', but also by rewarding them 'with everlasting life'.[130] In January 1553, when the City authorities urged parishioners to 'enlarge their devotions', they reminded them that 'the Lord of glory whose work this is shall yield unto you the reward that He hath promised to His faithful servants'.[131] Perhaps the theologically impeccable distinction between faith and works did not accord always with the way most Londoners lived and believed: there were threats and

[128] B. Tierney, *Medieval Poor Law: A Sketch of Canonical Theory in England* (Berkeley, 1959), pp. 46 ff.; W. K. Jordan, *Philanthrophy in England, 1480–1660* (1959), pp. 54, 146–7.

[129] Brinklow, 'Complaynt', p. 81; *Select Works of Robert Crowley*, ed. J. M. Cowper (EETS, extra series, xxii, 1874), pp. 14–16.

[130] Mercers' Company, Acts of Court, ii, fo. 224ᵛ, (25 Oct. 1548).

[131] CLRO, Journal 16, fo. 256ᵛ.

promises in the here and now as well as promises and threats for the world to come. But at the last, in their wills, very many Londoners did reveal the theological choices they had made at the Reformation.

LONDON HALF CONVERTED

Bishop Gardiner scorned the reformers' zeal and their certainty in their election: 'they are', he expostulated, 'up with the "living God" as though there were a dead God'; they thought that to enter Heaven 'needs no works at all, but only belief, only, only, nothing else'.[132] The tenor of many of the wills written in London during Edward's reign supports Gardiner's prejudices. 'Dying to the world', testators committed their souls, their friends and families to the 'living God', sure that He will 'send us Heaven at our latter end'.[133] In 'man's ways, invented of man, to work any mean to the salvation of man other wise than God in His Word doth allow' they reposed no trust. 'I *only* doth cleave and commit my whole man . . . unto the mercy of God in the deserving of Thy son Jesus Christ', avowed John Horsley in 1551.[134] 'I do detest all other sacrifice to be made for sin, but the sacrifice of Christ's death and passion', wrote Bartholomew Gibbs, a clothworker.[135] Those of the new faith steadfastly shared the same conviction: 'when we were all lost by Adam's transgression' the wrath of the Almighty was 'fully pacified through the blood shedding of our saviour Christ', whom 'I believe suffered death for my sins and rose again for my justification'. Reading the prophet Job (ch. 19) in the dark days of the 'sweat' in July 1551, Edward Fison found comfort: 'I believe that my redeemer liveth and that on the last day I shall rise out of the earth and that in my flesh shall see my saviour. This my hope is laid up in my bosom.'[136] Salvation was reserved, they clearly believed, only for the faithful, 'meaning thereby' 'the faithful *only* which have their faith only fixed and knit in the merit of [Christ's] blood'.

---

[132] *Narratives of the Reformation*, pp. 179–80; *Letters of Gardiner*, pp. 168–9. Scudamore customarily commended Hoby to the 'living Lord': NLS, Adv. MS 34.2.14.

[133] PRO, PCC, Prob. 11/34, fos. 115^r, 157^r.

[134] Ibid., fo. 138^r.

[135] Ibid., fo. 73^r.

[136] Guildhall, MS 9171/12, fo. 128^r.

The reformers wrote of their hope and expectation to be 'of the number of His elect', among those 'that shall be saved and live forever after this transitory life finished', 'children of salvation', 'saved souls at the Day of Judgement'.[137] It was clear by now, as not quite before, that they believed the unregenerate of the opposing faith could never become 'His elect vessel'.

Asserting the sufficiency of Christ's passion for their salvation, they renounced all their 'good works called good deeds', and asked that their goods be distributed 'by the grace of God', not to work for them but 'as fruits of faith'.[138] Friends and neighbours were called upon to 'rejoice' at their funerals; 'for that it hath pleased Almighty God to call me out of this vale of misery whereby I shall not afterward any more sin and offend Him'.[139] No 'worldly pomp' or 'pride of the world' was to attend these funerals; rather there should be a simple burial, with no ceremony, 'after a godly sort'. No dirge nor mass sung, no knells or lights, but 'say the blessing of God be with the soul and body'.[140] Godly psalms might be sung, and bequests made for edifying funeral sermons to be preached by men learned in Scripture—Dr Maydwell, Rogers, Hooper, Horne, Coverdale, Mackbray, and Lever, the leading evangelical preachers of Edwardian London.[141] Richard Canesby (buried, fittingly, by the pulpit at St Mary Woolchurch) asked for a sermon 'upon the dead and last day to put the people in a remembrance of all that only'.[142] William Catesby, a draper of St Martin Orgar, required a 'godly learned man . . . sad, discreet and sober' to preach a funeral sermon, not only that God should have mercy on him, but also that 'with his preaching through the might of Almighty God some of the people there present may turn from their evil vices and that the Holy Ghost may enter into their hearts'. But even seemingly convinced reformers had their doubts. Of the soul's fate in the hereafter

[137] See, for example, PRO, PCC, Prob. 11/34, fos. 73ʳ, 99ʳ, 103ᵛ, 117ʳ, 123ʳ, 136ʳ, 198ʳ, 207ʳ, 258ʳ; 11/36, fos. 7ʳ, 9ʳ; 11/37, fos. 91ʳ, 173ᵛ, 208ᵛ; Guildhall, MS 9171/12, fo. 15ʳ.

[138] PRO, PCC, Prob. 11/34, fos. 138ʳ, 149ʳ, 198ʳ.

[139] Ibid., fo. 158ᵛ.

[140] Ibid., fos. 73ʳ, 160ʳ; Prob. 11/36, fos. 8ʳ, 26ᵛ; Guildhall, MS 9171/12, fos. 19ᵛ, 50ᵛ, 80ʳ, 109ʳ, 110ᵛ, 119ᵛ, 157ᵛ.

[141] PRO, PCC, Prob. 11/34, fos. 46ᵛ, 149ʳ, 163ᵛ, 188ᵛ, 196ʳ, 198ʳ, 216ʳ, 263ʳ, 266ʳ; 11/35, fo. 111ʳ; 11/36, fos. 8ʳ, 10ʳ, 11ᵛ, 27ʳ; 11/37, fo. 128ᵛ; Guildhall, MS 9171/12, fos. 32ᵛ, 80ʳ, 84ᵛ, 85ᵛ, 98ᵛ.   [142] PRO, PCC, Prob. 11/34, fo. 149ʳ.

Catesby was uncertain: 'And Lord, if it be Thy will that I may pray for them in another world, or else not. . .'.[143]

Others also had not yet clearly crossed the religious divide between the reformed and Catholic faiths. John Maddock in March 1549 bequeathed his soul to his 'maker and redeemer with whose precious blood I knowledge myself to have been redeemed and bought'. But he besought too the aid of the Blessed Virgin and all saints to 'pray with me and for me, that God of His infinite mercy will take pity upon me and not reject me, most wretched sinner, being the price of His blood'.[144] An analysis of the preambles of all the wills registered in the Commissary Court in Edwardian London reveals a stark statistic: that nearly half of testators (44 per cent) bequeathed their souls to God alone, asserting neither a belief in saving faith nor one in the power of intercessory prayers (this, compared with only a fifth who had done likewise in the last years of Henry's reign).[145] Of the testators whose wills were proved in Register Bucke (1551) a third made a similar testimony.[146] Various explanations, all inconclusive, might be given for this ambiguity. To avow a certain belief in uncertain times may have seemed unnecessary to some; others were doubtless, and understandably, confused. In the grim days of July 1551, when many of these wills were made, testators fearing that the 'sweat' would or had already struck them down had literally no time to contemplate their soul's passage.

Fewer people now avowed their conviction in the old faith. Many of the traditional forms of piety were outlawed, it is true. No longer could the Catholic faithful endow chantries, bedeck images in their churches, join fraternities, decorate altars. All over London they had witnessed the desecration of their churches, and now no gifts were forthcoming. Alone in all the Commissary Court wills of the reign there is a bequest of 20d. for the old works of St Paul's, nothing else.[147] Yet the Virgin and saints still remained in Heaven, many thought, looking down upon their votaries, and ready to intercede for those who called

143 Guildhall, MS 9171/12, fo. 149ʳ.
144 Ibid., fo. 27ʳ.
145 Guildhall, MS 9171/12.
146 PRO, PCC, Prob. 11/34.
147 Guildhall, MS 9171/12, fo. 41ʳ.

upon them. Few did, at least in their wills. Only 8 per cent of testators in Register Bucke remembered the 'holy company of Heaven'; and only a quarter of Londoners whose wills were proved in the Commissary Court. A third of the wills in that register avowed distinctly Protestant convictions in the justificatory powers of Christ's passion, and so did more than half of the wills in Register Bucke.[148]

Late in 1552 a new Prayer Book was introduced which removed all the ambiguities which had allowed Catholics to worship according to the 1549 rite. The intent was to remove any ground for the ancient 'superstitions' of the Catholic past. The words of administration in the communion were changed to 'Take and eat this in remembrance that Christ died for thee, and feed on Him in thy heart by faith, with thanksgiving'; 'Drink this in remembrance'. All elements of transubstantiation were lost; spiritual manducation remained. In place of the Mass wafers ordinary bread was used. No longer were there prayers and Mass for the dead, for there was no Purgatory. Signing with the cross disappeared, except after baptism. The priest could not now anoint the sick and dying, and exorcism of the unborn child was abandoned.[149] This was the removal of idolatry for which the godly had hoped, but many found the changes abhorrent.

In December 1548 news had come to the country: 'as for the Mass, it is in London as it is in the country, some of the old-fashioned and some of the new'.[150] How much had changed by 1553? Was London half converted? To know the truth of the religious faith behind the numbers it is essential to understand the course in politics which had brought about the conversion. London may have been considered the stronghold of the reformed faith by the end of Edward's reign, but the Catholics were very many and anxious to restore the old faith, and perhaps even the converts would prove to be of less than adamantine conscience. At the end of August 1553, as Lady Jane Grey awaited her fate in the Tower, she asked: 'I pray you . . .

[148] PRO, PCC, Prob. 11/34; Guildhall, MS 9171/12.
[149] F. Procter and W. H. Frere, *A New History of the Book of Common Prayer* (1901); Dugmore, *The Mass and the English Reformers*, pp. 159 ff.; Brightman, *The English Rite*, i, pp. cl–clvi.
[150] Cited in Jordan, *Edward VI: The Young King*, p. 317.

have they Mass in London?' The reply which came could not
have comforted a Protestant: 'Yea, forsooth . . . in some
places.'[151]

[151] *Chronicle of Queen Jane*, p. 25.

XII London and High Politics in the Reign of Edward VI

LATE IN January 1547, as Henry VIII lay dying, men of desperate ambition had conspired to subvert the King's will for the rule of England during the minority of his young son. Henry's death on 28 January was kept secret from the people, Parliament met still, while behind the locked doors of the Privy Chamber leading councillors and courtiers schemed for three days to overturn the King's provisions for a Regency Council of sixteen, and contrived to grant instead the governorship of Edward and the protectorship of the realm to the Earl of Hertford alone. By doctoring the royal will to reward the conspirators with land and title the loyalty of some and the silence of others was bought, for a time.[1] Not until 31 January was the King's death declared to the people and Edward proclaimed.[2] The secret moves of Henry's last days left a dangerous political legacy. The conspirators looked always for further favours and for a share in the power they had handed over. 'Remember what you promised me in the gallery at Westminster, before the breath was out of the body of the King that dead is . . .', so Paget would remind a heedless Somerset in July 1549.[3] Lisle, Seymour, Hertford, Wriothesley were men who would stop at nothing—even, perhaps, fratricide—to

---

[1] For the overturning of Henry VIII's will, see D. E. Hoak, *The King's Council in the reign of Edward VI* (Cambridge, 1976), especially pp. 41–4, 231–6; D. B. Starkey, *The Reign of Henry VIII: Personalities and Politics* (1985), ch. 8; J. J. Scarisbrick, *Henry VIII* (1968), pp. 488–98; H. Miller, 'Henry VIII's unwritten will: grants of lands and honours in 1547', in *Wealth and Power in Tudor England: essays presented to S. T. Bindoff*, ed. E. W. Ives, R. J. Knecht, and J. J. Scarisbrick (1978).

[2] CLRO, Journal 15, fos. 303ᵛ, 305ᵛ. W. K. Jordan, *Edward VI: The Young King* (1968), p. 52. Still on 31 Jan. the Bishop of Rochester was eulogizing the generosity of Henry, as though he were still alive; Wriothesley, *Chronicle*, i, p. 177.

[3] Strype, *Ecclesiastical Memorials*, ii/2, p. 430.

save themselves or vanquish their rivals.[4] The ruling faction was always insecure; fearful of the consolidation of a rival party of disaffected lords, terrified that their opponents would win popular support.

No rulers ever felt their position more imminently threatened than those of Edward's reign, for they lacked the essential resource of power in early modern England: legitimacy. Some looked then to the Wars of the Roses to see the dangers of regency government. Somerset contrasted his own beneficent guardianship with the evil ambitions of Richard III, and sided with Duke Humphrey of Gloucester, the good uncle of Henry VI, against Cardinal Beaufort.[5] Only possession of the King's person could sanction their authority, and for control over Edward the Seymour uncles fought. In October 1549, as rival lords in the Council intrigued against Protector Somerset, who was holding the King at Hampton Court, a tract urged the commons to 'stick fast unto your most godly and Christian prince and King, for though they traitorously call them the body of the Council, yet they lack the head, then may we call it a monstrous Council'.[6] With government so insecure, the allegiance of London was more vital than ever. Londoners, witnessing the faction struggles, were reminded of the consequences of their rebellion against Henry III, and warned that they must choose their partisanship with great care for fear of future reprisals.[7] As during the Wars of the Roses, the capital's declaration for one group or the other in a quarrel might determine the outcome. But politics in Edward's reign were even more fraught than before, because now the rival lords struggled not for political control alone, but for the power to institute reform or reaction in religion; so at least the people believed, and it was for reasons of faith that they might consign their support to one or other faction. And in religion London was,

---

[4] The story of the politics of Edward's reign is told best in Hoak, *King's Council*, ch. 7; and in W. K. Jordan, *Edward VI: The Young King* and *Edward VI: The Threshold of Power* (1970).

[5] 'An Eye-Witness's account of the *coup d'état* of October, 1549', *EHR* lxx (1955), p. 606; BL, Additional MS 48216, fos. 6ʳ, 10ʳ. Foxe, too, compared Somerset with the 'Good Duke' Humphrey of Gloucester: *Acts and Monuments*, vi, p. 296.

[6] PRO, SP 10/9, fo. 12ʳ.

[7] Grafton, *Chronicle*, ii, p. 533.

of course, divided, and upon these divisions unscrupulous politicians would play.

THE 'STIRRING TIME'

Protector Somerset determined to rule alone, withdrawing increasingly from his fellow councillors and scorning their advice.[8] The Scottish campaigns would at last bankrupt England, bring political ruin to Somerset, and foster rebellion. Not even the income from the expropriated chantries could finance the prodigious expenditure of perhaps £200,000 annually in Scotland alone.[9] The military ambitions could be financed in two ways: by taxes from subjects and from royal finances. Under Somerset subjects paid twice: they paid subsidies, but also the Crown paid for its share by manipulating the currency, further bedevilling an economy which suffered enough without inept government interference. Extraordinary inflation followed the great debasement, and in London, where there was no food to be had unless it were bought, the suffering was acute. Between 1544 and 1551 prices in London rose by 89 per cent.[10] Londoners looked with loathing upon the high financiers.

It was a terrible irony that the commons suffered so under Somerset's protectorate, for Somerset surrounded himself by reformers to whose advice he listened, at first, and was persuaded to reform.[11] Yet the Protector's social measures, especially for agrarian reform, aroused expectations among the commons which could never be fulfilled, and meanwhile engendered disturbance in the country which, seemingly, everyone save Somerset had predicted. Lord Rich, in a speech before

[8] 'The Letters of William, Lord Paget of Beaudesert, 1547–1563', ed. B. L. Beer and S. M. Jack (Camden Miscellany, xxv, 1974), 12–14, 16, 19–25; Strype, *Ecclesiastical Memorials*, ii/2, pp. 427–37; Foxe, *Acts and Monuments*, vi. pp. 290–1; Hoak, *King's Council*, pp. 15–23, 114–18.

[9] For the Scottish campaigns and their influence upon the making of policy, see M. L. Bush, *The Government Policy of Protector Somerset* (1975); Jordan, *Edward VI: The Young King*, ch. 9.

[10] C. E. Challis, *The Tudor Coinage* (1978); S. R. Rappaport, 'Social Structure and Mobility in Sixteenth-Century London' (Columbia University, Ph.D. thesis, 1983).

[11] For Somerset as the 'Good Duke', see A. F. Pollard, *England under Protector Somerset* (1900); Jordan, *Edward VI: The Young King*, pp. 416–38. For the reformers around him, see Bush, *Government Policy*, pp. 58–73. See also, G. R. Elton, 'Reform and the "Commonwealth-Men" of Edward's reign', in his *Studies*, iii, pp. 234–54.

London's Mayor and Aldermen in October 1549, accused Somerset of creating disorder 'under pretence of [reform of] such matters such as all men desired might be redressed more gladly than he [Somerset], but in a more quiet and settled time'.[12] Paget warned Somerset constantly of the consequences of his 'lenity', his 'softness', his 'opinion to be good to the poor', knowing that his warnings, like Cassandra's, went unheard. In July 1549 he wrote that Somerset's vain desire to hear the benisons of the poor—'there was never man had the hearts of the poor as you have. Oh! the commons pray for you, sir, they say, God save your life!'—had occasioned universal disorder.[13] Dr Bush has suggested that Somerset's social policy and intentions towards the commons were, in truth, little different from those of any other sixteenth-century magnate, that there was nothing unusually liberal about his attitude—rather the contrary.[14] Yet what matters is that the people and his peers thought that he did intend reform and generosity to the poor, and that in time his peers chose this as the most credible accusation to make against him and the most convincing way to bring him down. Moreover, there is strong evidence that he did boast to provide justice for the poor.

Is Paget to be discounted? His was personal advice, given to ward off disaster, and not meant as propaganda. In London Somerset had plied the Court of Aldermen with successive suits and petitions, often, it seems, in favour of disadvantaged supplicants. These the Court began to turn down by early in 1549. For example, when in January the Duke wrote on behalf of William Yokens 'to continue in his dwelling house from the which the owner goeth about to put him out', no order was taken.[15] Latimer urgently requested the Protector in his Lenten sermons of 1549 that 'your grace would hear poor men's suits yourself', and maybe Somerset listened.[16] A charge against him later was that 'you had and held, against the law,

---

[12] Cited in Bush, *Government Policy*, p. 77.

[13] Strype, *Ecclesiastical Memorials*, ii/2, p. 430.

[14] Bush, *Government Policy*, pp. 73–83.

[15] CLRO, Repertory 12, fos. 38^v, 39^v, 40^r, 76^r; Yokens was, moreover, reformed in religion: GLRO, DL/C/614, fo. 30^r.

[16] Latimer, *Sermons*, p. 127; M. L. Bush, 'Protector Somerset and Requests', *HJ* xvii (1974), pp. 451–64.

in your own house, a Court of Requests'.[17] Maybe this Court
of Requests was no tribunal for the poor, but it is clear that
plaintiffs frequently did address their causes to Somerset, albeit
not always particularly poor ones.[18] Foxe remembered that
Somerset used to hear 'the causes and supplications of the
poor' as a father to his children.[19] Even if Somerset were not a
champion of the poor, and did not have the interests of the
commons close to his heart, there were very many who looked
to him as such. This is why there were so many suits to him to
act as 'good lord' before the Courts of Aldermen and Requests.
In London, as elsewhere, it may have been Somerset's ostens-
ible support from the poor which lost him the support of the
rich and powerful. In July 1549 Paget reported to Somerset the
rumour, dangerous to him, that 'you have some greater enterprise in your head that lean so much to the multitude'.[20]

'Society in a realm', wrote Paget, 'doth consist and is maintained by fearful love to God and the prince, which proceedeth
by religion and law.' If one or other lacked, then 'farewell
Kings'.[21] By the end of 1548 it was evident that the removal of
restraining laws (for example, for treason) had occasioned
instability, and reform in religion, disunity. A Council, finan-
cially bankrupt and with bankrupt policies, feared the coales-
cence of religious with social and economic disaffection to bring
disturbances worse and more widespread than those of the
previous summer. In London perennially unstable groups—the
dispossessed poor, servants and apprentices, maybe religious
radicals—might make common cause. In September 1548
preaching had been suspended, in London earlier than elsewhere, lest reformers urge reform more radical than the
Council could contemplate, and even exacerbate social tensions. Servants and apprentices were gated, prevented from
running to sermons at Paul's Cross, 'a gazing and a gaping' at
the empty pulpit.[22] But the subversive preaching of the radicals
continued. In January 1549, and again in May, John Hunting-

---

[17] Foxe, *Acts and Monuments*, vi, p. 290.
[18] Bush, 'Protector Somerset and Requests'; *Government Policy*, p. 79.
[19] Foxe, *Acts and Monuments*, vi, p. 295.
[20] Strype, *Ecclesiastical Memorials*, ii/2, p. 431; cited in Bush, *Government Policy*, p. 98.
[21] PRO, SP 10/8, no. 4; cited in Hoak, *King's Council*, p. 181.
[22] *TRP* i. 313; Mercers' Company, Acts of Court, ii, fo. 224^r.

don, from ardent Papist turned ardent Protestant, was reported for his 'lewd and slanderous' railing from the pulpit against the Mayor and Aldermen.[23] In April the Council procured the removal of John Goodale from the stewardship of St Martin Westminster on account of his 'naughty conversation'.[24] The plays and interludes performed in the City were regarded with suspicion, and rightly, for they often took subversive, heretical themes, and attracted to them the susceptible youth of London. On 23 May John Wilkinson was forbidden to hold plays in his house in future, and on the same day that ban was made universal.[25] The wardens of the City companies were commanded 'as they would answer at their peril' to impose until Michaelmas a nightly curfew between 10 p.m. and 4 a.m. upon all their 'servants and youth within their houses', forbidding any May games or attendance at games and interludes on holy days.[26] Sir William Pickering, boon companion to the late Earl of Surrey, was discovered once again at his old night games to disturb the City.[27]

The London authorities were preparing for trouble. Proclamations had been made in London on 30 April for all captains and soldiers to 'prepare themselves Northwards',[28] but soon they would be needed nearer home. On 4 June precepts came for a nightly watch in the City.[29] Within days the new Prayer Book was imposed, and in the West provoked open rebellion.[30] This rebellion the Protector delayed in suppressing, unwilling to take up the sword against subjects or to deploy troops elsewhere than in Scotland or France, and soon it was followed in East Anglia by another rebellion as serious, but stirred for reasons of economic distress and local political grievance rather than religion. Would London rebel? Between 18 June and 2 July 1549 there are no records of the

---

[23] CLRO, Repertory 11, fo. 327^v; Repertory 12, fo. 79^r.

[24] Hoak, *King's Council*, p. 216.

[25] CLRO, Repertory 12, fo. 91^r.

[26] Ibid., fo. 90^v; Mercers' Company, Acts of Court, ii, fo. 231^r.

[27] CLRO, Repertory 12, fo. 92^r.

[28] Ibid., fo. 76^r.

[29] Ibid., fo. 94^r.

[30] F. Rose-Troup, *The Western Rebellion of 1549* (1913); B. L. Beer, *Rebellion and Riot: Popular Disorder in the Reign of Edward VI* (1982): J. Cornwall, *Revolt of the Peasantry, 1549* (1977).

Court of Aldermen: evidence of the panicked preoccupation of London's governors. On 2 July orders came for a close watch of the City; the Aldermen and their deputies, according to the 'high charge that they are burdened withall', should ride nightly to survey the watches during 'this unquiet time'.[31] All householders were warned to keep their menservants at home between 9 p.m. and 5 a.m., and to forbid them to 'go abroad at their liberty on holy days during the time of this unquietness'.[32] Again, orders came for the prevention of interludes (had they discovered the incendiary Anabaptist play we know to have been written?), and buckler playing (such as had begun the Evil May Day riots) was forbidden.[33] Not even the watches were trusted: on 5 July the Aldermen summoned the ward constables and charged them to ensure that only 'sad persons and substantial householders' should guard the City; not 'boys and naked men'.[34] Five hundred troops were to be provided by the City companies. The City gates were fortified, but such was the insecurity that a false drawbridge was made for London Bridge 'in case need should require by reason of the stirring of the people (which God defend) to cast down the other' to prevent the rebels inside from inviting in those beyond the walls.[35]

Such precautions were usual in kind, but not in their extent. The Council's concern for the capital's security, its alarm that it would either be overrun by invading rebels or turn to rebellion itself, was manifested by the declaration of martial law in London on 18 July.[36] There were agitators at work to raise Londoners to support the rebels and, hardly surprisingly, they looked for supporters among the most volatile elements of the City population, especially servants and apprentices, those who happened to be most easily won to Protestantism also. For this last reason, in part, it was Kett's rebellion which was more likely to win adherents than the Prayer Book rebellion in the

---

[31] CLRO, Repertory 12, fo. 97ᵛ. For London and the rebellions of 1549, see Beer, *Rebellion and Riot*, ch. 7; Jordan, *Edward VI: The Young King*, pp. 445–6.

[32] CLRO, Repertory 12, fo. 98ʳ; Mercers' Company, Acts of Court, ii, fo. 232ʳ.

[33] CLRO, Repertory 12, fo. 99ʳ; see above, pp. 443–4.

[34] CLRO, Repertory 12, fo. 102ᵛ.

[35] Ibid., fo. 104ᵛ.

[36] Wriothesley, *Chronicle*, ii, p. 15; *Grey Friars Chronicle*, p. 60; *TRP* i. 341; Penry Williams, *The Tudor Regime* (Oxford, 1979), pp. 387–8.

West. Even the benchers at the Inns of Court were ordered to govern their unruly pupils during 'this stirring time'.[37] On 5 August John Wheatley, a saddler, was committed to Newgate 'for that he enticed men's servants and apprentices . . . to go with him to the rebels at Norwich'.[38] The following day the Aldermen were to 'charge every householder for the good governance and forthcoming of as many of his menservants as he will undertake for', and to ascertain who had gone missing.[39] Anthony Roberts of Tonbridge, Kent, was apprehended as a suspected rebel; some soldiers from the retinue of Captain Drury were expelled from the City; and two men committed for their 'slanderous and seditious words'.[40] An apprentice carpenter set upon the watch at Tower Hill and another fired arrows at it; there were windows broken at St Botolph Aldersgate when shots were fired.[41] The activity of the capital's most effective demagogues—the radical preachers—gave cause for alarm. On 11 July the Aldermen tried to discover in which churches Stevens, Huntingdon, Maydwell, Chamberlaine, Hooper, and Reynolds had been preaching, and who had preached at Westminster the previous Sunday. All these men were known zealots; at least three of them had been in trouble before for their sedition.[42] What they had been preaching is unknown, but Sir Stephen (Stevens?) had certainly been in touch with a rebel. At the gibbet at Aldgate, about to be executed for high treason, the bailiff of Romford appealed to his audience:

Good people, I am come hither to die, but know not for what offence except for words by me spoken yesternight to Sir Stephen, curate and preacher of this parish . . . He asked me, 'What news in the country?'. I answered, 'Heavy news'. 'Why?', quoth he. 'It is said', quoth I, 'that many men be up in Essex, but thanks be to God, all is in good quiet about us', and this was all as God be my judge.[43]

Whether or not there was treasonable connection between the

37 CLRO, Repertory 12, fo. 108ᵛ.
38 Ibid., fo. 122ʳ.
39 Ibid., fo. 123ʳ.
40 Ibid., fos. 110ʳ, 111ʳ, 115ʳ, 116ʳ.
41 Ibid., fo. 118ʳ; Walters, *London Churches*, p. 203.
42 CLRO, Repertory 12, fo. 104ᵛ.
43 Stow, *Survey*, i, p. 144.

rebels and City preachers, clearly the Council feared the subversive content and consequence of their preaching.

From 20 July the watch was kept by day as well as night, with eight bedells to search for vagabonds.[44] Soon the City was prepared to withstand siege. One hundred and forty labourers worked to clear the ditch which had once defended the City; all arms and gunpowder were accounted for to ensure that none fell into the wrong hands.[45] The citizens must be fed, although the harvest had failed; each Alderman lent £100 to buy wheat, and each householder laid in provisions to last a month.[46] So tense was the City that the proclamation of 2 July to stabilize food prices was stayed for a time.[47] St Bartholomew's fair was held as usual, but the watch was doubled and the traditional wrestling forbidden.[48] By the end of August the crisis had passed; the rebellions in the West and East Anglia had been put down, and Dussindale ran with the blood of slaughtered rebels. The toll was great, politically as otherwise. In London they had already informed the Council on 13 July that so great was the expense of relieving the poor and defending the City ('and toward all these charges there is no treasure in the City') that they could pay no subsidy for the King's war.[49] A loan to the Crown on 2 July was made only upon the personal securities of the councillors.[50] But the failure to pay loans and subsidies in August and September marked the City's failure in confidence in the Protector and his policies. While the country grew more impoverished Somerset's desire to 'enrich himself without measure' had been manifest. None saw better than the Londoners how 'he buildeth three, four or five palaces most sumptuously and leaveth the poor soldiers unpaid'.[51] The governors of London, like the councillors, bitterly blamed Somerset for the social policy which had engendered the rebellions and for his fainéance in failing to suppress them. By mid-September,

[44] CLRO, Repertory 12, fo. 112^{r-v}.
[45] Beer, *Rebellion and Riot*, p. 176; CLRO, Repertory 12, fos. 117^v, 119^r, 121^r, 121^v.
[46] Ibid., fos. 121^v, 122^{r-v}.
[47] *TRP* i. 336; Beer, *Rebellion and Riot*, p. 176.
[48] CLRO, Repertory 12, fo. 130^v.
[49] Ibid., fos. 106^v–107^r.
[50] Ibid., fo. 98^v.
[51] CLRO, Journal 16, fo. 36^r.

and maybe earlier, Somerset's rivals in Council were conspiring to remove him.[52]

THE FALL OF THE PROTECTORATE

In early October Protector Somerset waited at Hampton Court with the young King and a force of four thousand 'peasants' who had answered his call for a general array to save the King from a 'most dangerous conspiracy'.[53] In London the rival lords in the Privy Council waited. Both sides were waiting to see which way the country and the capital would commit their support. The King had the City's loyalty, of course. In Common Council on 8 October George Tadlowe (maybe tutored by Richard Grafton for this speech) reminded the citizens of how when the capital had stood with the barons against Henry III: 'was it forgotten? No, surely, nor forgiven during the King's life. The liberties of the City were taken away, strangers appointed to be our heads and governors.'[54] But the London Lords were hard to deny. The first that Londoners had known of the conflict between the Protector and the Lords was on Sunday, 6 October, when 'suddenly, of what occasion many marvelled and few knew, every lord and councillor went through the City weaponed and had their servants likewise weaponed, attending upon them in new liveries, to the wondering of many'.[55] A messenger to Somerset reported 'such a sort of horsemen in the streets that he could not number them . . . two thousand at the least', which 'seemed then to be such as would have no repulse'.[56] A powerful reason for co-operation. But the City's rulers had to make a decision knowing that to give it one way—to the Lords—would be to bring a force

[52] J. Berkman, 'Van der Delft's letter: a reappraisal of the attack on Protector Somerset', *BIHR* liii (1980).

[53] *TRP* i. 351; PRO, SP 10/9, fos. 1^r, 2^r; *CSP Sp.* ix, pp. 456–9; 'Eye-witness's account'. For accounts of the coup and Somerset's fall, see P. F. Tytler, *England under the reigns of Edward VI and Mary*, 2 vols. (1839), i, pp. 206–54; Jordan, *Edward VI: The Young King*, pp. 494–523; Pollard, *England under Protector Somerset*, ch. 9; R. R. Sharpe, *London and the Kingdom*, 3 vols. (1894), i, pp. 433–9.

[54] Grafton, *Chronicle*, ii, p. 523; Foxe, *Acts and Monuments*, vi, pp. 289–90; J. A. Kingdon, *Richard Grafton, Citizen and Grocer of London* (1901), pp. 36 ff.

[55] Grafton, *Chronicle*, ii, p. 522.

[56] 'Eye-witness's account', p. 605; Foxe, *Acts and Monuments*, vi, p. 289.

against Protector and King, and to outrage Somerset's feared popular following. To give it the other way would provoke the military might of the London Lords against London. To commit the City either way was to risk future reprisals in the likely event of another reverse in high politics. At last, after much discussion and dissension, the Common Council granted five hundred men, or a thousand if need be, and its support to the Lords.[57] How did it happen that the Earl of Warwick and the others came to command the support of the City governors while the poor and Protestant seemed to be for Somerset?

During the conspiracy to bring down the Protector, Somerset's cause was being zealously proclaimed by those who feared with his downfall the end of patronage for the poor and a halt to the furtherance of the reformed faith. On 9 October, as the Common Council committed the City to support the Lords, John Cardmaker, most charismatic of reforming preachers, rallied his congregation to the embattled Protector in the name of Protestantism: 'though he [Somerset] had a fall he was not undone, and that men should not have their purpose; and also he said that men would have up again their popish Mass; wherefore, good masters, stick unto it'.[58] His message was clear: if the new regime were to succeed Somerset then the present good laws in religion would be repealed and the Roman rites restored. Yet while many were convinced of Somerset's favour to the Gospel, others were not, and popular opinion on this matter was crucial and might be manipulated. So, Bonner's deprivation, so ruthlessly sought, had been part of a political campaign by Somerset to prove his own reforming credentials and to save himself.[59] But others could play that game: at Hampton Court on 6 October Somerset had heard that his rivals 'are not ashamed' to broadcast the false news that he was in the Tower already, and if freed 'would deliver the Bishops of London and Winchester out of the prisons and bring in again the old Mass'.[60] Both factions knew that this threat stirred passions, for and against, and used it in their

---

[57] CLRO, Journal 16, fos. 37^{r-v}; Letter Book R, fos. 40^v, 41^r–42^r.

[58] *Grey Friars Chronicle*, p. 64.

[59] See above, pp. 448–52.

[60] *Troubles connected with the Prayer Book of 1549*, ed. N. Pocock (Camden Society, xxxvii, 1884), p. 83.

propaganda. On 10 October and again on the 30th the London Lords issued proclamations against those who cast letters abroad in the City urging the fallen Protector's cause.[61]

Those handbills survive.[62] They urged the people to rise to save the Duke, whose authority was legitimate, and his cause, which was theirs. The threat to Protestantism was easily associated with a threat to the poor, and Somerset was—or so it was said—popularly believed to be the champion of both. His enemies, conversely, were accused both of conservatism in religion and oppressive social policies. It may have been that all this was 'black' propaganda, circulated not by Somerset's supporters but by the London Lords, intended to blacken Somerset's name further among those who feared social dislocation. Even so, this would be further proof of the Protector's reputation among the poor. One 'Henry A' warned lest people be 'carried away with the painted eloquence of a sort of crafty traitors' to turn from Somerset, who would have 'redressed things in the court of Parliament . . . to the intent that the poor commons might be godly eased'. Somerset's opponents, who had duplicitously worked with the Protector all this while, were moved by two damnable motives: 'insatiate covetousness and ambition', and the desire to 'plant again the doctrine of the Devil and Antichrist of Rome'. That London had been warned of this débâcle long ago gave the message urgency and conviction: Merlin had foretold that in London 'called Troy untrue . . . twenty three Aldermen of hers shall lose their heads on a day'. To which 'Henry A' added the private prayer: 'God grant to be shortly'.[63] Clearly London's rulers were seen as especially responsible for Somerset's fall. So they were, and one in particular: Sir John Yorke, master of the Southwark mint.[64] Somerset had been so vehemently opposed to Yorke's influence that in August 1549, in flagrant contempt of the City's liberties, he attempted to repudiate Yorke's election as Sheriff.[65] He failed, but the insult was not forgiven. It was

[61] CLRO, Journal 16, fos. 37ʳ, 38ᵛ, 39ʳ; PRO, SP 10/9, fo. 70ʳ; *TRP* i. 352.

[62] PRO, SP 10/9, fos. 12ʳ–14ʳ.

[63] Ibid. SP 10/9, fo. 12ʳ. On 26 Dec. a bill against prophecies passed in the Commons: NLS, Adv. MS 34.2.14, fo. 21ᵛ.

[64] For Yorke and his part in the coinage operations, see Challis, *Tudor Coinage.*

[65] CLRO, Repertory 12, fo. 120ᵛ; B. L. Beer, 'London and the Rebellions of 1548-9', *JBS* xii (1972), pp. 34–5.

Yorke who became the main intermediary between the City rulers and the London Lords during the coup, and to his house that Warwick moved on 6 October as he sought to win over the City, and he would go there again in January 1550 'to have the City on his side'.[66] Somerset's followers did not forget his traduction by Yorke, and their resentment grew as Yorke, Warwick's agent, manipulated the coinage, further threatening the estate of poverty.

Somerset's cause was not lost while he still lived. After his submission at Windsor and the handing over of Edward, the Duke 'with other traitors' was led through London to the Tower on 14 October. The City streets were ostentatiously decked to mark the return of the King and vaunt the loyalty of the citizens.[67] But when the rumour spread in mid-January 1550 that Somerset would be released, two thousand people, reportedly, waited outside the Tower in expectation of his happy deliverance.[68] The downfall of the Protector was far from concluding the savage factional struggles in the Council. Now it had to be decided whether Wriothesley and his conservative adherents would seize control or whether Warwick could prevail.[69] London observers reported the signs of division. Scudamore ended his letter to Hoby on 22 November, 'beseeching God . . . to keep an amity amongst our magistrates'.[70] At the end of November the two principal protagonists, Wriothesley and Warwick, lay sick in their Holborn houses.[71] 'Howsoever the world shall go' the prisoners in the Tower were unwary, 'indifferent merry', expecting a reverse in their fortunes.[72] In truth, the conservatives sought Somerset's life and, with his, Warwick's: they were 'traitors both; and both is worthy to die', advised Wriothesley.[73] Warwick knew that his fate and Somerset's were bound together, and now began to show Somerset unusual favour. On Christmas day Somerset

---

[66] *APC* ii, pp. 331–2; *CSP Sp.* x, p. 13.
[67] *Grey Friars Chronicle*, p. 64; Wriothesley, *Chronicle*, ii, pp. 27–8.
[68] *CSP Sp.* x, p. 7.
[69] For this faction struggle, see Hoak, *King's Council*, pp. 241 ff.
[70] NLS, Adv. MS 34.2.14, fo. 27^r.
[71] Ibid., fo. 17^r.
[72] Ibid., fo. 31^v.
[73] BL, Additional MS 48216, fo. 15^v; Hoak, *King's Council*, p. 255.

was allowed the freedom of the Tower, and his Duchess visited him. Hooper, minatory, preached a jeremiad in the Tower: taking a psalm of David, he spoke of 'governors that misordered their vocations, and persuading that God punished rulers for their sins'. Urging patience, he forbade the prisoners to 'seek revengements', for God would surely punish them with 'double plagues'. It was widely believed that Edward would free the prisoners as a New Year's gift.[74] Fearful of his life, Warwick moved once more to Yorke's house.[75] In a dramatic confrontation during the first week in January Warwick struck before he was himself struck down: Somerset would live, he told Wriothesley, laying his hand on his sword, for 'My lord, you seek his blood, and he that seeketh his blood would have mine also'.[76] Wriothesley and Arundel were banished from Court. 'What is laid to their charge is not openly known', reported Scudamore on 11 January, 'but some imagineth they were about the subversion of religion', and perhaps there were plans to seize the King, for Arundel had tampered with the locks of the Privy Chamber.[77] Somerset, too dangerous and too powerful to exclude, soon returned to political eminence—to the Privy Council and to intervene in London affairs in March— and his ambition to return to the principal place was always obvious.[78] His freedom remained the most serious of all the dangers to the government of Warwick. Acute social distress gave grounds for fears of insurrection, but worse still was the knowledge that the loyalty of the poor was with Somerset, and if they rose again it would be in his name.[79] Scudamore warned in January 1550 that 'by the division of the great the mad rage of the idle commoners is much provoked thereby to follow their naughty pretences, so that (unless God show His mercy over us) this year to come is like to be worse that any was yet'.[80]

---

[74] NLS, Adv. MS 34.2.14, fo. 21^r.
[75] *CSP Sp.* x, p. 13.
[76] BL, Additional MS 48216, fo. 16^r.
[77] NLS, Adv. MS 34.2.14, fo. 23^r.
[78] *APC* iii, pp. 235 ff.; CLRO, Repertory 12, fo. 217^r.
[79] Jordan, *Edward VI: The Threshold of Power*, pp. 56–7, 78, 105.
[80] NLS, Adv. MS 34.2.14, fo. 12^r.

CONSPIRACIES

Tudor government rested upon consent and upon popular support, but the regime of the Earl of Warwick possessed neither, and knew it. One way to win popular approbation might be to follow the path in religion which the people favoured, whatever that might be. While the councillors struggled among themselves for a new ascendancy after the October coup everyone speculated about the course the new regime would follow—for or against the Reformation—and everyone expected reaction. Hooper wrote to Bullinger on 7 November that 'we are greatly apprehensive of a change in religion . . . the papists are hoping and earnestly struggling for their kingdom'.[81] Gardiner looked forward eagerly to his imminent release.[82] But by 17 November Hilles, writing from London, could express the cautious hope that 'Christ may yet remain with us'; though a month before they had thought He would 'depart from us, even beyond sea'.[83] And on Christmas day 1549, in a move to end uncertainty, the King and Council issued an order to the Bishops that their clergy were to bring in all books of the old, Latin services and to enforce the new order in religion.[84] The papists had failed. 'God in His providence holds the helm, and raises up more favourers of His Word in his majesty's councils', reported Hooper on 27 December.[85] The motives which drew Warwick, whom everyone had taken for a Catholic, to the reformed faith, even to become 'a most holy and fearless instrument of the Word of God',[86] have been explained by Dr Hoak in terms of balance of power within the Council, of attempts to prevent the establishment of Princess Mary as Regent, of control over the King, who was determinedly Protestant and determined to rule, of Warwick's adherence to whatever form of religion was political orthodoxy.[87] But there was perhaps another pressure: if Warwick intended to prevent popular

[81] *Original Letters*, i, p. 69.
[82] Stow, *Annales*, p. 600; *Letters of Gardiner*, pp. 440-1.
[83] *Original Letters*, i, p. 268.
[84] *TRP* i. 353.
[85] *Original Letters*, i, p. 71.
[86] Ibid. i, p. 89.
[87] Hoak, *King's Council*, pp. 243 ff.

disturbance, as he needed to do 'in order to make his reign permanent',[88] he might adopt, or at least publicly support, what he perceived to be the prevailing faith. In London in 1549 that faith may have seemed to be reformed. The rapturous attendance at sermons, the iconoclasm, the patronage of the Protestant book trade, appear to have led Warwick to think that Londoners would resolutely resist any reversion to the old faith.[89]

If the government could not win popularity nor command obedience it must use repression. The character of the new regime was marked by the passage in December 1549 of the draconian act for the 'punishment of unlawful assemblies and risings of the King's subjects'.[90] On 1 February 1550, having heard reports of a new 'commotion' in the country, Parliament was dissolved so that its members could return to prevent a new rebellion.[91] (Moreover, the Commons was sent away because groups within it favoured Somerset.)[92] In London, the Imperial ambassador reported that 'the people murmur a good deal against the government of the Earl of Warwick',[93] and there was widespread disaffection in the capital, which was suffering from a shortage of work and food, and rising inflation. Some who might turn grudge into action were the soldiers returning from the less than glorious campaign in France, after Warwick's ignominious restitution of Boulogne. The government tried repeatedly to rid the capital of the disturbing presence of these armed men,[94] but the unfocused fears of what they might do became real in September 1550 when the City rulers heard that they planned to form 'several companies in London and come out of several lanes and streets . . . and set upon the citizens and their houses and take there such booty and spoil as they can lay hand upon'. Their vow that they would 'turn all England up so down at their pleasures', together

---

[88] *CSP Sp*. x, p. 301.

[89] See above, p. 425.

[90] 3 & 4 Edward VI c. 5; *Statutes of Realm*, iv/1, pp. 104–8; Jordan: *Edward VI: The Threshold of Power*, pp. 37–8.

[91] BL, Cotton MS Caligula E iv, fo. 207ʳ; Hoak, *King's Council*, pp. 198–9.

[92] Jordan, *Edward VI: The Threshold of Power*, pp. 75, 79–80; Hoak, *King's Council*, p. 75.

[93] *CSP Sp*. x, p. 47.

[94] CLRO, Journal 16, fos. 64ᵛ, 66ᵛ; *TRP* i. 363.

with fears that the City's pullulating vagabond population would join common cause with them, engendered alarm.[95]

A general and implacable hostility had soon grown against the Earl of Warwick and the band of followers with whom he swiftly packed the Privy Council. By January 1551 it was said that Warwick governed 'absolutely'; that he was 'hated by the commons and more feared than loved by the rest'.[96] The two men most bitterly detested by the populace were Warwick and Sir John Yorke, and not just for their traduction of Somerset.[97] As the effects of the debasement grew more savage and the penury caused by inflation was exacerbated by poor harvests, the City was desperate to provide food and fuel for the poor, and the suffering looked for someone to blame. There was not far to look. The commons certainly believed Warwick and his followers guilty of peculation (in a way they had not of Somerset).[98] Even someone in government noted: 'How after Somerset was deprived all the Lords had base money coined in the Tower, which was a great destruction'.[99] Certainly the identity of interest between those who controlled the state and those who served the mint gave rise to suspicion, especially as the man who benefited most was Yorke. His support of Warwick had been rewarded first with a knighthood and then with a series of profitable promotions and favours in the mint, and during the coinage operations between October 1550 and July 1551 Yorke was being paid the unprecedented sum of 10s. per oz. for silver for supplying a government desperate to maintain mint production.[100] Something of the transactions did reach the people—beyond the conviction that since they were not living well *someone* must be. In a tract of July 1553 one man's financial scrupulosity was contrasted with the venality of all the others: '*he* [Sir Edmund Peckham] never robbed his grace when he had all the rule of his treasure, *he* used not to buy silver for

---

[95] CLRO, Journal 16, fo. 91^r; Repertory 12, fo. 271^v. (On 25 Sept. a nightly watch was ordered.) Hoak, *King's Council*, pp. 194–5.

[96] *CSP Sp.* x, pp. 47, 216.

[97] Ibid. x, p. 279.

[98] See below, pp. 505–6, 523–4.

[99] BL, Additional MS 48023, fo. 351^v.

[100] Challis, *Tudor Coinage*, pp. 177–82; *Edward VI Chronicle*, pp. 48–9, 54, 58, 60, 63, 66.

4s. an oz. and make the King pay 5s. 4d. as other false traitors did'.[101]

Warwick's illness and withdrawal from Court during early 1550, and his tendency anyway to do business secretly, created the impression that his government's policies were all to promote his private gain.[102] In 1551 he determined to reform the coinage, so desperately debased, and inaugurated a deflationary policy, but even if anyone had understood his laudable intention, the damage was already done to his reputation. In a series of proclamations between April and July, disastrously mishandled, base coin was 'called down'; the shilling teston revalued first at 9d., then at 8d., finally at 6d.; the groat at 3d.[103] The poor suffered worst, as usual, for base money of low denomination was all they had. 'The richer sort' were safe-guarded, because they saved 'good gold and old [undebased] silver'. Moreover, they were forewarned, 'partly by friendship, understanding the thing beforehand', and between the proclamation of 30 April and its coming into effect in July 'did daily enhance and increase the prices of both wares and victuals, most miserably oppressing the poor', and protecting themselves from the effects of revaluation.[104] The Carpenters' Company even fined an officer for not having had the foresight to pay the company's alms to the hospitals before the fall in money.[105] On 8 July the Mercers were required to explain why the prices of their merchandise were so high: the answer came that English goods were so dear that they would not sell abroad, and that the Mercers were forced to demand high prices for imports to make up for losses, 'and yet scarce able to live'.[106] But it was the poor who suffered, and naturally they blamed their governors. The rumour circulating in late 1551 that Dudley had placed his own badge of the bear and ragged staff upon the King's coin and created his own mint at Dudley castle was only one among many of the same sort.[107] Such was

---

[101] *Chronicle of Queen Jane*, p. 119.
[102] Jordan, *Edward VI: The Threshold of Power*, p. 48.
[103] *TRP* i. 372, 373, 376, 378, 379; Challis, *Tudor Coinage*, pp. 103–11.
[104] Cooper, *Chronicle*, p. 351; *TED* ii, pp. 186–7; *Original Letters*, i, p. 108.
[105] Guildhall, MS 4326/2 (without folio numbers).
[106] Mercers' Company, Acts of Court, ii. fo. 250ʳ.
[107] *APC* iii, p. 462; cf. PRO, SP 10/15, fos. 7ʳ, 50ʳ; SP 10/14, fo. 72ʳ.

the suspicion of his government and his motives that the people
found stories like this not in the least incredible, and were
always prepared to believe the worst of him.

While Warwick's regime was so unpopular Somerset's
restoration would always be looked for. His presence in the
Privy Council and in the environs of London gave the expectation of his return to supreme authority. In March and April
1550 there were reports that everyone paid the deposed Protector the highest respect: 'my Lord of Somerset lieth at the
Court, and all men seeketh upon him', 'that there was no
doubt that Somerset will be back in his foremost place', and,
surprisingly, that this would be 'brought about by the hand of
the Earl of Warwick'.[108] Sir John Mason forecast in early 1550
that Somerset would be restored 'because there is no one else to
take his place'.[109] Other people were talking about the deposed
Protector's return to ascendancy, but was Somerset himself
working towards it? There was soon evidence that he was.
Warwick was worried. The marriage of Somerset's daughter to
Warwick's son in June 1550 gave the appearance of friendship,
but Warwick 'suspecteth he should have been betrayed there
and therefore came not thither'.[110] Later that month Warwick
interviewed Richard Whalley concerning his reckless support
for the scheming Somerset, who was once again 'taking private
ways by himself and attempting such perilous causes . . . he
will so far overthrow himself as shall pass the power of his
friends to recover'.[111] Somerset and his adherents were reportedly laying their confidence in Parliament that summer, where
they would represent that 'the people [were] oppressed with
fresh taxes, the King poorer than ever, and those in power
governed simply after their own caprice without respecting the
laws and customs of the realm'.[112] By the autumn of 1550 there
was a flood of rumours of discord and a change in government.
The Earls of Derby and Shrewsbury had quarrelled with

[108] NLS, Adv. MS 34.2.14, fo. 39ʳ; *CSP Sp.* x, pp. 62–3.

[109] *CSP Sp.* x, p. 62.

[110] BL, Additional MS 48023, fo. 350ʳ; NLS, Adv. MS 34.2.14, fo. 51ʳ.

[111] PRO, SP 10/10, fo. 9ʳ (26 June 1550); Tytler, *England under Edward VI and Mary*,
ii, pp. 21–4; Jordan, *Edward VI: The Threshold of Power*, pp. 75–6.

[112] John Haywarde, *The Life and Raigne of King Edward The Sixt* (1636), p. 129;
Jordan, *Edward VI: The Threshold of Power*, p. 76.

Warwick, and their loyalty was suspected.[113] In the Council all pretence of amity had disappeared and the opposing factions had polarized: 'some take Warwick's side, other my lord of Somerset's. The last is trying to win over the people, which he had not tried to do before.'[114] While those in authority speculated so did the people; and any perpetrators of rumours—or any who could be discovered—were severely punished. In August 1550 a miller's servant from Southwark was pilloried in Cheap and lost his ears for spreading a story that Somerset had proclaimed himself King. John Wilson presented a seditious bill against 'the head officers of the City' to the preacher at Paul's Cross in October.[115] What truth lay behind the rumours that Somerset was trying to marshal the popular support that had shown itself at his fall?

It was not until 7 October 1551 that Somerset was charged with conspiring to bring down his rivals. On that day Sir Thomas Palmer revealed to Warwick a plot: on St George's day that year (23 April) the Duke of Somerset 'went to raise the people'. 'A device was made to call the Earl of Warwick to a banquet with the Marquis of Northampton and divers other and to cut off their heads.' Palmer declared further that Sir Ralph Vane had two thousand men at the ready, Sir Thomas Arundel had assured the Tower, Miles Partridge was to raise London with cohorts of apprentices and take the Great Seal, Alexander Seymour and Laurence Hammond would attend upon them, and the King's own band of gendarmerie would be slain.[116] If the Duke were to be overthrown he would 'run through London and cry, "liberty, liberty" to raise the apprentices and, if he could, he would go to the Isle of Wight or to Poole'. The allegation was that Somerset's old political coterie and 'the idle people which took his part' would join common cause to bring down the ruling junta.[117] That this conspiracy had been conceived and planned the King, for one, never doubted, but historians since have claimed that the charges

[113] Ibid., pp. 76–80; CSP Sp. x, pp. 161–6, 167.

[114] Ibid. x, p. 186.

[115] BL, Additional Charter 981; Stow, *Annales*, p. 604; Wriothesley, *Chronicle* ii, p. 42; Haywarde, *Life and Raigne*, p. 129; CLRO, Journal 16, fo. 95ʳ.

[116] *Edward VI Chronicle*, pp. 87–8.

[117] Ibid., p. 89.

were without foundation, unjustly framed by Warwick to remove his rival once and for all, to assure his own ascendancy.[118]
To say that the charges were groundless is, however, to ignore
the evidence, and to declare Somerset guiltless as much an
article of faith as to believe Warwick 'the subtlest and most
devious disciple of Machiavelli'.[119] Was a conspiracy planned,
and if planned, were attempts made to carry it out?

'A conspiracy tending to rebellion' *was* discovered in the
capital in April 1551. Of the shadowy plot which appeared
around 12 April little is known beyond its existence, but the
Council was warned of worse to come. Orders came to the
Mayor and Aldermen to have 'a vigilant regard to the order of
the City; first, for their nightly watch; then for the correction of
vagabonds; thirdly for the repulsion of strangers coming into
the realm; fourthly for the reformation of disorder in churches
that an unity may be had'.[120] There was in London that April a
constant fear of conspiracy and rebellion: it was marked by the
measures taken for the City's security. As usual, all potentially
disruptive elements and all occasions for trouble were overseen:
a nightly watch was held, no buckler playing was permitted,
those who stayed in taverns during divine service were rounded
up, seventeen vagabonds were whipped through the City,
'suspect persons' were arrested, and the gates kept shut.[121]
The King and Council were reportedly raising armed bands to
the number of seven or eight hundred horse.[122] The manifest
uncertainty in high politics disturbed the citizenry, who saw
signs everywhere. 'There was seen in Lent beside St Martin's
Abbey, many men in harness sitting in the air, and so came
down again unto the ground, and faded away again.'[123] There
were, moreover, serious threats to stability. Since London
already suffered from a shortage of work and food, the influx of
literally countless foreigners—sectaries seeking refuge from
religious persecution and artisans seeking employment—was

---

[118] For such interpretations of Somerset's fall, see Jordan, *Edward VI: The Threshold
of Power*, pp. 73 ff.; Pollard, *England under Protector Somerset*, ch. 9; Tytler, *England under
Edward VI and Mary*, ii, pp. 34 ff.

[119] Pollard, *England under Protector Somerset*, p. 284.

[120] *APC* iii, pp. 256–7.

[121] *Original Letters*, i, p. 108; CLRO, Repertory 12, fos. 315^r–326^v.

[122] *CSP Sp.* x, p. 279.

[123] *Grey Friars Chronicle*, p. 69.

dangerous to the security of the traditionally xenophobic capital. There were rumours that there were forty thousand strangers; Scheyfve thought ten thousand; Dryander five, and Ochino four thousand. Ochino was most nearly right, although the repeated searches through the alien population for their age, status, and lodgings by the Court of Aldermen reveal both the difficulty and the importance of knowing.[124] Rumours circulated that all foreigners would be massacred, but some thought that the slaughter would go beyond that; that the dispossessed would attack 'also English merchants and rich burgesses, not forgetting the Council, who would not escape, especially my lord of Warwick and Mr Yorke . . .'.[125] Edward VI noted in his chronicle a plan by Essex men, discovered on 15 April, to attack strangers and to 'spoil the rich men's houses if they could'; though to this he added disparagingly: 'Woodcock'.[126] But was it?

The widespread conviction that London would rise on May Day to massacre foreigners was mistaken: the rising came before May Day and its purpose was more to destroy Warwick and his adherents than the aliens. On 16 April a gathering of 'ruffians and serving men' was discovered in London, their intention to 'excite the people to revolt'.[127] Insurrectionary peasants were to move first against their local gentry and then to march to the capital to aid the others (or so Scheyfve reported). From the North the disaffected Earls of Derby and Shrewsbury would come in support.[128] The report of a credulous ambassador and to be discounted? On that very night, 16 April, at midnight a constable of the watch discovered John Story, 'who had in his hand a naked sword all bloody, and said he was at a fray at Holborn bridge'. Story admitted that he and his master, William Drury, had been imprisoned by the Mayor and Sheriff earlier, and now 'trusted to see the day he should be even with him therefore, and said my Lord Privy Seal bade his master arrest master Sheriff and lay him

[124] *CSP Sp.* x, pp. 216, 265; *Original Letters*, i, pp. 336, 337; CLRO, Repertory 12, fos. 315^r, 321^r, 325^r.

[125] *CSP Sp.* x, p. 279.

[126] *Edward VI Chronicle*, p. 59; Haywarde, *Life and Raigne*, p. 122.

[127] *CSP Sp.* x, p. 279.

[128] Ibid.; C. A. Haigh, *Reformation and Resistance in Tudor Lancashire* (Cambridge, 1975), p. 141.

where he was laid before'.[129] On 18 April the City rulers requested the Privy Council that their officers be allowed to go armed in daylight.[130] What was Story doing with a blood-stained sword and so near to Warwick's residence? That he was servant to Drury, one of Warwick's captains, would suggest that he was there to defend rather than to attack. Why were Russell, Lord Privy Seal, and Drury so opposed to a London Sheriff, and why had Drury been imprisoned the previous week? There is clear evidence of a concatenation of conspiracy and popular disturbance in London that April.

That there was a political motive behind the disruption and collusion with those in high places was further suggested by the discovery of seditious bills. On 19 April came the City Recorder to the Privy Council to deliver 'certain books and bills of slanderous devices against the Council to move the people to rebellion'.[131] Another proclamation against those who 'sow, spread abroad and tell from man to man false lies, tales, rumours' against King and Council was issued on 28 April, and yet another set forth in May.[132] Somerset's earlier skill in declaring his cause to the people by rabble-rousing propaganda, and his use of the press, may suggest his involvement in these intrigues to raise the people against the regime. Probably the Tracy interrogated for unnamed treason in May 1551 was Richard Tracy, the writer of Protestant tracts.[133] More telling evidence of a conspiracy to bring down Warwick was the raising of armed men. In late March Sir Ralph Vane, now an ardent follower of Somerset having left Warwick's affinity, threatened Warwick's possession of Posterne Park in Kent, with forces of between 160 and 180 men, and boasted of raising three or four hundred.[134] But did he risk so much and raise so many men simply to challenge Warwick over a property? The disaffected Earls of Shrewsbury and Derby, always resolutely opposed to

---

[129] CLRO, Repertory 12, fo. 326ᵛ. This deposition was reported to the Council on 18 Apr.: ibid., fo. 327ʳ.

[130] Ibid., fo. 327ʳ.

[131] Ibid.; *APC* iii, p. 262.

[132] *TRP* i. 371, 374.

[133] *APC* iii, pp. 272–4; PRO SP 10/14, no. 78. Tracy was named among Somerset's accomplices in Oct. 1551.

[134] *APC* iii, pp. 244, 245–6, 279, 296; Jordan, *Edward VI: The Threshold of Power*, pp. 64–5.

Warwick's radical religious policies, were now very strongly suspected of intriguing to remove him. There were rumours that Somerset was prepared to reverse the Reformation, that if Gardiner were freed from the Tower it would be by 'Somerset's means', that Somerset would seize London, the Catholic Earls would raise the North, and Princess Mary would flee to Shrewsbury. The pact was sealed by the marriage of Derby's heir, Lord Strange, to another of Somerset's daughters. Early in 1551 the Imperial ambassador had heard that Warwick intended a pre-emptive strike against these northern earls, and indeed Warwick did send nobles loyal to him to the North to watch them.[135] It seems that Derby and Shrewsbury were involved in a plot with Somerset, but that they prudently abandoned the plan before it could be carried out; maybe because their intentions had been discovered. They asked for permission to be absent from the fateful St George's day feast, at which the massacre was to take place, and when summoned to Court refused to go.[136] In London the councillors banqueted together on four successive days from 23 to 26 April 'to show agreement among them, whereas discord was bruited';[137] indeed at this time there was a quarrel between Warwick and Somerset from which there was no going back. If Somerset's plan had been to seize London and assassinate Warwick it was unsuccessful.

THE TRIAL AND DEATH OF SOMERSET

There was a plot to bring down the government in April 1551 and Somerset was in it somewhere. On 28 April the Imperial ambassador reported that 'a great personage in England had rebelled against the governors and councillors of the King':[138] this could only be Somerset. 'Some whisper that my lord of Warwick harbours some suspicions of the Duke of Somerset . . . and is meditating revenge.'[139] The question is: if Somerset

---

[135] *APC* iii, pp. 264, 266–7; NLS, Adv. MS 34.2.14, fo. 45ʳ (21 June 1550); *CSP Sp.* x, pp. 279–82; B. L. Beer, *Northumberland: The Political Career of John Dudley, Earl of Warwick and Duke of Northumberland* (1973), p. 117.
[136] Haigh, *Reformation and Resistance*, pp. 141–2.
[137] *Edward VI Chronicle*, p. 60.
[138] *CSP Sp.* x, p. 285.
[139] Ibid., p. 280.

was manœuvring for power and stirring popular support in April, and Warwick suspected him, why was nothing done about it then and no moves made against him until Palmer's revelations in October? Somerset was arrested on 16 October and his accomplices shortly thereafter.[140] Warwick claimed then that 'this evil plot had long been in preparation, and the Council had suspected it, but their great zeal for the repose of the realm had caused them to wink at it for the time'.[141] There may have been some truth in this, for the summer of 1551 was one of extraordinary distress with the 'sweat' decimating the citizenry, and prices higher than ever. Yet the delay in charging Somerset has been seen as one of the most damning proofs of Warwick's cold-blooded determination permanently to remove his rival, and with the least regard to legal propriety. Warwick did confess upon the scaffold that he had framed the charges against Somerset (according to a French reporter), though Warwick's guilt would not necessarily exculpate Somerset.[142] It seems that Somerset once more in the autumn of 1551 was desperately attempting another counter-coup, perhaps on 2 October as one set of indictments noted.[143] In the first week of October Warwick managed to strike just ahead of Somerset.

Was the Duke of Northumberland (as the Earl of Warwick became—significantly enough—just after Somerset's arrest) guilty of judicial murder in January 1552 when Somerset went to the block, or had Somerset been conspiring treasonably?[144] Edward VI, after Somerset's execution, 'would often sigh and let fall tears' that he had allowed his uncle to 'lose his head for felony: for a felony neither clear in law and in fact weakly proved'.[145] The evidence against Somerset has often been discredited because it was given by such men as Sir Thomas ('busking') Palmer, 'hating the Duke and hated by him' (according to Edward VI), and Crane, who 'having consumed his own estate had armed himself to any mischief', but

[140] *APC* iii, p. 390.

[141] *CSP Sp.* x, pp. 384–5.

[142] Hoak, *King's Council*, pp. 74, 294 n. 188.

[143] PRO, KB/8/20, m. 19, 25.

[144] For views somewhat divergent from the account which follows, see Jordan, *Edward VI: The Threshold of Power*, pp. 81 ff.; Hoak, *King's Council*, pp. 73–6.

[145] BL, Harleian MS 2194, fo. 21^{r-v}.

much of it has the ring of truth:[146] for example, that Somerset would raise the apprentices, and the 'idle people' that took his part, or his intention that 'before the apprehension of the Lords he would have Sir John Yorke because he would tell many pretty things concerning the Mint'.[147] Warwick and the rest of the Council were convinced that Somerset and his adherents had been conspiring, but it was harder to prove it. They could, and did, torture Somerset's lesser followers, but not their equals.[148] The Earl of Arundel admitted to Northumberland and Northampton that he and Somerset had ' "determined to have apprehended you, but by the passion of God", quoth he, [disingenuously] "for no harm to your bodies" '. When Arundel's formidable interrogators challenged him of conspiring with Somerset not 'but once', as he had said, but several times at Syon and at Somerset Place, Arundel sighed, and 'lifting up his hands from the board said they knew all'. Arundel might capitulate, but Somerset did not.[149] The questions put to Somerset were specific, accusing him of definite actions and intentions, questioning him about known meetings with known accomplices—Palmer, Grey, Arundel, Partridge, Vane (who sent Somerset messages from the Tower), Paget, and Stanhope. Some of the questions, like this one, surely had evidence behind them: 'with how many have you conferred for the setting forth of the proclamation to persuade the people to mislike the government?'[150] But the indictments against Somerset contain fewer charges than the accusations in his interrogation, and appear rushed and contrived, as they were.

The Middlesex grand jury returned two indictments against Somerset:[151] in one he was charged with having upon 20 April 1551 plotted with others at his house at Somerset Place to deprive the King of his royal dignity and to seize his person. In preparation for this treason he with Stanhope, Partridge, Holcroft, Newdigate, and a hundred others assembled in order

---

146 PRO, SP 10/13, no. 65; BL, Harleian MS 2194, fo. 20ᵛ; Jordan, *Edward VI: The Threshold of Power*, pp. 83, 85, 91.

147 PRO, SP 10/13, no. 65; Tytler, *England under Edward VI and Mary*, ii, pp. 38–41.

148 *APC* iii, p. 407; Hoak, *King's Council*, pp. 229–30.

149 PRO, SP 10/13, no. 67: fo. 130ʳ.

150 BL, Cotton Cart. x. 6.

151 *DKPR, 4th report*, pp. 228–30. The commissions of oyer and terminer to the two Chief Justices were issued on 29 Oct. 1551; PRO, SP 10/13, no. 64, fo. 124ʳ.

to imprison Warwick and to seize the Great Seal and the Tower. They stirred the London citizenry to rebellion by sounding trumpets and drums and crying 'liberty, liberty'.[152] The other indictments differ slightly as to the time and place of the conspiracy, but not as to the intent and treason.[153] These indictments closely resemble Palmer's initial testimony to Warwick on 7 October in the names of the accused and the purpose of raising London, seizing the Tower and the Great Seal, but nothing further was said of the Northern conspiracy.[154] The charges appear almost formulaic: for example, that Somerset met with accomplices 'ad numerum centum' has the appearance of convenience rather than truth. Any rebels discovered would have been named, and that Vane was named as a conspirator on 20 May when he was not freed from the Tower until early June might put the veracity of the rest into doubt.[155]

But it was upon these indictments that Somerset was brought to his 'travesty of a trial' before a packed jury of peers on 1 December 1551.[156] There Somerset denied all the charges against him. We have Edward VI's account of his defence:

For London: he meant nothing for hurt of any lord, but for his own defence.

For the gendarmery: it were but a mad matter for him to enterprise, with his 100 against 900.

For having men in his chamber at Greenwich . . . it seemed he meant no harm, because, when he could have done harm, he did it not. . . . he did not intend to raise London . . .

His assembling of men was but for his own defence. He did not determine to kill the Duke of Northumberland, the Marquis etc., but spoke of it and determined afterward the contrary; and yet seemed to confess he went about their death.[157]

---

152 PRO, KB 8/19, m. 27.

153 The second indictment stated that at Somerset Place on 20 May the Duke procured Partridge and others to rise against Edward, and to capture and imprison Northumberland, Pembroke, and Northampton (PRO, KB 8/19, m. 26). The jury of the City of London found that the conspiracy took place in St Andrew Holborn on 20 Apr. (PRO, KB 8/19, m. 19), and the Kent jury placed the plot at Greenwich on 20 Apr. (PRO, KB 8/19, m. 12).

154 *Edward VI Chronicle*, pp. 86–8.

155 PRO, KB 8/20, m. 26.

156 Jordan, *Edward VI: The Threshold of Power*, pp. 92–8.

157 *Edward VI Chronicle*, p. 99.

These answers do not necessarily prove the innocence of a wronged man, but they are ambiguous. Certainly the jury was not prepared to find him guilty of treason according to these fudged indictments, and found only for felony, a lesser offence. This was a considerable rebuff to Northumberland. No less an authority than Sir Edward Coke declared that neither the charge of treason nor the commutation to felony was just under the law.[158]

Nothing revealed so well the fragility of the power base of the factions in Edward VI's reign or how swiftly the wheel of fortune might turn for them than the scenes in London at Somerset's fall and execution. To the very last there were hopes that the Duke would be pardoned and revenged upon his enemies. Thomas Holland came down from London to Bath and forecast 'thou shalt see another world ere Candlemas: the Duke of Somerset shall come forth of the Tower and the Duke of Northumberland shall go in'.[159] Northumberland and his Council lived in terror lest this should come about: Somerset's swift expedition was imperative, but fears of popular reaction gave them pause.

The utmost care was taken to assure the capital at Somerset's arrest. It was no coincidence that the very day after Palmer revealed Somerset's conspiracy the King's gendarmerie was mustered.[160] Bands of horse were moved to control the approaches to the City, a watch begun that did not cease until 24 November, and the Tower was fortified.[161] On 19 October the Privy Council summoned the City and gild officials to inform them of Somerset's machinations to bring down the government, and all householders were told to ensure the good order of their families and servants.[162] On 20 October the Mercers forbade any of their members to discuss Somerset's

[158] Jordan, *Edward VI: The Threshold of Power*, p. 91.
[159] *APC* iii, p. 462.
[160] Hoak, *King's Council*, pp. 200–1; *Edward VI Chronicle*, p. 86; BL, Royal MS 18 C xxiv, fo. 138ʳ; *APC* iii, pp. 293, 399; Jordan, *Edward VI: The Threshold of Power*, p. 87.
[161] CLRO, Repertory 12, fo. 401ʳ⁻ᵛ; Wriothesley, *Chronicle*, ii, pp. 57–8, 62; *Diary of Machyn*, pp. 10–11; *Grey Friars Chronicle*, p. 71.
[162] *APC* iii, p. 390; Wriothesley, *Chronicle*, ii, pp. 57–8; CLRO, Repertory 12, fos. 401ᵛ, 409ʳ.

captivity.[163] The allegation that Somerset would have 'destroyed the City of London and the substantial men of the same' did not convince the citizens.[164] At the end of October the Imperial ambassador reported that people were saying that 'it would be better for the King's security that he should be under the protection of the Sheriffs and the City authorities of London than in the hands of Northumberland and his following'.[165] Even the principal merchants who had supported Northumberland in 1549 were now calling him 'a tyrant, hating him, and declaring that Somerset was innocent'.[166] Maybe the strange 'fray' in the mayoral procession to St Paul's on 1 November when 'the Duke of Northumberland's servants would have taken the chain from the King's Sheriff's neck' suggested a wider hostility between the City oligarchs and the regime.[167]

The Council was on guard against London's potential disloyalty on the day of Somerset's trial, 1 December. City officers were placed in strategic positions, their arms evident but not drawn, and that night a double watch was kept.[168] Like the Duke of Buckingham in 1521, Somerset was taken from the Tower to his trial early in the morning and by river to avoid popular clamour, but even so such crowds gathered that 'one or two drowned by the way in the Thames'.[169] As the trial proceeded a throng of people stood outside Westminster Hall awaiting the verdict and when the axe was turned away, showing that the charge of treason was defeated, they took this for total acquittal, and 'made such a shriek, casting up of caps . . . that their cry was heard to the Long Acre behind Charing Cross, which made the Lords astonished'.[170] But popular jubilation was not universal: as the Duke returned through the City 'the one cried for joy that he was acquitted, the other cried out that he was condemned'.[171] If so much fury had been

[163] Mercers' Company, Acts of Court, ii, fo. 251r.
[164] APC iii, p. 390.
[165] CSP Sp. x, p. 389.
[166] Ibid. x, p. 393.
[167] Grey Friars Chronicle, pp. 71–2.
[168] CLRO, Repertory 12, fo. 426r.
[169] Stow, Annales, p. 604; Grafton, Chronicle, ii, p. 526.
[170] Stow, Annales, p. 607; Chronicle of Edward VI, pp. 99–100; Diary of Machyn, p. 12.
[171] Grafton, Chronicle, ii, p. 526

stirred by his trial more was feared at his execution. To intimidate the capital, as in October 1549, the councillors assembled
with their bands of horse: seven or eight hundred horsemen
'passed twice about St James's field and compassed it round
and so departed'.[172] More difficult was to silence the muttering
and the circulating propaganda. From 10 December the Mayor
sent an officer every Sunday to warn the Paul's Cross
preachers not to read out the bills presented to them.[173] The
writer and printer of a seditious tract set forth as a New Year's
gift were punished.[174] Diversions were created. That Christmas
there was a ceremonial visit to the City by the King's Lord of
Misrule and his entourage. The revels were to be grander by
far than usual—the Lord of Misrule's courtiers were bril-
liantly apparelled, not 'torchbearerlike'—for behind them lay a
political purpose. These revels were 'of the Council's appointment', devised to distract a troubled Edward from thoughts of
his uncle's imminent execution, and 'to remove fond talk out of
men's mouths', surmised Grafton, but the play concerned a
prison and a beheading, its props a 'pillory, gibbet, beheading
axe, a block, stocks and little ease'.[175]

Somerset went to the block on 22 January 1552. Early in the
morning crowds gathered to watch the spectacle, despite
attempts to prevent them.[176] Somerset's was the traditional
gallows speech: a penitent sinner, he was glad to have fore-
warning of his death that he might pay attention to last things,
but always 'a true and faithful man as any was unto the King's
majesty'. He died happy that he had brought religion 'nighest
to the word of the primitive Church'.[177] Suddenly, as the
sentence was about to be carried out, there was a great roar
'above in the element', like 'gun-powder set on fire in a close
house' or thundering hooves. The crowd scattered this way and

172 *Edward VI Chronicle*, p. 100; *Diary of Machyn*, pp. 12–13; *Two London Chronicles*,
p. 25; Stow, *Annales*, p. 607; *CSP Sp.* x, p. 396; *Grey Friars Chronicle*, pp. 72–3; Hoak,
King's Council, pp. 199–201.
173 CLRO, Repertory 12, fo. 434^r.
174 Ibid., fo. 437^r.
175 Grafton, *Chronicle*, ii, pp. 526–7; S. Anglo, *Spectacle, Pageantry and Early Tudor
Policy* (Oxford, 1969), pp. 302–9; Jordan, *Edward VI: The Threshold of Power*, pp. 98–9.
176 Antonio de Guaras, *The Accession of Queen Mary* (1892), p. 83.
177 BL, Cotton Cart. iv. 17; Harleian MS 2194, fo. 21^r; L. B. Smith, 'English
Treason Trials and Confessions in the Sixteenth Century', *Journal of History of Ideas*, xv
(1954), pp. 471–98.

that in 'wondrous fear', crying 'Jesus save us, Jesus save us'. Sir Anthony Browne appeared on the skyline, as though a saviour, and rode towards the scaffold. The people cried 'pardon, pardon, pardon, hurling up their caps and cloaks . . . God save the King'. But there was no pardon, and the Duke did not take this 'ruffle' as his opportunity to escape. As he was beheaded the people groaned; one of the very rare occasions of grief at a Tudor execution.[178] Around the block a crowd gathered, as after the death of Buckingham, 'talking about the Duke and bewailing his death', washing their hands in his blood and dipping their handkerchiefs in it.[179] Northumberland would never be forgiven for his rival's death.

'At this time many that bore affection to the Duke [of Somerset] talked that the young King was now to be feared.'[180] No longer to be manipulated by overmighty councillors, Edward had become a power in politics. The most radical reformation in religion yet was effected, in part, because he willed it. His determination to 'advance the profit of the commonwealth' was more than rhetoric. To the poor of London he gave Bridewell palace, but the rich he threatened. Most imperative of all the entries in his *Chronicle* is that for 8 June 1552: he recorded how the Lords of the Council sat at the Guildhall, and before a thousand citizens declared to the Mayor and Aldermen 'their slothfulness in suffering unreasonable prices of things, and to the craftsmen their wilfulness', warning them that if they did not amend 'I was wholly determined to call in their liberties as confiscate and to appoint officers that should look to them'.[181] The preachers called upon an unthankful people to repent, saying that the advance of the true religion was due entirely to God's mercy in giving them a godly prince, and that what He gave He could take away. In secular politics also Northumberland's power depended always upon the King's survival. Should Edward die the cause of reform and Northumberland's Council were both lost.

Rumours spread early in the capital of Edward's illnesses

---

[178] BL, Cotton Cart. iv. 17; Harleian MS 353, fo. 122ᵛ.

[179] *CSP Sp.* x, p. 453.

[180] Cooper, *Chronicle*, p. 355.

[181] *Edward VI Chronicle*, pp. 129, 167; Mercers' Company, Acts of Court, ii, fo. 256ᵛ; Wriothesley, *Chronicle*, ii, pp. 70–1; Jordan, *Edward VI: The Threshold of Power*, pp. 362–3, 367.

and imminent death. In July 1549 Edward had recorded laconically: 'because there was a rumour that I was dead, I passed through London'.[182] When in November 1550 he fell suddenly so ill that his life was despaired of the news was kept 'very secret'.[183] Alarming stories spread in early 1553 that he was already dead, and that he had died by poison.[184] The Council spread counter rumours of the royal recovery, propped Edward at a window that he might be seen alive, and early in May savagely punished three Londoners who had reported the King's death.[185] Once again, desperate councillors sought to pervert the succession, though this time with the dying King's approval. They swiftly concluded a plot to exclude Princess Mary from the throne and to crown instead Lady Jane Grey, the newly-wed wife of Northumberland's son, Guildford Dudley. When the King finally died on 6 July the news was withheld while Northumberland fortified the Tower and planned for Jane's accession. Not until two days later were leading citizens summoned to Greenwich, told of the death of the King and 'how he did ordain for the succession of the Crown . . . to the which they were sworn, and charged to keep it secret'.[186] Now, as before in 1547, the dependence of Tudor government upon the allegiance of London was compellingly shown by the infinite care to keep the King's death secret from the people until the succession and control of government in the next reign had been assured. Northumberland and his adherents feared the intervention of the people. That summer a 'marvellous strange monster' was born: girl twins, joined at the waist, looking east and west. To many this signified the two Queens proclaimed at Edward's death.[187] Which one would succeed?

---

[182] *Edward VI Chronicle*, p. 13.

[183] *CSP Sp.* x, p. 186.

[184] CLRO, Repertory 13, fo. 42ᵛ; *Diary of Machyn*, p. 35; 'Robert Parkyn's Narrative of the Reformation' in A. G. Dickens, *Reformation Studies* (1982), p. 306.

[185] *CSP Sp.* xi, p. 40; *APC* iv, pp. 263, 266.

[186] Jordan, *Edward VI: The Threshold of Power*, ch. 14; M. Levine, *Tudor Dynastic Problems, 1460–1571* (1973).

[187] Cooper, *Chronicle*, p. 356.

XIII   'Troy Untrue' and the Rebellions of Queen Jane and Queen Mary

ALL THE events of 1553 and 1554 had been foretold long before, by John of Bridlington in 1009; or so it was claimed by those who in April 1554 discovered this prophecy:

A 𝕵 who shall stand a small time, for in the right is not there the right set, for a horse [Arundel], a wolf, and a boar shall knit their tails together to bring 𝔐 to her prosperity, which 𝔐 shall be marvellous. She shall put the Fox [Gardiner] at liberty in her park, and make him ruler over her deer. Merlin saith that the Rose female will be the destruction of Troy [London], for France and Flanders shall play at the base, and Spain shall have the entry.[1]

The revelation of this 'ancient' prophecy was used as emotive and dangerous propaganda. London was of old feared to be disloyal—'Troy untrue'—and its citizens remembered its history and the consequences of past disloyalties. Yet in these years the Londoners were faced with disturbing and complex choices, not always sure where their loyalty lay. Queen Jane stood for the cause of reform, Queen Mary for the old faith. There could be no doubt of their fixed, and opposed, purposes in religion. Did conscience dictate a higher loyalty to a divine rather than to a secular power, a duty to a Protestant rather than a Catholic Queen, and what did prudence direct?

Queen Jane, Northumberland's puppet, made her entry into the capital on 10 July, but it was very far from triumphant. Only the Sheriff, the Herald, and some of the guard cried 'God save her!' at her proclamation. The citizens did not rejoice, but stood silent with 'sorrowful and averted countenances'.[2] Gilbert Potter stood out from the crowd to protest against

---

[1] BL, Harleian MS 559, fos. 12ʳ–13ʳ.

[2] *Grey Friars Chronicle*, p. 79; P. V., *Narratio Historica*, sig. Cviiʳ; *CSP Sp.* xi, pp. 80, 106; Antonio de Guaras, *The Accession of Queen Mary* (1892), p. 88.

the usurpation.[3] 'Thousands' of 'consenters' heard the proclamation, 'yet durst they not once move their lips to speak'.[4]
But there were some Londoners who had marked their acquiescence. Thirty-two City governors had signed their names to
Edward's letter patent for the succession on 8 July, and the
following day 'all the head officers' were sworn to Jane as
Queen.[5] On 14 July the Goldsmiths' Company held a general
assembly, the date marked 'A° R. Jane the first' in their court
book, and the company was warned that none should 'talk nor
meddle with the Queen's affairs but that it is honest and faithful'.[6] The Clothworkers were similarly dutiful to their 'sovereign lady Queen Jane',[7] and other companies too; while the
City duly collected contributions for the customary present for
a new monarch. A watch was ordered to prevent arms leaving
the City.[8]

A noble conspiracy had put Lady Jane Grey upon the
throne, a noble conspiracy would deprive her of it; the one
originating in the Council, the other in the country. In both
conspiracies the capital acquiesced, nothing more. Northumberland and his Council learnt on 12 July that Mary was
gathering support in East Anglia, for word had reached her of
Edward's death, the news brought by Reyns, her London goldsmith, upon the instructions of Sir Nicholas Throckmorton.[9]
The following day a reluctant Northumberland left his accomplices behind him as he rode from London to apprehend Mary.
Rightly suspecting the fidelity of men whom he had persistently
humiliated, he warned them that God would not 'count you

---

[3] *Grey Friars Chronicle*, p. 79; *Diary of Machyn*, pp. 35–6; Wriothesley, *Chronicle*, ii,
pp. 86–7; Guaras, *Accession of Queen Mary*, p. 88; P. V., *Narratio Historica*, sig. Cvii^r;
CLRO, Repertory 13, fo. 65^v. On 12 July apprentices and servants were ordered 'to
beware their talk and to take example by one which lost both his ears yesterday in
Cheapside'; Drapers' Company, Repertory B, fo. 15^r.

[4] 'Epistle of Poor Pratte', *Chronicle of Queen Jane*, p. 117.

[5] *State Trials*, i, p. 760; *Diary of Machyn*, p. 35.

[6] Goldsmiths' Company, Court Book I, fo. 177^r; Drapers' Company, Repertory B,
fo. 14^v.

[7] Clothworkers' Company, Court Book 1, fo. 247^v; Drapers' Company, Repertory
B, fos. 14^v–15^r.

[8] CLRO, Repertory 13, fo. 66^r; Journal 16, fos. 255^v, 257^v, 258^r, 261^r; Drapers'
Company, Repertory B, fo. 15^r.

[9] 'Vita Mariae', pp. 251, 294; *Chronicle of Queen Jane*, p. 2 n. The best account of
Mary's accession is found in J. Loach, *Parliament and the Crown in the Reign of Mary
Tudor* (Oxford, 1986), ch. 1.

innocent of our blood' if they should break their oaths and betray him, 'hoping thereby of life and promotion'.[10] A force of six hundred men accompanied the Duke as he marched against Mary, not men freely thronging to his cause, nor levied from the City, but rather his personal retinue and those pressed to his service. 'The drum is being beaten here to raise troops, and they are to have a month's pay in advance.'[11] As he rode eastwards through Shoreditch, Northumberland noted ominously: 'The people press to see us, but not one sayeth God speed us.'[12]

The machinations among Jane's dwindling adherents in the Tower are not properly our subject, but London politics and national politics were during these days almost indistinguishable. Northumberland's suspicions of the duplicity of the councillors were warranted: as word of each new proclamation for Mary reached them, so they began to 'pluck in their horns'. On 14 July Pembroke and Cheney 'sought to go out of the Tower to consult in London, but could not as yet'.[13] The Tower was fortified; not simply to prevent assailants entering, but to prevent treachery within.[14] At seven o'clock in the evening of 16 July the Tower gates were suddenly locked and the keys taken to Jane. No one knew why, but the loyalty of the Lord Treasurer was suspect. He was brought from his house in London to the Tower at midnight.[15] This arrest proved that some still remained loyal to Jane, but for how long? Between 16 and 19 July the Council in London was subverted and prepared to declare for Mary. The Earl of Arundel's biographer reveals him as the prime mover, winning over Pembroke, and describes how the nobles communed together at Pembroke's house in Baynard Castle under cover of a pretended conference with the French ambassador.[16] By the afternoon of 19 July the Imperial ambassadors knew that Mary would be proclaimed in London that night, and so she was.[17] All the chronicles report

[10] *Chronicle of Queen Jane*, p. 6.
[11] *CSP Sp*. xi, p. 87.
[12] *Chronicle of Queen Jane*, p. 8.
[13] Ibid., p. 9.
[14] *CSP Sp*. xi, pp. 91–2.
[15] *Chronicle of Queen Jane*, p. 9.
[16] *CSP Sp*. xi, p. 108; J. G. Nichols (ed.), 'The Life of Henry Fitz Allen, last Earle of Arundell', *Gentleman's Magazine*, 104 (1833), pp. 118–20.
[17] *CSP Sp*. xi, pp. 95, 96, 108.

the same story; of the City, 'which for some days had mortified itself as if astounded with silent grief', now jubilant.

Great was the triumph here at London . . . The bonfires were without number, and what with shouting and crying of the people, and ringing of the bells, there could no one hear almost what another said, besides banquetings and singing in the street for joy.[18]

The Hanse merchants gave wine for the street parties. The servants of the nobility paraded in new liveries emblazoned with the letter M, in ostentatious show of loyalty.[19] Only one man went to the pillory for slandering the 'good Queen Mary'.[20] Why were the Londoners so desperate to reverse Northumberland's coup, so delighted to repudiate Jane and proclaim Mary? Fear had probably made them comply in the first place: terror of Northumberland had compelled the City rulers to acquiesce, as it had done the councillors, or so they said. A conflict between prudential compliance and conscience had faced every citizen upon Jane's accession, but for none so urgently as for the City fathers. London's rulers had plainly been dreading Edward's death, suspecting the worst about Northumberland's intentions, especially after Jane's marriage to Guildford Dudley. On 10 June Alderman Jervis had appealed to his company, the Mercers, to save him from serving his turn as Mayor 'until the world be better established'.[21] Under these circumstances all the other petitions for exemption from City service, all the diplomatic absences from the capital in these months, may have stemmed from the same understandable desire to escape the toils of high politics.[22] Hatred of Northumberland, old suspicions of his motives, were enough to discredit Jane, Queen, on Northumberland's own admission, 'by our enticement'.[23] On the day of Mary's proclamation it was clear what Londoners would do with the Duke and his confederates,

[18] BL, Harleian MS 353, fo. 139^r; *Chronicle of Queen Jane*, pp. 11–12; P. V., *Narratio Historica*, sig. Dii^v–iii^r; *Two London Chronicles*, p. 27; *Grey Friars Chronicle*, p. 80; Guaras, *Accession of Queen Mary*, p. 96; *CSP Sp.* xi, pp. 108, 115; 'Robert Parkyn's Narrative of the Reformation', in A. G. Dickens. *Reformation Studies* (1982), p. 307; *Original Letters*, i, p. 369.

[19] P. V., *Narratio Historica*, sig. Diii^r.

[20] *Diary of Machyn*, p. 37.

[21] Mercers' Company, Acts of Court, ii, fo. 261^r.

[22] See, for example, CLRO, Journal 16, fos. 254^v, 261^r; Repertory 13, fo. 65^v.

[23] *Chronicle of Queen Jane*, p. 6.

given the chance. As Sir John Yorke rode through Leadenhall, oblivious of the reason for the wild excitement, he barely escaped a vigilante mob. That night 'there were many people outside the Sheriff's house, saying that they wanted Yorke'.[24] There were fears that if Northumberland were brought into London by daylight he would be massacred, so he and his accomplices were escorted by four thousand men to the Tower on 25 July. Londoners had watched his departure in grim silence; not so his return. Thousands lined the streets; some waved handkerchiefs soaked in the blood of Somerset, and the crowd pressed to mock and jeer.[25]

Hatred of the usurpation had never been disguised in London. Indeed, there was amazement that there had been no protest against Jane's proclamation, 'no demonstration against the Lady Mary's right'.[26] When on 16 July Ridley declared at a Paul's Cross sermon the two royal princesses to be bastards 'all the people were sore annoyed with his words, so uncharitably spoken by him in so open an audience'.[27] On the same day, with Jane still beleaguered in the Tower, a letter against her and her Council was found in St Paul's, and there were fears of a 'popular tumult'.[28] Yet, although Northumberland was acutely aware of his constant unpopularity in the City, still he hoped to find support for his cause there. How so? He always alleged the preservation of the reformed faith to be the first reason for the alteration of the succession: 'God's cause, which is the preferment of His Word and the fear of papistry's re-entrance, hath been . . . the original ground.'[29] This was the only cause he thought could win him London's partisanship, so he admitted later.

But in London, as elsewhere, it may well have been alarm at the prospect of continued reformation which inspired the people to resist Jane and rally to Mary's cause.[30] Northumber-

---

[24] *CSP Sp.* xi, p. 108.

[25] CLRO, Repertory 13, fo. 68ʳ; *Diary of Machyn*, pp. 37–8; Guaras, *Accession of Queen Mary*, p. 99; 'Vita Mariae', p. 268; Foxe, *Acts and Monuments*, viii, pp. 592–3.

[26] *CSP Sp.* xi, p. 83.

[27] *Grey Friars Chronicle*, p. 78; Wriothesley, *Chronicle*, ii, p. 88; *Two London Chronicles*, pp. 26–7.

[28] *CSP Sp.* xi, p. 92.

[29] *Chronicle of Queen Jane*, pp. 6–7.

[30] See Loach, *Parliament and the Crown*, pp. 7–10.

land had been mistaken. Not all Londoners were reformed in religion: far from it: many looked eagerly for the restoration of Catholicism. Mary's staunch adherence to the old faith and her defiance of Northumberland and her brother were well known, and admired. In 1551 she and her retinue had ridden through London wearing forbidden beads. When she first entered her capital after her accession images of the saints and of the Virgin, rescued from their hiding places, could be seen in the windows of the citizens' houses. The Grey Friars chronicler, albeit a Catholic witness, related how at the return of Northumberland to London 'all the people reviled him and called him traitor and *heretic*'.[31] Moreover, those of the new religion as well as of the old recognized that there had been a perversion of justice, and feared that divine vengeance would be visited upon the people for 'their wicked complaisance in allowing the Duke to cheat her [Mary] of her right'.[32] They had long been warned of impending punishment for their infidelity: Grindal, Lever, Haddon, and Bradford, with 'lamentable voice and weeping tears', had called upon the people to remember 'iudicium domini, iudicium domini', but were scorned as 'prating knaves'.[33] The *Pistel . . . sent to Gilbard Potter*, which circulated in July 1553, warned that judgement was at hand. The preachers' warnings 'that our King shall be taken away from us, and a tyrant shall reign; the Gospel shall be plucked away, the right heir shall be dispossessed' had gone unheeded. 'And thinkest thou not (Gilbard) the world is now come? Yea, truly.' Like the Ninevites who 'cast dust upon their heads, repented and bewailed their manifold sins', Londoners should 'not cease praying to God to send us quietness, and that the Lady Mary might enjoy the Kingdom'.[34] Since the present judgement seemed, at the time, far worse than anything Mary could do, the consequences for the Gospel, should she succeed, were not profoundly considered. For example, for all Mary's known Catholicism, 'the citizens were in good hope' that Mary

---

31 H. F. M. Prescott, *Mary Tudor* (1953), p. 190; *Grey Friars Chronicle*, p. 81.

32 'Epistle of Poor Pratte', *Chronicle of Queen Jane*, pp. 116–21.

33 John Knox, *An admonition or warning that the faithful Christians in London, Newcastel, Barwycke & others may auoide Gods vengeaunce*, (1554; *RSTC* 15059), sig. Biiii[r].

34 'Epistle of Poor Pratte', *Chronicle of Queen Jane*, p. 118.

would give a 'helping hand' to London's hospitals.[35] Though Edward Underhill wrote a ballad against the papists at Mary's accession, and even though he feared for the faith, still he had refused to honour his commission as Gentleman Pensioner and ride with Northumberland against Mary.[36] Despite his Protestantism, Throckmorton had sent word to Mary of Edward's death and of Northumberland's plot.[37] Later, when London's faithful had come to regret their loyalty to a Catholic Queen, Knox reminded them of the voices which had spoken in the wilderness against her 'when fires of joy and riotous banquetings were made at the proclamation of Mary . . . the stones and timber of those places shall cry in fire, and bear record that the truth was spoken'.[38]

THE ACCESSION OF QUEEN MARY

London was penitent at first. The City rulers did all in their power to make amends for the capital's earlier infidelity. A watch was ordered on 21 July, and any suspected of being 'adherents' of Northumberland stopped at the gates.[39] Men of dubious loyalty were apprehended.[40] As the Duke and fellow traitors were conveyed to the Tower 'every sad householder' of the eastern wards was 'in a readiness in harness'.[41] The City prepared its customary offering to a new sovereign, the second in a month, and there were hasty collections among the companies for the present given on 29 July by London's chastened Mayor (cash which it proved very difficult to recover from the company members).[42] The City scribes hastily sought to erase all signs of disloyalty, obliterating Jane's name from the records or, rather, trying to. For all these shows of loyalty, the capital still needed the Queen's explicit forgiveness, but this

---

[35] John Howes, *Contemporaneous Account in Dialogue-form of the Foundation and Early History of Christ's Hospital and of Bridewell and St Thomas' Hospitals* (1889), p. 23.
[36] *Narratives of the Reformation*, pp. 141–4.
[37] *Chronicle of Queen Jane*, p. 2 n.
[38] Knox, *Admonition*, sig. Aiiii[r].
[39] CLRO, Repertory 13, fo. 67[v]; Journal 16, fo. 257[r].
[40] Ibid., Repertory 13, fo. 67[v].
[41] Ibid., Repertory 13, fo. 68[r].
[42] Ibid., Journal 16, fo. 261[r]; Repertory 13, fo. 69[r]; Goldsmiths' Company, Court Book I, fo. 175[r]; Wriothesley, *Chronicle*, ii, pp. 91–2.

she declined to give until she had entered her City. Entry to London would signal Mary's accession to her kingdom, but it was potentially perilous, because she was 'among people so inconstant and so easily led astray'.[43] Her Council was divided, blaming each other for past events, evading Mary's attempts to discover the truth of the plots, divided too over the advice she sought, 'whether she ought to hasten entry into London, or put it off'.[44] She chose the bolder course, entering London on 3 August with a huge retinue amidst scenes of great rejoicing, elaborate display, 'the sounds of bells so long disused'.[45] The Queen was gracious and magnanimous, save to the children of Christ's hospital: 'she cast her eye another way . . . but if they had been so many Grey friars she would have given them better countenance'.[46]

Mary granted pardons to all who had sworn oaths to Jane (except to those who had entered the field against her): she could hardly have done otherwise, for among the disloyal were most of the ruling order. But, as Grafton wrote feelingly, the general pardon was 'interlaced with so many exceptions as they that needed the same the most, took smallest benefit thereby'.[47] Ninety individual Londoners received pardons for offences unknown; most probably sued through regret for supporting the illegitimate regime. The guilty Goldsmiths' Company sued for a corporate pardon.[48] Yet the stigma of having once stood for Jane was not soon forgotten. Mary rebuked Winchester, Arundel, and Pembroke, and was highly suspicious even when Pembroke had in train fifteen retainers instead of ten.[49] Thomas Mountain had ridden with Northumberland against Mary, and now he suffered the vindictiveness of Gardiner. Told that Mountain was among those exempt from the general pardon, Gardiner replied disingenuously, 'Is it even so? . . . Fetch me

---

[43] *CSP Sp*. xi, pp. 116–17.

[44] Ibid. xi, pp. 131–2.

[45] 'Vita Mariae', p. 271; Grafton, *Chronicle*, ii, p. 535; Wriothesley, *Chronicle*, ii, pp. 93–5; *CSP Sp*. xi, p. 151; *Two London Chronicles*, p. 29.

[46] Howes, *Foundation of Christ's Hospital*, p. 23; but see also *CSP Sp*. xi, p. 151.

[47] Grafton, *Chronicle*, ii, p. 536. (Grafton was himself imprisoned for printing Jane's proclamation.) Foxe, *Acts and Monuments*, vi, p. 540.

[48] *CPR P&M* i, pp. 410 ff. Very many were, in addition, known Protestants. Goldsmiths' Company, Court Book I, fo. 193[r].

[49] *CSP Sp*. xi, pp. 151–2; BL, Harleian MS 353, fo. 140[v].

the book that I may see it.' He looked in the book, Mountain remembered, 'as one ignorant . . . and yet he being the chief doer himself thereof'.[50]

Vox Populi Vox Dei proclaimed the congratulatory placards, and so Mary believed, but it soon became apparent that if many had at first supported her, many others had not, or were now moving swiftly into opposition. As soon as she reached London Mary 'had daily Mass and Latin service said before her in the Tower'.[51] She was persuaded against burying Edward with full Roman obsequies, but even so Gardiner said Mass in the Tower, and there were rumours that Edward had a popish burial. There was 'a good deal of murmuring about this, even among the Queen's bodyguard'.[52] The new Queen was naturally generous to her friends as well as to her erstwhile enemies. Some of those she greeted, 'these are my prisoners', and restored to liberty and authority—Gardiner, Norfolk, and Bonner[53]—were those whom many Londoners had been glad to see locked away, but others rejoiced to see them freed. As Bonner walked through the City streets on 5 August 'as many women as might kissed him', and the bells of his cathedral were pealed in exultation.[54]

The wisdom of the warnings about her unfaithful and volatile capital was soon evident to Mary. The City had appeared quiet enough for the watch to be abandoned on 6 August, and on the 8th Mary assured the Mayor and Aldermen that she would not 'compel or strain other men's consciences', at least for the time.[55] But soon there was trouble. On 11 August an old priest sung Mass at St Bartholomew's; 'after that Mass was done the people would have pulled him in pieces'.[56] A more concerted attempt was made the following Sunday. As Dr Bourne preached at Paul's Cross of his master's sufferings under Edward VI— 'unjustly cast into the vile dungeon . . . among thieves'—his

---

[50] *Narratives of the Reformation*, pp. 180–1.

[51] *CSP Sp.* xi, p. 134; Grafton, *Chronicle*, ii, p. 535.

[52] *CSP Sp.* xi, pp. 122, 131, 156–7, 169; Wriothesley, *Chronicle*, ii, pp. 96–7; *Diary of Machyn*, pp. 39–40; *APC* iv, p. 306; *Two London Chronicles*, p. 29; Guaras, *Accession of Queen Mary*, p. 101.

[53] BL, Harleian MS 353, fo. 140ʳ; Stow, *Annales*, p. 613; Wriothesley, *Chronicle*, ii, pp. 95–6; *Two London Chronicles*, p. 29.

[54] *Grey Friars Chronicle*, p. 82.

[55] *CSP Sp.* xi, pp. 157–9; *APC* iv, p. 317.

[56] BL, Harleian MS 353, fo. 141ʳ.

audience screamed 'thou liest', 'papist'. 'Great uproar and shouting . . . like mad people, what young people and women' ensued as Bourne inveighed against anarchic Protestants, who 'have given a cloak to their libertines' religion, which they cunningly dignified with the name of Gospel freedom'.[57] A dagger was thrown at Bourne, which 'as God would' missed him, and he fled to St Paul's school. Only the intervention of John Bradford could calm the enraged crowd, and but for the presence of the Mayor and Edward Courtenay 'there had been great mischief done'.[58] The outrage provoked swift retribution.

The Privy Council summoned the Mayor and Aldermen to give assurances that they could keep control of the City, 'without seditious tumults', to provide an account of how they proposed to do so, and 'if they be not able; then the Mayor to yield up his sword'. Rarely had London's liberties been so seriously compromised. The Common Council was called immediately to confirm that 'they would stick to my Lord Mayor and his brethren . . . or else their liberties would be taken away'.[59] Loyalty was promised on 16 August, but was easier to pledge than to prove. On the 15th an opposition pamphlet—the first of a flood—was broadcast in the City streets. 'Noblemen and gentlemen favouring the Word of God' were urged to band together with all their 'power and . . . following', and to abandon Mary because 'hardened and detestable papists' were her followers—Rochester, Hastings, Waldegrave, Pembroke, Arundel, Stourton, Peckham, Englefield, Drury, and Weston. Everyone else 'will assuredly prove tractable as we have seen by experience during the last seven years'. But first of all, 'Winchester, the great Devil, must be exorcized with his disciples . . . before he can poison the people and wax strong in his religion'.[60] In the Council they were jealous and jockeying for power, their retainers brawling in the City.[61]

[57] Grafton, *Chronicle*, ii, p. 536 (Grafton's account is followed by Foxe; *Acts and Monuments*, vi, pp. 391-2); *CSP Sp.* xi, pp. 169-70; Guaras, *Accession of Queen Mary*, pp. 104-5; Wriothesley, *Chronicle*, ii, pp. 97, 99; 'Vita Mariae', p. 272; *Diary of Machyn*, p. 41; *Grey Friars Chronicle*, p. 83.

[58] 'Vita Mariae', p. 272; Wriothesley, *Chronicle*, ii, p. 98; *Original Letters*, i, p. 369; *CSP Sp.* xi, p. 170.

[59] Wriothesley, *Chronicle*, ii, pp. 98-9; CLRO, Repertory 13, fos. 71ᵛ, 72ʳ; *APC* iv, p. 319.

[60] *CSP Sp.* xi, pp. 173-4.

[61] Ibid. xi, p. 172; *APC* iv, p. 323.

London's rulers did their utmost to calm the capital. The act against 'unlawful assemblies and rebellion' was proclaimed on 16 and 18 August, and read from the pulpit on Sundays.[62] London clergy were to forbid anyone else to preach in their churches, and all interludes and plays were halted.[63] 'Every man should take heed of himself and all his servants that they should beware of what communication they have where any commotion might rise.'[64] Children, apprentices, and servants must attend their parish churches and keep the peace; on no account were apprentices to wear 'dagger nor other weapon'.[65] On 18 August a royal proclamation followed, banning preaching, the expounding of Scripture, and the printing of 'false found books, ballads, rhymes and other lewd treatises in English'.[66] Any disruption was always blamed—usually with some justice—upon the reformed preachers: Becon, Veron, and Bradford were sent straight to the Tower.[67] For 'heinous words and seditious words' against the Queen, John Day, rector of St Ethelburga, went to the pillory on 21 August, and again on the 24th for 'more words'.[68] At the next Paul's Cross sermon no chances were taken: around the cross and in the churchyard stood 'almost all the guard', armed.[69]

But in London many Catholics had ached for the overthrow of heresy and now delighted at the chance to worship in the old way. On 23 August 'began the Mass at Saint Nicholas Cole Abbey, goodly sung in Latin', with candles on the altar and a cross.[70] The following day and the next, St Bartholomew's day, 'the old service in Latin tongue with the Mass was begun and sung in Paul's in the Shrouds' and in four or five other City churches '*not* by commandment but of the people's devo-

---

[62] CLRO, Repertory 13, fo. 71^v; Journal 16, fo. 261^v.

[63] Ibid., Journal 16, fo. 262^r.

[64] Goldsmiths' Company, Court Book I, fo. 181^r.

[65] CLRO, Journal 16, fos. 261^v–262^r; Foxe, *Acts and Monuments*, vi, p. 392; Guildhall, MS 5177/1, fo. 89^v (Bakers' Company); Goldsmiths' Company, Court Book I, fo. 181^r; Clothworkers' Company, Court Book 1, fo. 247 B^r.

[66] *TRP* ii. 390; Foxe, *Acts and Monuments*, vi, pp. 390–1; *Chronicle of Queen Jane*, p. 24.

[67] *Original Letters*, i, p. 369; *APC* iv, pp. 321–2; *CSP Sp*. xi, p. 175.

[68] *Diary of Machyn*, p. 42; Wriothesley, *Chronicle*, ii, pp. 100–1; CLRO, Repertory 13, fo. 73^r.

[69] BL, Harleian MS 353, fo. 141^r; *Chronicle of Queen Jane*, p. 18; Grafton, *Chronicle*, ii, p. 536; *Diary of Machyn*, p. 41; Wriothesley, *Chronicle*, ii, pp. 99–100.

[70] *Diary of Machyn*, p. 42.

tion': 'to see it is to be in a new world'.[71] That this coincided with the public conversion and execution of Northumberland and his confederates was perhaps more than coincidental. 'Like as his life was wicked and full of dissimulation, so was his end thereafter', said Lady Jane when she heard of his traduction of 'God's cause'. Northumberland received the Mass 'with elevation over the head, the pax giving, blessing and crossing on the crown . . . and all other rites and accidents', and leading citizens, some of them reformers who lived to regret their casuistry, were called upon to witness.[72] At his execution 'there were a great number turned with his words', for his catalogue of the divine punishments visited upon a sinful nation—the deaths of Henry VIII and Edward VI, 'then with rebellions and after with the sweating sickness, and yet ye would not turn'—was telling.[73]

Catholic joy at the restoration of their rites was deeply galling to the godly. By 5 September William Dalby was reporting contemptuously from London that 'the Mass is very rife'; 'there is no news but candlesticks, books, bells, censers, crosses and pipes'. The high altar of St Paul's 'is up again elevated . . . but for making haste the work fell. I hope it will be a token of some ill chance to come again, which God send quickly.'[74] In the revived general processions and parish processions the citizenry once again appeared as a Catholic community. On St Katherine's day (25 November) 'they of Paul's went a procession about Paul's steeple with great lights, and [before them] Saint Katherine, and singing with five hundred lights'.[75] These were not perhaps all spontaneous expressions of popular piety, for the prebends, City clergy, Mayor and Aldermen attended, and on 8 December Bonner ordered that every parish

---

[71] Wriothesley, *Chronicle*, ii, p. 101; Guaras, *Accession of Queen Mary*, p. 110; *Two London Chronicles*, p. 29.

[72] *Chronicle of Queen Jane*, pp. 18–19; Wriothesley, *Chronicle*, ii, p. 100; BL, Harleian MS 353, fo. 141ᵛ.

[73] BL, Harleian MS 284, fos. 127–128ʳ; Harleian MS 353, fo. 141ᵛ; *Chronicle of Queen Jane*, p. 21; Guaras, *Accession of Queen Mary*, pp. 106–8, 145–8.

[74] BL, Harleian MS 353, fo. 143ᵛ; *Grey Friars Chronicle*, p. 84; Wriothesley, *Chronicle*, ii, p. 104. The Mercers' Company gave orders on 6 Sept. for its altar to be set up again: Mercers' Company, Acts of Court, ii, fo. 264ʳ.

[75] *Diary of Machyn*, p. 49; Wriothesley, *Chronicle*, ii, p. 104; *Grey Friars Chronicle*, p. 85; *Two London Chronicles*, p. 31.

should go in procession every Sunday and Wednesday and Friday 'to pray for fair weather through London'. The battle lines over the issue of clerical celibacy were drawn before ever the proclamation in December 1553 that no married priest could minister or say Mass.[76] A defamation case was brought in September against one who had said a year earlier (when it would never have been punished): 'Tush man, for thou has married a Bishop's whore and thy wife is a Bishop's whore.'[77] At its most elevated the contention was over the nature of the priestly function, but the subject fascinated ordinary salacious minds. 'Parson Chicken', Thomas Sowdley, of the Catholic parish of St Nicholas Cole Abbey, was paraded around London in a cart on 24 November 'for he sold his wife to a butcher'.[78]

By September the Protestant establishment was moved to action as it became ever clearer how gravely the Gospel was threatened. Gardiner had found a way with the more faint-hearted of the brethren: simply summoning them to see him was enough to drive them into exile.[79] The 'painful peregrination' of the exiles began. Terentianus, driven from the Catholic bastion of Oxford, found London little safer, for his friends there were 'exposed to the greatest peril themselves'.[80] Yet some preachers were daring to announce that the *vox populi* which had acclaimed Mary was not *vox Dei*, rather *vox Diaboli*.[81] Cranmer, Lever, and Hooper were 'together in disputation daily'. In early September Cranmer had bills posted 'all over London', offering to prove in debate that Edwardian doctrine was 'sound, agreeable to Scripture . . . and approved by the authority of the ancient fathers'. This stand brought him before the Council on 14 September, and thence to the Tower.[82] But the 'placards of the Archbishop had rallied the spirits of the gospellers', or so a Protestant optimist thought, and Renard feared the 'many secret practices on foot for upholding the new

[76] *Diary of Machyn*, p. 50.

[77] GLRO, DL/C/628 (without folio numbers).

[78] CLRO, Repertory 13, fos. 96ᵛ, 99ᵛ; *Diary of Machyn*, p. 48; GLRO, DL/C/331, fos. 155ʳ–157ʳ; *CPR P&M* i, p. 466.

[79] *CSP Sp.* xi, p. 217.

[80] *Original Letters*, i, p. 370.

[81] *CSP Sp.* xi, p. 217.

[82] Foxe, *Acts and Monuments*, vi, p. 394; viii, p. 37; *Original Letters*, i, pp. 369–71; ii, p. 505; *Chronicle of Queen Jane*, p. 27; *Grey Friars Chronicle*, p. 84.

religion'.[83] Mary had been crowned in her capital with due pomp, despite the ambassador's dark fears that the citizenry were plotting and collecting arms in their houses, but there were clear signs of disaffection.[84] The Drapers imprisoned George Hopton for refusing to be a 'whiffler' at the coronation and there were also dissenters punished in the Bakers' Company.[85] Certain Londoners were refusing to pay up for the Queen's accession present; these were usually Protestants.[86]

Convocation met as Parliament met in early October, and on the 18th 'began the disputation in the long chapel in Paul's between the new sort and the old'. To the debate 'many . . . of the Court and of the City also' came to listen (though, thought the Grey Friars chronicler, 'they were never the wiser' for it).[87] Yet it hardly needed a theologian to understand the taunts and gratuitous interruptions of prolocutor Weston, or the impassioned denial of the Mass by Philpot, who would disprove transubstantiation or 'let me be burned with as many faggots as be in London, before the Court gates'.[88] At the end of the fifth day Weston appealed to the audience whether the Protestants were answered, to which a few cried 'yea', 'but they were not heard at all for the great multitude which cried "No, No"; which cry was heard and noised almost to the end of Paul's'.[89] While on 25 October Weston dilated upon the evils of clerical marriage, in another part of St Paul's, before 'a great multitude', John Stor was adducing damning evidence against him: 'I marvel much that there should be so much talk against married priests among them . . . for Mr Weston, who is the chiefest doer amongst them, had to do with a poor man's wife . . . in his own parish and burnt [infected] her.' This Stor could prove, and 'five hundred with me', for he could produce a witness: Weston's agent was sent to find a surgeon to heal the woman.[90]

---

[83] *Original Letters*, i, p. 371; *CSP Sp.* xi, p. 233.

[84] Ibid. xi, p. 238. The companies perforce paraded their loyalty by attending her coronation; Guildhall, MSS 5177/1, fo. 90ᵛ; 5174/2, fo. 36ᵛ; 5442/3; 16988/2, fo. 67ᵛ; 7086/2, fo. 166ᵛ; 15333/1, fo. 184ʳ; Clothworkers' Company, account book.

[85] Drapers' Company, MB 6, p. 3; Guildhall, MS 5177/1, fo. 91ᵛ.

[86] CLRO, Repertory 13, fos. 82ʳ, 86ᵛ.

[87] *Grey Friars Chronicle*, p. 85; *Two London Chronicles*, p. 30; *The Examination and Writings of John Philpot*, ed. R. Eden (Parker Society, Cambridge, 1842), pp. 179–214.

[88] *Examination of John Philpot*, p. 194; Foxe, *Acts and Monuments*, vi, p. 402.

[89] *Examination of John Philpot*, p. 206.

[90] GLRO, DL/C/628 (without folio numbers).

The campaign to ruin Weston reveals more than the hatred of this particular cleric (though that too): the reformers had determined to disgrace a leading member of the hierarchy, a chaplain to the Queen and candidate for a bishopric. The Convocation was so discordant, and gave such opportunity to Protestant opponents, that it was swiftly halted by royal command.[91] As Weston told Philpot, 'ye have the Word, and we have the sword'.[92]

WYATT'S REBELLION

Another royal policy threatened to stir trouble in the capital. 'In the beginning of November was the first notice among the people touching the marriage of the Queen to the King of Spain.'[93] The news was 'very much misliked', especially when a Parliamentary petition against the marriage was presented and rebuffed on 16 November.[94] Mary's determination to ally England to Spain, to restore England to Rome, created a dangerous association in opposition: those who hated foreign tyranny and those who hated priestly tyranny might make common cause. It is difficult to write the history of London for the last months of 1553 without remembering that that alliance was made. Though Londoners could not have known of the conspiracies in a nobleman's house in St Gregory's parish at the end of November,[95] religious dissidents were evident. John Huntingdon was before the Council, once again, for his 'railing rhyme against . . . the blessed sacrament'.[96] Mocking priests was the favourite pastime of London boys: by December the companies were ordering their servants and apprentices that 'none of them should mock or scorn with hemming and hissing no priests passing by the way', 'nor suffer their servants . . . so to do, as in throwing at them crabs'. Certain nobles 'fell out in

---

[91] Foxe, *Acts and Monuments*, vi, p. 411; 'Vita Mariae', p. 277; *Two London Chronicles*, p. 30.

[92] Grafton, *Chronicle*, ii, p. 537.

[93] *Chronicle of Queen Jane*, p. 32.

[94] Ibid., p. 35; D. M. Loades, *The Reign of Mary Tudor* (1981), p. 120; Loach, *Parliament and the Crown*, pp. 79–80.

[95] D. M. Loades, *Two Tudor Conspiracies* (Cambridge, 1965).

[96] *APC* iv, pp. 369, 375.

the night with a certain priest in the street'.[97] Assaults upon
priests were bound to happen once the Mass was restored and
all Protestant services outlawed. Catholic orthodoxy became
official on 15 December.[98]

London Protestants were too many and too determined to be
easily cowed. When on 3 January 'the churchwardens and sub-
stantiallest' of thirty parishes of London were summoned by
Gardiner to explain 'why they had not the Mass and service in
Latin in their churches, as some of them had not, as St [Mary
Magdalen] in Milk Street, and others', their answer was
ambiguous: 'they had done what lay in them'.[99] Either they
could not, or would not, do more. Resistance was often more
than passive. For 'framing a lewd bill' Richard and Thomas
Trendall were in the Tower by late December, but they were
unlucky to be discovered: most Protestant pamphleteers never
were.[100] On Christmas Eve a new English translation of
Gardiner's *De Vera Obediencia*, with Bonner's preface, was
published—'as it is said at Rouen', by Michael Wood (but not
so).[101] For Gardiner's justification of royal supremacy to be
broadcast just as the Queen was seeking ways of divesting
herself of that authority was embarrassing; so were the attacks
upon the clerical hierarchy, 'turning like weathercocks, ersy
versy as the wind bloweth', and the *ad hominem* insults of 'blink-
ing coxcomb Standish' and 'drunken Dr Weston (with his
burned breech)'.[102] But more dangerous by far was the propaganda which associated the arrival of the Spaniards with the
destruction of Protestantism.

On 2 January as the 'retinue and harbingers' of the King of
Spain 'came riding through London, the boys pelted at them
with snowballs; so hateful was the sight of their coming in to

[97] Goldsmiths' Company, Court Book I, fo. 188ʳ; Drapers' Company, MB 6, p. 6;
Clothworkers' Company, Court Book 1, fo. 250ʳ; *Chronicle of Queen Jane*, p. 33.

[98] *Diary of Machyn*, p. 50; CLRO, Repertory 13, fos. 105ᵛ, 106ᵛ; Wriothesley,
Chronicle, ii, p. 105; F. Youngs, *The Proclamations of the Tudor Queens* (Cambridge,
1976), p. 186.

[99] *Chronicle of Queen Jane*, p. 34.

[100] *APC* iv, p. 380.

[101] *Chronicle of Queen Jane*, p. 33; *De vera obediencia . . . and nowe translated into English
and printed by Michael Wood: with the preface and conclusion of the translatour. Marke the notes in
the margine* (from Roane, xxvi. of Octobre, MD.l.iii; *RSTC* 11585).

[102] *De Vera Obediencia*, sigs. Aiiiᵛ, Aviʳ.

them'.[103] Evil May Day had shown what London boys thought of foreigners, but there is a suspicion that these snowballing children had been put up to their game. At the ceremonial reception of the ambassadors 'the people, nothing rejoicing, held down their heads sorrowfully'.[104] By 5 January a gentleman was imprisoned for saying that 'the upshot of this match would not be as the Queen expected', and on the 7th Renard reported how 'the heretics' had posted slanderous placards against Mary, alleging that Philip was betrothed already.[105] The rumour discovered in the Queen's own Wardrobe on the 12th—that Edward lived—was more disturbing still, for it revealed Protestant hopes, and was to persist throughout Mary's reign as a mark of her unpopularity.[106] Known Protestants were arrested, often at church during services, but the excess zeal of the Mayor's officers was provocative, and they were restrained.[107] When on the 15th the City rulers swore to respect the Queen's betrothal to Philip, the rumours of the marriage were 'not only credited, but also heavily taken of sundry men'.[108] 'Foreign heretics' were visiting citizens' houses, declaring that 'the preachers spoke the truth when they announced that the kingdom would fall into foreign hands and the Gospel and religion would be altered'.[109] We know that steps were taken to prevent this—by conspiracy and rebellion: Wyatt's rebellion.[110]

There were rumours in London on 21 January that the Carews and others 'were up in Devonshire resisting of the King of Spain's coming', but these rumours were false, probably spread by the conspirators themselves.[111] Still in London

---

[103] *Chronicle of Queen Jane*, p. 34.

[104] Ibid.; Bodleian, MS Rawlinson B 102, fos. 83ʳ–85ʳ (printed in 'The Imperial Embassy of 1553/4 and Wyatt's rebellion', *EHR* xxxviii, 1923); CLRO, Journal 16, fo. 283ʳ.

[105] *CSP Sp*. xii, p. 17.

[106] *APC* iv, p. 384; *CSP Sp*. xii, p. 40; Loades, *Reign of Mary Tudor*, p. 166.

[107] CLRO, Repertory 13, fos. 110ʳ, 111ᵛ.

[108] *Chronicle of Queen Jane*, p. 35; *Diary of Machyn*, p. 51; *CSP Sp*. xii, pp. 31, 43.

[109] Ibid. xii, p. 31.

[110] For Wyatt's rebellion, see Loades, *Two Tudor Conspiracies*; P. Clark, *English Provincial Society from the Reformation to the Revolution: Religion, Politics and Society in Kent, 1500–1640* (Hassocks, 1977); W. B. Robison, 'The National and Local Significance of Wyatt's rebellion in Surrey', *HJ* xxx (1987), pp. 769–90.

[111] *Chronicle of Queen Jane*, p. 35.

preparations were being made for Philip's triumphal entry; pageants were devised, and on the 19th orders given for all imagery on the bridge to be painted.[112] By the 25th it was clear that London Bridge would be needed for purposes defensive rather than decorative. The Common Council which sat that day was afforced by all the 'head commoners' of the City, and all were warned to look 'to their own households and their neighbours'.[113] Watches were ordered, arms made ready, and the following day the Mayor issued the first warrant to the companies for the provision of armed men, such as they had equipped during the 'stirring time' in 1549.[114] Six hundred men were swiftly marshalled at Leadenhall on the evening of the 27th.[115] The civic spirit was lost whereby every householder armed for defence: now the citizens paid others. Perhaps few of the soldiers who collected at Leadenhall were free of the companies that sent them—or so it appears from the scanty evidence—rather they were pressed men, paid at the rate of 1s. to 3s. 4d., and armed from the company stores.[116] On 28 January the London Whitecoats, under the command of their captains, Brett, Pelham, Fitzwilliam, and the Duke of Norfolk, set out against Wyatt and the Kentish rebels.[117]

But their cause was already Wyatt's and not the Queen's. As the company marched towards the rebel-held Rochester Bridge, the London bands in the vanguard, Captain Brett drew his sword, turned in his saddle, and addressed the Whitecoats: 'Masters, we go about to fight against our native countrymen of England and our friends in a quarrel unrightful and partly wicked', for these rebels were assembled to prevent English-men from becoming 'slaves and villeins', to protect them against Spanish designs 'to spoil us of our goods and lands, ravish our wives before our faces, and deflower our daughters in our presence'; and so patriots should join their enterprise 'much less by fighting to withstand them'. At which the

---

[112] CLRO, Repertory 13, fos. 112ᵛ, 113ʳ, 118ʳ⁻ᵛ.

[113] Ibid., Repertory 13, fo. 116ʳ.

[114] Ibid., Repertory 13, fo. 116ʳ⁻ᵛ; Guildhall, MSS 5174/2, fo. 42ᵛ; 5606/1; 7086/2, fo. 168ᵛ; 15333/1, fo. 187ᵛ; Drapers' Company, Repertory B, fo. 22ᵛ.

[115] CLRO, Repertory 13, fo. 117ʳ⁻ᵛ; Wriothesley, *Chronicle*, ii, p. 107.

[116] Guildhall, MSS 16988/2, fo. 62ᵛ; 15333/1, fo. 187ᵛ; 5174/2, fo. 42ᵛ.

[117] *Chronicle of Queen Jane*, p. 37; Bodleian, MS Rawlinson B 102, fo. 84ʳ.

Londoners cried, 'We are all Englishmen', and vowed with their captains to die in that quarrel.[118] Only a quarter of the company returned home, bedraggled and disarmed; the London gilds thereby subsidizing the rebels at huge cost with men, money, and arms. The sight of the few returning troops who had not forsaken the Queen limping home was for the Londoners, no less than for the Queen and Council, 'a heart-sore and very displeasing'.[119]

The defection of the London Whitecoats to the rebels was no sudden decision; their rallying cry at Rochester Bridge had been a prearranged signal, their treachery part of a wider conspiracy. Wyatt knew in advance that the London bands would desert to his cause, for their captains had already been suborned, apparently by French agents, and Norfolk had been forewarned.[120] A letter discovered upon Wyatt's emissary 'wherein is declared that both pensioners, guard and Londoners would take such part as he [Wyatt] did' was sent to the Duke on the 29th.[121] Whether the Whitecoats were converted by persuasive captains on the way down to Kent or were already for the rebels when they joined at Leadenhall remains a mystery. No doubt some were mercenary in the full sense, for London was full of discharged soldiers, men who would go to the highest bidder, the sort who had been pressed to join Northumberland six months earlier. Yet Proctor's cynical assertion that they went over to the rebels with the hope of despoiling rich citizens' houses was never substantiated.[122] Wyatt himself certainly construed the Whitecoats' defection as an expression of the citizens' will that his cause should prevail, and was encouraged to believe 'that the City had been at his commandment'.[123] In

---

[118] *Chronicle of Queen Jane*, pp. 38–9; *Tudor Tracts*, pp. 229–31; Bodleian, MS Rawlinson B 102, fo. 84[r].

[119] *Chronicle of Queen Jane*, p. 39. Of the 28 men sent to Kent against Wyatt by the Clothworkers' Company only six returned. The Bakers' Company sent eight men; none returned, and all the armour was lost. From the Ironmongers' Company fourteen men went forth: the one man who returned was rewarded: Clothworkers' Company Court Book 1, fo. 251[r]; account book (without folio numbers); Guildhall, MSS 5174/2, fo. 42[v]; 16988/2, fo. 62[r–v]; 15333/1, fo. 187[v].

[120] *Tudor Tracts*, p. 230; Harbison, 'French Intrigues at Queen Mary's Court', *AHR* xlv (1940), p. 548; Loades, *Two Tudor Conspiracies*, p. 60.

[121] PRO, SP 11/2, fos. 48[r], 47[r].

[122] *Tudor Tracts*, p. 235.

[123] Bodleian, MS Rawlinson B 102, fo. 84[r].

London's partisanship lay the greatest chance for success, a chance which Wyatt must quickly exploit, so his captains urged: 'London longed sore for their coming; which they could by no means protract without breeding great peril and weakness to themselves.'[124] The Council too feared for the City's loyalty: 'it is thought no good policy to suffer him to come near unto London where it is thought there is expectation of meeting his friends', wrote Southwell on 25 January.[125] The rebel forces, perhaps three thousand strong, set out for London on the 30th.[126]

The Queen and the capital were beleaguered: threatened by rebels from without, but also in imminent danger from unknown rebels within. The intense anxiety about the City's loyalty, and how powerless the government was to prevent defection, was revealed by the chilling proclamation of 31 January. 'Wyatt and all his company were rank traitors, and all such as was gone to Wyatt', *but* 'as many as did take his part or spoke in his cause, and that all his well wishers should go through Southwark to him, and they should have free passage, &c'.[127] The drawbridge across London Bridge was cut down and cast into the Thames, not just through fear of rebel incursion, but in case 'some light headed citizens' should join the rebel cause. Such was the panic, that when the Queen rode from Westminster to the Guildhall on 1 February 'a number of Londoners, fearing lest they should be entrapped and put to death' fled before she could enter the City.[128] The Aldermen were ordered to search their wards daily and to have a 'vigilant eye . . . to the inhabitants and specially to the young'. Those who had departed already were noted.[129] The liveries of all the companies were informed by Winchester that two thousand men were needed for the defence of the City: every householder shall 'raise for his family . . . on pain of death', and arm

---

[124] *Tudor Tracts*, p. 235.

[125] PRO, SP 11/2, fo. 33[r].

[126] *Tudor Tracts*, p. 236; Bodleian, MS Rawlinson B 102, fo. 84[r]; *Diary of Machyn*, p. 52; *Chronicle of Queen Jane*, p. 40; *CSP Sp.* xii, p. 77.

[127] *Chronicle of Queen Jane*, p. 40.

[128] Grafton, *Chronicle*, ii, p. 541; *Two London Chronicles*, p. 32.

[129] CLRO, Repertory 13, fo. 119[v]. On 18 Jan. the companies had been ordered to report which of their members were absent; on 23 Jan. they were given three weeks to present lists of resident members: CLRO, Repertory 13, fos. 112[r], 113[v].

immediately for the defence of London 'and not elsewhere at their peril'.[130] Now the citizenry perforce revived the civic virtues, arming themselves. Their journeymen and apprentices could not be trusted on these days. Shrove Tuesday was the traditional day for apprentice misrule: now rebellion must not provide the ultimate occasion for disorder. The Drapers' apprentices were forbidden to 'flock together in councils'.[131] By 2 February, Candlemas, the 'most part of the householders of London' were in arms, accoutred in white coats emblazoned with the red cross of the City.[132] The Mayor and Aldermen dined in armour that evening in the Guildhall, and the lawyers at the Inns of Court 'pleaded in harness'.[133]

The ambassadors had already fled London, and many counselled that the Queen should follow. Mary determined to stay.[134] On 1 February the Queen rode from Westminster to the Guildhall. Before a large crowd, 'with her sceptre in her hand in token of love and peace', Mary promised never to act without her subjects' consent, to call a Parliament to discuss her marriage, and now that Wyatt was at London's gates she appealed for the citizens' allegiance.[135] The crowd threw caps in the air and 'cried out loudly that they would live and die in

[130] CLRO, Repertory 13, fos. 119^r, 120^v; PRO, SP 11/2, fo. 62^r; Guildhall, MS 16988/2, fo. 188^r.

[131] Drapers' Company, MB 6, p. 10; Mercers' Company, Register of Writings, ii, fo. 151^v.

[132] *Two London Chronicles*, p. 32. Mercers' Company, Register of Writings, ii, fos. 151^r–152^v. This time the companies sent their own members to guard the City: of twenty men sent to watch London Bridge by the Pewterers' Company sixteen are to be found in the quarterage lists (Guildhall, MS 7086/2, fo. 169^r). The Clothworkers sent 56 men 'of our own company' (Clothworkers' Company, account book). The Drapers sent 60 men (Drapers' Company, MB 6, fo. 10^r; Repertory B, fo. 24^r). Half of the Vintners' guard were, however, 'prest men' (Guildhall, MS 15333/1, fos. 187^v–188^r). The costs were considerable: the Grocers' Company accounts record military expenses of £108. 12s. 6d. between June 1553 and June 1554 (including the contribution to the garrison?): (Guildhall, MS 11570A, fo. 202^r). In providing watches to guard the City against Wyatt's incursion the Bakers' Company spent £9. 12s.; the Brewers £13. 10s.; the Carpenters £14. 5s.; Guildhall, MSS 5174/2, fo. 42^v; 5442/3; 4326/2. The total sum spent by the Vintners' Company 'at the insurrection' was £39. 8s.: Guildhall, MS 15333/1, fo. 188^v.

[133] *Chronicle of Queen Jane*, p. 40; Stow, *Annales*, pp. 618–19; Wriothesley, *Chronicle*, ii, p. 109.

[134] Bodleian, MS Rawlinson B 102, fo. 84^v; *CSP Sp*. xii, pp. 78–9.

[135] *Diary of Machyn*, pp. 52–3; *CSP Sp*. xii, p. 79; *Tudor Tracts*, pp. 239–40; Grafton, *Chronicle*, ii, pp. 539–40; Bodleian, MS Rawlinson B 102, fo. 84^v; *Grey Friars Chronicle*, p. 86; Wriothesley, *Chronicle*, ii, pp. 108–9.

her service, and that Wyatt was a traitor'; save one, a hosier and heretic, who protested: 'Your Grace may do well to make your foreward in battle of your Bishops and priests: for they be trusty, and will not deceive you.'[136] But most of those who came to hear her would most likely have been those who supported her anyway. Her cause was still far from sure.

Wyatt with 'five ancients', about two thousand men, marched from Deptford through Southwark on 3 February. On the City side of the river panic set in:

Then should ye have seen taking in wares of the stalls . . . there was running up and down . . . to weapons and harness . . . many women wept for fear; children and maids ran into their houses, shutting the doors for fear . . . so terrible and fearful at the first was Wyatt and his army's coming to the most part of the citizens.[137]

In Southwark, where discretion may have been the better part of valour, the scene was very different. Sympathy for Wyatt was clear. The rebels entered peaceably, and were entertained. Troops raised to ride against Wyatt by Lord William Howard, quartered in Southwark inns, went over to the rebels. For all Proctor's allegations that the rebels were only there for spoil, 'all men said that there was never men behaved themselves so honestly as his [Wyatt's] company did there'.[138] The fears in Southwark were not of Wyatt, rather of the cannons of the Tower ranged against their houses.[139]

If Wyatt and his bands' forbearance in not sacking Southwark marked their resolution, then the places they *did* choose to attack reveal much about their purpose. As soon as the rebels arrived in Southwark they stormed the Bishop of Winchester's palace, and they ransacked his library, 'so that men might have gone up to the knees in leaves of books, cut out and thrown under feet'.[140] This onslaught showed a particular grudge against Gardiner, and the despoiling of his library suggests a Protestant vendetta. Though Alexander Brett, the Whitecoat captain, had

136 *Tudor Tracts*, p. 240.
137 Ibid.; *Chronicle of Queen Jane*, p. 43.
138 Ibid., pp. 43, 45.
139 Ibid., pp. 44, 46.
140 *Tudor Tracts*, p. 241; Stow, *Annales*, p. 619.

spurred his bands to the rebel cause by alleging Spanish atro-
cities, his own reformed sympathies were clear, for he had once
stood surety for Anne Askew.[141] Wyatt proclaimed on his entry
that 'his coming was to resist the coming in of the Spanish King
&c', but his actions proved this not to be his sole motive.[142] He
sent 'one of his chaplains', who marched with the rebels, to the
Marshalsea, where 'they came in daily thick and threefold for
religion'. Wyatt's chaplains made them an offer, so Thomas
Mountain related: 'he would set us at liberty so many as lay for
religion: with the rest he would not meddle'. They answered
that they would wait on providence for their release: 'whether
it be by life or death we are content, His will be done upon us!
And thus fare you well.'[143] Wyatt also sent a message to Edwin
Sandys saying that he would be 'glad of his company and
advice', but Sandys, like Mountain, declined for if Wyatt's
mission was God's will it would succeed anyway.[144]

Wyatt and his followers 'were disappointed in that they
looked most certainly for': the union of the Kentish rebels with
their partisans in the City. Entry to London was across the
river, but three hundred men guarded the bridge night and
day, and boatmen were ordered to the north bank of the
Thames on pain of death.[145] A desperate foray over the roofs
of the bridge proved to Wyatt that 'this place is too hot for
us'.[146] To a passing merchant he gave a message:

Cousin Dorrell, I pray you commend me unto your citizens the
Londoners, and say unto them from me, that when liberty and
freedom was offered them they would not receive it . . . and therefore
they are the less to be moaned hereafter, when the miserable tyranny
of strangers shall oppress them.

Still he pressed on to London, hoping for support, and vowed:
'If I knock the third time I will come in, by God's grace.'[147]

[141] Guildhall, MS 9531/12, fo. 109ʳ.
[142] *Chronicle of Queen Jane*, p. 43. For the motives of the rebel leaders, see M. R.
Thorp, 'Religion and the Wyatt Rebellion of 1554', *Church History*, 47 (1978),
pp. 363–80.
[143] *Narratives of the Reformation*, pp. 185–6.
[144] Foxe, *Acts and Monuments*, viii, p. 594.
[145] *Chronicle of Queen Jane*, p. 44; Wriothesley, *Chronicle*, ii, pp. 109–10; *Tudor Tracts*,
p. 243.
[146] Ibid., pp. 243–4.
[147] *Chronicle of Queen Jane*, pp. 46–7.

The rebels marched fast and by night from Southwark to Kingston, crossing the river there, and reaching the outskirts of London by midnight. They intended to reach the Court gate by dawn, but exhaustion and the discovery that Pembroke's forces were already in the field led them to rest the night.[148] The capital was awakened by the news that the rebels were almost upon it. At dawn 'drums went through London'.[149] A pitched battle had been avoided before, but now seemed imminent. The failure to set upon the rebels as soon as they reached Southwark made some suspect treason. 'By God's mother', swore Sir John Abridges, 'I fear there is some traitor abroad that they be suffered all this while.'[150] It may have been that Pembroke was a more subtle tactician, planning to lure Wyatt into a trap,[151] but there were reasons (after Rochester Bridge) for the Queen's commanders to distrust their troops, reason too for the Queen to doubt her commanders, especially Pembroke. The night that Wyatt lay at Knightsbridge the Gentlemen Pensioners at Court were set to watch. Every man was needed, but the Council had already been warned to distrust the Pensioners, and Edward Underhill was sent home.

'May I depart in deed?'

'I tell you true . . . Mr Norres hath stricken you out of the book saying these words, "That heretic shall not watch here" '.[152]

On the morning of Ash Wednesday Wyatt refused the last offer of a pardon: 'they would fight it out to the uttermost'.[153] At about midday Wyatt entered London, coming down by St James's. Pembroke's horsemen 'hovered all this while without moving', while Wyatt's forces were 'suffered to pass, as was before their coming determined'.[154] From St James's a band led by Vaughan, the Knevetts, and Cobham turned down towards Westminster. The Court gates at Whitehall were

148 *Tudor Tracts*, pp. 246–7; *Chronicle of Queen Jane*, p. 48.
149 *Two London Chronicles*, p. 33.
150 *Chronicle of Queen Jane*, p. 44.
151 *Tudor Tracts*, p. 245.
152 *Narratives of the Reformation*, p. 162.
153 Bodleian, MS Rawlinson B 102, fo. 85^r.
154 *Chronicle of Queen Jane*, pp. 48–9; *Tudor Tracts*, p. 249; Bodleian, MS Rawlinson B 102, fo. 85^r.

already shut against them, but they futilely and 'most traitorously shot at them [so] that the arrows stuck there long after'.
Deciding that they were 'too few to do any feat there', they set
off to join Wyatt.[155] Yet had they known it, their very presence
had thrown the guards of the Chamber into complete disarray.
They fled in panic, and left the Queen virtually unprotected.
No one was left to hold the Court but the porters, no one to
protect Mary but Underhill and his fellows.[156]

Wyatt's own bands passed from St James's to Charing
Cross, encountering no resistance. At Charing Cross the Lord
Chamberlain's troops fired a salvo, but then turned tail and
fled for safety to the Court.[157] Failure to engage the rebels
confirmed fears of treachery within; that all was lost. In the
Court they cried 'Treason! treason!', thinking that Pembroke
had gone to Wyatt, and everywhere was terror and confusion.[158]
Again, meeting a band of the Lord Treasurer's troops, the
rebels passed safely by 'without any whit saying to them. Also,
this is more strange, nor did the citizens accost them.' No
one would make the first move since everyone feared the defection of their neighbour, the treachery of the royal troops and
commanders almost as much as they did the rebels. As Wyatt
and his 'daggle tails' marched past Temple Bar, along Fleet
Street, the citizens stood armed in their doorways. Wyatt's
delusion in believing that the citizens were on his side was only
equalled by the confusion of the citizens, who thought that
because Wyatt was unopposed Mary had pardoned the rebels.
Believing they had the support of 'some citizens, as like was',
Wyatt and his company marched through Fleet Street singing,
'God save the Queen, we are her friends'.[159] The rebels'
presence in the heart of the capital must prove that their pleas
were heard, their cause succeeding, they thought, 'and divers

[155] *Chronicle of Queen Jane*, pp. 48–9; *Tudor Tracts*, p. 251; *Grey Friars Chronicle*, p. 87;
Grafton, *Chronicle*, ii, pp. 541–2.

[156] *Narratives of the Reformation*, pp. 166–7.

[157] *Chronicle of Queen Jane*, p. 49; *Tudor Tracts*, p. 251.

[158] *Chronicle of Queen Jane*, p. 49. For Proctor, 'Pembroke's faith hath not been
wavering in his Catholic religion nor his truth and service doubtful at any time
towards his prince', but he must have been almost alone to think so. Cf. *Chronicle of
Queen Jane*, p. 62; *CSP Sp*, xi, pp. 151–2.

[159] *Chronicle of Queen Jane*, p. 50; Bodleian, MS Rawlinson B 102, fo. 85ʳ; *Two
London Chronicles*, p. 33.

of his men took the Queen's men by the hand as they went toward Ludgate'.[160]

As Wyatt advanced down Fleet Street as far as the Bell Savage at Ludgate, the Mayor and Aldermen stood paralysed, 'as men half out of their lives', sure that the failure to repulse Wyatt at St James's meant that 'all had not gone well with the Queen's side', but while they trembled 'many hollow hearts rejoiced in London at the same'.[161] Only the warning of John Harris, a merchant taylor of Watling Street, ensured that the Ludgate was locked at all. 'I know that these be Wyatt's ancients', he insisted. Others knew it too; 'some were very angry with him because he said so, but at his words the gates were shut'.[162] On this chance, seemingly, rested Wyatt's exclusion from the City. Despairing at last of the City's help, and thus of his enterprise, Wyatt and his bands turned down Fleet Street to be captured at Temple Bar, but before he departed he knocked, as he had promised: 'I have kept touch.'[163]

'THE QUEEN LOVETH NOT THE CITY'

Wyatt had kept his promise, but the Londoners had failed him. As in July 1553, the citizens would not support an usurper against their rightful Queen. Wyatt's purpose had never been clear: he knew that to have claimed openly that his cause was the reformed faith and the coronation of Princess Elizabeth would have lost him as much support as it won. But to proclaim no clear cause brought even greater confusion. Londoners had allowed usurpation in the past—of Richard III—and had come to regret it. They had lived through the political confusion of Edward's reign, and feared the return of such instability. The City's gospellers, although hating Mary's religion, would not oppose her. It was Underhill and the Gentlemen Pensioners, many of them 'good Protestants', who had guarded Mary, and Gregory Newman, a gospeller, who had held Newgate against

---

[160] *Diary of Machyn*, p. 54.
[161] *Chronicle of Queen Jane*, p. 51.
[162] *Grey Friars Chronicle*, pp. 87–8.
[163] Ibid.; Bodleian, MS Rawlinson B 102, fo. 85^r; *Chronicle of Queen Jane*, p. 50; BL, Additional MS 34176, fo. 24^r; Loades, *Two Tudor Conspiracies*, p. 74.

the rebels. But loyal Protestants did not expect to be rewarded by empty thanks and persecution. The Queen summoned her Gentlemen Pensioners and gave them 'great thanks' and promised favour, yet Underhill recorded that 'none of us got anything, although she was very liberal to many others that were enemies to God's Word'.[164] Their bitterness was understandable, and dangerous. And in the aftermath of rebellion Mary, in turn, suspected the Londoners of regretting not having joined Wyatt's cause, and knew that many opposed her 'proceedings in religion'. She proposed to call her next Parliament in loyal, and Catholic, Oxford. Only embassies from the Mayor and Aldermen prevented this dishonour to the City. To ensure London's security through the summer the Queen marshalled a garrison, for which the citizens paid.[165]

On 11 February Gardiner preached before the Queen, urging that she 'now be merciful unto the body of the commonwealth . . . which could not be unless the rotten and hurtful members thereof were cut off and consumed'.[166] Everyone then expected an effusion of rebel blood. The first to die were Lady Jane Grey and Guildford Dudley; Lady Jane penitent for her unwilling treachery, but resolute in her faith.[167] On the same day, 12 February, on twenty gibbets in the City, fifty of the most guilty of the rebels were hanged, drawn, and quartered, 'for a terror to the common sort'.[168] Two days later a portent appeared in the sky—an upturned rainbow and two red suns—and the Aldermen rushed out of the Guildhall to see it.[169] Searches for the rebels continued through London and Southwark, and those who had disappeared during the rebellion were summoned.[170] The City prisons were filled, and the City churches

---

[164] *Narratives of the Reformation*, pp. 163–8.

[165] Grafton, *Chronicle*, ii, p. 544; *Two London Chronicles*, p. 35; CLRO, Repertory 13, fos. 127ʳ, 128ᵛ, 130ʳ, 133ʳ, 139ʳ; *CSP Sp.* xii, p. 228; Drapers' Company, MB 6, p. 13; Repertory B, fos. 26ᵛ–28ʳ; Guildhall, MSS 7091/1 (26 Feb. 1554); 5174/2, fo. 42ʳ; 11570A, fo. 202ʳ.

[166] *Chronicle of Queen Jane*, p. 54.

[167] Ibid., pp. 55–9; Grafton, *Chronicle*, ii, p. 543.

[168] Grafton, *Chronicle*, ii, p. 543 (50 executed); *Diary of Machyn*, p. 55 (49 executed); Foxe, *Acts and Monuments*, vi, p. 415; *Two London Chronicles*, p. 34 (56 executed).

[169] Foxe, *Acts and Monuments*, vi, pp. 543–4.

[170] *Chronicle of Queen Jane*, p. 62. On 15 Feb. certificates of the names of citizens who had disappeared from Cheap and Bread Street wards during the rebellion had been read out: CLRO, Repertory 13, fo. 124ʳ.

too, 'by eighty, on a heap'. On 22 and 25 February the Queen graciously pardoned about 240 prisoners who were brought before her: at least, she forgave the Kentish rebels, but not the Londoners. 'Of the common sort very few were executed, save only of the White Coats; that, to say truth, deserved it trebly', wrote Proctor.[171]

We know the names of sixty-four London rebels, of forty-two from Southwark; we know their occupations and, for some, their parishes. Apart from the leaders, six were gentlemen, fourteen were yeomen, and they followed thirty-five, mostly lowly, trades. Seven can, with some certainty, be identified in the freedom lists of 1551–3. That little more than this can be known about them suggests what might anyway be expected: that most of the rebels were not of London's freedom or estab-lishment.[172] The quartered rebels hung on gibbets through that summer.

In the aftermath of rebellion a fifth column of dissidents was active in the capital, resisting still both the Spanish marriage and the restoration of the old religion. Queen and Council became more convinced that strict uniformity was essential for preventing further religious discord. Easter 1554 would provide the first test of faith, and as it loomed both Catholic authorities and the gospellers organized. The royal injunctions, issued on 4 March, demanded Catholic orthodoxy, although royal com-mand alone could not countermand statute, nor deny the royal supremacy, nor outlaw the practices in worship enshrined in Henrician and Edwardian statute. The 'articles to the clergy, wherein the chiefest points were the supremacy to be left out, *auctoritate regia*, and the dissevering of married priests from their wives, &c', were soon implemented.[173] Within weeks, London's married clergy were ousted.[174] Bonner had already sent a reminder to all his diocesans of their Catholic duty to confess to their curate and to receive the Mass: those who neglected this

[171] *Chronicle of Queen Jane*, pp. 59, 62; Bodleian, MS Rawlinson B 102, fo. 85ᵛ; *Diary of Machyn*, p. 56; Grafton, *Chronicle*, ii, p. 544; *Two London Chronicles*, p. 34; *Tudor Tracts*, p. 253.

[172] PRO, KB 8/26, m. 2, 4, 13, 25, 30; 8/29 m. 13; 8/32, m. 15, 16; 2, 3: 27/1169.

[173] *TRP* ii. 406; *Visitation Articles*, ii, pp. 355–6; Foxe, *Acts and Monuments*, vi, pp. 427–9; Youngs, *Proclamations of the Tudor Queens*, pp. 186–7; *Chronicle of Queen Jane*, p. 67.

[174] See below, pp. 570, 575–6.

sacred obligation and provided a wretched example to the young were to be reported to the Bishop's Chancellor by 6 April.[175] A mayoral order came to all householders 'both rich and poor' on 7 March to 'well and honestly use and behave themselves . . . in matters of religion and conservation of the Queen's peace', reporting 'all such as they do espy or know to disobey'.[176] No one over the age of twelve was to depart the City 'until this holy time of Easter be over'.[177] The orders were deeply provocative.

In the first week of March three hundred children gathered in Finsbury Fields to play a new game; 'some took Wyatt's part and some the Queen's', and they fought until some were wounded.[178] A game on such a scale had surely been stage-managed. A week later London was stirred by the strange spectacle of the 'bird that spake in the wall'. 'Seventeen thousand' Londoners, so Renard heard, had gathered to hear a voice issuing from the City wall in Aldersgate, the voice of an angel. What it said, and what it did not say, was alarming:

When they said to it, 'God save Queen Mary', it answered not; but when they said, 'God save the lady Elizabeth', the voice replied 'So be it'. Then was asked another question, 'What is the Mass?', and the reply came, 'Idolatry'.[179]

The 'white bird' was a young maiden, Elizabeth Crofts, who had been given a whistle by one Drake, and been persuaded to speak 'divers things of the Queen and Prince of Spain, of the Mass and confession', as she was directed by the curate of the parish (a known reformer), a player, and Hill, a weaver. The intention had been 'to benefit the prisoners', Princess Elizabeth and Courtenay.[180] At Whitsun 1554 a defamation case was

[175] Foxe, *Acts and Monuments*, vi, pp. 426–7; Guildhall, MS 9531/12, fo. 341ᵛ (Bishop Bonner's letter of 6 Mar. to the Archdeacon of Essex); G. M. V. Alexander, 'Bonner and Marian Persecutions', *History*, lx (1975) p. 377; Youngs, *Proclamations of the Tudor Queens*, p. 187 n. 12.

[176] CLRO, Repertory 13, fo. 131ʳ.

[177] Foxe, *Acts and Monuments*, vi, p. 429. Foxe's transcript of the mayoral order is appended: this no longer exists among the Corporation of London records.

[178] *Chronicle of Queen Jane*, p. 67; *CSP Sp.* xii, p. 146; CLRO, Repertory 13, fos. 131ʳ, 132ᵛ.

[179] *CSP Sp.* xii, pp. 154–5, 172; *Diary of Machyn*, p. 58.

[180] Wriothesley, *Chronicle*, ii, pp. 117–18; M. Hogarde, *The displaying of the protestantes* (1556; *RSTC* 13557), sig. Oviiᵛ; Stow, *Annales*, p. 624; *Two London Chronicles*, pp. 36–7.

brought in the King's Bench: Drake had allegedly promised that 'he would prepare five hundred men for master Peter Carew if he would take Exeter'.[181] Was Drake implicated in Wyatt's original conspiracy, and the bird in the wall part of a grander subversive design?

Once conspiracies are sought, every act of sedition may appear part of a master plan. The allegation of the man imprisoned for 'saying that the whole City were traitors and rebels' seemed alarmingly true that spring.[182] The Queen's opponents had 'discovered' in April an 'ancient' prophecy (they said in the cell of a nun of Syon) which foretold the destruction of the City rulers, Mary, and the Pope:

The streets [of London] shall run blood as red as any rose . . . As for the rulers of the City, twenty three shall lose their heads because of their untruth . . . The time of Ⓜ shall be very short. And in her days then shall the Pope be removed into England . . . but it shall not long continue. She shall take her flight to the castle of Julius Caesar, where she thinketh to be saved or succoured, but a dragon [Wyatt] will come to seek her destruction, which dragon's head shall be at Charing Cross, and his tail at Westchester. There shall go out of the gates of the City twenty thousand of the Pope's knights, but the dragon shall smite them down so fast with his tail that there shall be scant three thousand to return again.[183]

Rumours were circulating: that Paget and the 'white horse' (Arundel) were in the Tower. A seditious bill against Gardiner passed from hand to hand.[184] 'Misruled persons' were apprehended: since they claimed to be soldiers serving with Lord Clinton, maybe even the garrison was not to be trusted.[185] On 8 March arrows were fired at the priest celebrating Mass in St Dunstan in the East.[186] As the City churches were stripped of their Scriptural messages, and prepared for the Catholic celebration of Easter, further outrages were likely. Many parishes restored 'bearing of palms, and creeping

---

[181] PRO, KB 27/1175 r. 119. See D. M. Loades, 'Popular Subversion and Government Security in England during the Reign of Queen Mary I' (University of Cambridge, Ph.D. thesis, 1961), p. 33.
[182] CLRO, Repertory 13, fo. 129^v.
[183] BL, Harleian MS 559, fos. 12^r–13^r.
[184] CLRO, Repertory 13, fos. 129^r, 132^r.
[185] Ibid. Repertory 13, fo. 131^r.
[186] Ibid. Repertory 13, fo. 134^v.

to the cross on Good Friday, with the sepulchre lights and the resurrection on Easter day'.[187] For many of the new religion this was sacrilege, more than they could bear, and at least four hundred Londoners kept away from their parish churches that Easter. At St Pancras there was a different sort of sacrilege: as 'the priest rose to the resurrection that Easter, and . . . put his hand into the sepulchre, and said very devoutly, "surrexit; non est hic", he found his words true, for He was not there indeed'.[188]

'A plot started by the heretic element' was reported by Renard on 7 April.[189] Opposition tracts were being broadcast: a bill against the Council, another 'in favour of the Lady Elizabeth', and one insisting, 'Stand firm and gather together, and we will keep the Prince of Spain from entering the kingdom'.[190] The rumours that Edward still lived continued.[191] A ghastly joke was perpetrated on Sunday, 8 April: hanging on a gallows by St Matthew Friday Street was a dead cat, its crown shaven, dressed in a vestment, 'and a piece of white paper like unto a singing cake [sacramental bread]' between its paws. Despite the offer of a huge reward—£6. 13s. 4d.—the offenders were never found:[192] such was the network of support in London's evangelical underground.

Wyatt went to the block on 11 April, penitent because divine will had judged that his cause should not prevail. He had a last message for the crowd. He swore upon his death that Elizabeth and Courtenay never knew of the conspiracy: 'I cannot accuse them (God I take in witness).'[193] But at this Dr Weston, ever intemperate (he had preached in armour on Ash Wednesday), interrupted: 'Mark this, my masters.' Wyatt had shown written evidence to the Council which implicated Elizabeth and Courtenay.[194] Wyatt died a traitor's death, but many regarded

---

[187] Wriothesley, *Chronicle*, ii, p. 113.

[188] See below, pp. 558, 562ff; Foxe, *Acts and Monuments*, vi, p. 548. George Marsh, the former and deprived incumbent of the parish, was suspected.

[189] *CSP Sp*. xii, p. 212.

[190] CLRO, Repertory 13, fo. 145ᵛ; *CSP Sp*. xii, p. 213; see *APC* vi, pp. 8–9.

[191] CLRO, Repertory 13, fo. 146ʳ.

[192] Ibid., Journal 16, fo. 287ᵛ; Repertory 13, fo. 147ʳ; Wriothesley, *Chronicle*, ii, pp. 114–15; *Diary of Machyn*, pp. 59–60; Hogarde, *The displaying of the protestantes*, sig. Oviiiʳ.

[193] BL, Harleian MS 559, fo. 73ʳ; *Chronicle of Queen Jane*, pp. 73–4.

[194] Grafton, *Chronicle*, ii, p. 547; BL, Harleian MS 419, fo. 131ʳ.

him otherwise. Within a week his head was stolen away, the relic of a martyr.[195] On the evening of Wyatt's execution the news was brought to the Mayor as he dined: how Wyatt had exculpated Elizabeth and Courtenay and how Weston had denied it. 'Is this true? . . . said Weston so? In sooth, I never took him otherwise but for a knave.' Soon the City members of Parliament arrived, bringing the news current in the House: that Wyatt had urged Courtenay to confess.[196] The scaffold speech was soon the common talk of London. In the Rose tavern Robert Farrer, drunk as usual, inveighed against Elizabeth: 'that jill hath been one of the chief doers of this rebellion of Wyatt; and before all be done, she and all the heretics her partakers shall well understand of it'.[197] In another inn a grocer's apprentice, Richard Cutt, sat drinking, and his talk turned to Wyatt's gallows speech. Like many other apprentices, Cutt was probably sympathetic to the rebel cause, and he supposed that Wyatt had been forced by the Council to accuse Elizabeth and Courtenay. He soon found himself before Gardiner, and then on the pillory.[198] 'Lest the apprentices make any disturbance', London and the Queen's guards were to be in arms on May Day.[199] The City rulers were in disgrace: Gardiner warned the Mayor: 'My lord, take heed to your charge! The City of London is a whirlpool and sink of evil rumours, where they be bred, and from thence spread to all parts of the realm.'[200] The news that Wyatt had cleared the prisoners was even whispered to Sir Nicholas Throckmorton by sympathizers as he came to be arraigned at the Guildhall on 17 April.[201]

Throckmorton was charged with high treason: that he had been a prime mover of the rebellion and had plotted the capture

---

[195] Loades, *Two Tudor Conspiracies*, p. 115; *Grey Friars Chronicle*, p. 89; *Diary of Machyn*, pp. 59–60; *CSP Sp*. xii, p. 221; Wriothesley, *Chronicle*, ii, p. 115.

[196] Foxe, *Acts and Monuments*, vi, p. 431. Wyatt had visited Courtenay before his execution: *Chronicle of Queen Jane*, p. 72.

[197] Foxe, *Acts and Monuments*, viii, pp. 622–4.

[198] Ibid. vi, p. 431; CLRO, Repertory 13, fo. 153^r; *Chronicle of Queen Jane*, p. 75.

[199] *CSP Sp*. xii, p. 228.

[200] Foxe, *Acts and Monuments*, vi, p. 432.

[201] Bodleian, MS Rawlinson C 408, a contemporary account of the trial, is printed in Raphael Holinshed, *Chronicles of England, Scotland and Ireland*, ed. H. Ellis, 6 vols. (1807–8), iv, pp. 31–55, and *State Trials*, i, pp. 869–99.

of the Tower and the deposition of the Queen.[202] The jury sat from seven in the morning until five at night: finally their unanimous verdict was 'not guilty'. As Throckmorton was led away, the axe averted, 'many people rejoiced'. It was on that same night that Wyatt's head was stolen. Mary recognized the magnitude of the defeat: 'she was ill for three days, and not yet herself'.[203] On 21 April the jury were sent to prison, where they remained, despite their pleas for clemency, despite their disingenuous claims that they were 'poor merchant men', despite petitions in July from the Court of Aldermen for their release, until they were finally freed in December 1554, and then only with Philip's intercession and upon payment of crippling fines.[204] For a state prisoner to escape the toils of Tudor justice was rare indeed, and Throckmorton's defence and acquittal became a *cause célèbre*. John Bradford wrote of this 'memorable trial . . . worthy to be had in print'; of the accused who came as 'a little David with his sling'.[205] Although the jury had a 'papistical reward', a 'greater honour never came to the City of London than by these twelve men'.[206]

This jury, for whom conscience had prevailed over prudence, was specially chosen. Throckmorton, who knew the 'statutes, laws and chronicles of this realm', knew too the defendant's right to challenge the jury.[207] Ten jurors were presented to Throckmorton, but these ten challenged and removed. Only two were challenged for the Queen—Thomas Bacon and Geoffrey Walkden—and even then Throckmorton objected, 'Ah, ah, master Cholmley, will this foul packing never be left?'[208] How Throckmorton's final jury was found is uncertain—but Thomas Offley was one of the Sheriffs for that year. It was Offley who had 'saved many who should have died' after Wyatt's rebellion, and would save his brother Hugh by send-

---

[202] *State Trials*, i, p. 869.

[203] *CSP Sp.* xii, p. 221.

[204] *State Trials*, i, pp. 901–2; *Two London Chronicles*, pp. 35, 39, 40, 41; CLRO, Repertory 13, fos. 184ᵛ, 190ᵛ; Foxe, *Acts and Monuments*, vi, p. 549; Wriothesley, *Chronicle*, ii, p. 115.

[205] *The Writings of John Bradford*, ed. A. Townsend (Parker Society, Cambridge, 1848), pp. 405–7. In 1555 Bradford wrote the preface to an account of the Throckmorton trial.

[206] Ibid., p. 406.

[207] *State Trials*, i, p. 871.

[208] Ibid. i, p. 871.

ing him into exile.[209] Renard was right to suspect 'collusion
and malice', but were Throckmorton's jurors—Whetstone,
Banks, Martin, Lowe, Calthorp, Beswick, Kightley, Young,
Baskerville, Lucar, Pointer, Cater—'heretics to a man'?[210]

During his trial Throckmorton admitted a conversation with
Cuthbert Vaughan in St Paul's which revealed a motive very
similar to that attributed to Wyatt: 'We mind only the restitution of God's word'.

We talked of the incommodities of the marriage of the Queen with the
prince of Spain . . . Vaughan said, that it would be very dangerous
for any man that truly professed the Gospel to live here, such was the
Spaniards' cruelty, and especially against Christian men. Whereunto
I answered it was the plague of God justly come upon us; and now
Almighty God dealt with us as he did with the Israelites, taking from
them for their unthankfulness their godly Kings . . . Even so he
handled us Englishmen, which had a most godly . . . prince . . .
under whom we might . . . lawfully profess God's Word, which . . .
we handled so irreverently, that to whip us for our faults, he would
send us strangers . . . to exercise great tyranny over us . . . for every
man of every estate did colour his naughty affections with a pretence
of religion, and made the Gospel a stalking horse to bring their evil
desires to effect.[211]

Protestants' complacency had lost them the Gospel, but now
this jury dared oppose the judges, for they could not find 'him
guilty contrary to their consciences'.[212] They knew well the
consequences of thwarting the royal will. Whetstone had already
sued for a general pardon; Lucar was doubtless suspect for
having witnessed Edward's will; and Baskerville had been
summoned to attend the Duke of Northumberland's Mass of
reconciliation.[213] They would conform no longer. Among the
leaders of London's reforming establishment, they were intimately connected with each other, and known to Offley and
Throckmorton. Three of the jurors—Lowe, Lucar, and Young
—were in Offley's company, the Merchant Taylors. Humphrey
Baskerville had married Jane Packington of one of London's

---

209 Thomas Fuller, *The Worthies of England* (1662), Cheshire, p. 291; C. Garrett, *The
Marian Exiles: The Origins of Elizabethan Puritanism* (Cambridge, 1938).
210 *CSP Sp.* xii, pp. 221, 228.
211 *State Trials*, i, pp. 881–2.
212 'Recollections', p. 100.
213 *CPR P&M* i, p. 465; *State Trials*, i, p. 760; *Chronicle of Queen Jane*, p. 19.

first reforming dynasties.[214] Calthorp and Young were early benefactors of Christ's Hospital. That Offley, Young, Baskerville, Cater, and Banks were all governors of the City hospitals again reveals their association, and suggests their commitment to the ideals of the Edwardian commonwealth.[215] Perhaps some of these jurors were linked with those who would soon take a stand braver than theirs: the martyrs. Maybe John Calthorp was related to Master Bartram Calthorp, to whom Bartlet Green would look to relieve the prisoners for the faith, and maybe William Beswick was related to Roger Beswick, who protected his brother-in-law, Ridley.[216] Certainly these jurymen abetted their brethren in the Fleet. When Anthony Hickman and Thomas Locke joined them in prison, suspected of aiding the exiles and of heretical beliefs, the jurors, 'having liberty of the prison', used to stand under the window of their cell 'and talk aloud one to another . . . what they heard . . . and so by that means gave them light of many things before their examinations'.[217] Despite the dispatch of the Throckmorton jurors to prison on 25 April (or perhaps because of it) the jury which tried Throckmorton's accomplice, Sir James Croftes, on 28 April would not convict. The Council therefore sent for Robert Hartop, a man loyal to Mary and staunch in the old religion, to head the jury in order to get the right verdict, and Croftes was duly condemned.[218]

THE ARRIVAL OF THE SPANIARDS

In the summer of 1554, while London.was garrisoned, a 'merry fellow' came into St Paul's, and curtseying low, greeted the newly painted rood, as though the image of Christ was the Queen's man:

---

[214] *HSR*, St Michael Bassishaw, p. 108.

[215] Holinshed, *Chronicle*, iii, p. 1062; Howes, *Foundation of Christ's Hospital*, p. 13; Guildhall, MSS 12819/1, fos. 1ʳ–2ʳ, 7ᵛ, 8ʳ, 14ᵛ, 95ʳ; 12819/1, fo. 132ᵛ.

[216] Foxe, *Acts and Monuments*, vii, pp. 148, 260, 263, 265, 731, 743, 764.

[217] 'Recollections', p. 100.

[218] *Chronicle of Queen Jane*, p. 76; Wriothesley, *Chronicle*, ii, p. 115. Hartop's will shows him to have been a friend of Davy Woodruff, London's persecuting Sheriff for 1555, and of Richard Reyns, who had brought Mary the news of Edward's death: PRO, PCC, Prob 11/37, fo. 175ʳ.

Sir, your mastership is welcome to town. I had thought to have talked further with your mastership, but that ye be here clothed in the Queen's colours. I hope that ye be but a summer's bird, in that ye be dressed in white and green.[219]

The joyous proclamation of Mary had become a bitter memory for London's godly, as Knox had foretold. Yet many others rejoiced in the long awaited restoration of the old faith. It was the other policy dearest to Mary's heart which was universally lamented in London—the Spanish marriage. It was most deeply resented by the 'heretic element', for they feared that with the Spaniards would come their Inquisition.

London must show itself conspicuously loyal at Philip's entry if it was to escape further royal displeasure. Whether it would was in doubt. On 22 April Mary showed Renard a note threatening her, Gardiner, the Spaniards, and Philip; 'he would run risks when he came to England'.[220] A 'letter of sedition was let fall about the Parliament house', and an order came to the City companies on 10 May 'to beware of their talk concerning the Queen's majesty and the prince of Spain . . . And not to say the Queen loveth not the City, for her grace saith she loveth the City.' Moreover, they were told that the rumours that Lord Clinton would govern the City were false, but that 'he shall lie within twenty miles of the City to aid the City if any need should be'.[221] An alarming prospect. Gardiner judged it better for Mary to leave London for Windsor, and for Philip to enter at Southampton: 'the people were Catholics and obedient there and at Winchester; and there was no danger anywhere but in London'.[222]

In the capital preparations for Philip's arrival, interrupted by the rebellion, began again at the end of May. The Court of Aldermen deliberated what 'pageants and other open demonstrations of joy' there should be and, whether innocently or perversely, they appointed Richard Grafton to organize the celebrations.[223] The bodies of the rebels were tactfully removed

[219] Foxe, *Acts and Monuments*, vi, pp. 558–9.

[220] *CSP Sp.* xii, pp. 224–5.

[221] Drapers' Company, MB 6, p. 17.

[222] *CSP Sp.* xii, p. 239.

[223] CLRO, Repertory 13, fo. 162^v; see also fos. 165^r, 166^v, 184^r, 190^v. J. A. Kingdon, *Richard Grafton, Citizen and Grocer of London* (1901), pp. 63 ff.

from the gibbets on 1 June, the cross in Cheapside was newly gilded, and triumphal arches, giants, monuments to the Nine Worthies and to other noble Philips of history were erected along the path the King would take through the City.[224] Mary wrote on 14 August thanking the citizens for their 'good wills and forwardness in making of shows of honour and gladness' but she knew what bitterness lay behind the pretended jubilation; in the same letter the companies were ordered to 'use well the Spaniards'.[225] An order came from the Mercers the following day: they must 'be of honest behaviour . . . towards the strangers and to leave their mocking of them both in words and signs, and also the pick of quarrels, nor give any occasion of evil unto them nor cast nor throw neither in the streets nor out of their windows any manner of filth'.[226] Two bakers had gone to the Counter for 'misordering a Spaniard going in the street' on 31 July.[227] Philip's entry to the capital on 18 August went without mishap: indeed, many rejoiced at his coming.[228] Grafton made his point: Henry VIII was depicted with a Bible in his hand—*Verbum Dei*. But Gardiner made the painter cover the Gospel with gloves.[229]

'The English hate us Spaniards worse than they hate the Devil, and treat us accordingly'; so the Spanish visitors wrote home. Grumbling that the Londoners refused to provide them with lodgings, charged exorbitant prices, mocked them and their priests, they thought it was always open season for the lawless English to rob Spaniards and that malcontents were fostering rebellion against them.[230] They did have a point: the visitors were accommodated in the halls of the City companies since nowhere else was either safe or available.[231] Attacks upon the strangers were many, both in the Court and in the streets;

---

[224] 'John Elder's letter', *Chronicle of Queen Jane*, appendix x, pp. 146–52; *Two London Chronicles*, pp. 36, 38.

[225] CLRO, Repertory 13, fo. 191ʳ; Drapers' Company, MB 6, pp. 38–9.

[226] Mercers' Company, Acts of Court, ii, fo. 268ᵛ; Drapers' Company, MB 6, pp. 38–9.

[227] CLRO, Repertory 13, fo. 190ʳ.

[228] *Chronicle of Queen Jane*, p. 151; *Grey Friars Chronicle*, p. 91.

[229] *Chronicle of Queen Jane*, pp. 78–9. Some citizens refused to pay towards the fifteenth for the decoration of the City: CLRO, Journal 16, fo. 289ʳ.

[230] *CSP Sp*. xiii, pp. 3, 5, 26, 37, 56, 60, 72, 73.

[231] *Chronicle of Queen Jane*, p. 81.

moreover, the English aggressors were often pardoned.[232] Ruy Gomez, Philip's closest adviser, was set upon.[233] But the Londoners were sorely provoked: in a City already so over-crowded it seemed that 'there was so many Spaniards . . . that a man should have met in the streets for one Englishman above four Spaniards'.[234] The Spanish, for their part, were not above attacking the English.[235] The volatile citizenry, incited by the anti-Spanish literature, were prepared to believe anything of their cuckoo Spanish visitors. Rumours abounded: in early September there was 'a talk that the bishopric of Canterbury and metropolitanship . . . was given to a Spanish friar', and worse, 'talk of twelve thousand Spaniards coming more into the realm, they said to fetch the Crown'.[236] Orders on 20 September that innholders 'give a good ear to all such talk as their guests . . . shall have', and the proclamations against 'spreaders of rumours and false and seditious bills', could hardly silence the scaremongering.[237] A disturbing story was told of a dying Spaniard at the sign of the Pie by Aldgate revealing that his countrymen had come over to destroy London.[238] A death-bed confession had a special credibility. The worst fears of the Spanish coming—that it augured the Inquisition[239]—had been fulfilled that September: just as the Spaniards arrived Bishop Bonner began a quest for heresy throughout his diocese.

The visitation was at Bonner's initiative, before any formal reconciliation with Rome and without royal or conciliar mandate. 'It was a matter', said Bonner, 'pertaining to his own post.' Moreover, he knew that there would be opposition in Council, and that 'in religious matters it was meet to proceed firmly and without fear'.[240] Clergy and laity were to be

---

232 See, for example, *CPR P&M* ii, p. 242; iii, p. 112; Loades, *Reign of Mary Tudor*, pp. 215–16, 220.

233 *APC* v, p. 65.

234 *Chronicle of Queen Jane*, p. 81.

235 For example, *CPR P&M* ii, p . 243; Wriothesley, *Chronicle*, ii, p. 123.

236 *Chronicle of Queen Jane*, pp. 82, 81. On 18 Sept. Renard reported the rumour that Philip was sending for 10,000 Germans and 10,000 Spaniards: *CSP Sp.* xiii, p. 49.

237 CLRO, Repertory 13, fos. 201ᵛ, 204ʳ, 205ᵛ, 207ʳ, 212ᵛ; Journal 16, fos. 302ʳ, 303ʳ; *CPR P&M* ii, p. 103.

238 CLRO, Repertory 13, fo. 212ᵛ.

239 *CSP Sp.* xiii, p. 64. 'What gave offence was the use of the term "Inquisition".'

240 *CSP Sp.* xiii, p. 66.

examined upon 126 articles inquiring who had spoken against
Catholic ceremonies, who had asserted the doctrines of predestination and justification by faith alone, who refused to
attend church, to confess in Lent, or to attend processions or
Latin services.[241] (Bale soon produced his own, derogatory,
version of the articles.)[242] The Bishop toured his diocese, with
'sermons in every parish and place where he sat'. Opposition
was soon apparent, and 'many' were arrested.[243] At St Mary
Axe Roger Whaplett prevented the priest from reading the
visitation articles.[244] Three London parishes sent deputations
to the Bishop on 7 and 8 October protesting that the articles
could never be enforced. At the Guildhall 'strange words' were
spoken about the articles, and about King, Queen, and
Chancellor.[245] Throughout the City that autumn reformist
consciences were challenged openly; Catholic consciences also,
for the accusations made then against over four hundred of
London's reformist community came from those who thought
it their Christian duty to inform against the heterodox. Between
November 1554 and March 1555 Vicar-General Nicholas
Harpsfield sat in judgement upon London's Protestant community.[246] But between their delation and the passage of
judgement the laws had changed, because in December 1554
the medieval heresy laws were revived, enjoining the death
penalty upon those judged of doubtful faith. In England, as on
the Continent, an *auto-da-fé* began.

---

[241] *Visitation Articles*, ii, pp. 330–59; *Chronicle of Queen Jane*, p. 82; *Original Letters*, i,
p. 177; *CSP Sp*. xiii, p. 64; Guildhall, MSS 1586/1, fo. 67ʳ; 4570/1, fo. 125ᵛ. There are
no visitation records for the City, but those for Essex survive. Bonner was visiting
Essex in October; ibid., MS 9537/1.

[242] *A declaration of Edmonde Bonners articles concerning the cleargye of London dyocese whereby
that excerable Antychriste is reueled* (1554; *RSTC* 1289).

[243] Wriothesley, *Chronicle*, ii, p. 122; *CSP Sp*. xiii, p. 52.

[244] GLRO, DL/C/614, fos. 58ᵛ, 30ʳ.

[245] *CSP Sp*. xiii, p. 68. Which these parishes were is unknown.

[246] See ch. 14(i) below.

XIV   'Partly for Love, Partly for Fear, to Conformity'

I. THE GODLY IN BABYLON

TO THE reformers London had become Babylon, idolatrous capital of an ungodly Queen, and by staying there they faced many dangers. Once the Mass was restored in December 1553, 'with public profession of popery', and while the Catholics once more processed through the City streets on the festivals of the reinstated saints, the godly who stayed in London risked not only arrest but also damnation, through contamination. John Philpot warned his sister, 'you are . . . in the confines . . . of Babylon, where you are in danger to drink of the whore's cup, unless you be vigilant in prayer'.[1] How should the godly behave; could they live under a papistical regime attending popish services while somehow keeping their consciences undefiled? The letters they wrote to their ministers, now exiled or imprisoned, reveal their soul-searching. Might 'professors of the Gospel . . . prosecute their right and cause in any papistical court', or answer its summons or abide by its judgement?[2] A woman desired to know John Bradford's 'mind whether she, refraining from the Mass, might be present at the popish matins or evensong or not'.[3] 'A certain godly man' asked 'whether ye may lawfully in outward deeds and words resemble the papists' as long as 'ye retain the true profession of Christ inwardly in your heart?'.[4] The judgement was uncompromisingly against such Nicodemism. Ridley condemned the casuistry of a woman who asked 'what blame is in her if she use the religion here as

---

[1] Foxe, *Acts and Monuments*, vii, p. 695.

[2] BL, Harleian MS 416, fos. 62ᵛ–63ᵛ.

[3] Foxe, *Acts and Monuments*, vii, p. 221.

[4] *An answer to a certain godly mannes lettres, desiring iudgement, whether it be laufull to be present at the popishe masse* (*RSTC* 658), sig. Aiiʳ. See also, W. Musculus, *The temporysour* (*RSTC* 18312); Peter Martyr Vermigli, *A treatise of the cohabitacyon of the faithfull with the vnfaithfull* (*RSTC* 24673.5); A. Pettegree, *Foreign Protestant Communities in sixteenth-century London* (Oxford, 1986), pp. 124–5.

she may, though it be not as she would?'[5] Bradford could give
no worldly consolation: 'this Latin service is a plain mark of
Antichrist's Catholic synagogue': to attend was to be 'cut off
from Christ and His Church'.[6] Biblical warnings of the suffering endured 'in the fiery lake among the hypocrites, where there
is weeping and wailing and gnashing of teeth', were uttered
often. 'The worm of conscience' never died; and tales were told
of the pitiful ends of those who, like Judge Hales, recanting their
Protestantism, then repented, and in their despair and shame
drowned themselves.[7] 'Oh, what should it profit a man to have
this whole world at will, and to lose his own soul?'[8] The crisis
of conscience was recognized to be suffered most acutely by
'the good brethren and sisters in London'.[9]

In a 'little treatise', written in her old age, Rose Hickman
described the agonies of conscience she suffered during Mary's
reign. She had 'prayed earnestly to God to take either her or
me forth of the world', but in all that time, she boasted, she had
never attended a popish Mass, though she had been tempted.
Troubled, she had contacted the Bishops imprisoned in
Oxford's Bocardo. Should she allow her child to be baptized
'after the popish manner?' The answer came that she might,
for the 'sacrament of baptism was the least corrupted'. 'To
avoid the popish stuff' so far as she could, Rose put sugar
instead of the symbolic salt of the Catholic rite into the handkerchief she gave to the 'popish priest' at the baptism. But such
casuistry was itself corrupting. The Bishops had also told her:
'I might have been gone out of England before that time, if I
had done well.' Soon after, Rose and her husband Anthony
fled to Antwerp, 'delivered' from idolatry and 'blessed' with
peace of mind. Not so their in-laws, the Lockes. Thomas Locke
would have followed the Hickmans, but his wife refused to
leave. Rose warned her: 'Sister, you stay here for covetousness
and love of your husband's life and goods, but I fear the Lord's
hand will be upon you for it.' So it proved.[10]

[5] BL, Harleian MS 416, fo. 32ʳ.

[6] Foxe, *Acts and Monuments*, vii, pp. 244–6.

[7] Foxe, *Acts and Monuments*, vi, pp. 585, 710–712; *Narratives of the Reformation*, pp. 31,
192; *CSP Sp.* xii, p. 243; *Original Letters*, i, p. 177.

[8] *Narratives of the Reformation*, p. 192.

[9] John Bradford, *The hurte of hering masse* (*RSTC* 3494), sig. Aiiʳ; BL, Harleian
MS 422, fos. 104ʳ–132ʳ.   [10] 'Recollections', pp. 97–102.

Exile for conscience's sake was the only way to keep the faith inviolate and to live freely during Mary's reign. The ministers advised flight rather than conformity.[11] The first to depart London were those who had once sought there a religious haven: the stranger congregations. À Lasco, Utenhove, and Micronius, all forbidden now to preach, left for a second exile on 17 September 1553 with 175 of their most resolute followers (or as many as would fit on the boats).[12] Gardiner had a special way with others of the preachers, like Peter Alexander: his summoning of them hastened their departure.[13] Soon the more resolute or wealthy of London's godly and their households followed to an uncertain exile on the Continent, to towns where expatriate communities were established: Frankfurt, Strasburg, Emden, Basle, Zurich, Aarau, Weinhem, and Wesel. About one hundred Londoners, their wives and dependants, went abroad rather than conform, though their losses and insecurity were grave.[14] In exile, London's Protestant printers—Singleton, Day, Crowley, Scholoker, Banks, Bodley, Hall, Purfoot, Gibson—published works to aid the reformed cause.[15] Wealthy 'sustainers' helped their poorer brethren to escape or sent them money for subsistence abroad. Anthony Hickman, the Lockes, Richard Chambers, Burcher, Richard Springham, John Abell, Thomas Heton, John Ade, and, at first, Richard Hilles all aided those who remained abroad while Mary lived.[16] The story of Edwin Sandys's release from the Marshalsea, with the connivance of the Knight Marshal Sir Thomas Holcroft, and his desperate flight to the Continent reveals the risks the godly took. Pursued by all the constables of London, and with a £5 price upon his head, Sandys was moved from one hiding place to to the next by his friends, all 'good Protestants'. From William Banks's house he went to the house of another friend, he stayed for a week with Bartly, a 'stranger' in Mark Lane, and then

11 'Recollections', p. 99; *The Works of Bishop Ridley*, ed. H. Christmas (Parker Society, Cambridge, 1851), p. 419.

12 Pettegree, *Foreign Protestant Communities*, pp. 113–15.

13 *CSP Sp*. xi, p. 217.

14 This figure is derived from the census of exiles compiled by C. H. Garrett, *The Marian Exiles: The Origins of Elizabethan Puritanism* (Cambridge, 1938). It is difficult to define with certainty which of the exiles were Londoners.

15 P. M. Took, 'The Government and the Printing Trade, 1540–1560' (University of London, Ph.D. thesis, 1978), pp. 227–38.

16 Strype, *Ecclesiastical Memorials*, iii/1, pp. 224–5.

was hidden by Hurlestone, a skinner of Cornhill. He was conveyed, in disguise, out of London and fled to Antwerp.[17]

Leaving London, the exiles removed themselves from contamination, but thought always of the time when they could return. The lapse of time, which might cool the ardour of those left behind, inflamed that of the exiles. Letters from their persecuted ministers strengthened their resolve. In November 1553 John Bradford wrote to Cuthbert and Anne Warcup, Mrs Wilkinson, and 'others of his godly friends' exiled in Frankfurt.[18] 'In voluntary exile for the true religion of Christ', Joan Wilkinson made a will in December 1556 which revealed the networks of friends and the quality of a faith tempered by adversity.[19] With her in Frankfurt were her cousins Cuthbert Warcup and William Holland, City mercers, and her 'loving friend' Richard Chambers, a 'sustainer'. Her will was witnessed also by Edward Isaacs, Sandys's patron, Robert Horne, the preacher, and John Binks.[20] To the 'poor English congregations' of Frankfurt, Emden, Geneva, and Wesel she left £100, and to the French and Dutch exile congregations in Frankfurt £2 each. Back in London she had lent her books to Hooper 'during his life', and these, reclaimed from England, she entrusted to Holland 'to use to the profit of Christ's Church'. To Elizabeth Brown and Elizabeth Kelke, old friends of hers and favourers of the Gospel, she left £4 each, and £20 to Daniel Hooper 'towards his virtuous education'. 'In respect of the fidelity, old familiarity and friendship' of the Warcups and Chambers she charged them with the moral guidance of her daughter Jane. On pain of disinheritance, Jane might marry only 'a man fearing God . . . in religion of godly conversation utterly abhorring papistry'. Otherwise the 'painful peregrination' would have been all in vain.

For those who were left behind in London Easter 1554 was the first real trial of faith. In each City church the Mass was sung according to the old rite, with Latin mass books, with processions, with sacrament—or sacrifice—at restored altars.

[17] Foxe, *Acts and Monuments*, viii, pp. 590–8. For Holcroft's aid also to Thomas Mountain, see *Narratives of the Reformation*, p. 210.

[18] Foxe, *Acts and Monuments*, vii, pp. 219, 251.

[19] PRO, PCC, Prob. 11/42B, fos. 233ʳ–235ʳ. For the wills of Cuthbert and Anne Warcup, see ibid., fo. 351ᵛ; Prob. 11/53, fo. 320ʳ.

[20] Garrett, *Marian Exiles*, 51, 82, 211, 223, 431.

Rose Hickman had described, revealingly, that she had found safety in Antwerp not because of 'any more liberty of the Gospel given there', but because there were not parish churches but only a cathedral, 'wherein though the popish service was used, yet it could not easily be known who came to church, and who not'.[21] The contrast was with London, where all worshipped in the parish churches, and all were noticed. The most resolute of London's gospellers now refused to attend the antichristian services. Nearly two hundred people—half of the offenders haled before Harpsfield in visitation that autumn—were charged with staying away from their parish churches upon Sundays and holy days.[22] Absence from church was, in truth, no new offence, and might be excused by illness, or explained by the exigencies of trade, or simply by laziness. Yet most of these offenders in 1554 combined this offence with others which were doctrinally objectionable, and they were often, and significantly, discovered at Easter. It was the restoration of Catholicism which had driven these offenders from church: they said so explicitly.[23] Those who had gladly participated in the English services of Edward's reign now abstained from the Latin ones of Mary's. The dissident refused to attend church themselves, and like Christopher Fitzjohn of St Margaret Lothbury, 'seducer of the people', tried to deter others.[24] 'Since the restoring of the Latin service', James Golyver of St Mary Magdalen Old Fish Street 'hath not suffered nor would not suffer his apprentices and . . . servants to serve and sing in the choir'.[25] The visitation articles demanded who had stopped singing in the choirs, and seven were charged.[26] Henry Blakeham of St Alphage claimed that he was no singer; not that this had ever prevented him from singing the English psalms.[27] The Latin offended: as Robert Southam said, 'it was a fond question to ask a simple man, whether the Latin service be

[21] 'Recollections', p. 101.
[22] GLRO, DL/C/614.
[23] My conclusions, drawn from a detailed study of the individual offenders, differ from those of M. Jagger, 'Bonner's Episcopal Visitation of London, 1554', *BIHR* xlv (1972), pp. 306–11, and G. M. V. Alexander, 'Bonner and the Marian persecutions', *History*, lx (1975), p. 385.
[24] GLRO, DL/C/614, fo. 25^r.
[25] Ibid., fo. 22^v.
[26] *Visitation Articles*, ii, pp. 351–2; GLRO, DL/C/614, fos. 22^v, 26^r, 36^v, 52^v, 59^v.
[27] Ibid., fo. 52^v.

good and lawful'.[28] Candles, holy bread and holy water, palms
on Palm Sunday, ashes on Ash Wednesday, lights and
creeping to the cross were an abomination to the godly, and
nearly seventy of those charged had refused to countenance
them. John Kele at Christchurch 'the day he was married he
did blow out the light about the altar and would suffer no lights
to burn'.[29] From the processions held on Sundays and holy
days—the most remarkable sign of the old faith returning to
the City during the first year of Mary's reign—the gospellers
stayed away in disgust. More than a hundred people charged
before Harpsfield, including eight from St Martin Ludgate,
were accused of failing to attend. Christopher Mountford
asserted, with some truth, that it was old people not the young
who were 'accustomed to go in procession'.[30] John Payne,
William Cowper, and John Partridge had been persuaded by
London's Edwardian preachers that 'it is idolatry to go in procession', and would not now be convinced otherwise.[31]

It was conscience which prevented conformity with popish
services. Thomas Warbysshe of Aldermanbury refused to come
to church 'until such times as certain abuses be taken away, for
his conscience will not suffer him'.[32] Neither could John Warne
attend St Olave Silver Street 'saving as the service is in English
until he be otherwise persuaded by Scripture'. He never would
be: later he was accused of boasting that he was unrepentant
for staying away from church: 'thou art glad because thou hast
not therewith defiled thy conscience'.[33] Dorothy Griffin could
not bear to come to St Andrew by the Wardrobe because 'her
conscience is troubled', and George Edway protested that his
'conscience will not suffer him to hear Mass'.[34] Even to attend
church, to conform outwardly, might be to mask a deeper contempt. George Clerk of All Hallows London Wall, accused of
laughing as he received the Easter sacrament, confessed that

[28] Foxe, *Acts and Monuments*, viii, p. 471; GLRO, DL/C/614, fo. 19^v.
[29] Ibid., fo. 44^r.
[30] For the return of processions, see below, pp. 583–5. GLRO, DL/C/614, fo. 7^v.
[31] Ibid., fo. 8^r.
[32] Ibid., fo. 65^r.
[33] Ibid., fo. 48^v; Foxe, *Acts and Monuments*, vii, p. 80.
[34] GLRO, DL/C/614, fos. 37^v, 64^r. Mrs Griffin seems to have been a sustainer of
the prisoners: Foxe, *Acts and Monuments*, viii, p. 529. Edway had denounced a
conservative priest in 1538: *L&P* xiii/1. 1492.

'at Easter . . . upon occasion that some there present did laugh' he 'did a little smile and showed a merry countenance'. He was not alone in his parish to take the sacred mystery so lightly.[35]

To the godly the sacrament of the Mass was a delusion and perversion. 'Christ is ascended into Heaven', asserted James Golyver, and he 'therefore did call the sacrament an idol'.[36] The dissidents in the congregation were easily discovered by their efforts to evade witnessing the elevation of the host.[37] At St Stephen Walbrook they 'did use either to hang down their heads at the sacring time of the Mass, or else to sit in such a place of the church they cannot see the sacring'.[38] Some would walk around the churchyard; at St Andrew by the Wardrobe, Richard Smith sedulously read his English books during divine service.[39] Ninety Londoners were directly accused during the visitation of denying the cardinal doctrine of transubstantiation. For Thomas Williamson, James and Alice Golyver, Henry Pemerton, and Elizabeth Judde of St Mary Magdalen Old Fish Street 'Christ cannot be in Heaven and the sacrament at one time'.[40] Others, denying the Real Presence, could not believe that there was anything in the sacrament other than bread and wine. In St Bride Fleet Street there were several who believed 'Christ's very body not to be present really and truly in the said sacrament'.[41] Their doubts were profound, and some suffered for them, in time.[42]

Once, when all had shared a common faith, the Mass had bound the community. Now instead of reconciliation it brought contention. Catholics, believing in the saving power of the sacrament, and the godly, condemning it as idolatrous, could never be 'in charity'. Those of both confessions believed that anyone who took the sacrament while 'out of charity', mendaciously, must beware lest the Devil entered him as he

---

[35] GLRO, DL/C/614, fo. 63ᵛ.

[36] Ibid., fo. 22ᵛ.

[37] *Interrogatories vpon which . . . churchwardens shal be charged* (1558; *RSTC* 10117), sig. Aiiiᵛ. Cf. Foxe, *Acts and Monuments*, v, pp. 444–5; viii, p. 256.

[38] GLRO, DL/C/614, fo. 27ʳ.

[39] Ibid., fo. 46ᵛ.

[40] Ibid., fos. 21ᵛ, 22ᵛ, 23ʳ, 23ᵛ.

[41] Ibid., fos. 3ᵛ, 4ʳ, 4ᵛ, 5ᵛ.

[42] Thomas Fust, Thomas Browne, Isobel Foster, John Cardmaker.

had entered Judas.[43] Some were fearful of the threat. Anne
Williamson alleged as her reason for failing to go to church
throughout 1554 her conviction that the good might receive the
sacrament to their salvation, but the unworthy to their damnation.[44] At Easter 1553 John Walcott had refused to receive
holy communion because, being slandered by his neighbour in
Chancery Lane, he was 'out of charity'.[45] The transformations
in religious orthodoxy brought compelling reasons for being
'out of charity' and new justifications for not attending church:
not only might the service be abhorrent but also the prospect of
celebrating with parishioners who were gleeful because the
official faith was theirs, not yours, might have been more than
many could bear. William Morris had refused the Easter communion in 1553 because 'he thinketh and judgeth that it is not
lawful for him to receive the communion with him or them that
be of another faith than he is of'.[46] Within the Mass was a
symbol or ceremony of reconciliation: the kiss of peace or pax.
Bonner's visitation articles demanded to know whether the pax
was passed among the congregation during Mass, to remind
them of 'that peace, which Christ would have'.[47] Now because
of conflicting faiths the pax might be scorned. Ellen Morgan
refused to kiss the pax at St Anne and St Agnes, and she was
not alone.[48] For those who valued religious truth more highly
than social unity there could be little concord during Mary's
reign.

Though to attempt the redemption of sinners was a mark of
charity, the detection of four hundred Londoners for heresy
might itself seem the surest sign of a community divided, 'out
of charity'. But closer investigation reveals that Catholic neighbours had little reason to protect and every reason to report
many of those detected, for they had hardly been silent in the
congregation. Among these dissidents were the most adamant
and disruptive of the City's reformed community. Doubtless

---

[43] Foxe, *Acts and Monuments*, vi, p. 366; F. E. Brightman, *The English Rite*, 2 vols.
(1915), ii, pp. 650–8, 671–81.

[44] GLRO, DL/C/614, fo. 22ʳ. Cf. A. G. Dickens, *The Marian Reaction in the Diocese of
York*, 2 vols. (York, 1957), ii, pp. 139–40.

[45] Guildhall, MS 9168/10, fos. 191ᵛ–192ʳ.

[46] Ibid., fo. 251aᵛ.

[47] *Visitation Articles*, ii, p. 343.

[48] GLRO, DL/C/614, fos. 50ʳ, 53ʳ.

many of the four hundred detected had worshipped in the new ways peaceably enough, but others had profoundly shocked their fellow citizens by their iconoclastic outrages and subversive behaviour, or infuriated them by an excess of godly zeal. Now when the time came to report such extremists the orthodox were ready with evidence to indict them. Anne Williamson was accused in the autumn visitation of not going on procession in St Mary Magdalen Old Fish Street, nor witnessing the elevation of the host, and she confessed her scruples about receiving the eucharist.[49] But this was not the first time that she had been in trouble. Just before Christmas 1553 she had, 'contrary to womanhood', dared to enter the church without churching after childbirth, thereby breaking an old taboo and despising a Catholic rite. She refused to go from the church, to the horror of the 'most devout and worthy of the parish', would not offer chrism for her child's baptism, and only left, and then contemptuously, at the sacring of the Mass. In January 1554 the Vicar-General ordered her to undergo purification with suffrages and prayers, to sit in the churching pew, to offer tapers and chrism, and to do public penance. Whether she was 'very penitent', as was alleged, may be doubted.[50] Her husband Thomas was a confessed sacramentary, asserting that Christ could not be in the sacrament and in Heaven; he refused to attend Mass or go in procession. He was, moreover, an Anabaptist. Williamson alone was detected as an Anabaptist at the visitation, but he was not the only one guilty.[51] Michael Thumbe of St Peter Paul's Wharf was charged, with his wife, with not attending church nor receiving the sacraments. But Thumbe had recanted in 1549 the heresy, graver by far, that 'Christ took no flesh of Our Lady; and the baptism of infants is not profitable, because it goeth before faith'. Maybe he had truly recanted; maybe nothing could be proved against him in 1554.[52]

Others had old histories of heresy. Roger Whaplett (Whaplode) had failed to come to St Mary Axe, and he despised the

---

[49] Ibid., fo. 22[r].

[50] GLRO, DL/C/331, fos. 160[v], 162[r].

[51] GLRO, DL/C/614, fo. 23[r].

[52] Ibid., fo. 36[v]; Strype, *Memorials of Cranmer*, i, p. 257; W. K. Jordan, *Edward VI: The Young King* (1968), p. 228; I. B. Horst, *The Radical Brethren: Anabaptism and the English Reformation to 1558* (The Hague, 1972), pp. 111–12.

Mass and ceremonies of the Church; so much so that he buried his wife without the last rites or dirge. Not surprising, for Whaplett was, surely, Richard Hunne's son-in-law.[53] Thomas Tilseworth, accused of opening his shop in Christchurch parish during divine service, was of an old Lollard dynasty.[54] In the late 1530s the brethren had sent special advice to William Pittes for 'avoiding of the worst'; now he was in trouble again, for evading processions and despising holy bread and water.[55] George Edway had opposed popery in Henry's reign and did so still in Mary's.[56] During the first quest under the Six Articles Andrew Kempe had been indicted for disturbing services 'with brabbling of the New Testament'; now older and quieter, he simply failed to come to church.[57] John Warne was already a 'rank sacramentary' in 1546; nothing had persuaded him otherwise by 1554, nor ever would.[58]

One group detected in the visitation were the churchwardens of Edward's reign, who, having presided over the desecration of the parish churches, were held particularly responsible. At least fourteen of those convented before the Vicar-General had been churchwardens.[59] In no parish in London had there been unanimous support for the iconoclasm, and often the church-wardens had acted without the parish's consent. Now Catholics who had watched in horror, powerless, as parish treasures were destroyed or sold to the highest bidder, could take revenge upon the iconoclasts. In St Botolph Aldgate the battles between conservative and reformed coteries had been particularly bitter, settled only by mayoral arbitration. Now the most ardent reformers in the vestry were called to answer: John Frank, Anthony Anthony, and John Farminger were charged with patronage of reforming preachers and with holding heretical doctrine. Under investigation, they broke: 'before the Queen's

[53] GLRO, DL/C/614, fo. 58ᵛ; see above, pp. 121, 172–3.

[54] GLRO, DL/C/614, fo. 43ᵛ; see above, pp. 104–5.

[55] See p. 328 above; GLRO, DL/C/614, fo. 49ᵛ.

[56] See above, n. 34.

[57] Foxe, *Acts and Monuments*, v, p. 443; GLRO, DL/C/614, fo. 59ᵛ. He had been a churchwarden under Edward.

[58] CLRO, Repertory 11, fo. 300ʳ; and see below, p. 608.

[59] Cuthbert Beston, William Bodell, John Bullock, Dissell, Ellis, Finch, George Harrison, Robert Hilton, Andrew Kempe, Gregory Newman, Robert Shanks, John Sowle, Turner, Stephen Walden, William Warde, Henry Williamson: Walters, *London Churches*.

reign that now is they were . . . favourers of such doctrine as then was set forth, but not since'.[60] So they said.

Not all of the godly were those 'faithful professors' lauded by Foxe for their invisible faith and extreme moral effort. Some of those detected in 1554 were notorious disturbers of the peace. John Wheatley, accused in the visitation of reading heretical books and holding false beliefs, had been in grave trouble in 1549 for inciting City apprentices to join Kett's rebellion.[61] During a rising more dangerous still to London, Wyatt's, Richard Allen, a haberdasher of St Bride, was arrested for 'lewd and evil behaviour'. That Easter he failed to attend church or to receive communion.[62] In February 1549 William Abraham had ridden as a penitent through the City to the pillory at Cheap. A paper proclaimed his offence: that he had procured a man to murder his lover's husband. Such a man would be watched, and through 1554 he failed to attend church at Christchurch.[63] Robert Maior, charged with publicly denying the Mass, had earlier stood upon the pillory for cursing the Lord Mayor at his gate.[64] When George Tankerville, accused of sacramentarianism in the visitation, was re-arrested in February 1555, betrayed by a promoter, his wife hardly showed the spirit of Christian resignation. She picked up a spit and would have run it through Beard, the Judas, had she not been restrained.[65] That it was the most disruptive of London's reformed community, those who had been the iconoclasts in Edward's reign, those who had enemies, who were prosecuted is not surprising. But what was as yet unclear was what would happen once the new laws come in and thoroughgoing persecution began.

Reconciliation to the Catholic Church was the path of least resistance; it was also the 'broad pathway' which led to everlasting pain. Nevertheless at the turn of 1554 the prospects for those who persisted in opposing the regime were grim enough to make it seem the only course. Those convented before Harpsfield had already resisted for a year, and these were

[60] GLRO, DL/C/614, fo. 64ᵛ.
[61] Ibid., fo. 45ᵛ; CLRO, Repertory 12, fo. 122ʳ.
[62] Ibid., Repertory 13, fo. 173ᵛ; GLRO, DL/C/614, fo. 4ᵛ.
[63] CLRO, Repertory 12, fo. 53ʳ; GLRO, DL/C/614, fo. 43ʳ.
[64] Ibid., fo. 20ᵛ; CLRO, Repertory 12, fo. 89ʳ.
[65] GLRO, DL/C/614, fo. 21ʳ; Foxe, *Acts and Monuments*, vii, p. 344.

dangerous times. Offenders from eleven of the parishes had already seen their ministers deprived for marriage, and sent to exile or prison, to be replaced by priests staunch in their popery.[66] Twenty-two from St Sepulchre and twenty-four from St Bride waited the outcome of the trials of their ministers, Rogers and Cardmaker.[67] Some of London's godly were already recanting. On 1 July 1554 John Hill, a cutler from St Gregory who held obstinately against the sacrament of the altar, had been convented before the Dean of St Paul's. 'And so revoked his opinion openly before all the parish, and asked them mercy and forgiveness for his evil example.'[68] Two laymen and three married ex-friars, including John Lawe, rector of St Margaret Moses, stood penitent at St Paul's Cross on 4 November in white sheets and bearing candles, while John Harpsfield preached.[69] Precedents for accommodating the conscience with the new orthodoxy were becoming evident; so was the need to do so.

The authorities were vigilant. Almost sixty people had been rounded up early in October, at the same time as the visitation but not as part of it: their offence, importing forbidden books from the preachers fled to Germany. Masters Beston, a girdler, Spark, a draper, Randall Tirer, the stationer, and Brown, with other householders, apprentices, and servants were thrown into prison.[70] Such was the general suspicion and alarm that when Davy Cromwell, a poor priest, died in St Bartholomew's hospital at the turn of the year his books were seized and sent to the Vicar-General to examine. They were only books of astronomy.[71] In Strasburg John Banks lamented to Bullinger on 9 December that the 'godly men' who had supported the exiles

[66] John Rogers of St Sepulchre; John Harleton of St Bride; Thomas Cottisford of St Martin Ludgate; Thomas Brigotte of Christchurch; Thomas Chipping of St Mary Magdalen Old Fish Street; Henry Reed of St Margaret Lothbury; Thomas Becon of St Stephen Walbrook; Thomas Sowdley of St Mary Mounthaw and St Nicholas Cole Abbey; Edward Heydon of St Benet Paul's Wharf; Robert Crome of St Leonard Foster Lane; John Veron of St Alphage: E. L. C. Mullins, 'The Effects of the Marian and Elizabethan Religious Settlements upon the Clergy of London, 1553–1564' (University of London, M.A. thesis, 1948), appendix.

[67] Rogers and Cardmaker had been arrested in Aug. 1553: *APC* iv, p. 321.

[68] *Grey Friars Chronicle*, p. 90.

[69] Ibid., p. 92; *Diary of Machyn*, pp. 73–4. See also W. H. Frere, *The Marian Reaction in its relation to the English Clergy* (1896), pp. 64, 65, 83.

[70] Foxe, *Acts and Monuments*, vi, p. 561; *APC* v, pp. 84–5.

[71] St Bartholomew's Hospital, IIa 1/1, fos. 113^v–114^r.

hitherto were now imprisoned 'on our account', or if still free 'so carefully watched by the papists' that they could send no help 'without the greatest danger'.[72] The worst suspicions of what happened to the prisoners for faith if they fell into Bonner's clutches were soon confirmed. The news reached Anne Hooper in Frankfurt on 10 November that 'the hand of an individual had been burnt off, because he refused to hear Mass, and chose rather to be brought to the stake'. This was Thomas Tomkins, the weaver of Shoreditch, whose faith, according to Foxe, was tested by an ever more irascible Bonner, who held his hand over a candle flame until the sinews snapped. Anne Hooper heard too that 'some godly persons had lately been thrown into prison for the sake of religion'.[73]

Even while the visitation proceeded the gospellers continued to meet together secretly 'as their conscience did enforce them', but in ever greater danger. On 1 January 1555 at midnight a secret congregation of thirty and more was found in Bow churchyard by Gardiner's men. They had held an English service (an Edwardian service), with prayers and a reading.[74] Hooper heard that they 'prayed for the magistrates and estates of this realm'; Renard thought the worst, as usual: that they held 'heretic rites' and prayed for Elizabeth's 'freedom and prosperity'.[75] Thomas Rose, their minister, went to the Tower; the rest to the two Counters. In Bread Street Counter Hooper's informant found Chamber, Monger, and Sh*** 'strong', rejoicing that 'for well doing they are imprisoned'. 'Remember', wrote Hooper on 14 January in consolation; 'God and all his angels' looked down, and 'be ready always to take you up into Heaven, if ye be slain in this fight'.[76] At least one of the congregation did go to the stake: Elizabeth Warne.[77]

It was in these perilous times that those before Harpsfield, awaiting judgement, had to choose whether or not to conform. Only one chose exile: John Bartholomew of St Stephen Walbrook

---

[72] *Original Letters*, i, pp. 306–7.

[73] Ibid., i, p. 113; Foxe, *Acts and Monuments*, vi, pp. 718–19 (NB illustration).

[74] *Diary of Machyn*, p. 79.

[75] Foxe, *Acts and Monuments*, vi, p. 579; *CSP Sp.* xiii, p. 135.

[76] Foxe, *Acts and Monuments*, vi, pp. 585–7. Thomas Monger is mentioned in Sir Thomas Gresham's Day Book (in the possession of the Mercers' Company), and was probably a substantial merchant.

[77] Foxe, *Acts and Monuments*, vii, pp. 342–3.

left to join the English community in Basle.[78] But he had the money and freedom, as well as the courage, to leave: others had not. His fellow parishioner at St Stephen, Thomas Locke, had failed to attend church since the restoration of Latin services, but now he promised to reconcile himself. Soon after, he and seven of his children died: a divine judgement upon his casuistry, Rose Hickman thought.[79] Most of the four hundred chose the 'broad pathway'. Thirty were ordered to stand, white-sheeted penitents, at the choir of their parish churches after the offering of the Mass, while the priest declared their penance to the people, showing them turning from error to truth. Another sixty-nine were instructed to confess to their priest, and to receive the Mass. Nine were told to purge their offences. Forty-five were simply ordered to attend church and amend their behaviour, and another forty dismissed upon condition that they should 'conduct themselves in a Catholic manner as Christians should'.[80] In the French Church also fifty of the congregation stayed behind in London and heard Mass.[81]

Richard Hilles had once told Thomas Cromwell that he would never 'turn from Christ'. In 1541 he had counselled a backslider against that 'wickedness of falling away from the truth'. But by the end of 1554 his servant urged Bullinger to use his old influence and write 'a few words upon fleeing from the abomination of the Mass', for my 'master is now placing his soul in jeopardy'.[82] Even Ridley's own chaplains had apostasized, he feared. Nicholas Grimald had been released from the Marshalsea 'not without some becking or bowing, alas, of his knee to Baal'.[83] Now he was helping in the publication of More's works.[84] West, too, lacked his master's courage, 'said Mass against his conscience, and soon after

[78] GLRO, DL/C/614, fo. 26[r]; Garrett, *Marian Exiles*, 35.

[79] GLRO, DL/C/614, fo. 26[r]; PRO, PCC, Prob. 11/38, fo. 180[r]. He died in Oct. 1556: *Diary of Machyn*, p. 117.

[80] GLRO, DL/C/614.

[81] Pettegree, *Foreign Protestant Communities*, pp. 125, 131–2.

[82] *L&P* vi. 99; *Original Letters*, i, pp. 345, 347. Hilles was particularly named as a temporizer in *An answer to a certain godly mannes lettres*, sig. Aii[r].

[83] Foxe, *Acts and Monuments*, vii, p. 435; vi, p. 619. For Grimald's career and literary significance, see *The Life and Poems of Nicholas Grimald*, ed. L. R. Merrill (New Haven, 1925), and J. N. King, *English Reformation Literature: The Tudor Origins of the Protestant Tradition* (Princeton, 1982), pp. 242–4, 283, 297–9, 351.

[84] Thomas More, *Dialogue of Comfort*, *CW* 12, pp. xlix–l.

died'.[85] When in January 1555 Gardiner summoned those eighty still imprisoned for the faith, attempting to persuade them by 'promises, rewards and threatenings' to sign recantations, the rumour put about was that even Barlow and Cardmaker's constancy had failed.[86] The godly were 'in a marvellous dump and sadness'. But Cardmaker had 'beguiled as well the vain boasting of the Bishop as the light belief of the inconstant people'.[87] Soon after he went to the stake.

The godly waited anxiously to see how their preachers would stand the ultimate test of faith. Between 28 and 30 January 1555 Gardiner and other Bishops, with a commission from Cardinal Pole, sat in judgement upon the heretics, ready but reluctant to implement the new penalties. Bradford, Rogers, Saunders, Taylor, Hooper, and Cardmaker were condemned to death.[88] The first to suffer was John Rogers, who burnt at Smithfield on 4 February with heroic fortitude. Even Catholic opponents said so. The godly who had gathered wept and prayed God to give him courage to bear the pain and not to recant. Rogers's ashes were collected, a martyr's relics, and some seeing birds fly over as he expired thought this a sign of the Holy Ghost.[89] To die for the faith was a special benediction and a glorious victory over the idolaters. Ridley wrote, a few days later: 'And yet again I bless God in our dear brother and of this time protomartyr, Rogers . . .'.[90] Ridley, Bradford, and others condemned waited in prison: 'what shall come of them, God knoweth', wrote Scudamore on 2 March.[91] Within months Ridley and the most stalwart suffered. But such fortitude was given to few. In prison, Bradford lamented the prevailing casuistry of 'our Mass gospellers and popish Protestants'.[92] 'Not the tenth part abode in God's ways'; rather they had become

[85] Foxe, *Acts and Monuments*, vii, p. 406.

[86] *Original Letters*, i, p. 171; Foxe, *Acts and Monuments*, vi, p. 587; Wriothesley, *Chronicle*, ii, p. 126.

[87] BL, Harleian MS 421, fos. 67^r–68^v; Foxe, *Acts and Monuments*, vii, p. 82.

[88] BL, Harleian MS 421, fo. 36^r; Wriothesley, *Chronicle*, ii, p. 126. The wardens of the City companies were summoned to attend the arraignment: Drapers' Company, MB 6, fo. 48^r.

[89] *CSP Sp.* xiii, p. 138; Miles Hogarde, *The displaying of the protestantes, with a description of diuers their abuses* (*RSTC* 13557), sig. Gviii^r.

[90] BL, Harleian MS 416, fo. 16^r.

[91] NLS, Adv. MS 34.2.14, fo. 59^r.

[92] Bradford, *The hurte of hering masse*, sig. Cii^r.

'mangy mongrels', neither papist nor Protestant. Though they 'pretended popery outwardly' at the Mass, 'in their hearts, they said, and with their spirits they served the Lord'. Thereby 'they saved their pigs, I mean their worldly pleasures', while hoping still to be counted among the gospellers.[93] In darker moments the reformers feared that the Church's attempt at universal reconciliation was succeeding.

'THE CATHOLIC VISIBLE CHURCH'

By June 1555 Dr Story, the 'bloody Nimrod' of the Marian reaction, could report that London—'the spectacle of the whole realm'—was 'daily drawing partly for love, partly for fear, to conformity'. Through 'discreet severity' he hoped for 'universal unity in religion'.[94] The sight of the burnings in these days—for there had been twenty-eight taken from London to the fires by the end of June[95]—had persuaded some to an unwelcome conformity, but for very many more the return of the old faith had evidently been ardently awaited, and the announcement of reconciliation with Rome brought general rejoicing. In November 1554 Cardinal Pole had returned to England, the Pope's emissary, promising, 'I come to reconcile, not to condemn. I come not to compel, but to call again.' His long exile over, he preached in London before huge crowds—'Brothers, know that now is the hour for us to awake from sleep'—and his sermon 'moved a great number of the audience with sorrowful sighs and weeping tears to change their cheer'.[96] Gardiner preached at Paul's Cross on 2 December, urging the crowds joyfully to receive remission, and the bells of the City churches rang exultantly drowning his words.[97] In Parliament a bill passed at the end of December which reunited England with Rome.[98] Promising hope for the future was the announcement

[93] *The Writings of John Bradford*, ed. A. Townsend (Parker Society, Cambridge, 1848), p. 390.
[94] Foxe, *Acts and Monuments*, viii, p. 745; *CSP Ven.* vi/1, pp. 110–11; *A declaration of the life and death of John Story* (1571).
[95] Strype, *Ecclesiastical Memorials*, iii/2, pp. 554–6.
[96] 'John Elder's Letter', in *Chronicle of Queen Jane*, pp. 152–66; BL, Harleian MS 419, fo. 132^r; W. Schenk, *Reginald Pole: Cardinal of England* (1950).
[97] *Two London Chronicles*, pp. 40–1.
[98] J. Loach, *Parliament and the Crown in the Reign of Mary Tudor* (Oxford, 1986), pp. 104–15.

late in November that Mary had conceived a child.[99] On 25 January, the feast of the conversion of St Paul, processions and bonfires in every parish marked England's reconciliation with Catholic Christendom.[100]

During the Easter days of 1555 the people 'behaved in a very obedient manner. An incredible number of them took the holy sacrament.' So Renard reported from London.[101] As Lent began, Bonner had written to the laity of his diocese, promising that though they had wandered 'as lost sheep', had been punished by plague and rebellion, still God had 'cast his eye of mercy' upon them and would absolve the penitent from heresy and schism. All must seek absolution. Knowing how many were 'in scruple or doubt', Bonner promised that should their own priest in confession fail to satisfy their disturbed consciences they should seek resolution from his Archdeacons, his chaplains, or even himself.[102] Robert Bocher, for refusing ashes on Ash Wednesday, was called before the Bishop, and finally received absolution from a royal chaplain.[103] This was the form of absolution:

Our Lord Jesu Christ absolve you, and by the apostolic authority to me granted . . . I absolve you from the sentence of excommunication and from all other censures and pains into the which you be fallen by reason of heresy and schism . . . I restore you unto the unity of our holy mother Church and to the communion of all sacraments.
God save the King and Queen.[104]

Still, some in London tried to avoid the Easter Mass. Henry Locke and his family left the City to receive communion without Mass from a priest wearing only a vestment.[105] But that they had to leave London to worship as they wished is significant of the conformity of the majority of their fellow citizens and of the resolution and vigilance of the City clergy.

By around Easter 1554 one-third (41 or 42 of 129) of London benefices had been emptied by the deprivation, resignation, or

99 Wriothesley, *Chronicle*, ii, p. 124; *RSTC* 17561.
100 Wriothesley, *Chronicle*, ii, pp. 124, 126; *Two London Chronicles*, p. 41.
101 *CSP Sp.* xiii, p. 166.
102 Guildhall, MS 9531/12, fo. 372^{r-v}; Foxe, *Acts and Monuments*, vi, p. 710.
103 Hale, *Precedents*, ccccxxxiii.
104 Guildhall, MS 9531/12, fo. 372^v; *RSTC* 3280.3.
105 BL, Harleian MS 353, fo. 145^r.

imprisonment of their Edwardian reformist incumbents.[106]
The way was clear for stalwart Catholics to take up these
livings, or for those who had conformed during the schism now
openly to profess the Catholic faith. Ultra-Catholics, who had
left a schismatic England, now returned to work for the new
regime. Back to City benefices and to key positions in the
diocese also came uncompromising Catholic clergy. Most, it is
true, had never left. Underhill claimed: 'I think if you look
among the priests in Paul's, ye shall find some old mumpsimus
there.'[107] As Ridley prepared for martyrdom he bade farewell
to the prebendaries of St Paul's, but he wondered whom there
was left to salute: 'all those who loved God's Word' were exiled
or, like himself, awaiting 'cruel death for Christ's Gospel sake'.
'As for the rest of them, I know they could never brook me
well, nor could I ever in them delight.'[108]

Among the London clergy of Mary's reign were leading
Catholic theologians and apologists, prime movers in the
English Catholic Reformation. From Louvain returned Richard
Smith and Nicholas Harpsfield, who had spent their time in
exile writing for the Catholic cause: Smith upon the sacrament,
Harpsfield his *Life and Death of Sir Thomas Moore.* Smith now
became vicar of St Dunstan in the East and chaplain extraordinary to the King and Queen; Harpsfield a prebendary of St
Paul's, and later Archdeacon of Canterbury.[109] His brother,
John Harpsfield, became Bonner's chaplain, and in April 1554
was collated Archdeacon of London.[110] Henry Cole, who had
been pardoned in 1544 for his treasonable mission to Rome
and Cardinal Pole, had prudently, though reluctantly, conformed during Edward's reign, but now showed himself again
in his popish colours. He became, in turn, prebend and Dean
of St Paul's, Vicar-General in spirituals to Pole, and Dean of
the Arches.[111] Cole succeeded Feckenham, another of the
Bishop's chaplains, as Dean of St Paul's when Feckenham was
instituted first Abbot of the restored Benedictine monastery of

[106] Mullins, 'Effects', pp. 94–126, 291.
[107] *Narratives of the Reformation*, p. 141.
[108] Foxe, *Acts and Monuments*, vii, p. 558.
[109] Harpsfield, *Life*, pp. clxxv–ccxiv; *DNB*; *CPR P&M* ii, pp. 72, 195.
[110] *DNB*. The Harpsfields had University scholarships from their father's
company: Mercers' Company, Acts of Court, ii, fo. 267ʳ.
[111] Strype, *Ecclesiastical Memorials*, iii/2, pp. 27–8.

St Peter's Westminster.[112] Like Feckenham, William Chedsey, a leading conservative preacher of Henry VIII's last years, had been imprisoned under Edward for opposition to reform. Chedsey became vicar of All Hallows Bread Street, Archdeacon of Middlesex, prebend in the cathedral, and canon of St George's Windsor.[113] Most uncompromising of all the London divines was Thomas Darbishire, Bonner's nephew, who became Chancellor of the London diocese.

When the world changed again, at the accession of Elizabeth, these men proved once more their commitment to the old faith. Darbishire was sent to the Council of Trent for the answer to a question which perplexed Catholics as well as their confessional opponents: would attendance at heretical services be a sin? It would. Darbishire left London for Rome, and in 1563 entered the Society of Jesus.[114] Rather than subscribe to the Elizabethan Oath of Supremacy, Smith, Cole, Chedsey, the Harpsfields, Darbishire, Boxall, and Feckenham chose prison or exile. Fifteen other London clerics were deprived, almost certainly because of their romanist convictions; seventeen resigned, and twelve more were somehow removed.[115] So nearly half of the beneficed clergy of London in Mary's reign were so committed to the Catholic cause that they could not accommodate their consciences to serve the times. Others, certainly, were of more fleeting loyalty. Alan Percy survived in his parish of St Mary at Hill through four reigns (1521–60), his convictions now, and possibly then, obscure.[116] In Henry's reign John Standish had been a conservative champion, but with each successive regime, through politic conformity, he adroitly found favour. On Mary's accession he abandoned the wife and the reformed principles he had espoused under Edward (whose chaplain he had been), and advocated the burning of English Bibles. In 1558 he lost his prebend at St Paul's, but in 1559 was restored.[117] Henry Pendleton, Bonner's chaplain, and Marian rector successively of St Martin Outwich and St Stephen Walbrook, had preached

---

112 Knowles, *Religious Orders*, iii, pp. 423–4, 428–30.
113 *CPR P&M* i, pp. 223, 323.
114 *DNB*.
115 Mullins, 'Effects', pp. 293–4.
116 Venn, *Alumni*.
117 *BRUO*.

against the new faith in Henry's reign, but converted to it under Edward. In Mary's reign he reverted. Foxe claimed that Pendleton died regretting that ever he had 'yielded to the doctrine of the papists as he did', but such allegations were repudiated in 'A declaration of H. P., D.D. in his sickness of his faith in all points as the Catholic Church teacheth, against sclanderous reports against him'.[118] For the most part the clergy instituted to City benefices were learned and committed, fit for their task of reconciling and converting the community most effectively evangelized by the reformers.[119]

'Partly by love, partly by fear' the Londoners were being drawn back into the Catholic fold. Alongside the persecution of the obdurate, Bonner and his leading clergy mounted a campaign of instruction and re-education. On 1 July 1555 thirteen Homilies were published. Composed by the Bishop and his chaplains, they were concerned with the central issues of faith: of man's fall and redemption, Christian love and charity, the nature and authority of the Church, the centrality of the sacrament of the altar, and an assertion of the doctrine of transubstantiation. Every Sunday and holy day one homily must be read in the churches, should no licensed preacher be present. In September Bonner published *A profitable and necessarye doctryne*, a new formulary of the Catholic faith, based closely upon the King's Book of 1543. This formulary of the 'Catholic trade and doctrine of the Church' was designed to instruct the clergy and educated laity who had been taught 'in the late . . . pestiferous schism' to call this by 'a new, envious and odious term . . . Papistry', and it laid special emphasis upon the sacrifice of the Mass.[120] For children—those most urgently in need of instruction—Bonner prepared a little catechism, printed in red and black to help them read: *An honeste, godlye instruction*.[121] Londoners of the previous generation, so Pole reminded them, had

[118] *BRUO*; Foxe, *Acts and Monuments*, viii, p. 635.

[119] The conclusions of Dr Alexander differ: 'Bonner and the Marian Persecutions', p. 387.

[120] *RSTC* 3281.5–3285. Strype, *Ecclesiastical Memorials*, iii/1, pp. 440–1. The campaign for re-education is described in Alexander, 'Bonner and the Marian Persecutions', pp. 386–8, and J. Loach, 'The Marian Establishment and the Printing Press', *EHR* c (1986), pp. 135–48.

[121] Edmund Bonner, *An honestly godlye instruction, and information for bringinge vp of children* (*RSTC* 3281). P. Tudor, 'Religious Instruction for Children and Adolescents in the Early English Reformation', *JEH* xxxv (1984), pp. 391–413.

been vouchsafed special 'miracles' to strengthen them in their faith. When 'the realm was falling from the unity of the Church, when the foundation began to move', there were 'marvellous examples' of constancy: the martyrdom of Thomas More, of Fisher, and of the Carthusians.[122] Now More's circle returned from exile and were preparing the great edition of his *English Works*, which Rastell published in 1557.[123]

The people might best learn of perfect Catholic spirituality, as before, by witnessing the austere and virtuous life of the religious. The epitome of Catholic hopes for the restoration of religion was the return of the religious houses and of the religious life. Yet the despoliation had gone so far; the lands had been sold and re-sold; the new owners had proved markedly reluctant to cede them in the name of religion. London's religious houses now housed courtiers, or stabled their horses, or sheltered the indigent, the sick, and the orphaned. London's citizens and courtiers had benefited most from the despoliation: how would they ever be persuaded to divest themselves of land and wealth? But the Queen had close to her heart the cause of restoring the religious houses which her father had despoiled.[124] King Philip had brought with him 'Franciscans and other monks to make a beginning'.[125] Sixteen Benedictine monks, clad in the order of their habit, led by Dean Feckenham, appeared at Court and informed a tearful Mary that they would renounce their preferments in order to return to the monastic life.[126] In November 1556 a new monastic community was founded at Westminster.[127] On Palm Sunday 1555 Observant Franciscans returned to their house at Greenwich, led by Henry's old opponents, Elstow and Peto. Some were the survivors of the English exiles, others were Spanish friars, but in time there were some new recruits.[128] Exiled Dominicans returned to London with the Franciscans, and that summer established a house at West

---

[122] Strype, *Ecclesiastical Memorials*, iii/2, pp. 490 ff.

[123] Harpsfield, *Life*, p. 100; *RSTC* 18076.

[124] For the Marian restoration of monasticism, see Knowles, *Religious Orders*, iii, pp. 421–43; K. Brown, 'The Franciscan Observants in England, 1482–1559' (University of Oxford, D.Phil. thesis, 1987), pp. 222–38; D. M. Loades, *The Reign of Mary Tudor* (1979), pp. 352–5.

[125] *CSP Sp.* xiii, p. 23.

[126] *CSP Ven.* vi/1. 32.

[127] Knowles, *Religious Orders*, iii, pp. 423–4; *Diary of Machyn*, pp. 118–19.

[128] *CSP Sp.* xiii, p. 146; Wriothesley, *Chronicle*, ii, p. 128; *Grey Friars Chronicle*, p. 95.

Smithfield with William Peryn as their Prior.[129] None had suffered so terribly as the London Carthusians; now, the faith restored, they—or those who were left, including John Foxe and Maurice Chauncy—returned. (Sir Robert Rochester, controller of Mary's household, was the brother of one of their martyrs—dan John.) The House of the Salutation in the City was now a noble mansion; the returned monks lodged first at the Savoy, before refounding their house at Sheen in November 1555.[130] The community of Bridgettine nuns at Syon was once more enclosed in August 1557, charged 'never to go forth as long as they do live'.[131]

The initiative for the refoundation came from the religious themselves; the main financial support from the Crown. Though the Church courts imposed donation to the religious houses as a penance—the heretic Gertrude Crockhay was ordered to pay 6s. 8d. to the Friars Minor in Smithfield[132]—the Church had no money for the religious. Hardly a single owner of the monastic lands made so much as a gesture to return his spoils. The one Londoner to return his land was Thomas Bowes, who had leased the Observant friary at Southampton. But he surrendered his lease only in return for a royal licence to export 40,000 woollen kerseys, not through penitential fervour.[133] Landlords were bound to think not only of their heirs, but also of likely confiscations by Mary's successors. But Londoners may have given during their lives to the religious, for the cause was urgent. Certainly, some left bequests in their wills to aid 'poor religious persons'.[134] As early as April 1554 Thomas Lewen looked to restoration. Though he left his property to find a priest to celebrate Mass at Ṣt Nicholas Olave, this bequest was to stand only 'until such time as it shall fortune' a religious house to be refounded at Sawtre.[135] Benet Jackson gave £4 in August 1555 to the 'maintenance of the house of the

---

[129] BRUO.

[130] E. M. Thompson, The Carthusian Order in England (1930), pp. 500–9.

[131] Diary of Machyn, p. 145.

[132] Guildhall, MS 9168/11, fo. 165ʳ.

[133] CPR P&M iii, pp. 360–1; Loades, Reign of Mary Tudor, p. 352.

[134] See, for example, PRO, PCC, Prob. 11/37, fos. 222ʳ, 263ᵛ, 272ʳ; Prob. 11/40, fos. 76ᵛ, 84ᵛ, 106ᵛ, 107ʳ, 194ʳ, 291ᵛ. I have found no bequests to the religious in the wills proved in the Commissary Court during Mary's reign.

[135] PRO, PCC, Prob. 11/37, fos. 222ʳ–223ʳ.

poor friars of Greenwich . . . to the continuance of prayer and God's service there'.[136] To the friars of St Bartholomew West Smithfield, Richard Bartlett gave £6 in his will of January 1557, and more in reversion.[137] In 1558 Robert Reynolds left 40s. each to the houses at Sheen and Syon, to the Dominicans, and to the Observants at Greenwich, and Elizabeth Nott gave £5 each to the Black Friars and the Observants 'to pray for her soul and all Christian souls'.[138] Monks and friars had as their first spiritual purpose the duty to pray for souls departed, but this essential Catholic doctrine had not survived the religious changes unscathed.

Belief in Purgatory had lain at the heart of late medieval religion. No one doubted that a stay in Purgatory awaited souls departed, and everyone lived in fear of future suffering there. The living knew their ceaseless duty to the dead: by endless prayer to release the souls in Purgatory. Much of the time and money of the Catholic community had been expended in that endeavour. When in the 1530s the reformers turned to deny the existence of Purgatory and to deride the rites of intercession for the dead, their attacks—'newest and most strange to the people'—were most resented. Surely the belief that prayers could avail the departed was not easily abandoned, nor quickly, but by Mary's reign the doctrine had been profoundly, perhaps irrevocably, undermined. William Wolcar, a goldsmith, offered his company in September 1556 £100 to pay for a beadsman to pray for his soul and all Christian souls forever.[139] But he was almost alone among the citizens to make such a bequest. Fear of future sequestration doubtless dissuaded many, but the reluctance went deeper. The undermining of belief in Purgatory was of massive social consequence.

Fraternities had been a vital and enduring part of the religious and community life of London, until the Edwardian Chantries Act outlawed them along with Purgatory. The conviction that the rites of intercession could win salvation from Purgatory was only the first among many reasons for the creation of the gilds. Acts of mutual charity were more than

---

136 Ibid., fo. 263^r.
137 PRO, PCC, Prob. 11/40, fo. 194^r.
138 Ibid., fos. 107^r, 291^v.
139 Goldsmiths' Company, Court Book I, fo. 253^r.

ever needed in the economically and politically volatile 1550s; the desire for commensality, for charity, and for prayer had surely lived on after the dissolution of the gilds. Would there be a proliferation of restored fraternities upon Mary's accession? Thomas Lewen confidently bequeathed 10*s*. in April 1554 to Jesus brotherhood in St Paul's.[140] That fraternity was restored —although not until July 1556—but none of the others were: so, at least, it seems from the marked silence about gilds in the City wills and churchwardens' accounts. Fraternities were not returned to the parish churches along with the images of patronal saints: nervousness about re-endowing them is understandable, but the absence of spontaneous associations, of gild meetings, among the parishioners is mysterious. For the gospellers there were, of course, voluntary associations of another kind, quite different in the beliefs that led them to worship together, but perhaps not so different in organization.[141]

The one London fraternity which was restored was that of the Name of Jesus in the Shrouds of St Paul's. It was re-established by royal grant on 28 July 1556 upon petition of the rector and parishioners of St Faith's. This gild was, under Mary, as it had been before its dissolution, the gild for the City's most devout Catholic establishment. Its president was Cook, Dean of the Court of Arches, its masters Richard Lyell and Thomas Argall, advocate and registrar of the Arches. Robert Johnson, Bonner's registrar, and John Cawood, the royal printer, were members. Darbishire, Marian commissioner and Chancellor of London, and future Jesuit, was its first vice president. Strype thought, wrongly but reasonably, that this gild was newly founded 'after the old popish custom', and named after the Society of Jesus. This was London's first Counter Reformation confraternity, and its last.[142]

Jésus gild buried its brethren with full Catholic rite: with processions of mourners in black carrying lighted candles, and in attendance all the brotherhood in black satin hoods, decorated

---

[140] PRO, PCC, Prob. 11/37, fo. 222ʳ.

[141] A. G. Rosser, 'Communities of Parish and Guild in the Late Middle Ages', in S. J. Wright (ed.), *Parish, Church and People: Local Studies in Lay Religion, 1350–1750* (1988), pp. 29–55.

[142] *CPR P&M* iii, p. 274; Strype, *Ecclesiastical Memorials*, iii/2, p. 108. J. J. Scarisbrick, *The English Church and the English People* (Oxford, 1984), p. 37.

'IHS', and all the priests of Paul's.[143] Once more leading citizens were interred with great splendour and elaborate ritual: for the richest, company halls and the City streets were hung with black, the heralds and the Mayor and Aldermen attended in violet gowns, the poor, carrying torches and banners painted with images, came as mourners and were rewarded with great doles. Prayers were said for the soul of the departed, and dirige and requiem masses sung.[144] At the funeral of the redoubtable Catholic Judge Morgan, Darbishire preached; others had sermons by the Observant friars or by leading City clergy.[145] Henry Machyn, the undertaker, recorded these elaborate City funerals in professional detail. Now even those Edwardian City governors, whose sympathies had seemed to be for reform, were buried with Catholic pomp and obsequies. Sir Richard Dobbes, celebrated by Ridley as founding father of the hospitals, had dirige and requiem sung; one for the Trinity in pricksong and another of Our Lady.[146] The funeral masses were followed once again by month's minds for the departed soul, and Sir Harry Amcottes founded chantries, one in London, one in Lincolnshire, as he thought, for ever.[147]

The most visible sign of the old faith returning was the restoration of processions through the City. Under Edward 'for lack of devotion', it seemed, they had been abandoned, but upon Mary's accession triumphantly they returned; at first in a wave of popular enthusiasm, thereafter by command. In November 1553 St Katherine's image was processed around the steeple of St Paul's, with five hundred lights, on her patronal feast, and on St Andrew's day there was 'a general procession . . . in Latin with *ora pro nobis*'.[148] Devotional processions were encouraged in sermons, and sanctioned on 8 December by Bonner's order that every parish church provide for a cross, a staff, and a cope for processions.[149] Machyn chronicled the calendar of City processions through Mary's

---

[143] *Diary of Machyn*, pp. 166, 172, 178–9.
[144] See, for example, *Diary of Machyn*, pp. 68, 70–1. Machyn recorded all the important City funerals of Mary's reign.
[145] Ibid., pp. 106, 91, 98.
[146] Ibid., p. 106.
[147] Ibid., pp. 70–1.
[148] *Grey Friars Chronicle*, p. 77; *Two London Chronicles*, p. 31.
[149] *Diary of Machyn*, pp. 49–50.

reign, and they were many. Beside the processions of parishes on the festivals of their patronal saints, there were the processions of the Parish Clerks, of the Fishmongers, of the Court on St George's day, and of the parishes of St Clement and Islington.[150] After Mass at the Guildhall on 7 May 1554, sung by the Queen's choristers, every parish clerk of the City went in procession, two by two, each wearing a cope and garlanded, and after them were borne eighty banners. Then came the waits playing, and four choirs singing *Salva festa dies*, and the blessed sacrament carried under a canopy, up St Laurence Lane, through Cornhill to Leadenhall, down Bishopsgate to St Ethelburga.[151] On days for special rejoicing the whole City processed in thanskgiving: to celebrate the defeat of Wyatt's rebels, or victory over the French, or the Queen's quickening with child. In hard times there were propitiatory processions.[152] Because 'many would have it', an order came in November 1554 that the ancient festival of St Nicholas the Boy Bishop, abrogated in 1538, be celebrated once more on the next Childermas day. The order was promptly countermanded, through fears of misrule, but even so 'there went about those St Nicholases in divers parishes'.[153] Still in 1556 and 1557 the boy bishops paraded from house to house, and, while rejected by some, were 'received with many good people into their houses, and had as much good cheer as ever they had'.[154]

On 1 July 1556 precise orders came for parishes to process every Monday, Wednesday, Friday, Sunday, and holy day. First should come the youth of the parish, being least accustomed and most reluctant: the parish schoolmaster with his scholars, bareheaded before the cross and carrying their prayer books; then apprentices and serving men with their primers 'if they can read, or else with their beads, every one devoutly praying and using himself godly'. Then came the choir, fol-

---

[150] *Diary of Machyn*, pp. 49, 51, 55, 61, 62–3, 64, 65, 66, 75, 76–7, 78, 80–1, 82, 85, 87, 89, 92, 98, 106, 107, 113, 119, 121, 137, 138, 139, 140, 141, 147, 150–1, 159, 160, 165; *Grey Friars Chronicle*, pp. 85, 86, 89, 94–5, 97; Wriothesley, *Chronicle*, ii, pp. 104–5, 106, 115.

[151] *Diary of Machyn*, pp. 62–3.

[152] Ibid., pp. 55, 147, 150–1, 152; *Grey Friars Chronicle*, p. 93.

[153] *Diary of Machyn*, pp. 75, 78. At St Mary at Hill they paid 8*d.* in 1555 for a book for St Nicholas: *St Mary at Hill*, ii, p. 399.

[154] *Diary of Machyn*, pp. 121, 160. For opposition from one Protestant extremist, see Foxe, *Acts and Monuments*, viii, pp. 579–80.

lowed by householders, their wives, widows, and lastly, their 'maidens with books or beads'. Penalties for absence were severe.[155] Parishes took heed, buying 'books for the order of going of procession', but the demands were too great: the command was 'but little looked upon, and the more pity'.[156] What had begun with spontaneous enthusiasm ended in apathy.

Processions had once signalled a shared faith and the communal spirit of the City. Most sacred of the festival processions was that of Corpus Christi, where the host was born ascendant. In 1554 there were 'many godly processions in many parishes . . . and staff torches burning, and many canopies'.[157] But to the godly this was idolatry. As the St Sepulchre's procession passed by, John Street drew his dagger, attacking the priest and trying to seize the sacrament.[158] What had once brought reconciliation now engendered dissension. But all those Catholic ceremonies and games which the reformers thought derogated from the true faith and had outlawed as superstitious now returned to the parishes of the City: the blessing of palms and reading of the Passion on Palm Sunday, the veiling of the rood during the Easter days, the hallowing of candles at Candlemas, creeping to the cross, the ringing of bells on All Saints' day, and processions at Rogationtide.[159]

THE RETURN OF THE OLD FAITH
TO THE PARISHES

An Elizabethan churchwarden of St Andrew Holborn, looking back through his predecessor's accounts for the first year of Mary's reign, was horrified by the zeal of the idolaters, and the pusillanimity of the rest. 'Without a great varying, by commandment', the parishioners had raced to 'set up all manner of superstitious things again': 'so ready they were to maintain idolatrous service, and forward to further superstition, and in

---

[155] Guildhall, MS 9531/12, fo. 403^v; Wriothesley, *Chronicle*, ii, p. 136; *Grey Friars Chronicle*, p. 97. The companies too punished those who failed to attend: Guildhall, MS 5177/1, fo. 108^r.

[156] Guildhall, MSS 2596/1, fo. 118^r; 4570/1, fo. 116^v; *Grey Friars Chronicle*, p. 97.

[157] *Diary of Machyn*, p. 63.

[158] Ibid., p. 64; *Grey Friars Chronicle*, p. 89.

[159] R. Hutton, 'The Local Impact of the Tudor Reformations', in C. A. Haigh (ed.), *The Reformation Revised* (Cambridge, 1987), pp. 128–31.

so short a space'.[160] 'Not by commandment but of the people's devotion' had the Mass first been restored to London.[161] So it was for the restoration of church treasures, in some parishes. Treasures had been hidden, safe from the despoilers, during Edward's reign. Now they were returned. At St Stephen Walbrook they still had a little 'fosser' in 1558 containing the sacred relics of the parish: a relic of the place where God appeared to St Mary Magdalen, relics of St Stephen, of Mount Calvary, a piece of the rock where God spoke to Moses, of the stone upon which Christ's body was washed after the deposition, a piece of bone of one of the eleven thousand virgins, a finger of one of the holy innocents.[162] Such treasures were not to be acquired in the market-place, even in Mary's reign.

The restoration of the Mass and of Catholic ceremonies demanded the restitution of all the service books, pyxes, paxes, chalices, censers, chrismatories, albs, altar cloths necessary for their celebration. But church treasures which had been donated by devout parishioners over generations had been desecrated, seized by the Crown, or auctioned under Edward. 'Popish books' had not only been removed but defaced.[163] The altars and the great roods had been replaced by communion tables and the royal arms. The walls of the City churches were decorated now not with dooms and images of saints, but with scriptural messages (graffiti, thought Bonner). 'Children of iniquity' had scrawled 'Scripture wrongly applied', intended to 'uphold their liberty of the flesh, and marriage of priests, and destroy . . . the reverent sacrament of the altar'.[164] By Easter 1554 the Scriptures and royal arms had been obliterated in most of the churches in London diocese, six months in advance of Bonner's order.[165] Retaining the biblical writings was now taken as a token of continuing resistance. Visiting the Marshalsea late in 1553, Brook 'beheld a piece of Scripture painted above the door', and warned the prisoners in Gardiner's name to remove

[160] Edward Griffiths, *Cases of supposed exemption from Poor Rates . . . with a Preliminary Sketch of the Ancient History of the Parish of St Andrew Holborn* (1831), appendix, 'Bentley's Book', p. xviii.

[161] Wriothesley, *Chronicle*, ii, p. 101.

[162] C. Pendrill, *Old Parish Life in London* (1937), pp. 21–2.

[163] See above, pp. 428–32, 457.

[164] *Documentary Annals*, i, p. 135; Foxe, *Acts and Monuments*, vi, pp. 565–6.

[165] Wriothesley, *Chronicle*, ii, pp. 113–14.

it: if I find it here, my lord shall know of it, by the holy Mass!'[166] At Gardiner's command also the parish of St Benet Gracechurch whitewashed the Scriptures in the church. They 'whited' the royal arms over the altar at St Mary Magdalen Milk Street and all the writings and pictures on the walls, and so too at St Matthew Friday Street, St Botolph Aldgate, St Mary at Hill, and the other churches.[167] But would the damaged, whitewashed churches be restored to their former splendour? The Imperial ambassador was pessimistic, writing in May 1554 that 'most of the churches here are in ruins, such is this people's faithlessness'.[168]

Between Mary's accession in July and the official restoration of the Mass in December 1553 there was a period of relative religious liberty. Some parishes—like Thomas Mountain's St Michael Paternoster—used that freedom to continue worshipping according to the Edwardian service, but others—like St Nicholas Cole Abbey—instantly reverted to Catholic rites. At St Dunstan in the West the altar and old service were restored by the end of September.[169] In March 1554 royal orders came for the restoration of 'all laudable and honest ceremonies', and Bonner was already alarmed that in his diocese, most vigilantly observed by the Council and commissioners, the orders were not obeyed promptly enough.[170] Since the orders came so swiftly for the alteration of religion and for the provision of all things necessary for the Mass and Catholic sacraments it is difficult to discover how far popular enthusiasm or government command spurred the restitution. The re-endowment of the City churches was an act of faith in more ways than one, for there remained always the possibility, the likelihood, that the government and religion might change again, and the treasures be sequestrated once more. There is some evidence of a prudential reluctance to re-invest. Until June 1557 the parish of St Martin Orgar had lent out copes, crosses, books, canopy cloths, and chalices to parishes poorer or less compliant. At St Botolph Aldgate service

166 *Narratives of the Reformation*, p. 185.

167 Guildhall, MSS 2596/1, fo. 113[r]; 1016/1, fo. 15[v]; 9235/1 (unfoliated); *St Mary at Hill*, ii, p. 397.

168 *CSP Sp.* xii, p. 243.

169 R. Hutton, 'Local Impact of the Tudor Reformations', p. 128.

170 *TRP* i. 390, 407; Guildhall, MS 9531/12, fos. 341[r], 362[r]; Strype, *Ecclesiastical Memorials*, iii/1, p. 34.

books were hired rather than bought for Easter and Whitsun 1554, and the Mercers' Company continued to borrow a chalice from St Paul's school until the summer of 1556.[171] But some enthusiasts had soon begun to bequeath ornaments to their parish churches. In March 1554 Thomas Claydon left two goblets to St Mary at Hill to be used daily at the high altar, and 'that I may be daily prayed for'; the following month Thomas Lewen gave 20s. for the repair of ornaments at St Nicholas Olave.[172] Of 345 of the London wills made during Mary's reign, thirty left money for ornaments or for the edification of their churches (but then perhaps sometimes rather to support their fellow parishioners in their great charges than as Catholic benefaction).[173] But though wills generally provide evidence of individuals' devotion and of their hope that the old faith would continue, they do not reveal the extent of spontaneous donation during Mary's reign. Then the need for restitution was urgent: the gifts must be made during life in order for services to be properly provided and the royal will obeyed.

The gilded images of Christ with Mary and John which had once dominated the interior of every City church had been pulled down as idols. But for Catholic worshippers the rood was of central significance, especially during the Easter days. Commands came for the roods to be restored. In most parishes the roods were not returned by Easter 1554, nor yet by Easter 1555, for in October 1555 Pole appointed Dr Story with other commissioners to visit every church in London and Middlesex to ensure that the roods were rebuilt, adorned with images of Mary and John.[174] Accordingly, on 14 October the vestry of St Martin Orgar met and decided to set up two new altars and a rood. At St Pancras an order came from the commissioners not only that they provide a rood, but that it be five foot long and completed within six weeks. At St Margaret Pattens they paid an 'earnest' of 20s. for their rood and veil cloth, and spent a further £3, but this was hardly enough to restore what had been

[171] Guildhall, MSS 959/1, fo. 18ʳ; 9235/1; Mercers' Company, Acts of Court, ii, fos. 278ʳ, 284ᵛ.

[172] PRO, PCC, Prob. 11/37, fos. 169ᵛ, 222ʳ–223ʳ.

[173] The wills studied were registered in Guildhall, MS 9171/13 and PRO, PCC, Prob. 11/37 and 40. For a testator expressing a reformist intent also leaving money to edify his church, see Prob. 11/37, fo. 173ᵛ.

[174] Wriothesley, *Chronicle*, ii, p 131.

one of the most famous roods in England, an object of veneration.[175] Restoring the roods was extremely, even prohibitively,
expensive. In April 1554 Robert Chertsey left £10 to his
wealthy parish of St Laurence Jewry to rebuild the rood, but
few parishes were similarly endowed.[176] Christchurch, Dr
Story's own parish, sought to make the City pay, but Story's
request 'seemeth to the Court [of Aldermen] very unreasonable'. The parishioners' petition for aid was in July 1556
referred to the governors of Christ's Hospital, itself almost
bankrupt.[177] Most parishes did restore their roods, because
they were bound to, but slowly, and in the knowledge that they
might not last for long. Not until 1557 did the parish of St
Mary Magdalen Milk Street rebuild its rood, paying the carver
£4. 6s. 8d. for the 'Rood Mary and John' and 25s. for the
image of their patronal saint, St Mary Magdalen. Within two
years they gave 16d. in drink to the workmen to 'pull down the
rood loft and burn the idols'.[178]

The images restored were the visual expression of the return
of the old faith and obedience. These were 'lay men's books',
possessing, wrote Bonner, a greater power than Scripture in
revealing the mysteries of the faith: 'they are the most apt to
receive light that are more obedient to follow ceremonies than
to read'.[179] But, if in ways quite different from Protestantism,
the Catholic faith depended upon written texts. Catholic liturgy
was precisely enshrined in the service books, according to the
various uses, which defined the rites and prayers to be followed. Restoration of the sung Mass required a provision of—as
Bonner ordered—'a legend, an antiphoner, a grail, a psalter,
an ordinal, to say or solemnize divine office, a missal, manual
and processional'.[180] Yet all such 'popish books' had been destroyed. Catholic service books were now urgently reprinted, in

175 Guildhall, MS 959/1, fo. 17ʳ; J. P. Malcolm, *London Redivivum*, 4 vols. (1802–7),
ii, p. 169; Guildhall, MS 4570/1, fos. 114ᵛ, 115ʳ.
176 PRO, PCC, Prob. 11/37, fo. 247ʳ; Guildhall, 2590/1, fo. 6ᵛ.
177 CLRO, Repertory 13, fos. 332ᵛ, 406ᵛ.
178 Guildhall, MS 2596/1, fos. 119ʳ, 122ʳ. For the rebuilding of roods in other
parishes, see ibid., MSS 1016/1, fo. 19ʳ; 1586/1, fo. 76ʳ; 'Bentley's Book', pp. xix–xx;
Pendrill, *Old Parish Life*, p. 29.
179 Bonner, *A profitable and necessary doctryne*, cited in J. Phillips, *The Reformation of
Images: The Destruction of Art in England, 1535–1660* (Berkeley and Los Angeles, 1973),
pp. 104–5.
180 Guildhall, MS 9531/12, fo. 368ʳ.

590 'Partly for Love, Partly for Fear'

many editions; and not through direct government patronage. In 1554 there were five separate editions of the manual of the Mass printed, and in 1555 27 editions of liturgical works were produced in London.[181] The City parishes compliantly purchased the necessary service books. On 2 September 1553—the very day that their errant governor, Richard Grafton, printer of the Edwardian Prayer Books, was released from prison—St Bartholomew's Hospital ordered a Mass book, manual, and psalter, and the following July purchased four processionals and another manual.[182] The order for the restoration of Latin services came on 15 December 1553, and on the 24th the parishioners of St Margaret Pattens bought a manual and processional at Paul's churchyard for 6s. 8d., and paid their erstwhile curate 4s. for his Mass book.[183] St Matthew Friday Street bought their service books in January 1554, selling meanwhile all their outlawed English books. So it was in the other parishes.[184] And having bought the books, they were reluctant to destroy them again. Upon his visitation to St Martin Orgar in 1560 the Archdeacon found still all the Catholic service books which the Marian regime could have desired. It was not until 1577 that the parishioners of St Christopher le Stocks, having invested so much in the Catholic books, determined to burn 'certain old popish books' which remained.[185]

Those who had gained from the despoliation under Edward were now bound to make restitution. Revenge was sought upon the most triumphant iconoclasts. Sir Thomas Cawarden, the despoiler of St Anne Blackfriars, was forced to provide the parishioners with a new church. The greatest beneficiary had been the Crown, and Mary, abhorring the sacrilege, determined

[181] P. M. Took, 'The Government and the Printing Trade, 1540–1560' (University of London, Ph.D. thesis, 1978), pp. 253–60.

[182] St Bartholomew's Hospital, Ha 1/1, fos. 91ʳ, 98ʳ, 102ᵛ.

[183] Guildhall, MS 4570/1, fo. 124ʳ.

[184] Guildhall, MS 1016/1, fos. 15ʳ, 16ʳ; St Michael Cornhill spent £7. 10s.; St Botolph Aldgate £8. 15s.; St Stephen Walbrook £11. 5s.; St Mary at Hill £6. 12s.; St Mary Woolnoth £6. 2s.: *The accounts of the churchwardens of the parish of St Michael Cornhill*, ed. W. H. Overall (1871), pp. 114–16, 123, 129, 130, 135, 140; Guildhall, MSS 9235/1; 593/2 (unfoliated); *St Mary at Hill*, ii, pp. 395, 397; *Transcript of the registers of St Mary Woolnoth and St Mary Woolchurch Haw*, ed. J. M. S. Brooke and A. W. C. Hallen (1886), p. xix.

[185] Guildhall, MS 959, fo. 19ʳ; W. Jenkinson, *London Churches before the Great Fire* (1917), p. 139.

to 'let every parish for to have' its treasures again 'by her coming to the Crown'. On 14 September 1553 the Edwardian commissioners for London were instructed to return the spoils to the churches whence they came.[186] The City parishes hastened to stake their claims. St Mary Woolnoth sent a boat to Westminster for the parishioners to collect their ornaments, and paid Mr Sturton, the royal receiver of goods, for searching 'divers parishes' for two tabernacles. St Benet Gracechurch spent £4. 13s. 4d. attempting to 'get the specialities belonging to the parish out of the Queen's Council's hands'. These parishes were fortunate, but 'Trinity parish had not their cope of cloth of gold again'.[187] In May 1554 the Court of Aldermen mediated between St Mary Abchurch and St Katherine Creechurch in a dispute over a cope.[188] Treasures sold by the churches were, where possible, bought back. In September 1553 the churchwardens of Christchurch asked the Treasurer of St Bartholomew's Hospital to recompense the 'stranger' to whom Grafton had sold the altar stone for £3, and they were looking for their vestments.[189]

Catholic memories were long regarding who had seized which treasures. In St Stephen Walbrook Thomas Maugham, a reformer, sold back in 1556, for the same price he had paid, the banners he had acquired in 1548.[190] In 1555 Thomas Devon gave to St Botolph Aldgate a cloth to hang before the rood; 'nevertheless in times past it did belong to the church'.[191] The most eager iconoclasts had not been forgotten, nor forgiven. In the 1554 visitation the Vicar-General ordered parishioners to return the church property still in their possession.[192] Gregory Newman and John Hilton, Edwardian churchwardens of St Sepulchre, were called before the Court of Aldermen and then before the Vicar-General, to answer for the destruction of the rood loft. And when in March 1555 Newman was found

---

[186] Stow, *Survey*, i, p. 341; *APC* iv, p. 348.

[187] Pendrill, *Old Parish Life*, p. 28; Guildhall, MS 1568/1, fo. 97^r; *Diary of Machyn*, p. 166.

[188] CLRO, Repertory 13, fos. 163^v, 169^v.

[189] St Bartholomew's Hospital, Ha 1/1, fos. 91^r, 92^r.

[190] Guildhall, MS 593/2 (unfoliated); GLRO, DL/C/614, fo. 27^r.

[191] Guildhall, MS 9235/1 (unfoliated).

[192] GLRO, DL/C/614, fos. 5^v, 6^v, 9^r, 34^r, 42^r, 53^v, 55^r, 62^v.

guilty of quite another offence his penance was—fittingly—to give 20s. to edify St Sepulchre.[193]

The story of the Marian restitution was different for each parish, and the stories are now obscure. The churchwardens' accounts record the payments, but rarely the deliberations which preceded them. Yet throughout London, from parish to parish, the restoration of Catholic artefacts revived deeper contention about the nature of worship. Old scores might now be settled. In St Botolph Aldgate the old guard hastened to restore the church and the old ways. 'Certain women' of the parish, in the traditional way, collected money on Hock Monday 1555 and bought a cross cloth for 20s. For the 'good will and zeal he hath borne ever unto the church' William Green, the farmer, restored the sepulchre and gave over £10. But more was needed. In 1555 the commissioners came to oversee the repair of the church 'which was fallen in great ruin and decay'. Through natural justice, or revenge, the desecrators were made to pay for the re-edification. At St Botolph it was those reforming parishioners, who had come before the Vicar-General in the visitation and who were still under suspicion, who were persuaded to return their ill-gotten gains, and to re-endow the church. Anthony Anthony, John Farminger, and John Frank, who had brought reforming preachers to the pulpit, were now made to contribute £20 towards the rood. The rood, which cost £60, stood as a reproach to the reformers and as a Catholic victory.[194]

Yet still in November 1557 Pole castigated the citizens for allowing their parish churches to lie in ruins; though never 'cast down by colour of authority' as the abbeys were, nevertheless they had been allowed to fall down through neglect.[195] For all the energy and money expended upon re-endowment, the churches' treasures were cast on to bonfires at Elizabeth's accession, with no struggle to save them. According to Wriothesley, church goods worth £2,000 went to the flames; doubtless their value was even greater.[196] The first despoli-

---

[193] CLRO, Repertory 13, fo. 179ʳ; GLRO, DL/C/614, fos. 2ᵛ, 8ᵛ; DL/C/331, fos. 188ʳ–189ʳ.

[194] Guildhall, MS 9235/1 (unfoliated).

[195] Strype, *Ecclesiastical Memorials*, iii/2, p. 482.

[196] Wriothesley, *Chronicle*, ii, p. 146.

ations, of Henry's and Edward's reigns, had been hardest to bear; with time the citizens had grown resigned. Some, of course, rejoiced as the idols were pulled down. Yet the iconoclasm of Mary's reign, even in London, was limited. When attacks came they were directed at images which were most conspicuous and most venerated. In Cheapside at the Mercers' chapel the image of Becket stood once more. At dead of night in February 1554 the head of the image was knocked off: 'with great shame it was done'. Every time this image was restored determined iconoclasts 'shamefully mangled it', for it was a special symbol of London's idolatry.[197] But most idolatrous of all was the restoration of the Mass, and it provoked the gravest sacrilege. At Easter 1555 William Flower, seeing the people at Westminster Abbey fall down before that 'most shameful and detestable idol'—the sacrament—and 'moved with extreme zeal for my God', stabbed the ministering priest. He wore a paper proclaiming 'Deum time, Idolum fuge'.[198]

HERESY AND DISORDER

For the Queen all heresy was sedition. And so, often, it was. When priests were attacked or, as in March 1555, blood drawn in church in protest against the visitation,[199] the godly could not be regarded otherwise than rebellious and profane. Yet by thinking and treating the gospellers in the City as by nature her opponents Mary might make them so, in time.

One episode in the spring of 1555—among others similar—reveals why the government feared that heretical and subversive forces made common cause, and that even City officers might be in complicity with them. About Shrovetide John Tooley, citizen and poulterer, had robbed a Spaniard at Westminster, and was condemned to hang. At the gallows on 26 April Tooley avowed his reformed faith: he trusted to be saved only by the merits of Christ's passion, not by masses or trentals, images or saints. He cried out: 'from the tyranny of

---

[197] Mercers' Company, Acts of Court, ii, fo. 271ʳ; *Diary of Machyn*, p. 82; *Two London Chronicles*, p. 42; Wriothesley, *Chronicle*, ii, p. 127; *CSP Ven.* vi/1, p. 28; Foxe, *Acts and Monuments*, vi, p. 705.

[198] Ibid. vii, p. 70. Strype, *Ecclesiastical Memorials*, iii/1, pp. 336–7.

[199] Ibid. iii/1, p. 334.

the Bishop of Rome . . . from false doctrine and heresy . . .
good lord deliver us!' Demanding of the crowd, 'All you that
be true Christian men, say with us, Amen', three hundred
responded, three times in unison. As he read aloud a bill,
'Beware of Antichrist', an apprentice, one of the youthful
reformers of the City, richly dressed in violet coat and white
ruff, handed around copies. In the crowd were Haukes, a future
martyr, and Robert Bromley, an old friend of Tooley's and an
officer of Sheriff Chester. Bromley thought Tooley 'died well',
and even in Chester's own kitchen Bromley, Chester's son, his
butler and steward read together 'Beware of Antichrist'. The
authorities suspected that this gallows scene was stage-
managed. To Newgate prison came 'so strait a commandment
that no man might come to us'; because, so Richard Smith
heard, 'Tooley cursed the Pope at the gallows. They thought it
to be our counsel.' On 7 May Tooley was posthumously condemned as a 'lewd heretic', and his body was exhumed and
burnt.[200] Thus, for Mary's enemies, Tooley died a hero: not
only anti-Spanish but also a Protestant martyr.

Pamphlets circulated in the City 'in favour of the Lady
Elizabeth', and attacking the Spaniards, 'justly hated like dogs
all the world over'.[201] Such tracts were dangerous; their
authors 'traitors and heretics'. One of such 'knaves is able to
undo a whole city', warned Southwell.[202] Though most of the
tracts were printed abroad, they were usually sponsored from
London. A London sailor, Broder, ordered the printing in
Danzig of a bill urging Elizabeth's cause—*Certaine Questions
demaunded and asked by the noble Realme of Englande, of her true
naturall chyldren and subjectes*—and Bartlet Green, a young lawyer
of the Temple, was probably behind its circulation. Green had
written to the exiled Christopher Goodman, telling him of 'certain questions which were cast abroad in London', and was
arrested when his reply to Goodman's question, 'Was the

---

[200] Foxe, *Acts and Monuments*, vii, pp. 90–7, 368; *Two London Chronicles*, pp. 42–3;
Wriothesley, *Chronicle*, ii, p. 128.

[201] The best discussions of the seditious literature of Mary's reign are found in
Took, 'Government and the Printing Trade', pp. 271–93; D. M. Loades, 'The
Theory and Practice of Censorship in Sixteenth-Century England', *TRHS* 5th series,
xxiv (1974), pp. 141–57; J. Loach, 'Pamphlets and Politics, 1553–8', *BIHR* 48 (1975),
pp. 31–44.

[202] *Narratives of the Reformation*, p. 188.

Queen dead?'—that she was not, yet—was intercepted.[203]
Thomas Mountain had a copy of a libel of the King and Queen
('a certain description was made of his person, Queen Mary
being joined in the same'), written by Thomas Stoning, which
had been smuggled to the Marshalsea by the curate of St
Bride.[204] The pamphlets and broadsheets had common
themes, more political than overtly religious: that the entry of
the Spaniards augured England's ruin, that Mary had usurped
the Crown, and that Elizabeth would be England's saviour.[205]
In May 1555 a thousand copies of *The Dialogue*, a pamphlet
supporting Elizabeth's claims, were confiscated. A further
proclamation against seditious literature followed.[206] Most
subversive of all was *A Warnyng for England, conteynyng the horrible
practises of the kyng of Spayn in the Kyngedome of Naples*. With the
refrain, 'Beware of had I wist', it dilated upon the depredations
of the Spaniards in their other vassal states, and also of 'certain
close practices for the recovery of abbey lands'. It warned that
'the faggots be already prepared' for those who had disobeyed
the 'Pope's laws . . . nothing wanteth but a day to kindle the
fire'. 'Fond beggarly afterwit is my lord Forewit's fool.' Every
City company was promptly charged on 1 December to find
who had seen or heard of the book; 'whom they know to have
come late from beyond the sea'. The Drapers admitted that
some of the books had been discovered among them, and on 27
December the whole company was ordered to 'beware of pernicious books let fall abroad'.[207]

The attacks upon Philip, the opposition to his coronation,
and the support for Elizabeth were the more dangerous, and
the more bitter, for Mary as she hoped against hope for a child.
In the birth of an heir rested Catholic hopes for the future.
News of the Queen's quickening at the end of November 1554

---

[203] *Foreign Calendar, Mary*, p. 105; Foxe, *Acts and Monuments*, vii, pp. 734–5; Took, 'Government and the Printing Trade', pp. 279–81.

[204] *Narratives of the Reformation*, pp. 187–8.

[205] See, for example, *CSP Sp.* xiii, pp. 102, 135.

[206] *CSP Ven.* vi/1, pp. 70, 475; CLRO, Repertory 13, fo. 288ʳ; Foxe, *Acts and Monuments*, vii, pp. 127–8.

[207] *RSTC* 10023.7. *CSP Ven.* vi/1, pp. 269–70, 272; Drapers' Company, MB 6, fos. 82ʳ, 84ʳ; Guildhall, MS 7090/1, fo. 42ʳ. Reference is made to this tract in John Bradford, *The copye of a letter* (*RSTC* 3504.5), sig. Fiʳ, and *The lamentacion of England* (*RSTC* 10014), sig. Aiiʳ.

had been celebrated in London.[208] But at the end of March 1555 Alice Perwiche, the wife of a merchant taylor in St Botolph Aldersgate, voiced a dangerous suspicion: that 'the Queen's grace is not with child', and that she would desperately substitute another woman's baby at her confinement and call it her own. Some about the Queen, she said, plotted against Mary's life: the deed would be done by 'a certain lady' to be 'nameless at this time'. Anne Shackleton too spoke 'heinous and seditious words'.[209] As the royal birth grew closer more apocalyptic rumours spread. At the sign of the Hedgehog in Paul's churchyard a woman bore a son who lived only a week. The dying child began to talk, urging the people to 'Rise and pray', 'the Kingdom of God is at hand'.[210] News came of the birth of a prince, on 30 April, and again early in June; bells rang; the *Te Deum* was sung; the priest at St Anne extolled the prince's handsomeness. But the news was false. Royal advisers, so it was said, now stooped to that duplicity of which Perwiche had warned: to Isabel Malt, lying in childbed in Horn Alley, Aldersgate, came Lord North, asking her to part with her child.[211] Philip would now look for a new wife, wrote the author of *A Warnyng for England*.[212] The rumours continued that Edward still lived, and would come again. A pretender rode around Kent claiming to be a 'messenger from Edward'. He was brought to the capital to be publicly shamed 'riding in a cart in a fool's coat', was whipped and banished north, and finally executed.[213] If not Edward, then Elizabeth would reign. Mary's unpopularity in the City grew. In September 1555 Richard Daye sought a new lease from St Bartholomew's Hospital, but said revealingly, 'he is the Queen's servant, but he is no more beloved than the Queen's highness is in the City of London'.[214]

---

[208] *Grey Friars Chronicle*, p. 93; *Diary of Machyn*, pp. 76–7; Foxe, *Acts and Monuments*, vi, pp. 567–8; Wriothesley, *Chronicle*, ii, p. 124.

[209] *CPR P&M* iii, pp. 184–5; CLRO, Repertory 13, fo. 276ᵛ.

[210] BL, Harleian MS 353, fo. 145ʳ; *Diary of Machyn*, p. 88; Wriothesley, *Chronicle*, ii, pp. 128–9.

[211] Guildhall, MS 1586/1, fo. 76ʳ; *CSP Sp.* xiii, p. 16; Foxe, *Acts and Monuments*, vii, pp. 125–6.

[212] *A Warnyng for England*, sig. Aviiʳ.

[213] CLRO, Repertory 13, fos. 162ʳ, 164ʳ; *APC* v, pp. 122, 126, 221, 228; Wriothesley, *Chronicle*, ii, p. 129; *Diary of Machyn*, p. 88; *CSP Ven.* vi/1, pp. 324, 339.

[214] St Bartholomew's Hospital, Ha 1/1, fo. 127ᵛ. See also Guildhall, 12071/1, fo. 532ʳ.

It was not only Protestants who hated the Spaniards, so it was politic for Protestant opponents of Mary and her marriage to hide their religious motives in order not to lose Catholic support. The author of the vituperative anti-Spanish *The copye of a letter sent by J. Bradford* claimed to be moved by love of the Queen and her faith. Yet the association between hatred for the Spaniards and hatred for their faith was a natural one. So it was on Corpus Christi that a mob of Londoners five hundred strong gathered to attack the Spanish visitors in 1555, attempting another Sicilian Vespers.[215] There were rumours that Londoners planned to rise against the Spaniards at midsummer, and watches were ordered.[216] In mid-July 1555 a suspicious gathering of gentlemen, adherents of Elizabeth, was rounded up in London. One of them, Edward Randall, had informed Noailles of a conspiracy for 'recovering their liberties' by the end of August.[217] That winter, as Parliament met, a group of malcontents plotted in Arundel's tavern. The conspirators, led by Dudley, aimed to overthrow the government by domestic insurrection and by invasion from France.[218] (In aim, as in personnel, the plot had much in common with Wyatt's conspiracy.) Later it was claimed that the conspirators 'did very sore mislike such Catholic proceedings as . . . the Queen . . . went about', and 'did declare themselves to be right Protestants'.[219] While some of the malcontents, like Peckham, were moved by personal grudge rather than by religious zeal, there is no doubting the reforming inclinations of such as Cawarden, Kingston, Throckmorton, and Arnold.[220] At the execution of Peckham and Daniel in London there were public demonstrations of sorrow, and it was said that the Queen's heart hardened against her opponents.[221]

In London there were perennially subversive elements, easily rallied to a cause. The fear was that the Queen's enemies knew

---

[215] *CSP Ven.* vi/1, pp. 85, 126; D. M. Loades, *Two Tudor Conspiracies* (Cambridge, 1965), p. 143.

[216] *CSP Ven.* vi/1, p. 133; CLRO, Journal 16, fo. 338ᵛ.

[217] Loades, *Reign of Mary Tudor*, p. 228.

[218] Loades, *Two Tudor Conspiracies*, chs. 8, 9.

[219] PRO, SP 2/8/35; cited in Loades, *Two Tudor Conspiracies*, p. 185.

[220] See Loach, *Parliament and the Crown*, ch. 10; *House of Commons, 1509–1558* (Cawarden, Kingston, Throckmorton, Arnold); W. B. Robison, 'The National and Local Significance of Wyatt's Rebellion in Surrey', *HJ* xxx (1987), pp. 769–90.

[221] *CSP Ven.* vi/1, p. 447.

how to appeal to their susceptibilities. So it was that John Bradford's warning of the consequences of Spanish invasion had a special resonance for the young: implacable Spaniards would massacre all those aged under twenty, he alleged, 'for we were born out of the faith, and so, say they, we shall die'.[222] Catholic preachers and writers who sought to explain the pernicious spread of heresy invariably blamed the 'younger and carnal sort' who, in search of 'parasite liberty', gave their allegiance to the new faith. In the doctrines of Protestantism, it was said, the rebellious young found the perfect excuse and legitimation for their natural tendency to disobedience, to follow baser instincts.[223] Everyone in a position of authority was now scorned. Children, claimed Bishop Christopherson, made a 'merry mockery' of their parents, and said: 'My father is an old doting fool and will fast upon the Friday; my mother goeth always mumbling on her beads.'[224] The clergy could no longer command respect, and 'disobedience to magistrates and aged men was at no time more practised'. Such insubordination led to anarchy, it was said: servants and apprentices became 'not only odious to the world but also unthrifty to their masters, and in manner became masters themselves'.[225] All this was, of course, propaganda: not even novel propaganda; but the behaviour of the vocal young people of London in Mary's reign made it increasingly credible. These were youth, as Pole lamented, 'brought up in a contrary trade'.[226]

Mocking priests had become a favourite pastime of London boys. Attacks upon priests became commoner as the Reformation progressed. 'Ye shall see that every boy in the street shall spit in the priests' faces and hurl stones at them', so it had been foretold.[227] But it was leading Catholic clerics who were the targets of irreverent young radicals, and the old faith which was pilloried. In May 1554 a draper's apprentice was whipped

---

[222] Strype, *Ecclesiastical Memorials*, iii/2, p. 347.

[223] See S. E. Brigden, 'Youth and the English Reformation', *P&P* 95 (1982), pp. 37–67.

[224] John Christopherson, *An exhortation to all menne to take hede and beware of rebellion* (*RSTC* 5207), sig. Tii[v]; cited in K. V. Thomas, 'Age and Authority in Early Modern England', *Proceedings of the British Academy*, lxii (1976), p. 45.

[225] Hogarde, *Displaying of the protestantes*, sigs. Lv[v], Lvii[r].

[226] Strype, *Ecclesiastical Memorials*, iii/2, pp. 497 ff.

[227] PRO, SP 1/220, fo. 68[r] (*L&P* xxi/1. 1027); *TRP* i. 292.

for having shorn a little boy's head like a priest's.[228] When Rowland Taylor was forced to don popish vestments, he postured and jested: 'If I were in Cheap, should I not have boys enough to laugh at these apish toys and toying trumpery?' That Bonner was persistently insulted by juvenile opponents is suggested by Foxe's story of the Bishop's furious and supposedly unprovoked assault upon a group of boys bathing in the Thames.[229] The restored monks provided new targets for derision, or worse. In October 1555 a boy offered a grey friar a baby's bottom to kiss.[230] For sport or through outraged conviction it was the young who were the most reckless and eager iconoclasts. At St Faith's on Ascension day 1559 a 'rascally lad servant' of a printer snatched the cross at the head of the procession and tore away the figure of Christ, and waving it at the women said that it was 'the Devil's guts'. The processions of the young were likely to be mocking masquerades of monks.[231]

Young Protestants went along to the sermons of conservative preachers to heckle and jeer. When Dr Bourne preached in the summer of 1553 there was 'great uproar . . . what young people and women [as] ever was heard, as hurly burly and casting up of caps'.[232] A 'stripling' was whipped through the streets in November 1555 for deriding a Bishop who preached.[233] But while the City youth taunted the Catholic leaders, the evangelists were their heroes. Crome, Latimer, and Cardmaker had gathered crowds of devotees: perhaps the same crowds from whom Weston, Bourne, Gardiner, and Watson had to be protected. In prison, John Philpot was sustained, as Anne Askew had been, by apprentices. Bonner heard that prisoners in his 'coal house' with a 'meany' (retinue) of apprentices had climbed upon the roof and stood 'gazing abroad as though they had been at liberty'. And he threatened Philpot: 'I will cut off your resort; and as for the prentices, they were as good not to come

---

[228] CLRO, Repertory 13, fo. 157^r; Drapers' Company, MB 6, fo. 16^r.

[229] Foxe, *Acts and Monuments*, vi, p. 691; viii, p. 526.

[230] CLRO, Repertory 13, fo. 335^r; *APC* v, p. 169; Hogarde, *Displaying of the protestantes*, sigs. Miiii^v–v^r.

[231] *CSP Ven.* vii. 71; Pendrill, *Old Parish Life*, p. 70; King, *English Reformation Literature*, pp. 179–80.

[232] *Diary of Machyn*, p. 41.

[233] Ibid., pp. 97–8; Grafton, *Chronicle*, ii, p. 536; *APC* iv, p. 317.

to you, if I take them.'[234] As the bands of martyrs came to suffer, their youthful supporters—often their contemporaries—gathered at the stake to comfort them, to the alarm of the persecutors.

Any gathering might seem subversive to a government constantly vigilant for heresy and sedition. Under persecution the godly went underground, seeking clandestine ways and secret places to continue their worship. At the Reformation plays and interludes had been turned to evangelical purposes, for the Protestant faith was not yet at war with music and drama: on the contrary. Now plays and players were regarded with particular suspicion, especially in the capital, and regularly banned.[235] The discovery of 'naughty plays' in June 1557 brought complete censorship. But in September the play *A Sack full of News* was being performed at the Boar's Head, Aldgate: a play either satirical or heretical, probably both.[236] Minstrels travelled around singing 'Scripture songs'. At an Essex wedding the ballad *News out of London*, concerning the Mass and the Queen's 'mis-proceedings' was sung. Not only were the themes of the plays subversive, but the audiences might have ulterior motives for attending. At the Saracen's Head in December 1557, under cover of seeing a play, the Islington congregation met to celebrate the Protestant communion.[237]

As soon as the Mass was restored the godly began to profess the gospel in secret conventicles, always watched and always in danger. Yet Edward Underhill remembered how 'some were preserved still in London, that in all the time of persecution never bowed their knees unto Baal, for there was no such place to shift in in this realm as London, notwithstanding there great spyall and search'.[238] Foxe records the providential preservation of the London congregation: the godly met secretly in

---

[234] *The Examinations and Writings of John Philpot*, ed. R. Eden (Parker Society, Cambridge, 1842), p. 70; Foxe, *Acts and Monuments*, vii, p. 639.

[235] King, *English Reformation Literature*, ch. 6; CLRO, Journal 16, fos. 287ᵛ, 328ʳ; Repertory 13, fo. 484ᵛ. I have profited greatly from reading Professor Collinson's third Birkbeck lecture: 'The English Reformation and the English People: the Rise and Fall of Protestantism as a Popular Cause' (1982).

[236] *APC* vi, pp. 102, 119–20, 168–9.

[237] Foxe, *Acts and Monuments*, viii, pp. 435, 578, 444–5.

[238] *Narratives of the Reformation*, p. 149.

Cawarden's mansion at Blackfriars, at Aldgate, in a cloth-workers' loft by the great conduit, aboard a ship at Billingsgate, and aboard another, named the Jesus, which was moored between Ratcliffe and Rotherhithe, in a cooper's house in Pudding Lane, in Thames Street. Abhorring the Catholic services, they would go into the fields to sing English psalms and read the Bible.[239] From late in 1553 until their exile the Hickmans received into their house 'well disposed Christians'. 'And we . . . did table [take communion?] together in a cham-ber, keeping the doors close shut for fear of the promoters, as we read in the Gospel the disciples of Christ did for the Jews.'[240] The letters and depositions and testimonies recorded by Foxe reveal something of the world of the conventicles. The under-ground congregations remained loyal to the Edwardian com-munion and service book. Ministers remained to celebrate the services for the godly in private houses, and as long as they did so the London congregations might remain theologically and liturgically grounded in the traditions of Edwardian Protestant-ism.[241] At Easter 1554 and 1555 Bartlet Green received the communion—'sacramental bread and sacramental wine'—in the chamber of the minister John Pulline at Cornhill, Pulline alone reading 'the words of the institution'. Christopher Goodman and Michael Reniger were communicants there also in 1554; a year later they were exiled.[242] Hating the other Catholic sacraments also, the godly adhered to the reformed use. At Pentecost 1557 Gertrude Crockhay was godmother at the baptism of Thomas Saunders's child in the house of a mid-wife in Mincing Lane. Daniel performed the baptism in English 'after the form used in the time of the schism'.[243] The following year, as she lay dying, Gertrude refused to receive the last rites of the Church; she would be sick, and would Dr Mallet 'have

---

[239] Foxe, *Acts and Monuments*, viii, pp. 558–9, 470, 474; *Zurich Letters*, ed. H. Robinson (Parker Society, Cambridge, 1845), pp. 29–30; B. R. White, *The English Separatist Tradition* (Oxford, 1971), pp. 10 ff.

[240] 'Recollections', p. 99.

[241] See Foxe, *Acts and Monuments*, viii, pp. 142, 445, 458; *Diary of Machyn*, p. 79.

[242] Foxe, *Acts and Monuments*, vii, p. 738; Garrett, *Marian Exiles*, 171, 334, 344; GLRO, DL/C/331, fos. 158^r, 159^r, 187^r.

[243] Guildhall, MS 9168/11, fo. 165^r; Foxe, *Acts and Monuments*, viii, p. 579. Roger Holland's child too was baptized at home: ibid. viii, p. 474.

her to cast up their god again?' She scorned the threat that she would be denied Christian burial: 'how happy am I, that I shall not rise with them, but against them'.[244]

A window upon the 'secret society . . . of God's children' was opened when the congregation at the Saracen's Head in Islington was betrayed. Led by their minister, John Rough, and their deacon, Cuthbert Simpson, the London congregation moved from one 'safe house' to another. They met at the house of Frogg, a Dutch shoemaker, in the liberty of St Katherine's, at the King's Head at Ratcliffe, at the Swan at Limehouse. A group of about twenty—Dutchmen, Frenchmen, and young English merchants—came twice to Alice Warner's house at the King's Head in Stepney, demanding 'good cheer', and retired to a back room. There she saw them reading: a Bible and another book, she could not tell which. They called each other 'brother', she said. If a minister were present there was an English service—'wholly as it was in the time of King Edward . . . neither praying for the King nor the Queen'—and preaching. One Sunday after evensong Coste read three English psalms. The deacons collected money for the prisoners for the faith, and for their own poor members. (Their betrayer thought that some came 'more for money than for ought else'.) There was talk of the iniquities of the Mass and of the Pope, and of the corruption of attending popish services.[245] When Gertrude Crockhay had succumbed and performed the penance imposed by the Commissary she was overcome by remorse, fearing that she would be damned. Rough comforted her with sentences from Scripture and advised her to confess her fault to the congregation that she be 'received into their fellowship again'.[246] Like Christians of the apostles' time they saluted one another 'with the holy kiss'.[247]

The most subversive cells of gospellers in London were found in the City prisons. At least two hundred were incarcerated for their faith and there they had nothing left to lose. Philpot signed his letters, 'dead to the world and living to Christ, your own brother, sealed up in the verity of Christ for ever'.[248] In prison

[244] Foxe, *Acts and Monuments*, viii, pp. 727–8.
[245] Ibid. viii, pp. 458–60.
[246] Guildhall, MS 9168/11, fo. 165ʳ; Foxe, *Acts and Monuments*, viii, p. 728.
[247] Ibid. viii, p. 449.
[248] Ibid. vii, p. 701.

they had, at last, freedom of worship. In the Nun's Bower and
Little Ease the Edwardian services were daily celebrated, and
the reformers preached freely.[249] The prisoners spent their time
in prayer, singing psalms, writing exhortatory letters, or
holding services. Ending a letter abruptly, Bradford explained
that 'our common prayer time calleth me'.[250] In Newgate,
Richard Gibson, a prisoner not for heresy but for debt, evangelized energetically 'to the pernicious and evil example of the
inhabitants of the City of London and the prisoners'.[251] The
authorities sent in their own agents to prevent the conversions.
The visitor of Newgate appointed in Edward's last days to read
Scripture was swiftly replaced by a priest to celebrate the
Catholic sacraments.[252] It was perhaps partly to relieve the
'infection' of heresy, as well as overcrowding, that in March
1557 the keeper of Newgate was financed to ransom prisoners
in debt for fees alone.[253]

City gospellers daringly visited the prisoners. Thomas Green,
apprentice to Wayland the printer, went to Newgate to 'eat
and drink together with our friends, in the fear of God'. He
met a Frenchman there who sold him a book 'as spoke both
treason and heresy': *Antichrist*. Edward Benet of St Bride was
discovered smuggling Coverdale's New Testament to Tingle in
Newgate.[254] Bartlet Green's lawyer friends came to visit him
in prison, and for a remembrance brought him books to auto-
graph. In Thomas Hussey's book he inscribed:

> Behold thyself by me, such one was I, as thou:
> And thou in time shall be, even dust as I am now.[255]

Robert Smith sent a letter to his wife, with news, and money
bequeathed for her by the martyrs. If she wished to reply she
should deliver her letters to Mrs Tankerville in Chancery Lane
—'ye shall always hear of me at Tankerville's house'.[256] Sympathetic gaolkeepers, traducing their trust, allowed the prisoners

---

[249] Ibid. vii, p. 145.
[250] Ibid. vii, p. 234.
[251] Ibid. viii, p. 436.
[252] St Bartholomew's Hospital, Ha 1/1, fos. 76ʳ, 97ᵛ, 102ʳ, 105ʳ.
[253] CLRO, Repertory 13, fo. 491ʳ.
[254] Foxe, *Acts and Monuments*, viii, pp. 521–2, 561–2.
[255] Ibid. vii, p. 742; *House of Commons, 1509–1558* (Hussey).
[256] Foxe, *Acts and Monuments*, vii, p. 368.

access to their well-wishers. In close prison, 'as a sparrow alone in a house top', Bradford had no news: he asked Augustine Bernher, former servant to Latimer, to 'cause Mrs Pierrepoint to learn of the Sheriff, master Chester, what they purpose to do with me'.[257] The network of sustainers in London was very extensive—over sixty are known by name—and included leaders of the City establishment. William Chester, Sheriff for 1554–5, had been Laurence Saunders's master, and seeing his zeal for the Gospel, had freed him from his apprenticeship to study Hebrew and divinity. Now as he led the martyrs to the stake he 'wept apace'.[258] Other City officers who were favourers of the Word did more to aid the persecuted.[259]

To die for the faith was a special benediction. 'Many are called, but few are chosen.' The London congregation rallied to aid those who took up the cross, some wishing that they had been summoned to join them. 'Me thinketh I see you desiring to be under the same [cross]', wrote Philpot to Robert Harrington.[260] The martyrs wrote letter upon letter of counsel to those whom they left behind, exhorting them never to succumb. 'Abide the trial, dear friends', wrote Thomas Whittle 'to all the true professors' in London. The sight of the martyrs' suffering 'in this sinful Sodom, this bloody Jerusalem, this unhappy City of London' should strengthen them in their faith.[261] The godly, in turn, sent presents to the martyrs and tokens. Mrs Marler made Bradford a new shirt for his burning; Lady Vane sent Philpot a scarf 'that ye may present my handy work before your captain [Christ]'.[262] The Londoners now in exile never forgot the prisoners, and the generosity of the Warcups, of Mrs Brown and Wilkinson, was constantly praised.[263] In Oxford's Bocardo Ridley, Latimer, and Cranmer received 'meat, money and shirts' from London.[264] All this support was in open defiance of

[257] Foxe, *Acts and Monuments*, vii, p. 262.

[258] Ibid. vi, pp. 612–13; vii, pp. 194–5, 262; Drapers' Company, apprenticeship records.

[259] See, for example, *Narratives of the Reformation*, pp. 196, 210; Foxe, *Acts and Monuments*, vii, pp. 685, 742; viii, pp. 409, 523–4, 594–5; CLRO, Repertory 13, fo. 512[r].

[260] Foxe, *Acts and Monuments*, vii, p. 699.

[261] Ibid. vii, pp. 726–7.   [262] Ibid. vii, pp. 147, 726–7.

[263] Ibid. vi, p. 673; vii, pp. 212, 251, 255; viii, pp. 100, 186; PRO, PCC, Prob. 11/42B, fos. 351[v], 233[r]; 11/53, fo. 320[r].

[264] Foxe, *Acts and Monuments*, vii. p. 427.

a government which wished the martyrs to be universally condemned or forgotten.

As the martyrs came to suffer their supporters gathered to celebrate their courage and their election. When a band of martyrs was brought from Colchester to London a crowd met them at Stratford and it increased until it was a thousand strong in Cheapside. Bonner feared the support the heretics gained 'ex promiscua plebe'; not least because it was often the youthful friends of the martyrs who rallied.[265] Orders came to the Court of Aldermen from the Council on 14 January 1556 for householders to keep their apprentices and servants at home at the time of the burnings. On 22 January a new order was issued for a curfew for 'young folk', for that morning at Smithfield 'the greatest [number] was as has been seen at such a time'. The following Monday, 27 January, the beadle of every ward was to stand armed to prevent any 'servant or child' from venturing forth before 11 o'clock, for Bartlet Green and his fellows were to be burnt.[266] In September 1557 a prisoner was rescued from the Alderman's deputy in Blackfriars.[267] Thomas Bentham, the minister of the London congregation, wrote to the exiled Lever of the 'fervent zeal of so many, and such increase of our congregation in the midst of this cruel and violent persecution . . . *a domino factum est*'. He told of how when seven of his congregation went to Smithfield on 27 June 1558 they were 'comfortably taken by the hands and so godly comforted', despite the 'fearful proclamation' that none should approach the martyrs, and despite the Sheriff's warnings.[268] No one doubted the horrors of death by burning. The martyrs signed their letters as a 'burnt offering' or a 'sacrifice' to the Lord.[269] Meeting Rough in September 1557, Farrer asked him whence he came. 'I have been where I would not for one of my eyes but I had been'; to Austoo's burning 'to learn the way'.[270]

[265] Ibid. viii, p. 307.
[266] *APC* v, p. 224; Strype, *Ecclesiastical Memorials*, iii/2, p. 501; Foxe, *Acts and Monuments*, vi, pp. 728, 740; *Diary of Machyn*, pp. 99–100; CLRO, Journal 16, fo. 367ʳ.
[267] CLRO, Repertory 13, fo. 539ᵛ.
[268] BL, Harleian MS 416, fo. 63ʳ; Strype, *Ecclesiastical Memorials*, iii/2, pp. 133–4; Foxe, *Acts and Monuments*, viii, pp. 479, 559.
[269] Ibid. vii, p. 262.
[270] Ibid. viii, p. 448.

MARTYRDOM

'Glorious is the course of the martyrs of Christ at this day', wrote
Philpot. 'Never had the elect of God a better time for their glory
than this is.'[271] And to London God had given a special providence. For as Cardinal Pole told the Londoners in November 1557,
meaning to shame them, 'a greater number of these brambles
and briars [heretics] were cast into the fire here among you than
in any place beside'.[272] Within less than four years seventy-five
were burnt in the fires of Smithfield, Islington, Stratford le Bow,
Westminster, Uxbridge, and Brentford, with London crowds
to witness.[273] Deaths for the faith less glorious were suffered in
the 'ugsome holes' of the City prisons by men 'hunger pined',
hanging on chains, or caught in stocks given placatory names,
'the widow's alms', the 'Lord Chancellor's alms'.[274] Not all
the seventy-five who suffered in and around the City came from
London; nor were all the Londoners who suffered burnt in
London. How are London's martyrs to be counted? Numbers
involve judgements. Ridley, born in Northumberland, is
usually counted as an Oxford martyr, but he was Bishop of
London after all. Laurence Saunders and George Marsh were
sent home to Coventry and Chester to suffer, but their cures of
souls had been in the capital, their evangelical preaching there,
and Saunders had been apprenticed in the Drapers' Company.
John Leaf, William Hunter, and Roger Holland were apprentices in London, although not born there. For some of the
martyrs little is known, beyond their membership of the
London congregation. This is here taken to mean that they
were resident in London. Upon such assumptions rests the list
in Table 6 of those from London who died for the faith.

Why did the martyrs die? Because the Queen, her Bishops,
her Council, her commissioners, and a substantial number of
the lay governors and the people believed that wilful disobedience to the Catholic Church—heresy—was the worst of sins
and to be extirpated, lest it 'infect' more. Pole believed, as he
told the citizens, 'there cannot be a greater work of cruelty

---

271 Foxe, *Acts and Monuments*, vii, p. 699.
272 Strype, *Ecclesiastical Memorials*, iii/2, p. 490.
273 Ibid. iii/2. lxxxv.
274 Ibid. iii/1, pp. 400–1; iii/2, p. 428; Foxe, *Acts and Monuments*, viii, p. 204.

against the commonwealth than to nourish or favour any such [heretics]. For be you assured, there is no kind of treason to be compared with theirs.'[275] Some persecutors claimed that they acted from love, and that love might return schismatics to the fold of the universal Church. Bonner wrote only one of the *Homilies* of 1555 himself: that upon Christian love. Charity, the first of Christian virtues, he wrote, must be qualified: 'to love our enemies is the proper condition only of them that be children of God, the followers of Christ'. Charity commands that all governors should correct offenders within their jurisdictions.[276] Bonner was to entreat the heretics 'charitably withal' to amend. But where persuasion failed there was no way but one: as Bonner told John Mills, 'as truly as thou seest the bodies of them in Smithfield burnt, so truly their souls do burn in Hell, because they err from the true Church'.[277] Story wrote in 1555 of how by 'discreet severity' they hoped to regain uniformity in religion,[278] but with time any discretion would be undermined by the tenacity of the martyrs. The martyrs died because there were those either so aware of their duty or so hating heresy that they reported or presented them: but the main reason for their deaths was their absolute refusal to recant.

The persecutors did not want the gospellers to die but to be reconciled. Every martyrdom was, for them, a defeat. Alphonsus, a Spanish grey friar and Philip's confessor, preached against the persecuting Bishops: 'they learned it not in Scripture to put any death for conscience, but rather to let them live and be converted'.[279] Bonner and his officers tried endlessly to persuade the errant to reconcile: this was why Philpot was examined fifteen times, Gibson three times, Richard Woodman six times, Elizabeth Young nine times, and why Cuthbert Simpson was tortured.[280] To Roger Holland's examinations were brought old family friends from Lancashire to persuade

---

[275] Strype, *Ecclesiastical Memorials*, iii/2, p. 487. For the Marian persecution in general, see P. Hughes, *The Reformation in England*, 3 vols. (1951–4), ii, part 3.

[276] Alexander, 'Bonner and the Marian Persecutions', pp. 376–7.

[277] Foxe, *Acts and Monuments*, vii, p. 86; viii, p. 485.

[278] *CSP Ven.* vi/1, pp. 110–11.

[279] Strype, *Ecclesiastical Memorials*, iii/1, pp. 332–3.

[280] Foxe, *Acts and Monuments*, vii, pp. 605–80; viii, pp. 436–40, 338–74, 536–48, 455.

TABLE 6: *London's Martyrs*

Name	Status or Occupation	Parish or Ward	Age at death	Heresy	Place and date of execution
John Rogers	prebend of St Pancras; rector of Holy Trinity the Less, 1532–4; vicar of St Sepulchre; rector of St Margaret Moses		c.55	denial of the Mass	Smithfield, 4 Feb. 1555
Laurence Saunders	rector of All Hallows Bread Street; once apprentice to Sir William Chester, draper			denial of the Mass	Coventry, 8 Feb. 1555
Thomas Tomkins	weaver	Shoreditch		denial of the Mass	Smithfield, 16 Mar. 1555
William Hunter	apprentice to a silk weaver	St Stephen Coleman Street	19	denial of the Mass	Brentwood, 26 Mar. 1555
William Flower	monk of Ely; thereafter priest, schoolteacher, surgeon	Lambeth	c.50	denial of the Mass; antinomianism	Westminster, 24 Apr. 1555
John Tooley	citizen and poulterer			(?)	body exhumed and burnt, 7 May 1555
John Cardmaker	Observant friar; prebendary and chancellor of Wells; vicar of St Bride Fleet Street			denial of the Mass	Smithfield, 30 May 1555
John Warne	upholsterer; member of the Clothworkers' Company	St Olave Silver Street	29	denial of the Mass; refusal to confess	Smithfield, 30 May 1555

Name	Description	Parish	Age	Charge	Place and date
John Bradford	deputy paymaster at Boulogne, 1544; lawyer of Inner Temple; prebend of Kentish Town		c.45	denial of the Mass	Smithfield, 1 July 1555
John Leaf	apprentice to a tallow chandler	Christchurch Newgate	19	denial of the Mass	Smithfield, 1 July 1555
George Marsh	schoolmaster; curate at St Pancras		40	denial of the Mass	Chester, 24 Apr. 1555
Robert Smith	painter; once retainer to Sir Thomas Smith			denial of the Mass	Uxbridge, 8 Aug. 1555
Elizabeth Warne	widow	St Olave Silver Street		denial of the Mass; despising ceremonies	Stratford le Bow, 23 Aug. 1555
George Tankerfield	cook	St Dunstan in the West	27 or 28	denial of the Mass	St Alban's, 26 Aug. 1555
Stephen Harwood	brewer			denial of the Mass	Stratford le Bow, Aug. 1555
Thomas Fust	hosier	St Bride Fleet Street		denial of the Mass	Ware, 1555
Nicholas Ridley	quondam Bishop of London		c.55	denial of the Mass	Oxford, 16 Oct. 1555
Bartlet Green	lawyer of the Inner Temple	born in Basinghall	25	denial of the Mass	Smithfield, 27 Jan. 1556
John Tudson	apprentice	St Mary Botolph	27	denial of the Mass	Smithfield, 27 Jan. 1556
John Went	artificer		27	denial of the Mass	Smithfield, 27 Jan. 1556
Thomas Browne		St Bride Fleet Street		denial of the Mass	Smithfield, 27 Jan. 1556

Table 6, cont.

Name	Status or Occupation	Parish or Ward	Age at death	Heresy	Place and date of execution
Isabel Foster		St Bride Fleet Street		denial of the Mass	Smithfield, 27 Jan. 1556
Joan Warne, alias Lashford		born in All Hallows the Less		denial of the Mass	Smithfield, 27 Jan. 1556
Hugh Laverack	painter	Barking; dwelling in Seething Lane		denial of the Mass	Stratford, 15 May 1556
John Apprice	clothworker	St Thomas Apostle			Stratford, 15 May 1556
Lyon Cawch (Lion à Coise)	merchant from Flanders		28		Stratford, 27 June 1556
	barber	Lime Street			Smithfield, 6 Apr. 1557
	goodwife at sign of the Crane; inn keeper	Crossed Friars by the Tower			Smithfield, 6 Apr. 1557
man + wife		St Dunstan in the East			Islington, 17 Sept. 1557
James Austoo		All Hallows Barking		denial of the Mass	Islington, 17 Sept. 1557
William Sparrow				selling blasphemous ballads; calling the Mass abominable	Smithfield, 30 Nov. 1557

Name	Description	Location	Age	Charge	Place and date
John Hallingdale				refusing the Mass; christening his child according to the Edwardian service	Smithfield, 30 Nov. 1557
Richard Gibson	gentleman; son of the royal sergeant-at-arms, swordbearer, and bailiff of Southwark; grandson of Sir William Bayly, Lord Mayor for 1524-5			denial of the Mass	Smithfield, 30 Nov. 1557

The Islington Congregation

Name	Description	Location	Age	Charge	Place and date
Margaret Mearing		Mark Lane, Aldgate			Smithfield, 22 Dec. 1557
John Rough	quondam black friar of Stirling; appointed preacher on the borders by Somerset; beneficed priest at Hull under Edward VI; minister of the London congregation		c.30		Smithfield, 22 Dec. 1557
Cuthbert Simpson	a wealthy tailor; deacon of the London congregation			denial of the Mass	Smithfield, 21 or 28 Mar. 1558
Hugh Foxe	hosier	Wood Street		denial of the Mass	Smithfield, 21 or 28 Mar. 1558
John Devenish				denial of the Mass	Smithfield, 21 or 28 Mar. 1558
Henry Pond				denial of the Mass	Smithfield, 27 June 1558

Table 6, cont.

Name	Status or Occupation	Parish or Ward	Age at death	Heresy	Place and date of execution
Reinald Eastland	an alien?			denial of the Mass	Smithfield, 27 June 1558
Robert Southam		St Dunstan in the West		denial of the Mass; calling it an idol	Smithfield, 27 June 1558
Matthew Ricarby				denial of the Mass	Smithfield, 27 June 1558
John Floyd				denial of the Mass; asserting that prayer to saints is idolatrous	Smithfield, 27 June 1558
John Holiday				denial of the Mass	Smithfield, 27 June 1558
Roger Holland	merchant taylor; apprenticed to Kempton at the Black Boy in Watling Street; born in Lancashire of a gentry family		late 20s	denial of the Mass	Smithfield, 27 June 1558

him. 'And as I mean thee good', said Bonner, 'so, Roger, play the wise man's part and come home with the lost son . . . and the Church shall . . . kill the fatling to make thee good cheer withal.'[281] Friendly persuasion, or moral blackmail, was employed. 'Wouldst thou not be at home with thy children with a good will?'[282] Sometimes the persuasions were successful, or nearly so, though Foxe is here usually discreetly silent. As Rough knew, 'the Devil is very busy to persuade, the world to entice with promises and fair words'. Even Cardmaker was close to recantation.[283] The largest audience ever seen at a parish church came, in hope or dread, to hear Richardson preach, for 'he came thither to have recanted, but he would not'.[284] Sir John Cheke's recantation in July 1556 was a triumph for the government and profoundly demoralizing for the Protestant cause. (By his copy of the recantation Matthew Parker wrote: *Homines sumus*: 'we are but men'.) Within weeks prisoners were submitting. On 20 October about thirty prisoners in the Lollards' Tower were freed on promise of future good and Catholic behaviour.[285] These prisoners may have been won over by Story and Feckenham (iron hand and velvet glove). At his treason trial in 1571, seeking to vindicate himself of the charge of cruelty, Story told of how he and Feckenham 'persuaded' with 'twenty eight condemned to the fire', and found them to be 'very tractable'. They, with the Cardinal, 'did sue to the Queen, and laid both our swords together', and all the prisoners, save an old woman living by Paul's churchyard, were pardoned and received absolution.[286] In the autumn of 1557 the Cardinal intervened again to pardon twenty-two Essex heretics whom Bonner had condemned.[287]

With time Bonner, who had at first needed to be urged to his task, became more relentless: the 'bloody Bonner' of legend. The wilfulness and obduracy and teasing of the godly goaded

[281] Foxe, *Acts and Monuments*, viii, pp. 473–9.
[282] Ibid. viii, p. 546.
[283] Ibid. viii, p. 449; Strype, *Ecclesiastical Memorials*, iii/1, p. 433.
[284] *Diary of Machyn*, p. 91.
[285] *CSP Ven*. vi/2, p. 769; Strype, *Ecclesiastical Memorials*, iii/1, pp. 504, 514–17; *Diary of Machyn*, p. 118.
[286] *State Trials*, i, p. 1094.
[287] Strype, *Ecclesiastical Memorials*, iii/2, pp. 29–30.

him. Richard Gibson drew up in October 1556 a submission to the Church, but a deeply ambiguous one. Gibson was learned, adroit, a gentleman and grandson of a Lord Mayor of London. He knew how to taunt Bonner. He sent the Bishop nine articles of faith: 'what the gorgeous and glittering whore is, that sitteth upon the beast, with a cup of gold in her hand full of abhomination?' At last Gibson went to the stake on 30 November 1557.[288] Ralph Allerton also had reconciled once before finally suffering martyrdom.[289] To all the gospellers the chance was given, and given again, for them to reconcile and live. But they prayed that they would be given the strength to resist. John Leaf was sent two bills in the Counter: one a confession of heresy, the other a reconciliation. It was said that these were read out to him, because he was illiterate, and that upon the confession he sprinkled his own blood as a token.[290]

The martyrs died because they would not recant opinions which the Marian Church anathematized. Precisely what these convictions were is difficult to determine. Some of the martyrs left behind full confessions of faith.[291] Most did not. It is within John Foxe's great *Acts and Monuments* that the martyrs' trials and deaths are principally recorded and, for all that work's remarkable completeness and accuracy, suspicions remain that where the martyrs deviated from the right beliefs of the godly band of professors a discreet veil is drawn over their utterances.[292] Sometimes; not always. Foxe tells of William Flower who attacked the priest at Westminster at Easter 1555. His fellow prisoners condemned the sacrilege, and Flower too lived to regret the act, if not the belief that inspired it. Flower gave an account of his action: how 'compelled by the Spirit . . . being in mind fully content to die for the Lord, [I] gave over my flesh willingly, without all fear, I praise God'. 'I humbly beseech you to judge the best of the Spirit, and condemn not God's doings.' This suggests antinomianism, and surely here

---

288 Strype, *Ecclesiastical Memorials*, iii/2, pp. 46–61.

289 Foxe, *Acts and Monuments*, viii, p. 411.

290 Ibid. vii, p. 193.

291 For example, Strype, *Ecclesiastical Memorials*, iii/2. lxi, lxiii.

292 P. Collinson, 'Truth and Legend: the Veracity of John Foxe's Book of Martyrs' in *Clio's Mirror: Historiography in Britain and the Netherlands*, ed. A. Duke and C. A. Tamse (Zutphen, 1985), pp. 31–54.

was a fanatic who would have been condemned by any government, Catholic or reformed.[293]

The authorities knew that strange ideas were held about
the incarnation of Christ, that some denied original sin, the
descent of Christ into Hell, and the baptism of infants. Some
were accused of Anabaptism. Elizabeth Young was charged:
'she hath a certain spark of the Anabaptists, for she refuseth to
swear upon the four evangelists before a judge'.[294] To Saunders
it was put that Joan Bocher was 'of your church', but this was
vehemently denied. Thomas Spurdance was said to be an Anabaptist.[295] Joan Bocher had triumphantly asserted at her death
that there were a thousand Anabaptists in London diocese: she
must have exaggerated, but certainly they were an established
presence.[296] On 12 April 1557 five people ('out of Essex') were
burnt at Smithfield:

One holding that Christ is not yet come. Another, He was not yet
ascended. Another, that He was not yet equal with the Father in godhead. The fourth that a child begotten between a Christian man and a
Christian woman was christened in the mother's womb and ought no
otherwise to be baptized. The five held that all men's wives ought to
be in common.

These were extremists. Not one of them was mentioned by
Foxe.[297] The sectaries who had so alarmed Hooper in Edward's
reign may have proliferated in the chaos under persecution,
and their extreme views were anathematized by the 'true
professors' whose underground organization they thought to
penetrate. 'I hear say', said Dr Martin, 'that you have divers
churches and faiths in the King's Bench.' This was true: there
was fierce contention between good Protestants and the 'free
willers' who denied God's predestining providence which was
the gospellers' orthodoxy.[298] Cuthbert Simpson, deacon of the
London congregation, assured his wife that 'there is nothing

---

[293] Foxe, *Acts and Monuments*, vii, pp. 68–74, 368. See Hughes, *The Reformation in
England*, ii, pp. 261–2.
[294] Foxe, *Acts and Monuments*, viii, p. 539.
[295] Ibid. vi, p. 619; vii, p. 631; viii, pp. 432, 475.
[296] See above, p. 444.
[297] BL, Harleian MS 357, fo. 112ʳ; Wriothesley, *Chronicle*, ii, p. 137; *Diary of
Machyn*, p. 130.
[298] Foxe, *Acts and Monuments*, viii, p. 164; Strype, *Ecclesiastical Memorials*, iii/2. xliii.

that cometh unto us by fortune or chance, but by our heavenly Father's providence'.[299]

Among the congregations was the sect of the 'free willers', led by Henry Hart. Hart had drawn up thirteen articles of faith for the brotherhood, and it was said that none could join 'except he be sworn'. It followed that the sect of free willers was very small. Yet the gospellers saw its dangers. One who had been of 'the free will men' and come to regret it blamed the 'calamity of this realm of England' upon the sins of the gospellers themselves: 'and one cause was because we were not sound in the predestination of God'. Hart lived at the foot of London Bridge with John Kempe, another free willer, in the house of Curle, a cutler. Robert Coles and John Ledley, lodgers at the Bell Inn in Gracechurch Street, 'great counsellors' among the prisoners, had also for a time denied God's predestinate grace.[300] In prison Philpot wrote an 'Apology for spitting on an Arian; with an invective against the Arians'. The brethren were not to marvel at contention among the prisoners, 'for . . . it is necessary that heresies should be, that the elect might be tried'.[301] When Feckenham accused a group of Essex heretics that 'there were sixteen of them and that they were of sixteen sundry opinions' they responded by issuing a confession of sound reformed doctrine.[302] It was a false rumour, Stephen Cotton assured his brother John, that in Newgate 'we are of contrary sects and opinions'.[303] William Punt, Ridley's messenger, and author of 'Mistress Missa' tracts in the previous reign, 'had in his bosom a certain book against the sect of the Anabaptists'.[304] The urgency and vehemence with which the gospellers denounced the sectaries reflected their horror of heresies and the threat of confusion and separation.

Denial of the Mass and adherence to the doctrines enshrined in the Edwardian Book of Common Prayer of 1552 was suffi-

[299] Foxe, *Acts and Monuments*, viii, p. 460.

[300] Strype, *Ecclesiastical Memorials*, iii/2. xliii; Foxe, *Acts and Monuments*, viii, p. 384; J. W. Martin, 'English Protestant Separatism: Henry Hart and the Freewillers', *Sixteenth Century Journal*, vii (1976), pp. 55–75.

[301] Strype, *Ecclesiastical Memorials*, iii/2. xlviii.

[302] Ibid. iii/1, p. 588; iii/2. lxii, lxiii; Foxe, *Acts and Monuments*, viii, pp. 155–6. Three were pardoned, but thirteen were burnt at Stratford on 27 June 1556.

[303] Inner Temple, Petyt MS 538/47, fo. 3^r; Foxe, *Acts and Monuments*, viii, p. 525; Strype, *Ecclesiastical Memorials*, iii/1, p. 588.

[304] Foxe, *Acts and Monuments*, viii, p. 384; vii, p. 426.

cient to condemn someone for heresy. Where there is testimony of the beliefs of the martyrs and of the conduct of their hidden congregations it seems that they were loyal to the teachings of the Edwardian Church. In October 1553 Thomas Mountain was brought before Gardiner for celebrating the 'schismatical' service still: 'Take and eat this, &c', and 'Drink this, &c'.[305] It was for holding communion according to the Edwardian use that Rose and the congregation in Bow churchyard were arrested in January 1555.[306] Rough was accused of ministering according to the Edwardian Book of Common Prayer and of knowing others who 'yet do keep books of the said communion and use the same in private houses'.[307] To follow the 1552 Prayer Book was, of course, implicitly to deny transubstantiation, and most of the martyrs condemned the doctrine explicitly also. It was for this that they died. Roger Holland's confession of faith might stand for that of others:

I say and believe . . . that the sacrament of the supper of Our Lord, ministered in the holy communion, according to Christ's institution, I being penitent . . . in perfect love and charity, do there receive by faith the body and blood of Christ. And though Christ in His human person sit at the right hand of the Father, yet (by faith, I say) His death, His passion, His merits are mine, and by faith, I dwell in Him and He in me. And as for the Mass, transubstantiation, and the worshipping of the sacrament, they are mere impiety and horrible idolatry.[308]

Like Holland, many who died for the faith were recent converts to it. Tankerfield had been 'a very papist' until Mary's accession; Green had been a riotous young lawyer until his conversion; and Holland had only been converted late in Edward's reign, by attending the daily lectures at All Hallows and the sermons at St Paul's, and by reading Scripture and the Prayer Book.[309] Others had long held sacramentarian views: Flower was converted as early as 1529; Robert Southam had refused the sacrament of the altar since 1548 at least; Warne and Tomkins were certainly committed sacramentaries by the end

[305] *Narratives of the Reformation*, p. 178.
[306] *Diary of Machyn*, p. 79.
[307] Foxe, *Acts and Monuments*, viii, pp. 444–6.
[308] Ibid. viii, p. 478.
[309] Ibid. vii, p. 343; viii, p. 473.

of Henry's reign.[310] But some of the martyrs were so young
that they had never known England under papal authority;
and, reaching the age of confirmation after the Edwardian
changes, had never received the Mass. William Cecil, looking
back, specially lamented those who had never 'heard of any
other kind of religion, but only that which by their blood and
death, in the fire, they did as true martyrs testify'.[311] The
Venetian ambassador could even claim in 1557 that there were
no Catholics in England under the age of thirty-five, for anyone
younger could know only heresy and schism.[312] This was true
in law, if not in fact; and truer in the south-east than it was of
the North and 'dark corners' of the land. The Marian interrogators sought to discover what the faith of the accused had
been at the age of fourteen—the age of spiritual discretion—in
order to know whether or not they were responsible for their
heresy.[313] Although some had been brought up in the reformed
faith—like Joan Warne, who 'from the time that she was eleven
years of age, she hath misliked the sacrifice of the Mass'[314]—
they were martyred none the less. Their examiners at first
pitied their ignorance and confusion, even offering William
Hunter, an apprentice, the opportunity to be set up in mastership if he would be circumspect. Bishop Bird pleaded with
Haukes: 'ye are a young man, and I would not wish you to go
too far, but learn of your elders to bear somewhat'.[315] But few
were persuaded to emulate the worldly mutability of perjured
ecclesiastical officials who changed faith with the regime. Calling these young protesters 'peevish boy', 'stout boyly heretic',
and 'malapert', the Bishops sent them to the fires.[316] The
youth of the martyrs may have made the burnings seem to
some more poignant, less explicable.

'Those that be evil, of love, we ought to procure their correction.' So Bonner had written. But as the months passed it
became clearer that the persecution would fail, not only in

[310] Foxe, *Acts and Monuments*, vii, p. 73; viii, p. 471.
[311] William Cecil, *Execution of Justice in England* (1583), cited in Strype, *Ecclesiastical Memorials*, iii/2, pp. 153–4. See Hughes, *The Reformation in England*, ii, pp. 289–93.
[312] *CSP Ven.* vi/2, pp. 1074–5.
[313] Foxe, *Acts and Monuments*, vi, pp. 730, 738; vii. p. 716.
[314] Ibid. vii, p. 717; viii, p. 142.
[315] Ibid. vi, p. 727; vii, pp. 104–5; see also vii, pp. 44, 101.
[316] Ibid. vii, pp. 111, 119, 151, 353; viii, pp. 145, 157, 161, 467, 477, 483.

charity but also in its effectiveness, if they chose the wrong victims. The persecutors realized that persecution must be tempered, for there were degrees of guilt and degrees of heretical pravity. In a speech to the Commons in 1559 Story regretted that he and his fellow commissioners had 'laboured only about the young and little twigs, whereas they should have struck at the roots'; that they had concentrated upon those who knew only heresy, or did not know what was heresy and what was not.[317] In October 1556 Boswell had complained to Bonner how in Essex 'the sworn inquest for heresies do . . . let the arch heretics go; which is one great cause that moveth the rude multitude to murmur, when they see the simple wretches (not knowing what heresy is) to burn'.[318] But in that month also, at the behest of Story and Feckenham, Pole had pardoned twenty-seven City heretics because they were 'nescientes quid fecerunt'. Story had stormed before the Bishops and Council, denouncing the fainéance of allowing heretics of high rank to go free. Perhaps he knew that the London congregation met in Sir Thomas Cawarden's house in Blackfriars.[319] Time was when the heresy quests had looked first for the influential, but no more.

An aversion developed to the campaign against the gospellers: not because of sympathy with their beliefs, but because of the manner in which it was conducted. A gospeller, admittedly partial, wrote to Bonner in 1556 that every child knew and could say that 'Bloody Bonner is Bishop of London', and that he would never succeed in his purpose 'so long as he went this way to work. And that . . . they had lost the hearts of twenty thousand that were rank papists within this twelvemonth.'[320] The burnings at Smithfield were halted after June 1558, 'for I saw well', claimed Story, 'that it would not prevail'.[321] The City officers grew less and less amenable to their task. When, at the end of Mary's reign, Julian Livings came before Dr Darbishire it was the constable of St Bride Fleet Street, the

[317] Foxe, *Acts and Monuments*, vi, p. 554; *House of Commons, 1509–1558* (Story).

[318] Foxe, *Acts and Monuments*, viii, p. 388; cited in Hughes, *The Reformation in England*, ii, p. 273.

[319] *State Trials*, i, p. 1094; *House of Commons, 1509–1558* (Story); Foxe, *Acts and Monuments*, viii, p. 558.

[320] Strype, *Ecclesiastical Memorials*, iii/1, p. 471.

[321] *State Trials*, i, p. 1094.

parish which had sent more martyrs for burning than any other, who stood surety for her. Darbishire challenged him: 'You be constable and should give her good counsel.' 'So do I,' the constable replied, 'for I bid her to go to Mass, and to say as you say. For, by the Mass, if you say the crow is white, I will say so too.'[322] But the aversion to the persecution was not all to do with religion, for with the Marian reaction had come judicial changes and social policies which were, in the circumstances, seen as equally oppressive. Even at the stake it was not for his papistry that Ridley condemned Bonner, but for violating the leases given by Ridley to poor men.[323]

FAITH AND CHARITY

The persecution was conducted against a background of social and economic misery. Through the winter of 1555 it had rained as though without end, flooding Westminster, Southwark, and Lambeth. The harvests in the following years failed disastrously, leaving food in very short supply and prohibitively expensive. Late in September 1556 there was uproar in the grain markets, and the City governors intervened to provision the capital.[324] Wheat was imported from France.[325] Torrential rains came again that autumn.[326] By the winter of 1556 the situation was desperate: by Christmas malt was sold in Gracechurch Street market for 40s. a quarter; at New Year 1557 for 44s., and in the spring for 46s. Not until the following year's harvest did prices fall, and then dramatically. The penny loaf, which weighed in 1556 only 11 oz., weighed 56 oz. after the 1557 harvest. For demand, too, had fallen, and for terrible reasons. 'The poor died for hunger in many places' in 1556, and, according to grim precedent, dearth was followed by pestilence.[327] There began in 1556 'hot burning fevers and other strange diseases', which over the next two years became epidemic. Plague was the most awful threat facing the citizens,

[322] Foxe, *Acts and Monuments*, viii, p. 530.

[323] Ibid. vii, p. 550.

[324] Strype, *Ecclesiastical Memorials*, iii/1, p. 502; CLRO, Repertory 13, fos. 404^v, 424^v–425^r.

[325] Ibid., Repertory 13, fo. 476^r.

[326] *Two London Chronicles*, p. 43.

[327] *Cooper's Chronicle*, fo. 375^v; Stow, *Annales*, pp. 631, 634.

compared with which persecution for religion might even seem distant.[328]

Charity was more than ever needed. Some in their dire poverty were now abandoning their children in rich men's doorways: at the houses of Philip Gunter, John Sowle, and Sir John Leigh, founding fathers of the City hospitals, and of a Spaniard, Granado.[329] Babies were found in church pews, and the charitable thought of them as 'godsendings'.[330] Londoners began to remember adopted foundlings in their wills—'Roger, a man child which was laid and found at my gate in his infantery and swaddling clowts'; 'Alice, the poor child, a foundling'—or they might even bequeath them in their turn, as 'my little child William, whom I keep of alms, I give as freely as he was given me'.[331] Late in November 1556 came a proclamation that no one should leave children in the City streets or at citizens' houses.[332] On 17 December a woman was whipped at Bridewell for abandoning her child.[333] Christ's Hospital, the home for foundlings and orphans, was full.[334] The charity of Londoners was, in fact, increasing in these years: a third of the testators whose wills were proved in the Prerogative Court of Canterbury remembered the poor and the hospitals in Mary's reign.[335] But all their benefactions and all their private charity could not satisfy the growing and desperate need. When Londoners looked to the government for a solution to this problem of poverty they found little help, for its concerns were quite otherwise: to provision an unpopular war with France, to restore the old faith, and to persecute, unavailingly. In the City the actions of the authorities seemed to justify the charges that the poor were specially, often unjustly, persecuted. In March 1557 Peter Bartly and his wife were imprisoned for breaking the Lenten fast. But, it was objected, all they had done was to eat a cockerel which they had found dead on a dunghill; 'they be

[328] P. Slack, *The Impact of Plague in Tudor and Stuart England* (1985), pp. 56–8.
[329] Guildhall, MS 12806/1, fos. 4[r], 5[r], 6[r], 10[v], 14[r]; CLRO, Repertory 12, fo. 171[r]; Repertory 13, fos. 72[v], 197[v], 236[v], 361[r], 462[v].
[330] Guildhall, MS 12806/1, fos. 11[r], 12[r], 13[r], 18[r].
[331] Guildhall, MS 9171/13, fos. 1[r], 32[v], 46[r]; 9171/15, fo. 110[v].
[332] CLRO, Repertory 13, fo. 453[v]; *Diary of Machyn*, p. 119.
[333] Guildhall, MS 12806/1, fo. 3[v]; CLRO, Repertory 13, fo. 462[v].
[334] Ibid., Repertory 13, fos. 442[r], 487[r].
[335] Guildhall, MS 9171/13; PRO, PCC, Prob. 11/37 (Register More, 1554–5); PCC, Prob. 11/40 (Register Noodes I, 1557–8).

very poor folks, and did eat it of very necessity and not of any contempt'.[336]

The achievements of the Edwardian commonwealth in solving the problem of poverty in London were now systematically challenged. The City hospitals stood as a reproach to good Catholics in Mary's reign, for their very creation marked the loss of the religious houses. The story of the perilous survival of the hospitals under Mary was told by John Howes, Richard Grafton's servant, in the form of a dialogue between Duty and Dignity. The hospitals' records prove Howes's story to be largely true, if partial. Asked Dignity disingenuously: 'Although God took away the good King, yet this good work could never die, for men's devotions continue for ever . . .?' Not so, replied Duty; 'the change of religion had almost overturned all'. 'In Queen Mary's time the City had much to do to keep them from suppressing.'[337] The hospitals' supporters were fearful for their survival. Robert Toy left Christ's Hospital 20s. annually for ten years in 1555, but only 'if the poor so long continue' there.[338] The hospitals were a perpetual reminder not only of lands lost to the Church but of power lost too, for control of the hospitals belonged to the City. In Edward's reign Bishop Ridley and the City governors had worked in unprecedented and harmonious partnership to establish 'Christ's holy hospitals and *truly* religious houses', but in Mary's reign there was jealousy and mutual recrimination.[339] 'Could not the Pope's clergy and Bridewell be friends?' asked Dignity. Duty alleged that because 'Bridewell did somewhat abridge the ecclesiastical courts of their jurisdiction' the 'governors were never in quiet'.[340] The Repertories of the Court of Aldermen record this contention over jurisdiction.[341] The governors of Bridewell had usurped some of the powers of the Bishop and his officials to try

---

[336] CLRO, Repertory 13, fo. 486ʳ. There may have been another motive behind the accusation: was this Master Bartly, one of the godly, who had sheltered Edwin Sandys? See above, p. 561.

[337] John Howes, *Contemporaneous Account in Dialogue-form of the Foundation and Early History of Christ's Hospital and of Bridewell and St Thomas' Hospital* (1889), pp. 23–5.

[338] PRO, PCC, Prob. 11/37, fo. 304ᵛ.

[339] Paul Slack, 'Social Policy and the Constraints of Government, 1547–1558', in J. Loach and R. Tittler (eds.), *The Mid-Tudor Policy, c.1540–1560* (1980), pp. 110–13.

[340] Howes, *Foundation*, p. 25.

[341] CLRO, Repertory 13, fos. 456ᵛ, 457ʳ, 463ʳ, 478ᵛ, 480ᵛ, 509ᵛ, 569ʳ.

moral offences: these powers Bonner was determined to regain. By 1558 the quarrels between the Bishop and the Communalty had become so intractable that the City gave aid for the defence of citizens 'sued and troubled' in the spiritual courts over evidence that they had given in the trial of a bawd.[342] The Bishop and his clergy used any means to discredit the governors, Duty alleged, 'what with preaching and what with process', and Duty may have been right. In April 1558 the outraged Court of Aldermen appealed to the Lord Chancellor for 'reformation' against the 'rash and slanderous' accusation made by a Paul's Cross preacher that an apprentice had been beaten so severely at Bridewell that 'he should die thereof'.[343]

The London hospitals had been transformed from godly houses in which the Word of God was set forth into Catholic strongholds. At Christ's Hospital, Gardiner had intervened to ensure that chapels were built and ornaments restored, and 'to have a Mass priest to be their hospitaller and to have daily Mass said'.[344] To the premises of Richard Grafton, the printer of the Bible and the Edwardian Prayer Books, now came Robert ('Robin Papist') Caly, returned from exile. There he printed works of Catholic propaganda and spied upon the gospellers.[345] Dr Story also settled in the Grey Friars. The hospital governors spent £8 on refurbishing a house for him there, and so from an enemy 'made a friend', but Story used this house for interrogating suspects.[346] Instead of being houses for the godly upbringing of the young the hospitals became houses of correction for the sort of young Protestant they had formerly nurtured. So Thomas Green, an apprentice, was sent to Christ's Hospital for distributing a pamphlet, *Antichrist*, and had there 'for the time the correction of thieves and vagabonds', and was whipped before Story.[347] Now even 'the Bible would breed heresies', and the Word was confiscated from two young reformers, Gie and Waterson: 'a bibble-babble were more fit

[342] Ibid., Repertory 14, fo. 49[r].
[343] Ibid., Repertory 14, fos. 25[v], 29[v].
[344] Howes, *Foundation of Christ's Hospital*, p. 25.
[345] Took, 'Government and the Printing Trade', p. 250; Foxe, *Acts and Monuments*, viii, pp. 438, 551, 552, 724, 792.
[346] CLRO, Repertory 13, fo. 135[r]; Guildhall, MS 12819/1, fo. 123[r]; Howes, *Foundation of Christ's Hospital*, p. 24.
[347] Foxe, *Acts and Monuments*, viii, pp. 521–4.

for thee', said Story. They were sent to Bridewell to be whipped upon the cross of correction. But Bridewell's governor was Grafton, the Bible's printer, to whom their punishment would be anathema, so Story was sent to see that 'the matter should not be lightly handled'.[348] The Marian persecution took many forms, and consciences were tested in different ways.

Thomas More had feared, even when the new faith first entered the City, that complacent Catholics did not cast out the heretics in their midst. Had they, a generation later, after all the transformations in religion, learnt to tolerate the heterodoxy of their reformist neighbours; these neighbours who had for a time been allowed to worship freely while they themselves had not? The homily against brawling of 1547 had reminded all those who now slandered each other, 'he is a papist, he is a gospeller, he is of the new sort, he is of the old faith' that 'St Paul could not abide to hear among the Corinthians these words of discord . . . I hold of Paul, I of Cephas'.[349] This homily, and all the proclamations forbidding those of rival beliefs to taunt each other, had a special resonance for London where such contention was commonplace. But now, when the dangers of disobedience to the Catholic Church were so great, did they still openly revile each other? What were the social consequences for this community of the Marian reaction and persecution, during which those who were not reconciled were to be cast out, and terribly?

London saw more burnings than anywhere else in England. Some, maybe many, rejoiced to see the martyrs die. As Tankerfield suffered in St Albans he found more well-wishers than opponents, according to Foxe, but after his martyrdom 'some superstitious old women' said that 'the Devil was so strong with him as all such heretics as he was, that they could not feel pain almost'.[350] When the promoters, Cluny and Avales, tried to force John Lithall to kneel before the rood in St Paul's and say a paternoster and ave in worship of the five wounds, a crowd gathered. Some spat at him, he remembered, and said 'it was a

---

[348] Foxe, *Acts and Monuments*, viii, pp. 724–5.
[349] *Certain Sermones or Homilies appoynted by the Kynges Maiestie* (1547; *RSTC* 13639), sig. Siii[r].
[350] Foxe, *Acts and Monuments*, vii, pp. 346–7.

pity I was not burned already'.[351] But for the most part, the evidence from both the persecutors and the persecuted is of sympathy for the victims; of the authorities' fears and the gospellers' pride at the numbers who gathered to give support. That is the public response; the private response is hidden. Knowing who delated the gospellers to the authorities, as they were commanded, would reveal much about the citizens' sense of their social and religious duty, and their attitude to heresy. In the autumn and winter of 1554–5 almost four hundred Londoners had appeared before the Vicar-General. Dutiful churchwardens and right-thinking parishioners had reported them. But this had been before the restoration of the medieval heresy laws, which returned the initiative in heresy persecution to the clergy, and before the first martyr suffered. Once the persecution began all but a few of the four hundred reconciled themselves. Thereafter even Catholic parishioners, knowing the fate of heretics, may have been more reluctant to inform against their neighbours. And still there were some who kept their faith secret. When the fiercely Catholic wives of Stepney conspired to murder Underhill, the 'hot gospeller', he was warned by a woman 'whose good will to the Gospel was unknown to the rest'.[352]

The initiative in detection of heresy in London came from the commissioners, of whom Dr Story was the most determined. In Lent 1557 they demanded all the clergy and churchwardens in the diocese to report all those who did not attend church, confess, and receive the sacrament that Easter. The following year the commissioners may have conducted another visitation in the City; at least they published a book of *Interrogatories*, inquiring into orthodoxy, to be answered by all churchwardens, which must have been preparatory.[353] Certainly the commissioners knew the names of many of the City's gospellers: Darbishire showed Julian Living 'a roll of certain names of citizens'.[354] Who had given them the names remains doubtful. If the clergy were the main providers of information, neither

[351] Ibid. viii, p. 533.

[352] *Narratives of the Reformation*, p.160, see also p. 148.

[353] *Interrogatories upon which . . . the churchwardens . . . shalbe charged* (1558); Alexander, 'Bonner and the Marian Persecutions', p. 379.

[354] Foxe, *Acts and Monuments*, viii, p. 529.

Foxe nor any other source names them. It seems clear that even Bonner and his officials acted, as duty bound, upon information given to them rather than searching out and arresting suspects themselves.[355] According to the Protestant chroniclers of the tribulations of the godly—Foxe, Mountain, Underhill—London's brethren were almost all detected by Judas-like promoters, by Caly, Cawood, Banbery, Beard, Avales, and arrested by Bonner's apparitor, Cloney, by the Lord Chancellor's men, or by Aldermen's deputies or sergeants. It was through fear of the promoters that Underhill, the Hickmans, and Mountain hid.[356] Sometimes the gospellers suspected traitors in their midst: once Rough had even doubted Margaret Mearing, who was to be martyred with him.[357] Underhill had fled from the Catholic wives of Stepney.[358] But of Londoners delating their fellow citizens there is almost no evidence. When Protestants were reported, as during the 1554 visitation, it was often the trouble-makers, disliked for other reasons than their religion alone, who were named, as we have seen. As the persecution began, the reformist community were sometimes sheltered by those who, while not sharing their convictions, did not wish to see them suffer for them. They might be protected through family loyalty or simple humanity. Stephen Mylady admitted that 'he hath coming and going to his house his own unmarried sister, which sometime was married to a priest, and that he doth for charity help and succour her'.[359] When two women offered to stand surety for Elizabeth Young, Darbishire accused them: 'do ye not smell a little of heresy also?' But they denied it, and perhaps not untruthfully: they helped because Elizabeth's children 'were like to perish', and if Elizabeth died the upbringing of the children would fall to them.[360] The persecution touched more than the martyrs alone, and cast their families to the care of the community.

Citizens might draw back from accusing their neighbours of

[355] Alexander, 'Bonner and the Marian Persecutions', pp. 380, 391.

[356] This assertion rests upon an analysis of all the arrests recorded in Foxe's *Acts and Monuments*, and on the accounts of their experiences by Underhill, Mountain, and the Hickmans.

[357] Foxe, *Acts and Monuments*, viii, pp. 450–1.

[358] *Narratives of the Reformation*, p. 160.

[359] GLRO, DL/C/614, fo. 30ʳ.

[360] Foxe, *Acts and Monuments*, viii, pp. 547–8.

heresy, but the old suspicions and grievances harboured against confessional enemies did not go away. Catholic parishioners found other ways of persecuting the gospellers. At St Botolph Aldgate, where the divisions were bitter, the reformist coterie was forced to finance the restoration of the church. In St Sepulchre, a parish notorious for its papistry in Edward's reign, the old guard awaited the chance to punish the iconoclasts. Gregory Newman, a Protestant stalwart, to whom Underhill had resorted during Wyatt's rebellion, had reconciled to the Catholic Church in December 1554, but his fellow parishioners waited with other charges to accuse him. In March 1555 he was delated, *publica fama*, for his adultery with his maidservant, Anne Savage, another gospeller. His horror at the prospect of public penance may have stemmed from his knowledge that many looked forward to seeing him shamed. Knocking himself upon the breast, he proclaimed himself deeply contrite, and his penance was commuted to a donation towards that church he had desecrated.[361] Grudges remained.

Those who had suffered a 'painful peregrination' during Mary's reign later found it hard to forgive those who, still pretending to be gospellers, had joined the idolaters. There were bitter divisions within the stranger churches during Elizabeth's reign between those who had gone into exile and those who had remained behind. In the French Church all who had celebrated Mass were ordered to make an act of public penitence.[362] Deep resentments remained in the Drapers' Company after the Throckmorton trial: one juror, Pointer, escaped gaol, and another, Calthorp, who had languished in the Fleet, could not forgive him. Still in 1559 the Drapers' Court sought reconciliation between them, 'for the good maintenance of brotherly love between brother and brother of this house, all malice ceasing'.[363] Hardest of all was to forgive those who claimed falsely that they had been steadfast in faith. John Bolton, a silk weaver of Long Lane, boasted that he had never attended Mass in Mary's reign, but John Moyar knew better:

And when I refused to be partaker of their idolatrous Mass, they said I was mad, and was sent to the stocks, and there cruelly handled till

[361] GLRO, DL/C/331, fos. 188ʳ–189ᵛ.
[362] Pettegree, *Foreign Protestant Communities*, pp. 124–5. ·
[363] Drapers' Company, Repertory C, fos. 159ʳ, 164ʳ.

Monday morning, and then you dissemblingly, as I may charitably say, received at their Mass before the whole parish.[364]

The Marian persecution, and its aftermath, left much bitterness and much guilt. Bishop Ridley had recognized the unhappiness. Bidding the citizens of London farewell before his martyrdom, he wrote: 'I do not doubt but that in that great City there may be many privy mourners, which do daily mourn for that mischief, the which never did nor shall consent to that wickedness.'[365] He was too optimistic in hoping that the faith was kept inviolate. Harmony had come to the parishes, as the reformed among the citizens chose overwhelmingly to reconcile and, outwardly at least, to conform.

LAST THINGS

In the years of persecution fear compelled most of the godly to bear with the times, and to become those 'popish Protestants' whom the martyrs condemned. They hoped somehow to preserve the faith inwardly and to be forgiven. After his famous, or notorious, recantation Sir John Cheke bequeathed his soul to God, trusting that 'it cannot perish that is committed to Him'.[366] The divergence between the ways people believed privately and worshipped publicly might be revealed in their wills. But not always. In the will of Alderman Hynde the reformed convictions which he originally expressed in the preamble were anxiously crossed out. At his funeral in August 1554 there was nothing to displease the Catholic authorities.[367] It was not only fear which led the godly to conform, but perhaps an inclination not to oppose the clear will of the majority of the citizens. In the first months of Mary's reign Londoners had hastened to return to the old ways. The Mass was 'very rife' and altars and masses had been restored, 'faster than ever they

---

[364] Strype, *Ecclesiastical Memorials*, iii/2, pp. 427–30. Later, as a member of the Plumbers' Hall congregation of Elizabethan London, Bolton recanted once more, suffered excommunication from the congregation, and hanged himself: P. Collinson, *The Elizabethan Puritan Movement* (1967), p. 90.

[365] Foxe, *Acts and Monuments*, vii, p. 559.

[366] PRO, PCC, Prob. 11/40, fo. 12r.

[367] Ibid. Prob. 11/37, fo. 45r; *Diary of Machyn*, p. 67.

were put down'.[368] The wills written by Londoners in Mary's reign reveal a remarkable obedience to the Catholic Church; they believed, as one wrote, 'according as the holy mother Church of Rome doth command me'.[369] Very nearly half of testators whose wills were proved in the Commissary Court now committed their souls to the protection of the Virgin and saints in Heaven, fully twice as many as had done so in the previous reign. A third bequeathed their souls to God alone, maker and redeemer, and only a fifth asserted the reformed conviction that belief in the redeeming passion of Christ alone sanctified sinners.[370] And so, too, of the wills proved in the Prerogative Court of Canterbury in the last years of Mary's reign nearly half looked for the intercession of saints.[371] Once more, though far, far more rarely than before, testators began to ask for prayers for souls: Robert Warner required the Drapers to say three times during their dinner after his funeral, 'God have mercy on my soul and all Christian souls.'[372] When, on the eve of the feast of the conception of the Virgin, Cuthbert Thompson's charitable bequest of coals was distributed, the poor were required to pray for his soul and all Christian souls.[373] Joan Baker's '*whole* trust' for the remission of her sins lay in the 'merits of Christ's blessed passion', but still she provided for a priest to sing for souls in St Laurence Jewry.[374] Catholics might now insist, more overtly than before, that the Mass was an essential application of the merits of Christ's passion: the symbol of that passion here on earth.

To suppose simply that in their wills, and in their lives, most citizens prudentially followed the prevailing orthodoxy may be to ignore significant changes in the ways they expressed their faith. For in the wills written in the last years of Mary's reign is revealed a spirit which helps to answer the perplexing question of how citizens of opposed convictions could worship together, 'in charity'. Again and again, bequests were made which juxtapose the conventions and the spirit of the reformed and the Catholic faiths. What had once been rare now became

[368] BL, Cotton MS Titus B vi, fos. 198ᵛ, 196ʳ.
[369] PRO, PCC, Prob. 11/37, fo. 234ᵛ.
[370] Guildhall, MS 9171/13.
[371] PRO, PCC, Prob. 11/40 (Will Register Noodes).
[372] Ibid. Prob. 11/37, fo. 219ʳ.   [373] Ibid. Prob. 11/40, fo. 185ʳ.
[374] Guildhall, MS 9171/13, fo. 32ᵛ.

common: testators came to avow a certainty in the merits of Christ's passion to save, while at the same time desiring the Virgin and saints in Heaven to intercede and pray for them.[375] Not that Catholics had ever doubted the sanctifying power of the Passion; rather, as the reformers had alleged, this saving message had been obscured in the welter of superstition, 'man's traditions', and the multiplicity of laws. Now the wills of Londoners reveal that the central evangelical truth—of justification by faith—had reached them. But this belief might be held, as it had always been held, alongside other beliefs which the evangelicals scorned, and not just for the purpose of pacifying the Catholic authorities. In November 1553 Joan Pecke of St Sepulchre bequeathed her soul to God, 'trusting to be saved *only* by the death and resurrection of . . . Jesus Christ, whose mercy is my trust and hope, and in His holy mother Mary . . . and the congregation of all saints in Heaven to pray for me'.[376] And at the end of Mary's reign, John Vawdry, citizen and vintner, committed his soul to

Jesus Christ who with His most precious blood and bitter passion redeemed me and the whole world from the captivity of the Devil and the danger of sins, most humbly beseeching Him . . . to receive me, wretched sinner, to his inestimable mercy wherein *singly* and above all things I have my confidence and trust . . . I also most heartily beseech the most glorious Virgin Mary and all the holy company of Heaven to pray with me and for me.[377]

On Good Friday 1561 Richard Alington thought upon the passion of Christ and made his will.[378] He left a fortune in trust to establish 'some lively remembrance of the passion and death of Christ'. For Alington had had since his childhood 'ineffable, peculiar and special comfort by a picture of the cross and Christ nailed thereunto'. Yet even this grace had not reformed him, and he was haunted by his sins. 'A monstrous fowl' like a raven appeared before him, bearing a death's head which he thought his own. Months later, believing himself about to die, he summoned his neighbours around him to hear his confession:

375 Guildhall, MS 9171/13, fos. 11^r, 53^r, 73^r, 84^v; PRO, PCC, Prob. 11/40, fos. 19^r, 24^r, 83^r, 107^r, 166^v, 185^v, 230^v, 234^v, 283^v.
376 Guildhall, MS 9171/13, fo. 11^r.
377 PRO, PCC, Prob. 11/40, fo. 166^v.
378 Ibid. Prob. 11/45, fo. 46^v.

'Good masters, for Christ's passion's sake give good ear unto me, and pray continually.' Alington had grown rich through usury, and thinking himself 'condemned eternally', was prevented by his conscience from going to church 'at any time of common prayer'. But Christ in His mercy sent apparitions 'like puppets' to torment him, and to show him accounts of his usurious gains and whom he must repay.[379] Once, Alington had been intrigued by reform. John Louthe had 'lessoned with him about the sacrament . . . touching the sense of *hoc est corpus meum*'.[380] But he resisted, and became associated with the most fervent of all the Marian Catholic establishment. He chose as his executors John Feckenham, Henry Cole and Dr Scott (quondam Abbot of Westminster, Dean of St Paul's and Bishop of Chester), Boxall and William Roper. Yet, as he thought himself dying, Alington conceded the common Christianity which those of opposed faith shared. He told those gathered about him: 'And masters, I cannot tell of what religion you be that be here, nor I care not, for I speak to tell you the truth, and to accuse mine adversary the Devil.' Awareness that in the end those of the old faith and the new alike shared a common saviour and a common enemy urged some peace between them. And on his last will Richard Alington had drawn a crucifixion.

---

[379] Stow, 'Memoranda', pp. 117–21.
[380] *Narratives of the Reformation*, p. 46.

'The Crucifixion drawn on Richard Alington's Will'

Epilogue

ON THE eve of Corpus Christi 1561 the citizens of London were visited by a providential punishment. Lightning struck St Paul's and burnt the steeple. Had it not 'pleased God to turn and calm the wind' the whole City might have been ablaze. Everyone sought the reason. Some found material causes: 'it was the negligence of the plumbers'. But most saw the divine hand at work, for 'could not the Devil have done it without God's permission and to some purpose'. Though God's judgements were strictly 'unsearchable', this catastrophe was clearly sent to punish the Londoners for their sins. But for which particular sins? That the lightning struck on the eve of this sacred Catholic festival, once more suppressed, was surely a sign; so was the fact that only the cross from the steeple was saved and only the communion table in the chancel destroyed. Here was God's judgement upon the iconoclasts, the reformers in religion. The cathedral had been the 'beauty of the City, the beauty of the whole realm', built and edified by the devotion of past generations of Catholics, and by all those still living who had given to the 'old works of St Paul's'. Now it was ruined, as so many other ancient, beautiful, and venerated religious houses were ruined; their desecration, for Catholics, a monument to the heresy and sacrilege of their City. It became very plain in 1561 how many citizens lamented the devastation of their churches, which they had stood by and watched, and how many mourned the loss of that faith which they had celebrated there. Corpus Christi had once been a festival of reconciliation, but never again.

The explanation which the Londoners found for this affliction alarmed the newly established Protestant Church of Elizabeth, for it revealed how many had been won back to the Catholic Church under Mary, or indeed had never left it. Bishop Pilkington preached at Paul's Cross two days after the fire, telling the citizens that it must not be taken as a sign of God's anger against the reformers, rather as a judgement upon their own sinful lives.[1] Protestants warned constantly against superstition; they had predicated their own Church upon its abandonment, but they had their own superstitions. They, too, looked to blame natural disaster upon the people's persistence in false belief. When in 1563 plague once again brought fearful casualty to the City it was said that 'the cause thereof was the superstitious religion of Rome . . . so much favoured by the citizens'. God would send worse: fire and the sword.[2] So profoundly had the Reformation divided the citizens that any calamity was attributed to the perversity and irreligion of their neighbours, and taken as a warning of divine vengeance.

There began then that almost atavistic fear among English Protestants of Catholic retribution and plot, a fear which was strongly present in London and which would have such alarming political consequences through the Civil War, the Popish Plot, the Exclusion Crisis, and beyond.[3] Henceforth, Catholics who wished to worship in London were forced to do so in secret, finding priests to celebrate Mass behind the closed doors of the houses of the citizens and nobility. And they, in their clandestine congregations, were hunted down, as they had hunted the godly. When discovered, these recusant Catholics were followed through the City streets and taunted, *Dominus vobiscum*, because He was not.[4] The Catholic faith which survived in London, or was created anew under persecution in Elizabeth's reign, was of a different kind than before. Because the Mass and transubstantiation, Purgatory, the authority of Rome, the saints and their images, remained at the heart of the faith, but nowhere in the liturgy or institutions of the Church of

---

[1] 'The true report of the burning of the steeple and church of St Paul's, London', *Tudor Tracts*, pp. 401–8; C. Barron, C. Coleman, and C. Gobbi (eds.), 'The London Journal of Alessandro Magno, 1562', *London Journal*, ix (1983), pp. 142–3; Stow, 'Memoranda', pp. 116–17; P. Collinson, *Archbishop Grindal, 1519–1583* (1979), pp. 154–5.

[2] Stow, 'Memoranda', p. 128.

[3] C. Hibbard, *Charles I and the Popish Plot* (1983); J. Miller, *Popery and Politics in England, 1660–1688* (Cambridge, 1973); R. Clifton, 'Fear of Popery', in C. Russell (ed.), *The Origins of the English Civil War* (1975), pp. 144–67; 'The Popular Fear of Catholics during the English Revolution', in P. Slack (ed.), *Rebellion, Popular Protest and the Social Order* (Cambridge, 1984), pp. 129–61; T. Harris, *London Crowds in the Reign of Charles II: Propaganda and Politics from the Restoration until the Exclusion Crisis* (Cambridge, 1987).

[4] Stow, 'Memoranda', pp. 121–2.

England, to partake of the Mass became an act of separation and a denial of public duty. The dissent was perhaps plainest in London where, because of the presence of the godly and the proximity of government, the old forms of worship could hardly be 'winked at' by a sympathetic clergy and magistracy. Yet from the secrecy of the capital the Catholic missionary priests were sent out into the countryside to convert, as the brethren had been before.[5]

And London also was the centre for the campaign to bring true reformation to a Church 'but halfly reformed'. There, in Elizabeth's reign, a 'church within a Church' was created: of curates, preachers, lecturers, and the 'brethren of London', who determined upon further reformation. Here the leaders of the radical, even revolutionary, puritan movement preached and gathered support.[6] 'The people resort unto them as in popery they were wont to run on pilgrimage.'[7] For London was 'no grange'. The godly, as so often before, crossed parish boundaries and traversed the City to hear sermons in those parishes with radical religious traditions: St Stephen Coleman Street, Holy Trinity Minories, St Botolph Aldgate, All Hallows Honey Lane, St Antholin. In time this growing voluntaryism might undermine the solidarity of the Christian community of the territorial parish.[8] Persecuted again by a monarch and an episcopate determined upon conformity and uniformity in religion, the godly went underground once more to worship in houses, in fields, aboard ships, as the London congregation had done in Mary's reign. They recognized their association with the godly who had survived persecution and exile under

[5] C. Haigh, 'The Continuity of Catholicism in the English Reformation', in C. Haigh (ed.), *The English Reformation Revised* (Cambridge, 1987), pp. 176–208; J. J. Scarisbrick, *The Reformation and the English People* (Oxford, 1984), ch. 7; G. de C. Parmiter, *Elizabethan Popish Recusancy in the Inns of Court* (*BIHR* supplement, xi, 1976); J. Bossy, *The English Catholic Community, 1570–1850* (1975).

[6] P. Collinson, *The Elizabethan Puritan Movement* (1967); *The Religion of Protestants: The Church in English Society, 1559–1625* (Oxford, 1982); T. Liu, *Puritan London: A Study of Religion and Society in the City Parishes* (Delaware, 1986).

[7] Cited in Collinson, *Elizabethan Puritan Movement*, p. 150.

[8] D. A. Williams, 'London Puritanism: the Parish of St Botolph without Aldgate', *Guildhall Miscellany*, ii (1960), pp. 24–38; D. A. Kirby, 'The Radicals of St Stephen's Coleman Street, London', ibid. iii (1970), pp. 98–119; H. G. Owen, 'A Nursery of Elizabethan Nonconformity, 1567–72', *JEH* xvii (1966), pp. 65–76; V. Pearl, *London and the Outbreak of the Puritan Revolution* (1961), pp. 163–5.

Mary, and they too were strengthened by adversity.[9] And 'persecution grew so fast as that it brought many a hundred to know one another that never knew before'.[10] Abhorring a Church of England which was still polluted by 'popish dregs', a tiny band of the godly moved at last to separatism,[11] but by far the greater part did not, staying within the Church and hoping that it would become the 'true Church' in time.

Within London, even within Protestant London, these 'known men' in their godly communities were marked apart, separated from conventional society by the very fervour and strength of their own exclusive fellowship. Sometimes it seemed as though the band of the elect, the 'saints' of later generations, must remain a very small one; that they expected it to be so. John Bradford had written from prison to one of his brethren: 'I trust you be of that little flock which shall inherit the kingdom of Heaven.'[12] One woman had even hoped that at the General Resurrection she would not rise with her confessional opponents, but against them.[13] Few can have felt so strongly, so 'uncharitably'. Yet it was hard for the godly to contemplate communion with the ungodly, for the faithful to worship with the unfaithful, especially when so many had suffered so much for the faith. The Church of England established by Elizabeth aspired to contain within itself a variety of religious convictions. Experience had shown that it must be a broad Church. And even the godly, who hated the moral and spiritual inertia of all those who 'quietly enjoy the world; they care not what religion comes',[14] still customably went to Church with the worldlings.

What can ever be known of the private faith of that great majority who did their public duty by conforming through all these reformations? Some thought that the conformity of the mass of the people would be inevitable. For Paget, the 'master

9 Collinson, *Elizabethan Puritan Movement*, pp. 84–9; C. Burrage, *Early English Dissenters*, 2 vols. (1912), ii. pp. 9–13. Stow, 'Memoranda', p. 143; *Zurich Letters*, ed. H. Robinson (Parker Society, Cambridge, 1845), p. 29; *The Remains of Edmund Grindal*, ed. W. Nicholson (Parker Society, Cambridge, 1843), pp. 201–16.

10 Cited in Collinson, *Elizabethan Puritan Movement*, p. 88.

11 B. R. White, *The English Separatist Tradition* (Oxford, 1971); M. Tolmie, *The Triumph of the Saints: the Separate Churches of London, 1616–1649* (Cambridge, 1977); P. S. Seaver, *Wallington's World: A Puritan Artisan in Seventeenth-Century London* (1985).

12 Cited in Collinson, *Religion of Protestants*, p. 255.

13 See above, pp. 601–2.

14 Cited in Collinson, *Elizabethan Puritan Movement*, p. 26.

of practices', 'what countenance soever men make outwardly was to please them in whom they see the power resteth'.[15] Private faith was another matter. Were Londoners such time servers? The godlessness of cities was a contemporary truism. London was Babylon, Nineveh, a sink of sin and full of worldly distractions; its covetous citizens thought upon 'greedy gain'; its feckless poor in their deprivation and distress had literally no time for the Word. So it was said. Comparisons were made between the simple piety of the countryside and the venality and vanity of cities.[16] It was a matter of simple observation that had every Londoner suddenly decided to come to church their parish churches could never have held them all by the end of Elizabeth's reign. Yet these same churches had been the objects of the citizens' continuing devotion and veneration: even until the desecration of the Reformation began they had given of their wealth and labour towards edifying them. It seems as though the citizens of late Elizabethan and Jacobean London did, conformably, attend their Easter communion and take their 'rights' almost universally; the more faithful, surely, side by side with the less faithful.[17] And had the poor really so little spirituality? They had been called upon in their thousands by their Catholic neighbours who often knew them to pray for souls, as though their imprecations gave a special benediction. Even humbler members of City society were found to be thinking often and deeply upon matters of faith. Religion could give, and gives, to the dispossessed a special consolation in their deprivation.

In London the experience of Reformation had been most intense and most immediate. It was here that the first evangelicals began their campaign to bring a whole nation from darkness to light. While the people of remoter regions, 'the dark corners of the land', could linger long in ignorance of Scriptural truth and remain bound to their Catholic, even pre-

---

[15] See above, p. 425.
[16] See above, pp. 448–9. *A dispraise of the life of a courtier*, translated by Sir Francis Bryan (1548; *RSTC* 12431), sig. Fiii[r].
[17] Collinson, *Religion of Protestants*, pp. 203–4; J. P. Boulton, 'The Limits of Formal Religion: The Administration of the Holy Communion in Late Elizabethan and Early Stuart London', *London Journal*, 10 (1984), pp. 135–54; *Neighbourhood and Society: A London Suburb in the Seventeenth Century* (Cambridge, 1987), pp. 275–8.

Christian, 'country divinity',[18] the Londoners could hardly be unaware of the doctrinal debates which raged around them. Though many wished never to listen to the new heresies, or, worse, the new orthodoxies taught, they could not easily forget what they had once, however unwillingly, heard. Without preaching there was strictly no doctrine, but in London there had been so much preaching; too much preaching, some said. With preaching had come confusion and division. A variety of faiths and of religious experience was the consequence. In these heady days of the early Reformation the force of events had carried many to a sometimes unlikely, unwelcome prominence and led hitherto anonymous men and women to take stands for conscience. 'Good Catholic men'—as far from the fervent evangelicals as 'frost from fire'—were moved to defend a Church which they had never dreamed would need defending. Men and women converted to the new faith in the first revolutionary generations of reform believed that they could transform religion and society. And they succeeded, at great cost. Although, in the way of things, the citizens chose overwhelmingly the anonymity of their households and the peaceful obscurity of their shops to the great sacrifices which the pursuit of a cause or the resistance to authority demanded, and though domestic life asserted, for most, a higher claim, nevertheless there was no one left untouched by the great transformations effected by the Reformation.

A world was lost which could never be recovered. The altars and images, the dooms and roods of the parish churches, the towering spires and peaceful cloisters of the religious houses were destroyed and profaned. But even the loss of such treasures was perhaps of little moment compared to the shattering of the beliefs which they had symbolized. The desecration threatened the end of mediation, propitiation, and spiritual solace, and many were left bewildered, bereft. The thousand thousand prayers which had been said for souls lingering in Purgatory were, the people were told, otiose. But if there was no intermediary world in the hereafter, and no possibility of the intercession for the dead by the living, then a starker judgement of election or reprobation awaited the Christian. Now, for the

---

[18] Collinson, *Religion of Protestants*, pp. 201–2; Christopher Hill, 'Puritans and "the Dark Corners of the Land" ', *TRHS*, 5th ser. xiii (1963), pp. 77–103.

first time, the believer faced choices between diverging paths to salvation, and must examine his conscience to decide whether the godly and the ungodly, the Catholic and the Protestant, could live together 'in charity'. The world of shared faith was broken by the Reformation and the Christian community divided.

Manuscript Sources

LONDON

BRITISH LIBRARY

Additional MSS	5,758	21,946	37,664
	5,809	24,515	45,131
	5,843	26,748	45,359
	5,853	28,585	48,022
	10,617	33,271	48,126
	12,462	33,376	
	18,783	33,923	

Additional Charter   981

C 18 e 2

Cottonian MSS   Caligula   B vii
Cleopatra   E iv–vi
F i, ii
Galba   B v, ix, x
Nero   B iv, vi
C x
Titus   B i, ii, vi
Vespasian   F xiii
Vitellius   B ix, x, xiii

Cottonian Charters   iv, 17; x, 6.

Egerton MSS   1995
2148
2603
2623
2815

Hargrave MS   134

Harleian MSS	133	372	424
	249	416	425
	283	417	530
	284	419	537
	298	421	540
	353	422	544
	367	423	559

	601	1197	2344
	643	2194	2383
	651	2252	6148
	660	2342	

Lansdowne MSS	56
	762
	lxxviii
	ciii
	cxiv 3

Royal MSS	17 B xxxv
	18 C xxiv

Salisbury MSS	volumes	150	microfilms	M/485/39
		198		M/485/52
		200		M/485/53
		332		M/485/76

Sloane MSS	156
	1066
	1710
	1786
	2442

Stowe MSS	142
	147
	1066

CLOTHWORKERS' COMPANY

Court Book 1	Court Orders, 1536/7–58.

CORPORATION OF LONDON RECORD OFFICE

Repertories of the Court of Aldermen, 2–14.
Journals of the Court of Common Council, 2–17.
Letter Books, N–Q.

DRAPERS' COMPANY

A vii, 4, 7, 266	Grants.
Repertory 7(i)	Court Minutes, 1516–29.
Repertory B	Court Minutes, 1552–7.
MB6	Court Minutes and Records, 1553–5.
Repertory C	Court Minutes, 1557–61.

GOLDSMITHS' COMPANY

Court Books D–K   Wardens' Accounts and Court Minutes.

GREATER LONDON RECORD OFFICE

Consistory Court

Act Book, Office
DL/C/614   Nov. 1554–June 1555.

Act Books, Instance
DL/C/1   Oct. 1496–Oct. 1505.
DL/C/3   Oct. 1540–Apr. 1545.
DL/C/4   Apr. 1545–Feb. 1547/8.
DL/C/5   Mar. 1547/8–Apr. 1551.
DL/C/649   Mar. 1551/2–Jan. 1552/3.
DL/C/607   Apr. 1553–May 1555.
DL/C/608   June 1555–May 1557.
DL/C/609   May 1557.
DL/C/6   June 1557–Nov. 1559.
DL/C/610   Oct. 1557–May 1559.

Deposition Books
DL/C/206   Oct. 1510–May 1516.
DL/C/207   Jan. 1520/1–June 1524.
DL/C/208   Easter 1529–Feb. 1532/3.
DL/C/209   Dec. 1552–Feb. 1552/3.
DL/C/628   Apr. 1553–Nov. 1554.

Vicar-General's Books
DL/C/330   'Foxford' 1520/1–1538/9.
DL/C/331   'Crooke' 1546–1560.

St Thomas's Hospital
H1/ST/A1/1   Court Minutes, 1556–64.

GUILDHALL LIBRARY

Diocesan Records

Bishops' Registers
MS 9531/9   Fitzjames 1506–22.
MS 9531/10   Tunstall 1522–29/30.
MS 9531/11   Stokesley 1530–9.
MS 9531/12   Bonner   1540–9.
                         Ridley   1550–3.
                         Bonner   1553–8.

Commissary Court

MSS 9064/9–11	Acta quoad correctionem delinquentium, 1502–16.
MSS 9168/5–11	Act Books, Probate and Administration, 1516–58.
MSS 9171/10–13	Registers of Wills, 1522–58.
MS 9172/1	Original Wills, 1523–51.

Consistory Court

MS 9065A	Deposition Book in Testamentary Causes.
MS 1231	Account of the trial and execution of Sir Thomas More and of the suppression of the Charterhouse and the execution of Prior Houghton.

Company Records

Armourers and Brasiers' Company

MS 12071/1	Court minute book, 1413–1559.
MS 12073	Yeomanry Court minute book, 1552–1604.
MS 12140	Chantry estates.

Bakers' Company

MS 5197	Ordinances and oath book.
MS 5177/1	Court minute book, 1537–61.
MS 5179/1	Quarterage book, 1518–56.

Barber Surgeons' Company

MS 5257/1	Court minute book, 1557–86.

Brewers' Company

MS 5445/1	Court minute book, 1531–54.
MS 5445/2	Court minute book, 1557–63.

Carpenters' Company

MS 4329/1	Court minute book, 1533–73.

Coopers' Company

MS 5603/1	Rough court minutes, 1552–67.

Grocers' Company

MS 11570A	Memorandum, ordinance and account book, 1463–1557.
MS 11158/1	Court minute book, 1556–91.

Ironmongers' Company

MS 16,967/1	Court minute book, 1555–1602.

MS 16,981/1	Apprentices' book of signed oaths.
MS 16,987/1	Yeomanry quarterage book, 1524–59.
MS 16,988/2	Wardens' accounts, 1539–92.

Merchant Taylors' Company

| 298(2) | Accounts, 1545–57. |
| 323(27) | Freemen list. |

Parish Clerks' Company

| MS 4889 | Bede roll of the Fraternity of St Nicholas, 1450–1521. |

Pewterers' Company

| MS 7090/1 | Court minute book, 1551–61. |
| MS 7086/2 | Wardens' accounts, 1530–72. |

Vintners' Company

| MS 15333/1 | Wardens' accounts, 1522–82. |

Wax Chandlers' Company

| MS 9481/1 | Renters Wardens' account book, 1531–97. |

Weavers' Company

| MS 4645 | 'Ancient Book'. |

Parish Records

Churchwardens' Accounts

All Hallows Staining

| MS 4956/1 | 1491–1550. |
| MS 4956/2 | 1533–1628. |

St Alphage

| MS 1432/1 | 1527–53. |
| MS 1432/2 | 1553–80. |

St Andrew Hubbard

| MS 1279/1 | 1454–1524. |
| MS 1279/2 | 1525–1621. |

St Benet Gracechurch

| MS 1568/1 | 1548–1723. |

St Botolph Aldersgate

| MS 1454 | 1466–1635. |

St Botolph Aldgate

| MS 9235/1 | 1547–85. |

St Dunstan in the East

| MS 4887 | 1494–1509. |

St Dunstan in the West
MS 2968/1   1516–1608.

St James Garlickhithe
MS 4810/1   1555–1627.

St Laurence Pountney
MS 3907/1   1530–1681.

St Margaret Pattens
MS 4570/1   1506–57.
MS 4570/2   1558–1653.

St Martin Outwich
MS 6842   1508–28 & 1537–45.

St Mary Magdalen Milk Street
MS 2596/1   1518–1606.

St Mary Woolnoth
MS 1002/1   1539–1641.

St Matthew Friday Street
MS 1016/1   1547–1678.

St Michael Cornhill
MS 4071/1   1455–1608.

St Michael le Querne
MS 2895/1   1514–1604.

St Peter Westcheap
MS 645/1   1435–1601.

St Stephen Walbrook
MS 593/1   1474–1538.

Vestry Minutes

St Laurence Jewry
MS 2590/1   1556–1669.

St Martin Orgar
MS 959/1   1557–1643.

Christ's Hospital
MS 12806/1   Court minute book, 1556–63.
MS 12819/1   Treasurer's account book, 1552–8.

INNER TEMPLE LIBRARY

Petyt   MS 538, vol. 47.

LAMBETH PALACE LIBRARY

Register of Archbishop Cranmer.
Carte Miscellane, VIII, 2a–e.

MERCERS' COMPANY

Acts of Court, ii, 1527–60.
Sir Thomas Gresham's Day Book.
Renters Wardens' Accounts, 1525–47.
Register of Writings, ii.

PUBLIC RECORD OFFICE

The following classes of records have been used:

C 1	Early Chancery Proceedings.
C 85	Chancery, Significavit of Excommunication.
Ct Req	Court of Requests.
E 36	Exchequer, Treasury of Receipt, Miscellaneous Books.
E 135	Exchequer, King's Remembrancer, Ecclesiastical Documents.
E 159	Exchequer, King's Remembrancer, Memoranda Rolls.
E 301	Exchequer, Court of Augmentations, Certificate of Colleges and Chantries, Henry VIII and Edward VI.
KB 27	Placita Coram Rege.
KB 29	Controllment Rolls of the Court of King's Bench.
KB 9	Ancient Indictments in King's Bench.
KB 8	Baga de Secretis.
PCC Prob. 11	Prerogative Court of Canterbury, Will Registers.
SP 1	State Papers, Henry VIII.
SP 2	State Papers, Henry VIII, folio volumes.
SP 3	Lisle Papers.
SP 6	Theological Tracts.
SP 10	State Papers, Edward VI.
SP 11	State Papers, Mary I.
SP 46	Johnson Papers.
Sta Cha	Star Chamber Proceedings, Henry VIII–Philip & Mary.

ST BARTHOLOMEW'S HOSPITAL, WEST SMITHFIELD

Ha 1/1	Minutes of the Board of Governors, 1549–61; Churchwardens' accounts of Christchurch, Newgate, 1546–8.

Ha 13   Papers concerning Governors. Rules and
                               Orders for Hospitals, 1552, 1557.

OUTSIDE LONDON

TRINITY COLLEGE, DUBLIN

MS 775   Archbishop Ussher's transcript of a lost
                               register of Bishop Fitzjames.

BALLIOL COLLEGE, OXFORD

MS 354   The Commonplace Book of Richard Hill.

BODLEIAN LIBRARY, OXFORD

MS Don C 42   William Latimer, 'Treatyse' on Anne
                               Boleyn.
Tanner MS 221   Collection of letters patent, ordinances and
                               accounts of the Gild of the Name of
                               Jesus, St Paul's.
Tanner MS 304
MS Rawlinson C 102
MS Rawlinson C 408   Account of the trial of Sir Nicholas Throck-
                               morton, 1554.
Ashmole 861   The Chronicle of Anthony Anthony.
Fol. Δ 624   The Chronicle of Anthony Anthony
                               (marginal notes).
Jesus MS 74
MS Gough London 10

NATIONAL LIBRARY OF SCOTLAND

Adv. MS 34.2.14   Letters of Richard Scudamore to Sir Philip
                               Hoby, 1549–54.

Index